Communications in Computer and Information Science 2857

Series Editors

Rationale
The CCIS series is devoted to the publication of proceedings of computer science conferences. Its aim is to efficiently disseminate original research results in informatics in printed and electronic form. While the focus is on publication of peer-reviewed full papers presenting mature work, inclusion of reviewed short papers reporting on work in progress is welcome, too. Besides globally relevant meetings with internationally representative program committees guaranteeing a strict peer-reviewing and paper selection process, conferences run by societies or of high regional or national relevance are also considered for publication.

Topics
The topical scope of CCIS spans the entire spectrum of informatics ranging from foundational topics in the theory of computing to information and communications science and technology and a broad variety of interdisciplinary application fields.

Information for Volume Editors and Authors
Publication in CCIS is free of charge. No royalties are paid, however, we offer registered conference participants temporary free access to the online version of the conference proceedings on SpringerLink (http://link.springer.com) by means of an http referrer from the conference website and/or a number of complimentary printed copies, as specified in the official acceptance email of the event.

CCIS proceedings can be published in time for distribution at conferences or as post-proceedings, and delivered in the form of printed books and/or electronically as USBs and/or e-content licenses for accessing proceedings at SpringerLink. Furthermore, CCIS proceedings are included in the CCIS electronic book series hosted in the SpringerLink digital library at http://link.springer.com/bookseries/7899. Conferences publishing in CCIS are allowed to use our online conference service (Meteor) for managing the whole proceedings lifecycle (from submission and reviewing to preparing for publication) free of charge.

Publication process
The language of publication is exclusively English. Authors publishing in CCIS have to sign the Springer CCIS copyright transfer form, however, they are free to use their material published in CCIS for substantially changed, more elaborate subsequent publications elsewhere. For the preparation of the camera-ready papers/files, authors have to strictly adhere to the Springer CCIS Authors' Instructions and are strongly encouraged to use the CCIS LaTeX style files or templates.

Abstracting/Indexing
CCIS is abstracted/indexed in DBLP, Google Scholar, EI-Compendex, Mathematical Reviews, SCImago, Scopus. CCIS volumes are also submitted for the inclusion in ISI Proceedings.

How to start
To start the evaluation of your proposal for inclusion in the CCIS series, please send an e-mail to ccis@springer.com

Mahault Albarracin · David Benrimoh ·
Christopher L. Buckley · Pablo Lanillos ·
Riddhi J. Pitliya · Hideaki Shimazaki ·
Ivilin Peev Stoianov · Tim Verbelen ·
Martijn Wisse
Editors

Active Inference

6th International Workshop, IWAI 2025
Montreal, QC, Canada, October 15–17, 2025
Revised Selected Papers

Editors
Mahault Albarracin
VERSES
Los Angeles, CA, USA

David Benrimoh
McGill University
Quebec, QC, Canada

Christopher L. Buckley
University of Sussex, Brighton, UK
and VERSES
Los Angeles, CA, USA

Pablo Lanillos
Cajal Neuroscience Center, Spanish National Research Council
Madrid, Spain

Riddhi J. Pitliya
University of Oxford, UK and VERSES
Los Angeles, CA, USA

Hideaki Shimazaki
Graduate School of Informatics Kyoto University
Kyoto, Japan

Ivilin Peev Stoianov
Institute of Cognitive Sciences and Technologies (ISTC), National Research Council (CNR)
Roma, Italy

Tim Verbelen
VERSES
Los Angeles, CA, USA

Martijn Wisse
Delft University of Technology
Delft, The Netherlands

ISSN 1865-0929 ISSN 1865-0937 (electronic)
Communications in Computer and Information Science
ISBN 978-3-032-16954-9 ISBN 978-3-032-16955-6 (eBook)
https://doi.org/10.1007/978-3-032-16955-6

This Springer imprint is published by the registered company Springer Nature Switzerland AG
The registered company address is: Gewerbestrasse 11, 6330 Cham, Switzerland

Preface

Over 170 students and researchers from academia and industry joined us for the 6th International Workshop on Active Inference (IWAI), held in the beautiful Montreal, Canada, from Oct 15–17, 2025. This year's IWAI was the biggest yet and had a packed and exciting three-day program featuring leading experts in active inference, including a tutorial by Ryan Smith, an opening keynote by Christoph Mathys, an exciting presentation on free energy by Dalton Sakthivadivel, an inspiring session on robotics by Jun Tani, a lively fireside chat with Gary Marcus and Karl Friston, capped off with a closing keynote by Friston which offered both an excellent retrospective on active inference and a look forward into the future.

We encouraged submissions from three different streams: Computational Theory and Simulations (mathematical and computational developments); Cognitive, Philosophical, and Neural Models (modeling biological, neural, and cognitive phenomena, and integrating applied philosophy); and Empirical, Clinical, and Real-World Applications (data-driven studies applied to robotics, IoT, clinical, industrial, or societal challenges). This volume presents the 24 full papers that were accepted for publication and presented as contributed talks at the workshop. The papers were selected out of 76 submissions through a double-blind review process in which submissions received three reviews each on average process, and cover a wide range of disciplines and domains: from computer science and neuroscience to applications in self-driving cars and robotics.

The overarching theme of the conference was the ability of Active Inference to serve as a unifying framework across domains and levels of explanation. What makes IWAI unique is that it is a true testament to the power of principled computational modelling. In the same conference where we heard about how active inference can be used to explain mental health conditions, we also heard about how it can be used in robot navigation. And despite coming from completely different contexts and functioning at different levels of abstraction, the computational psychiatrist and the roboticist--so long as they both 'speak' active inference--can come together and exchange ideas on an equal footing, share ideas, and find commonalities. Herein lies the power of active inference--its ability to act as a universal language of organized systems.

We would like to thank the Program Committee for their valuable work, all authors for their contributions, all speakers for their inspiring keynotes, and, of course, all the attendees. We would also like to thank our sponsors, VERSES, Noumenal, Waymo, and NEXT, whose support made this event possible.

October 2025

David Benrimoh
Mahault Albarracin
Ivilin Peev Stoianov
Martjin Wisse

Organization

General Chairs

David Benrimoh	McGill University, Canada
Mahault Albarracin	VERSES, USA

Technical Program Chairs

Ivilin Peev Stoianov	Institute of Cognitive Sciences and Technologies (ISTC), National Research Council (CNR), Italy
Martijn Wisse	Delft University of Technology, Netherlands

Advancement Chair

Susie Kim	VERSES, USA

Organizing Committee

Mahault Albarracin	VERSES, USA
David Benrimoh	McGill University, Canada
Christopher L. Buckley	University of Sussex, UK; VERSES, USA
Pablo Lanillos	Donders Institute for Brain, Cognition and Behaviour, Netherlands; Instituto Cajal (CSIC), Spain
Riddhi J. Pitliya	University of Oxford, UK; VERSES, USA
Hideaki Shimazaki	Kyoto University, Japan
Ivilin Peev Stoianov	Institute of Cognitive Sciences and Technologies (ISTC), National Research Council (CNR), Italy
Tim Verbelen	VERSES, USA
Martijn Wisse	Delft University of Technology, Netherlands

Reviewers

Mahault Albarracin	VERSES, USA
David Benrimoh	McGill University, Canada
Marc A. Broberg	Two Trees Physical Therapy, USA
Matthew Brown	ThoughtForge Inc., USA
Christopher L. Buckley	University of Sussex, UK; VERSES, USA
Victor Casamayor Pujol	Universitat Pompeu Fabra, Spain
Ozan Catal	VERSES, USA
Peter Chin	Dartmouth College, USA
Leo Christov-Moore	Institute for Advanced Consciousness Studies, USA
Poppy Collis	University of Sussex, UK
Lancelot Da Costa	VERSES, USA
Peter Dayan	Max Planck Institute, Germany
Daria de Tinguy	Ghent University, Belgium
Bert de Vries	Eindhoven University of Technology, The Netherlands
Bart Dhoedt	Ghent University, Belgium
Kenji Doya	Okinawa Institute of Science and Technology Graduate University, Japan
Johan Engstrom	Waymo, USA
Lukas J. Fiderer	Universität Innsbruck, Austria
Karl Friston	University College London, UK
Emma Graham	Dartmouth College, USA
Richard Granger	Dartmouth College, USA
Yusuke Hayashi	AI Alignment Network, USA
Conor Heins	VERSES, USA
Catherine F. Higham	University of Glasgow, UK
Nicolás Hinrichs	Okinawa Institute of Science and Technology, Japan
Yufei Huang	Rutgers University, USA
David Hyland	University of Oxford, UK
Takuya Isomura	RIKEN, Japan
Mohsen A. Jafari	Rutgers University, USA
Farzad Kamrani	Swedish Defence Research Agency, Sweden
Kazuharu Kidera	Honda, Japan
Markus Klar	University of Glasgow, Scotland
Karlo Koledić	University of Zagreb, Croatia
Georgia Koppe	Heidelberg University, Germany
Magnus Koudahl	VERSES, USA

Wouter M. Kouw	Eindhoven University of Technology, The Netherlands
Pablo Lanillos	Spanish National Research Council, Spain
Andrea Matta	Polytechnic Institute of Milan, Italy
Tadayuki Matsumura	Hitachi, Japan
Shingo Murata	Keio University, Japan
Roderick Murray-Smith	University of Glasgow, UK
Wouter W. L. Nuijten	Eindhoven University of Technology, The Netherlands
Joséphine Pazem	Universität Innsbruck, Austria
Corrado Pezzato	VERSES, USA
Marco Perin	VERSES, USA
Ivan Petrovic	University of Zagreb, Croatia
Riddhi J. Pitliya	VERSES, USA
Marc Pritsch	Heidelberg University, Germany
Maxwell Ramstead	Noumenal, USA
Francesca Rossi	Scuola Superiore Meridionale, Italy
Giovanni Russo	University of Salerno, Italy
Adam Safron	Johns Hopkins University School of Medicine, USA
Tommaso Salvatori	VERSES, USA
Whitney Sales	ThoughtForge Inc., USA
Dalton Sakthivadivel	City University of New York, USA
Zahra Sheikhbahaee	MILA, Canada
Hideaki Shimazaki	Kyoto University, Japan
Sebastian Stein	University of Glasgow, UK
Ivilin Peev Stoianov	National Research Council of Italy, Italy
Jun Tani	Okinawa Institute of Science and Technology, Japan
Theodore J. Tinker	Okinawa Institute of Science and Technology, Japan
Alexander Tschantz	VERSES, USA
Tim Verbelen	VERSES, USA
Toon Van de Maele	VERSES, USA
Thijs van de Laar	Eindhoven University of Technology, The Netherlands
Kirstin Wagner	University of Birmingham, Uk
Ran Wei	Waymo, USA
John H. Williamson	University of Glasgow, UK
Martijn Wisse	Delft University of Technology, The Netherlands
Jeong Hwan Yoon	University of California, Los Angeles, USA

Contents

Models of Decision Making

On the Variational Costs of Changing Our Minds

David Hyland[1(✉)] and Mahault Albarracin[2]

[1] University of Oxford, Oxford, United Kingdom
[2] VERSES AI Research Lab, Los Angeles, CA 90016, USA

Abstract. The human mind is capable of extraordinary achievements, yet it often appears to work against itself. It actively defends its cherished beliefs even in the face of contradictory evidence, conveniently interprets information to conform to desired narratives, and selectively searches for or avoids information to suit its various purposes. Despite these behaviours deviating from common normative standards for belief updating, we argue that such 'biases' are not inherently cognitive flaws, but rather an adaptive response to the significant pragmatic and cognitive costs associated with revising one's beliefs. This paper introduces a formal framework that aims to model the influence of these costs on our belief updating mechanisms.

We treat belief updating as a motivated variational decision, where agents weigh the perceived 'utility' of a belief against the informational cost required to adopt a new belief state, quantified by the Kullback-Leibler divergence from the prior to the variational posterior. We perform computational experiments to demonstrate that simple instantiations of this resource-rational model can be used to qualitatively emulate commonplace human behaviours, including confirmation bias and attitude polarisation. In doing so, we suggest that this framework makes steps toward a more holistic account of the motivated Bayesian mechanics of belief change and provides practical insights for predicting, compensating for, and correcting deviations from desired belief updating processes.

Keywords: Belief Change · Motivated Reasoning · Active Inference · Cognitive Effort

"*[H]uman reason is both biased and lazy. Biased because it overwhelmingly finds justifications and arguments that support the reasoner's point of view, lazy because reason makes little effort to assess the quality of the justifications and arguments it produces.*" (Mercier and Sperber, 2017, p. 9) [35].

1 Introduction

Humanity faces an increasingly paradoxical epistemic problem. Never before have people been able to obtain so much information so quickly, yet at the

M. Albarracin et al. (Eds.): IWAI 2025, CCIS 2857, pp. 3–23, 2026.
https://doi.org/10.1007/978-3-032-16955-6_1

same time, many societies have become increasingly polarised and paralysed by conflicting narratives. A feature that is common to several manifestations of this predicament, including public health crises and conspiracy theorising, is the presence of actors who tenaciously defend beliefs long after the balance of evidence has shifted. What, exactly, makes changing our minds so hard?

A natural approach to answering this question may begin by supposing a normative standard or benchmark against which to compare the actual processes of human belief change. The primary normative model for rational belief updating is Bayesian reasoning. According to this model, probabilistic beliefs should be adjusted proportionally to the strength of evidence according to Bayes' rule. However, the persistent discrepancy between the Bayesian standard and actual human belief updating raises questions about whether our epistemic processes are inherently irrational or if something is missing from the traditional picture [34].

The Bayesian paradigm largely remains silent on *why* humans fall short of its ideals, primarily due to its assumptions that belief-revision is cost-free and that the driver of epistemic processes should be probabilistic coherency [26]. In practice, revising one's beliefs incurs metabolic costs, cognitive effort, and crucially, pragmatic risks and opportunities. A scientist retracting a cherished hypothesis, a politician breaking ranks with their party, or a public figure admitting error each pay tangible costs that a cost-free Bayesian calculus does not account for. Without a principled way to model the effect of such costs on belief revision, apparent "irrationalities" including confirmation bias [36], motivated reasoning [29], attitude polarisation [33], and belief persistence [15] seem like fundamental flaws in human cognition.

We argue that such apparent deviations from Bayesian norms are adaptive responses of agents operating under motivational considerations and real resource constraints. In particular, we formalise cognitive belief-change costs using the KL divergence to quantify informational distances between belief states, representing the 'informational work' required for belief state transitions. Our approach also integrates social and pragmatic factors. Beliefs are influenced by identity, social status, and interpersonal dynamics; fears of ostracism or admitting errors can increase resistance to change. Our hope is that through such modelling, we can take steps toward developing general frameworks that explain not just isolated sources of non-Bayesian belief updating, but also the inherent trade-offs between competing considerations associated with changing our minds.

1.1 Contributions and Paper Structure

Our primary contribution is the proposal of a variational cost functional for belief revision that models the influence of pragmatic affordances and cognitive costs on human belief updating. Secondly, we present results from simplified computational experiments demonstrating how varying conservatism and likelihood weighting parameters qualitatively exhibit phenomena such as confirmation bias, evidence search asymmetries, and attitude polarisation. Finally, we discuss the implications of our model and promising future directions.

2 Related Literature

"There is considerable evidence that people are more likely to arrive at conclusions that they want to arrive at, but their ability to do so is constrained by their ability to construct seemingly reasonable justifications for these conclusions. (Kunda, 1990) [29].

2.1 Decision-Theoretic Models of Belief Updating

Drawing on frameworks of decision making, human belief revision is increasingly being treated as a value-based decision [28,47,51]. On this view, beliefs are updated not purely based on their accuracy, but are associated with a *utility*. The utility of holding a belief is derived from the outcomes it leads to, which can be internal (emotional comfort, positive feelings) or external (acceptance within a community, job opportunities) [51]. This is supported by several arguments highlighting the centrality of affect in decision-making and belief-updating [13,27,50]. Certain beliefs may give rise to utility in proportion to how well they track or predict reality, in which case, there is an incentive for the agent to seek truthful beliefs. Other beliefs may demand that the agent confabulates an elaborate yet tenuous narrative that coincidentally supports their desired conclusion. In other words, a belief's usefulness can be orthogonal to its truthfulness.

2.2 Cognitive Costs

In addition to the pragmatic incentives that shape belief updating, there are unavoidable costs that any agent must pay to change their beliefs. These costs have been studied from the perspective of bounded/resource/computational rationality, where the presence of some form of cost associated with cognition is explicitly modelled in an agent's decision-making [21,30,31,39–41,54,64]. Belief updating can be understood as a thermodynamic process involving transitions between mental states, where each transition incurs unavoidable dissipative costs [17]. These costs arise from fundamental physical principles governing information processing in biological systems at the level of neural computation and metabolic energy expenditure [62,63].

The transition between belief states involves both work-like and heat-like components. The work-like component corresponds to the directed shift of the belief state, while heat dissipation occurs in the form of entropy production during the transitions between belief states in finite amounts of time [3]. The total dissipation produced by belief updating can be quantified as the difference between the reversible work theoretically possible and the actual work captured during the transition process. This represents the unavoidable cost of finite-time belief changes [43].

This thermodynamic perspective helps to explain why, other considerations being equal, rapid belief changes tend to be more costly and inefficient compared to gradual updates. The system must balance the speed of belief revision against the increased dissipative costs of rapid change. This intuition can be made more precise using concepts from finite-time thermodynamics [3]. The total entropy production in a sequence of step-equilibrations is bounded by $\Delta S^u \geq \frac{L^2}{2K}$, where L is the thermodynamic length of the belief change pathway and K is the number of intermediate equilibration steps [48]. Thus, increasing the number of steps decreases the lower bound on total entropy production, permitting more efficient pathways of belief change. The brain appears to possess several remarkable features that aid in minimising these costs. For instance, the efficient coding hypothesis suggests that neural representations of sensory information are structured to minimise the number of neuronal spikes required to transmit a given signal [4].

Understanding these fundamental thermodynamic constraints provides insight into why belief change can be so difficult even in the presence of contradictory evidence. The brain must carefully balance the energetic and informational costs of updating against the potential benefits. It is unclear precisely how significantly the thermodynamic costs of belief change contribute to this effect, and it would be worthwhile empirically investigating how such considerations can contribute to and explain belief inertia.

2.3 Social Costs

Human beliefs serve not only as internal models of the world but also as social signals and commitments. In active inference and variational learning frameworks, agents update beliefs to minimise surprise or prediction error, yet these updates occur in a social context where beliefs fulfil both epistemic (truth-seeking) and social-coordination functions [2,6,9,35,57,58,61]. Believing (or disbelieving) certain propositions can grant individuals emotional comfort or group acceptance, independent of the belief's accuracy [51]. This dual role means that an agent's posterior after observing new evidence is not determined by epistemic considerations alone, but also by the expected social and personal utilities associated with holding particular beliefs [1,22,29,51]. Consequently, standard Bayesian updates, which are focused purely on data and prior likelihood, are often tempered by an additional motive: to align with valued identities and norms that confer utility on the belief state [9,22,35,61]. This insight echoes the idea that *all thinking is "wishful" thinking* to some extent, with motivational imperatives modulating inferential processes [28]. The free energy minimised during belief updating thus implicitly includes not just accuracy-related (surprisal) terms but also pragmatic terms capturing the work required to overcome cognitive inertia and social repercussions [2,8].

Changing one's mind can threaten group affiliations and invite real or perceived social sanctions (e.g., loss of status, trust, or membership) [6]. Beliefs often function as markers of group identity, so revising a key belief may signal disloyalty or value misalignment, incurring social costs like ostracism or ridicule. Anticipation of such costs creates a strong deterrent to belief revision, especially for identity-linked beliefs maintained by tight-knit communities and normative expectations [2,35]. Indeed, social norms enforce a kind of epistemic conformity: individuals internalize the expectation that they "ought" to hold certain beliefs to remain in good standing [6,22]. From a decision-analytic perspective, the utility of a belief therefore includes not only its truth-tracking benefits but also its social payoff. A false or unfounded belief might persist if it brings social acceptance or emotional relief, whereas a truthful belief might be resisted if it carries stigma or existential dread. Accordingly, belief change in social contexts resembles a form of motivated reasoning: agents are inclined to arrive at the conclusions they want (or need) to reach, as long as they can justify them to themselves and others [29]. Here, "wants" are not arbitrary whims but structured by social identity and normative pressures—what one wants to believe is often what one's group wants one to believe. An agent will unconsciously search for justifications to retain beliefs that serve valued social goals (e.g. solidarity, consistency, pride) and discount evidence that threatens those goals. The free energy landscape is warped by social potential energy. Certain directions of belief change appear steep (costly) due to the interpersonal consequences associated with them.

Empirical research supports these principles. For instance, people consistently overestimate the severity of the social sanctions they will face for changing a politically charged belief, leading to excessive self-censorship and public conformity [55]. In one set of their studies, U.S. partisans expected far more backlash from their in-group if they voiced a dissenting opinion than what actually materialised, with an average overestimation effect size of $d \approx 0.87$. These inflated expectations of ostracism or punishment (sometimes stemming from an egocentric bias in social perspective-taking) make belief revision seem riskier than it truly is. Accordingly, individuals often stick to publicly defending their prior attitudes, even when privately grappling with contrary evidence. Social psychologists refer to this pattern as identity-protective cognition, wherein reasoning processes bend to protect the agent from the social identity costs of admitting error. The effect can become self-reinforcing. If everyone fears speaking up or changing their mind, the apparent unanimity of belief within the group remains unchallenged, further raising the perceived cost of dissent. Yet research also shows that these perceived social costs are malleable. Prompting individuals to reflect on their past loyalty and contributions to the group can reassure them that a change of mind will not irrevocably brand them as "disloyal," thereby significantly reducing their concern about sanction and encouraging more open expression of revised beliefs.

Beliefs are multi-functional cognitive tools that balance accuracy, utility, and inertia. They must at once represent the world (epistemic accuracy), support our emotional needs and moral values, and coordinate with our social milieu (utility), all while minimising drastic revisions that incur cognitive and social "work" (inertia). This perspective prepares us to interpret classic phenomena—confirmation bias, selective exposure to information, and attitude polarisation, not as inexplicable failures of rationality, but as strategic trade-offs given the agent's objectives. An agent facing high costs for belief change will rationally exhibit a kind of conservatism. Agents will favour information that confirms existing beliefs and avoids provoking costly updates. Indeed, a confirmation bias in information-seeking can be seen as an adaptive strategy to preserve high-utility beliefs by selectively attending to congruent evidence and filtering out challenges. Experimental studies of selective exposure document that people spend more time with news and arguments that align with their preexisting attitudes than with those that contradict them, even when source credibility is controlled [60]. By skimming "friendly" evidence, individuals reduce the likelihood of encountering data that would demand painful social readjustments or internal value conflicts. Similarly, communities may become polarised when each side's beliefs carry their own social rewards—members of opposing groups double down on group-consistent narratives, bolstering internal cohesion at the expense of cross-group accuracy. Over time, this self-reinforcing selection and interpretation of evidence drives group attitudes further apart, as each group lives in a bubble where maintaining their version of reality is pragmatically advantageous [2]. The following sections will explore how confirmation bias in evidence appraisal, asymmetrical information search, and polarisation dynamics emerge naturally once we acknowledge that changing one's mind is not "free." It incurs variational costs, paid in both cognitive effort and social capital, which a resource-rational mind navigates by carefully weighing when belief change is truly worth the price.

2.4 Confirmation Bias

Confirmation bias manifests through selective attention mechanisms, as shown in recent experimental work [46,56]. Westerwick et al. demonstrated that when selecting political information online, participants spent more time with content matching their existing views, regardless of source quality [60]. This bias emerged from participants' choices rather than the content itself. Building on this, [56] revealed that making a categorical choice selectively enhanced sensitivity to subsequent evidence consistent with that choice, similar to attentional cueing effects. [46] proposed a neural mechanism for this bias, suggesting that choices direct feature-based attention to amplify processing of choice-consistent sensory evidence while suppressing inconsistent information. Together, these findings indicate that confirmation bias operates through early attentional selection rather than solely in later-stage decision processes.

2.5 Motivated Reasoning

Confirmation bias is a type of motivated reasoning, a process where information processing is biased toward achieving desired outcomes rather than accuracy alone [29]. Motivations can be accuracy-driven, encouraging unbiased reasoning, or directional, prompting strategies that reinforce existing beliefs, identity, or preferred conclusions. However, motivated reasoning remains constrained by plausibility; people select cognitive processes, such as memory retrieval and interpretation, that justify favoured conclusions rather than inventing implausible beliefs [14,24,44].

Individuals revise beliefs asymmetrically, giving more weight to confirmatory or emotionally favourable evidence than to equally informative negative evidence [32]. Motivated reasoners also selectively trust or avoid sources based on alignment with their views, effectively assigning lower reliability to disconfirming information. This acts like a biased Bayesian filter, reducing the impact of contradictory evidence on belief updating [45].

Consequently, shared evidence can polarise rather than unify groups with opposing and even similar priors. When interpreting balanced evidence through a biased lens, individuals' initial beliefs often become more extreme, exacerbating attitude polarisation [2,5].

2.6 Biased Reasoning

Two recent frameworks have been proposed to explain some of the biases that occur in reasoning: coherence-based reasoning (CBR) [53] and belief-consistent information-processing (BCIP) [37]. Coherence-based reasoning posits that individuals strive to maintain a consistent and interconnected set of beliefs, minimising cognitive dissonance. More specifically, in CBR, a constraint-satisfaction network settles into an attractor by bidirectionally reshaping both beliefs and incoming information to maximise overall coherence. Crucially, strongly activated priors are harder to dislodge, so they often anchor the attractor state. Similarly, belief-consistent information processing describes the tendency to favour information that aligns with pre-existing beliefs, a process that is less cognitively demanding than evaluating and integrating contradictory evidence. BCIP is a special case of CBR's coherence construction under conditions of dominant priors [38].

These two frameworks can be reconciled with our account through the lens of cognitive economy. Our variational cost framework formalises this principle by suggesting that altering one's beliefs incurs a cognitive cost, quantified by the KL divergence, which measures the informational distance between prior and posterior beliefs. Moreover, the weight of other firmly held beliefs can be explained by the presence of costs associated with revising more strongly held beliefs. For example, the expected cost associated with modifying a fundamental belief such as "I make correct assessments of the world" would be increased levels of doubt about the reliability of one's assessments, potentially leading to greater general levels of uncertainty and the accompanying negative affect that often arises.

In this sense, both coherence-based reasoning and belief-consistent information processing can be viewed as cognitive strategies that minimise the costs that we describe. By maintaining coherence and selectively processing information, individuals reduce the "informational work" required to update their mental models of the world, thereby avoiding the significant pragmatic and cognitive expenditures associated with belief revision. In essence, these frameworks highlight different facets of the same underlying drive to manage cognitive resources efficiently, where the perceived utility of a belief is weighed against the inherent costs of mental reorganisation.

3 A Motivated Variational Belief Change Model

As a starting point, we take inspiration from *variational inference* [7,59], which underpins the mathematical formalism of the Free-Energy Principle and Active Inference [18,19]. In the standard Bayesian paradigm, the goal is to infer a posterior belief $p(s \mid o)$ about the state of the world s given an observation o. According to Bayes' rule, finding this posterior requires one to compute the model evidence or marginal likelihood $p(o)$, which is an intractable problem in general. Variational inference aims to reformulate this problem by recasting it as an *optimisation problem* over a variational family $\mathcal{Q}$ of probability distributions. The objective function of this optimisation problem is the negative *evidence lower bound* (ELBO) or *variational free energy* (VFE), and is given by

$$F[q(s), o] = -\underbrace{\mathbb{E}_{q(s)}[\log p(o \mid s)]}_{\text{Accuracy}} + \underbrace{D_{\mathrm{KL}}\left[q(s) \mid\mid p(s)\right]}_{\text{Complexity}} \tag{1}$$

$$= \underbrace{-\mathbb{E}_{q(s)}[\log p(s, o)]}_{\text{Energy}} - \underbrace{H[q(s)]}_{\text{Entropy}}. \tag{2}$$

The decomposition in Eq. 1 highlights a key tension between the expected log likelihood of the observation (accuracy) and the KL divergence from the prior to the variational posterior (complexity). In particular, this complexity acts as a *regulariser* on the agent's posterior beliefs, penalising models that differ more from the prior.

Core to the description of any agent is a description of its boundary, also commonly known as its *Markov blanket.* The Markov blanket of an object describes the interface via which it is coupled to its environment. According to the Free Energy Principle (FEP), the internal paths of systems possessing a Markov blanket can be viewed as probabilistic beliefs about external paths, and the internal and active paths of the system appear to minimise its VFE [20]. When moving to descriptions of agents, however, the system's internal states embody not only a predictive model of the world, but also *preferences* over possible configurations of the agent. This is where *active inference* (AIF) comes into the picture (Fig. 1).

AIF extends the FEP to recognise the role of the *actions* that agentic systems can perform to influence the environment and, vicariously, their observations [10, 12,42]. Cast here in the variational perspective, the objective function that is posited to drive decision-making in AIF is the *expected free energy* (EFE), which is a functional of a policy (sequence of actions) π, and is given by

$$G(\pi) = -\underbrace{\mathbb{E}_{q(s,o|\pi)}\left[D_{\mathrm{KL}}\left[q(s \mid o, \pi) \mid\mid q(s \mid \pi)\right]\right]}_{\text{Epistemic value}} - \underbrace{\mathbb{E}_{q(o|\pi)}[\log \tilde{p}(o)]}_{\text{Pragmatic value}}, \tag{3}$$

where $\tilde{p}(o)$ is a probability distribution representing the agent's preferences over their own observations, commonly known as a *prior preference* [10]. Here, we propose to extend this picture by considering the implications of assuming that

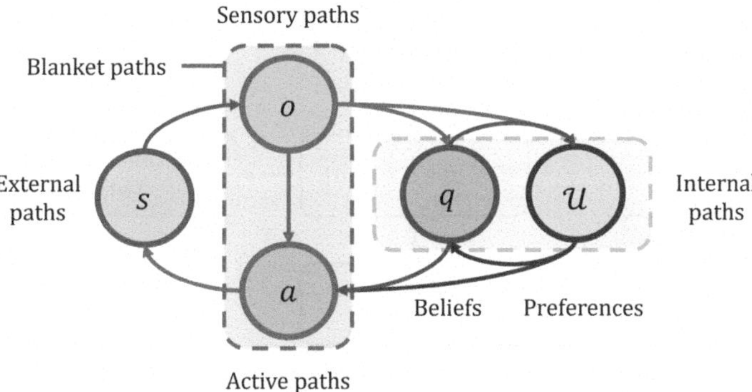

Fig. 1. A depiction of the causal influences of different components of the agent-environment pair on each other. External paths s represent states of the world external to a system or agent under consideration. This system possesses a Markov blanket, which separates internal from active paths and can itself be divided into sensory and active paths. We further assume that internal paths consist of two distinct components: beliefs and preferences, which mutually influence each other, are influenced by sensory paths, and in turn influence active paths.

agents have *preferences about their own beliefs*, and develop a mathematical framework for describing the ensuing implications for belief updating. In other words, we suggest that the C matrix, which is used in the active inference literature to parameterise the preference prior [10,23,42], can be extended to be defined over the agent's own beliefs as well, rather than only observations/states.

For the purposes of this study, we focus on the mechanisms that drive belief change in agents. In particular, we are interested in the mapping from sensory states to belief states. We investigate the consequences of assuming that this mapping is comprised of two key components. The first component is a preference satisfaction component, represented here by an"expected utility" term, which can be related to prior preferences through a softmax transformation [11]. The second component is a direct cost for belief updating, which is quantified by the KL divergence from the agent's prior beliefs to their posterior beliefs.

We will further assume that agents' preferences are grounded only in particular paths, and not directly on external paths, which is in concordance with an affect-driven view on motivation [49,52]. Under these assumptions, an agent's preferences can be described mathematically by a *utility functional* $\mathcal{U} : \mathcal{Q} \times \mathcal{O} \to \mathbb{R}$ defined over observations and *beliefs*, but not external states (Fig. 2).

Given this, we model an agent's belief updating processes as a variational optimisation process that maximises the following functional of beliefs and observations:

$$\mathcal{F}[q(s), o] = \underbrace{\mathcal{U}[q(s), o]}_{\text{belief utility}} - \lambda \underbrace{D_{\mathrm{KL}}\left[q(s) \mid\mid p(s)\right]}_{\text{complexity}}, \tag{4}$$

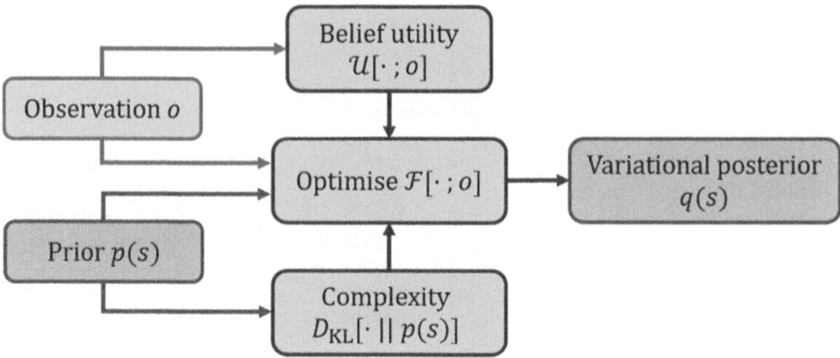

Fig. 2. Schematic depicting the components involved in our proposed model of belief updating. Prior beliefs and observations act as inputs to a process that optimises a balance between a belief utility functional and a complexity term, measuring the KL divergence from prior to the ultimate posterior beliefs.

where $\lambda \geq 0$ is a parameter determining the relative strength of the cost of belief updating to the belief utility term. The belief utility term may or may not depend on the accuracy of the model, and depending on its form, can give rise to different belief updating behaviours. Under this model, taking $\lambda \to 0$ induces a belief update that is purely driven by belief utility, which is akin to assuming that the agent is able to instantaneously and effortlessly convince themselves of whatever they wish to believe. On the other hand, taking $\lambda \to \infty$ increases the cost of updating to the point that the agent is no longer able to change their mind, under any circumstances.

Importantly, one particular form for the belief utility that we investigate is a linear combination of what we term an *affective utility* and a weighted expected log-likelihood or *accuracy* term, which takes the following form:

$$\mathcal{U}[q(s), o] = \underbrace{U[q(s), o]}_{\text{affective utility}} + \alpha \underbrace{\mathbb{E}_{q(s)}[\log p(o|s)]}_{\text{accuracy}}, \tag{5}$$

where $\alpha \geq 0$ is a *likelihood weighting* parameter, which determines the extent to which the agent's final belief distribution explains the data it has observed. A higher value of α can be interpreted as a stronger desire to arrive at beliefs that explain the observed data well. Moreover, for constant affective utility functions and $\alpha = \lambda = 1$, we recover the VFE as a special case of "accuracy-motivated" belief updating. In the following section, we study the predictions made by adopting the belief utility functional given in Eq. 5.

4 Experiments and Results

To study the implications of our proposed model on how motivated agents update their beliefs, we conducted a series of minimal experiments using categorical distributions[1]. In all simulations, we consider how a single piece of evidence presented in the form of a likelihood distribution may be selected and subsequently influence the belief updating process. In particular, we demonstrate that under our model, several key features of human belief updating are qualitatively recovered. Moreover, our model can serve as a framework to generate testable predictions and simulations of human behaviour in various scenarios.

4.1 How Do Different Agents React to Differing Degrees of Good vs Bad News?

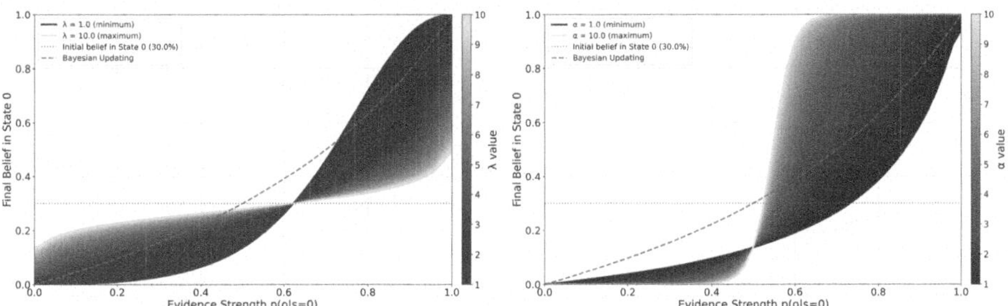

Fig. 3. Plots depicting how agents with different conservatism parameters λ and likelihood weight parameters α respond to evidence that confirms or contradicts their belief preferences to varying degrees. Left: Final belief of the probability $q(s = 0)$ of state 0 occurring as the evidence strength (in the form of a likelihood $p(o|s = 0)$) varies from 0 to 1 for different values of λ. Right: Final belief $q(s = 0)$ as evidence strength varies for different values of α.

In our first set of experiments, we aim to understand and illustrate the effects of varying the strength of evidence, the conservatism parameter λ and the likelihood weight parameter α on belief updating. In this scenario, an agent begins with an initial prior over the outcomes of a Bernoulli random variable (i.e., a biased coin flip) specified by $p(s = 0) = 0.3$, and receives evidence of varying strengths in the form of a likelihood $p(o \mid s)$. For Bernoulli random variables, we will assume that the hidden state s may take the values 0 or 1. The agent updates their beliefs to minimise the objective in Eq. 4 under the belief utility given in Eq. 5.

In Fig. 3, we plot the final belief in state 0 as we vary the evidence strength $p(o \mid s = 0)$ between 0 and 1 along the x axis and the values of λ (left) and α (right) as a spectrum for $\lambda \in [1, 10]$ and $\alpha \in [1, 10]$, along with the Bayesian update. From the left plot, we observe that higher values of λ lead to updates that are closer to the prior, whereas lower values of λ lead to updates that are

[1] The code for generating the experimental results can be found at https://github.com/dkhyland/motivated-variational-belief-updating.

more sensitive to the affective utility. From the right plot, the opposite effect is observed – higher likelihood weights lead to more sensitivity to the evidence, and lower likelihood weights increase sensitivity to the affective utility.

4.2 How Do the Relative Strengths of Belief Utility and Conservatism Affect the Selection of Evidence?

In this study, we demonstrate the presence of a form of confirmation bias in our model, and seek to understand how different components of the model affect the selection of evidence in our motivated agent. In particular, recall that a crucial tenet within active inference is that agents are active sense-makers, selecting evidence to resolve uncertainty in both specific and non-specific manners in order to develop a better model of the world and achieve their ultimate objectives. In this experiment, we extend this notion to include the motivated selection of evidence to either confirm or disconfirm an agent's preferences.

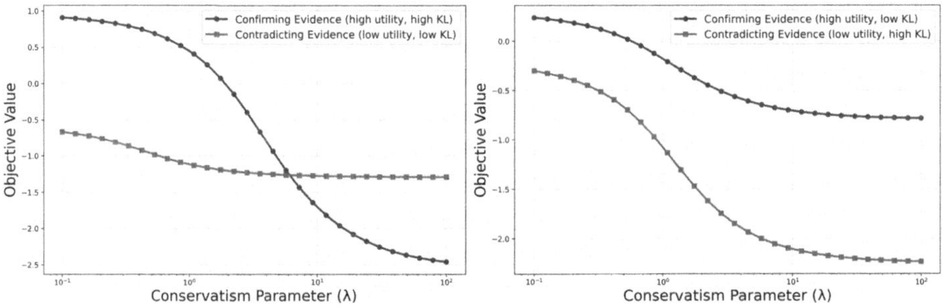

Fig. 4. Plots of the objective value for Scenarios 1 and 2 with different combinations of evidence. In both scenarios, we fix $\alpha = 2.0$ and use a linear affective utility functional with $U[q(s), o] = q(s = 0)$. Left: Scenario 1, where Evidence A has high utility but high KL from the prior and Evidence B has low utility but low KL from the prior. Right: Scenario 2, where Evidence A has high utility and low KL, whereas Evidence B has low utility and high KL.

We consider what happens when an agent with a linear affective utility who prefers to believe that $p(s = 0) = 1$ is presented with two pieces of evidence. We studied two different scenarios for what these pieces of evidence may be. In Scenario 1, the first piece of evidence (Evidence A) is *'confirmatory'*, in the sense that it provides evidence for the agent's desired belief, but is further (induces updates with a larger KL divergence) from the agent's prior compared to the second piece of evidence (Evidence B). Evidence B is *'contradictory'*, in the sense that it is evidence against the agent's desired belief but is closer to the agent's prior. In Scenario 2, Evidence A has both a higher affective utility and induces updates with a lower KL from the prior to the posterior. In Fig. 4, the objective value is plotted as we sweep across values of $\lambda \in [0.1, 100]$. In

Scenario 1, we observe a threshold at which the agent switches from selecting confirmatory evidence to selecting contradictory evidence, whereas this does not occur in Scenario 2. Intuitively, this is because for low values of λ, the utility term dominates belief updating, but for higher values of λ, the cognitive cost term dominates. In contrast, when both the utility component is higher and cognitive costs are lower for one piece of evidence over the other, there is never a reason for the agent to choose to observe disconfirmatory evidence.

4.3 How Do Belief Conservatism and Likelihood Weighting Affect Attitude Polarisation?

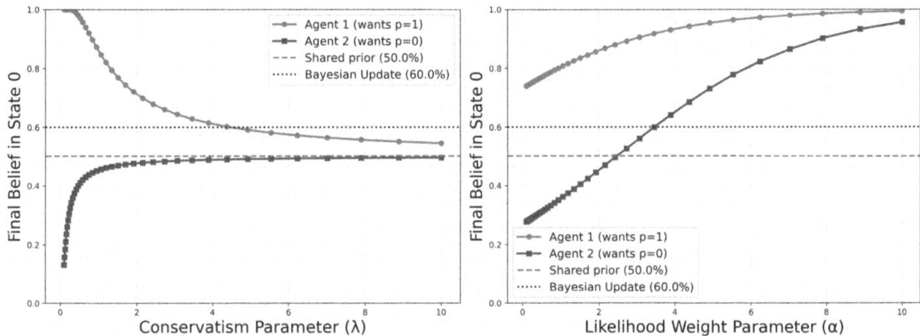

Fig. 5. Plots of attitude polarisation effects between two agents who begin with the same prior beliefs and observe the same evidence, but have different affective utilities. Agent 1 linearly prefers to believe that $q(s = 0) = 1$, whereas Agent 2 linearly prefers to believe that $q(s = 0) = 0$. Left: Final beliefs as we vary λ from 0 to 10. Right: Final beliefs as we vary α from 0 to 10.

In our final experiment, we simulated a basic attitude polarisation scenario, where two agents, Agents 1 and 2, began with the same prior belief about the probability $q(s = 0)$, and observed the same evidence in the form of a likelihood. Both agents were endowed with linear affective utility functions, but Agent 1 had a preference for believing that $q(s = 0) = 1$ and Agent 2 had a preference for believing that $q(s = 0) = 0$. Plotting the final beliefs after updating according to our model as we varied λ and α independently in Fig. 5, we observe that for low values of both parameters, the two agents' final beliefs differed significantly, demonstrating a basic form of attitude polarisation. However, as we increase the two parameters, the agents' final beliefs converged toward similar (though not necessarily Bayes rational) beliefs.

5 Discussion

Though much work needs to be done in empirically validating instantiations of the framework, our findings lend plausibility to the idea that realistic belief updating is subject to significant inertia and bias, driven largely by the constraints imposed on human agents by both internal cognitive limitations and external structures.

Realistic belief revision is rarely drastic, especially in the presence of cognitive costs for updating. From a cognitive standpoint, rapid updates necessitate more complex neural rewiring and a higher cognitive load, which can overwhelm limited cognitive resources. Agents would tend to avoid such costly leaps. Incremental steps across belief space reduce immediate costs but may also cumulatively result in lower total energetic and informational expenditure.

Under this view, we hypothesise that gradual transitions are typically more sustainable and preferable overall. Such considerations could explain why individuals are naturally inclined to resist abrupt changes in their belief systems despite potentially strong contradictory evidence, reinforcing conservative patterns of information integration.

Strategies for Effective Belief Updating. Our basic model suggests several strategic insights for promoting more effective belief updating. Given the high cost of large leaps in belief space, strategies should prioritise incrementalism. This involves structuring information exposure in manageable segments that progressively lead individuals towards desired beliefs, thereby reducing the energetic, cognitive, and social resistance to dramatic changes. Social networks should be leveraged strategically: encouraging cross-cutting social ties and diversity in informational environments can reduce the perceived social risks associated with belief change.

5.1 Future Directions

In considering future avenues for research based on our current findings, several promising directions are worth exploring.

Completing the Action-Perception Loop. So far, our model has focused on the processes involved in the updating of beliefs, taking into account sensory evidence. Our preliminary data selection investigation takes this a step further by demonstrating how decisions about what data to observe can influence decision-making. However, further work is required to fully integrate motivation into the perception-action loop.

Addition of Temporal Considerations. So far, our model has not explicitly incorporated the temporal aspect of belief updating, which we believe to be significant in modelling the various costs that must be taken into consideration. Indeed, several works have posited a central role of *rates of change* in free energy/prediction errors as crucial to understanding affect [16,25]. Extending the model to account for the role of time would allow a more detailed analysis of how belief trajectories could be optimised, rather than single updates.

Extensions to Group Dynamics. Further work could more explicitly incorporate group dynamics, particularly focusing on how social networks influence belief inertia and revision costs. Future work could explore the degree to which group identity and perceived social costs shape belief stability, potentially replicating frameworks similar to those presented by [2] on epistemic communities. By examining how belief updates propagate through structured networks and assessing how identity-protective reasoning reinforces certain belief states, we can quantify the inertia inherent within closely knit communities. Moreover, evaluating the relative weight of belief confidence levels and their susceptibility to drift could provide deeper insights into the dynamics of belief evolution in social contexts. Such extensions may also clarify how networked beliefs reinforce each other, creating feedback loops that stabilise misinformation.

Further Empirical Validation. Empirical validation remains essential for confirming and refining our theoretical propositions. Future empirical work will rigorously test model predictions using controlled laboratory experiments, field studies, and simulation analyses. For instance, quantifiable predictions derived from our framework—such as the relationship between KL divergence, belief revision speed, and associated cognitive or social costs—could be tested experimentally by monitoring physiological or neural responses during belief updating tasks. Longitudinal field studies examining belief trajectories within real-world social groups could also provide valuable validation, providing insights into how incremental versus rapid belief changes correlate with tangible social and cognitive outcomes.

Acknowledgments. The authors would like to thank Lancelot Da Costa and Tomáš Gavenčiak for helpful discussions and feedback.

A Analytical Solution for Optimal Belief Updates in the Linear Affective Utility Case

In this appendix, we derive the closed–form optimal posterior that minimises the variational objective introduced in 3. Throughout, let S be a finite set of latent states $s \in S$, $q(s)$ the candidate posterior, and $p(s)$ the fixed prior. Observed data are denoted by o with likelihood $p(o \mid s)$. The objective functional to be minimised is

$$\mathcal{F}[q(s), o] := \underbrace{U[q(s), o]}_{\text{affective utility}} + \alpha \underbrace{\mathbb{E}_q\big[\log p(o \mid s)\big]}_{\text{accuracy}} - \lambda \underbrace{D_{\mathrm{KL}}\left[q(s) \,||\, p(s)\right]}_{\text{complexity}}. \quad (6)$$

Here $U[q(s), o]$ represents the *affective* utility of a belief state $q(s)$ and observation o, and α, λ modulate respectively the weight assigned to the epistemic evidence and the inertia (or cost) of deviating from the prior.

For the case of linear affective utilities, we have

$$U[q(s), o] = \sum_{s \in S} c_s \, q(s), \quad (7)$$

with coefficients $c_s \in \mathbb{R}$ capturing the valence of believing state s.

We maximise (6) under the normalisation constraint $\sum_s q(s) = 1$. Introducing a Lagrange multiplier $\eta \in \mathbb{R}$ gives the augmented Lagrangian

$$\mathcal{L}(q, o, \eta) = \sum_{s \in S} c_s\, q(s) + \alpha\, \mathbb{E}_q[\log p(o \mid s)] - \lambda\, D_{\mathrm{KL}}\, [q(s) \;||\; p(s)] + \eta\Big(1 - \sum_s q(s)\Big). \tag{8}$$

Stationarity with respect to each $q(s)$ yields

$$0 = \frac{\partial \mathcal{L}}{\partial q(s)} = c_s \;+\; \alpha\, \log p(o \mid s) \;-\; \lambda\left[1 + \log \frac{q(s)}{p(s)}\right] \;-\; \eta. \tag{9}$$

Solving (9) for $q(s)$ and exponentiating, we obtain

$$q(s) = p(s) \exp\left[\frac{1}{\lambda}\Big(c_s + \alpha\, \log p(o \mid s) - \eta\Big) - 1\right] \tag{10}$$

$$\propto p(s) \exp\Big[\tfrac{1}{\lambda}\big(c_s + \alpha\, \log p(o \mid s)\big)\Big]. \tag{11}$$

Normalising with the partition function

$$Z(o) \;:=\; \sum_{s' \in S} p(s')\, \exp\Big[\tfrac{1}{\lambda}\big(c_{s'} + \alpha\, \log p(o \mid s')\big)\Big], \tag{12}$$

we arrive at the optimal variational posterior

$$q^{\star}(s) = \frac{p(s) \exp\big[\lambda^{-1}\big(c_s + \alpha\, \log p(o \mid s)\big)\big]}{Z(o)}. \tag{13}$$

B Additional Figures

See (Figs. 6 and 7)

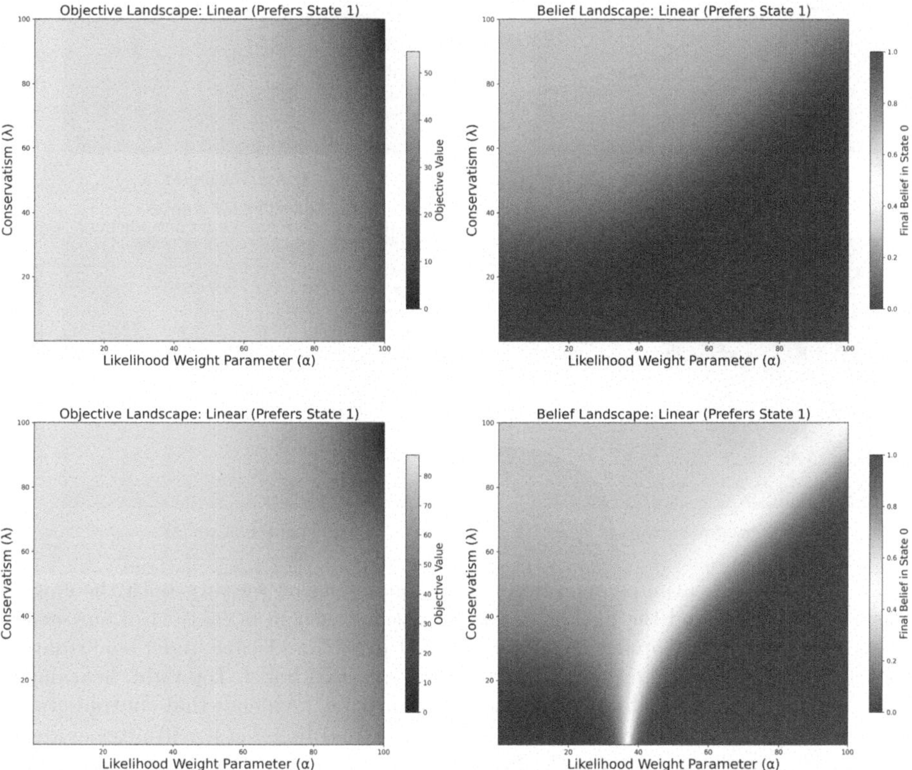

Fig. 6. Heatmaps depicting the variational objective and final belief landscapes for different (λ, α) pairs. The upper two panels depict the objective and belief landscapes (left and right, respectively) for evidence in the form of a likelihood where $p(o|s = 0) = 0.3$, and the bottom two represent the same but for evidence $p(o|s = 0) = 0.7$. For disconfirmatory evidence (top).

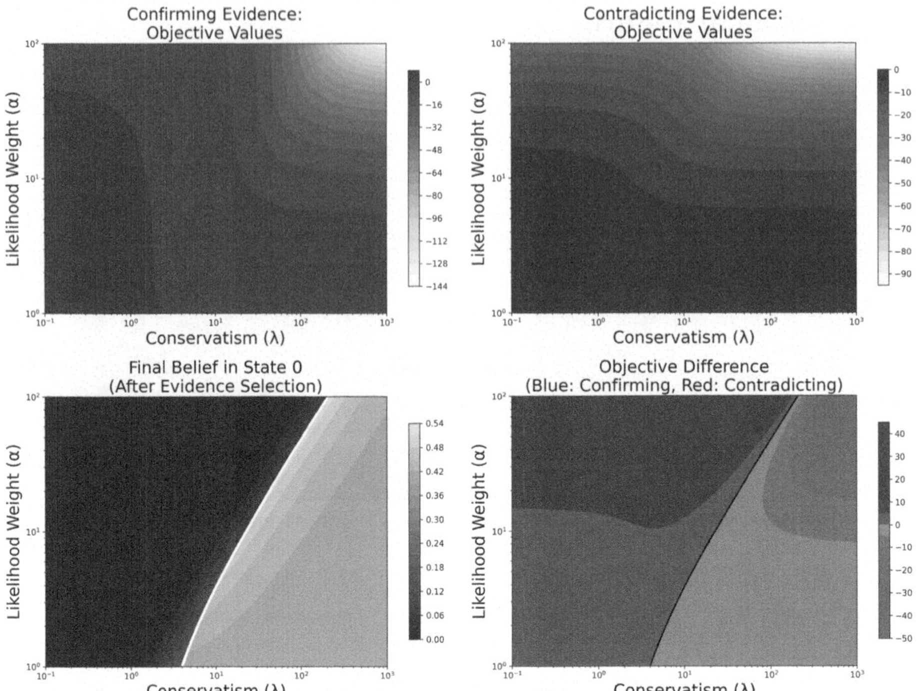

Fig. 7. Contoured heatmaps depicting various quantities as we vary both the conservatism parameter λ and the likelihood weight parameter α in Scenario 1 of our second experiment (Sect. 4.2). Top left: heatmap of the objective landscape under confirmatory evidence, i.e., evidence that aligns with the agent's desired belief. Top right: heatmap of the objective landscape under contradictory evidence, i.e., evidence that contradicts the agent's desired belief. Bottom left: heatmap of the final belief $q(s = 0)$ after evidence selection. The white line depicts the boundary at which the agent selects Evidence A (left of boundary) over Evidence B (right of boundary). Bottom right: heatmap showing the difference in the objectives for confirmatory and contradictory evidence. The black line again depicts the boundary between choosing Evidence A over Evidence B.

References

1. Albarracin, M., Bouchard-Joly, G., Sheikhbahaee, Z., Miller, M., Pitliya, R.J., Poirier, P.: Feeling our place in the world: an active inference account of self-esteem. Neurosci. Consciousness **1**, niae007 (2024)
2. Albarracin, M., Demekas, D., Ramstead, M.J., Heins, C.: Epistemic communities under active inference. Entropy **24**(4), 476 (2022)
3. Andresen, B.: Current trends in finite-time thermodynamics. Angew. Chem. Int. Ed. **50**(12), 2690–2704 (2011)
4. Barlow, H.B., et al.: Possible principles underlying the transformation of sensory messages. Sensory Commun. **1**(01), 217–233 (1961)
5. Bartels, L.M.: beyond the running tally: partisan bias in political perceptions. Polit. Behav. **24**(2), 117–150 (2002)

6. Bicchieri, C., Mercier, H.: Norms and Beliefs: How Change Occurs. Oxford University Press, Oxford (2014)
7. Blei, D.M., Kucukelbir, A., McAuliffe, J.D.: variational inference: a review for statisticians. J. Am. Stat. Assoc. **112**(518), 859–877 (2017)
8. Bouizegarene, N., Ramstead, M.J.D., Constant, A., Friston, K.J., Kirmayer, L.J.: Narrative as active inference: an integrative account of cognitive and social functions in adaptation. Front. Psychol. **15**, 1345480 (2024)
9. Constant, A., Ramstead, M.J.D., Veissière, S.P.L., Friston, K.J.: regimes of expectations: an active inference model of social conformity and human decision making. Front. Psychol. **10**, 679 (2019)
10. Da Costa, L., Parr, T., Sajid, N., Veselic, S., Neacsu, V., Friston, K.: Active inference on discrete state-spaces: a synthesis. J. Math. Psychol. **99**, 102447 (2020)
11. Da Costa, L., Sajid, N., Parr, T., Friston, K., Smith, R.: Reward maximization through discrete active inference. Neural Comput. **35**(5), 807–852 (2023)
12. Da Costa, L., Tenka, S., Zhao, D., Sajid, N.: Active inference as a model of agency. arXiv preprint arXiv:2401.12917 (2024)
13. Deane, G., Mago, J., Fotopoulou, A., Sacchet, M., Carhart-Harris, R., Sandved-Smith, L.: The computational unconscious: adaptive narrative control, psychopathology, and subjective well-being (2024)
14. Ditto, P.H., Pizarro, D.A., Tannenbaum, D.: Motivated moral reasoning. Psychol. Learn. Motiv. **50**, 307–338 (2009)
15. Ecker, U., et al.: The psychological drivers of misinformation belief and its resistance to correction. Nat. Rev. Psychol **1**, 13–29 (2022)
16. Fernandez Velasco, P., Loev, S.: Affective experience in the predictive mind: a review and new integrative account. Synthese **198**(11), 10847–10882 (2021)
17. Fields, C., Goldstein, A., Sandved-Smith, L.: Making the thermodynamic cost of active inference explicit. Entropy **26**(8), 622 (2024)
18. Friston, K.: The free-energy principle: a unified brain theory? Nat. Rev. Neurosci. **11**(2), 127–138 (2010)
19. Friston, K., et al.: The free energy principle made simpler but not too simple. Phys. Rep. **1024**, 1–29 (2023)
20. Friston, K., et al.: Path integrals, particular kinds, and strange things. Phys. Life Rev. (2023)
21. Gershman, S.J., Horvitz, E.J., Tenenbaum, J.B.: Computational rationality: a converging paradigm for intelligence in brains, minds, and machines. Science **349**(6245), 273–278 (2015)
22. Guénin-Carlut, A., Albarracin, M.: On embedded normativity: an active inference account of agency beyond flesh. In: Active Inference, vol. 1630 of Communications in Computer and Information Science. Springer Nature, Cham, Switzerland, pp. 91–105 (2024). https://doi.org/10.1007/978-3-031-47958-8_7
23. Heins, C., et al.: PYMDP: a Python library for active inference in discrete state spaces. arXiv preprint arXiv:2201.03904 (2022)
24. Jain, S.P., Maheswaran, D.: Motivated reasoning: a depth-of-processing perspective. J. Consum. Res. **26**(4), 358–371 (2000)
25. Joffily, M., Coricelli, G.: Emotional valence and the free-energy principle. PLoS Comput. Biol. **9**(6), e1003094 (2013)
26. Jones, M., Love, B.C.: Pinning down the theoretical commitments of Bayesian cognitive models. Behav. Brain Sci. **34**(4), 215–231 (2011)
27. Kiverstein, J., Miller, M., Rietveld, E.: Desire and motivation in predictive processing: an ecological-enactive perspective. Rev. Philosop. Psychol., 1–21 (2024)

28. Kruglanski, A.W., Jasko, K., Friston, K.: All thinking is 'wishful' thinking. Trends Cogn. Sci. **24**(6), 413–424 (2020)
29. Kunda, Z.: The case for motivated reasoning. Psychol. Bull. **108**(3), 480 (1990)
30. Lewis, R.L., Howes, A., Singh, S.: Computational rationality: linking mechanism and behavior through bounded utility maximization. Top. Cogn. Sci. **6**(2), 279–311 (2014)
31. Lieder, F., Griffiths, T.L.: Resource-rational analysis: understanding human cognition as the optimal use of limited computational resources. Behav. Brain Sci. **43**, e1 (2020)
32. Little, A.T.: How to distinguish motivated reasoning from Bayesian updating. Political Behavior, pp. 1–25 (2025)
33. Lord, C., Ross, L., Lepper, M.: Biased assimilation and attitude polarization: the effects of prior theories on subsequently considered evidence. J. Person. Soc. Psychol. **37**, 2098–2109 (1979)
34. Mandelbaum, E.: Troubles with bayesianism: an introduction to the psychological immune system. Mind Lang. **34**(2), 141–157 (2019)
35. Mercier, H., Sperber, D.: The Enigma of Reason. Harvard University Press (2017)
36. Nickerson, R.S.: Confirmation bias: a ubiquitous phenomenon in many guises. Rev. Gen. Psychol. **2**(2), 175–220 (1998)
37. Oeberst, A., Imhoff, R.: Toward parsimony in bias research: a proposed common framework of belief-consistent information processing for a set of biases. Perspect. Psychol. Sci. **18**(6), 1464–1487 (2023)
38. Oeberst, A., Mischkowski, D., Imhoff, R.: Belief-consistent information processing or coherence-based reasoning: integrating two parsimonious frameworks for biases. Eur. J. Soc. Psychol. (2025)
39. Ortega, P.A., Braun, D.A.: Thermodynamics as a theory of decision-making with information-processing costs. Proc. Royal Soc. A: Math. Phys. Eng. Sci. **469**(2153), 20120683 (2013)
40. Ortega, P.A., Braun, D.A., Dyer, J., Kim, K.-E., Tishby, N.: Information-theoretic bounded rationality. arXiv preprint arXiv:1512.06789 (2015)
41. Parr, T., Holmes, E., Friston, K.J., Pezzulo, G.: Cognitive effort and active inference. Neuropsychologia **184**, 108562 (2023)
42. Parr, T., Pezzulo, G., Friston, K.J.: Active Inference: The Free Energy Principle in Mind, Brain, and Behavior. MIT Press (2022)
43. Parrondo, J.M., Horowitz, J.M., Sagawa, T.: Thermodynamics of information. Nat. Phys. **11**(2), 131–139 (2015)
44. Patterson, R., Operskalski, J.T., Barbey, A.K.: Motivated explanation. Front. Hum. Neurosci. **9**, 559 (2015)
45. Pilgrim, C., Sanborn, A., Malthouse, E., Hills, T.T.: Confirmation bias emerges from an approximation to Bayesian reasoning. Cognition **245**, 105693 (2024)
46. Prat-Ortega, G., de la Rocha, J.: Selective attention: a plausible mechanism underlying confirmation bias. Curr. Biol. **28**(19), R1151–R1154 (2018)
47. Priniski, J.H., Solanki, P., Horne, Z.: A Bayesian decision-theoretic framework for studying motivated reasoning (2022)
48. Salamon, P., Andresen, B., Nulton, J., Roach, T.N., Rohwer, F.: More stages decrease dissipation in irreversible step processes. Entropy **25**(3), 539 (2023)
49. Sennesh, E., Ramstead, M.: An affective-taxis hypothesis for alignment and interpretability. arXiv preprint arXiv:2505.17024 (2025)
50. Sharot, T., Garrett, N.: Forming beliefs: why valence matters. Trends Cogn. Sci. **20**(1), 25–33 (2016)

51. Sharot, T., Rollwage, M., Sunstein, C.R., Fleming, S.M.: Why and when beliefs change. Perspect. Psychol. Sci. **18**(1), 142–151 (2023)
52. Shenhav, A.: The affective gradient hypothesis: an affect-centered account of motivated behavior. Trends Cognit. Sci. (2024)
53. Simon, D., Read, S.J.: Toward a general framework of biased reasoning: coherence-based reasoning. Perspect. Psychol. Sci. **20**(3), 421–459 (2025)
54. Simon, H.A.: Theories of bounded rationality. In: Decision and Organization, pp. 161–176 (1964)
55. Spelman, T., Elnakouri, A., Kteily, N., Finkel, E.J.: Overestimating the social costs of political belief change. J. Exp. Soc. Psychol. **105**, 104115 (2023)
56. Talluri, B.C., Urai, A.E., Tsetsos, K., Usher, M., Donner, T.H.: Confirmation bias through selective overweighting of choice-consistent evidence. Curr. Biol. **28**(19), 3128–3135 (2018)
57. Vasil, J., Badcock, P.B., Constant, A., Friston, K.J., Ramstead, M.J.D.: A world unto itself: human communication as active inference. Front. Psychol. **11**, 417 (2020)
58. Veissière, S.P.L., Constant, A., Ramstead, M.J.D., Friston, K.J., Kirmayer, L.J.: Thinking through other minds: a variational approach to cognition and culture. Behav. Brain Sci. **43**, e90 (2020)
59. Wainwright, M.J., Jordan, M.I., et al.: Graphical models, exponential families, and variational inference. Found. Trends® Mach. Learn. **1**(1–2), 1–305 (2008)
60. Westerwick, A., Kleinman, S.B., Knobloch-Westerwick, S.: Confirmation bias in online searches: impacts of selective exposure before an election on political attitude strength and shifts. J. Commun. **67**(4), 660–684 (2017)
61. Williams, D.: Socially adaptive belief. Philos. Stud. **178**(3), 785–804 (2021)
62. Wolpert, D.H.: The free energy requirements of biological organisms; implications for evolution. Entropy **18**(4), 138 (2016)
63. Wolpert, D.H., Korbel, J., Lynn, C.W., Tasnim, F., Grochow, J.A., Kardeş, G., Aimone, J.B., Balasubramanian, V., De Giuli, E., Doty, D., et al.: Is stochastic thermodynamics the key to understanding the energy costs of computation? Proc. Natl. Acad. Sci. **121**(45), e2321112121 (2024)
64. Zhu, J.-Q., Sanborn, A., Chater, N., Griffiths, T.: Computation-limited Bayesian updating. In: Proceedings of the Annual Meeting of the Cognitive Science Society, vol. 45 (2023)

Cognitive Effort in the Two-Step Task: An Active Inference Drift-Diffusion Model Approach

Álvaro Garrido-Pérez[1](✉), Viktor Lemoine[1], Amrapali Pednekar[1], Yara Khaluf[2], and Pieter Simoens[1]

[1] IDLab, Department of Information Technology, Ghent University - imec, Gent, Belgium
alvaro.garridoperez@ugent.be

[2] Wageningen University and Research, Wageningen, The Netherlands

Abstract. High-level theories rooted in the Bayesian Brain Hypothesis often frame cognitive effort as the cost of resolving the conflict between habits and optimal policies. In parallel, evidence accumulator models (EAMs) provide a mechanistic account of how effort arises from competition between the subjective values of available options. Although EAMs have been combined with frameworks like Reinforcement Learning to bridge the gap between high-level theories and process-level mechanisms, relatively less attention has been paid to their implications for a unified notion of cognitive effort. Here, we combine Active Inference (AIF) with the Drift-Diffusion Model (DDM) to investigate whether the resulting AIF-DDM can simultaneously capture the effort arising from both habit violation and value discriminability. To our knowledge, this is the first time AIF has been combined with an EAM. We tested the AIF-DDM on a behavioral dataset from the two-step task and compared its predictions to an information-theoretic definition of cognitive effort based on AIF. The model's predictions successfully accounted for second-stage reaction times but failed to capture the dynamics of the first stage. We argue the latter discrepancy likely stems from the experimental design rather than a fundamental flaw in the model's assumptions about cognitive effort. Accordingly, we propose several modifications of the two-step task to better measure and isolate cognitive effort. Finally, we found that integrating the DDM significantly improved parameter recovery, which could help future studies to obtain more reliable parameter estimates.

Keywords: Active inference · Drift-diffusion model · Cognitive effort

1 Introduction

Building upon the Bayesian Brain Hypothesis (BBH), numerous studies in the past decade have tried to formalize cognitive effort using information-theoretic principles [25]. According to the BBH, humans maintain an internal world model

M. Albarracin et al. (Eds.): IWAI 2025, CCIS 2857, pp. 24–44, 2026.
https://doi.org/10.1007/978-3-032-16955-6_2

encoded in prior beliefs, which they continuously update as they interact with the environment. Within this context, cognitive effort arises from the conflict between a pre-existing belief about how to act (a habit) and an updated belief about the optimal policy [10].

In a decision-making task, cognitive effort may also arise from the competition between the subjective values of the available choices. When these values are closer together—that is, when value discriminability is low—reaction times (RTs) tend to increase (e.g., [3,7]). Although RTs are not a direct measure of cognitive effort, they are often used as a proxy (e.g., [24]), based on the assumption that slower responses reflect more information processing.

From an information-theoretic perspective, the effect of value discriminability on RTs can be understood as the additional cognitive effort required to resolve increased choice uncertainty [6,8,13]. Yet, information theory provides no explicit account of the underlying deliberation process, which complicates the task of linking its predictions to specific neural signatures. In contrast, evidence accumulator models (EAMs) offer a mechanistic explanation for how value competition shapes decision speed, which may be empirically tested (e.g., [18]).

A major limitation of using EAMs to study cognitive effort is that they are agnostic to how beliefs are formed. This limitation is often addressed by combining EAMs with Reinforcement Learning (RL) [11]. However, in recent years, Active Inference (AIF) has emerged as a powerful alternative to RL, offering a first-principles perspective on perception, learning, and decision-making [5].

Here, we investigate whether integrating a Drift Diffusion Model (DDM), a prominent class of EAM, with AIF can simultaneously capture the influence of both value discriminability and habit violation on cognitive effort. To our knowledge, this is the first attempt to combine AIF with an EAM to model human behaviour. We evaluate the integrated AIF-DDM using the two-step task, a version of the multi-armed bandit in which participants must plan two steps ahead [2]. Furthermore, we compare its predictions with a recently proposed definition of cognitive effort in AIF [10].

Our work builds directly on a recent study that developed an AIF model of the two-step task [4]. The study demonstrated that AIF outperformed a Hybrid Reinforcement Learning (HRL) model (which combines model-free and model-based strategies [16]) in two out of four datasets, while achieving comparable performance in the remaining two. Moreover, the authors provided compelling evidence for directed exploration—a key differentiator between AIF and HRL. Despite these achievements, the study could not determine which specific AIF learning mechanisms participants were using. Given prior evidence that integrating a DDM into an HRL model improved the reliability of model-based estimates [16], we tested whether the combined AIF–DDM could resolve this ambiguity.

2 Methods

2.1 Participants and Behavioural Task

In this section, we will briefly describe the two-step task and the behavioural dataset that we used to fit the computational models. For more details on the experimental procedure, we encourage reading the paper that made the dataset publicly available [17].

As the name suggests, each trial of the two-step task consists of two stages (Fig. 1). In the first stage, participants choose between two actions. Each action leads to one of the two second-stage states through a probabilistic transition that is either common ($p = 0.7$) or rare ($p = 0.3$). Importantly, the two first-stage actions have opposite most-likely transitions. These transition probabilities remain constant throughout the task. After transitioning to a second-stage state, participants make a final choice between two actions. Each of these second-stage actions results in a monetary reward or no reward, depending on its current *outcome* probability. In contrast to the fixed transitions, these outcome probabilities fluctuate independently over time, following Gaussian random walks.

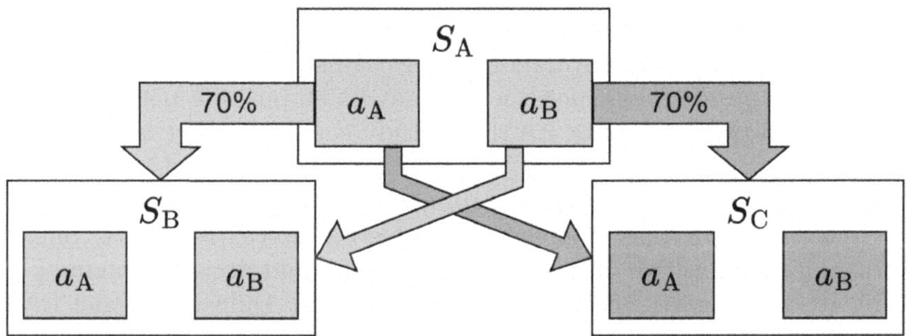

Fig. 1. Abstract representation of the two-step task. At the first stage (top), a choice leads to one of two second-stage states (bottom). Transitions are either common (p=0.7, thick arrows) or rare (p=0.3). The two initial actions have opposing common transitions. A second-stage choice may result in a monetary reward with a probability that fluctuates over time.

We analysed data from the "Magic Carpet" dataset, which was made available by [17] and originally comprised 24 participants. In this experiment, participants had a 2-second deadline to respond at each stage. Every time this deadline was surpassed, the trial was labelled as a *missed* trial and the participant moved on to the next stage or trial. Before conducting any analysis, we identified and removed these trials. We also considered invalid trials those with at least one RT smaller than 100 ms (to exclude anticipatory responses that are too fast to reflect genuine deliberation [9]). Finally, any participant with more than 10% invalid trials was removed from the dataset. One participant was excluded under this criterion, with 44.1% of their trials being either missed or too fast. Therefore, the original sample size was reduced to n=23. In total, 6.53% of the data was excluded from the analysis (including all the data from the removed participant).

2.2 An Active Inference Drift-Diffusion Model of the Two-Step Task

In this section, we introduce the AIF model developed by [4] and discuss how we integrate it with a DDM. Further explanations of the equations are provided in Appendix A. To establish a comparative benchmark, we implemented an HRL-DDM. Since the latter model is secondary to our main analysis, its mathematical formulation is only detailed in Appendix B.

In what follows, we will use the same notation as in [4]. For a given trial t, the first-stage state and chosen actions are denoted by $s_{1,t}$ and $a_{1,t}$, respectively. Likewise, the second-stage state and chosen action are denoted by $s_{2,t}$ and $a_{2,t}$, respectively. Note that the first-stage state can only be s_{A}, but the second-stage state may be s_{B} or s_{C}, depending on the transition (see Fig. 1). Outcomes $o_t \in \{0, 1\}$ represent the observed reward (or absence of it) after choosing a second-stage action.

According to the AIF model developed by [4], an agent performing the two-step task is equipped with a generative model (i.e., a set of beliefs about how the task environment evolves given its actions). In the two-step task, agents must learn the outcome probabilities of each of the four second-stage actions. We will refer to the set of these four probabilities as θ, and the belief distribution over θ, at a given trial t, $\pi_t(\theta)$. In addition, the agent must learn the four transition probabilities $p(s_{2,t}|s_{1,t}, a_{1,t})$. However, as pointed out by [4], these transitions are accurately learned in a few trials and therefore, action-selection is only sensitive to information regarding outcome probabilities.

Thus, an agent performing the two-step task must optimally balance getting as many rewards as possible (exploitation) and learning the four outcome probabilities (exploration). According to AIF, this balance is achieved by selecting actions that minimize a quantity known as Expected Free Energy (EFE), which in this case is given by the following equation:

$$G_t(a) = \underbrace{-\mathbb{E}_{p(o_t;\pi_t(\theta)|a)}[\ln p(o_t|C)]}_{\substack{\text{Realising preferences} \\ \text{(Exploitation)}}} - \underbrace{\mathbb{E}_{p(o_t;\pi_t(\theta)|a)}[D_{\mathrm{KL}}(\pi_t(\theta)|o_t, a\|\pi_t(\theta))]}_{\substack{\text{Model parameter exploration} \\ \text{(Active Learning)}}} \tag{1}$$

where $G_t(a)$ is the EFE of a given action at a given trial, $p(o_t|C)$ is the distribution over prior preferred observations (Eq. 5, Appendix A), which depends on a free parameter λ. Note that in the two-step task, the *preferred* observation is to get the reward after a second-stage choice. D_{KL} is the Kullback-Leibler divergence between prior beliefs about second-stage outcome probabilities $\pi_t(\theta)$ and posterior beliefs after selecting an action and observing its outcome. $p(o_t; \pi_t(\theta)|a)$ is a distribution equal to $p(o_t|\theta, a)\pi_t(\theta)$. Note that, unlike traditional EFE formulations, there is no *hidden state exploration* term because in the real experiment participants can always see the state in which they are [15].

Equation 1 can be used to calculate EFEs for second-stage actions. However, for first-stage actions, the level of optimality depends upon the EFEs of the final states and the transition probabilities. Therefore, for the first-stage actions, the EFE equation is given by:

$$G(a_j) = p(s_B|s_A, a_j) \sum_{a_2 \in \mathcal{A}_B} G(a_2) \; + \; p(s_C|s_A, a_j) \sum_{a_2 \in \mathcal{A}_C} G(a_2) \quad (2)$$

where $\mathcal{A}_B$ and $\mathcal{A}_C$ are the sets of available actions in the second-stage states s_B and s_C respectively, and $G(a_2)$ are the second-stage EFEs given by Eq. 1.

Previous studies have reported a tendency for participants to repeat initial-stage actions, regardless of the outcome history [4]. Within AIF, this behaviour is formalised through a habit term E. In the model of [4], E is parameterised by κ, which modulates the extent of an agent's reliance on habits (see Eq. 6, Appendix A).

At the first stage, an agent will select an action that minimizes the EFE corrected by the habitual bias $G_t^{\text{net}}(s_1, a_1) = G_t(s_1, a_1) + E(a_1)$. At the second stage, no habitual bias is assumed, and therefore $G_t^{\text{net}}(s_2, a_2) = G_t(s_2, a_2)$.

For the original AIF model [4], the choice probabilities of an agent at trial t, stage $p = \{1, 2\}$ and state s, will be given by the softmax distribution of the corresponding net EFEs parametrized by an inverse temperature parameter, γ_p (one for each stage). However, for the AIF-DDM the choice will be determined by a drift-diffusion process, with a non-decision time t_{nd}, boundary separation a_{bs} and a drift rate $v_{p,s,t}$, that depends on the difference in G^{net} of the two available actions such that:

$$v_{p,s,t} = v_p^{\text{mod}} [G_t^{\text{net}}(s, a) - G_t^{\text{net}}(s, a')] \,, \quad a, a' \in \mathcal{A}_s \quad (3)$$

where $\mathcal{A}_s$ is the set of available actions at state s, and v_p^{mod} is a free parameter that regulates the sensitivity to G^{net} differences (which can also be interpreted as the agent's information-processing speed). Following the original AIF model [4], we fit a separate v_p^{mod} parameter for each stage, an approach analogous to the use of stage-specific inverse temperatures.

In the classical DDM, a starting-point bias parameter z is often included. However, in the two-step task, the symbols representing the different actions were randomly displayed either on the left or the right of the screen [17]. Therefore, we set $z = 0.5$ (effectively cancelling the bias in the drift-diffusion process).

After completing every trial, the agent updates its beliefs about the second-stage outcome probabilities. In [4], the updating rules for the prior distribution over outcome probabilities (Eqs. 8 and 9 in Appendix A), depend on four free parameters: the learning rate l, the prior volatility, v_{PS}, which modulates the influence of surprise on the agent's beliefs, volatility of sampled actions v_{SD}, which modulates how beliefs over chosen actions *decay*, or are forgotten, and the volatility of unsampled actions v_{UD}, similar to v_{SD} but for unchosen actions.

The study by [3] tested four AIF variants distinguished by their learning rules: a No Unsampled-Decay model (AIF_{NUD}, $v_{\text{UD}} = 0$); a No Sampled-Decay model (AIF_{NSD}, $v_{\text{SD}} = 0$), a No Predictive Surprise model (AIF_{NPS}, $v_{\text{PS}} = 0$) and a model including all the learning mechanisms (AIF_{FULL}). Although solid evidence was found favouring AIF over HRL, the best-fitting AIF variant could not be determined [4]. Consequently, we fitted all four models in our analysis.

2.3 Cognitive Effort and Active Inference

A recent formalisation of cognitive effort (ξ) based on AIF defines it as the KL divergence between context-sensitive beliefs about how to act $P_G(\pi)$ and context-insensitive prior beliefs, or habits $P_E(\pi)$ [10]:

$$\underbrace{\xi \triangleq D_{\mathrm{KL}}[P_G(\pi)||P_E(\pi)]}_{\text{Effort}} = \underbrace{\mathbb{E}_{P_G}[\ln P_G(\pi)]}_{\text{Context sensitive}} - \underbrace{\mathbb{E}_{P_G}[\ln P_E(\pi)]}_{\text{Context insensitive}} \tag{4}$$

$$P_E(\pi) = \mathrm{Cat}(\sigma(-\mathbf{E})) \qquad P_G(\pi) = \mathrm{Cat}(\sigma(-\mathbf{G} - \mathbf{E}))$$

where $\mathbf{G}$ and $\mathbf{E}$ are vectors comprising the context-sensitive EFEs and the context-insensitive priors (or habits) of the available actions, respectively.

From Eq. 4, two key predictions can be derived. First, high effort will be experienced when there is an incongruence between $\mathbf{G}$ and $\mathbf{E}$. Second, when the elements of $\mathbf{G}$ are of similar magnitude to one another, cognitive effort is minimal regardless of $\mathbf{E}$.

For the first stage of the two-step task, the AIF-DDM aligns with the first prediction of Eq. 4. For a choice between actions a_A and a_B, with $\mathbf{G} = [G_A, G_B]$, and $\mathbf{E} = [E_A, E_B]$, RTs should increase as the magnitude of the drift rate decreases, which is proportional to $|\Delta G^{\mathrm{net}}| = |G_A + E_A - G_B - E_B|$. Therefore, higher RTs may occur when, for example $|G_A| \gg |G_B|$, $|E_A| \ll |E_B|$ and $|G_A| \approx |E_B|$. In other words, when there is an incongruence between $\mathbf{G}$ and $\mathbf{E}$.

However, our model contradicts the second prediction of Eq. 4, because for a constant $\mathbf{E}$, RTs should increase as $|G_A|$ gets closer to $|G_B|$. Interestingly, for the second stage, Eq. 4 predicts minimal effort if we assume no habits (i.e. $E_A = E_B = 0$), in contrast to AIF-DDM, where effort can still be high if the context-sensitive EFEs are closely matched.

2.4 Model Fitting and Comparison Procedures

We followed the same Maximum Likelihood Estimate (MLE) procedure as in [4], to fit the models' free parameters to the behavioural data[1] For each participant and model, we found the parameter set with the highest likelihood using Scipy's 'L-BFGS-B' algorithm [19]. This step was repeated 35 times for each participant, with different (uniformly) randomized initializations for all parameters. After completing all the runs, we selected the parameter set with the maximum likelihood and used its value for model comparison.

For the pure AIF and HRL models, we computed the likelihoods using the choice probability distributions of each model. For AIF-DDM and HRL-DDM

[1] **Code** available at https://github.com/decide-ugent/aif-ddm.

however, the likelihoods were given by the Wiener's First-Passage Time Distribution (WFPT) provided by the HDDM Python package [22]. Further details can be found in Appendix C.

For model comparison, we relied on the Statistical Parametric Mapping (SPM) [21] and Variational Bayesian Analysis (VBA) [1] toolboxes, using MATLAB R2022b. We performed a group-level random-effect Bayesian model selection (BMS) procedure, which requires the log-model evidence (LME) of each model for each participant. We used two LME approximations: the Bayesian Information Criterion (BIC) and the Akaike's Information Criterion (AIC) scores [12] (see Appendix C). Using the BMS procedure, we estimated four model comparison metrics: the Expected Posterior Probability (EPP); the Estimated Model Frequency (EMF), or the proportion of participants best fit by the model; the Exceedance Probability (EP), which quantifies the likelihood that the model is more frequent than competing models across participants; and the Protected Exceedance Probability (PEP), similar to EP but accounting for the possibility that apparent differences in model frequencies arise due to chance [14].

The SPM toolbox allows comparing the performance of model *families* (i.e., models that share a common feature). As in [4], we use this feature to compare the overall AIF performance (considering its four variants) to HRL. Since PEPs are unavailable for family-level inference, we only calculate EPs for these analyses.

2.5 Model and Parameter Recovery Procedures

To perform the parameter recovery analysis, we simulated a dataset for each model using parameter values sampled from uniform distributions with ranges equal to those used to fit the real dataset. Next, we fitted the synthetic dataset and obtained new (recovered) parameters. Finally, we checked if the original and recovered parameters were correlated by calculating Pearson's correlations. This process was repeated 23 times (equal to the number of participants in the real experiment) for each model.

For the model recovery analysis, each model was first used to simulate a full experimental dataset with the same parameter sets as in the parameter recovery analysis. We then fitted all models to each simulated dataset and calculated the proportion of times each model was identified as best-fitting. Results were summarized in a confusion matrix (Fig. 2).

3 Results

3.1 Model Recovery Analysis

Model recovery analysis indicates that AIF-DDM_{FULL} is frequently misclassified as either AIF-DDM_{NSD} or AIF-DDM_{NSD} (Fig. 2). The poor recovery likely stems from the full model encompassing all update rules present in the other two models [4]. The rest of the models had a better but modest recovery (except HRL-DDM, which had a perfect recovery). Thus, combining AIF with a DDM could not help substantially to differentiate between different learning variants as we hypothesised. Overall, both AIC and BIC scores show similar results. Although the AIC score seems slightly more reliable since it could achieve a better recovery for both AIF-DDM$_{\mathrm{FULL}}$ and AIF-DDM$_{\mathrm{NUD}}$.

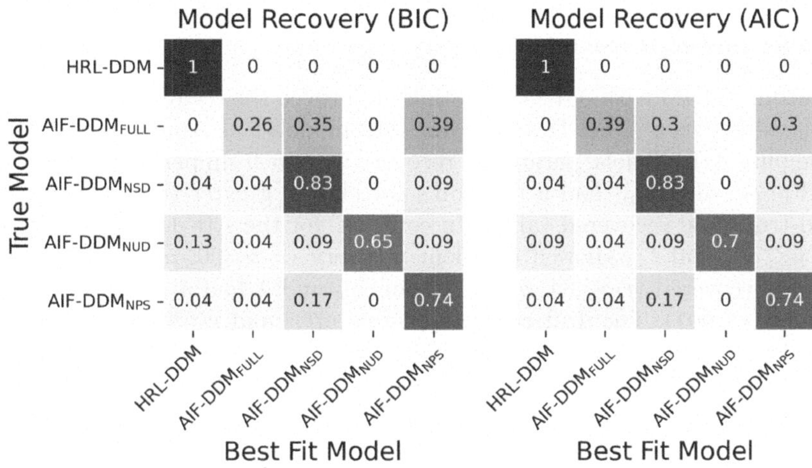

Fig. 2. Model recovery results. Each cell contains the fraction of simulated experiments from a given true model, for which a corresponding model was classified as the best-fitting, according to the BIC score (left) or the AIC score (right).

3.2 Model Comparison Results

The BMS analysis revealed different results depending on the score used. According to the BIC score, AIF-DDM$_{\mathrm{NSD}}$ and AIF-DDM$_{\mathrm{NPS}}$ dominated moderately across participants with Expected Posterior Probabilities, EPP $= [0.39, 0.31]$, Protected Exceedance Probabilities, PEP $= [0.63, 0.29]$ and Estimated Model Frequencies, EMF $= [0.44, 0.35]$, respectively (see Fig. 5, first row). However, according to the AIC scores AIF-DDM$_{\mathrm{FULL}}$ was the best-fitting model with EPP $= 0.57$, PEP $= 0.97$ and EMF $= 0.7$ (see Fig. 5, second row). The disagreement between AIC and BIC scores was also reported for the model comparison between pure HRL and AIF models [4], and likely stems from the fact that BIC penalizes complexity more than AIC.

Even though the two BMS procedures produced conflicting results, we selected AIF-DDM$_{\mathrm{FULL}}$ as the best-fitting model, and focus exclusively on its results in the subsequent sections. This decision was based on the model recovery analysis, which showed that the full model is frequently misclassified as either AIF-DDM$_{\mathrm{NSD}}$ or AIF-DDM$_{\mathrm{NPS}}$. Nonetheless, we acknowledge that the BIC-based comparison favoured AIF-DDM$_{\mathrm{NSD}}$ and AIF-DDM$_{\mathrm{NPS}}$, so we replicated our analysis for these models and show their results in Appendix D. We excluded AIF-DDM$_{\mathrm{NUD}}$ from further analysis as it performed poorly in both

BMS procedures. Finally, the model family selection analysis revealed similar results to those reported in the study that compared pure AIF and HRL [4]. We found that the AIF-DDM family outperformed the HRL-DDM model according to both BIC (EP = 1, EMF = 0.9, EPP = 0.88) and AIC (EP = 1, EMF = 0.92, EPP = 0.9) scores.

3.3 Parameter Recovery Analysis

Integrating a DDM with AIF substantially improved the recovery of the parameters shared between AIF-DDM and pure AIF models (see Appendix E, Table 1). For the pure AIF models, parameter recovery was far from perfect. For example, for AIF_{FULL}, only v_{DU} had a Pearson's correlation greater than .6 between its ground-truth and recovered values. In contrast, for the AIF-DDM models, v_{DU}, v_{DS}, v_{PS}, a_{bs} and t_{nd} showed excellent recovery ($r > .90$, $p < .001$), l and κ were well recovered ($r > .84$, $p < .001$) and p_r and λ had a moderate recovery ($r > .67$, $p < .001$). The drift rate parameters v_1^{mod} and v_2^{mod} had moderate-to-poor recovery, depending on the specific model. Nonetheless, their recovery was still superior to that of the inverse temperature parameters (γ_1 and γ_2) from the pure AIF model.

3.4 Expected Free Energy Discriminability Affects Second Stage, but Not First Stage Reaction Times

To evaluate the models' goodness-of-fit and predictive accuracy, we compared model-simulated behaviour on the two-step task to the observed behaviour of participants. This was done using a decile-binned analysis based on the absolute difference in the net EFE between the two actions, $|\Delta G^{\text{net}}|$. For each AIF-DDM model (except $\text{AIF-DDM}_{\text{NUD}}$), we first computed the trial-by-trial $|\Delta G^{\text{net}}|$ for each participant for both stages, based on their best-fit model parameters, choice history, and the experienced sequence of rewards and transitions. From the resulting distribution of $|\Delta G^{\text{net}}|$ values, for each participant and stage, we identified the decile boundaries (10th-90th percentiles). We then binned each trial's observed RT and choice into one of ten decile bins according to its $|\Delta G^{\text{net}}|$ value. For example, a trial with a $|\Delta G^{\text{net}}|$ value below the 10th percentile was placed in the first bin (0th-10th). Within each bin, we calculated two metrics for each participant: (1) the mean RT, and (2) the probability of choosing the action with the lower net EFE, $P(\text{choose } G_{\text{min}}^{\text{net}})$. These participant-level metrics were then averaged across participants for each decile bin. Finally, for each trial stage, we generated 100 simulations to estimate the model's predicted RT and choice probability. These simulated metrics were then binned and averaged in the same way as the observed data.

The simulation analysis results were qualitatively similar across all models, therefore we only display the predictions of the $\text{AIF-DDM}_{\text{FULL}}$ model as an example in Fig. 3. Equivalent plots for the remaining models are presented in the Appendix F. In both stages, participants were more likely to select the action with the minimum net EFE as the value of $|\Delta G^{\text{net}}|$ increased. This trend was well-captured by all the AIF-DDM models (see Appendix F).

The AIF-DDM models also predict that the mean RT should decrease as a function of EFE discriminability (i.e., as $|\Delta G^{\text{net}}|$ increases) for both stages. This effect can be observed for the second stage; however, it is almost negligible for the first one (Fig. 3, bottom left).

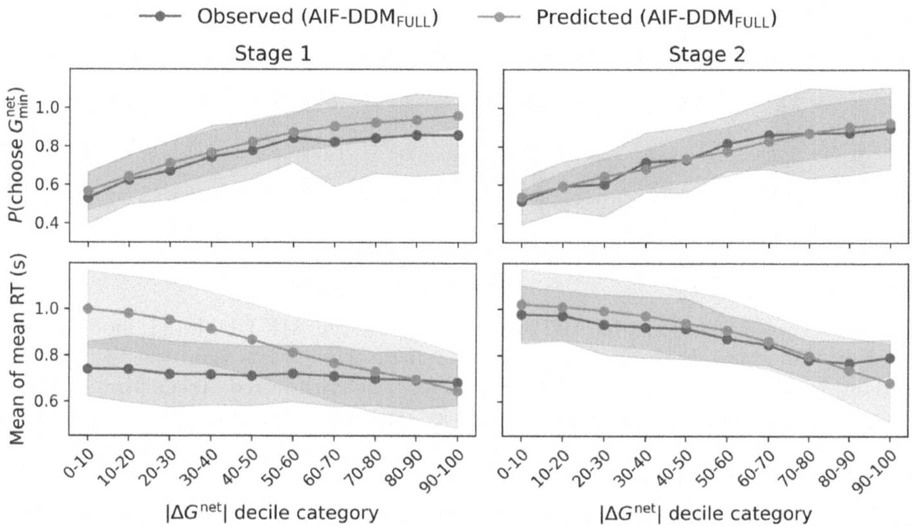

Fig. 3. Comparison of observed (blue) and model-predicted (orange) behaviour. Lines show the mean choice probability (top) and mean reaction time (bottom) across participants, binned by net Expected Free Energy ($|\Delta G^{\text{net}}|$) deciles. Shaded areas are ± 1 standard error of the mean.

Multiple explanations may account for the mismatch in first-stage RTs. The first concerns the 2-second deadline. As described in Sect. 2.1, all missed trials were excluded from the analysis, effectively truncating the observed RT distribution. In contrast, the AIF-DDM models do not impose this time restriction, resulting in higher predicted RT means (see Appendix G for further details). However, this effect alone is unlikely to fully explain the misfit, since the second stage had the same deadline, yet a far less pronounced mismatch.

A second possible explanation stems from the well-documented tendency for participants to repeat their previous first-stage choice [4]. In both AIF(-DDM) and HRL(-DDM), a habit term is included only in the first-stage equations. It is therefore plausible that, for most first-stage choices, the habit strength of the previously selected option is sufficient to produce fast responses—even in trials falling into the lowest $|\Delta G^{\text{net}}|$ deciles. In contrast, with the lack of a habit term in the second-stage equations, $|\Delta G^{\text{net}}|$ values may be smaller or more variable. This effect was particularly notable for the participant with the largest fitted κ

value (where κ regulates the tendency to repeat the previous first-stage choice). However, it was not observed for the rest of the participants (see Fig. 4).

A third potential explanation is that a substantial portion of the cognitive effort in the initial stage may arise from mental simulation or partial exploration of the decision tree, a process not accounted for by the assumption that RTs depend solely on $|\Delta G^{\text{net}}|$. Nevertheless, the limited complexity of the task (involving only four possible branches) constrains the extent of such planning.

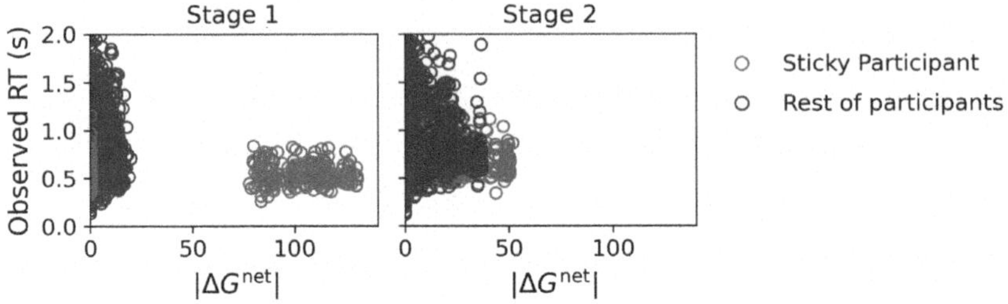

Fig. 4. Observed reaction time (RT) vs. $|\Delta G^{\text{net}}|$ for the AIF-DDM$_{\text{FULL}}$ model, separated by task stage. Each circle represents a single trial. Trials from the 'sticky' participant, identified by the highest fitted κ value, are shown in red. Trials from all other participants are shown in blue. For the sticky participant, notice the cluster of first-stage trials with high $|\Delta G^{\text{net}}|$ values compared to the second-stage trials that are more concentrated at low $|\Delta G^{\text{net}}|$ values.

3.5 The Lack of Net EFE Discriminability Effect on First-Stage Reaction Times Might Be Explained by the Experimental Design

A final potential explanation for the lack of $|\Delta G^{\text{net}}|$ effect on first-stage RTs is the experimental design. In the trial sequence of the two-step task, the elapsed time from the moment participants make the second-stage choice of the previous trial until they are presented with the first-stage options of the next trial is variable and long (3.2–3.8 s). This time interval could allow participants to engage in "pre-thinking" about their upcoming first-stage choice before the options are presented. As a result, the recorded first-stage RT may only capture the final execution of an already made decision, rather than the full deliberation process. In contrast, the interval between the first- and second-stage choices is fixed and shorter (2 s), leaving less room for pre-thinking and making the second-stage RT a more reliable measure of cognitive effort.

The difference in inter-choice intervals could also explain why mean first-stage RTs are shorter than mean second-stage RTs (see Fig. 3), which may seem counterintuitive since, according to AIF, participants in the first stage should engage in planning, a more costly cognitive process than the one-shot second-stage decision.

4 Discussion

In this study, we introduced a novel AIF-DDM and evaluated its predictions in the two-step task. We discussed the difference between the model predictions and a recently developed information-theoretic formalism of cognitive effort based on AIF (Eq. 4, [10]).

Our model predicts that both the need to overcome a pre-existing habit and the difficulty of discriminating between options with similar subjective values (EFEs) increase cognitive effort during the first stage of the two-step task. We argue that for the first stage, both effects can be regarded as a single competition process between the 'habit-corrected' or net EFEs of the available options. For the second stage, habits are assumed to play no role, and therefore, the model predicts that cognitive effort only depends on the discriminability between pure context-sensitive EFEs.

We found empirical evidence that EFE discriminability affects second-stage RTs of the two-step task as predicted by the AIF-DDM model – an effect that is not captured by the information-theoretic formalism. However, we found that the discriminability between the habit-corrected EFEs had a negligible effect on the first-stage RTs, contrary to the AIF-DDM predictions. We argue that the latter observation could be a result of the long inter-trial intervals that may allow participants to "pre-think" their choice before being presented with the options. Thus, first-stage RTs may only reflect motor execution, not the preceding cognitive effort.

Lastly, we found that integrating a DDM with AIF did not help to distinguish between different learning mechanisms proposed by [4]. However, the DDM integration improved parameter recovery compared to pure AIF. The latter result was expected, since the same effect has been reported for HRL in the two-step task [16]. Nonetheless, the result might be relevant for computational psychiatry studies interested in finding the link between certain AIF parameters and specific pathologies.

Our study has multiple limitations. Firstly, we focused on comparing the predictions of AIF-DDM with a single information-theoretic definition of cognitive effort. Our intention, however, is not to disprove this particular formalism, but to illustrate that a more complete theory of cognitive effort should include both the effect of subjective value competition and the violation of prior expectations. Secondly, we focused on studying how cognitive effort may result from the competition between the EFEs and prior beliefs of the different available options. We did not, however, address the costs associated with learning or with *generating* EFEs, which are likely to differ between the two stages. Finally, cognitive effort is a complex phenomenon that may depend upon multiple factors, such as the objective cognitive demand of a task or the participant's information-processing speed. Some of these factors may be captured by different AIF-DDM parameters (e.g., a_{bs} and κ for impulsiveness and motivation, respectively); however, this analysis was beyond the scope of this paper.

A future study could evaluate whether the pre-thinking effect is responsible for the AIF-DDM misfit for the first-stage RTs by reducing the inter-trial intervals. This could be achieved by, for example, reducing the amount of time that the rewards are displayed. Moreover, the problem of ruling out missed trials could be mitigated by increasing the available time to make a choice, using a likelihood function that can take into account the probability of missing a deadline, or introducing collapsing thresholds in the DDM [20]. Implementing these changes could move us closer to a better understanding of cognitive effort in the two-step task.

Acknowledgments. This work was supported by European Union's Horizon 2020 FET research program under grant agreement No. 964464 (ChronoPilot). We gratefully acknowledge the authors of [17] for making their dataset publicly available and the authors of [4] for their open-source code.

Disclosure of Interests. The authors have no competing interests to declare that are relevant to the content of this article.

A Further Details on Active Inference in the Two-Step Task

In [4], the distribution over prior preferred observations is given by the following equation:

$$P(o_t|C) = \frac{1}{Z(\lambda)} e^{o_t \lambda} e^{-(1-o_t)\lambda} \tag{5}$$

where λ is a free parameter that regulates how much an agent prefers to realize prior preferences (in this case, getting a reward) over gaining new information. Thus, λ regulates the exploration-exploitation trade-off.

As mentioned in Sect. 2.2, agents are assumed to tend to repeat first-stage actions independent of the observed outcomes. This behaviour is modelled in AIF by introducing a *habit* term. In [4], this quantity is described by:

$$E(a_j) = \frac{1}{Z(\kappa)} e^{\delta_{a_{t-1},a_j} \kappa_e - (1-\delta_{a_{t-1},a_j})\kappa} \tag{6}$$

where κ is a precision parameter regulating the agent's reliance on habits.

After every trial, the AIF agent updates its beliefs about the second-stage outcome probabilities. Since these probabilities are mutually independent, the overall distribution equals the product of four beta distributions, each describing the believed outcome probability of its respective action and parameterized by its corresponding α and β.

$$\pi_t(\theta) = \prod_{s=1}^{2} \prod_{a=1}^{2} \mathcal{B}e(\alpha_{s,a,t}, \beta_{s,a,t}) \tag{7}$$

According to the learning rule implemented by [4], after observing the outcome of a chosen action, its corresponding α and β parameters are updated following these rules:

$$\alpha_{s,a,t} = (1 - \chi_t)\alpha_{s,a,t-1} + \delta_{a_t,a}o_t l + (1 - \nu_{SD})\alpha_{s,a,t-1} + \nu_{SD}\alpha_0 \tag{8}$$

$$\beta_{s,a,t} = (1 - \chi_t)\beta_{s,a,t-1} + \delta_{a_t,a}(1 - o_t)l + (1 - \nu_{SD})\beta_{s,a,t-1} + \nu_{SD}\beta_0 \tag{9}$$

$$\chi_t = \frac{m\ PS}{1 + m\ PS} \tag{10a}$$

$$m = \frac{\nu_{\mathrm{PS}}}{1 - \nu_{\mathrm{PS}}} \tag{10b}$$

Where $\chi_t \in [0, 1]$ is the surprise-modulated adaptation rate for trial t which depends on the predictive surprise $PS = -\ln p(o_t; \pi_t(\theta))$, and the prior volatility parameter $\nu_{\mathrm{PS}} \in [0, 1]$ which modulates the influence of surprise on the agent's beliefs. In Eqs. 8 and 9, l is the learning rate parameter, ν_{SD} is the volatility parameter of chosen actions and ν_{SD} modulates how over time α and β decay towards their prior values α_0 and β_0, respectively.

Evidence from multi-armed bandit tasks suggests a bias in prior outcome probabilities. To model this effect, a free parameter, p_r, is fitted to each participant, which determines the initial (or uninformed) α and β prior values, such that $\alpha_0 = 2(1 - p_r)$ and $\beta_0 = 2p_r$ [4].

The decay effect of the chosen actions captured by ν_{SD}, may also affect unchosen actions, and therefore, for these, the updating rules reduce to:

$$\alpha_{s,a,t} = (1 - \nu_{\mathrm{UD}})\alpha_{s,a,t-1} + \nu_{\mathrm{UD}}\alpha_0 \tag{11}$$

$$\beta_{s,a,t} = (1 - \nu_{UD})\beta_{s,a,t-1} + \nu_{UD}\beta_0 \tag{12}$$

where ν_{UD} is the volatility of unchosen actions.

The surprise update mechanism for chosen actions and the decay mechanisms for chosen and unchosen actions may not all be simultaneously operating [4]. Therefore, four variants of the AIF models can be implemented, each with a different combination of update rules as pointed out in Sect. 2.2.

In addition to the outcome probabilities, agents must also learn the transition probabilities. As pointed by [4], before starting the experiment, participants completed an instruction phase as well as 50 practice trials and therefore, they were aware that the transition probabilities of initial-stage actions were mirrored and could either be $p(s_B|s_A, a_A) = p(s_C|s_A, a_B) = 0.7$ or $p(s_B|s_A, a_A) = p(s_C|s_A, a_B) = 0.3$ (with $p(s_B|s_A, a_A) = 1 - p(s_C|s_A, a_A)$ and $p(s_B|s_A, a_B) = 1 - p(s_C|s_A, a_B)$, respectively). In [4], this transition learning is modelled by having agents count transitions, and on each trial, choosing the most likely transition structure based on the observed frequencies.

B Hybrid Reinforcement Learning Drift-Diffusion Model

According to the HRL-DDM model, an agent performing the two-step task in state s, and trial t, will select an action a that maximizes the expected discounted future reward (or state-action value), denoted by $Q_t(s,a)$. Traditionally, state-action values can be computed using either a model-free Q_t^{MF} or a model-based strategy Q_t^{MB}.

In the two-step task, after completing each trial, model-free state-action values are updated following the SARSA(λ) algorithm. For the chosen second-stage action a_2 and the final state (either s_B or s_c, depending on the transition), its corresponding state-action value is updated as follows:

$$Q_{t+1}^{\text{MF}}(s_2,a_2) = Q_t^{\text{MF}}(s_2,a_2) + \alpha_2\delta_{2,t} \tag{13}$$

where $\alpha_2 \in [0,1]$ is the second-stage learning rate parameter and $\delta_{2,t}$, is the second-stage prediction error, defined as the difference between the observed outcome and its state-action value, such that $\delta_{2,t} = o_t - Q_t^{\text{MF}}(s_2,a_2)$

Since there is no outcome after choosing a first-stage action, and its optimality ultimately depends on the outcome observed in the second stage, the updating rule for the model-free state action value of the first-stage action will depend on the second-stage prediction error $\delta_{2,t}$ and will be given by the following equation:

$$Q_{t+1}^{\text{MF}}(s_{\text{A}},a_1) = Q_t^{\text{MF}}(s_{\text{A}},a_1) + \alpha_1[\delta_{2,t} + \lambda\delta_{1,t}] \tag{14}$$

where $\alpha_1 \in [0,1]$ is the first-stage learning rate parameter, $\delta_{1,t}$, is the first-stage prediction error, defined as $\delta_{1,t} = Q_t^{\text{MF}}(s_2,a_2) - Q_t^{\text{MF}}(s_{\text{A}},a_1)$, and $\lambda \in [0,1]$ is the eligibility parameter, which determines how much the second-stage prediction error influences the first-stage state-action value.

For the model-based strategy, agents are assumed to have a generative model of the task, which in this case means that they know (or learn) the state transition probabilities and use these to calculate state-action values. For the first-stage actions, the model-based values depend on the second-stage model-free values, such that:

$$Q_t^{\text{MB}}(s_{\text{A}},a_1) = \sum_{s_2 \in \{s_{\text{B}},s_{\text{C}}\}} P(s_2|s_{\text{A}},a_1) \max_{a_2 \in \mathcal{A}_{s_2}} Q_t^{\text{MF}}(s_2,a_2) \tag{15}$$

In the first stage, the agent will follow a combination of model-free and model-based strategies, such that the total net expected reward of a given action a_j, at the first-stage state, s_A is given by a linear combination of the trial's model-free Q_t^{MF} and model-based Q_t^{MB} expected rewards:

$$Q_t^{\text{net}}(s_{\text{A}},a_j) = wQ_t^{\text{MB}}(s_{\text{A}},a_j) + (1-w)Q_t^{MF}(s_{\text{A}},a_j) + \rho\,\text{rep}_t(a_j) \tag{16}$$

where $w \in [0,1]$ is a weight parameter that regulates the relative influence of model-based and model-free strategies in the decision-making. The last term in the equation is added to model the tendency observed in participants to repeat the first-stage choice of the previous trial. The level of response stickiness is regulated by the parameter ρ, which multiplies the function rep(a) – equal to 1 if the first-stage action is the same as the one chosen in the previous trial and 0 otherwise.

At the second stage, the state-action values are computed using only a model-free strategy, and it is assumed that there is no tendency to repeat the previous action. Therefore, for the second stage, the net state-action value reduces to $Q_t^{\text{net}}(s_2, a_2) = Q_t^{\text{MF}}(s_2, a_2)$.

Once state-action values are computed, the HRL-DDM agent, will pick an action a at trial t, state s and stage $p = \{1, 2\}$ through a drift-diffusion process with a drift rate given by the following equation:

$$v_{p,s,t} = v_p^{\text{mod}}[Q_t^{\text{net}}(s, a) - Q_t^{\text{net}}(s, a')] , \quad a, a' \in \mathcal{A}_s \tag{17}$$

where $\mathcal{A}_s$ is the set of available actions at state s.

C Further Details on the Model Fitting and Comparison Procedures

For the two-step task, the MLE procedure consists of finding the participant's set of parameter values of the model m, $\hat{\theta}_m^{\text{MLE}}$, that maximizes the likelihood of the first-stage and second-stage behavioural data ($d_{1:T}^1$ and $d_{1:T}^2$, respectively), given the parameters set, $p(d_{1:T}^1, d_{1:T}^2 | \theta_m, m)$, or equivalently, log-likelihood, $LL = \log p(d_{1:T}^1, d_{1:T}^2 | \theta_m, m)$. This quantity can be expressed as the sum of the individual trial LLs [23], such that:

$$LL = \sum_{t=1}^{T} \log p(d_t^1, d_t^2 \mid d_{1:t-1}^1, d_{1:t-1}^2, \theta_m, m) \tag{18}$$

To use a standard minimization procedure, instead of maximizing the LL, we minimize the negative of it (NLL). Moreover, we can express the LL of each trial's data as the product of the first-stage and second-stage data, giving the following expression for the NLL:

$$NLL = -\sum_{t=1}^{T} \log p(d_t^1 \mid d_{1:t-1}^1, d_{1:t-1}^2, \theta_m, m) + \log p(d_t^2 \mid d_{1:t}^1, d_{1:t-1}^2, \theta_m, m) \tag{19}$$

where $p(d_t^1 \mid d_{1:t-1}^1, d_{1:t-1}^2, \theta_m, m)$ and $p(d_t^2 \mid d_{1:t}^1, d_{1:t-1}^2, \theta_m, m)$ are the probabilities of each choice (or choice and RT) of the first and second stages, respectively, given the parameters of the model and the information available up to that moment.

For AIF and HRL models, d_t^p, corresponds to the choice made at stage p and trial t, so Eq. 19 can be evaluated using the corresponding choice probability distributions of each model. For AIF-DDM and HRL-DDM, however, d_t^p, corresponds to the choice and RT tuple of stage p and trial t. Hence, to compute the likelihood of each trial and phase, we use the WFPT distribution as implemented by the HDDM Python package. The WFPT distribution takes as input the parameters a_{bs}, t_{nd}, as well as the drift rates, which for HRL-DDM and AIF-DDM are calculated using Eqs. 17 and 3, respectively.

To perform the BMS procedure for model comparison, we require an approximation for the LME of each model for each participant. Instead of calculating this quantity, we use two approximations; one based on the BIC score ($LME \approx -BIC/2$) and another based on the AIC score ($LME \approx -AIC/2$) [12], where the BIC and AIC scores are computed using the following equations:

$$\text{BIC} := k \ln(n) - 2\hat{LL} \tag{20}$$

$$\text{AIC} := 2k - 2\hat{LL} \tag{21}$$

where k is the number of free parameters in the model, n is the number of trials, and $\hat{LL}$ is the maximum LL value, for a given participant and model.

D Model Comparison Plots

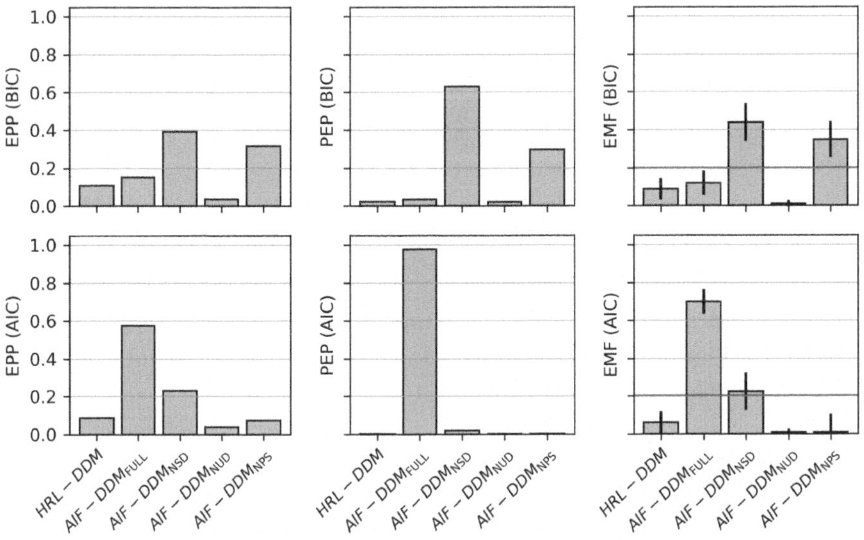

Fig. 5. Bayesian model selection results comparing the five candidate models. The analysis was performed separately using log-model evidence approximations based on BIC (top row) and AIC (bottom row) scores. Each column represents a different metric for model comparison: (left) Expected Posterior Probabilities (EPP), (middle) Protected Exceedance Probabilities (PEP), and (right) Estimated Model Frequencies (EMF). In the EMF plots, the red horizontal line indicates the chance level (0.2), and error bars represent the standard deviation. The AIC-based analysis consistently favours the AIF-DDM$_{\text{FULL}}$ model, while the BIC-based results are more distributed across AIF-DDM$_{\text{NSD}}$, and AIF-DDM$_{\text{NPS}}$.

E Parameter Recovery Results

Table 1. Parameter recovery results for the four AIF variants. The Pearson's correlation and p-values between original and recovered parameters for the pure AIF models are denoted by r and p, respectively. For the AIF-DDM models, the Pearson's correlations and p-values are denoted by r_{ddm} and p_{ddm}, respectively.

(a) AIF(-DDM)$_{\text{FULL}}$

Parameter	r	p	r_{DDM}	p_{DDM}
l	.41	.054	.89	< .001
v_{DU}	.91	< .001	> .99	< .001
v_{DS}	.42	.043	.99	< .001
v_{PS}	.52	.012	.98	< .001
λ	.51	.014	.71	< .001
κ	.53	.009	.97	< .001
p_r	.28	.190	.93	< .001
γ_1	.17	.428		
γ_2	.38	.073		
v_1^{mod}			.46	.027
v_2^{mod}			.73	< .001
a_{bs}			.98	< .001
t_{nd}			>.99	< .001

(b) AIF(-DDM)$_{\text{NSD}}$

Parameter	r	p	r_{DDM}	p_{DDM}
l	.62	.002	.96	< .001
v_{DU}	.90	< .001	.97	< .001
v_{PS}	.76	< .001	.98	< .001
λ	.49	.017	.68	< .001
κ	.68	< .001	.85	< .001
p_r	.83	< .001	.97	< .001
γ_1	.10	.644		
γ_2	.31	.143		
v_1^{mod}			.68	< .001
v_2^{mod}			.59	.003
a_{bs}			.99	< .001
t_{nd}			>.99	< .001

(c) AIF(-DDM)$_{\text{NUD}}$

Parameter	r	p	r_{DDM}	p_{DDM}
l	.47	.025	.84	< .001
v_{DS}	.69	< .001	.93	< .001
v_{PS}	.57	.005	.94	< .001
λ	.66	< .001	.79	< .001
κ	.82	< .001	.93	< .001
p_r	.32	.131	.69	< .001
γ_1	.06	.798		
γ_2	.22	.315		
v_1^{mod}			.55	.007
v_2^{mod}			.60	.003
a_{bs}			.95	< .001
t_{nd}			>.99	< .001

(d) AIF(-DDM)$_{\text{NSL}}$

Parameter	r	p	r_{DDM}	p_{DDM}
l	.69	< .001	.93	< .001
v_{DU}	.94	< .001	.91	< .001
v_{DS}	.81	< .001	.97	< .001
λ	.56	.005	.67	< .001
κ	.68	< .001	.87	< .001
p_r	.94	< .001	.93	< .001
γ_1	.18	.413		
γ_2	.31	.146		
v_1^{mod}			.25	.246
v_2^{mod}			.58	.004
a_{bs}			> .99	< .001
t_{nd}			> .99	< .001

F AIF-DDM$_{NSD}$ and AIF-DDM$_{NPS}$ Simulation Results

See Fig. 6.

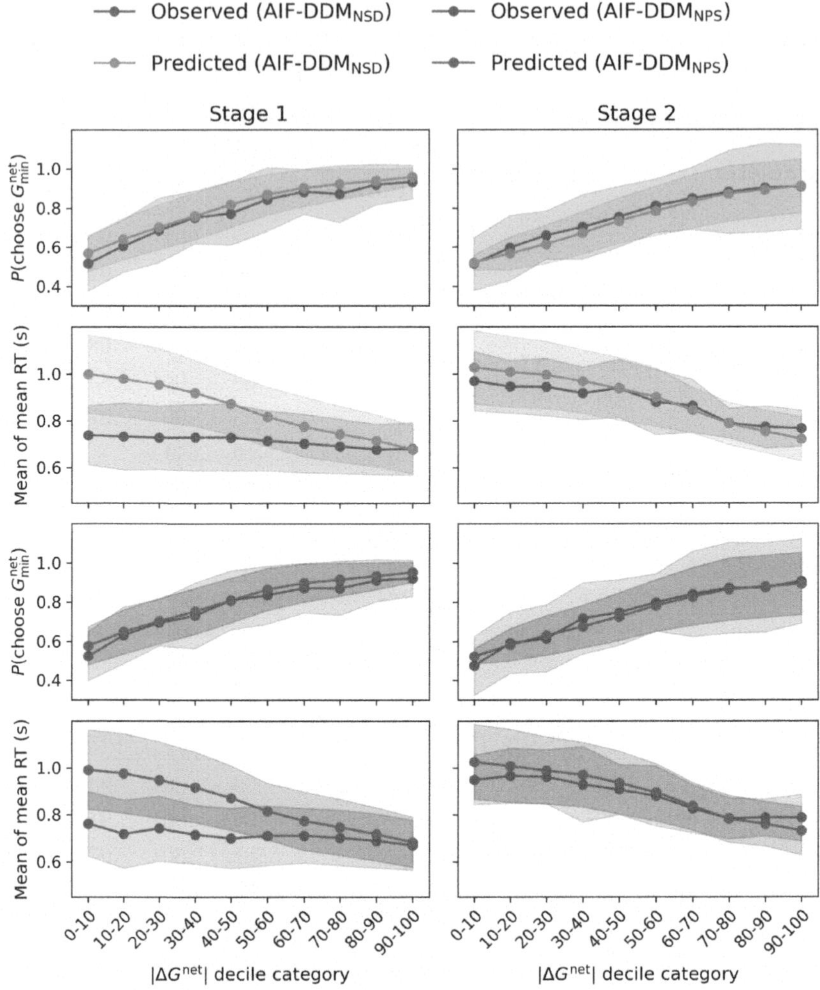

Fig. 6. Comparison of observed and model-predicted behaviour for AIF-DDM$_{NSD}$ and AIF-DDM$_{NPS}$ models.

G Reaction Time Q-Q Analysis

To further investigate the reaction time (RT) misfit mentioned in Sect. 3.4, we generated Quantile-Quantile (Q-Q) plots comparing observed versus simulated within-participant RT deciles ($10^{th} - 90^{th}$). Simulated RT deciles were produced for each participant by running 50 simulated experiments with 1000 trials each (with the same reward volatility as the real experiment) using the best-fitted parameters. As shown in Fig. 7, the plots reveal that all models consistently overestimate the higher RT quantiles for both stages. This overestimation was expected, as the models' predictions are unconstrained, whereas the participants' observed RTs were truncated by a 2-second response deadline, with missed trials excluded from the analysis.

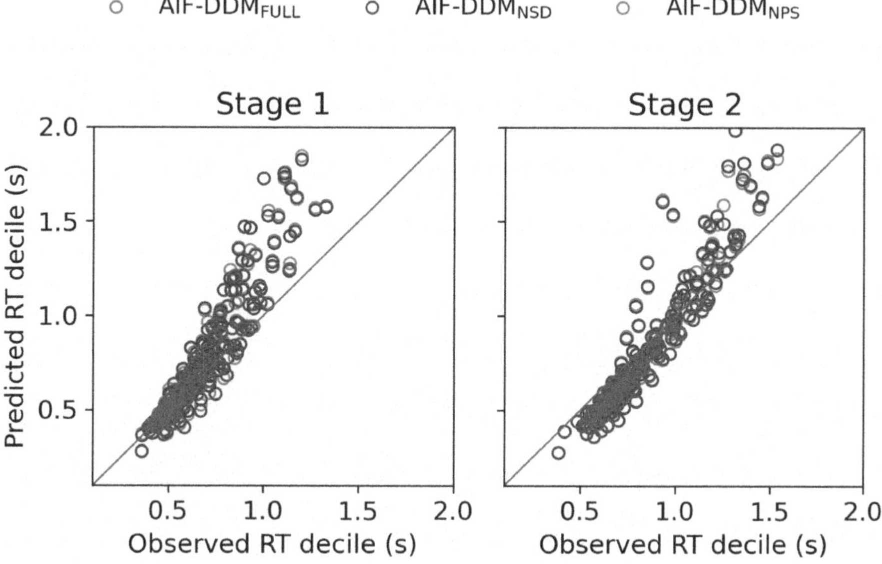

Fig. 7. Systematic overestimation of higher reaction time (RT) deciles by the AIF-DDM models. The Q-Q plots compare observed versus predicted RT deciles for Stage 1 (left) and Stage 2 (right). Points lying above the identity line (red) indicate that the model's predicted RT is higher than the observed RT. This effect is most pronounced for the longest reaction times, where predictions deviate furthest from the identity line. Note that the purple markers indicate an overlap of red and blue points.

References

1. Daunizeau, J., Adam, V., Rigoux, L.: Vba: a probabilistic treatment of nonlinear models for neurobiological and behavioural data. PLoS Comput. Biol. **10**(1), e1003441 (2014)
2. Daw, N.D., Gershman, S.J., Seymour, B., Dayan, P., Dolan, R.J.: Model-based influences on humans' choices and striatal prediction errors. Neuron **69**(6), 1204–1215 (2011)
3. Fiedler, S., Glöckner, A.: The dynamics of decision making in risky choice: an eye-tracking analysis. Front. Psychol. **3**, 335 (2012)

4. Gijsen, S., Grundei, M., Blankenburg, F.: Active inference and the two-step task. Sci. Rep. **12**(1), 17682 (2022)
5. Hodson, R., Mehta, M., Smith, R.: The empirical status of predictive coding and active inference. Neurosci. Biobehav. Rev. **157**, 105473 (2024)
6. Lai, L., Gershman, S.J.: Human decision making balances reward maximization and policy compression. PLoS Comput. Biol. **20**(4), e1012057 (2024)
7. Lee, D.G., d'Alessandro, M., Iodice, P., Calluso, C., Rustichini, A., Pezzulo, G.: Risky decisions are influenced by individual attributes as a function of risk preference. Cogn. Psychol. **147**, 101614 (2023)
8. McDougle, S.D., Collins, A.G.: Modeling the influence of working memory, reinforcement, and action uncertainty on reaction time and choice during instrumental learning. Psychon. Bulletin & Rev. **28**(1), 20–39 (2021)
9. Myers, C.E., Interian, A., Moustafa, A.A.: A practical introduction to using the drift diffusion model of decision-making in cognitive psychology, neuroscience, and health sciences. Front. Psychol. **13**, 1039172 (2022)
10. Parr, T., Holmes, E., Friston, K.J., Pezzulo, G.: Cognitive effort and active inference. Neuropsychologia **184**, 108562 (2023)
11. Pedersen, M.L., Frank, M.J., Biele, G.: The drift diffusion model as the choice rule in reinforcement learning. Psychon. Bulletin & Rev. **24**, 1234–1251 (2017)
12. Penny, W.D.: Comparing dynamic causal models using aic, bic and free energy. Neuroimage **59**(1), 319–330 (2012)
13. Proctor, R.W., Schneider, D.W.: Hick's law for choice reaction time: a review. Quart. J. Exp. Psychol. **71**(6), 1281–1299 (2018)
14. Rigoux, L., Stephan, K.E., Friston, K.J., Daunizeau, J.: Bayesian model selection for group studies—revisited. Neuroimage **84**, 971–985 (2014)
15. Schwartenbeck, P., Passecker, J., Hauser, T.U., FitzGerald, T.H., Kronbichler, M., Friston, K.J.: Computational mechanisms of curiosity and goal-directed exploration. elife **8**, e41703 (2019)
16. Shahar, N., et al.: Improving the reliability of model-based decision-making estimates in the two-stage decision task with reaction-times and drift-diffusion modeling. PLoS Comput. Biol. **15**(2), e1006803 (2019)
17. Feher da Silva, C., Hare, T.A.: Humans primarily use model-based inference in the two-stage task. Nat. Human Behav. **4**(10), 1053–1066 (2020)
18. Steinemann, N., Stine, G.M., Trautmann, E., Zylberberg, A., Wolpert, D.M., Shadlen, M.N.: Direct observation of the neural computations underlying a single decision. Elife **12**, RP90859 (2024)
19. Virtanen, P., et al.: Scipy 1.0: fundamental algorithms for scientific computing in python. Nat. Methods **17**(3), 261–272 (2020)
20. Voskuilen, C., Ratcliff, R., Smith, P.L.: Comparing fixed and collapsing boundary versions of the diffusion model. J. Math. Psychol. **73**, 59–79 (2016)
21. Wellcome Centre for Human Neuroimaging: satistical parametric mapping software package. https://www.fil.ion.ucl.ac.uk (1991). Accessed 10 May 2025
22. Wiecki, T.V., Sofer, I., Frank, M.J.: Hddm: Hierarchical bayesian estimation of the drift-diffusion model in python. Front. Neuroinform. **7**, 14 (2013)
23. Wilson, R.C., Collins, A.G.: Ten simple rules for the computational modeling of behavioral data. Elife **8**, e49547 (2019)
24. Yao, Z.F., Yang, M.H., Hsieh, S.: Neural correlates of span capacity during visual discrimination under varying cognitive demands. Sci. Rep. **15**(1), 31071 (2025)
25. Zenon, A., Solopchuk, O., Pezzulo, G.: An information-theoretic perspective on the costs of cognition. Neuropsychologia **123**, 5–18 (2019)

Addressing the Subsumption Thesis: A Formal Bridge Between Microeconomics and Active Inference

Noé Kuhn(✉)

Department of Philosophy, Carnegie Mellon University, Pittsburgh, USA
nkuhn@andrew.cmu.edu

Abstract. As a unified theory of sentient behaviour, active inference is formally intertwined with multiple normative theories of optimal behaviour. Specifically, we address what we call the subsumption thesis: The claim that expected utility from economics, as an account of agency, is subsumed by active inference. To investigate this claim, we present multiple examples that challenge the subsumption thesis. To formally compare these two accounts of agency, we analyze the objective functions for MDPs and POMDPs. By imposing information-theoretic rationality bounds (ITBR) on the expected utility agent, we find that the resultant agency is equivalent to that of active inference in MDPs, but slightly different in POMDPs. Rather than being strictly resolved, the subsumption thesis motivates the construction of a formal bridge between active inference and expected utility. This highlights the necessary formal assumptions and frameworks to make these disparate accounts of agency commensurable.

Keywords: Active Inference · Expected Utility · Information-Theoretic Bounded Rationality · Microeconomics

1 Introduction

Since the middle of the previous century, expected utility has formed the bedrock of the agency underwriting microeconomics. With early implementations dating back to Bernoulli in 1713 [38], expected utility has undergone many augmentations in order to reflect realistic deliberate decision processes. The comprehensive start of this lineage can be traced to the classic utility theorem [8,25], with earlier applications found in [32]. Subsequent accounts include Bayesian Decision Theory [6,35], Bounded Rationality [27,37], Prospect Theory [20], and many more flavours. The algorithmic implementation of expected utility theory is found in the Reinforcement Learning literature [3]. While seemingly disparate, practically all expected utility accounts of agency depict an agent making decisions in a probabilistic setting to attain optimal reward – to pursue utility [5].

Coming from the completely different background in neuroscience, Active Inference is a comparatively new account of agency [13], positioning itself as

M. Albarracin et al. (Eds.): IWAI 2025, CCIS 2857, pp. 45–61, 2026.
https://doi.org/10.1007/978-3-032-16955-6_3

"a unifying perspective on action and perception [...] richer than the common optimization objectives used in other formal frameworks (e.g., economic theory and reinforcement learning)" [30, pg. 1;4]. Here, the agent seeks to minimize information-theoretic surprisal expressed as free energy (See Definition 3, 4). Active inference allows for a realistic modeling of the very neuronal processes underwriting biological agency [28].

Given the breadth of successful applications [9] combined with its strong fundamental first principles [14], some proponents of active inference posited what we call the subsumption thesis: Expected utility theory as seen in economics is subsumed by active inference – it is an edge case. A formulation in the same vein posits: "Active inference [...] englobes the principles of expected utility theory [...] it is theoretically possible to rewrite any RL algorithm [...] as an active inference algorithm" [11]. So how does the subsumption thesis hold up in the given examples? Is it possible to formally delineate how expected utility and active inference differ? This paper then establishes a firm connection between microeconomics and active inference, which has scarcely been explored before [18].

To formally compare the two accounts of agency, we require a commensurable space for agent-environment interactions: MDPs and POMDPs (Definition 1 and 2). These agent-environment frameworks are the bread and butter of expected utility applications [3,4,21]. Active inference agency has more recently also been specified for the same frameworks [9–11,29]. As such (PO)MDPs provide a theoretical arena for the subsumption thesis to be evaluated.

What exactly is at stake that motivates this inquiry into the subsumption thesis? Firstly, expected utility and active inference rest upon different first principles to substantiate their respective account of agency [11]. Analysis of the formal relationship between these two accounts could provide insights into how the first principles of one account might be a specification the other's first principles. Secondly, this inquiry will shed light on how each account handles the exploration-exploitation dilemma [7]: How should an agent prioritise between exploring an environment versus exploiting what they already know about the environment for utility? Finally, if active inference truly subsumed expected utility, then there are large ramifications for welfarist economics: Currently, the formal mainstream understanding of welfare which informs economic policy [31] is based on aggregating individual agents acting according to expected utility [24, pg. 45] [12,33] within the hugely influential school of revealed preference theory. The subsumption thesis challenges foundations of 'optimal' economic policy if expected utility only captures a sliver of 'optimal' behaviour.

To investigate the subsumption thesis, the rest of the paper is structured as follows. In Sect. 2 the agent-environment frameworks are defined alongside the relevant accounts of agency and basic concepts in microeconomics. In Sect. 3, some examples are investigated which challenge the subsumption thesis. In Sect. 4, the formal bridge between expected utility and active inference is established via Information Theoretic Bounded Rationality (ITBR) [27]. Finally Sect. 5 provides some concluding and summarizing remarks.

2 Preliminary Definitions and Microeconomics

2.1 Agent-Environment Frameworks

A finite Markov Decision Process (MDP) is a mathematical model that specifies the elements involved in agent-environment interaction and development [3]. This formalization of sequential decision making towards reward maximization originates in dynamic programming, and currently enjoys much popularity in model-based Reinforcement Learning (RL). Although potentially reductive, employing MDPs and POMDPs allows for formal commensurability between different accounts of agency.

Definition 1: (Finite Horizon MDP). *An MDP is defined according to the following given tuple:* $(\mathbb{S}, \mathbb{A}, P(s'|a, s), R(s', a), \gamma = 1, \mathbb{T})$

- $\mathbb{S}$ *is a finite set of states.*
- $\mathbb{A}$ *is a finite set of actions.*
- $P(s'|a, s)$ *is the transition probability of posterior state* s' *occurring upon the agent's selection of action* a *in the prior state* s.
- $R(s', a) \in \mathbb{R}^+$ *is the reward function taking as arguments the agent's action and resulting state. For our purposes, the action taken will be irrelevant to the resulting reward:* $R(s', a) = R(s')$.
- γ *denotes the discount factor of future rewards. This is set to* 1 *as this parameter is not commonly used in the cited active inference literature.*
- $\mathbb{T} = \{1, 2, \ldots, t, \ldots, \tau, \ldots, T\}$ *is a finite set for discrete time periods whereby* $t < \tau$ *and the horizon is* T.

Note that time period subscripts e.g., s_τ *are sometimes omitted when unnecessary.*

In a single-step decision problem, an expected reward-maximizing agent would evaluate the optimal action a^* as follows:

$$a_t^* = \arg\max_{a \in \mathbb{A}} E_{P(s_\tau|a_t, s_t)} R(s_\tau) \tag{1}$$

Further, a Partially Observable Markov Decision Process (POMDP) generalizes an MDP by introducing observations o that contain incomplete information about the latent state s of the environment [3,21]. The agent can only infer latent states via observations. Thus, POMDPs are ideal for modeling action-perception cycles [14] with the cyclical causal graphical model $a \to s \to o \ldots$

Definition 2: (Finite Horizon POMDP). *A finite horizon POMDP further adds two elements to the previously given MDP tuple:* $(\mathbb{O}, P(o|s))$

- $\mathbb{O}$ *is a finite set of observations.*
- $P(o|s)$ *is the probability of observation* o *occuring to the agent given the state* s.

2.2 Active Inference Agency

With the environment-agent frameworks established, we can proceed to define how an active inference agent approaches a (PO)MDP. Although fundamental and interesting, the Variational Free Energy objective crucial to perception in active inference will not be examined here; Inference on latent states is assumed to occur through exact Bayesian inference [10, pg. 16]. The central objective function for agency in active inference is the Expected Free Energy (EFE), the formulation of which for (PO)MDPs we will take from [9–11,29]. Essentially, the agent takes the action trajectory $\pi = \{a_\tau, \ldots, a_T\}$ that minimizes the cumulative expected free energy G, which is roughly the sum of the single-step EFEs G_τ. By inferring the resultant EFE of policies through $Q(\cdot)$, the optimal trajectory π^* corresponds to the most likely trajectory – the path of least action. [14]. Formally:

$$
\begin{aligned}
\pi^* &= \arg\min_{\pi} G(\pi) \\
G(\pi) &\approx \sum_{\tau}^{T} G_\tau(\pi) \\
G_\tau(\pi) &= G_(a_t)
\end{aligned}
\tag{2}
$$

We can then define the EFE for single-step for MDPs and POMDPs. Note that this could also be scaled up to trajectories/vectors of the relevant elements e.g. $s_{t:T}$. For simplicity we will look at single-step formulations for the remainder of the paper.

Definition 3: (EFE on MDPs). *For an agent in an MDP with preference distribution $P(s|C)$, the Expected Free Energy of an action for some given current state s_t is defined as follows:*

$$
\begin{aligned}
G_\tau(a_t) &= D_{KL}[P(s_\tau|a_t, s_t)||P(s|C)] \\
&= - \underbrace{\mathfrak{H}[P(s_\tau|a_t, s_t)]}_{\text{Entropy of future states}} \underbrace{-E_{P(s_\tau|a_t)}[logP(s|C)]}_{\text{Expected Surprise}}
\end{aligned}
\tag{3}
$$

As seen in the rearranged objective function of the second line, the agent seeks to keep future options open while meeting preferences; The entropy of future possible states is to be maximized while the information-theoretic surprisal according to the preference distribution is to be minimized. The conditionalisation on C specifies a parameterized preference distribution [29].

Definition 4: (EFE in POMDPs). *For an agent in a POMDP with preference distribution $P(s|C), P(o|C)$, the Expected Free Energy of an action for some given current state s_t is defined as follows:*

$$G_\tau(a_t) = \underbrace{E_{P(s_\tau|o_t,a_t)}\mathfrak{H}[P(o_\tau|s_\tau)]}_{\text{Ambiguity}} + \underbrace{D_{KL}[P(s_\tau|a_t,o_t)|P(s|C)]}_{\text{Risk}} \tag{4}$$

$$= -\underbrace{E_{P(o_\tau|a_t)}[D_{KL}[P(s_\tau|o_\tau)||P(s_\tau|a_t)]]}_{\text{Intrinsic Value}} - \underbrace{E_{P(o_\tau,s_\tau|a_t)}[logP(o|C)]}_{\text{Extrinsic Value}}$$

With some auxiliary assumptions [11, pg. 10] which are admissible for our purposes, the two formulations of EFE in a POMDP are equivalent, and both contain a curiosity inducing term and an exploitation term [16]. The first formulation motivates the agent to minimize the expected entropy of observations given unknown states and to minimize the divergence between actual states and preferred states. The second formulation motivates the agent to maximize the expected informational value of observations while also maximizing the expected log probability of preferred observations – note how the underbrace does not include the minus.

2.3 Microeconomics

As this paper investigates an intersection between fields which are generally not in direct contact, a brief introduction to risk attitudes and lotteries in microeconomics is provided. The origin of these studies can be traced back to the gambling houses of the 18th century; As early as 1713, Bernoulli employed marginally decreasing utility functions to resolve the famous St. Petersburg Paradox [38]. This paradox asks what amount a rational agent would be willing to pay to enter lottery with an infinite expected value. To answer this question, we utilize lotteries [8] and risk-attitudes [1] from microeconomics:

Definition 5: (Lottery). *A (monetary) lottery is a probability distribution over outcomes x that are the argument of the utility function. Therefore, a lottery L can be modelled as an integrable random variable defined by the probability space triplet consisting of a sample space, sigma algebra, and probability measure:* $(\Omega, \mathfrak{F}, \mu)$

A decision maker then evaluates their preference over a set of lotteries according to their utility function $U(x) \in \mathbb{R}^+$ where $x \in \mathfrak{F}$. The expected utility of each lottery then induces a preference ordering over lotteries. For example, the strong preference relation $L_1 \succ L_2$ means that lottery L_1 is more preferable to lottery L_2. Classically, this ordering is in line with the von Neumann-Morgenstern axioms of completeness, transitivity, continuity, and independence [25]. Any such preference ordering is also maintained for any positive affine transformation of $U(x)$ [8,25]. By juxtaposing the expected utility $E[U(L)]$ of a lottery against the utility of the expectation of the same lottery $U(E[L])$, risk aversion can be defined.

Definition 6: (Risk Aversion). *An agent with some utility function $U(\cdot)$ is considered risk averse if for some lottery the following preference relation holds:* $U(E[L]) \succ U(L)$

This preference relation occurs if an agent's utility function is concave, i.e. the marginal utility is decreasing. A risk loving agent conversely acts according to a convex utility function, and a risk neutrality is associated with a linear utility function. Accordingly, Bernoulli used lotteries and a log-utility function to resolve the St. Petersburg paradox, the solution of which is relegated to Appendix A for readers unfamiliar with the problem – the pertinent point is that concave utility functions on set rewards are extensively studied in economics.

3 Subsumption Examples

Equipped with an understanding of marginal utility and lotteries, we can now tackle two manifest exhibits of the subsumption thesis by proponents of active inference. Further, an illustrative MDP demonstrates the divergence in behaviour between active inference and expected utility. The results of the simulated behaviour are directly taken from the discussed papers. These exhibits then motivate the bridging in Sect. 4 later.

The first [34] and second [11] exhibit both concern agency in a classical T-maze: A simple forked pathway in which the agent can either go left or right (See Fig.1 below). This environment is also called "Light" in POMDP literature [21]. The agent-environment dynamics are modeled using a POMDP; Unbeknownst to the agent, the reward is either in the left or right arm. The agent can also go down to observe a cue indicating the definite location of the reward. Going down the 'wrong' arm of the fork leads to a punishment equal to the negative reward, say -1. The performance of the agency is evaluated by the reward attainment of the agent within a two period horizon. At this point however, the setup of the first and second exhibit diverge crucially.

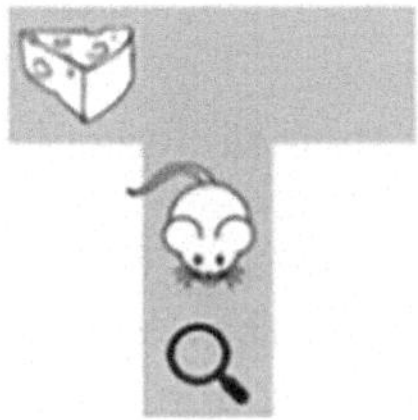

Fig. 1. An agent in a T-Maze with unknown context. Illustration from [34].

In the first exhibit [34], the right and left fork are absorbing states – the agent cannot leave them upon entry. As such, the agent cannot correct going down the wrong arm in the first period by the second period. Given this setup, the expected utility agent performs very poorly, while the active inference is cue-seeking and therefore performs optimally [34, pg. 138]. The expected utility agent performs so poorly because supposedly "the agent does not care about the information inferred and is indifferent about going to the cue location or remaining at the central location" [34, pg. 137]. This appears reductive, as an expected utility agent facing two lotteries will behave the same as the active inference agent. Consider the risky lottery L_1 which is the result of a gambling and non-information seeking strategy. Contrast this lottery with L_2, which is the degenerate lottery of investigating the cue first and going to the reward in the second period. Assuming even just a linear utility function $U(R) = R(s)$, then $U(L_1) = 0.5 \cdot 1 + 0.5 \cdot -1 = 0$ and $U(L_2) = 1 \cdot 1$. Clearly, the expected utility agent holds a preference which motivates cue-seeking behaviour: $L_2 \succ L_1$.

Regarding the second exhibit [11], there is a slight difference in the setup. The arms of the fork are no longer absorbing states, which allows for mistake correction and a cumulative reward of 2 over two periods. Now, the focus of [11] isn't anymore on performance comparison but instead achieving the desiderata of risk-aversion and information sensitivity [11, pg. 10]. While the agency according to active inference meets the desiderata, the expected utility agent does not. However, risk aversion and the resulting information sensitivity can easily be induced by using a concave utility function. Consider again the risky lottery L_1 and a cue-seeking lottery L_2. Assuming a utility function taking the reward as argument $U(R) = R(s)^c$ where $c \in \mathbb{R}^+$, then $U(L_1) = 0.5 \cdot 0 + 0.5 \cdot 2^c$ and $U(L_2) = 1^c$. Accordingly if $c < 1$, then $L_2 \succ L_1$, and only if $c = 1$, then indeed the agent is indifferent $L_1 \sim L_2$. As is evident, it is the risk-neutral agent who does not meet the desiderata.

Finally, consider the following single-step MDP created for illustrative purposes. A paraglider stands at the foot of two steep mountains s_1, s_2 separated by a chasm s_3 and must decide which one to climb. While still risky, the path up mountain 1 is far more secure than the path up mountain 3. However, mountain 2 is taller than mountain 1 and therefore allows for a more enjoyable flight. This decision process can aptly be modeled in an MDP (See Fig. 2 below). Note that the subscript here does not relate to the period. Taking a_1 gives $\{P(s_1|a_1), P(s_2|a_1), P(s_3|a_1)\} = \{0.6, 0, 0.4\}$, and a_2 gives $\{P(s_1|a_2), P(s_2|a_2), P(s_3|a_2)\} = \{0, 0.4, 0.6\}$. The height in kilometers gives the reward function $\{R(s_1), R(s_2), R(s_3)\} = \{1, 1.5, 0\}$. With the MDP sufficiently specified, we can compare the agency of an active inference agent and an expected utility agent. See Appendix B for details on the resulting expected utility and free energy.

The active inference agent is indifferent between the two actions as both actions result in the same EFE – Eq. (3). For expected utility however, only the linear utility function agent is indifferent; The risk averse agent prefers the safer mountain and the risk loving agent prefers the riskier mountain due to the concavity or convexity of the utility function respectively. On the contrary, the canonical softmax preference distribution of active inference has no straight

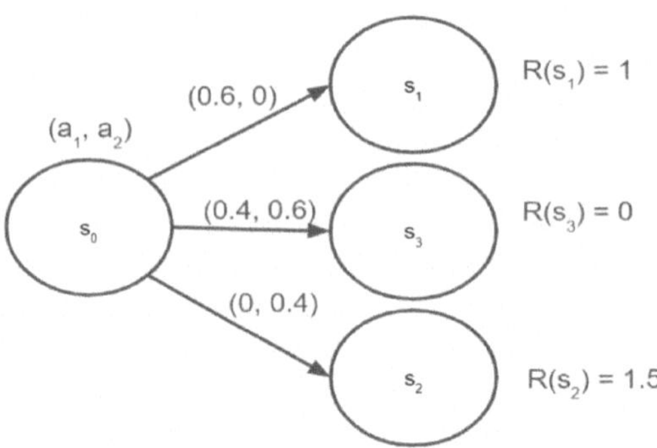

Fig. 2. 'Paraglider' MPD with states, actions, transition probabilities, and rewards.

forward way of encoding risk aversion to our knowledge. As such, this simple but valid MDP provides a setup in which specified expected utility may better meet the desiderata than the active inference agent.

It should be clear by now that wrapping a utility function around the rewards is a well-studied and principled approach which differs from simply including "ad-hoc exploration bonuses in the reward function" [11, pg. 2]. Introducing non-linearity over the rewards seems to lead to an impasse in the comparison between expected utility. The most direct case for comparing and subsuming expected utility [10] only considers a linear utility function ($U(\cdot) = R(\cdot)$) for the expected utility agent. Recently, it was shown that the intrinsic value of EFE (4) approximates the value of information for linear expected utility [39].

To bridge the two agencies, we would like to draw attention to physical and biological constraints on agents which have motivated active inference. For example, tractability is a central concern for active inference, as evidenced by the appeal to variational Bayes. Luckily, there already exists an account of agency which imbues expected utility with such constraints: ITBR [17,27]. The connection between ITBR and active inference in an MDP has briefly been explored before [26]. We seek to now clearly establish this conceptual bridge between microeconomics and active inference for both MDPs and POMDPs.

4 From Expected Utility to Active Inference via ITBR

4.1 In MDPs

Let us first establish the bridge between expected utility and active inference in an MDP. Essentially, both objective functions can both be transformed into the "Divergence Objective" [22]:

$$a^* = \underset{a \in \mathbb{A}}{\arg\min} D_{KL}[P(s_\tau|a_t)||P^*(s)] \tag{5}$$

where a^* is the optimal action and $P^*(s)$ is a preference distribution over states, for example, a softmax or Gibbs distribution with temperature parameter. Note the immediate similarity to the EFE objective function for MDPs (3) – here conditionalisation on the current state s_t is omitted for brevity as we consider a single step.

To get there from expected utility, we can consider the following Lagrangian constraints on the utility objective function [17, pg. 3]. Let $P(\cdot)$ be the prior distribution over relevant elements of the MDP, and $Q(\cdot)$ the posterior distribution after a limited search or 'bounded deliberation; see [27] for details. The deliberation bound is given as an information-theoretic quantity e.g. *nats* or *bits*; hence the name information-theoretic bounded rationality. Let $K \in \mathbb{R}^+$ *nat* – although the information theoretic unit base *nat* is arbitrary:

$$D_{KL}[Q(s_\tau|a_t)||P(s_\tau|a_t)] \leq K \tag{6}$$

The constraint of Eq. 6 can be interpreted as a bound on the search for the optimal action. The second constraint of (6) means that the agent is uncertain about the 'true' transition probabilities in the MDP. This constraint gives us the following ITBR free energy objective functional [26]:

$$F_{ITBR}(Q) = \sum_s Q(s|a)\left(U(s,a) - \frac{1}{\beta}log\frac{Q(s|a)}{P(s|a)}\right) \tag{7}$$

This functional is to be maximized $(Q^*(s|a))$ with given parameter $\beta \in \mathbb{R}^+$. See Appendix C for how the maximizing solution is derived. We can now use the maximizing argument of the objective function (7) as a 'goal' for the agent, or a preference distribution over states $P^*(s)$. Like in active inference, we assume that the preference distribution over states is independent of the action taken to get there. This preference is given by the Gibbs distribution:

$$P^*(s|a) = \frac{P(s|a) \cdot e^{\beta U(s,a)}}{Z_\beta} \rightarrow P^*(s) \tag{8}$$

We can now solve (8) for $U(s,a)$ and input this into (7) to obtain the divergence objective (5):

$$a^* = \underset{a \in \mathbb{A}}{\arg\min} - D_{KL}[Q(s_\tau|a_t)||P^*(s)] + constant \tag{9}$$

where the constant is irrelevant for optimization purposes. The details of this derivation relegated to Appendix D. Evidently, the same optimal agency arises in an MDP for an active inference and ITBR agent. Next, let us bridge expected utility to active inference in a POMDP.

4.2 In POMDPs

Analogously to the MDP setting, we can transform the ITBR objective to get to the divergence objective function for POMDPs. Fortunately, this divergence objective has previously been formulated as the "Free Energy of the Expected Future" (FEEF): [23, pg. 10]. Again, this objective function motivates a minimal posterior divergence from a preference distribution, now jointly over states and observations:

$$a^* = \arg\min_{a \in \mathbb{A}} D_{KL}[P(o_\tau, s_\tau | a_t) || P^*(o, s)] \tag{10}$$

To attain this expression, we can formulate a new ITBR objective in the POMDP framework [17] and transform it analogously to the MDP case before. We can again consider the information-theoretic bound $V \in \mathbb{R}^+ nat$:

$$D_{KL}[Q(s_\tau, o_\tau | a_t) || P(s_\tau, o_\tau | a_t)] \leq V \tag{11}$$

Considering these constraints, we can express the ITBR Free energy objective functional again:

$$F_{ITBR}(Q) = \sum_s Q(o, s|a) \left(U(o, s, a) - \frac{1}{\beta} log \frac{Q(o, s|a)}{P(o, s|a)} \right) \tag{12}$$

where the solution is again the Gibbs distribution:

$$P^*(o, s|a) = \frac{P(o, s|a) e^{\beta U(o,s,a)}}{Z_\beta} \tag{13}$$

By combining (13) and (12) we get the resultant minimization objective, where the resultant optimal agency is of course the same as that of the divergence minimization objective (10):

$$a^* = \arg\min_{a \in \mathbb{A}} - D_{KL}[Q(o_\tau, s_\tau | a_t) || P^*(o, s)] + constant \tag{14}$$

Which again intuitively motivates the agent to have the inferred posterior distribution given the action be as close as possible to the prior preference distribution over states. It is crucial to note however that this is not the same objective function as EFE in POMDPs (4)! To get from the divergence objective (10) for POMDPs to EFE (4), we can follow the steps taken in [23]; for a detailed discussion of the relationship between the divergence objective and EFE, the reader should also consult [22,23]. Essentially, the divergence objective can also be decomposed into an exploitative and explorative term. However, while the explorative term is equal to that of active inference, the divergence objective additionally further encourages the agent to increase posterior entropy of observations given latent states – to keep options open. Note that in the formulation below, both objective functions below (15), (4) are to be minimized.

$$\begin{aligned} - F_{ITBR} &= D_{KL}[Q(o, s|a) || P^*(o, s)] \\ &= \underbrace{E_{Q(s|a)} [\ D_{KL}[Q(o|s) || P^*(o)]\ ]}_{\text{Extrinsic Value}} - \underbrace{E_{Q(o|a)} [\ D_{KL}[Q(s|o) || Q(s|a)]\ ]}_{\text{Intrinsic Value}} \end{aligned} \tag{15}$$

$$G = -\underbrace{E_{Q(o,s|a)}[logP(o|C)]}_{\text{Extrinsic Value}} - \underbrace{E_{Q(o|a)}\left[\ D_{KL}[Q(s|o)||Q(s|a)]\ \right]}_{\text{Intrinsic Value}} \tag{4}$$

where by the relationship between G and $-F_{ITBR}$ is as follows:

$$G - E_{Q(o|a)}\mathfrak{H}[Q(o|s)] = -F_{ITBR} \tag{16}$$

Comparing then the decomposed divergence objective to active inference, in the pursuit of extrinsic value the boundedly rational utility agent seeks to additionally keep posterior options open compared to the active inference agent – similarly to the agency in an MDP.

4.3 The Bridge Summarized

Let us reconsider the entire journey from expected utility to active inference so as to not lose sight of the forest in front of all the trees. First, simply incorporate a utility function into the reward maximizing objective function (1) to get an expected utility agent. Then, impose information-theoretic deliberation constraints on the optimization process (6). Consequently, the agent faces a Lagrangian optimization problem (7). The solution to this optimization problem is taken as a preference distribution for the agent. Combining the preference distribution and the objective function results in the divergence objective [22], which can then be compared with the active inference objective function. In an MDP, the resultant agency is the exact same (5). However, in a POMDP, the objective functions differ (16).

Bar this difference, one key aspect must be elucidated for the extrinsic value terms in both MDPs and POMDPs. Although the intrinsic value term is the same for the different objective functions, the two prior preference distributions $P^*(s)$ of ITBR (8) and $P(s|C)$ of active inference (3) are not necessarily the same. For $\beta = 1$, if we consider $P(s|C)$ as a Gibbs distribution as per [10, pg. 9];[29, pg. 134], then the two preference distributions are only equal if either the utility function is linear, or if active inference admits agent-specific non-linear utility functions (17); An admission which prima facie seems irreconcilable with the physicalist/nonsubjectivist philosophy behind active inference. This larger discussion is, however, to be relegated to a later paper.

$$P^*(s) = \frac{e^{U(s)}}{\sum\limits_{s} e^{U(s)}} \quad \text{and} \quad P(s|C) = \frac{e^{R(s)}}{\sum\limits_{s} e^{R(s)}} \tag{17}$$

where optimal behaviour a^* in an MDP is the same for both accounts of agency only if $U(s)$ is a positive affine transformation of $R(s)$, such as when $R(s)$ is simply multiplied by some temperature parameter.

5 Conclusion

Having formalized the bridge from expected utility to active inference, we can re-evaluate the subsumption thesis. Simple reward-oriented agency ($U(\cdot) = R(\cdot)$) can be effectively subsumed by active inference in MDPs, and if exact Bayesian inference is used, also in POMDPs [10]. However, as shown in Sect. 2, expected utility in microeconomics uses utility functions that take rewards as arguments to reflect risk-attitudes. As seen in Sect. 3 then, there are various examples where the subsumption argument does not hold up; Expected utility acts the same as active inference, or under specific circumstances, may meet desiderata of agency even more. In Sect. 4, we establish the formal bridge between expected utility and active inference. By using ITBR [27], we can directly compare the objective functions of (bounded) expected utility and active inference. Upon considering agent-environment assumptions, the divergence objective [22] is used as a reference point to compare the two accounts of agency. It is demonstrated that in an MDP, ITBR and active inference lead to the same agency [26]. In a POMDP, ITBR is equivalent to the divergence objective, which however differs from the active inference objective function [23]. While the explorative/information-seeking terms are equal, the exploitative/reward-oriented term differs: $E_{Q(o|a)}\mathfrak{H}[Q(o|s)]$ must be subtracted from the active inference objective function, and the preference distributions are not necessarily equal. These results of course rely on the explicit form of EFE used here [9,10,29] and corresponding assumptions [11, pg. 10].

An area where expected utility cannot compete however is in the first principles which motivate agency [2,11,13–15]. Still, the debate on what objective function follows from the first principles is not yet sealed in this flourishing field [23]. Perhaps more intriguing links between brain function and the statistical-physical applications of information theory lurk underneath the bridge established here. Furthermore, computational simulations [17] and empirical studies [36] might flesh out the practical comparison between bounded expected utility and active inference; Computational efficiency has not remotely been addressed in this paper. The results of this paper will also interest the school of bounded rationality in microeconomics. Finally, it would be especially interesting for economics to understand how an economy could develop from multiple ITBR or active inference agents [19]. By integrating interdisciplinary approaches to agency, we aim to foster a holistic approach that enriches the roles of both human and artificial agents in society.

Acknowledgments. I would like to express my immense gratitude to my supervisor Martin Kolmar at the University of St. Gallen Institute for Business Ethics for allowing me to delve into this topic and lending his support along the way. Further, I want to thank the various researchers willing to so openly discuss the contents and concepts of the paper.

Disclosure of Interests. The author has no competing interests to declare that are relevant to the content of this article.

Appendix

A Resolving the St. Petersburg Paradox

Consider a lottery on the outcome of a fair coin toss. Starting at two dollars, the stake doubles with every subsequent outcome of heads. The game ends once tails comes up for the first time in the sequence. The expected payout $E[L]$ of the game is thus infinite:

$$E[L] = \sum_{i=1}^{\infty} \frac{1}{2^i} \cdot 2^i = \infty$$

How much would someone pay to participate in this game? Taking a linear utility function on the payout, the gambler should be willing to pay any amount to enter the game. Daniel Bernoulli suggested a logarithmic utility function $U(x) = ln(x)$. Assume the cost of entry is x. Then the expected utility of the lottery has a finite value; The amount the agent at most would be willing to enter the lottery:

$$E[U(L)] = \sum_{i=1}^{\infty} \frac{1}{2^i} \cdot ln(2^i) = 2 \cdot ln(2)$$

Therefore the agent expects finite utility from the payout of the lottery due to the concavity of the utility function. As such, only a finite amount will be paid to enter the game.

B Expected Utility and Active Inference for the 'Paraglider' MDP

The single-step MDP is specified as follows. Therefore note that the subscript does not pertain to the period:

$$\begin{aligned}
&\mathbb{S} = s_1, s_2, s_3 \\
&\mathbb{A} = a_1, a_2 \\
&\{P(s_1|a_1), P(s_2|a_1), P(s_3|a_1)\} = \{0.6, 0, 0.4\} \\
&\{P(s_1|a_2), P(s_2|a_2), P(s_3|a_2)\} = \{0, 0.4, 0.6\} \\
&\{R(s_1), R(s_2), R(s_3)\} = \{1, 1.5, 0\}
\end{aligned}$$

Consider an expected utility agent with utility function $U(R(s)) = R(s)^c$ where $c \in \mathbb{R}^+$. As such,

$$\begin{aligned}
&E[U(a_1)] = 0.6 \cdot 1^c \\
&E[U(a_2)] = 0.4 \cdot 1.5^c \\
&\text{For } c < 1 \rightarrow \arg\max_{a \in \mathbb{A}} E[U(a)] = a_1 \\
&\text{For } c > 1 \rightarrow \arg\max_{a \in \mathbb{A}} E[U(a)] = a_2
\end{aligned}$$

So a risk-averse expected utility agent will scale the smaller but safer mountain.

The active inference agent however is indifferent between the two actions. If we assume the preference distribution to be a softmax on the rewards, then we can ignore the normalizing denominator as it is constant w.r.t to action. Note that no temperature parameter is used here. Therefore we can write the relevant objective function as:

$$
\begin{aligned}
G(a_t) &= -\sum_s P(s_\tau|a_t)\cdot R(s_\tau) - \sum_s P(s_\tau|a_t) log \frac{1}{P(s_\tau|a_t)} \\
G(a_1) &= -0.6 - 0.3065 - 0.366 = G(a_2) \\
\rightarrow & \ \underset{a\in\mathbb{A}}{\arg\min} G(a) = \{a_1, a_2\}
\end{aligned}
$$

Therefore the optimal action of the risk-averse expected utility agent is a subset of the optimal active inference agency.

C Preference Distribution Derivation

We maximize the ITBR objective function (7) via first order condition.

$$
\frac{\delta F_{ITBR}}{\delta Q(s|a)} = U(s,a) - \frac{1}{\beta}\left(log\frac{Q(s|a)}{P(s|a)} + 1\right) \stackrel{!}{=} 0
$$

Solve for $Q(s|a)$, and normalize to attain the Gibbs distribution

$$
\begin{aligned}
Q(s|a) &= P(s|a)e^{\beta U(s,a)-1} \propto P(s|a)e^{\beta U(s,a)} \\
Q^*(s|a) &= \frac{P(s|a)e^{\beta U(s,a)}}{\sum_s P(s|a)e^{\beta U(s,a)}} = \frac{P(s|a)e^{\beta U(s,a)}}{Z_\beta}
\end{aligned}
$$

Which gives us (8)

D Getting from ITBR to the Divergence Objective via the Gibbs Distribution

Solve (8) for $U(s,a)$:

$$
\begin{aligned}
P^*(s|a) &= \frac{P(s|a)e^{\beta U(s)}}{Z_\beta} \\
\frac{1}{\beta} ln(P^*(s|a)\cdot Z_\beta) &= U(s)
\end{aligned}
$$

Plug this into the ITBR objective function (12) and consider the maximizing argument a:

$$\begin{aligned}
&\arg\max_{a\in\mathbb{A}} \frac{1}{\beta}E_{Q(s|a)}[lnP^*(s|a)+ln(Z_\beta)] - \frac{1}{\beta}ln\frac{Q(s|a)}{P(s|a)} \\
&= \arg\max_{a\in\mathbb{A}} E_{Q(s|a)}[lnP^*(s|a)+ln(Z_\beta) - lnQ(s|a) + lnP(s|a)] \\
&= \arg\min_{a\in\mathbb{A}} E_{Q(s|a)}[-lnP^*(s|a)-ln(Z_\beta) + lnQ(s|a) - lnP(s|a)] \\
&= \arg\min_{a\in\mathbb{A}} D_{KL}[Q(s|a)||P^*(s|a)]
\end{aligned}$$

Which is the divergence objective for MDPs (5). In a POMDP setting, the derivation proceeds analogously to obtain the Free Energy of the Expected Future (10).

References

1. Arrow, K.J.: Essays in the Theory of Risk Bearing. Markham Publishing Co, Chicago (1971)
2. Barp, A., Da Costa, L., França, G., Friston, K., Girolami, M., Jordan, M.I., Pavliotis, G.A.: Geometric methods for sampling, optimisation, inference and adaptive agents. Handbook of Statistics **46**, 21–78 (2022). https://doi.org/10.48550/arXiv.2203.10592, https://doi.org/10.48550/arXiv.2203.10592, arXiv:2203.10592v3 [stat.ML]
3. Barto, A., Sutton, R.S.: Reinforcement Learning: An Introduction. The MIT Press, 2nd edn. (2018)
4. Bellman, R.: A markovian decision process. J. Math. Mech. **6**, 679–684 (1957). https://doi.org/10.1512/iumj.1957.6.56038, https://doi.org/10.1512/iumj.1957.6.56038
5. Bentham, J.: An Introduction to the Principles of Morals and Legislation. Batoche Books, Kitchener, 2000 edn. (1781)
6. Berger, J.O.: Statistical Decision Theory and Bayesian Analysis. Springer Series in Statistics, Springer-Verlag, New York, 2nd edn. (1985). https://doi.org/10.1007/978-1-4757-4286-2
7. Berger-Tal, O., Nathan, J., Meron, E., Saltz, D.: The exploration-exploitation dilemma: A multidisciplinary framework. PLoS ONE **9**(4), e95693 (2014). https://doi.org/10.1371/journal.pone.0095693
8. Bonanno, G.: Decision making (2017). https://faculty.econ.ucdavis.edu/faculty/bonanno/PDF/DM_book.pdf
9. Da Costa, L., Parr, T., Sajid, N., Veselic, S., Neacsu, V., Friston, K.: Active inference on discrete state-spaces: a synthesis. J. Math. Psychol. **102447**, 36 (2021). https://doi.org/10.1016/j.jmp.2020.102447, https://doi.org/10.48550/arXiv.2001.07203, submitted on 20 Jan 2020 (v1), last revised 28 Mar 2020 (this version, v2)
10. Da Costa, L., Sajid, N., Parr, T., Friston, K., Smith, R.: Reward maximisation through discrete active inference **v4** 18 p. (2022). arXiv preprint arXiv:2009.08111
11. Da Costa, L., Tenka, S., Zhao, D., Sajid, N.: Active inference as a model of agency. arXiv preprint arXiv:2401.12917 (2024). https://doi.org/10.48550/arXiv.2401.12917, accepted in RLDM2022 for the workshop 'RL as a model of agency'

12. Fleurbaey, M., Mongin, P.: The utilitarian relevance of the aggregation theorem. Am. Econ. J. Microecon. **8**(3), 289–306 (2016). https://doi.org/10.1257/mic.20140308
13. Friston, K.: The free-energy principle: A rough guide to the brain? Trends Cogn. Sci. **13**(7), 293–301 (2009). https://doi.org/10.1016/j.tics.2009.04.005
14. Friston, K., et al.: The free energy principle made simpler but not too simple. Physics Reports **1024**, 1–29 (2023). https://doi.org/10.1016/j.physrep.2023.07.001
15. Friston, K., et al.: Path integrals, particular kinds, and strange things. Physics of Life Reviews **47** (2023). https://doi.org/10.1016/j.plrev.2023.08.016, https://doi.org/10.48550/arXiv.2210.12761
16. Friston, K., Rigoli, F., Ognibene, D., Mathys, C., Fitzgerald, T., Pezzulo, G.: Active inference and epistemic value. COGNITIVE NEUROSCIENCE **6**(4), 187–224 (2015). https://doi.org/10.1080/17588928.2015.1020053, http://dx.doi.org/10.1080/17588928.2015.1020053
17. Genewein, T., Leibfried, F., Grau-Moya, J., Braun, D.A.: Bounded rationality, abstraction, and hierarchical decision-making: An information-theoretic optimality principle. Frontiers Robot. AI **2**, 27 (2015). https://doi.org/10.3389/frobt.2015.00027, https://doi.org/10.3389/frobt.2015.00027, this article is part of the Research Topic Theory and Applications of Guided Self-Organisation in Real and Synthetic Dynamical Systems
18. Henriksen, M.: Variational free energy and economics: Optimizing with biases and bounded rationality. Frontiers Psychol. **11** (2020). https://doi.org/10.3389/fpsyg.2020.549187, https://doi.org/10.3389/fpsyg.2020.549187
19. Hyland, D., et al.: Free-energy equilibria: Toward a theory of interactions between boundedly-rational agents. In: ICML 2024 Workshop on Models of Human Feedback for AI Alignment (2024)
20. Kahneman, D., Tversky, A.: Prospect theory: An analysis of decision under risk. Econometrica **47**(2), 263–291 (1979)
21. Littman, M.: A tutorial on partially observable markov decision processes. J. Math. Psychol. **53**(2), 119–125 (2009)
22. Millidge, B., Seth, A., Buckley, C.: Understanding the origin of information-seeking exploration in probabilistic objectives for control. arXiv preprint arXiv:2103.06859 (2021). https://doi.org/10.48550/arXiv.2103.06859, submitted on 11 Mar 2021 (v1), last revised 24 Nov 2021 (this version, v7)
23. Millidge, B., Tschantz, A., Buckley, C.L.: Whence the expected free energy? Neural Comput. **33**(2), 447–482 (2021). https://doi.org/10.1162/neco_a_01354, https://doi.org/10.1162/neco_a_01354
24. Mongin, P.: A concept of progress for normative economics. Economics and Philosophy **22**, 19–54 (2006). https://doi.org/10.1017/S0266267105000696
25. von Neumann, J., Morgenstern, O.: Theory of Games and Economic Behavior. Princeton University Press, Princeton, NJ (1953)
26. Ortega, P.A., Braun, D.A.: What is epistemic value in free energy models of learning and acting? a bounded rationality perspective. Cognitive Neurosci. **6**(4), 215–216 (2015). https://doi.org/10.1080/17588928.2015.1051525, https://doi.org/10.1080/17588928.2015.1051525
27. Ortega, P.A., Braun, D.A., Dyer, J., Kim, K.E., Tishby, N.: Information-theoretic bounded rationality. arXiv preprint arXiv:1512.06789 (2015). https://doi.org/10.48550/arXiv.1512.06789, submitted on 21 Dec 2015
28. Parr, T., Markovic, D., Kiebel, S.J., Friston, K.J.: Neuronal message passing using mean-field, bethe, and marginal approximations. Sci. Rep. **9**(1), 1–18 (2019). https://doi.org/10.1038/s41598-019-50764-9

29. Parr, T., Pezzulo, G., Friston, K.J.: Active Inference: The Free Energy Principle in Mind, Brain, and Behavior. The MIT Press (2022)
30. Pezzulo, G., Parr, T., Friston, K.: Active inference as a theory of sentient behavior. Biol. Psychol. **186** (2024). https://doi.org/10.1016/j.biopsycho.2023.108741, under a Creative Commons license
31. Pigou, A.C.: The Economics of Welfare. Macmillan and Co., Limited, London (1920)
32. Ramsey, F.P.: Truth and probability. In: Braithwaite, R.B. (ed.) The Foundations of Mathematics and Other Logical Essays, chap. VII, pp. 156–198. Kegan, Paul, Trench, Trubner & Co. and Harcourt, Brace and Company, London and New York (1931), originally published in 1926
33. Ross, D.: Philosophy of Economics. Palgrave Philosophy Today, Palgrave Macmillan London, 1 edn. (2014). https://doi.org/10.1057/9781137318756, https://doi.org/10.1057/9781137318756
34. Sajid, N., Da Costa, L., Parr, T., Friston, K.: Active inference, bayesian optimal design, and expected utility. In: Cogliati Dezza, I., Schulz, E., Wu, C.M. (eds.) The Drive for Knowledge: The Science of Human Information Seeking, pp. 124–146. Cambridge University Press, Cambridge (2022)
35. Savage, L.J.: The Foundations of Statistics. Dover Publications, Inc., New York, N.Y., revised and enlarged edn. (1972), originally published by John Wiley & Sons in 1954
36. Schwartenbeck, P., FitzGerald, T.H.B., Dolan, R.J., Friston, K.J.: Evidence for surprise minimization over value maximization in choice behavior. Scientific Reports **5**, 16575 (2015). https://doi.org/10.1038/srep16575, https://doi.org/10.1038/srep16575
37. Simon, H.A.: Models of Man. John Wiley & Sons, New York (1957)
38. Szipro, G.: Risk, Choice, and Uncertainty: Three Centuries of Economic Decision-Making. Columbia University Press, New York City (2020)
39. Wei, R.: Value of Information and Reward Specification in Active Inference and Pomdps. arXiv preprint arXiv:2408.06542 (2024)

Forney Style Factor Graphs

Bethe Predictive Coding

Magnus Koudahl[1(✉)], Alexander Tchantz[1,2], Tommaso Salvatori[1,3], Hampus Linander[1], Lancelot Da Costa[1,4], Jeff Beck[1], and Christopher Buckley[1,2]

[1] VERSES AI Research Lab, Los Angeles, CA 90016, USA
magnus.koudahl@verses.ai
[2] Department of Informatics, University of Sussex, Brighton and Hove, UK
[3] Department of Informatics, Vienna University of Technology, Vienna, Austria
[4] ELLIS Institute Tübingen, Tübingen, Germany

Abstract. Predictive coding has recently gained momentum in the machine learning community as an energy based, biologically inspired method to train neural networks via local computations only. While its energy functional has an established interpretation as a variational free energy, its relationship to other variational inference algorithms has so far not been thoroughly explored. In this work, we formally show that the predictive coding energy objective can be re-derived through a careful choice of constraints on a Bethe Free Energy functional. Doing so confers a number of benefits: First, it provides a recipe for deriving closed-form updates for any variable that is part of the predictive coding network, opening the possibility of fully gradient free, Hebbian learning. Second, it allows us to combine the tools of predictive coding with advances from both conventional Bayesian deep learning as well as traditional Bayesian methods. Finally, making the constraints explicit allows them to be manipulated in order to derive new predictive coding based objectives and algorithms. We demonstrate this by introducing a novel predictive coding network employing Matrixnormal-Wishart priors over the weights which can be trained using closed-form updates.

Keywords: Bethe Free Energy · Predictive Coding · Variational Message Passing

1 Introduction

Predictive coding (PC) is an influential theory from computational neuroscience that has recently started to gain traction in machine learning (ML) [6,15,17]. It has been successfully used to train deep neural networks that perform comparably to standard deep learning models trained using backpropagation [13]. Predictive coding networks (PCNs) accomplish this using only local updates, while enjoying several other interesting properties [19]. Among these, is the ability to run supervised learning tasks in reverse [14,20] and to train models with arbitrary graph topologies [18]. However even though PC has an established

M. Albarracin et al. (Eds.): IWAI 2025, CCIS 2857, pp. 65–74, 2026.
https://doi.org/10.1007/978-3-032-16955-6_4

interpretation as a variational free energy minimisation procedure, its relationship to the wider field of variational inference has not yet been definitively established. In effect, this creates an artificial separation between PC and the wider worlds of approximate Bayesian inference and Bayesian deep learning.

In this paper we present a new derivation of predictive coding on arbitrary graphs using the framework of constrained Forney-style factor graphs (CFFGs) [8] and constrained Bethe free energy (CBFE). In particular we show that the classical PC objective—precision weighted prediction errors—can be obtained from minimisation of a Bethe free energy (BFE) under a naive mean field factorisation with Dirac δ-form constraints on the posterior q. This provides a rigorous foundation for combining PC and other approximate Bayesian inference methods as well as for understanding recent results on PC-graphs, whose energy function has not been theoretically justified [18]. Additionally it allows for a principled framework within which to explore alternative versions of PC by varying the choice of constraints and optimiser.

2 Deriving Bethe Predictive Coding

In this section we introduce the CBFE objective on Forney-style factor graphs (FFGs). Afterwards, we will extend the formalism to CFFGs and subsequently use it to derive the PC objective on free-form graphs.

Forney-Style Factor Graphs: An FFG [5] is a collection of nodes $\mathcal{V}$ and edges $\mathcal{E}$. FFGs are commonly used to represent factorised functions

$$f(\mathbf{s}) = \prod_{a \in \mathcal{V}} f_a(s_a) \tag{1}$$

where $\mathbf{s}$ denotes the set of all variables and $s_a \in \mathbf{s}$ denotes the subset that are arguments of the factor f_a. In the FFG representation, each node $a \in \mathcal{V}$ corresponds to a factor f_a and each edge $i \in \mathcal{E}$ to a variable. An edge is connected to a node iff the variable of that edge is an argument of the factor f_a.

In this work, we will use the following notation: we denote the edges neighbouring a node a by $\mathcal{E}(a)$, and use $s_{a \setminus i}$ to denote all s_a except s_i. We use d_i to denote the degree of the edge i. By design, d_i can only be 1 or 2, since an edge can connect to maximum two nodes. When a variable is the argument of more than two factors, "copies" of it are created using equality nodes [5].

PC Objective Derivation: We now derive a version of the PC objective for generic graphical models. We do so by casting it as a CBFE and applying results from [24,25]. Following [25], we define the BFE as

$$F[q, f] = \sum_{a \in \mathcal{V}} \int q_a(s_a) \log \frac{q_a(s_a)}{f_a(s_a)} \mathrm{d}s_a - \sum_{i \in \mathcal{E}} (1 - d_i) \int q_i(s_i) \log q_i(s_i) \mathrm{d}s_i, \tag{2}$$

where q is our approximate posterior. Notice that instead of having a global q, we have instead separated q into a set of local contributions associated with either nodes $q_a(s_a)$ or edges $q_i(s_i)$. This means that instead of working with a global free energy, we can work with local free energies for each node and local entropies for each edge. Each local contribution to q in addition satisfies the constraints:

$$q_i(s_i) = \int q_a(s_a) \mathrm{d}_{s_{a \setminus i}}, \quad \text{for all } a \in \mathcal{V} \text{ and all } i \in \mathcal{E}, \tag{3}$$

$$\int q_i(s_i) \mathrm{d}s_i = 1, \quad \text{for all } i \in \mathcal{E}, \quad \int q_a(s_a) \mathrm{d}s_a = 1, \quad \text{for all } a \in \mathcal{V}. \tag{4}$$

The constraints in Eq. (3) enforce consistency among edge and node local marginals while Eq. (4) ensures that all marginals are correctly normalised. From this point onward we omit subscripts on q and instead let its argument define the marginal we are referring to: i.e., $q(s_i) = q_i(s_i)$.

To continue our derivation, we now impose a full mean field factorisation on q such that

$$q(s_a) = \prod_{i \in \mathcal{E}(a)} q(s_i). \tag{5}$$

That is, the node marginals are the product of neighbouring edge marginals. We also enforce δ-form constraints on each $q(s_i)$ such that

$$q(s_i) = \delta(s_i - \hat{s}_i) \quad \text{for all } i \in \mathcal{E}. \tag{6}$$

which forces the edge marginal over any variable s_i to take the form of a Dirac δ centered at the point $\hat{s}_i$. Enforcing the δ-constraints of Eq. (6) causes the edge entropies $H[q_i]$ to be constant,[1] leaving only the node-local free energies in (2):

$$F[q, f] = \sum_{a \in \mathcal{V}} \int q(s_a) \log \frac{q(s_a)}{f_a(s_a)} \mathrm{d}s_a - \underbrace{\sum_{i \in \mathcal{E}} (1 - d_i) \int q(s_i) \log q(s_i) \mathrm{d}s_i}_{=C}, \tag{7}$$

[1] Either 0 or $-\infty$. In either case they are constant under optimisation and can therefore be disregarded.

Substituting in the results of enforcing our constraints and omitting constants, we obtain

$$F[q,f] = \sum_{a\in\mathcal{V}} \int \overbrace{\prod_{i\in\mathcal{E}(a)} q(s_i)}^{q(s_a)} \log \frac{\overbrace{\prod_{i\in\mathcal{E}(a)} q(s_i)}^{q(s_a)}}{f_a(s_a)} \mathrm{d}s_a \qquad \text{by mean field factorisation} \tag{8a}$$

$$= \sum_{a\in\mathcal{V}} \int \prod_{i\in\mathcal{E}(a)} \delta(s_i - \hat{s_i}) \log \frac{\prod_{i\in\mathcal{E}(a)} \delta(s_i - \hat{s_i})}{f_a(s_a)} \mathrm{d}s_{i\in\mathcal{E}(a)} \qquad \text{by } \delta\text{-form constraint} \tag{8b}$$

$$= -\sum_{a\in\mathcal{V}} \int \prod_{i\in\mathcal{E}(a)} \delta(s_i - \hat{s_i}) \log f_a(s_a) \mathrm{d}s_{i\in\mathcal{E}(a)} + C. \tag{8c}$$

In the last line we once again recognise that the entropy of a product of δ's is constant, and can be disregarded during optimisation. Now we can solve the integral in (8c) by applying the sifting property of the Dirac δ to recover the PC objective. To exemplify, we show the result of substituting in the definition for the node function f_a corresponding to a single MLP layer (with weights W and nonlinearity $g(\cdot)$) in a PCN [2,9] and writing out the appropriate definition for q for a single node:

$$\begin{aligned}
&\int \prod_{i\in\mathcal{E}(a)} \delta(s_i - \hat{s_i}) \log f_a(s_a) s_{i\in\mathcal{E}(a)} \\
&= \iiiint \delta(x-\hat{x})\delta(y-\hat{y})\delta(W-\hat{W})\delta(\Sigma^{-1}-\hat{\Sigma}^{-1}) \log \mathcal{N}(y \mid g(Wx), \Sigma^{-1}))\mathrm{d}x\mathrm{d}y\mathrm{d}\Sigma^{-1}\mathrm{d}W \\
&= -\log \mathcal{N}\left(\hat{y} \mid g(\hat{W}\hat{x}), \hat{\Sigma}^{-1}\right) .
\end{aligned} \tag{9}$$

Of note here is that the δ-form constraints effectively make the objective (8c) a hybrid between an energy based model and variational Bayes; equivalently, this objective combines variational inference with EM-as-message-passing [3]. Which marginals are treated in what way is then controlled by the application of δ-form constraints. The flexibility afforded by the CBFE approach allows one to mix and match which marginals are treated in what way, opening the door for hybrid approaches.

3 Extending PC by Constraint Manipulation

The main advantage of the CBFE approach to PC is that, by making the constraint set and optimiser of standard PC explicit, we can start to manipulate them to obtain new variants of PC. This is most easily illustrated using the CFFG notation introduced in [8] and detailed in Appendix A. In brief, CFFGs enhance the default FFG notation by also denoting the constraint set using beads on nodes and edges. Form constraints are denoted by symbols inside beads, and each bead corresponds to a marginal in q. Figure 1 shows the CFFG of the composite factor node of a single multilayer perceptron (MLP) layer in a standard PCN.

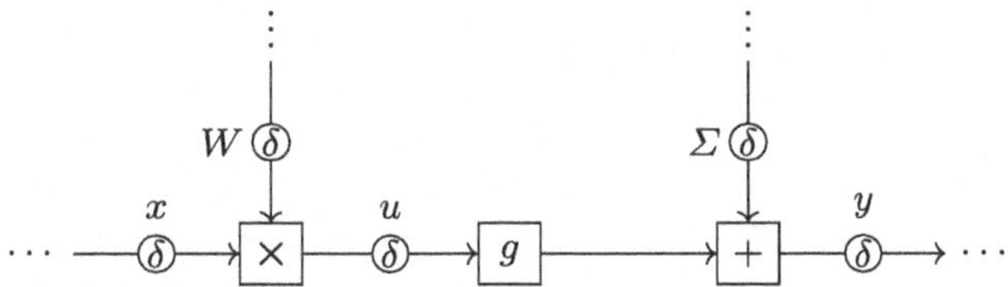

Fig. 1. CFFG of a single MLP layer with δ-constraints.

When used with a gradient descent optimiser, Fig. 1 recovers the standard algorithm for training a single layer of a PCN. However more generally, the CBFE provides a unified framework for deriving—through variational calculus—a variety of gradient-free (message passing) optimisation algorithms [24,25].

As an example, the constraint set we have investigated so far would correspond to a variational EM procedure over all variables in the PCN [25].

Without the δ-constraints in Fig. 1, we would instead recover the mean field variational message passing (VMP) updates [22]. Further relaxing the mean field factorisation in q would instead lead us to performing belief propagation (BP) [12]. Of note here is that we still need to actually be able to compute the requisite messages in order to get a practical algorithm.

In short, CBFE provides a direct recipe for obtaining closed form solutions to all relevant quantities in a PCN. The resulting closed form, gradient free update rules can then be freely mixed and matched with gradient based optimisers as desired.

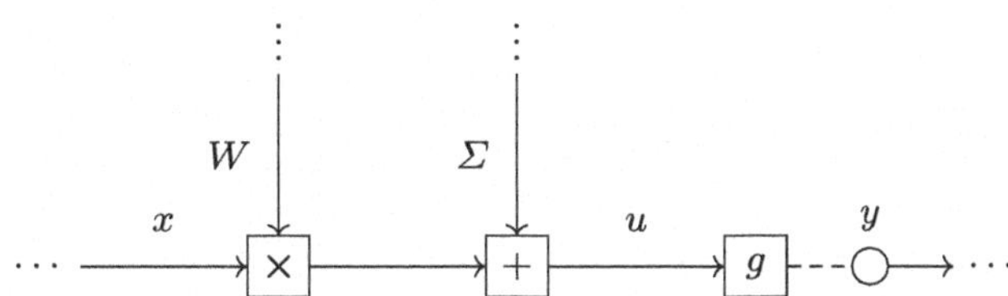

Fig. 2. CFFG of a DVI layer.

Another feature of the CBFE approach is that it allows us to unify PC with work done on traditional Bayesian Neural Networks. This has very concrete advantages such as allowing for principled uncertainty propagation through PCNs. For example, the DVI method of [23] can be succinctly expressed by rearranging some nodes on the CFFG and applying a moment matching constraint (dashed line) as illustrated in Fig. 2 instead of the mean field/δ-constraint combination. To get a bias term, one can assume that Σ is Gaussian with non-zero mean. A subtle difference compared to the PC of Fig. 1 is that uncertainty is also propagated through the non-linearity g.

4 Experiments

It follows that the CBFE approach affords new PC algorithms. For example, by separating the pre- and post-activation marginals shown in Fig. 1, we now have a CFFG that is locally linear around the weights. This allows us to add conjugate (MatrixNormal-Wishart; MNW) priors over both the weights and the noise, so that the closed form updates become straightforward. We illustrate this in Fig. 3, with the node function $\mathcal{L}$ denoting the linear transition $\mathcal{N}(u \mid Wx, \Sigma^{-1})$. All experiments are available at https://github.com/VersesTech/cbfe_pc.

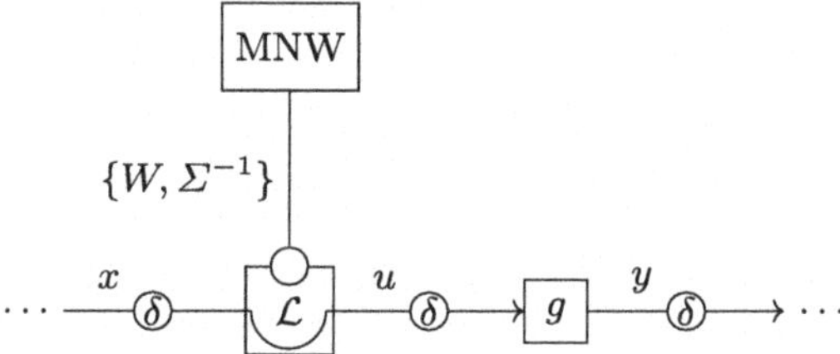

Fig. 3. CFFG of an MLP layer with conjugate MNW prior. Variables $\{x, u, y\}$ are optimised with gradient descent for experiments on the `UCI hardware` dataset.

Several methods exist for passing messages through a non-linearity, e.g. [4,7,16,21,23]. We choose to employ the PC solution of using δ-form constraints, where the difficult task of accurately passing messages through the non-linearity is solved at the cost of inverting the non-linearity and not propagating uncertainty. We plan to investigate other methods in future work. We test the resulting algorithm on the `two moons` data set with a two-layer MLP network. We used a tanh non-linearity and layers with hidden dimensions [32, 16]. We trained the network using exclusively closed-form updates and achieve 99% test accuracy within a single round of message passing. The resulting decision boundary is shown in Fig. 4.

While the `two moons` dataset remains a toy example, it provides proof of concept that the gradient-free optimisers made available from the CBFE approach to PC can afford efficient training of neural networks.

To illustrate the efficiency of our approach on real world datasets, as well as the ability to vary the choice of optimiser, we trained a similar model with layer dimensions [50, 50] using `ReLU` nonlinearities on the `UCI hardware` dataset available at https://archive.ics.uci.edu/. Here we chose a gradient descent optimiser for the marginals enclosed by the box in Fig. 3. Effectively, this extends

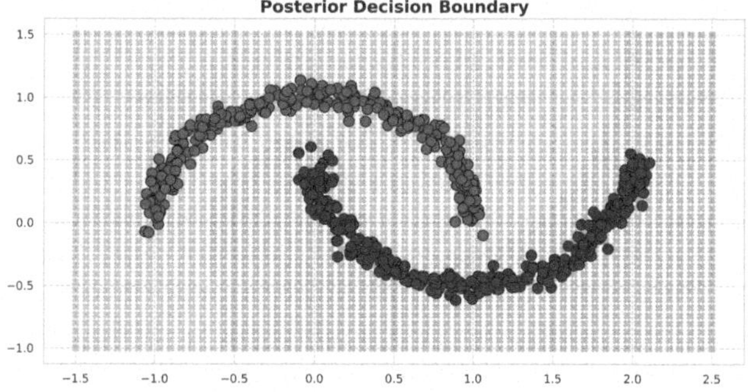

Fig. 4. Decision boundary on the two moons dataset.

the standard PCN with conjugate MatrixNormal-Wishart priors for uncertainty on the weights. We obtain comparable performance to Bayes-by-Backprop [1] with a slightly faster convergence. Results are shown in Fig. 5.

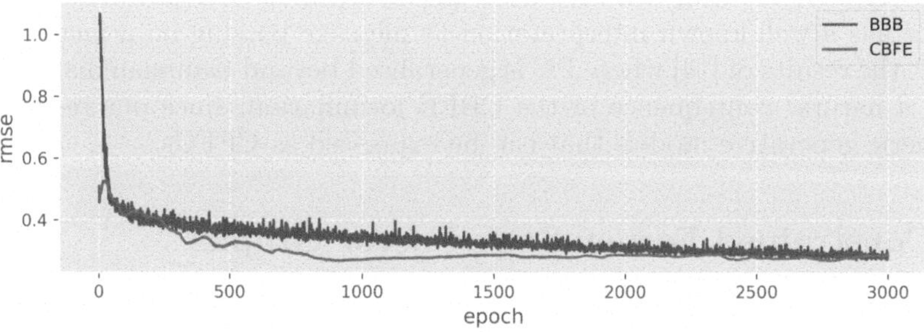

Fig. 5. Root mean squared error (RMSE) for CBFE vs Bayes-by-Backprop (BBB) [1] on UCI hardware dataset.

The goal of these experiments is mainly to illustrate the flexibility afforded by the CBFE approach. We explicitly manipulate the factorisation in q to use MNW conjugate priors, combine VMP updates with gradient descent (GD) optimisation and isolate the preactivation marginal through careful construction of the CFFG. We envision that scaling our approach to deeper models and more complicated architectures as well as more complex datasets, is going to present empirical challenges. We therefore limit the present exposition to proof-of-concept demonstrations and leave further experimentation to future work.

5 Conclusions

We have shown how to derive the PC objective as a special case of CBFE. This expands upon previous results showing the relation between PC and variational free energy (VFE) minimisation [2,6] by making explicit the VFE functional and the precise set of constraints required to obtain PC.

Having established this link immediately generalises PCNs from layered structures to arbitrary graph topologies. This has already been empirically investigated in [18]. Our results substantiates [18] and provides a principled grounding for their results, which have previously been mainly heuristically motivated.

The CBFE approach also greatly expands the set of candidate node functions we can consider when constructing PCNs since we now have the full expressive power of probabilistic generative models available. Broadening the scope of available building blocks for PCNs also means we are now able to construct networks such that they can—either fully or partially—be solved using gradient-free message passing methods. We demonstrated this experimentally by training a PCN with conjugate MNW priors over the weights without using a single gradient update. This was made possible by carefully attending to the underlying CFFG.

The CBFE approach also provides a principled foundation for understanding various peculiarities of PC, such as the ability to run classifiers in reverse [20]—since the underlying CFFG is undirected and the marginals can be optimised to equilibrium, it is not surprise that the model can be inverted. The close relationship to Kalman filtering demonstrated in [10,11] is also natural since Kalman filtering has a well-known interpretation as message passing on a factor graph. Finally, the results of [14] where PC is generalised beyond Gaussian distributions is also a natural consequence of the CBFE formulation, since our results hold for generic generative models that can be expressed as CFFGs.

A Constrained Forney-Style Factor Graphs

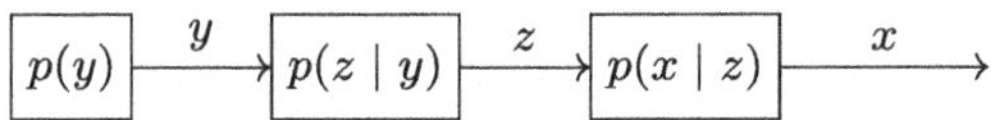

Fig. 6. An example FFG.

Forney-style factor graphs (FFGs) are a graphical formalism for describing probablistic generative models introduced in [5]. An FFG differs from a Bayesian network in that *edges* represent variables and nodes represent factors. For example, the model

$$p(x, y, z) = p(x|z)p(z|y)p(y) \tag{10}$$

would correspond to the FFG in Fig. 6. Constrained Forney-style factor graphs (CFFGs) are a generalisation of FFGs that introduce notation for constraints on the BFE implied by the graph. CFFGs denote marginals in q by circular beads with each bead corresponding to a marginal. By default, CFFGs only explicitly denote factorisations or constraints beyond normalisation and marginalisation, e.g. beyond (2) and (5). Beads for edge marginals are only included when they are explicitly modified, otherwise they are left implicit. Constraints on individual marginals are denoted by symbols inside beads. For example, if we wanted to: *(i)* add a local mean field factorisation around the node $p(x|z)$ in Fig. 6, *(ii)* leave a joint factorisation around $p(z|y)$ and *(iii)* add a δ-form constraint to y, we would use the CFFG in Fig. 7.

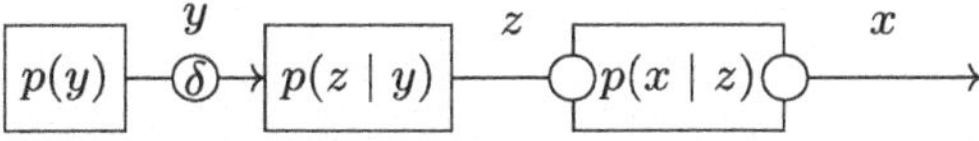

Fig. 7. Sample FFG updated with constraints, making it a CFFG.

This allows for specifying both the generative model p and the variational distribution q graphically. Together, these completely specify the inference problem at hand. For further details we refer to [8].

References

1. Blundell, C., Cornebise, J., Kavukcuoglu, K., Wierstra, D.: Weight Uncertainty in Neural Networks. arXiv:1505.05424 [cs, stat] (May 2015)
2. Buckley, C., Kim, C., McGregor, S., Seth, A.: The free energy principle for action and perception: a mathematical review. J. Math. Psychol. (2017)
3. Dauwels, J., Korl, S., Loeliger, H.-A.: Expectation maximization as message passing. arXiv preprint cs/0508027 (2005)
4. Eskin, E., Smola, A., Vishwanathan, S.: Laplace Propagation. In: Advances in Neural Information Processing Systems, vol. 16. MIT Press (2004)
5. Forney, G.D., Jr.: Codes on graphs: normal realizations. IEEE Trans. Inform. Theory **47**(2), 520–548 (2001)
6. Friston, K.: A theory of cortical responses. Philosop. Trans. Royal Soc. B: Biolog. Sci. **360**(1456) (2005)
7. Goulet, J.-A., Nguyen, L.H., Amiri, S.: Tractable approximate gaussian inference for bayesian neural networks. J. Mach. Learn. Res. **22**(251), 1–23 (2021)
8. Koudahl, M., van de Laar, T., de Vries, B.: Realising synthetic active inference agents, part i: Epistemic objectives and graphical specification language (2023)
9. Millidge, B., Seth, A., Buckley, C.L.: Predictive coding: a theoretical and experimental review (2021)
10. Millidge, B., Tang, M., Osanlouy, M., Harper, N.S., Bogacz, R.: Predictive coding networks for temporal prediction. PLoS Comput. Biol. **20**(4), e1011183 (2024)
11. Millidge, B., Tschantz, A., Seth, A., Buckley, C.: Neural kalman filtering (2021)
12. Pearl, J.: Reverend bayes on inference engines: a distributed hierarchical approach. In: Proceedings of the Second AAAI Conference on Artificial Intelligence, Aaai 1982, pp. 133–136. AAAI Press, Pittsburgh, Pennsylvania (1982)
13. Pinchetti, L., et al.: Benchmarking predictive coding networks–made simple. arXiv preprint arXiv:2407.01163 (2024)

14. Pinchetti, L., Salvatori, T., Millidge, B., Song, Y., Yordanov, Y., Lukasiewicz, T.: Predictive coding beyond Gaussian distributions. In: 36th Conference on Neural Information Processing Systems (2022)
15. Rao, R.P.N., Ballard, D.H.: Predictive coding in the visual cortex: a functional interpretation of some extra-classical receptive-field effects. Nat. Neurosci. **2**(1), 79–87 (1999)
16. Ribeiro, M.I.: Kalman and extended kalman filters: concept, derivation and properties. Instit. Syst. Robot. **43**(46), 3736–3741 (2004)
17. Salvatori, T., et al.: Brain-inspired computational intelligence via predictive coding. arXiv preprint arXiv:2308.07870 (2023)
18. Salvatori, T., et al.: Learning on arbitrary graph topologies via predictive coding. arXiv:2201.13180 (2022)
19. Song, Y., Millidge, B., Salvatori, T., Lukasiewicz, T., Xu, Z., Bogacz, R.: Inferring neural activity before plasticity as a foundation for learning beyond backpropagation. Nat. Neurosci. **27**(2), 348–358 (2024)
20. Tscshantz, A., Millidge, B., Seth, A.K., Buckley, C.L.: Hybrid predictive coding: inferring, fast and slow. PLoS Comput. Biol. **19**(8), e1011280 (2023)
21. Wan, E., Van Der Merwe, R.: The unscented Kalman filter for nonlinear estimation. In: Proceedings of the IEEE 2000 Adaptive Systems for Signal Processing, Communications, and Control Symposium (Cat. No.00EX373), pp. 153–158. IEEE, Lake Louise, Alta., Canada (2000)
22. Winn, J., Bishop, C.M.: Variational message passing. J. Mach. Learn. Res. **6**(4), 661–694 (2005)
23. Wu, A., Nowozin, S., Meeds, E., Turner, R.E., Hernández-Lobato, J.M., Gaunt, A.L.: Deterministic variational inference for robust bayesian neural networks (2019)
24. Zhang, D., Song, X., Wang, W., Fettweis, G., Gao, X.: Unifying message passing algorithms under the framework of constrained bethe free energy minimization. IEEE Trans. Wireless Commun. **20**(7), 4144–4158 (2021)
25. Şenöz, I., van de Laar, T., Bagaev, D., de Vries, B.: Variational message passing and local constraint manipulation in factor graphs. Entropy **23**(7), 807 (2021)

A Message Passing Realization of Expected Free Energy Minimization

Wouter W. L. Nuijten[1(✉)], Mykola Lukashchuk[1], Thijs van de Laar[1], and Bert de Vries[1,2]

[1] Eindhoven University of Technology, 5612, AP Eindhoven, The Netherlands
w.w.l.nuijten@tue.nl
[2] GN Hearing, 5612, AB Eindhoven, The Netherlands

Abstract. We present a message passing approach to Expected Free Energy (EFE) minimization on factor graphs, based on the theory introduced in [37]. By reformulating EFE minimization as Variational Free Energy minimization with epistemic priors, we transform a combinatorial search problem into a tractable inference problem solvable through standard variational techniques. Applying our message passing method to factorized state-space models enables efficient policy inference. We evaluate our method on environments with epistemic uncertainty: a stochastic gridworld and a partially observable Minigrid task. Agents using our approach consistently outperform conventional KL-control agents on these tasks, showing more robust planning and efficient exploration under uncertainty. In the stochastic gridworld environment, EFE-minimizing agents avoid risky paths, while in the partially observable minigrid setting, they conduct more systematic information-seeking. This approach bridges active inference theory with practical implementations, providing empirical evidence for the efficiency of epistemic priors in artificial agents.

Keywords: Active Inference · Epistemic Planning · Expected Free Energy · Factor Graphs · Message Passing

1 Introduction

Expected Free Energy (EFE) minimization, rooted in the Free Energy Principle, provides a framework for modeling intelligent behavior by unifying reward-seeking (pragmatic) and information-seeking (epistemic) drives [16,18]. While control-as-inference approaches have made significant advances in formulating decision-making as probabilistic inference problems [1,20], EFE minimization extends this paradigm by explicitly accounting for epistemic uncertainty [13], though its practical application faces computational challenges for extended planning horizons and high-dimensional state-spaces [30].

Traditional approaches to computing EFE often involve evaluating all possible action sequences, which becomes intractable for non-trivial problems. While various approximations have been developed to address this tractability issue,

M. Albarracin et al. (Eds.): IWAI 2025, CCIS 2857, pp. 75–98, 2026.
https://doi.org/10.1007/978-3-032-16955-6_5

traditional approaches typically use EFE as a cost function for evaluating policies, rather than as an objective functional for variational optimization of beliefs [8,19,29].

This paper provides empirical validation of the theoretical foundation presented in [37], which reformulates EFE minimization directly as a variational inference problem on factor graphs. By introducing appropriate epistemic priors, we show that minimizing EFE can be achieved through standard Variational Free Energy (VFE) minimization, making it consistent with the Free Energy Principle's core tenet that all processes are fundamentally based on variational free energy minimization.

We implement this approach through an iterative message passing algorithm on factorized state-space models. We evaluate its performance in environments with different uncertainty characteristics: a stochastic gridworld with perilous transitions and a partially observable Minigrid environment requiring active exploration for successful completion. Our results confirm that agents using our inference-based method exhibit the same characteristic advantages over KL-control agents as direct EFE computation, particularly in handling epistemic uncertainty. This validates our approach while providing a computationally efficient framework for planning under uncertainty.

The remainder of this paper is organized as follows:

- Sect. 2 provides background on necessary materials.
- Sect. 3 discusses related work in control as inference and active inference.
- Sect. 4 presents our methodology for reformulating EFE minimization as an inference problem.
- Sect. 5 describes our evaluation environments and experimental design.

2 Background

2.1 Variational Inference

Variational inference (VI) provides a principled framework for approximating complex posterior distributions in Bayesian models [6,7,24,38]. The central challenge in Bayesian inference is computing the posterior distribution $p(\boldsymbol{x}|\boldsymbol{y})$ of hidden state sequence $\boldsymbol{x}$ given an observed data sequence $\boldsymbol{y}$, which requires evaluating the model evidence $p(\boldsymbol{y})$ [11,21]. This normalization constant is typically intractable for complex models.

VI reformulates inference as an optimization problem by approximating the Bayesian posterior with a simpler, tractable distribution $q(\boldsymbol{x})$ from a family of distributions $\mathcal{Q}$ [7]. The functional we will minimize is the Variational Free Energy (VFE). The VFE is defined as $F[q] = D_{KL}(q(\boldsymbol{x})\|p(\boldsymbol{x}|\boldsymbol{y})) - \log p(\boldsymbol{y})$, making it clear that minimizing the VFE is equivalent to minimizing the KL divergence since $\log p(\boldsymbol{y})$ is constant with respect to q. The VFE also provides a tractable upper bound on the negative log evidence, with $F[q] \geq -\log p(\boldsymbol{y})$ [23].

2.2 Factor Graphs

Factor graphs are a specific type of probabilistic graphical model that explicitly represents the factorization structure of the model, where factors represent (conditional) probability distributions. In our work, we employ Forney-style factor graphs (FFGs) [15], which offer a specific representation approach with notation following [26].

An FFG represents a factorized function $f(\boldsymbol{s})$ as

$$f(\boldsymbol{s}) = \prod_{a \in \mathcal{V}} f_a(\boldsymbol{s}_a), \tag{1}$$

where $\boldsymbol{s}$ encompasses all variables in the model, and $\boldsymbol{s}_a \subseteq \boldsymbol{s}$ represents the subset of variables that participate in factor f_a.

In the FFG representation, nodes ($a \in \mathcal{V}$) correspond to factors in the model, while edges ($\mathcal{E} \subseteq \mathcal{V} \times \mathcal{V}$) represent variables. An edge connects to a node precisely when the variable appears as an argument in the corresponding factor. We denote the set of edges connected to node $a \in \mathcal{V}$ as $\mathcal{E}(a)$, and the nodes connected to edge $i \in \mathcal{E}$ as $\mathcal{V}(i)$.

To illustrate, the FFG representation of the factorized function

$$f(s_1, s_2, s_3, s_4) = f_a(s_1) f_b(s_1, s_2) f_c(s_3) f_d(s_2, s_3, s_4) \tag{2}$$

in shown in Fig. 1.

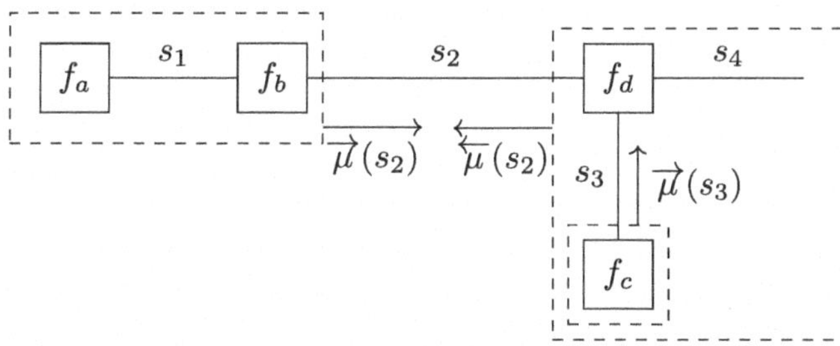

Fig. 1. A Forney-style factor graph representation of the factorization in (2).

A common approach to realizing efficient variational inference on factor graphs involves the Bethe assumption, which posits that the posterior distribution factorizes as a product of local marginals associated with the nodes and edges of the graph. This structural assumption on the posterior distributions enables the formulation of message passing algorithms that seek out stationary points of the Bethe free energy [14,39,40].

To illustrate the computational benefits of message passing, consider the generative model in (2), and assume we are interested in computing $p(s_2)$. This marginal distribution can be obtained by summing out all other variables from the joint

$$p(s_2) = \sum_{s_1} \sum_{s_3} \sum_{s_4} f(s_1, s_2, s_3, s_4), \tag{3}$$

which, when each s_i can take 10 values, contains about a thousand terms. However, taking into account the factorization of the generative model and the distributive law of the product, (3) can be rewritten as

$$p(s_2) = \underbrace{\left(\sum_{s_1} f_a(s_1) f_b(s_1, s_2) \right)}_{\overrightarrow{\mu}(s_2)} \cdot \underbrace{\left(\overbrace{\left(\sum_{s_3} f_c(s_3)\right)}^{\overrightarrow{\mu}(s_3)} \sum_{s_4} f_d(s_2, s_3, s_4) \right)}_{\overleftarrow{\mu}(s_2)}. \tag{4}$$

The computation in (4) requires only a few hundred summations and is preferable from a computational standpoint. In larger models, the number of computations scale linearly with the number of factor nodes, instead of exponentially. The intermediate results $\overrightarrow{\mu}(s_i)$ and $\overleftarrow{\mu}(s_i)$ afford an interpretation as local message in the FFG representation of the model, see Fig. 1. For comprehensive treatments of factor graphs and associated (variational) message passing algorithms, we refer readers to [14,26,27,39,40].

3 Related Work

Autonomous decision-making under uncertainty remains a central challenge in control theory and artificial intelligence. This section reviews key developments that contextualize our contribution.

3.1 Control as Inference

The pursuit of efficient and high-performing autonomous systems has driven significant research in control theory. Optimal control [3,4,32] provides a mathematical framework for determining the control inputs that minimize a predefined cost function for a given system. Building upon these foundations, Model Predictive Control (MPC) algorithms address the challenges of real-time control by incorporating a feedback loop and a receding horizon strategy [5,12,33,34]. This approach allows for online adaptation to disturbances and constraints.

A significant paradigm shift in recent years involves viewing control as an inference problem. This perspective allows the application of powerful probabilistic tools to address control challenges, particularly in complex and uncertain environments. Under deterministic dynamics, the sequential decision-making process in closed-loop receding horizon MPC can be elegantly mapped to inference on a factor graph [25,28].

When dealing with stochastic dynamics or the need for state estimation under uncertainty, stochastic optimal control methods can be reformulated using variational inference [20,22]. Here, the intractable posterior distribution over states and/or controls is approximated by a tractable variational distribution.

Active inference [10,13] addresses control under uncertainty by proposing that information gained about the system is also a form of reward. The framework suggests that variational inference naturally balances exploration and exploitation by optimizing the Expected Free Energy [18], which elegantly combines the drive to minimize uncertainty about the environment (information gain) with the need to achieve desired outcomes. However, a current limitation of active inference lies in the computational cost associated with computing the Expected Free Energy [18], which has spurred recent research into efficient algorithms [8,17,29,30].

Recently, [37] proposed an alternative approach to Expected Free Energy minimization by framing EFE minimization as a regular variational free energy minimization task. This approach is promising for scalable implementation of EFE-minimizing planning algorithms, but offers a theoretical account, without considering practical implementation or empirical validation. In the next section, we will propose a message passing realization of this approach.

4 Methodology

For the main contribution of this paper, we will elaborate on Theorem 1 from [37]. For convenience, we will repeat the theorem here, albeit without the inclusion of model parameters θ:

Theorem 1 (Expected Free Energy Theorem). *Consider an agent with generative model $p(\boldsymbol{y}, \boldsymbol{x}, \boldsymbol{u})$, and prior beliefs $\hat{p}(\boldsymbol{x})$ about future desired states.*

Consider the Variational Free Energy functional

$$F[q] \triangleq E_{q(\boldsymbol{y},\boldsymbol{x},\boldsymbol{u})}\left[\log \frac{\overbrace{q(\boldsymbol{y},\boldsymbol{x},\boldsymbol{u})}^{\text{posterior}}}{\underbrace{p(\boldsymbol{y},\boldsymbol{x},\boldsymbol{u})}_{\substack{\text{generative}\\\text{model}}}\ \underbrace{\hat{p}(\boldsymbol{x})}_{\substack{\text{preference}\\\text{prior}}}\ \underbrace{\tilde{p}(\boldsymbol{u})\tilde{p}(\boldsymbol{x})}_{\substack{\text{epistemic}\\\text{priors}}}}\right], \tag{5}$$

where the generative model in the denominator is augmented by both a preference prior $\hat{p}(\cdot)$ and epistemic priors $\tilde{p}(\cdot)$.

If the epistemic priors are chosen as

$$\tilde{p}(\boldsymbol{u}) \propto \exp(H[q(\boldsymbol{x}|\boldsymbol{u})]) \tag{6a}$$

$$\tilde{p}(\boldsymbol{x}) \propto \exp(-H[q(\boldsymbol{y}|\boldsymbol{x})]) \tag{6b}$$

then $F[q]$ decomposes as

$$F[q] = \underbrace{E_{q(\boldsymbol{u})}\left[G(\boldsymbol{u})\right]}_{\substack{\text{expected policy}\\\text{costs}}} + \underbrace{E_{q(\boldsymbol{y},\boldsymbol{x},\boldsymbol{u})}\left[\log \frac{q(\boldsymbol{y},\boldsymbol{x},\boldsymbol{u})}{p(\boldsymbol{y},\boldsymbol{x},\boldsymbol{u})}\right]}_{\text{complexity}} + \text{constant}, \tag{7}$$

where

$$G(\boldsymbol{u}) = \mathbb{E}_{q(\boldsymbol{y},\boldsymbol{x}|\boldsymbol{u})}\left[\log\left(\frac{q(\boldsymbol{x}|\boldsymbol{u})}{\hat{p}(\boldsymbol{x})} \cdot \frac{1}{q(\boldsymbol{y}|\boldsymbol{x})}\right)\right] \tag{8}$$

is the expected free energy as defined in [13]. In (6),

$$H[q(y|x)] = -\int q(y|x) \log q(y|x)\, dy \tag{9}$$

is the entropy functional.

Proof. The proof of (7) is given in [37, Appendix A]. □

While (7) shows that minimization of the $F[q]$ leads to minimization of (expected) $G(u)$, the proof of (7) is declarative and does not provide an explicit algorithm for minimizing $F[q]$.

In the following sections, we will describe a message passing algorithm on factor graphs that can be used as a practical approach to search for stationary points of the free energy functional.

4.1 Factorized Models and Factorized Posteriors

Theorem 1 is a general result, however, in practice, we are often interested in factorized state-space models of the form

$$p(\boldsymbol{y}, \boldsymbol{x}, \boldsymbol{u}) = p(x_0) \prod_{t=1}^{T} p(y_t|x_t) p(x_t|x_{t-1}, u_t) p(u_t) \tag{10}$$

We can make an additional assumption that the posterior distribution factorizes in the same way as the generative model:

$$q(\boldsymbol{y}, \boldsymbol{x}, \boldsymbol{u}) = q(x_0) \prod_{t=1}^{T} q(y_t|x_t) q(x_t|x_{t-1}, u_t) q(u_t)\,. \tag{11}$$

Note that this is consistent with making the Bethe assumption, which says that the variational posterior distribution can be decomposed into local contributions:

$$q(\boldsymbol{s}) = \prod_{a \in \mathcal{V}} q_a(\boldsymbol{s}_a) \prod_{i \in \mathcal{E}} q_i(s_i)^{-1} \tag{12}$$

for $\mathcal{G} = (\mathcal{V}, \mathcal{E})$ the underlying FFG. Under this assumption, we can derive a corollary to Theorem 1 that provides more specific expressions for the epistemic priors.

Corollary 1. *Consider an agent with Variational Free Energy functional as in* (5)*, comprising a generative model* (10)*, a posterior distribution factorized as in* (11)*, and a preference prior* $\hat{p}(\boldsymbol{x}) = \prod_{t=1}^{T} \hat{p}(x_t)$*. If the epistemic priors are chosen as*

$$\tilde{p}(u_t) \propto \exp(H[q(x_t, x_{t-1}|u_t)] - H[q(x_{t-1}|u_t)]) \tag{13a}$$
$$\tilde{p}(x_t) \propto \exp(-H[q(y_t|x_t)]) \tag{13b}$$

then the Variational Free Energy functional (5) *decomposes as*

$$F[q] = E_{q(\boldsymbol{u})}\big[G(\boldsymbol{u})\big] + E_{q(\boldsymbol{y},\boldsymbol{x},\boldsymbol{u})}\left[\log \frac{q(\boldsymbol{y}, \boldsymbol{x}, \boldsymbol{u})}{p(\boldsymbol{y}, \boldsymbol{x}, \boldsymbol{u})}\right] + \text{constant}. \tag{14}$$

While this corollary is a special case and a direct application of Theorem 1, an elaboration of the proof is given in Appendix A. This corollary states that the preference and epistemic priors can be reduced to local contributions. We will implement the preference and epistemic priors as factor nodes that act as prior distributions during the inference procedure. A timeslice of the augmented factor graph is shown in Fig. 2.

The benefit of this approach is that inference on factor graphs is well-understood and can be implemented efficiently using reactive message passing [2]. Effectively, this means that the computational complexity of Expected Free Energy minimization is the same as the computational complexity of variational inference on a factor graph.

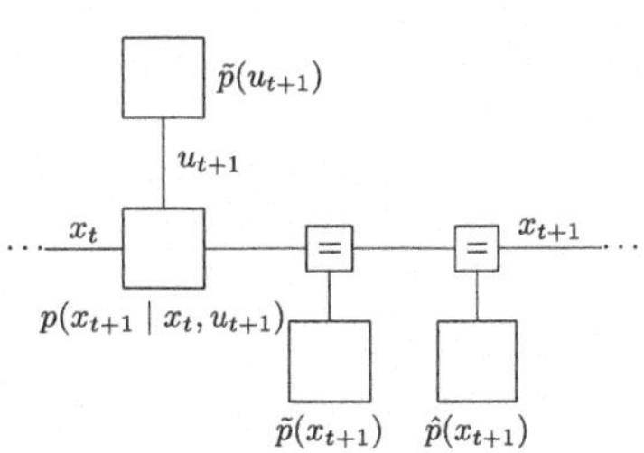

Fig. 2. Slice of the factor graph representation of the augmented generative model. The original generative model (10) is augmented with epistemic priors $\tilde{p}(u_{t+1})$ and $\tilde{p}(x_{t+1})$, and preference priors $\hat{p}(x_{t+1})$ for future timesteps.

4.2 Inferring a Policy Posterior

Corollary 1 introduces a circular dependency in the model definition: to define the VFE functional with epistemic priors (13), we need access to the variational posterior distribution, but the variational posterior can only be obtained by minimizing the VFE functional given the generative model.

This circular dependency can be resolved through an iterative variational inference procedure implemented as message passing on the factor graph. We first initialize the variational posterior and then iteratively update both the posterior beliefs and epistemic priors until convergence.

On a factor graph, we can implement variational inference using message passing algorithms that iteratively updates posterior distributions [31].

Each message passing iteration τ refines both the posteriors and priors simultaneously. To that extent, let $q_\tau(\cdot)$ be the variational posterior distribution at iteration τ, we then define the epistemic priors as

$$\begin{aligned}\tilde{p}_\tau(u_t) &= \sigma\big(H[q_{\tau-1}(x_t, x_{t-1}|u_t)] - H[q_{\tau-1}(x_{t-1}|u_t)]\big)\\ \tilde{p}_\tau(x_t) &= \sigma\big(-H[q_{\tau-1}(y_t|x_t)]\big)\,.\end{aligned} \tag{15}$$

Here, σ is the softmax function, which guarantees proportionality as in Eqs. 13a and 13b. A formal description of the algorithm is given in Algorithm 1. While this approach solves the initialization problem, there are some subtleties that need to be addressed. Specifically, although the subtraction of entropies in line Eq. 21a results in a constant when using the same variational distribution q for both the epistemic prior $\tilde{p}$ and the optimization, this property no longer holds when we use different distributions - namely, when we use $q_{\tau-1}$ to define $\tilde{p}_\tau$

but optimize with respect to q_τ. While this is not a problem if the inference procedure converges, this convergence is not guaranteed.

Algorithm 1. EFE minimization as VFE minimization

Input: Factorized generative model $p(\boldsymbol{y}, \boldsymbol{x}, \boldsymbol{u})$, preference prior $\hat{p}(\boldsymbol{x})$, number of VI iterations τ_{max}

Output: Policy posterior $q_{\tau_{max}}(\boldsymbol{u})$

$q_0(\boldsymbol{y}, \boldsymbol{x}, \boldsymbol{u}) \leftarrow$ Uninformative distribution

for $\tau \leftarrow 1$ to τ_{max} **do** ▷ Iterations of variational inference algorithm

  for each time step t **do**

    $\tilde{p}_\tau(u_t) \leftarrow \sigma(H[q_{\tau-1}(x_t, x_{t-1}|u_t)] - H[q_{\tau-1}(x_{t-1}|u_t)])$

    $\tilde{p}_\tau(x_t) \leftarrow \sigma(-H[q_{\tau-1}(y_t|x_t)])$

  end for

  $q_\tau(\boldsymbol{y}, \boldsymbol{x}, \boldsymbol{u}) \leftarrow \text{infer}(p(\boldsymbol{y}, \boldsymbol{x}, \boldsymbol{u}))$ ▷ Message passing (4)

end for

return $q_{\tau_{max}}(\boldsymbol{u})$

5 Evaluation

This section evaluates our EFE-minimizing policy inference method. In this section, we will evaluate the performance of the proposed method. The addition of preference priors is consistent with the literature on KL control [35,36], which means the main point of interest is the influence of the epistemic priors on the policy posterior. To this extent, we will execute the experiments both with and without the epistemic priors, which will correspond to a KL-control and an EFE-minimizing policy, respectively. KL-control is known to be prone to optimistic planning in the face of stochasticity and uncertainty [25,28], so we will explore partially observable Markov decision processes (POMDPs) with stochastic dynamics and observation noise.

For our experimental evaluation, we consider scenarios where the environment dynamics are completely known to the agent, though they may be stochastic or contain inherent uncertainty. This known-dynamics assumption allows us to isolate and evaluate the specific effects of epistemic priors on decision-making, without conflating them with model learning.

5.1 Experimental Design

We designed a stochastic grid environment that specifically challenges agents with uncertainty in dynamics and observations. Additionally, we evaluate our method on the Minigrid door-key environment [9], which tests how agents handle partial observability. Both environments highlight the differences between KL-control and EFE-minimizing policies in the presence of epistemic uncertainty.

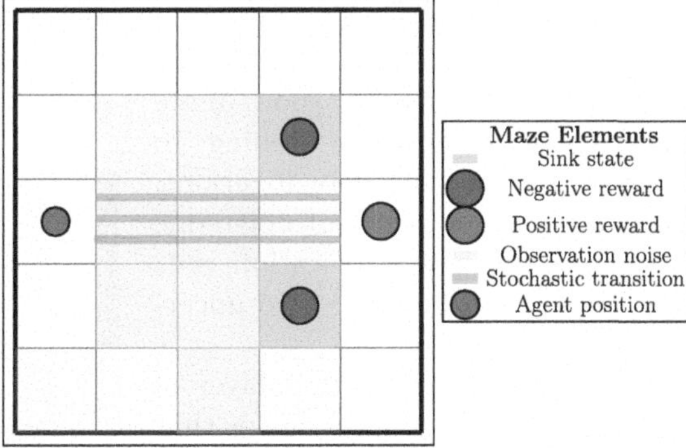

Fig. 3. The stochastic grid environment. The agent should traverse the grid with both stochastic transitions and observation noise. Cells with stochastic transitions appear on the shortest path, creating a risk-reward tradeoff. Opacity for observation noise is used to indicate the uncertainty in the environment.

Stochastic Grid Environment. For our first experiment, we focus on a stochastic grid environment. In this environment, the agent has to traverse the grid from one end to the other, with hazards and stochastic transitions. The key challenge is that on the shortest path from the start to the goal, there are cells in which the transition matrix is stochastic, with the risk that the agent will end up in a sink state. The stochasticity presents a direct test of how agents handle uncertainty in dynamics: the KL-control agent is expected to plan optimistically through these uncertain transitions, while the EFE-minimizing agent should recognize the epistemic risk and avoid these cells. This environment also features observation noise, adding another layer of uncertainty that forces the agent to maintain beliefs over possible states rather than having full observability.

A longer but safer path exists that avoids all stochastic transitions. The optimal policy for a risk-aware agent would be to take this safer path, despite it requiring more steps. A visualization of the environment is shown in Fig. 3.

The agent receives a reward of 1 for reaching the goal. When ending up in a sink state, the agent receives a penalty of -1. The full specification of the generative model can be found in Appendix B.

Minigrid Door-Key Environment. The second environment we consider is a Minigrid environment, specifically a 4×4 door-key environment. This environment tests a different aspect of epistemic uncertainty, namely, partial observability. The agent has a limited field of view, which means that the agent must actively explore to reduce uncertainty about the environment state.

The task requires the agent to locate and pick up a key, find and open a door, and finally reach the goal square. This multi-step process creates a natural exploration challenge that tests how agents handle partial observability. The agent location, key location, and door location are randomized in each episode, which means that the agent has epistemic uncertainty about the environment state.

The EFE-minimizing agent should show more directed exploration behavior, actively seeking to reduce uncertainty about the key and door locations. In contrast, the KL-control agent (without epistemic priors) might exhibit less efficient exploration patterns, as it lacks the intrinsic drive to resolve uncertainty.

The Minigrid environment adds another layer of complexity to the task, as the field of view means that the observations are relative to the agent, while the goals are formulated in an external frame of reference. This means that the observation space of the agent is much larger than the state space. The observation space is of size $\approx 5^{49}$, which makes algorithms like Sophisticated Inference [17] intractable. Furthermore, the planning horizon of 22 timesteps makes standard Expected Free Energy computation as policy evaluation intractable. The computational complexity of the door-key environment is where the benefits of the proposed method are most evident.

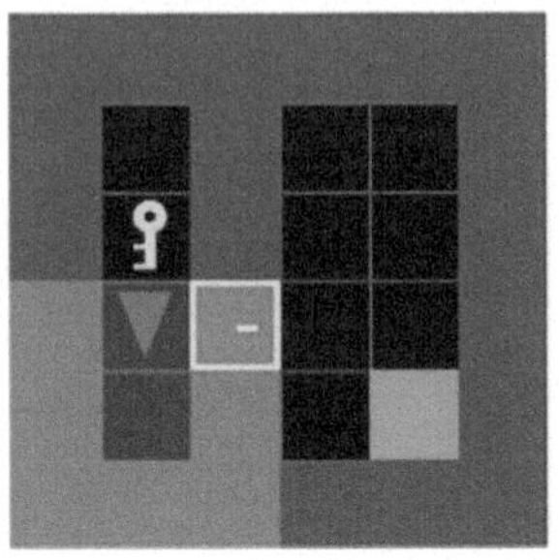

Fig. 4. An initial state of the Minigrid environment. The agent has a limited field of view, indicated by the highlighted cells.

A visualization of the initial state of the Minigrid environment is shown in Fig. 4. The agent receives a reward when reaching the goal, proportional to the number of steps taken. The full specification of the used generative model can be found in Appendix C. The source code and implementation details for all experiments presented in this paper are publicly available in our online repository[1].

5.2 Results

Stochastic Grid Environment. We evaluated the performance of both agents across 100 episodes, Table 1, left, summarizes the quantitative results.

This table suggests distincly different navigational patterns between both agents. The EFE-minimizing agent consistently chooses the longer but safer path around the stochastic transition cells, demonstrating risk-averse behavior that aligns with theoretical predictions. In contrast, the KL-control agent attempts the shorter path through cells with stochastic transitions, exhibiting the optimistic planning tendency typical of approaches that wrongly account for the system's aleatoric uncertainty. A more detailed visualization of the trajectories for both agents, as well as an empirical convergence analysis of our algorithm, is provided in Appendix D.

[1] https://github.com/biaslab/EFEasVFE.

Table 1. Performance comparison across environments (100 episodes for Stochastic Grid, 200 episodes for Minigrid).

Stochastic Grid			Minigrid Door-Key		
Metric	KL	EFE (ours)	Metric	KL	EFE (ours)
Success Rate	21%	**100%**	Success Rate	85.0%	**95.0%**
Avg. Reward	0.22 ± 0.77	$\mathbf{1.00 \pm 0}$	Avg. Reward	0.82 ± 0.35	$\mathbf{0.92 \pm 0.21}$
–	–	–	Avg. Time to Key Visibility	2.0 ± 3.65	$\mathbf{1.28 \pm 0.64}$

Minigrid Door-Key Environment. We evaluated both agents across 200 experimental episodes with a planning horizon of 25 steps. Table 1, right, presents the quantitative comparison between the EFE-minimizing and KL-control agents in the Minigrid door-key environment.

The EFE-minimizing agent demonstrates more effective exploration patterns, particularly in scenarios requiring active information seeking. This is especially evident in the reduced time needed to locate the key, confirming that epistemic priors enable more directed information-seeking in partially observable environments.

A more detailed visualization of the trajectories for both agents and an empirical convergence analysis of our algorithm is provided in Appendix E.

6 Discussion

Our experimental results demonstrate that agents using the proposed message passing approach for EFE minimization exhibit the characteristic behaviors of active inference: risk-averse path selection in stochastic environments and information-seeking exploration in partially observable settings. These behaviors emerge naturally from the inclusion of epistemic priors in the variational free energy objective, without requiring explicit computation of expected free energy.

The reformulation of EFE minimization as a variational inference problem provides several advantages: it maintains theoretical consistency with the Free Energy Principle's core tenet; transforms a combinatorial search problem into a tractable inference procedure using message passing on factor graphs; and eliminates the need for ad hoc policy pruning, replacing it with principled reactive processing where the agent minimizes VFE at each point in time. This approach is particularly valuable in complex environments where traditional EFE computation becomes intractable, as demonstrated in our Minigrid experiments.

While our implementation shows promising results, the convergence properties of our iterative approach to handling self-referential epistemic priors require further theoretical investigation. Future research should investigate the inclusion of additional parameters in the generative model, particularly those related to environment dynamics. A natural extension of our work would be to incorporate parameter learning within the epistemic priors. This would allow agents to infer policies that facilitate sample-efficient learning of model parameters. This concept has already been introduced in [37]. However, the exact functional form of the empirical prior has not yet been derived.

7 Conclusion

In this paper, we presented a message passing implementation of Expected Free Energy minimization on factor graphs. Our approach reframes EFE minimization as a variational inference problem, allowing us to use standard message passing algorithms for efficient policy inference. The key insight is that by introducing appropriate epistemic priors, we can transform the expected free energy objective into a modified variational free energy objective that can be optimized through standard inference techniques.

Our experimental results in both stochastic and partially observable environments demonstrate that this approach reproduces the characteristic behaviors of active inference: risk aversion in environments with hazardous stochasticity and information seeking in partially observable environments. The message passing implementation shows significant advantages in computational efficiency compared to traditional methods for computing expected free energy, particularly in complex environments with high-dimensional observation spaces and long planning horizons.

By reformulating EFE minimization as variational inference, our work contributes to unifying the theoretical frameworks of the Free Energy Principle and active inference with practical implementations for decision-making under uncertainty. This bridges the gap between theoretical accounts of intelligent behavior and efficient algorithms for artificial agents, offering a principled approach to balancing pragmatic and epistemic objectives in complex and uncertain environments.

Acknowledgements. This publication is part of the project "ROBUST: Trustworthy AI-based Systems for Sustainable Growth" with project number KICH3.LTP.20.006, which is (partly) financed by the Dutch Research Council (NWO), GN Hearing, and the Dutch Ministry of Economic Affairs and Climate Policy (EZK) under the program LTP KIC 2020-2023.

A Proof of Corollary 1

Proof. Proof of Corollary 1. This proof is an adjusted proof of the proof of Theorem 1, which is given in [37].

$$F[q] = E_{q(\boldsymbol{y},\boldsymbol{x},\boldsymbol{u})}\left[\log \frac{q(\boldsymbol{y},\boldsymbol{x},\boldsymbol{u})}{p(\boldsymbol{y},\boldsymbol{x},\boldsymbol{u})\hat{p}(\boldsymbol{x})\tilde{p}(\boldsymbol{u})\tilde{p}(\boldsymbol{x})}\right] \tag{16a}$$

$$= E_{q(\boldsymbol{u})}\left[\log \frac{q(\boldsymbol{u})}{p(\boldsymbol{u})} + \underbrace{E_{q(\boldsymbol{y},\boldsymbol{x}|\boldsymbol{u})}\left[\log \frac{q(\boldsymbol{y},\boldsymbol{x}|\boldsymbol{u})}{p(\boldsymbol{y},\boldsymbol{x}|\boldsymbol{u})\hat{p}(\boldsymbol{x})\tilde{p}(\boldsymbol{u})\tilde{p}(\boldsymbol{x})}\right]}_{C(\boldsymbol{u})}\right] \tag{16b}$$

$$= E_{q(\boldsymbol{u})}\left[\log \frac{q(\boldsymbol{u})}{p(\boldsymbol{u})} + \underbrace{G(\boldsymbol{u}) + E_{q(\boldsymbol{y},\boldsymbol{x}|\boldsymbol{u})}\left[\log \frac{q(\boldsymbol{y},\boldsymbol{x}|\boldsymbol{u})}{p(\boldsymbol{y},\boldsymbol{x}|\boldsymbol{u})}\right] + constant}_{C(\boldsymbol{u}) \text{ if (13) holds}}\right] \tag{16c}$$

$$= E_{q(\boldsymbol{u})}\left[G(\boldsymbol{u})\right] + E_{q(\boldsymbol{y},\boldsymbol{x},\boldsymbol{u})}\left[\log \frac{q(\boldsymbol{y},\boldsymbol{x},\boldsymbol{u})}{p(\boldsymbol{y},\boldsymbol{x},\boldsymbol{u})}\right] + constant \quad \text{if (13) holds} \tag{16d}$$

□

In the above derivation, we still need to prove the transition for $C(\boldsymbol{u})$ from (16b) to (16c), which we address next.

Lemma 1 (Proof of equivalence $C(\boldsymbol{u})$ in (16b) and (16c)).

$$C(\boldsymbol{u}) = \mathbb{E}_{q(\boldsymbol{y},\boldsymbol{x}|\boldsymbol{u})}\left[\log \frac{\overbrace{q(\boldsymbol{y},\boldsymbol{x}|\boldsymbol{u})}^{posterior}}{\underbrace{p(\boldsymbol{y},\boldsymbol{x}|\boldsymbol{u})}_{predictive}\underbrace{\hat{p}(\boldsymbol{x})}_{utility}\underbrace{\tilde{p}(\boldsymbol{u})\tilde{p}(\boldsymbol{x})}_{epistemic\ priors}}\right] \tag{17a}$$

$$= \underbrace{\mathbb{E}_{q(\boldsymbol{y},\boldsymbol{x}|\boldsymbol{u})}\left[\log\left(\underbrace{\frac{q(\boldsymbol{x}|\boldsymbol{u})}{\hat{p}(\boldsymbol{x})}}_{risk}\cdot\underbrace{\frac{1}{q(\boldsymbol{y}|\boldsymbol{x})}}_{ambiguity}\right)\right]}_{G(\boldsymbol{u})=Expected\ Free\ Energy} + \tag{17b}$$

$$+\mathbb{E}_{q(\boldsymbol{y},\boldsymbol{x}|\boldsymbol{u})}\left[\log\left(\underbrace{\frac{\hat{p}(\boldsymbol{x})q(\boldsymbol{y}|\boldsymbol{x})}{q(\boldsymbol{x}|\boldsymbol{u})}}_{inverse\ factors\ from\ G(\boldsymbol{u})}\cdot\underbrace{\frac{q(\boldsymbol{y},\boldsymbol{x}|\boldsymbol{u})}{p(\boldsymbol{y},\boldsymbol{x}|\boldsymbol{u})\hat{p}(\boldsymbol{x})\tilde{p}(\boldsymbol{u})\tilde{p}(\boldsymbol{x})}}_{factors\ from\ (17a)}\right)\right]$$

$$= G(\boldsymbol{u}) + \underbrace{\mathbb{E}_{q(\boldsymbol{y},\boldsymbol{x}|\boldsymbol{u})}\left[\log \frac{q(\boldsymbol{y},\boldsymbol{x}|\boldsymbol{u})}{p(\boldsymbol{y},\boldsymbol{x}|\boldsymbol{u})}\right]}_{=B(\boldsymbol{u})} + \underbrace{\mathbb{E}_{q(\boldsymbol{y},\boldsymbol{x}|\boldsymbol{u})}\left[\log \frac{q(\boldsymbol{y}|\boldsymbol{x})}{q(\boldsymbol{x}|\boldsymbol{u})\tilde{p}(\boldsymbol{u})\tilde{p}(\boldsymbol{x})}\right]}_{choose\ epistemic\ priors\ to\ let\ this\ vanish} \tag{17c}$$

$$= G(\boldsymbol{u}) + B(\boldsymbol{u}) + \mathbb{E}_{q(\boldsymbol{x}|\boldsymbol{u})}\left[\log \frac{1}{q(\boldsymbol{x}|\boldsymbol{u})\tilde{p}(\boldsymbol{u})}\right] + \mathbb{E}_{q(\boldsymbol{y}|\boldsymbol{x})}\left[\log \frac{q(\boldsymbol{y}|\boldsymbol{x})}{\tilde{p}(\boldsymbol{x})}\right]$$

Now here we can replace the general $q(\boldsymbol{y}|\boldsymbol{x})$ and $q(\boldsymbol{x}|\boldsymbol{u})$ with the factorised $\prod_t q(y_t|x_t)$ and $\prod_t q(x_t|x_{t-1}, u_t)$.

$$
\begin{aligned}
C(\boldsymbol{u}) &= G(\boldsymbol{u}) + B(\boldsymbol{u}) + \mathbb{E}_{q(\boldsymbol{x}|\boldsymbol{u})}\left[\log \frac{1}{\prod_t q(x_t|x_{t-1}, u_t)\tilde{p}(u_t)}\right] + \\
&\quad + \mathbb{E}_{q(\boldsymbol{y}|\boldsymbol{x})}\left[\log q(y_t|x_t) - \log \tilde{p}(x_t)\right] \qquad (18a)\\
&= G(\boldsymbol{u}) + B(\boldsymbol{u}) \\
&\quad + \sum_{\boldsymbol{x}} \mathbb{E}_{q(x_t, x_{t-1}|u_t)}\left[-\log q(x_t|x_{t-1}, u_t) - \log \tilde{p}(u_t)\right] + \\
&\quad + \sum_{\boldsymbol{y}} \mathbb{E}_{q(y_t|x_t)}\left[\log q(y_t|x_t) - \log \tilde{p}(x_t)\right]. \qquad (18b)
\end{aligned}
$$

Now we can recognize the following:

$$
\begin{aligned}
&\mathbb{E}_{q(x_t, x_{t-1}|u_t)}\left[-\log q(x_t|x_{t-1}, u_t)\right] \qquad (19a)\\
&\quad = \mathbb{E}_{q(x_t, x_{t-1}|u_t)}\left[-\Big(\log q(x_t, x_{t-1}|u_t) - \log q(x_{t-1}|u_t)\Big)\right] \qquad (19b)\\
&\quad = H[q(x_t, x_{t-1}|u_t)] - H[q(x_{t-1}|u_t)], \qquad (19c)
\end{aligned}
$$

and

$$
\mathbb{E}_{q(y_t|x_t)}\left[\log q(y_t|x_t)\right] = -H[q(y_t|x_t)]. \qquad (20)
$$

Which, when substituted into (18b), together with the definitions of $\tilde{p}(u_t)$ and $\tilde{p}(x_t)$, yields

$$
\begin{aligned}
&= G(\boldsymbol{u}) + B(\boldsymbol{u}) \\
&\quad + \sum_{\boldsymbol{x}} \underbrace{H[q(x_t, x_{t-1}|u_t)] - H[q(x_{t-1}|u_t)] - \log \tilde{p}(u_t)}_{=c_x \text{ if } \tilde{p}(u_t) \propto \exp(H[q(x_t, x_{t-1}|u_t)] - H[q(x_{t-1}|u_t)])} + \\
&\quad + \sum_{\boldsymbol{y}} \underbrace{H[q(y_t|x_t)] - \log \tilde{p}(x_t)}_{=c_y \text{ if } \tilde{p}(x_t) \propto \exp(-H[q(y_t|x_t)])} \qquad (21a)\\
&= G(\boldsymbol{u}) + \mathbb{E}_{q(\boldsymbol{y}, \boldsymbol{x}|\boldsymbol{u})}\left[\log \frac{q(\boldsymbol{y}, \boldsymbol{x}|\boldsymbol{u})}{p(\boldsymbol{y}, \boldsymbol{x}|\boldsymbol{u})}\right] + c_x + c_y, \quad \textit{if (13) holds.} \qquad (21b)
\end{aligned}
$$

B Generative Model for the Gridworld Environment

The generative model for the stochastic grid environment is defined as follows:

$$
\begin{aligned}
x_0 &\sim p(x_0) \qquad (22a)\\
x_t &\sim \mathrm{Cat}(x_t|x_{t-1}, u_t, B) \qquad (22b)\\
y_t &\sim \mathrm{Cat}(y_t|x_t, A) \qquad (22c)\\
x_T &\sim \hat{p}(x_T). \qquad (22d)
\end{aligned}
$$

Here, s_t represents the agent's state at time t, y_t is the observation, and u_t is the action. The transition dynamics are governed by B, and A represents the observation model. The agent starts with a prior belief $p(s_0)$ and aims to reach the goal state by the end of the planning horizon T.

In the case of the KL-control agent, the prior on the control is given by

$$u_t \sim \text{Cat}(u_t|\mathbf{1}/4) \quad \text{for } t = 1, \ldots, T\,. \tag{23}$$

The EFE-minimizing agent uses empirical priors on the control and the states, given by

$$u_t \sim \text{Cat}(u_t|\sigma(H[q(x_t, x_{t-1}|u_t)] - H[q(x_{t-1}|u_t)])) \qquad \text{for } t = 1, \ldots, T \tag{24a}$$

$$x_t \sim \text{Cat}(x_t|\sigma(-H[q(y_t|x_t)])) \qquad \text{for } t = 1, \ldots, T\,. \tag{24b}$$

C Generative Model for the Minigrid Environment

The generative model for the Minigrid environment uses a factorized state and observation space, which makes the model computationally tractable. Here, the location of the agent is denoted by l, the orientation by o, the key-door state by s, the door location by d, and the key location by k. The key-door state is a categorical variable with three possible values: $\{0, 1, 2\}$, where 0 indicates that the key is not picked up yet, 1 indicates that the key is picked up but the door is not opened yet, and 2 indicates that the key is picked up and the door is opened. For the observations, $y_{t,(x,y)}$ is the observation at time t for cell (x, y) of the field of view. The generative model for the Minigrid environment is defined as follows:

$$l_0 \sim p(l_0) \tag{25a}$$

$$o_0 \sim p(o_0) \tag{25b}$$

$$s_0 \sim p(s_0) \tag{25c}$$

$$d \sim p(d) \tag{25d}$$

$$k \sim p(k) \tag{25e}$$

$$l_t \sim \text{Cat}(l_t|l_{t-1}, o_{t-1}, k, d, s_{t-1}, u_t, B^l) \tag{25f}$$

$$o_t \sim \text{Cat}(o_t|o_{t-1}, B^o, u_t) \tag{25g}$$

$$s_t \sim \text{Cat}(s_t|s_{t-1}, l_{t-1}, o_{t-1}, k, d, u_{t-1}, B^s) \tag{25h}$$

$$y_{t,(x,y)} \sim \text{Cat}(y_{t,(x,y)}|l_t, o_t, k, d, s_t, A_{(x,y)}) \quad \forall (x, y) \in \{1, \ldots, 7\}^2 \tag{25i}$$

with terminal state goal priors

$$l_T \sim \hat{p}(l_T) \tag{26a}$$

$$s_T \sim \text{Cat}(s_T|[0, 0, 1]) = \hat{p}(s_T)\,, \tag{26b}$$

where B^l is the location transition tensor, B^o is the orientation transition tensor, B^s is the key-door state transition tensor, and $A_{(x,y)}$ are the observation tensors for each cell in the field of view.

In the case of the KL-control agent, the prior on the control is given by

$$u_t \sim \text{Cat}(u_t|\mathbf{1}/5) \quad \text{for } t = 1, \ldots, T\,. \tag{27}$$

The EFE-minimizing agent uses empirical priors on the control and the states, given by

$$\begin{aligned}
u_t \sim \text{Cat}(u_t|\sigma(& & \text{(28a)}\\
& H[q(l_t, l_{t-1}, o_{t-1}, k, d, s_{t-1}|u_t)] - H[q(l_{t-1}, o_{t-1}, k, d, s_{t-1}|u_t)] + & \text{(28b)}\\
& H[q(s_t, s_{t-1}, l_t, o_{t-1}, k, d|u_t)] - H[q(s_{t-1}, l_t, o_{t-1}, k, d|u_t)] + & \text{(28c)}\\
& H[q(o_t, o_{t-1}|u_t)] - H[q(o_{t-1}|u_t)])) \quad \text{for } t = 1, \ldots, T & \text{(28d)}
\end{aligned}$$

$$l_t \sim \text{Cat}\left(l_t|\sigma\left(\sum_{(x,y)} -H[q(y_{t,(x,y)}, o_t, s_t, k, d, |l_t)] + H[o_t, k, d, s_t|l_t]\right)\right)$$
$$\text{for } t = 1, \ldots, T \tag{28e}$$

$$o_t \sim \text{Cat}\left(o_t|\sigma\left(\sum_{(x,y)} -H[q(y_{t,(x,y)}, l_t, s_t, k, d, |o_t)] + H[l_t, s_t, k, d|o_t]\right)\right)$$
$$\text{for } t = 1, \ldots, T \tag{28f}$$

$$s_t \sim \text{Cat}\left(s_t|\sigma\left(\sum_{(x,y)} -H[q(y_{t,(x,y)}, l_t, o_t, k, d, |s_t)] + H[l_t, o_t, k, d|s_t]\right)\right)$$
$$\text{for } t = 1, \ldots, T \tag{28g}$$

$$k \sim \text{Cat}\left(k|\sigma\left(\sum_t \sum_{(x,y)} -H[q(y_{t,(x,y)}, l_t, o_t, s_t, d, |k)] + H[l_t, o_t, s_t, d|k]\right)\right) \tag{28h}$$

$$d \sim \text{Cat}\left(d|\sigma\left(\sum_t \sum_{(x,y)} -H[q(y_{t,(x,y)}, l_t, o_t, s_t, k, |d)] + H[l_t, o_t, s_t, k|d]\right)\right). \tag{28i}$$

D Additional Results for the Stochastic Grid Environment Experiments

In this section, we will provide further analysis of the results presented in Sect. 5.2. We will go into more detail on the convergence of the Bethe Free Energy over different iterations of the variational inference procedure, and we will elaborate on a trajectory in a specific episode.

D.1 Convergence Analysis

In Fig. 5, we plot the Bethe Free Energy over the iterations of the message passing procedure, along the state of the environment at which the inference procedure is being called.

As we can see, even though we have not provided a proof of convergence, in this specific example, the Bethe Free Energy converges to a constant value, indicating that our inference procedure has converged.

Note that the Bethe Free Energy is an approximation of the true Variational Free Energy, and can therefore not be used for model comparison [39]. Although RxInfer minimizes the Bethe Free Energy, this explains the upwards trend in the Bethe Free Energy curve, and we can only use the Bethe Free Energy as a sanity check to check convergence of the inference procedure.

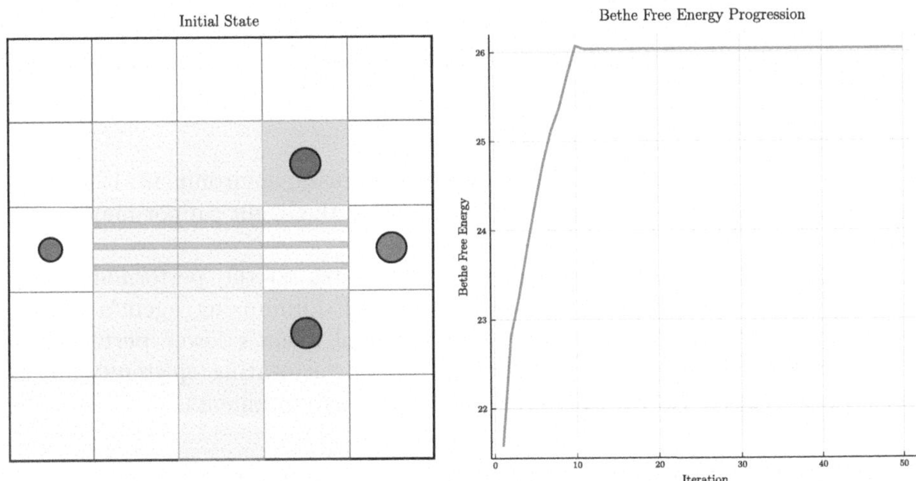

Fig. 5. Visualization of the inference results for the stochastic grid environment. On the left, the initial state of the environment is shown. On the right we show the Bethe Free Energy curve over the iterations of message passing. Convergence to a constant value indicates convergence of the inference procedure.

D.2 Trajectory

Figure 6 provides a frame-by-frame comparison of the trajectories taken by the EFE-minimizing agent (left) and the KL-control agent (right) in the stochastic grid environment. This visualization clearly demonstrates the differences in planning strategies between the two approaches.

The EFE-minimizing agent immediately chooses the longer but safer path, moving upward and around the cells with stochastic transitions. This risk-averse behavior is a direct result of the epistemic priors that penalize uncertainty in transitions. By frame $t = 8$, the agent has successfully reached the goal state without encountering any hazardous transitions.

In contrast, the KL-control agent attempts to optimize for the shortest path, moving directly through cells with stochastic transitions. This optimistic planning is characteristic of approaches that don't account for aleatoric uncertainty.

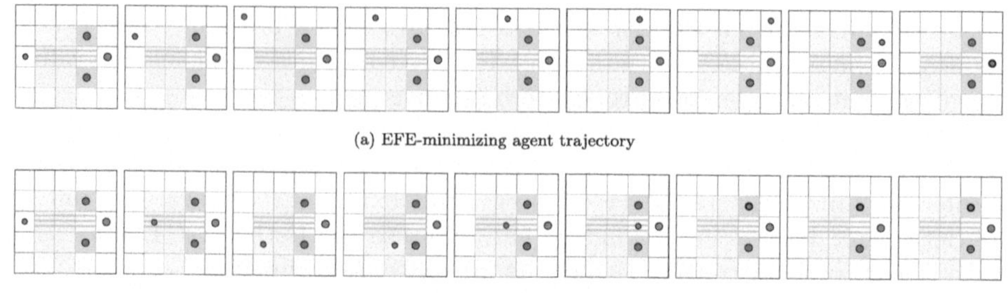

(a) EFE-minimizing agent trajectory

(b) KL-control agent trajectory

Fig. 6. Comparison of agent trajectories in a stochastic maze environment. Top: EFE-minimizing agent with epistemic priors. Bottom: KL-control agent without epistemic priors.

While this strategy would be optimal in a deterministic environment, it leads to potential failures in this stochastic setting because the agent cannot manipulate its own luck.

The difference in trajectories directly translates to the performance gap observed across the 100 trial episodes. The EFE-minimizing agent's perfect success rate (100%) compared to the KL-control agent's lower performance (21%) confirms the theoretical prediction that incorporating epistemic uncertainty leads to more robust planning in stochastic environments.

E Additional Results for the Minigrid Environment Experiments

E.1 Convergence Analysis

In Fig. 8, we perform inference on the initial state of the Minigrid environment shown in Fig. 7. The figure displays the Bethe Free Energy progression during the inference process, along with the agent's final beliefs about its current location, orientation, and the state of the key and door after the last iteration. We observe that the BFE stabilizes to a constant value, indicating that our inference procedure successfully converges.

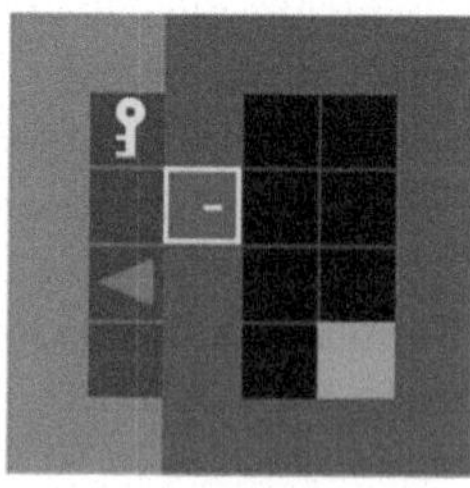

Fig. 7. Initial state of the Minigrid environment.

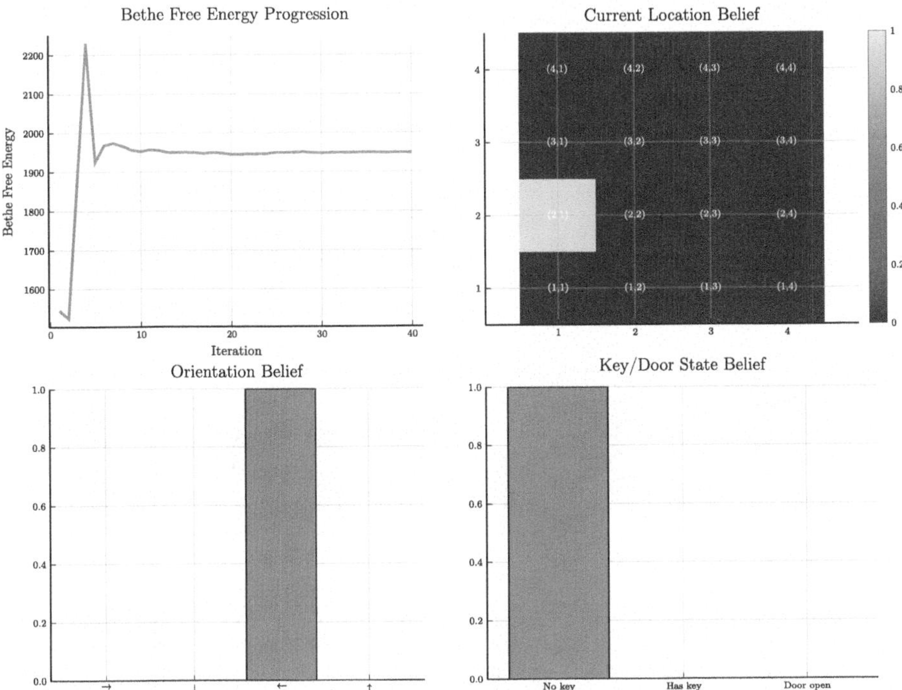

Fig. 8. Visualization of the inference results for the Minigrid environment. Top left: Bethe Free Energy curve over the iterations of message passing. Top right: Agent's belief of its current location after the last iteration. Bottom left: Agent's belief of its current orientation after the last iteration. Bottom right: Agent's belief of the state of the key and door after the last iteration.

E.2 Trajectory

Figures 9 and 10 provide a frame-by-frame comparison of the trajectories taken by the EFE-minimizing agent and the KL-control agent in the Minigrid environment. This visualization clearly demonstrates the differences in planning strategies between the two approaches, and highlights the shortcomings of the KL-control approach.

The EFE-minimizing agent is able to solve the task at hand, while the KL-control agent stays in the corner of the grid facing the wall. As shown in Fig. 9, the EFE-minimizing agent reaches the goal state, while the KL-control agent does not.

The difference in trajectories directly translates to the performance gap observed across the test episodes. The EFE-minimizing agent's superior performance confirms our theoretical prediction that incorporating epistemic uncertainty leads to more efficient planning in partially observable environments like Minigrid.

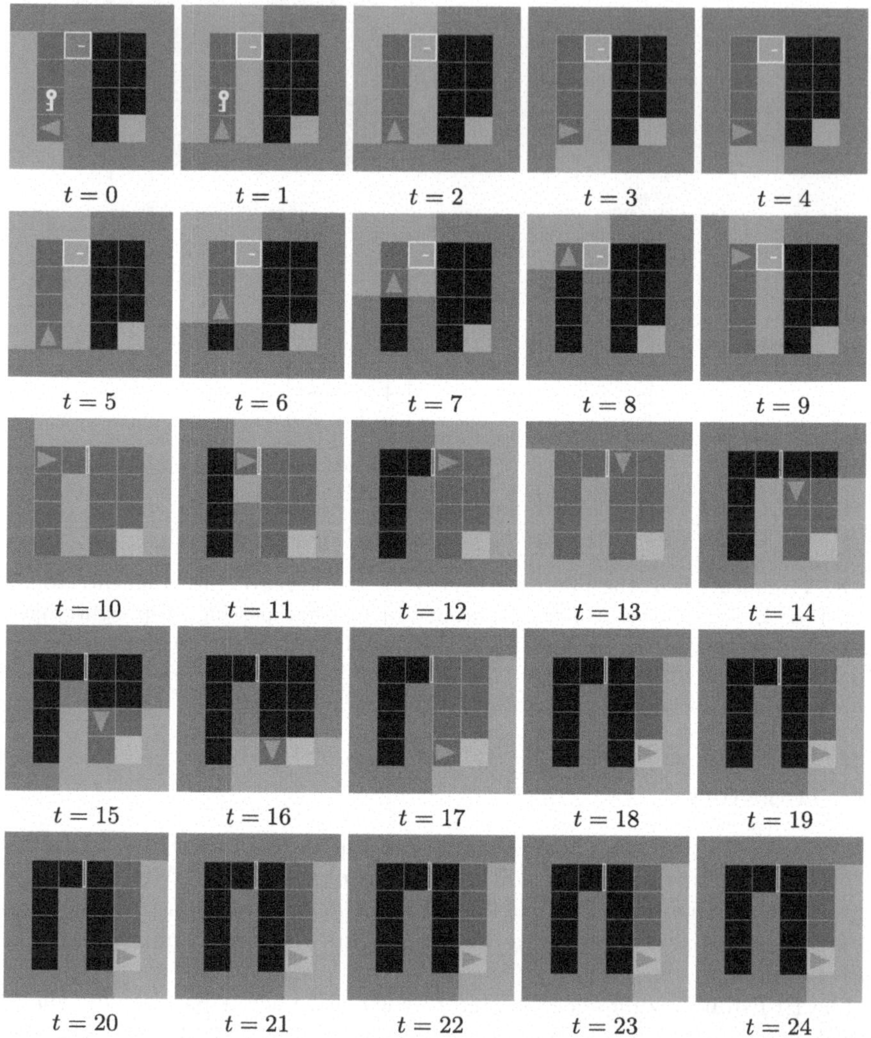

Fig. 9. Visualization of the agent's trajectory in the Minigrid environment using EFE-based control. The 5×5 grid shows the sequential frames of the agent's movement.

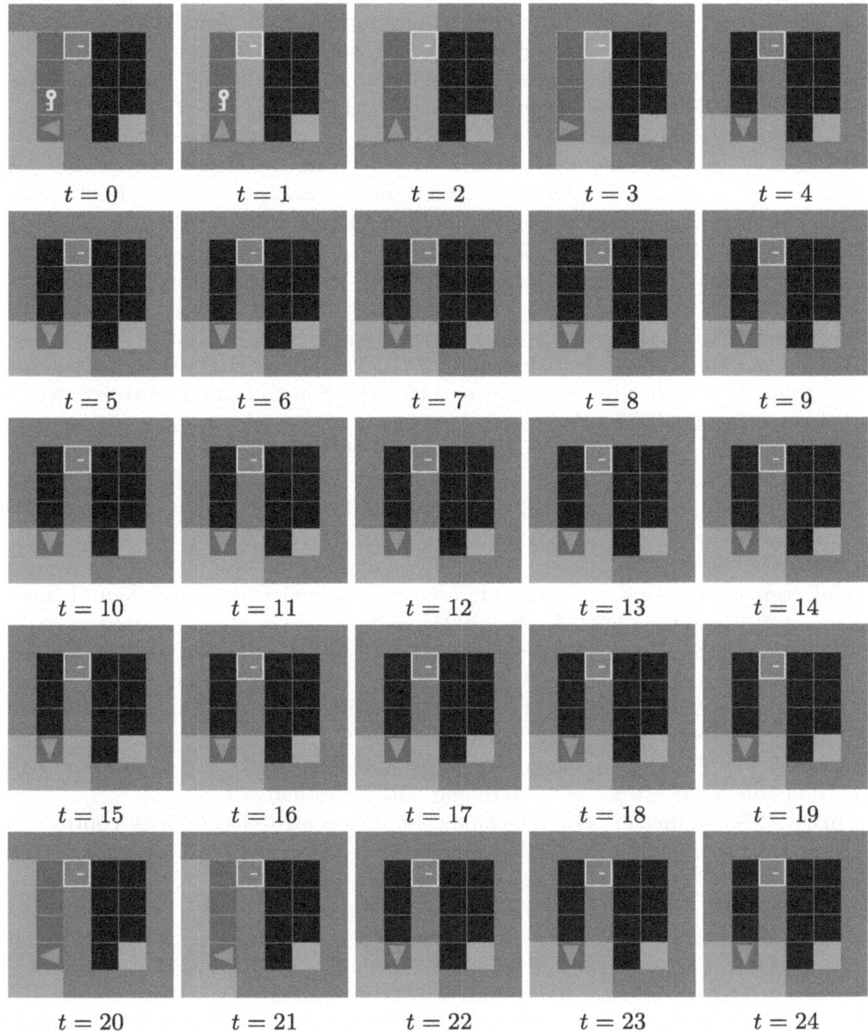

Fig. 10. Visualization of the agent's trajectory in the Minigrid environment using KL control. The 5×5 grid shows the sequential frames of the agent's movement throughout the episode.

References

1. Attias, H.: Planning by probabilistic inference. In: International Workshop on Artificial Intelligence and Statistics, pp. 9–16. PMLR (2003). https://proceedings.mlr.press/r4/attias03a.html
2. Bagaev, D., de Vries, B.: Reactive Message Passing for Scalable Bayesian Inference (Dec 2021). https://doi.org/10.48550/arXiv.2112.13251

3. Bellman, R.: The theory of dynamic programming. Bull. Am. Math. Soc. **60**(6), 503–515 (1954). https://doi.org/10.1090/S0002-9904-1954-09848-8, https://www.ams.org/bull/1954-60-06/S0002-9904-1954-09848-8/
4. Bellman, R.: Dynamic programming. Science **153**(3731), 34–37 (1966). https://www.jstor.org/stable/1719695
5. Bertsekas, D.: Dynamic programming and optimal control: Volume I, vol. 4. Athena scientific (2012). https://books.google.com/books?hl=en&lr=&id=qVBEEAAAQBAJ&oi=fnd&pg=PR1&dq=Dynamic+Programming+and+Optimal+Control&ots=x0bAav0O5n&sig=s3UxthkdnzR2UpqCUsUsQ7zKgLc
6. Bishop, C.M., Nasrabadi, N.M.: Pattern recognition and machine learning, vol. 4. Springer (2006). https://link.springer.com/book/9780387310732
7. Blei, D.M., Kucukelbir, A., McAuliffe, J.D.: Variational Inference: a Review for Statisticians. J. Am. Statist. Associat. **112**(518), 859–877 (2017). https://doi.org/10.1080/01621459.2017.1285773
8. Champion, T., Da Costa, L., Bowman, H., Grześ, M.: Branching Time Active Inference: the theory and its generality. Neural Netw. **151**, 295–316 (2022). https://doi.org/10.1016/j.neunet.2022.03.036, https://www.sciencedirect.com/science/article/pii/S0893608022001149
9. Chevalier-Boisvert, M., et al.: Minigrid & miniworld: modular & customizable reinforcement learning environments for goal-oriented tasks. Adv. Neural Inform. Process. Syst. **36**, 73383–73394 (2023). https://proceedings.neurips.cc/paper_files/paper/2023/hash/e8916198466e8ef218a2185a491b49fa-Abstract-Datasets_and_Benchmarks.html
10. Costa, L.D., Tenka, S., Zhao, D., Sajid, N.: Active Inference as a Model of Agency (Jan 2024). https://doi.org/10.48550/arXiv.2401.12917
11. Cox, R.T.: Probability, frequency and reasonable expectation. Am. J. Phys. **14**(1), 1–13 (1946). http://www.cs.toronto.edu/~ilya/cox1946.pdf
12. Cutler, R.R., Ramaker, B.L.: Dynamic matrix control-a computer control algorithm. In: Proc. Joint Automatic Control Conference, 1979 (1979). https://cir.nii.ac.jp/crid/1570291225777284224
13. Da Costa, L., Parr, T., Sajid, N., Veselic, S., Neacsu, V., Friston, K.: Active inference on discrete state-spaces: A synthesis. J. Math. Psychol. **99**, 102447 (2020). https://doi.org/10.1016/j.jmp.2020.102447, https://www.sciencedirect.com/science/article/pii/S0022249620300857
14. Dauwels, J.: On variational message passing on factor graphs. In: IEEE International Symposium on Information Theory, Nice, France, pp. 2546–2550 (2007). https://doi.org/10.1109/ISIT.2007.4557602
15. Forney, G.D.: Codes on graphs: normal realizations. IEEE Trans. Inform. Theory **47**(2), 520–548 (2001). https://ieeexplore.ieee.org/abstract/document/910573/
16. Friston, K.: The free-energy principle: a unified brain theory? N. Rev. Neurosci. **11**(2), 127–138 (2010). https://doi.org/10.1038/nrn2787, https://www.nature.com/articles/nrn2787 Publishing Group
17. Friston, K., Costa, L.D., Hafner, D., Hesp, C., Parr, T.: Sophisticated Inference (Jun 2020). https://doi.org/10.48550/arXiv.2006.04120
18. Friston, K., Rigoli, F., Ognibene, D., Mathys, C., Fitzgerald, T., Pezzulo, G.: Active inference and epistemic value. Cognitive Neurosci. **6**(4), 187–214 (2015). https://doi.org/10.1080/17588928.2015.1020053, http://www.tandfonline.com/doi/full/10.1080/17588928.2015.1020053
19. Friston, K.J., et al.: Active inference and intentional behavior. Neural Comput. **37**(4), 666–700 (2025). https://doi.org/10.1162/neco_a_01738

20. Ito, K., , Kashima, K.: Kullback–Leibler control for discrete-time nonlinear systems on continuous spaces. SICE J. Control Measurem. Syst. Integrat. **15**(2), 119–129 (2022). https://doi.org/10.1080/18824889.2022.2095827
21. Jaynes, E.T.: Probability Theory: The Logic of Science, 1 edn. Cambridge University Press(Apr 2003). https://doi.org/10.1017/CBO9780511790423, https://www.cambridge.org/core/product/identifier/9780511790423/type/book
22. Kappen, B., Gomez, V., Opper, M.: Optimal control as a graphical model inference problem. Mach. Learn. **87**(2), 159–182 (2012). arXiv:0901.0633 [math]
23. Kingma, D.P., Welling, M.: Auto-Encoding Variational Bayes (Dec 2022). https://doi.org/10.48550/arXiv.1312.6114
24. Koller, D., Friedman, N.: Probabilistic Graphical Models: Principles and Techniques. MIT Press (Jul 2009), google-Books-ID: 7dzpHCHzNQ4C
25. Levine, S.: Reinforcement Learning and Control as Probabilistic Inference: Tutorial and Review (May 2018). https://doi.org/10.48550/arXiv.1805.00909
26. Loeliger, H.A.: An introduction to factor graphs. IEEE Signal Proces. Mag. **21**(1), 28–41 (2004). https://doi.org/10.1109/MSP.2004.1267047
27. Loeliger, H.A., Dauwels, J., Hu, J., Korl, S., Ping, L., Kschischang, F.R.: The factor graph approach to model-based signal processing. Proc. IEEE **95**(6), 1295–1322 2007). https://doi.org/10.1109/JPROC.2007.896497
28. Lázaro-Gredilla, M., Ku, L.Y., Murphy, K.P., George, D.: What type of inference is planning? (Nov 2024). https://doi.org/10.48550/arXiv.2406.17863
29. Paul, A., Sajid, N., Costa, L.D., Razi, A.: On efficient computation in active inference. Expert Syst. Appli. **253**, 124315 (2024). https://doi.org/10.1016/j.eswa.2024.124315
30. Paul, A., Sajid, N., Gopalkrishnan, M., Razi, A.: Active inference for stochastic control. In: Kamp, M., et al. (eds.) Machine Learning and Principles and Practice of Knowledge Discovery in Databases, pp. 669–680. Springer International Publishing, Cham (2021). https://doi.org/10.1007/978-3-030-93736-2_47
31. Pearl, J.: Reverend bayes on inference engines: a distributed hierarchical approach. In: AAAI 1982 Proceedings, pp. 133–136. AAAI Press, Carnegie Mellon University, Pittsburgh PA (1982). https://books.google.com/books?hl=nl&lr=&id=e59kEAAAQBAJ&oi=fnd&pg=PA129&ots=qrs53bNhtS&sig=auOq_v4YW1fTTgNOvsmydDN6RPY,
32. Pontryagin, L.S.: Mathematical Theory of Optimal Processes. Routledge, London (May 2018). https://doi.org/10.1201/9780203749319
33. Richalet, J.: Algorithmic control of industrial processes. In: Proc. of the 4th IFAC Sympo. on Identification and System Parameter Estimation, pp. 1119–1167 (1976). https://cir.nii.ac.jp/crid/1570854174674016512
34. Richalet, J., Rault, A., Testud, J.L., Papon, J.: Model predictive heuristic control: applications to industrial processes. Automatica **14**(5), 413–428 (1978). https://doi.org/10.1016/0005-1098(78)90001-8, https://www.sciencedirect.com/science/article/pii/0005109878900018
35. Todorov, E.: Linearly-solvable Markov decision problems. In: Advances in Neural Information Processing Systems, vol. 19. MIT Press (2006). https://proceedings.neurips.cc/paper_files/paper/2006/hash/d806ca13ca3449af72a1ea5aedbed26a-Abstract.html
36. Todorov, E.: General duality between optimal control and estimation. In: 2008 47th IEEE Conference on Decision and Control, pp. 4286–4292 (2008). https://doi.org/10.1109/CDC.2008.4739438, https://ieeexplore.ieee.org/abstract/document/4739438, ISSN: 0191-2216

37. De Vries, B., et al.: Expected Free Energy-based Planning as Variational Inference (Apr 2025). https://doi.org/10.48550/arXiv.2504.14898
38. Winn, J., Bishop, C.: Variational Message Passing. J. Mach. Learn. Res. **6**, 661–694 (2005)
39. Yedidia, J.S., Freeman, W., Weiss, Y.: Constructing free-energy approximations and generalized belief propagation algorithms. IEEE Trans. Inform. Theory **51**(7), 2282–2312 (2005). https://doi.org/10.1109/TIT.2005.850085
40. Senöz, İ.: Message Passing Algorithms for Hierarchical Dynamical Models. Phd Thesis 1 (Research TU/e / Graduation TU/e), Eindhoven University of Technology, Eindhoven (Jun 2022), iSBN: 9789038655321

Spike-Timing-Dependent Plasticity for Bernoulli Message Passing

Sepideh Adamiat[1(✉)], Wouter M. Kouw[1], and Bert de Vries[1,2]

[1] Electrical Engineering Department, TU Eindhoven, Eindhoven, Netherlands
s.adamiat@tue.nl

[2] Lazy Dynamics B.V., Eindhoven, Netherlands

Abstract. Bayesian inference provides a principled framework for understanding brain function, while neural activity in the brain is inherently spike-based. This paper bridges these two perspectives by designing spiking neural networks that simulate Bayesian inference through message passing for Bernoulli messages. To train the networks, we employ spike-timing-dependent plasticity, a biologically plausible mechanism for synaptic plasticity which is based on the Hebbian rule. Our results demonstrate that the network's performance closely matches the true numerical solution. We further demonstrate the versatility of our approach by implementing a factor graph example from coding theory, illustrating signal transmission over an unreliable channel.

Keywords: Bayesian inference · factor graphs · message passing · leaky integrate-and-fire neurons · spiking neural networks · spike-timing-dependent plasticity

1 Introduction

Numerous perceptual and motor tasks carried out by the human nervous system can be effectively described using a Bayesian inference framework [7,15]. According to the Bayesian brain hypothesis, the brain updates its beliefs about the world by integrating sensory input with prior knowledge [4]. This concept aligns with the Free Energy Principle (FEP), which asserts that living systems adapt to their environment by maximizing evidence for an internal generative model of sensory observations [9]. This adaptation process is conceptually carried out by variational free energy minimization, physically realized by exchanging action potentials between neurons.

Recent studies have explored the information processing capacity of in vitro biological neurons cultured on top of multi-electrode arrays, demonstrating their potential for unsupervised learning, speech recognition, and decision-making tasks [3]. Some of these studies have employed the Free Energy Principle, including research by [14], which harnesses the inherent adaptive computation of neurons in a simulated game world, and [13], which investigates their application

M. Albarracin et al. (Eds.): IWAI 2025, CCIS 2857, pp. 99–112, 2026.
https://doi.org/10.1007/978-3-032-16955-6_6

in blind source separation. These studies are evidence supporting the FEP and Bayesian Brain hypothesis.

Bayesian inference can be computationally intensive due to the complexity of integrating and marginalizing over probability distributions. Graphical representations help to manage this complexity by breaking the problem into smaller, localized computations. These graphical models provide a structured framework for inference, where message-passing algorithms (also known as belief propagation) efficiently compute posterior probabilities by passing information along the graph's edges. This approach simplifies inference for large and sparsely connected systems, making Bayesian methods suitable for practical applications [19]. Interestingly, active inference, which is a corollary of FEP, can also be realized by message passing on a factor graph representation of the generative model [10].

On the other hand, from a biological point of view, all electrical communication in the brain is realized by spike-based message passing, and to understand the brain's behavior or communicate with cultured neurons, we need to encode and decode information in spike forms. Spiking Neural Networks (SNNs) are a class of artificial neural networks that provide a more accurate simulation of neural processes, making them significant for advances in computational neuroscience [20]. These biologically inspired neural models communicate with each other by discrete spike trains as input and output. There are several advances in using SNNs in machine learning tasks, including [29], which implemented reinforcement learning to control an inverted pendulum problem, and [17] that proposed a self-learning spiking network to control a mobile robot. These studies highlight the potential of SNNs for robotic control and autonomous decision making. SNNs are also recognized for their efficient energy consumption when running on neuromorphic devices [30].

Numerous studies have explored the biological plausibility of message passing algorithms [25]. In [21], an innovative approach to implementing sum-product message passing within spiking neural networks is introduced. This approach utilizes a network of interconnected liquid state machines. While their work offers valuable insights into bridging the gap between SNNs and message passing, it does not address the implementation of a learning mechanism for synaptic weights—a gap that this paper aims to fill.

This paper presents a method for training SNNs with spike-timing-dependent plasticity (STDP), a biologically inspired synaptic update rule widely used in neuromorphic computing, to implement sum-product message passing for Bernoulli-distributed messages. We demonstrate that the proposed networks, constructed with a minimal number of neurons and synaptic connections, yield results closely aligned with numerical results. Furthermore, we apply the proposed network to an example of a noisy signal transmission channel, demonstrating the principles and generality of our approach.

2 Background

In this section, we outline two key concepts we aim to connect: Bayesian inference with message passing and spiking neural networks. Readers already familiar with these topics may choose to skip this section.

2.1 Message Passing on Forney-Style Factor Graphs

Sum-product message passing, or belief propagation, in Forney-style Factor Graphs (FFGs) is a powerful algorithm for performing Bayesian inference. A large variety of algorithms in fields of machine learning, signal processing, coding, and statistics may be viewed as special cases of this method [18,19,24]. An FFG offers a graphical description of a factorized function [8]. In an FFG, nodes represent functions, and edges represent variables. An edge connects to a node if and only if it represents a variable that is an argument of the node's function. As an example, consider the factorized function

$$f(x_1, x_2, x_3, x_4) = f_a(x_1) f_b(x_1, x_2) f_c(x_2, x_3, x_4) f_d(x_2) f_e(x_3) f_f(x_4), \tag{1}$$

with its corresponding FFG is illustrated in Fig. 1. Note that in an FFG, each edge can maximally connect to two nodes. If a variable is an argument in more than two factors, we use an "equality" node $f(x, y, z) = \delta(z - y)\delta(z - x)$ that enforces the same beliefs across all variables (x, y and z) that connect to the equality node.

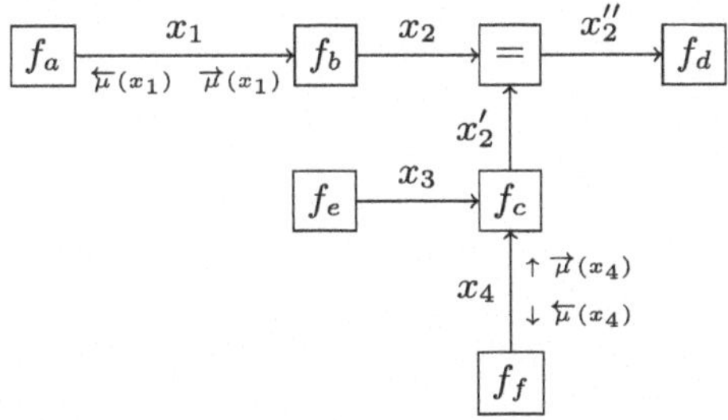

Fig. 1. The FFG corresponding to the example model defined in Equation (1).

The sum-product algorithm passes "messages", which are probabilistic distributions, along the graph's edges. This method is an efficient approach for conducting probabilistic inference within sparsely connected generative models. Message-passing-based inference is particularly scalable because it exploits the model's independence structure, significantly reducing the computational complexity compared to naive Bayesian inference, which requires summing over all possible configurations globally. The local and distributed nature of the sum-product algorithm allows it to handle large-scale problems efficiently, making it widely applicable in areas such as multi-agent trajectory planning [6], control of non-linear systems [1] and active inference in non-linear environments [16].

In FFG, messages are passed bidirectionally along the edges. For notational convenience, we assign a direction to each edge and denote messages propagating in the designated direction by $\overrightarrow{\mu}(\cdot)$, while those traveling in the opposite direction are represented by $\overleftarrow{\mu}(\cdot)$. In general, for any node $f(y, x_1, \ldots, x_n)$, the sum-product rule for an outgoing message over edge y is given by

$$\underbrace{\overrightarrow{\mu}_y(y)}_{\text{outgoing messages}} = \int \underbrace{\overrightarrow{\mu}_{x_1}(x_1)...\overrightarrow{\mu}_{x_n}(x_n)}_{\text{incoming messages}} \underbrace{f(y, x_1, ..., x_n)}_{\text{node function}} \mathrm{d}x_1...\mathrm{d}x_n. \tag{2}$$

By pre-calculating the update rules for common model components, one can efficiently construct and adjust inference algorithms without extensive computation. Certain studies have proposed message update rules [19,28], while there exist software toolboxes that provide the pre-computed message update rules for commonly used distributions and factors [2,22] to automate the inference process for making applications. Table 1 present some key update rules for processing incoming messages that carry a Bernoulli distribution. In this study, we aim to implement these computations using spiking neural networks, offering a biologically inspired approach to probabilistic inference.

Table 1. Sum-product message update rules for the equality and XOR nodes, for given input messages $\overrightarrow{\mu}_x(x) = \mathcal{B}er(x|p_x)$ and $\overrightarrow{\mu}_y(y) = \mathcal{B}er(y|p_y)$ [18].

Node function	Update rule
X → [=] → Z, Y ↑	$\overrightarrow{\mu}_z(z) = \mathcal{B}er(z \mid \frac{p_x p_y}{1-p_x+2p_x p_y-p_y})$
X → [XOR] → Z, Y ↑	$\overrightarrow{\mu}_z(z) = \mathcal{B}er(z \mid p_x - 2p_x p_y + p_y)$

2.2 Spiking Neural Networks

SNNs are a class of artificial neural networks that more closely mimic biological neural systems compared to traditional artificial neural networks. Unlike standard models that use continuous activation values, SNNs process and transmit information using discrete spike events over time. This event-driven paradigm makes SNNs well-suited for combining with MP, and also enables more energy-efficient computation.

Neuron Models. SNNs employ neuron models that describe the dynamics of membrane potential and spike generation. One of the most commonly used models is the Leaky Integrate-and-Fire (LIF) neuron, which approximates the behavior of a biological neuron with a differential equation. The membrane potential $V(t)$ evolves according to

$$\tau_m \frac{dV(t)}{dt} = -(V(t) - V_{\text{rest}}) + RI(t)\,,$$

where τ_m is the membrane time constant, V_{rest} is the resting membrane potential, R is the membrane resistance, $I(t)$ is the synaptic input current [11]. $I(t)$ represents the weighted sum of spikes from presynaptic neurons, capturing the total synaptic input to the neuron.

The leak mechanism refers to the gradual decay of the membrane potential $V(t)$ back toward its resting value V_{rest} in the absence of input. This models the natural tendency of neurons to return to a stable baseline and prevents indefinite accumulation of input. The integrating aspect represents the accumulation of incoming currents $I(t)$, which drive $V(t)$ upward. When the membrane potential reaches a certain threshold V_{th}, the neuron fires: it emits a spike (action potential), and the membrane potential is immediately reset to a lower value V_{reset}. This is the fire mechanism, mimicking the all-or-nothing nature of biological spikes. After firing, the neuron may enter a refractory period during which it is temporarily unable to spike, ensuring separation between consecutive spikes and limiting firing rates.

Synaptic Models. Synaptic models in SNNs determine how spikes from presynaptic neurons affect the membrane potential of postsynaptic neurons. In addition to shaping input currents, synapses can also adapt over time through learning mechanisms. One biologically inspired form of synaptic plasticity is STDP, which adjusts synaptic weights based on the relative timing of pre- and postsynaptic spikes. The change in synaptic weight Δw is typically modeled as

$$\Delta w = \begin{cases} A_+ \exp\left(-\frac{\Delta t}{\tau_+}\right) & \text{if } \Delta t \geq 0 \\ -A_- \exp\left(\frac{\Delta t}{\tau_-}\right) & \text{if } \Delta t < 0, \end{cases}$$

where $\Delta t = t_{\text{post}} - t_{\text{pre}}$ is the time difference between the postsynaptic and presynaptic spikes, A_+ and A_- are learning rates, and τ_+ and τ_- are time constants for potentiation and depression, respectively [26].

In summary, STDP strengthens synapses when the presynaptic neuron fires shortly before the postsynaptic neuron and weakens them when the order is reversed, thus encoding temporal correlations in spike activity.

3 Methodology

In this section, we introduce networks of LIF neurons trained with the biologically plausible synaptic learning rule, STDP, to simulate factor nodes in the message passing algorithm. This is achieved by encoding Bernoulli distributions into spike trains, passing them through the proposed networks, and decoding the output spike trains back into Bernoulli distributions. The output messages are compared with the numerical results derived from the sum-product rule (2). The code used to produce the results in this paper is available at https://github.com/biaslab/stdp-bernoulli-message-passing.

Since this study focuses on Bernoulli messages, we start with basic logical operations AND, OR, and NOT, shown in Table 2. Functionally complete sets, such as {AND, NOT} or {OR, NOT}, serve as the foundation of logical computation, allowing for the construction of any logical operation through combinations of these primitives [5]. As a further example, we implement the XOR factor node using a combination of the AND, OR, and NOT networks. The architecture of the proposed spiking neural network is illustrated in Fig. 2. Each circle represents a LIF neuron. The network consists of three layers: input, output, and a temporary training layer. The input layer comprises two neurons, each receiving encoded spike trains of the input Bernoulli messages. To encode Bernoulli distributions as spike trains, we sample them at a rate of 100 samples per second (i.e., every 10 ms). These input neurons are connected to the output neuron via trainable synaptic weights, denoted as ω_1 and ω_2, which are trained to produce the desired logical output.

Table 2. Logical Gates. The logical relationships defined here specify the desired output signals for training the networks using STDP.

S_1	S_2	OR	AND	NOT-S_1	XOR
0	0	0	0	1	0
0	1	1	0	1	1
1	0	1	0	0	1
1	1	1	1	0	0

To facilitate training, a training layer is used to activate the output neuron during learning; this layer is removed after training is complete. According to the STDP learning rule, synapses are strengthened when the output neuron fires shortly after the input spikes. During training, we generate the correct output spike trains based on the truth table shown in Table 2. These target spike trains are delayed by 1 ms and provided to the network via the positive training neuron, to activate the output neuron slightly after the input neurons. To weaken synapses when no spike is expected in the true output, we preemptively activate the output neuron 1 ms before the input spikes occur. This mechanism ensures that STDP decreases the synaptic strength in undesired scenarios.

The evolution of the synaptic weights ω_1 and ω_2 during training is shown in Fig. 4. This training approach is inspired by [23], and all neuron and synapse model parameters follow that study. The parameter values are summarized in Table 3. We used the Brian2 toolbox [12] to simulate neural activities.

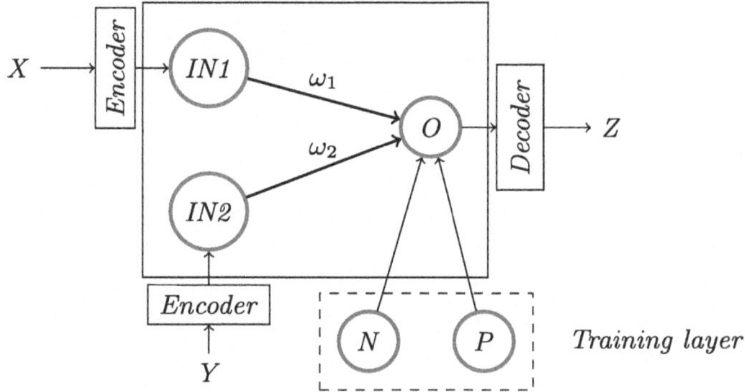

Fig. 2. SNN architecture for simulating the sum-product update rules of AND, OR, and NOT factor nodes, as represented in Table 4. Random variables X and Y are encoded into spike trains, and the resulting output spike train is decoded into Z. Each circle represents a LIF neuron. The training layer is removed after training the synaptic weights ω_1 and ω_2 using the STDP algorithm. For computing NOT of X, the variable Y is configured to emit constant spikes at every time step by defining its Bernoulli distribution as $\overrightarrow{\mu}_y(y) = \mathcal{B}er(y \mid p_y = 1)$.

Table 3. The parameter values used for LIF and STDP in Sect. 3.

V_{th}	V_{rest}	τ_m	R	w_{max}	w_{min}
−50 mV	−80 mV	5 ms	1	1	−1
$\mathbf{a}^+$	$\mathbf{a}^-$	τ^+	τ^-	$\mathbf{t}_{\mathbf{interval}}$	**lr**
0.005	$-a^+$	20 ms	20 ms	5 ms	0.005

For training, the input probabilities p_x and p_2 are initially set to 0.5. However, for the NOT operation, where the goal is to compute the negation of input 1, we fix $p_y = 1$, ensuring that the second input neuron fires at every time step. As shown in the figure, the network learns to strengthen the synaptic connection ω_2 while weakening ω_1. This effectively causes the output neuron to spike only when input 1 does not spike, correctly implementing the NOT function.

After the training phase, the training layer is removed, and the spikes from the output neuron are decoded into a Bernoulli distribution. The output message is defined as

$$\overrightarrow{\mu}_z(z) = \mathrm{Ber}(z \,|\, p_z), \qquad p_z = \rho/\tau \tag{3}$$

where ρ represents the total spike count observed at the output neuron, and τ is defined as the total number of samples drawn from the input messages. In this decoding process, $\mu_z(z)$ reflects the probability that the output neuron fires.

Figure 3 compares these decoded results with the numerical outcomes obtained from the sum-product update rule defined in (2), as summarized in Table 4. The derivation details are provided in Appendix A. As shown in Fig. 3, the proposed spiking-based method closely matches the numerical results.

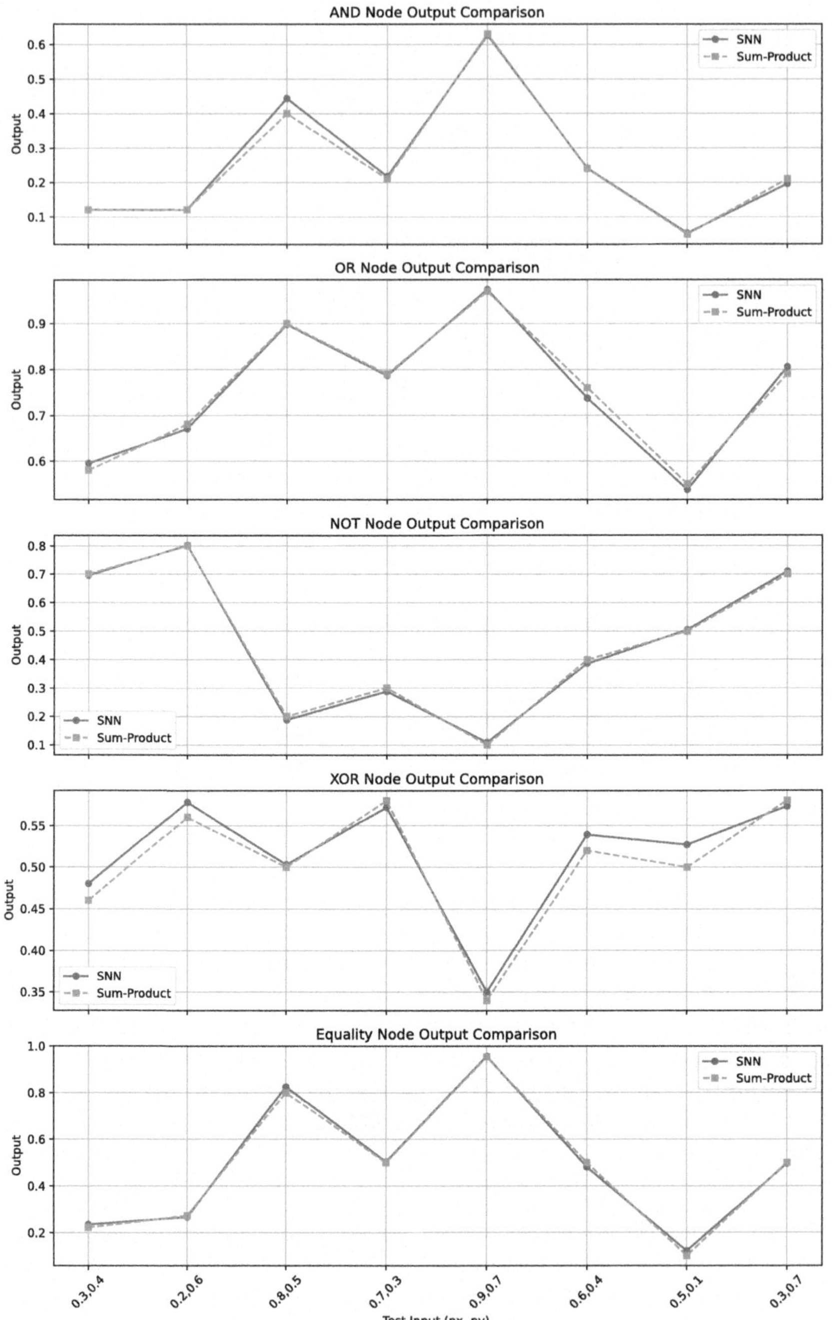

Fig. 3. Comparison of the results obtained from the proposed SNN-based nodes for passing Bernoulli messages with those produced by the sum-product algorithm. Validation was performed using eight random pairs of p_x and p_y.

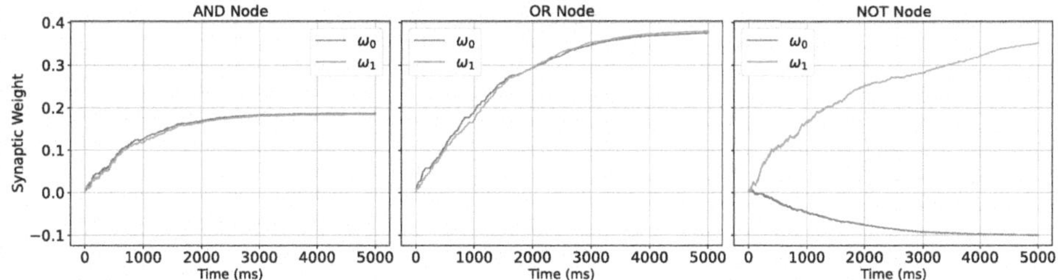

Fig. 4. Evolution of synaptic weights for AND, OR, and NOT nodes, from STDP-based training described in Sect. 3.

Table 4. Sum-product message update rules for the AND, OR, and NOT nodes, for given input messages $\overrightarrow{\mu}_x(x) = \mathcal{B}er(x|p_x)$ and $\overrightarrow{\mu}_y(y) = \mathcal{B}er(y|p_y)$. Detailed derivations are provided in Appendix A.

Node function	Update rule
X → AND → Z; Y ↑	$\overrightarrow{\mu}_z(z) = \mathcal{B}er(z \mid p_x p_y)$
X → OR → Z; Y ↑	$\overrightarrow{\mu}_z(z) = \mathcal{B}er(z \mid p_x - p_x p_y + p_y)$
X → NOT → Z	$\overrightarrow{\mu}_z(z) = \mathcal{B}er(z \mid 1 - p_x)$

The proposed networks can be used to implement more complex logical operations. As an example, we construct the XOR operation using combinations of the basic logical networks; its architecture is shown in Fig. 5. In Fig. 3, the output of the spiking-based XOR implementation is compared with the numerical results obtained from the sum-product update rule, as presented in Table 1.

The final operation we address in this paper is the equality constraint, a fundamental factor node in the sum-product algorithm. Although it is not a logical operation in the traditional sense, it can still be simulated using the proposed spiking networks. We expect a behavior consistent with the values reported in Table 5 for the equality constraint. As shown in the

Table 5. Equality Constraint. The table presents the expected output spike behavior of an equality node.

S_1	S_2	Equality
0	0	0
0	1	Not Defined
1	0	Not Defined
1	1	1

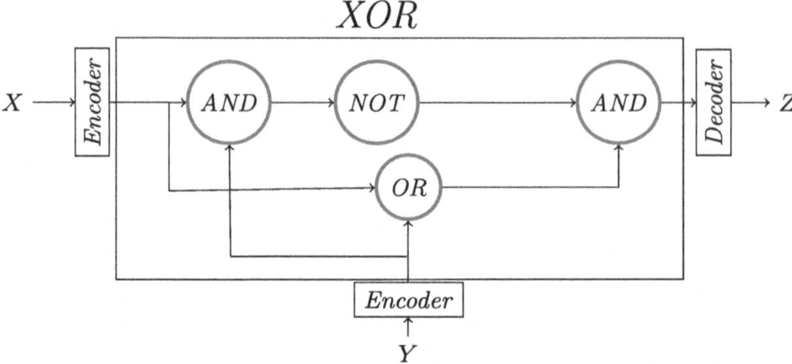

Fig. 5. SNN architecture for simulating the sum-product update rule of XOR factor node as defined in Table 1. The component networks reused from the AND, OR, and NOT implementations are shown in Fig. 2, and the same encoding and decoding methods are applied.

table, the total number of output spikes, denoted by ρ in (3), closely resembles the result of an AND operation applied to the input spike trains. To determine an appropriate value for τ in (3) under this constraint, cases labeled as 'Not Defined' are excluded from the total number of samples. Interestingly, these cases align with the spike count of the XOR logical operation. The final outcome is compared in Fig. 3 with the numerical sum-product result, obtained using the update rules presented in Table 1.

4 Example

In this section, we evaluate the inference capability of the proposed spiking network model using an example based on an unreliable binary communication channel. This signal transmission scenario has been used in prior work [18] [27] as a benchmark for evaluating Bernoulli message-passing algorithms. Consider a simple binary code $C = \{(0,0,0,0),\ (0,1,1,1),\ (1,0,1,1),\ (1,1,0,0)\}$ illustrated as a FFG in Fig. 6 (left). Each bit of a binary codeword $X = (x_1, x_2, x_3, x_4)$ is transmitted through an unreliable communication channel with a known bit-flip (cross-over) probability of $\varepsilon = 0.1$. The generative model is given by

$$f(x_1, \ldots x_4, z \mid y_1, \ldots y_4) \propto f_{xor}(x_1, x_2, z)\, f_{=}(x_3, x_4, z) \prod_{i=1}^{4} p(y_i \mid x_i).$$

where z is a latent variable and the y_i are noisy observations. Given the observed values $(y_1, y_2, y_3, y_4) = (0, 0, 1, 0)$, the corresponding Bernoulli messages can be obtain as

$$(\overrightarrow{\mu}_{x_1}(x_1), \overrightarrow{\mu}_{x_2}(x_2), \overrightarrow{\mu}_{x_3}(x_3), \overrightarrow{\mu}_{x_4}(x_4)) = (\mathcal{B}er(0.1), \mathcal{B}er(0.1), \mathcal{B}er(0.9), \mathcal{B}er(0.1))\,.$$

as detailed in Appendix B.

The goal is to compute the marginal posterior for each bit x_i. The table in Fig. 6 (right) compares all the messages produced by our SNN-based model with the ground-truth messages passed along each edge of the FFG. The results show that our method closely approximates the expected message values. Finally, the marginal posteriors are obtained via the normalized product

$$\overrightarrow{\mu}_{x_i}(x_i) \times \overleftarrow{\mu}_{x_i}(x_i) = Ber(0.109), Ber(0.109), Ber(0.168), Ber(0.175), \quad (4)$$

which closely match the messages produced by the sum-product rule $Ber(0.1)$, $Ber(0.1)$, $Ber(0.18)$, $Ber(0.18)$ [18].

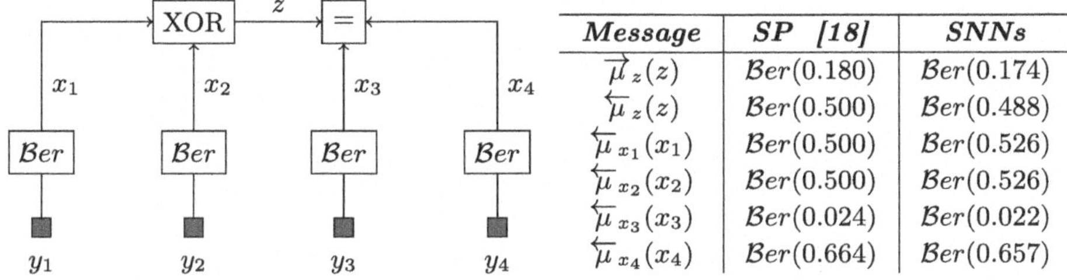

Message	*SP [18]*	*SNNs*
$\overrightarrow{\mu}_z(z)$	$\mathcal{B}er(0.180)$	$\mathcal{B}er(0.174)$
$\overleftarrow{\mu}_z(z)$	$\mathcal{B}er(0.500)$	$\mathcal{B}er(0.488)$
$\overleftarrow{\mu}_{x_1}(x_1)$	$\mathcal{B}er(0.500)$	$\mathcal{B}er(0.526)$
$\overleftarrow{\mu}_{x_2}(x_2)$	$\mathcal{B}er(0.500)$	$\mathcal{B}er(0.526)$
$\overleftarrow{\mu}_{x_3}(x_3)$	$\mathcal{B}er(0.024)$	$\mathcal{B}er(0.022)$
$\overleftarrow{\mu}_{x_4}(x_4)$	$\mathcal{B}er(0.664)$	$\mathcal{B}er(0.657)$

Fig. 6. (Left) The FFG corresponding to the unreliable binary channel example. (Right) The sum-product calculated messages versus messages produced by the SNNs.

5 Conclusions

We presented a biologically plausible network of spiking neurons trained using STDP to perform message-passing for Bernoulli messages. This work represents a potential step toward bridging the gap between theoretical models of brain function, such as the Bayesian Brain hypothesis and the FEP, and the spiking behavior of biological neurons.

Acknowledgments. This work was carried out in the context of the BayesBrain project. We gratefully acknowledge financial support from the Eindhoven Artificial Intelligence Systems Institute (EAISI) at TU Eindhoven.

Disclosure of Interests. The authors declare no conflict of interest.

A Derivation of Sum-Product Update Rules

We can derive the forward sum-product message $\overrightarrow{\mu}_z(z)$ according to (2), given the input messages $\overrightarrow{\mu}_x(x) = \mathcal{B}er(x|p_x)$ and $\overrightarrow{\mu}_y(y) = \mathcal{B}er(y|p_y)$.

$$\begin{aligned}
\overrightarrow{\mu}_z(z) &= \sum_x \sum_y \overrightarrow{\mu}_x(x)\, \overrightarrow{\mu}_y(y)\, f(z,x,y) \\
&= \mathcal{B}er(0|p_x)\, \mathcal{B}er(0|p_y))f(z,0,0) + \mathcal{B}er(1|p_x)\, \mathcal{B}er(0|p_y) f(z,1,0) \\
&\quad + \mathcal{B}er(0|p_x)\, \mathcal{B}er(1|p_y) f(z,0,1) + \mathcal{B}er(1|p_x)\, \mathcal{B}er(1|p_y) f(z,1,1) \\
&= (1-p_x)(1-p_y) f(z,0,0) + p_x(1-p_y) f(z,1,0) \\
&\quad + (1-p_x)p_y f(z,0,1) + p_x p_y f(z,1,1)\,.
\end{aligned}$$

Using the truth table, we can substitute z and evaluate the terms for different operations. As an example, the AND operation is

$$\begin{aligned}
\overrightarrow{\mu}_z(z) &= \begin{cases} p_x p_y & \text{if } z=1 \\ (1-p_x)(1-p_y) + p_x(1-p_y) + (1-p_x)p_y & \text{if } z=0 \end{cases} \\
&= \begin{cases} p_x p_y & \text{if } z=1 \\ 1-p_x p_y & \text{if } z=0 \end{cases} \quad = \quad \mathcal{B}er(z|p_x p_y)\,.
\end{aligned}$$

Similarly, we have the following computations for the OR factor node

$$\begin{aligned}
\overrightarrow{\mu}_z(z) &= \begin{cases} p_x p_y + p_x(1-p_y) + (1-p_x)p_y & \text{if } z=1 \\ (1-p_x)(1-p_y) & \text{if } z=0 \end{cases} \\
&= \begin{cases} p_x + p_y - p_x p_y & \text{if } z=1 \\ 1-p_x-p_y+p_x p_y & \text{if } z=0 \end{cases} \quad = \quad \mathcal{B}er(z|p_x + p_y - p_x p_y)\,.
\end{aligned}$$

B Derivation of Messages in the Example

Under the bit-flip channel model, the likelihood is given by

$$p(y_i|x_i) = \begin{cases} (1-\varepsilon)x_i + \varepsilon(1-x_i) & \text{if } y_i = 1 \\ \varepsilon x_i + (1-\varepsilon)(1-x_i) & \text{if } y_i = 0 \end{cases} = \mathcal{B}er(y_i|\,(1-\varepsilon)x_i + \varepsilon(1-x_i))\,.$$

According to the sum-product rule, the messages can be obtained by

$$\begin{aligned}
\overrightarrow{\mu}_{x_i}(x_i) &= \sum_{y_i\in\{0,1\}} \delta(y_i - \hat{y}_i)\mathcal{B}er(y_i|\,(1-\varepsilon)x_i + \varepsilon(1-x_i)) \\
&= \begin{cases} (1-\varepsilon)x_i + \varepsilon(1-x_i) & \text{if } \hat{y}_i = 1 \\ \varepsilon x_i + (1-\varepsilon)(1-x_i) & \text{if } \hat{y}_i = 0 \end{cases} \\
&= \begin{cases} \mathcal{B}er(x_i|1-\varepsilon) & \text{if } \hat{y}_i = 1 \\ \mathcal{B}er(x_i|\varepsilon) & \text{if } \hat{y}_i = 0 \end{cases}.
\end{aligned}$$

References

1. Adamiat, S., Kouw, W.M., van Erp, B., de Vries, A.B.: Message passing-based Bayesian control of a cart-pole system. In: International Workshop on Active Inference. Springer (2024)
2. Bagaev, D., de Vries, B.: Reactive message passing for scalable Bayesian inference. Sci. Program. **2023**(1), 6601690 (2023)
3. Cai, H., et al.: Brain organoid reservoir computing for artificial intelligence. Nat. Electron. **6**(12), 1032–1039 (2023)
4. Doya, K.: Bayesian Brain: Probabilistic Approaches to Neural Coding. MIT Press (2007)
5. Enderton, H.B.: A mathematical introduction to logic. Elsevier (2001)
6. van Erp, B., Bagaev, D., Podusenko, A., Şenöz, İ., de Vries, B.: Multi-agent trajectory planning with NUV priors. American Control Conference (2024, in press)
7. Fiorillo, C.D., Tobler, P.N., Schultz, W.: Discrete coding of reward probability and uncertainty by dopamine neurons. Science **299**(5614), 1898–1902 (2003)
8. Forney, G.D.: Codes on graphs: normal realizations. IEEE Trans. Inf. Theory **47**(2), 520–548 (2001)
9. Friston, K., Kilner, J., Harrison, L.: A free energy principle for the brain. J. Physiol. **100**(1–3), 70–87 (2006)
10. Friston, K.J., Parr, T., de Vries, B.: The graphical brain: belief propagation and active inference. Netw. Neurosci. **1**(4), 381–414 (2017)
11. Gerstner, W., Kistler, W.M.: Spiking Neuron Models: Single Neurons, Populations, Plasticity. Cambridge University Press (2002)
12. Goodman, D.F., Brette, R.: Brian: a simulator for spiking neural networks in Python. Front. Neuroinform. **2**, 350 (2008)
13. Isomura, T., Kotani, K., Jimbo, Y.: Cultured cortical neurons can perform blind source separation according to the free-energy principle. PLoS Comput. Biol. **11**(12), e1004643 (2015)
14. Kagan, B.J., et al.: In vitro neurons learn and exhibit sentience when embodied in a simulated game-world. Neuron **110**(23), 3952–3969 (2022)
15. Knill, D.C., Richards, W.: Perception as Bayesian Inference. Cambridge University Press (1996)
16. Van de Laar, T.W., De Vries, B.: Simulating active inference processes by message passing. Front. Robot. AI **6**, 20 (2019)
17. Lobov, S.A., Mikhaylov, A.N., Shamshin, M., Makarov, V.A., Kazantsev, V.B.: Spatial properties of STDP in a self-learning spiking neural network enable controlling a mobile robot. Front. Neurosci. **14**, 88 (2020)
18. Loeliger, H.A.: An introduction to factor graphs. IEEE Signal Process. Mag. **21**(1), 28–41 (2004)
19. Loeliger, H.A., Dauwels, J., Hu, J., Korl, S., Ping, L., Kschischang, F.R.: The factor graph approach to model-based signal processing. Proc. IEEE **95**(6), 1295–1322 (2007)
20. Maass, W.: Networks of spiking neurons: the third generation of neural network models. Neural Netw. **10**(9), 1659–1671 (1997)
21. Maass, W.: Liquid state machines: motivation, theory, and applications. Computability in Context, pp. 275–296 (2011)
22. Minka, T., Winn, J., Guiver, J., Zaykov, Y., Fabian, D., Bronskill, J.: /Infer.NET 0.3 (2018). Microsoft Research Cambridge. http://dotnet.github.io/infer

23. Mo, L., Wang, M.: Logicsnn: a unified spiking neural networks logical operation paradigm. Electronics **10**(17), 2123 (2021)
24. Palmieri, F.A., Pattipati, K.R., Di Gennaro, G., Fioretti, G., Verolla, F., Buonanno, A.: A unifying view of estimation and control using belief propagation with application to path planning. IEEE Access **10**, 15193–15216 (2022)
25. Parr, T., Markovic, D., Kiebel, S.J., Friston, K.J.: Neuronal message passing using mean-field, bethe, and marginal approximations. Sci. Rep. **9**(1), 1889 (2019)
26. Song, S., Miller, K.D., Abbott, L.F.: Competitive hebbian learning through spike-timing-dependent synaptic plasticity. Nat. Neurosci. **3**(9), 919–926 (2000)
27. Steimer, A., Maass, W., Douglas, R.: Belief propagation in networks of spiking neurons. Neural Comput. **21**(9), 2502–2523 (2009)
28. Winn, J., Bishop, C.M., Jaakkola, T.: Variational message passing. J. Mach. Learn. Res. **6**(4) (2005)
29. Wu, G., Liang, D., Luan, S., Wang, J.: Training spiking neural networks for reinforcement learning tasks with temporal coding method. Front. Neurosci. **16**, 877701 (2022)
30. Yamazaki, K., Vo-Ho, V.K., Bulsara, D., Le, N.: Spiking neural networks and their applications: a review. Brain Sci. **12**(7), 863 (2022)

Fundamental Active Inference Extensions

Diffusion-Generated Latent Spaces for Continuous Amortized Active Inference Agents: A Mathematical Framework

Zahra Sheikhbahaee[1(✉)], Adam Safron[1,2], Dalton A. R. Sakthivadivel[3,4], Mahault Albarracin[4,5], and Irina Rish[6,7]

[1] Institute for Advanced Consciousness Studies, 2811 Wilshire Blvd # 510, Santa Monica, CA 90403, USA
zahra.sheikhbahaee@iacstudies.org, adam.safron@tufts.edu
[2] Tufts University, 419 Boston Ave, Medford, MA 02155, USA
[3] VERSES Research Lab, Department of Mathematics, CUNY Graduate Centre, New York, NY, USA
dalton.sakthivadivel@cuny.edu
[4] Département d'informatique, Université du Québec à Montréal, Montréal, Canada
mahault.albarracin@uqam.ca
[5] Département d'informatique, Institut de Sante et Societe, Université du Quebec a Montréal (UdeM), Montréal H2L 2C4, Canada
[6] Mila, Quebec AI Institute, 6666 Rue Saint-Urbain, Montréal, QC H2S 3H1, Canada
irina.rish@mila.quebec
[7] Université de Montréal (UdeM), Montréal H3T 1J4, Canada

Abstract. We present a mathematical integration of latent diffusion models with active inference for continuous state and action spaces. Traditional active inference implementations rely on variational inference with parametric distributions, limiting their expressiveness in complex environments. We propose generating latent belief states through reverse diffusion processes, fundamentally altering how agents represent uncertainty about their environment (i.e. via diffusion-based posteriors). Our key contribution is a reformulation of the expected free energy (EFE) computation over diffusion-generated trajectories, providing a principled approach to action selection that naturally balances epistemic and pragmatic values. We establish the mathematical foundations connecting diffusion processes to the free energy principle and present algorithms for practical implementation.

Keywords: Active Inference · Diffusion Models · Free Energy Principle · Continuous Control · Variational Inference

1 Introduction

Active inference casts both perception and action as minimization of variational free energy [14,40]. Agents maintain their *structural integrity*—the preservation

M. Albarracin et al. (Eds.): IWAI 2025, CCIS 2857, pp. 115–136, 2026.
https://doi.org/10.1007/978-3-032-16955-6_7

of their characteristic organizational patterns and functional coherence—through their *Markov blanket* by minimizing surprise about sensory observations, leading to behavior that actively seeks preferred states while reducing uncertainty about the environment. Here, a Markov blanket defines the statistical boundary separating internal states from external environmental dynamics, where sensory and active states mediate this boundary as conditional independence relationships [20]. This formalism ensures that an agent's internal states remain statistically separated from external fluctuations except through sensory observations and motor actions. In most cases, active inference is implemented in discrete categorical state space. However, implementing active inference in continuous, high-dimensional spaces presents significant challenges, particularly in representing complex, multimodal belief distributions [4].

We propose a different approach to belief representation in active inference through diffusion processes. Rather than optimizing parametric variational distributions, our approach learns score functions that implicitly define beliefs through reverse diffusion dynamics. This enables the emergence of rich, multimodal belief states through iterative refinement conditioned on high-dimensional observations. Such an approach fundamentally expanding the representational capacity of active inference agents.

The key insight of our approach is that belief updating in continuous spaces can be reformulated as a reverse diffusion process, where the score function $\nabla_{\mathbf{s}} \log p_{\mathcal{T}}(\mathbf{s} \mid \mathbf{o})$ guides noisy samples toward regions of high posterior probability. This eliminates the need for explicit parametric distributions while maintaining mathematical rigor through the connection to variational free energy minimization.

Our contributions are threefold:

1. By mathematically connect diffusion processes and variational free energy minimization, we show how score-based models implement belief updates.
2. We derive a novel formulation of expected free energy over diffusion trajectories, enabling principled action selection in continuous spaces.
3. We present practical algorithms that implement these theoretical insights.

2 Background: Mathematical Framework

2.1 Active Inference in Continuous Spaces

In Active Inference, agents optimize behavior by performing approximate Bayesian inference over latent states, parameters, and actions in a generative model of their environment [5]. This generative model encodes how observations arise from latent causes and how those causes evolve over time, under the influence of action. The generative process of a standard agent decomposes the joint distribution over hidden states $\mathbf{s}$, observations $\mathbf{o}$, and actions $\mathbf{a}$, as [37]

$$p(\mathbf{o}_{0:T}, \mathbf{s}_{0:T}, \mathbf{a}_{0:T}) = p(\mathbf{s}_0) \prod_{t=1}^{T} p(\mathbf{s}_t \mid \mathbf{s}_{t-1}, \mathbf{a}_{t-1})\, p(\mathbf{o}_t \mid \mathbf{s}_t)\, \pi(\mathbf{a}_t \mid \mathbf{s}_t) \tag{1}$$

where $p(\mathbf{s}_0)$ is the initial state prior, $p(\mathbf{s}_t \mid \mathbf{s}_{t-1}, \mathbf{a}_{t-1})$ encodes the dynamics model, $p(\mathbf{o}_t \mid \mathbf{s}_t)$ represents the observation likelihood, and $\pi(\mathbf{a}_t \mid \mathbf{s}_t)$ denotes the agent's policy. This factorization is commonly used in discrete and finite spaces, where $\mathbf{s}_t$, $\mathbf{o}_t$, and $\mathbf{a}_t$ take values from categorical distributions defined over a finite state space.

We define sequences that span over time steps with the superscript $\sim$, i.e., for states and outcomes, therefore we have $\tilde{\mathbf{s}} = \{\mathbf{s}_1, \mathbf{s}_2, ..., \mathbf{s}_T\}$ and $\tilde{\mathbf{o}} = \{\mathbf{o}_1, \mathbf{o}_2, ..., \mathbf{o}_T\}$, respectively. The variational free energy for the belief distribution over trajectories $q(\tilde{\mathbf{s}} \mid \tilde{\mathbf{o}})$ relative to a generative model $p(\tilde{\mathbf{o}}, \tilde{\mathbf{s}})$ is [17]

$$F_{\pi,\theta}[q] = \mathbb{E}_{q(\tilde{\mathbf{s}}|\tilde{\mathbf{o}})}[\log q(\tilde{\mathbf{s}} \mid \tilde{\mathbf{o}}) - \log p(\tilde{\mathbf{o}}, \tilde{\mathbf{s}})] \tag{2}$$

$$= D_{\mathrm{KL}}[q(\tilde{\mathbf{s}} \mid \tilde{\mathbf{o}}) \| p(\tilde{\mathbf{s}})] - \mathbb{E}_{q(\tilde{\mathbf{s}}|\tilde{\mathbf{o}})}[\log p(\tilde{\mathbf{o}} \mid \tilde{\mathbf{s}})] \tag{3}$$

where $-\log p(\tilde{\mathbf{o}}) = -\int \log p(\tilde{\mathbf{o}}, \tilde{\mathbf{s}}) d\tilde{\mathbf{s}}$ plays the role of surprisal (the non-equilibrium steady-state potential) and minimizing F drives $q(\tilde{\mathbf{s}} \mid \tilde{\mathbf{o}})$ toward the system's steady-state attractor [15]. Traditionally, the posterior probability $q(\tilde{\mathbf{s}} \mid \tilde{\mathbf{o}})$ is parameterized explicitly (e.g., as Gaussian distributions). We instead propose generating samples from $q(\tilde{\mathbf{s}} \mid \tilde{\mathbf{o}})$ through a learned diffusion process.

2.2 Mathematical Foundation: From Stochastic Processes to Belief Generation

The theoretical foundation connecting diffusion processes to active inference rests on the observation that belief updating in complex environments requires mechanisms capable of representing multimodal, non-Gaussian uncertainty. Traditional variational approaches constrain posterior approximations to parametric families (typically Gaussian), limiting expressiveness in high-dimensional latent spaces with complex dependency structures.

Any continuous Markov process with time-indexed marginal $p_\tau(\mathbf{s} \mid \mathbf{o})$ can be expressed in the general form [15,33]

$$d\mathbf{s} = -[\mathbf{D}(\mathbf{s}) + \mathbf{Q}(\mathbf{s})] \nabla H(\mathbf{s}) d\tau + \mathbf{\Gamma}(\mathbf{s}) d\tau + \sqrt{2\mathbf{D}(\mathbf{s})} d\mathbf{W}_\tau \tag{4}$$

where $\mathbf{W}_\tau$ is a Wiener process and $H(\mathbf{s}) = -\log p_\tau(\mathbf{s} \mid \mathbf{o})$ is the information potential at diffusion time τ. Throughout this work, we distinguish between environmental time steps (denoted by subscripts, e.g., $\mathbf{s}_t$ for state at environment time t) and diffusion process time (denoted by τ, e.g., $p_\tau(\mathbf{s} \mid \mathbf{o})$ for the diffusion distribution at diffusion time $\tau \in [0, 1]$).

If we assume that the stationary distribution $p_\tau(\mathbf{s} \mid \mathbf{o})$ is Gaussian, then it implies that states evolve toward an equilibrium with harmonic potential[1]. $\mathbf{D}(\mathbf{s})$ is a positive semidefinite diffusion matrix controlling noise injection, $\mathbf{Q}(\mathbf{s})$ is a skew-symmetric curl matrix encoding *solenoidal flow*—divergence-free velocity fields that preserve phase space volume through rotational dynamics without net expansion or contraction. This component enables non-dissipative dynamics

[1] In the detailed-balance case with no curl term, the dynamics reduce to an Ornstein–Uhlenbeck (OU)/overdamped Langevin gradient flow in a quadratic potential.

such as oscillatory or conservative motion patterns, complementing the dissipative drift toward equilibrium imposed by the diffusion matrix $\mathbf{D}(\mathbf{s})$. The parameter $\boldsymbol{\Gamma}_i(\mathbf{s}) = \sum_j \frac{\partial}{\partial_{\mathbf{s}_j}}(\mathbf{D}_{ij}(\mathbf{s}) + \mathbf{Q}_{ij}(\mathbf{s}))$ emerges from Itô calculus and ensures mathematical consistency with the associated Fokker-Planck equation governing the probability density evolution. This divergence term compensates for the non-uniform diffusion and drift coefficients, preventing spurious probability flows that would violate conservation of probability mass.

In our framework, we consider homogeneity across states in $\mathbf{D}(\mathbf{s}) = \frac{1}{2}\beta(\tau)\mathbf{I}$ and set $\mathbf{Q}(\mathbf{s}) = \mathbf{0}$ (no momentum-induced/curl flow). Here $\beta(\tau)$ is the time-dependent noise schedule that drives both the linear drift and the diffusion: it scales the drift magnitude $-\frac{1}{2}\beta(\tau)\mathbf{s}$ and the diffusion coefficient $\sqrt{\beta(\tau)}$. As such, we obtain

$$d\mathbf{s} = -\tfrac{1}{2}\beta(\tau)\,\mathbf{s}\,d\tau + \sqrt{\beta(\tau)}\,dW_\tau. \tag{5}$$

This is the well-known Ornstein-Uhlenbeck (OU) process. The OU forward path is logarithmically concave (Gaussian), satisfying the standard log-Sobolev inequality (LSI) conditions. Therefore, the solution to this SDE is a transition kernel, given by:

$$q(\mathbf{s}_\tau \mid \mathbf{s}_0) = \mathcal{N}(\mathbf{s}_\tau; \alpha(\tau)\mathbf{s}_0, \sigma^2(\tau)I) \tag{6}$$

where $\alpha(\tau) = \exp\left(-\frac{1}{2}\int_0^\tau \beta(\tau')d\tau'\right)$ and $\sigma^2(\tau) = 1 - \alpha^2(\tau)$. The time-reversed process [1]

$$d\mathbf{s} = \left[-\frac{1}{2}\beta(\tau)s - \beta(\tau)\nabla_{\mathbf{s}}\log p_\tau(\mathbf{s} \mid \mathbf{o})\right]d\tau + \sqrt{\beta(\tau)}dW_\tau \tag{7}$$

generates beliefs by systematically removing noise through a denoising process. To understand this mechanism, consider that the forward diffusion progressively corrupts the initial state s_0 into pure Gaussian noise as $\tau \to 1$. The reverse process inverts this corruption by following the score function $\nabla_{\mathbf{s}}\log p_\tau(\mathbf{s} \mid \mathbf{o})$, which points toward regions of higher probability density in the data distribution.

At each reverse timestep, the score function provides the direction of steepest ascent in log-probability space, effectively guiding the noisy sample back toward the manifold of plausible belief states conditioned on observation $\mathbf{o}$. This process can be interpreted as iterative variational inference: each denoising step refines the belief state by incorporating both the learned prior over latent states (encoded in the score network) and the conditioning information from observations. The stochastic term $\sqrt{\beta(\tau)}dW_\tau$ maintains the correct marginal distributions throughout the reverse trajectory, ensuring that the final sample $\mathbf{s}_0$ is drawn from the desired posterior $q(\mathbf{s} \mid \mathbf{o})$.

The optimal score function $\nabla_{\mathbf{s}}\log p_\tau(\mathbf{s} \mid \mathbf{o})$ for the time-reversed diffusion process can be learned through denoising score matching. We distinguish between the true score function $\nabla_{\mathbf{s}_\tau}\log p_\tau(\mathbf{s}_\tau \mid \mathbf{o})$ and its learned approximation $s_\theta(\mathbf{s}_\tau, \tau, \mathbf{o})$, where θ denotes the neural network parameters.

The objective of achieving the de-noise score matching leverages the fact that, for the forward diffusion process $q(\mathbf{s}_\tau \mid \mathbf{s}_0) = \mathcal{N}(\mathbf{s}_\tau; \alpha(\tau)\mathbf{s}_0, \sigma^2(\tau)I)$, the score has a closed-form expression:

$$\nabla_{\mathbf{s}_\tau} \log q(\mathbf{s}_\tau \mid \mathbf{s}_0) = -\frac{\mathbf{s}_\tau - \alpha(\tau)\mathbf{s}_0}{\sigma^2(\tau)} = -\frac{\epsilon}{\sigma(\tau)} \tag{8}$$

where $\mathbf{s}_\tau = \alpha(\tau)\mathbf{s}_0 + \sigma(\tau)\epsilon$ with $\epsilon \sim \mathcal{N}(0, I)$. This crucial relationship allows us to reformulate the score matching problem as a noise prediction task. Instead of directly learning the score function, we can train a network $\epsilon_\theta(\mathrm{s}_\tau, \tau, \mathbf{o})$ to predict the noise ϵ, then recover the score as $s_\theta(s_\tau, \mathbf{o}) = -\epsilon_\theta(s_\tau, \tau, \mathbf{o})/\sigma(\tau)$. This noise parameterization is more stable during training and has become the standard approach in diffusion models.

This yields the practical training objective:

$$\mathcal{L}_{\text{score}}(\theta) = \mathbb{E}_{\tau, \mathbf{s} \sim q(\mathbf{s}|\mathbf{o}), \epsilon \sim \mathcal{N}(0,I)} \left[\lambda(\tau) \left\| s_\theta(\mathbf{s}_\tau, \tau, \mathbf{o}) - \left(-\frac{\epsilon}{\sigma(\tau)} \right) \right\|^2 \right] \tag{9}$$

where $s_\theta \colon \mathcal{S} \times [0, 1] \times \mathcal{O} \to \mathcal{S}$ denotes *the learned score approximator* parameterized by θ that approximates the score function $\nabla_{\mathbf{s}} \log p_\tau(\mathbf{s} \mid \mathbf{o})$ [28]. This network takes as input the current state $\mathbf{s}_\tau$, the diffusion time τ, and the observation o, and outputs an estimate of the score (the gradient of the log-probability density with respect to the state).

$$\mathcal{L}_{\text{score}}(\theta) = \int_{\lambda_{\min}}^{\lambda_{\max}} w(\lambda) \mathbb{E}_{\epsilon \sim \mathcal{N}(0,I)} \left[\left\| \epsilon - \epsilon_\theta(\mathbf{s}_\lambda; \lambda, \mathbf{o}) \right\|^2 \right] d\lambda \tag{10}$$

The network architecture processes three inputs: the noisy state $\mathbf{s}_\tau$, the diffusion timestep τ, and the conditioning on observation $\mathbf{o}$ [43]. The log signal-to-noise ratio (log-SNR) for time τ is defined by $\lambda(\tau) = \log\left(\frac{\alpha^2(\tau)}{\sigma^2(\tau)}\right)$ ensures uniform contribution across all noise levels, preventing the loss from being dominated by high-noise timesteps [47,51]. Conditioning the score function on observations $\mathbf{o}$ to generate belief states for active inference transforms the diffusion process into an implicit variational inference mechanism: rather than explicitly parameterizing a variational distribution $q(\mathbf{s} \mid \mathbf{o})$, we generate samples from this distribution through the learned reverse diffusion process. This approach creates an implicit variational posterior that minimizes variational free energy while maintaining the flexibility to represent complex, multimodal belief distributions.

In Eq. (10), the weighted diffusion loss is a weighted integral of ELBOs across noise levels; and for monotonic weights $w(\lambda)$ this collapses to an ELBO computed on a data distribution augmented with Gaussian noise [28]. Here the constant depends only on the endpoints $(\lambda_{\min}, \lambda_{\max})$ and is independent of θ.

Once we train using this denoising objective, the network ϵ enables flexible inference strategies. For belief generation, we can use DDPM sampling with the full T diffusion steps for maximum sample quality, or DDIM with the $K \ll T$ steps for accelerated inference [45]. Both methods use the same trained network but differ in their reverse dynamics-DDPM simulates a Markovian, stochastic reverse process that injects Gaussian noise at each step while DDIM follows a (deterministic, ODE-like) non-Markovian update which trades sample diversity for computational efficiency. This flexibility is crucial for real-time active inference applications where computational budgets vary.

2.3 Free Energy with Diffusion-Generated Beliefs

The key insight is that diffusion-generated beliefs implicitly minimize variational free energy through the following mechanism. The coupled forward-backward SDE framework can be considered as an implementation of variational principles underlying active inference, where the backward process aims to construct the belief dynamics that minimize variational free energy in self-organizing systems [16]. Let $q_\theta(\mathbf{s} \mid \mathbf{o})$ be the distribution induced by the reversed diffusion process. The optimal parameters satisfy

$$q_\theta^* = \arg\min_\theta F[q_\theta] \quad \text{where} \quad F[q_\theta] = \mathbb{E}_{q_\theta}[\log q_\theta(\mathbf{s} \mid \mathbf{o}) - \log p(\mathbf{o}, \mathbf{s})]. \tag{11}$$

This allows us to compute free energy through the diffusion objective

$$F[q_\theta] \approx \underbrace{\mathbb{E}_{q_\theta(\mathbf{s}_t|\mathbf{o}_t)}[-\ln p(\mathbf{o}_t \mid \mathbf{s}_t)]}_{\text{reconstruction loss}} + \underbrace{D_{\mathrm{KL}}[q_\theta(\mathbf{s}_t \mid \mathbf{o}_t) \| p_\psi(\mathbf{s}_t \mid \mathbf{s}_{t-1}, \pi)]}_{\text{complexity penalty}} + \underbrace{\mathcal{L}_{\text{score}}(\theta)}_{\text{score matching}} \tag{12}$$

where the observation-conditioned posterior over the next latent is computed by the reverse diffusion sampler driven by the learned score function s_θ and the transition prior is a latent dynamic network. We compute the Kullback–Leibler divergence via a Donsker–Varadhan variational representation which under optimization over a restricted function class yields a lower bound to the true KL [11]. Because both the diffusion-based posterior q_θ and the learned transition prior p_ψ are implicit, the KL terms in the sequential ELBO are intractable. We therefore use the Donsker–Varadhan variational representation to estimate the KL from samples. The score loss, $\mathcal{L}_{\text{score}}$, is a denoising score-matching regularizer that, under standard log–Sobolev (LSI) conditions along the OU forward path, controls an explicit upper bound on the posterior KL, thus tightening an explicit upper bound on free energy. This statement is motivated by showing that the KL divergence is upper bounded by half of this Fisher divergence [50]:

$$D_{\mathbf{KL}}\Big[q_\theta(\mathbf{s}_\tau \mid \mathbf{o}) \| p(\mathbf{s}_\tau)\Big] \le \frac{1}{2\nu} \int q_\theta(\mathbf{s}_\tau \mid \mathbf{o}) \left\| \nabla_{\mathbf{s}_\tau} \log \frac{q_\theta(\mathbf{s}_\tau \mid \mathbf{o})}{p(\mathbf{s}_\tau)} \right\|^2 d\mathbf{s}_\tau. \tag{13}$$

where $\nu > 0$ (e.g., we consider $\nu = 1$ for the standard Gaussian) and shows that $\mathcal{L}_{\text{score}}$ is a weighted integral over noise levels; it controls KLs at those levels. In practice, we propose minimizing a surrogate objective that combines (*i*) a reconstruction term, (*ii*) a complexity term comparing $q_\theta(\mathbf{s}_t \mid \mathbf{o}_t)$ with the transition prior $p_\psi(\mathbf{s}_t \mid \mathbf{s}_{t-1}, \pi)$, and (iii) the loss of diffusion score matching[2].

[2] If one uses the exact Donsker–Varadhan variational representation (sup over all test functions Φ) where $D_{\mathrm{KL}}(q \parallel p) = \sup_\Phi\{\mathbb{E}_q[\Phi] - \log \mathbb{E}_p[\exp(\Phi)]\}$, then it is equal to the KL [3]. While we obtain a lower bound on KL if we optimize over a restricted parametric family of Φ.

3 Expected Free Energy over Diffusion Trajectories

The fundamental innovation of our framework lies in reformulating expected free energy computation over trajectories generated through diffusion processes rather than parametric posterior approximations. This reformulation preserves the theoretical elegance of active inference while dramatically expanding representational capacity.

Active inference selects actions by minimizing expected free energy, which quantifies the expected surprise and uncertainty under a policy. We extend this to diffusion-generated beliefs.

For a policy π, the expected free energy over the time-horizon T is [36]

$$G(\pi) = \sum_{t=1}^{T} \mathbb{E}_{q(\mathbf{o}_t, \mathbf{s}_t \mid \pi)} \left[\underbrace{D_{\mathrm{KL}}\left[q(\mathbf{s}_t \mid \mathbf{o}_t) \| q(\mathbf{s}_t \mid \pi)\right]}_{\text{Epistemic value}} - \underbrace{\log p(\mathbf{o}_t)}_{\text{Pragmatic value}} \right] \tag{14}$$

With diffusion-generated beliefs, we reformulate this computation. First, we define the diffusion-based EFE as follows. Let beliefs be generated via the diffusion process $\mathbf{s}_t \sim \text{Diffusion}(\mathbf{o}_t, \theta)$ ($\mathbf{s}_t$ is generated by the SDE in (5)). The expected free energy becomes

$$G_{\text{diff}}(\pi) = \sum_{t=1}^{T} \mathbb{E}_{\tau \sim \pi} \left[\mathcal{E}_t + \mathcal{P}_t + \mathcal{C}_t\right] \tag{15}$$

where:

$$\mathcal{E}_t = I(\mathbf{o}_t; \theta \mid \mathbf{s}_t, \pi) \quad \text{(Epistemic value)} \tag{16}$$

$$\mathcal{P}_t = r(\mathbf{s}_t, \mathbf{a}_t)/T + \gamma \mathbb{E}_{\mathbf{s}_{t+1}}[V(\mathbf{s}_{t+1})] \quad \text{(Pragmatic value)} \tag{17}$$

$$\mathcal{C}_t = -H[\pi(\mathbf{a}_t \mid \mathbf{s}_t)] \quad \text{(Consistency)} \tag{18}$$

We present the full mathematical derivation in Appendix A.1, where we show how diffusion sampling can be integrated into epistemic value calculations through parameter uncertainty. One can compute value by applying soft Bellman backup operator, while accounting for stochastic transition dynamics [31].

The epistemic term $\mathcal{E}_t$ quantifies information-gain about model parameters [19], naturally emerging from the stochastic nature of diffusion sampling. We estimate it by

$$\mathcal{E}_t \approx H[p(\mathbf{o}_t \mid \mathbf{s}_t, \pi)] - H[p(\mathbf{o}_t \mid \mathbf{s}_t, \theta, \pi)] \tag{19}$$

where the first term captures total uncertainty and the second captures irreducible noise. In Appendix A.2 we guarantee control on the error of this estimation.

The epistemic value $I(\mathbf{o}; \theta \mid \mathbf{s}, \pi)$ can be estimated using Mutual Information Neural Estimation (MINE) [2]:

$$I(\mathbf{o}; \theta \mid \mathbf{s}) = \sup_{T \in \mathcal{T}} \mathbb{E}_{p(\mathbf{o}, \theta \mid \mathbf{s})}[T(\mathbf{o}, \theta)] - \log \mathbb{E}_{p(\mathbf{o} \mid \mathbf{s}) p(\theta)}[e^{T(\mathbf{o}, \theta)}] \tag{20}$$

where $T : \mathcal{O} \times \Theta \to \mathbb{R}$ is a deep learning network for approximating Bayesian inference. To operationalize epistemic value estimation in practice, we introduce *function-space features* $f_\theta(\mathbf{s}_t)$ as intermediate representations extracted from the decoder network layers. To operationalize epistemic value estimation, we use Neural Tangent Kernel (NTK)-inspired function-space features built from Jacobian–vector products (JVPs) of the decoder. These features capture the functional uncertainty inherent in the learned mapping from latent states to observations, providing a computational proxy for parameter uncertainty that enables tractable mutual information estimation through the MINE framework. In practice, we approximate this by:

$$\mathcal{E}_t = \mathbb{E}_{p(\mathbf{o}_t, f_\theta|\mathbf{s}_t)}\left[T_\xi\big(\mathbf{o}_t, f_\theta(\mathbf{s}_t)\big)\right] - \log \mathbb{E}_{p(\mathbf{o}_t|\mathbf{s}_t)p(f_\theta)}\left[e^{T_\xi\left(\mathbf{o}_t, f_\theta(\mathbf{s}_t)\right)}\right]. \tag{21}$$

In Appendix A.3, we give a full derivation following [2] and in Appendix A.4, we provide an algorithmic implementation.

3.1 Pragmatic Value Emergence from Expected Free Energy

We show that the pragmatic value component $\mathcal{P}_t = r(\mathbf{s}_t, \mathbf{a}_t)/T + \gamma V(\mathbf{s}_{t+1})$ emerges naturally from expected free energy (EFE) minimization. To encode agent preferences within the active inference framework, we adopt the maximum entropy principle from optimal control theory. Following the formulation established in reinforcement learning [31,53], preferences over observations are represented through the exponential family

$$p(\mathbf{o}_t) = \frac{1}{Z_t} \exp\left[r(\mathbf{o}_t)/T\right], \tag{22}$$

where $r(\mathbf{o}_t)$ represents the reward signal, $T > 0$ is the temperature parameter controlling the concentration of the preference distribution, and Z_t is the partition function ensuring normalization. Lower temperatures ($T \to 0$) concentrate probability mass on high-reward observations, creating sharp preferences, while higher temperatures induce more exploratory, diffuse preference distributions. This temperature-controlled preference encoding naturally bridges reinforcement learning objectives with free energy minimization.

The expected free energy is given by

$$G(\pi) = \sum_t \mathbb{E}_{q(\mathbf{o}_t, \mathbf{s}_t|\pi)}\left[D_{\mathrm{KL}}\big(q(\mathbf{s}_t \mid \mathbf{o}_t) \,\|\, q(\mathbf{s}_t \mid \pi)\big) - \log p(\mathbf{o}_t)\right]. \tag{23}$$

Substituting the preference model (22) yields $\log p(\mathbf{o}_t) = r(\mathbf{o}_t)/T - \log Z_t$. Since Z_t depends only on time (not on actions or the policy), we drop the additive baseline $\log Z_t$ without changing the optimal control solution. Because observations are generated from latent states and rewards in control tasks depend on state–action pairs, we adopt the standard identification $r(\mathbf{o}_t) \approx r(\mathbf{s}_t, \mathbf{a}_t)$ with $\mathbf{a}_t \sim \pi(\cdot \mid \mathbf{s}_t)$, which lets us express the pragmatic contribution $-\log p(\mathbf{o}_t)$ as $-r(\mathbf{s}_t, \mathbf{a}_t)/T$.

Using the law of iterated expectations together with Markovian dynamics $p(\mathbf{s}_{t+1} \mid \mathbf{s}_t, \mathbf{a}_t)$ and a discount factor $\gamma \in (0, 1)$ for infinite-horizon problems, we obtain the one-step decomposition

$$G_t(\pi) = \mathbb{E}_{\mathbf{s}_t, \mathbf{a}_t \sim \pi}\left[-\frac{1}{T}\, r(\mathbf{s}_t, \mathbf{a}_t) + \gamma\, G_{t+1}(\pi)\right] + \mathcal{E}_t, \tag{24}$$

where $\mathcal{E}_t := \mathbb{E}\big[D_{\mathrm{KL}}\big(q(\mathbf{s}_t \mid \mathbf{o}_t) \,\|\, q(\mathbf{s}_t \mid \pi)\big)\big]$ collects the epistemic (information-gain) term that rewards observations expected to reduce posterior uncertainty. Equation (24) makes explicit the separation between a pragmatic soft cost and an epistemic drive for information.

We define the value function as the negative optimal expected free energy, $V(\mathbf{s}_t) := -\min_\pi G_t(\pi \mid \mathbf{s}_t)$, and we minimize (24) over $\pi(\cdot \mid \mathbf{s}_t)$ yields a soft Bellman relation. The entropy term arises from the Gibbs/convex-dual identity [10]

$$\max_{\pi(\cdot|\mathbf{s}_t)} \left\{\mathbb{E}_{\mathbf{a}_t \sim \pi}[f(\mathbf{a}_t)] + H\left[\pi(\cdot \mid \mathbf{s}_t)\right]\right\} = \log \int \exp\left(f(\mathbf{a}_t)\right) d\mathbf{a}_t, \tag{25}$$

with maximizer $\pi^\star(\mathbf{a}_t \mid \mathbf{s}_t) \propto \exp(f(\mathbf{a}_t))$ (for discrete actions, the integral becomes a sum; for continuous actions, $H[\cdot]$ is differential entropy relative to the base measure) [38]. Applying this with $f(\mathbf{a}_t) = \frac{1}{T} r(\mathbf{s}_t, \mathbf{a}_t) + \gamma\, \mathbb{E}_{\mathbf{s}_{t+1}} V(\mathbf{s}_{t+1})$, and introducing the soft Q-function

$$Q(\mathbf{s}_t, \mathbf{a}_t) := \tfrac{1}{T}\, r(\mathbf{s}_t, \mathbf{a}_t) + \gamma\, \mathbb{E}_{\mathbf{s}_{t+1} \sim p(\cdot|\mathbf{s}_t, \mathbf{a}_t)}\big[V(\mathbf{s}_{t+1})\big], \tag{26}$$

where we obtain the maximum-entropy Bellman form as

$$V(\mathbf{s}_t) = \max_\pi \mathbb{E}_{\mathbf{a}_t \sim \pi(\cdot|\mathbf{s}_t)}\left[Q(\mathbf{s}_t, \mathbf{a}_t) + H\left[\pi(\mathbf{a}_t \mid \mathbf{s}_t)\right]\right] = \log \int_{\mathcal{A}} \exp\left(Q(\mathbf{s}_t, \mathbf{a}_t)\right) d\mathbf{a}_t \tag{27}$$

which is the standard maximum-entropy RL optimality condition [22,31,53]. In (27), the pragmatic value $\mathcal{P}_t$ emerges as the reward and value components, while the entropy term $H[\pi(\mathbf{a}_t \mid \mathbf{s}_t)]$ encourages exploration. Thus, maximum entropy RL naturally arises from active inference when preferences are encoded as a Boltzmann distribution over outcomes (observations mapped to state–action rewards), the EFE objective recovers the maximum-entropy RL form: the pragmatic term contributes $\frac{1}{T} r(\mathbf{s}_t, \mathbf{a}_t) + \gamma\, V(\mathbf{s}_{t+1})$(up to a policy-independent constant), while policy entropy and epistemic information-gain are added explicitly.

Previous work on the free energy principle has linked constraints to preferences in the sense that a control system maintaining preferred transitions in its own states is equivalent to constraining its representation of environmental dynamics to minimize variational free energy [44], such that evolving through desired configurations or phenotypes is the same as a constrained variational inference process [26]. The present derivation enhances this picture by showing that maximum entropy RL has a natural interpretation as minimizing expected free energy. Likewise, in the discrete setting similar principles were gestured at in [7]; the development here provides a detailed extension to the continuous setting.

4 Related Work

Recent advances have demonstrated the efficacy of diffusion models beyond generative tasks. [27] prove that the standard denoising diffusion training loss is exactly a weighted variational lower bound on the data likelihood, however, they do not introduce any action selection or expected free energy. [25] introduced Diffuser, applying diffusion models for trajectory planning in offline RL and they learn trajectory optimizer which can be used for goal-conditioning. [52] proposed Diffusion-QL, where they use a conditional diffusion model as an expressive policy class and train it with a behavior-cloning diffusion loss plus a Q-guided improvement term. The Diffusion World Model [9] extends this paradigm by learning world dynamics through diffusion processes, enabling long-horizon planning without step-by-step rollouts. Our work differs from these other approaches by focusing on belief state generation rather than trajectory or dynamics modeling. This procedure allows for hierarchal belief estimation, with the potential for representing beliefs (and confidence) over multiple levels of abstraction.

The free energy principle [14,18] provides a unifying framework for perception and action. Recent implementations include [13] for discrete state spaces and [35,49] for continuous control. Da Costa and Pavliotis proposed theoretical foundations for entropy-production in diffusion processes, with the time-irreversibility aspects of diffusion-based belief generation representing a promising area of expansion for our current framework [6]. However, these approaches typically rely on parametric distributions, limiting their expressiveness in complex, multimodal environments. Moreover, recently [12] proposed a score-based diffusion model of POMDP trajectories in small grid-worlds and then employ conditioned sampling to Monte-Carlo estimate Expected Free Energy to select actions with the lowest EFE.

Score-based models [46,47] have revolutionized generative modeling through their connection to stochastic differential equations. Notably, [23] demonstrated practical training procedures, and [8] showed superior performance over GANs. We further extend these approaches by conditioning score functions on observations for belief state generation in active inference.

Amortized variational inference [29,42] learns inference networks to approximate posteriors. [34] proposed iterative amortized inference for improved approximations. Our approach can be viewed as amortizing belief updates through diffusion, but with the flexibility to represent arbitrary multimodal distributions.

5 Method

5.1 Diffusion-Based Active Inference with Latent State Conditioning

We develop a diffusion framework that operates in learned latent spaces where belief states $\mathbf{s}_0$ are explicitly conditioned on sensory observations $\mathbf{o}$. This formulation extends the Variational Diffusion Model (VDM) [28] to hierarchical latent representations essential for active inference, where the diffusion prior captures generative models of state dynamics rather than raw observations.

Our approach differs from standard VDM in the role of conditioning. We introduce dual conditioning: the latent state $\mathbf{s}_0$ represents the clean belief state, while context $\mathbf{o}$ provides sensory and task information. This separation is critical for active inference architectures where perception ($\mathbf{o} \to \mathbf{s}_0$), generative modeling ($\mathbf{s}_T \to \mathbf{s}_0$), and action selection ($\mathbf{s}_0 \to \boldsymbol{\pi}$) must remain functionally distinct.

We start with the variational bound on negative log-likelihood:

$$\mathbf{E}[-\log p(\mathbf{o})] \leq \mathbf{E}_q\Bigg[\mathrm{D}_{KL}\big[q(\mathbf{s}_{T_H}|\mathbf{s}_0,\mathbf{o})||p(\mathbf{s}_{T_H})\big] + \sum_{\tau\geq 1}^{T_H} \mathrm{D}_{KL}\big[q(\mathbf{s}_{\tau-1}|\mathbf{s}_\tau,\mathbf{s}_0,\mathbf{o})||\ p(\mathbf{s}_{\tau-1}\mid \mathbf{s}_\tau)\big] - \log p(\mathbf{o}|\mathbf{s}_0)\Bigg]$$

where We use T_H (H for "horizon") to denote the number of discrete diffusion steps and the forward diffusion process conditions on the clean latent state is defined as:

$$q(\mathbf{s}_\tau|\mathbf{s}_0,\mathbf{o}) = \mathcal{N}(\mathbf{s}_\tau; \sqrt{\bar{\alpha}_\tau}\mathbf{s}_0, (1-\bar{\alpha}_\tau)\mathbf{I}), \tag{28}$$

with reparameterization $\mathbf{s}_\tau = \sqrt{\bar{\alpha}_\tau}\mathbf{s}_0 + \sqrt{1-\bar{\alpha}_\tau}\boldsymbol{\epsilon}$ where $\boldsymbol{\epsilon} \sim \mathcal{N}(\mathbf{0},\mathbf{I})$ [45]. The noise schedule is defined through $\alpha_{\tau_k} := 1-\beta_{\tau_k}$ and cumulative products $\bar{\alpha}_{\tau_k} := \prod_{j=1}^{k}\alpha_{\tau_j}$ for discrete timesteps τ_k.

The exact posterior that conditions on the initial latent state is:

$$q\big(\mathbf{s}_{\tau_{k-1}}|\mathbf{s}_{\tau_k},\mathbf{s}_0\big) = \mathcal{N}(\mathbf{s}_{\tau_{k-1}}; \tilde{\boldsymbol{\mu}}_{\tau_k}(\mathbf{s}_{\tau_k},\mathbf{s}_0), \tilde{\sigma}^2_{\tau_k}\mathbf{I}), \tag{29}$$

where the mean explicitly depends on the clean state $\mathbf{s}_0$:

$$\tilde{\boldsymbol{\mu}}_{\tau_k}(\mathbf{s}_{\tau_k},\mathbf{s}_0) = \frac{\sqrt{\bar{\alpha}_{\tau_{k-1}}}(1-\alpha_{\tau_k})}{1-\bar{\alpha}_{\tau_k}}\mathbf{s}_0 + \frac{\sqrt{\alpha_{\tau_k}}(1-\bar{\alpha}_{\tau_{k-1}})}{1-\bar{\alpha}_{\tau_k}}\mathbf{s}_{\tau_k}, \tag{30}$$

and variance $\tilde{\sigma}^2_{\tau_k} = \frac{(1-\alpha_{\tau_k})(1-\bar{\alpha}_{\tau_{k-1}})}{1-\bar{\alpha}_{\tau_k}}$. This posterior structure matches DDIM's non-Markovian formulation [45] but makes the $\mathbf{s}_0$ dependence explicit. This distinction is crucial for architectures where $\mathbf{s}_0$ emerges from an encoder network $q(\mathbf{s}_0|\mathbf{o})$.

The generative model replaces the true $\mathbf{s}_0$ with the network's prediction:

$$p_\theta(\mathbf{s}_{\tau_{k-1}}|\mathbf{s}_{\tau_k}) = q(\mathbf{s}_{\tau_{k-1}}|\mathbf{s}_{\tau_k}, \mathbf{s}_0{=}\hat{\mathbf{s}}_\theta(\mathbf{s}_{\tau_k},\tau_k,\mathbf{o}),\mathbf{o}), \tag{31}$$

where $\hat{\mathbf{s}}_\theta$ can be parameterized equivalently through noise prediction: $\hat{\mathbf{s}}_\theta(\mathbf{s}_{\tau_k},\tau_k,\mathbf{o}) = \frac{1}{\sqrt{\bar{\alpha}_{\tau_k}}}(\mathbf{s}_{\tau_k} - \sqrt{1-\bar{\alpha}_{\tau_k}}\boldsymbol{\epsilon}_\theta(\mathbf{s}_{\tau_k},\tau_k,\mathbf{o}))$.

The per-timestep KL divergence between posterior and model reduces to a signal-to-noise ratio (SNR) weighted denoising objective:

$$D_{\mathrm{KL}}[q(\mathbf{s}_{\tau_{k-1}}|\mathbf{s}_{\tau_k},\mathbf{s}_0,\mathbf{o})||p_\theta(\mathbf{s}_{\tau_{k-1}}|\mathbf{s}_{\tau_k})] = \frac{1}{2}(\mathrm{SNR}(\tau_{k-1}) - \mathrm{SNR}(\tau_k))\big\|\boldsymbol{\epsilon} - \boldsymbol{\epsilon}_\theta(\mathbf{s}_{\tau_k},\tau_k,\mathbf{o})\big\|^2, \tag{32}$$

where $\mathrm{SNR}(\tau) = \bar{\alpha}_\tau/(1-\bar{\alpha}_\tau)$. This yields the discrete-time training objective which is a surrogate for tightening the variational bound:

$$\mathcal{L}_{T_H}(\mathbf{o}) = \frac{T_H}{2}\mathbb{E}_{\epsilon\sim\mathcal{N}(\mathbf{0},\mathbf{I}),i\sim\mathcal{U}\{1,T_H\}}\left[(\mathrm{SNR}(\tau_{i-1}) - \mathrm{SNR}(\tau_i))\|\epsilon - \epsilon_\theta(\mathbf{s}_{\tau_i},\tau_i,\mathbf{o})\|^2\right]. \tag{33}$$

For continuous time ($T_H \to \infty$), the loss simplifies under log-SNR parameterization $\gamma(\tau) = -\log \mathrm{SNR}(\tau)$:

$$\mathcal{L}_\infty(\mathbf{o}) = \frac{1}{2}\mathbb{E}_{\epsilon\sim\mathcal{N}(0.1),\tau\sim\mathcal{U}(0,1)}\left[\gamma'(\tau)\|\epsilon - \epsilon_\theta(\mathbf{s}_\tau,\tau,\mathbf{o})\|^2\right], \tag{34}$$

here the weight $\gamma'(\tau)$ naturally emerges from the ELBO decomposition. This continuous-time view establishes equivalence with score-based SDEs:

$$\epsilon_\theta = -\sqrt{1-\bar{\alpha}_\tau}\nabla_{\mathbf{s}}\log q(\mathbf{s}_\tau|\mathbf{o}). \tag{35}$$

5.2 Algorithmic Implementation

The sampling procedure implements the non-Markovian family with explicit $\mathbf{s}_0$ prediction:

$$\mathbf{s}_{\tau_{k-1}} = \sqrt{\bar{\alpha}_{\tau_{k-1}}}\hat{\mathbf{s}}_\theta(\mathbf{s}_{\tau_k},\tau_k,\mathbf{o}) + \sqrt{1-\bar{\alpha}_{\tau_{k-1}} - \sigma^2_{\tau_k}}\epsilon_\theta(\mathbf{s}_{\tau_k},\tau_k,\mathbf{o}) + \sigma_{\tau_k}\mathbf{z}_{\tau_k}, \tag{36}$$

where $\mathbf{z}_{\tau_k} \sim \mathcal{N}(\mathbf{0},\mathbf{I})$ and σ_{τ_k} controls stochasticity. By setting $\sigma_{\tau_k} = 0$, it yields deterministic DDIM inference; $\sigma_{\tau_k} = \sqrt{\frac{1-\bar{\alpha}_{\tau_{k-1}}}{1-\bar{\alpha}_{\tau_k}}}\sqrt{1-\frac{\bar{\alpha}_{\tau_k}}{\bar{\alpha}_{\tau_{k-1}}}}$ recovers ancestral DDPM sampling.

Algorithm 1. Latent-Conditioned Diffusion for Active Inference

Require: Observation $\mathbf{o}$, noise network ϵ_θ, diffusion steps T_H, guidance step sizes $\{\eta_{\ell,\tau_k}\}$

1: Initialize $\mathbf{s}^{(0)} \sim \mathcal{N}(\mathbf{0},\mathbf{I})$
2: **for** $k = 1$ to T_H **do**
3: $\tau_k \leftarrow T_H - k$ {Discrete diffusion timestep}
4: $\hat{\epsilon} \leftarrow \epsilon_\theta(\mathbf{s}^{(k-1)},\tau_k,\mathbf{o})$ {Predict noise ϵ}
5: $\hat{\mathbf{s}}_0 \leftarrow \frac{1}{\sqrt{\bar{\alpha}_{\tau_k}}}\left(\mathbf{s}_{\tau_{k-1}} - \sqrt{1-\bar{\alpha}_{\tau_k}}\hat{\epsilon}\right)$ {Predict clean latent state}
6: $\mathbf{s}_{\tau_k} \leftarrow \sqrt{\bar{\alpha}_{\tau_{k-1}}}\hat{\mathbf{s}}_0 + \sqrt{1-\bar{\alpha}_{\tau_{k-1}}}\hat{\epsilon}$ {DDIM update (σ_{τ_k}=0)}
7: $\mathbf{s}_{\tau_k} \leftarrow \mathbf{s}_{\tau_k} + \eta_{\ell,\tau_k}\nabla_{\mathbf{s}}\log p_\phi(\mathbf{o} \mid \mathbf{s}_0)$ {Likelihood guided posterior step}
8: **end for**
9: **return** $\mathbf{s}_{\tau_k}$ {Final belief state for policy conditioning}

In Algorithm 1, we illustrate the implementation of this framework through discrete-time approximation. The likelihood guidance step (line 7) adds explicit posterior correction through the gradient of the observation model $\nabla_{\mathbf{s}} \log p_\phi(\mathbf{o} \mid \mathbf{s}_0)$ [8]. The score network is implemented as a Diffusion Transformer (DiT) [41] processing the noisy state $\mathbf{s}_{\tau_k}$, timestep τ_k, and observation $\mathbf{o}$. The generated belief states $\mathbf{s}_0$ serve as continuous latent representations for policy conditioning, enabling multimodal uncertainty representations that challenge traditional Gaussian variational methods while maintaining the computational efficiency of amortized inference.

Algorithm 2. Active Inference with Score-Based Diffusion Beliefs

Require: Environment $\mathcal{E}$, replay buffer $\mathcal{B}$
Initialize networks: score s_θ, policy π_ϕ, value V_ψ, dynamics f_ξ {All networks operate in latent space; gradients isolated at belief boundaries}
while training **do**
  Observe $\mathbf{o}_t$ from environment
  $\mathbf{s}_t \leftarrow$ ScoreBasedBelief$(\mathbf{o}_t, s_\theta)$ {Algorithm 1}
  Compute $G_{\text{diff}}(\pi) = \mathcal{E}_t + \mathcal{P}_t + \mathcal{C}_t$ {Diffusion-based EFE}
  Compute policy gradient: $g_\pi = -\nabla_{\pi_\phi} G_{\text{diff}}(\pi_\phi)|_{\mathbf{s}_t}$
  Sample action: $\mathbf{a}_t \sim \pi_\phi(\cdot \mid \mathbf{s}_t)$
  Update policy parameters: $\phi \leftarrow \phi - \eta g_\pi$ during training
  Execute $\mathbf{a}_t$, observe $\mathbf{o}_{t+1}, r_t$
  Store $(\mathbf{o}_t, \mathbf{a}_t, r_t, \mathbf{o}_{t+1})$ in $\mathcal{B}$
  if update step **then**
    Sample batch $\{(\mathbf{o}_i, \mathbf{a}_i, r_i, \mathbf{o}'_i)\}$ from $\mathcal{B}$
    Update score network: $\theta \leftarrow \theta - \eta \nabla_\theta \mathcal{L}_{\text{score}}$
    Update policy via EFE: $\phi \leftarrow \phi - \eta \nabla_\phi G_{\text{diff}}(\pi_\phi)$
    Update value/dynamics: Standard TD learning in latent space
  end if
end while

Now in Algorithm 2, we employ the inferred diffusion-generated beliefs to the complete active inference loop. At each timestep, rather than updating parametric belief distributions, the agent generates a new belief state through reverse diffusion (Algorithm 1), which naturally incorporates observation uncertainty through the conditional denoiser $\boldsymbol{\epsilon}_\theta(\cdot, \cdot, \mathbf{o})$. The policy update implements gradient-based optimization of the expected free energy objective, where the gradient is computed through the three-term decomposition. Specifically, the epistemic value gradient encourages information-seeking actions, while the pragmatic gradient drives goal-directed behavior. The training phase updates three key components: the diffusion network learns to predict injected noise via an ELBO-consistent diffusion loss, the policy network optimizes the EFE objective through gradient descent, and the value/dynamics networks are updated via standard temporal difference learning. Thus, we maintain theoretical grounding in the free energy principle while leveraging the expressiveness of diffusion models.

Algorithm 2 is a direct instantiation of the three-term decomposition of expected free energy introduced in Eqs. (16)–(18). With this mapping in place, we can formalize why the combined loss $\mathcal{G} = E_t + P_t + C_t$ recovers an active-inference agent whose behaviour is both *goal-directed* and *uncertainty-aware.*

The epistemic value computation employs function-space uncertainty estimation through Jacobian features of the decoder network, approximated via finite differences. This provides a computationally tractable alternative to parameter-space uncertainty while maintaining theoretical grounding in the neural tangent kernel framework [24,30,32].

6 Connection to Active Inference Principles

Our approach maintains core active inference principles while extending to continuous spaces. The diffusion process implicitly minimizes variational free energy by learning to denoise observations into likely latent states. Using the variance-preserving Ornstein-Uhlenbeck formulation, we train a noise prediction network $\epsilon_\theta(\mathbf{s}_\tau, \tau, \mathbf{o})$ which recovers the score function as $\boldsymbol{s}_\theta = -\epsilon_\theta/\sigma(\tau)$, and in the continuous-time limit this objective equals the ELBO up to constants. While traditional active inference uses variational updates of the form

$$q(\mathbf{s}_t \mid \mathbf{o}_{1:t}) \propto p(\mathbf{o}_t \mid \mathbf{s}_t) q(\mathbf{s}_t \mid \mathbf{o}_{1:t-1}), \tag{37}$$

our reverse-diffusion update implements *iterative refinement*: each denoising step computes $q(\mathbf{s}_{\tau-1} | \mathbf{s}_\tau, \mathbf{s}_0, \mathbf{o})$, thereby incorporating observational constraints through the conditional denoiser and converts KL divergence into a weighted quadratic of prediction error. Unlike mean-field approximations that collapse uncertainty into unimodal distributions, diffusion-generated beliefs preserve the full complexity of posterior geometry while remaining computationally tractable through amortized inference. For policy selection, our reformulation keeps the decomposition into epistemic and pragmatic components intact. The epistemic term naturally arises from parameter uncertainty in the stochastic diffusion sampling, while the pragmatic term integrates learned preferences through the reward function. The result is active inference that scales to high-dimensional continuous spaces without sacrificing the capacity to represent rich, structured uncertainty.

7 Discussion

Our integration of diffusion models with active inference addresses fundamental limitations in applying the free energy principle to continuous domains. We highlight deep connections between score-based generative models and variational inference, and show that diffusion processes naturally implement a principled belief-updating mechanism.

While this work establishes the mathematical foundations, empirical validation remains for future work. We believe that applying diffusion models to represent multimodal beliefs will be particularly beneficial for contact-rich manipulation tasks and partially observable environments, where traditional parametric approaches often fall short. We have reformulated the epistemic term in our EFE formulation, which guides exploration by quantifying the expected information gain about model parameters. This approach emerges naturally from the mathematical structure of minimizing expected free energy over future policies, rather than relying on ad hoc exploration bonuses.

Future theoretical work should establish tighter bounds on the approximation quality of diffusion-generated beliefs and investigate connections to predictive coding and hierarchical active inference. Practically, reducing the computational cost of diffusion sampling through distillation or few-step methods would improve real-time applicability. Our theoretical framework anticipates empirical validation across canonical continuous-control domains, particularly MuJoCo environments [48], where epistemic exploration capabilities are expected to demonstrate superior sample efficiency in scenarios with complex posterior geometries.

8 Conclusion

We have presented a mathematical framework unifying diffusion models with active inference for continuous state and action spaces, fundamentally advancing both theoretical understanding and practical implementation of uncertainty-aware autonomous agents. By generating beliefs through reverse diffusion processes, we overcome critical limitations of parametric variational inference while maintaining rigorous theoretical grounding in the free energy principle.

Our reformulation of expected free energy over diffusion trajectories provides a principled approach to action selection that naturally balances epistemic exploration with pragmatic goal attainment. This framework resolves the long-standing challenge of representing complex, multimodal belief distributions in continuous active inference, opening new avenues for scaling these principles to real-world applications.

The principled uncertainty quantification inherent in our approach addresses fundamental challenges in AI safety research. By explicitly modeling epistemic uncertainty through diffusion-generated beliefs, our framework enables multiple desirable features:

- The epistemic term naturally drives information-seeking behavior while maintaining awareness of model limitations, crucial for safe deployment in uncertain environments.
- Unlike point-estimate approaches, our method provides well-calibrated uncertainty estimates essential for human-AI collaboration and high-stakes decision-making.
- The multimodal representational capacity enables graceful handling of out-of-distribution scenarios, reducing brittle failure modes common in conventional approaches.

This work establishes foundations for several critical research directions: hierarchical active inference architectures for multi-scale decision-making, integration with formal verification methods for safety-critical applications, and extension to multi-agent scenarios where coordinated uncertainty-aware behavior becomes paramount.

The convergence of diffusion models with active inference represents more than a technical advancement—it provides a mathematically principled pathway toward AI systems that maintain human-interpretable uncertainty representations while scaling to complex real-world domains. The principled uncertainty quantification inherent in our approach may prove valuable for safe deployment of autonomous systems, as the epistemic term naturally drives information-seeking behavior while maintaining awareness of model limitations.

Acknowledgments. We gratefully acknowledge the Survival and Flourishing Fund and the Institute of Advanced Consciousness Studies for their generous support of Dr. Sheikhbahaee and Dr. Safron. We also thank VERSES AI for supporting Dr. Sakthivadivel and Dr. Albarracin, and Dr. Lancelot Da Costa for helpful discussions regarding diffusion models.

A Appendix

A.1 Expected Free Energy Decomposition

In this section, we prove expected free energy decomposes as prescribed in (15). We begin with the standard definition of EFE [36],

$$G(\pi) = -\sum_t \mathbb{E}_{q(\mathbf{o}_t,\mathbf{s}_t|\pi)}\left[D_{\mathrm{KL}}[q(\mathbf{s}_t \mid \mathbf{o}_t)\|q(\mathbf{s}_t \mid \pi)] + \log p(\mathbf{o}_t)\right] \tag{38}$$

$$= \underbrace{D_{\mathrm{KL}}\left[q(\mathbf{o}_t \mid \pi)\|p(\mathbf{o}_t)\right]}_{\text{risk}} + \underbrace{\mathbb{E}_{q(\mathbf{s}_t|\pi)}\left[H[\ln p(\mathbf{o}_t \mid \mathbf{s}_t)]\right]}_{\text{Ambiguity}} \tag{39}$$

The second line is obtained by using Bayes' rules which indicates that the ratio of posterior and prior ($q(\mathbf{s}_t|\pi)$) is equal to the corresponding likelihood ($q(\mathbf{o}_t|\mathbf{s}_t,\pi) \approx p(\mathbf{o}_t|\mathbf{s}_t)$) and marginal likelihood $\frac{p(\mathbf{s}_t|\mathbf{o}_t)}{q(\mathbf{s}_t|\pi)} = \frac{q(\mathbf{o}_t|\mathbf{s}_t,\pi)}{q(\mathbf{o}_t|\pi)}$ [39]. With diffusion-generated beliefs, $q(\mathbf{s}_t \mid \mathbf{o}_t)$ is implicitly defined by the reverse process parameterized by θ. The KL divergence decomposes into

$$D_{\mathrm{KL}}[q(\mathbf{s}_t \mid \mathbf{o}_t)\|q(\mathbf{s}_t \mid \pi)] = \mathbb{E}_{q(\mathbf{s}_t|\mathbf{o}_t)}[\log q(\mathbf{s}_t \mid \mathbf{o}_t) - \log q(\mathbf{s}_t \mid \pi)] \tag{40}$$

$$= H[q(\mathbf{s}_t \mid \mathbf{o}_t), q(\mathbf{s}_t \mid \pi)] - H[q(\mathbf{s}_t \mid \mathbf{o}_t)]. \tag{41}$$

The entropy $H[q(\mathbf{s}_t \mid \mathbf{o}_t)]$ depends on the diffusion parameters θ. Under parameter uncertainty, one has

$$H[q(\mathbf{s}_t \mid \mathbf{o}_t)] = H[p(\mathbf{s}_t \mid \mathbf{o}_t, \theta)] + I(\mathbf{s}_t; \theta \mid \mathbf{o}_t). \tag{42}$$

However, for planning, epistemic value must depend on the policy. This gives the definition of the epistemic term as $\mathcal{E}_t = I(\mathbf{o}_t; \theta \mid \mathbf{s}_t, \pi)$ after marginalization on trajectories.

For the pragmatic term, we use the standard result that $\log p(\mathbf{o}_t) = r(\mathbf{o}_t)/T - \log Z$ where Z is a normalization constant. The value function arises from the dynamic programming decomposition of future rewards.

The consistency term $\mathcal{C}_t = -H[\pi(\mathbf{a}_t \mid \mathbf{s}_t)]$ regularizes the policy, preventing premature convergence and maintaining exploration.

A.2 Estimation of Epistemic Value

In (19) an estimator for $\mathcal{E}_t = I(\mathbf{o}_t; \theta \mid \mathbf{s}_t, \pi)$, itself equal to

$$H[p(\mathbf{o}_t \mid \mathbf{s}_t, \pi)] - \mathbb{E}_\theta[H[p(\mathbf{o}_t \mid \mathbf{s}_t, \theta, \pi)]]$$

is given by the expression

$$H[p(\mathbf{o}_t \mid \mathbf{s}_t, \pi)] - H[p(\mathbf{o}_t \mid \mathbf{s}_t, \theta, \pi)].$$

The relation between these two expressions can be understood as replacing the second term, the expected aleatoric uncertainty under parameter beliefs, by a plug-in Monte Carlo estimator using a diffusion sample. If the observations are valued in a finite discrete set of length m and $H[p(\mathbf{o}_t \mid \mathbf{s}_t, \pi)]$ is known exactly then the following bound on the error is satisfied. Let us set

$$\mathcal{E}_t = H[p(\mathbf{o}_t \mid \mathbf{s}_t, \pi)] - \mathbb{E}_\theta[H[p(\mathbf{o}_t \mid \mathbf{s}_t, \theta, \pi)]] \tag{43}$$

and

$$\hat{\mathcal{E}}_t = H[p(\mathbf{o}_t \mid \mathbf{s}_t, \pi)] - H[p(\mathbf{o}_t \mid \mathbf{s}_t, \theta, \pi)]. \tag{44}$$

Firstly note that $\mathbb{E}_\theta[\hat{\mathcal{E}}_t] = \mathcal{E}_t$ and by the Bhatia–Davis bound, the variance of $H[p(\mathbf{o}_t \mid \mathbf{s}_t, \theta, \pi)]$ is bounded above by $(\log m)^2/4$. Hoeffding's inequality with one sample then gives

$$P(|\hat{\mathcal{E}}_t - \mathcal{E}_t| \geqslant \varepsilon) \leqslant 2 \exp\left\{-\frac{2\varepsilon^2}{(\log m)^2}\right\}.$$

As an immediate corollary the root mean squared error satisfies

$$\sqrt{\mathbb{E}[|\hat{\mathcal{E}}_t - \mathcal{E}_t|^2]} \leqslant \frac{\log m}{2}.$$

A similar bound is possible in the continuous case, making it applicable to our scenario.

A.3 MINE-Based Epistemic Value

The epistemic value in active inference quantifies the expected information gain about model parameters from future observations. In this section we estimate the mutual information $I(\mathbf{o}; \theta \mid \mathbf{s})$ between observations $\mathbf{o}$ and model parameters θ conditioned on latent state $\mathbf{s}$. We also derive the neural estimation of mutual information $I(\mathbf{o}; \theta \mid \mathbf{s})$.

Beginning from the KL divergence representation, we have

$$\begin{aligned} I(\mathbf{o}; \theta \mid \mathbf{s}) &= D_{\mathrm{KL}}[p(\mathbf{o}, \theta \mid \mathbf{s}) \| p(\mathbf{o} \mid \mathbf{s}) p(\theta \mid \mathbf{s})] && (45)\\ &= \mathbb{E}_{p(\mathbf{o}, \theta \mid \mathbf{s})}\left[\log \frac{p(\mathbf{o}, \theta \mid \mathbf{s})}{p(\mathbf{o} \mid \mathbf{s}) p(\theta \mid \mathbf{s})}\right]. && (46) \end{aligned}$$

However, directly computing this quantity is intractable for deep neural networks. Motivated by the neural tangent kernel (NTK) framework [24], we adopt a function-space perspective. The key insight is that in the infinite-width limit, neural networks evolve according to linear dynamics in function space, allowing us to characterize uncertainty through the network's functional behavior rather than its parameters directly. This motivates approximating the epistemic uncertainty through local geometric properties of the decoder function.

Let $f_\theta\colon \mathcal{S} \to \mathcal{O}$ denote our decoder network mapping from latent states to observations. We characterize the epistemic uncertainty through the decoder's local geometric properties captured by Jacobian-vector products. Given a latent state s, we first linearize the decoder at s to obtain the Jacobian operator $J_\theta(s)\colon \mathcal{S} \to \mathcal{O}$.

We sample N random perturbation directions $\{\delta_i\}_{i=1}^N$ where $\delta_i \sim \mathcal{N}(0, I)$ are normalized such that $\|\delta_i\| = 1$. For each direction, we compute the exact directional derivative using automatic differentiation:

$$J_i(s) = J_\theta(s) \cdot \delta_i = \lim_{t\to 0} \frac{f_\theta(s + t\delta_i) - f_\theta(s)}{t} \tag{47}$$

where $J_\theta(s) \cdot \delta_i$ denotes the Jacobian-vector product, which gives the instantaneous rate of change of f_θ at s in the direction δ_i. This quantity is computed exactly through automatic differentiation via the linearization:

$$f_\theta(s + \epsilon v) = f_\theta(s) + \epsilon J_\theta(s) \cdot v + O(\epsilon^2) \tag{48}$$

By extracting the linear term, we obtain $J_i(s) = J_\theta(s) \cdot \delta_i$ without numerical approximation errors. These directional derivatives characterize the local sensitivity of the decoder's output to perturbations in latent space, providing a principled measure of functional uncertainty.

While finite difference approximations $J_i^{\text{fd}}(s) = [f_\theta(s + \epsilon\delta_i) - f_\theta(s)]/\epsilon$ could be used for computational simplicity, they suffer from truncation error $O(\epsilon)$ and numerical instability when ϵ is too small. Our approach using automatic differentiation eliminates these issues while maintaining computational efficiency via forward-mode JVP computation. This provides a principled framework for uncertainty quantification grounded in NTK theory.

To estimate the epistemic value as mutual information $I(J; s)$ between Jacobian features and latent states, we employ the Mutual Information Neural Estimation (MINE) framework [2]. Using the Donsker-Varadhan representation:

$$I(J; s) = \sup_{T\in\mathcal{T}} \mathbb{E}_{p(J,s)}[T(J, s)] - \log \mathbb{E}_{p(J)p(s)}[e^{T(J,s)}] \tag{49}$$

We parameterize a statistics network $T_\phi : \mathcal{J} \times \mathcal{S} \to \mathbb{R}$ that takes as input the concatenation of processed Jacobian features and latent representations. The MINE objective provides a lower bound:

$$I(J; s) \geq \mathbb{E}_{p(J,s)}[T_\phi(J, s)] - \log \mathbb{E}_{p(J)p(s)}[e^{T_\phi(J,s)}] \tag{50}$$

where $J = \psi([J_1(s), ..., J_N(s)])$ represents the processed concatenation of Jacobian features, with ψ being a learned feature extractor (convolutional networks with spatial attention for pixel observations, or linear networks for state observations).

The epistemic value is then $\mathcal{I}_{\text{epistemic}} = I(J; s)$ which is a proxy for the epistemic value. This quantifies the expected information gain about the decoder's functional behavior from observing the Jacobian features given the latent state. The connection to active inference emerges as policies that lead to states with higher epistemic value are expected to reduce uncertainty about future observations.

A.4 Practical Implementation

Algorithm 3. MINE Epistemic Value Estimation via Automatic Differentiation

Require: Latent distribution $\mathcal{N}(\mu_s, \Sigma_s)$, decoder f_θ, statistics network T_ϕ
Sample M latent states: $\{s_j\}_{j=1}^{M} \sim \mathcal{N}(\mu_s, \Sigma_s)$
Set decoder to evaluation mode: f_θ.eval() {Disables dropout}
for each s_j **do**
  Linearize decoder at s_j: $(f_j, \text{jvp}_j) \leftarrow \text{linearize}(f_\theta, s_j)$
  Sample N directions: $\delta_i \sim \mathcal{N}(0, I)$, normalize: $\delta_i \leftarrow \delta_i / \|\delta_i\|$
  for $i = 1$ to N **do**
    Compute exact directional derivative: $J_{j,i} = \text{jvp}_j(\delta_i)$ $\{= J_\theta(s_j) \cdot \delta_i\}$
  end for
  Stack Jacobian features: $\mathbf{J}_j = [J_{j,1}, ..., J_{j,N}]$
  Extract features via encoder: $\mathbf{j}_j = \psi(\mathbf{J}_j)$ $\{\psi$ encodes JVP outputs$\}$
  Latent input: $\mathbf{h}_j = s_j$ {identity map; same normalization in joint and marginal}
end for
Compute joint statistics: $t_{\text{joint}} = \frac{1}{M} \sum_{j=1}^{M} T_\phi(\mathbf{j}_j, \mathbf{h}_j)$
Create marginal by permuting: $\sigma \sim \text{Permutation}(M)$
Compute marginal statistics with EMA stabilization:

$$t_{\text{marginal}} = \log \left(\frac{1}{M} \sum_{j=1}^{M} \exp(T_\phi(\mathbf{j}_{\sigma(j)}, \mathbf{h}_j)) \right) \tag{51}$$

Update running mean: $\bar{t} \leftarrow \alpha \cdot t_{\text{marginal}} + (1 - \alpha) \cdot \bar{t}$
return Epistemic value: $\mathcal{I}_{\text{epistemic}} = t_{\text{joint}} - t_{\text{marginal}}$

Our implementation computes epistemic value through Jacobian-vector products and mutual information estimation described in Algorithm 3. The key features of our implementation are as follows:

- Exact Jacobian-vector products via automatic differentiation: Rather than using finite difference approximations, we compute exact directional derivatives $J_\theta(s) \cdot \delta$. This eliminates truncation errors and numerical instability issues associated with finite differences.

- Decoder in evaluation mode: The decoder is explicitly set to evaluation mode, disabling all dropout layers. While dropout can be used for uncertainty estimation [21], we focus on function space sensitivity.
- Efficient computation through linearization: The linearization $(f, \text{jvp}) \leftarrow \text{linearize}(f_\theta, s)$ provides a function handle for computing Jacobian-vector products without using the full Jacobian matrix, maintaining memory efficiency even for high-dimensional outputs.
- Stable MINE estimation: The mutual information neural estimator uses exponential moving average (EMA) for the marginal term to reduce gradient variance.

This approach provides a principled estimation of epistemic uncertainty that captures the decoder's functional variability through its exact local linear approximation, without requiring stochastic inference-time operations or numerical approximations.

References

1. Anderson, B.D.: Reverse-time diffusion equation model. Stochast. Process. Appl. **12**, 313–326 (1982)
2. Belghazi, M.I., et al.: Mutual information neural estimation. In: International Conference on Machine Learning, pp. 531–540. PMLR (2018)
3. Birrell, J., Katsoulakis, M.A., Pantazis, Y.: Optimizing variational representations of divergences and accelerating their statistical estimation. IEEE Trans. Inf. Theory **68**(7), 4553–4572 (2022)
4. Champion, T., Grześ, M., Bowman, H.: Multimodal and multifactor branching time active inference. Neural Comput. **36**(11), 2479–2504 (2024)
5. Da Costa, L., Parr, T., Sajid, N., Veselic, S., Neacsu, V., Friston, K.: Active inference on discrete state-spaces: a synthesis. J. Math. Psychol. **99**, 102447 (2020)
6. Da Costa, L., Pavliotis, G.A.: The entropy production of stationary diffusions. J. Phys. A Math. Theor. **56**(36), 365001 (2023)
7. Da Costa, L., Sajid, N., Parr, T., Friston, K., Smith, R.: Reward maximization through discrete active inference. Neural Comput. **35**(5), 807–852 (2023)
8. Dhariwal, P., Nichol, A.: Diffusion models beat GANs on image synthesis. In: Advances in Neural Information Processing Systems, vol. 35 (2021)
9. Ding, Z., Zhang, A., Tian, Y., Zheng, Q.: Diffusion world model: future modeling beyond step-by-step rollout for offline reinforcement learning. preprint arXiv:2402.03570 (2024)
10. Donsker, M.D., Varadhan, S.R.S.: Asymptotic evaluation of certain Markov process expectations for large time–III. Commun. Pure Appl. Math. **29**(4), 389–461 (1976)
11. Donsker, M.D., Varadhan, S.R.S.: Asymptotic evaluation of certain Markov process expectations for large time–IV. Commun. Pure Appl. Math. **36**(2), 183–212 (1983)
12. Durand, E.A., Joffily, M., Khamassi, M.: A diffusion model-based approach to active inference. In: 2024 IEEE International Conference on Future of Machine Learning and Data Science (FMLDS 2024) (2024)
13. Fountas, Z., Sajid, N., Mediano, P., Friston, K.: Deep active inference agents using Monte-Carlo methods. Adv. Neural. Inf. Process. Syst. **33**, 11662–11675 (2020)
14. Friston, K.: The free-energy principle: a unified brain theory? Nat. Rev. Neurosci. **11**(2), 127–138 (2010)

15. Friston, K.: A free energy principle for a particular physics. arXiv preprint arXiv:1906.10184 (2019)
16. Friston, K., et al.: The free energy principle made simpler but not too simple. Phys. Rep. **1024**, 1–29 (2023)
17. Friston, K., FitzGerald, T., Rigoli, F., Schwartenbeck, P., Pezzulo, G.: Active inference: a process theory. Neural Comput. **29**(1), 1–49 (2017)
18. Friston, K., Kilner, J., Harrison, L.: A free energy principle for the brain. J. Physiol. **100**(1–3), 70–87 (2006)
19. Friston, K., Rigoli, F., Ognibene, D., Mathys, C., Fitzgerald, T., Pezzulo, G.: Active inference and epistemic value. Cogn. Neurosci. **6**(4), 187–214 (2015)
20. Friston, K.J., Wiese, W., Hobson, J.A.: Sentience and the origins of consciousness: from cartesian duality to Markovian monism. Entropy **22**(5), 516 (2020)
21. Gal, Y., Ghahramani, Z.: Dropout as a Bayesian approximation: representing model uncertainty in deep learning. In: International Conference on Machine Learning, pp. 1050–1059. PMLR (2016)
22. Haarnoja, T., Zhou, A., Abbeel, P., Levine, S.: Soft actor-critic: off-policy maximum entropy deep reinforcement learning with a stochastic actor. In: International Conference on Machine Learning, pp. 1861–1870. PMLR (2018)
23. Ho, J., Jain, A., Abbeel, P.: Denoising diffusion probabilistic models. Adv. Neural. Inf. Process. Syst. **33**, 6840–6851 (2020)
24. Jacot, A., Gabriel, F., Hongler, C.: Neural tangent kernel: convergence and generalization in neural networks. In: Advances in Neural Information Processing Systems, vol. 31 (2018)
25. Janner, M., Du, Y., Tenenbaum, J., Levine, S.: Planning with diffusion for flexible behavior synthesis. In: International Conference on Machine Learning (2022)
26. Kiefer, A.B.: Psychophysical identity and free energy. J. R. Soc. Interface **17**(169), 20200370 (2020)
27. Kingma, D., Salimans, T., Poole, B., Ho, J.: Variational diffusion models. Adv. Neural. Inf. Process. Syst. **34**, 21696–21707 (2021)
28. Kingma, D.P., Gau, R.: Understanding diffusion objectives as the elbo with simple data augmentation. In: Advances in Neural Information Processing Systems, vol. 36 (2023)
29. Kingma, D.P., Welling, M.: Auto-encoding variational Bayes. In: International Conference on Learning Representations (2014)
30. Lee, J., Bahri, Y., Novak, R., Schoenholz, S.S., Pennington, J., Sohl-Dickstein, J.: Deep neural networks as Gaussian processes. In: International Conference on Learning Representations (2018)
31. Levine, S.: Reinforcement learning and control as probabilistic inference: tutorial and review. preprint arXiv:1805.00909 (2018)
32. Liu, J., Lin, Z., Padhy, S., Tran, D., Bedrax-Weiss, T., Lakshminarayanan, B.: Simple and principled uncertainty estimation with deterministic deep learning via distance awareness. In: Advances in Neural Information Processing Systems, vol. 33, pp. 7498–7512 (2020)
33. Ma, Y.A., Chen, T., Fox, E.: A complete recipe for stochastic gradient MCMC. In: Advances in Neural Information Processing Systems, vol. 28, pp. 2917–2925 (2015)
34. Marino, J., Yue, Y., Mandt, S.: Iterative amortized inference. In: International Conference on Machine Learning, pp. 3403–3412. PMLR (2018)
35. Millidge, B.: Deep active inference as variational policy gradients. J. Math. Psychol. **96**, 102348 (2020)
36. Millidge, B., Tschantz, A., Buckley, C.L.: Whence the expected free energy? Neural Comput. **33**(2), 447–482 (2021)

37. Millidge, B., Tschantz, A., Seth, A.K., Buckley, C.L.: On the relationship between active inference and control as inference. In: Active Inference, First International Workshop. Communications in Computer and Information Science, vol. 1326, pp. 3–11. Springer (2020)
38. Nachum, O., Norouzi, M., Xu, K., Schuurmans, D.: Bridging the gap between value and policy based reinforcement learning. In: Advances in Neural Information Processing Systems, vol. 30 (2017)
39. Parr, T., Friston, K.: Generalised free energy and active inference. Biol. Cybern. **113**, 495–513 (2019)
40. Parr, T., Pezzulo, G., Friston, K.J.: Active inference: the free energy principle in mind, brain, and behavior. MIT Press (2022)
41. Peebles, W., Xie, S.: Scalable diffusion models with transformers. In: Proceedings of the IEEE/CVF International Conference on Computer Vision, pp. 4195–4205 (2023)
42. Rezende, D.J., Mohamed, S., Wierstra, D.: Stochastic backpropagation and approximate inference in deep generative models. In: International Conference on Machine Learning, pp. 1278–1286. PMLR (2014)
43. Ribeiro, F.D.S., Glocker, B.: Demystifying variational diffusion models. Found. Trends Comput. Graph. Vision **17**, 76–170 (2025)
44. Sakthivadivel, D.A.: A worked example of the Bayesian mechanics of classical objects. In: Active Inference: Third International Workshop. Communications in Computer and Information Science, vol. 1721, pp. 298–318. Springer (2023)
45. Song, J., Meng, C., Ermon, S.: Denoising diffusion implicit models. In: International Conference on Learning Representations (2021)
46. Song, Y., Ermon, S.: Generative modeling by estimating gradients of the data distribution. In: Advances in Neural Information Processing Systems, vol. 32 (2019)
47. Song, Y., Sohl-Dickstein, J., Kingma, D.P., Kumar, A., Ermon, S., Poole, B.: Score-based generative modeling through stochastic differential equations. In: International Conference on Learning Representations (2021)
48. Todorov, E., Erez, T., Tassa, Y.: Mujoco: a physics engine for model-based control. In: 2012 IEEE/RSJ International Conference on Intelligent Robots and Systems, pp. 5026–5033. IEEE (2012)
49. Tschantz, A., Millidge, B., Seth, A.K., Buckley, C.L.: Reinforcement learning through active inference. In: In "Bridging AI and Cognitive Science" Workshop at International Conference on Machine Learning (2020)
50. Vempala, S., Wibisono, A.: Rapid convergence of the unadjusted langevin algorithm. In: Conference on Neural Information Processing System, vol. 33 (2019)
51. Vincent, P.: A connection between score matching and denoising autoencoders. Neural Comput. **23**(7), 1661–1674 (2011)
52. Wang, Z., Hunt, J.J., Zhou, M.: Diffusion policies as an expressive policy class for offline reinforcement learning. In: International Conference on Learning Representations (2023)
53. Ziebart, B.D.: Modeling purposeful adaptive behavior with the principle of maximum causal entropy. Ph.D. thesis, Carnegie Mellon University (2010)

Deep Active Inference with Neural Stochastic Differential Equations

Marc Pritsch[1,2](✉) and Georgia Koppe[1,2,3]

[1] Interdisciplinary Center for Scientific Computing, Heidelberg University, Heidelberg, Germany
georgia.koppe@iwr.uni-heidelberg.de

[2] Hector Institute for AI in Psychiatry and Department of Psychiatry and Psychotherapy, Central Institute of Mental Health, Mannheim, Germany
marc.pritsch@iwr.uni-heidelberg.de

[3] Hertie Institute for AI in Brain Health, Tübingen University, Tübingen, Germany

Abstract. Active inference (AIF), grounded in the free energy principle, provides a unified framework for learning and decision-making under uncertainty. However, standard computational AIF formulations are restricted to finite state spaces and discrete time, limiting their applicability in environments with continuous space and time. To address this, we parameterise continuous-time state evolutions with neural stochastic differential equations (NSDEs), whose neural-network–based drift and diffusion can flexibly capture complex, nonlinear stochastic dynamics in high-dimensional state spaces. By embedding these NSDEs in a variational inference architecture, we derive a continuous-time expected free energy objective and show how to optimise it using established stochastic differential equation (SDE) solvers and Monte Carlo sampling. In particular, we show that continuous-time AIF naturally accommodates irregular or sporadic time series data, accurately recovering observed dynamics and preferred outcomes on a synthetic benchmark while outperforming fixed-step baselines. These results demonstrate that our NSDE-based AIF preserves the stochastic nature of AIF in continuous time while remaining grounded in mathematical theory and computational methods for SDEs.

Keywords: Active Inference · Neural Differential Equations · Variational Inference · Continuous-Time Control · Irregular Sampling

1 Introduction

Active inference (AIF) is a theoretical framework that explains how agents perceive, learn, and act in uncertain environments. Grounded in the free energy principle, it posits that agents minimise *expected free energy* (EFE) by continuously updating their beliefs about hidden states and selecting actions that reduce expected surprise in a stochastic world [2,11–14]. AIF has been used to explain

M. Albarracin et al. (Eds.): IWAI 2025, CCIS 2857, pp. 137–150, 2026.
https://doi.org/10.1007/978-3-032-16955-6_8

behaviours across a variety of domains, including substance use disorders [38], hallucinations [1], saccadic eye movements [33], and robotic control [4].

Common computational formulations of AIF are defined for finite state and action spaces and in discrete time [8,37], limiting its extension to large or continuous state spaces and time. Deep AIF overcomes part of these limitations by introducing neural networks (NNs) to approximate the required probability densities, thereby scaling the approach to continuous state spaces [3,7,10,17,30,39]. Nonetheless, most deep AIF methods still assume a discrete-time formulation with equidistant time steps.

Neural stochastic differential equations (NSDEs)[1] have recently emerged as a continuous-time extension of fixed-grid generative time series models, combining the flexibility of NNs with the theoretical foundations of differential equations [19,21,28]. In this work, we integrate NSDEs into the AIF framework, yielding a flexible, NN-based approach for principled decision-making in continuous time.

Notation. Throughout the paper, random variables are denoted by Roman italic uppercase letters, e.g., X, and their realisations by Roman italic lowercase letters, e.g., x. One exception is the policy, which is always denoted by π. Time is treated either in discrete steps, where we write Z_t for the value of a discrete-time stochastic process $(Z_t)_{t\in\{t_1,\dots,t_N\}}$ at time $t \in \{t_1, \dots, t_N\}$, or in continuous time, where we write $Z(t)$ for the value of a continuous-time stochastic process $(Z(t))_{t\in[0,T]}$ at time $t \in [0, T]$. Boldface symbols, e.g., $\mathbb{P}$ and $\mathbb{Q}$, are reserved for probability distributions, and we write $\mathbb{E}_{\mathbb{P}}[\cdot]$ for expectations taken with respect to $\mathbb{P}$. The distribution of a random variable X or of a stochastic process Z is denoted by $\mathbb{P}_X$ or $\mathbb{P}_Z$, respectively, unless stated otherwise. For a finite-dimensional distribution $\mathbb{P}_X$, the corresponding density or probability mass function (PMF) is denoted by $p(x)$. The set of real numbers is denoted by $\mathbb{R}$, with $\mathbb{R}^d$ denoting the d-dimensional Euclidean space. Calligraphic symbols and Greek uppercase letters, e.g., $\mathcal{S}$ or Π, are reserved for sets (with dimensionality $d_{\mathcal{S}}$).

1.1 Active Inference in Discrete Time and Space

In discrete time and for discrete state spaces, the AIF problem is well established and can be formalised as a partially observable Markov decision process (POMDP) (see Fig. 1) [8,34,37]. The agent is assumed to possess a generative model, specified by a joint PMF over trajectories of observations $o_{1:T} \in \mathcal{O}^T$ and hidden states $s_{1:T} \in \mathcal{S}^T$, and policies $\pi \in \Pi$, of the form

$$p(o_{1:T}, s_{1:T}, \pi) = p(\pi)p(s_1)\prod_{t=2}^{T} p(s_t \mid s_{t-1}, \pi)\prod_{t=1}^{T} p(o_t \mid s_t) \tag{1}$$

[1] A stochastic differential equation (SDE) defines a stochastic process through explicit functional forms for its drift and diffusion terms. An NSDE refers to an SDE in which the drift and/or diffusion functions are parameterised by NNs, allowing for flexible, data-driven modelling of complex, nonlinear dynamics [19].

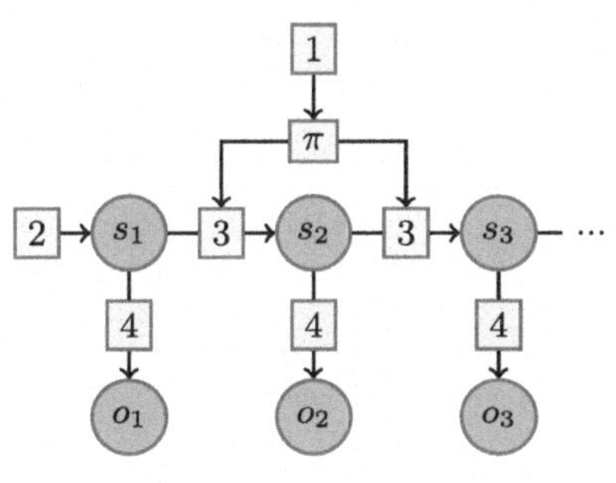

Generative model of the agent given by

1. a prior over policies $\pi \sim \mathbb{P}_\pi$,
2. an initial state distribution $S_1 \sim \mathbb{P}_{S_1}$,
3. a state transition model $S_t \mid S_{t-1}, \pi \sim \mathbb{P}_{S_t|S_{t-1},\pi}$ for all $t \in \{2, \dots, T\}$,
4. an observation model $O_t \mid S_t \sim \mathbb{P}_{O_t|S_t}$ for all $t \in \{1, \dots, T\}$.

Fig. 1. Schematic illustration of a partially observable Markov decision process.

as the corresponding joint distribution $\mathbb{P}_{O_{1:T},S_{1:T},\pi}$ *factorise* into the prior over policies $\mathbb{P}_\pi$ (encoding the agent's preferences or biases over possible action sequences), the initial state distribution $\mathbb{P}_{S_1}$, the state transition model $\mathbb{P}_{S_t|S_{t-1},\pi}$, $t \in \{2, \dots, T\}$, for a given policy $\pi \in \Pi$, and the observation model $\mathbb{P}_{O_t|S_t}$, $t \in \{1, \dots, T\}$, defining the dependence of observations given the hidden state.

Perception. As the agent receives observations from the environment $o_{1:t} \in \mathcal{O}^t$ up to $t \in \{1, \dots, T\}$, it seeks to infer the hidden (past and future) states of the world $s_{1:T}$ given its current policy π. Since inverting the generative model and computing the posterior over hidden states, $\mathbb{P}_{S_{1:T}|o_{1:t},\pi}$, is typically intractable, the agent employs variational inference (VI) to approximate this posterior by choosing a family of approximate posterior distributions over hidden states given a policy, denoted by $\mathcal{Q}$. Thus, the agent solves an optimisation problem given by

$$\underset{\mathbb{Q}_{S_{1:T}|\pi} \in \mathcal{Q}}{\operatorname{argmin}} \; F_\pi(\mathbb{Q}_{S_{1:T}|\pi}, o_{1:t}, t),$$

where $F_\pi(\mathbb{Q}_{S_{1:T}|\pi}, o_{1:t}, t)$ is the *variational free energy* (VFE) of $\mathbb{Q}_{S_{1:T}|\pi} \in \mathcal{Q}$ at time $t \in \{1, \dots, T\}$, defined as

$$F_\pi(\mathbb{Q}_{S_{1:T}|\pi}, o_{1:t}, t) = \mathrm{KL}(\mathbb{Q}_{S_{1:T}|\pi} \,\|\, \mathbb{P}_{S_{1:T}|\pi}) - \mathbb{E}_{\mathbb{Q}_{S_{1:T}|\pi}} \left[\log p(o_{1:t} \mid S_{1:T}) \right], \quad (2)$$

where $\mathbb{P}_{S_{1:T}|\pi}$ is the conditional distribution over hidden states obtained from the state transition model in Eq. 1, $p_{(o_{1:t}|s_{1:T})}$ is the PMF of the conditional distribution over observations up to time t obtained from restricting the observation model in Eq. 1, and KL is the Kullback–Leibler divergence between approximate posterior and prior [26].

Learning. In AIF, learning refers to the process of updating the parameters (e.g., state and transition probabilities) of the distributions that constitute the generative model (Eq. 1). This is achieved by minimising the VFE (Eq. 2) with respect to these parameters.

Planning. Given observations $o_{1:t} \in \mathcal{O}^t$, $t \in \{1, \ldots, T-1\}$, the agent maintains a posterior belief over hidden states $\mathbb{Q}_{S_{1:T}|o_{1:t},\pi}$ and selects policies by solving

$$\underset{\pi \in \Pi}{\operatorname{argmin}}\, G(\pi, t),$$

where $G(\pi, t)$ denotes the EFE of a policy $\pi \in \Pi$ at time $t \in \{1, \ldots, T-1\}$, defined as

$$\begin{aligned} G(\pi, t) = &\, \mathbb{E}_{\tilde{\mathbb{Q}}_{O_{t+1:T}, S_{t+1:T}}} \left[-\mathrm{KL}(\mathbb{Q}_{S_{t+1:T}|O_{t+1:T},\pi} \parallel \mathbb{Q}_{S_{t+1:T}|o_{1:t},\pi}) \right] \\ & + \mathbb{E}_{\tilde{\mathbb{Q}}_{O_{t+1:T}, S_{t+1:T}}} \left[-\log p^{\mathrm{pref}}(O_{t+1:T}) \right], \end{aligned} \tag{3}$$

where $\tilde{\mathbb{Q}}_{O_{t+1:T}, S_{t+1:T}}$ denotes the predictive distribution over future observations and states given by the product of the inferred latent belief $\mathbb{Q}_{S_{t+1:T}|o_{1:t},\pi}$ and the restricted observation model $\mathbb{P}_{O_{t+1:T}|S_{t+1:T}}$. The first term quantifies expected information gain as a divergence between current beliefs $\mathbb{Q}_{S_{t+1:T}|o_{1:t},\pi}$ and counterfactual beliefs based on imagined future observations $\mathbb{Q}_{S_{t+1:T}|O_{t+1:T},\pi}$, and the second encodes the agent's specified preferences via the expected likelihood of a preferred state distribution p^{pref}.

1.2 Extending AIF to Continuous State and Action Spaces

Deep AIF introduces NNs to approximate the generative model and posterior densities, as well as to parameterise the agent's policy [3,7,10,17,30,39]. Although usually formulated in discrete time, this approach scales inference and planning to much larger state spaces than traditional methods. In practice, deep AIF models are typically trained by rolling out simulated trajectories under candidate policies and updating the NNs to better match the true generative model and posterior statistics, while also learning purposeful action selection [3,17,30].

1.3 Extending AIF to Continuous-Time Environments

Neural Stochastic Differential Equations. Nevertheless, many real-world processes are inherently continuous in time. To model such dynamics, we now introduce NSDEs, a flexible framework for learning continuous-time, noise-driven systems with NNs [20,28,40]. Originally introduced for ordinary differential equations. [5], the idea was subsequently extended to stochastic differential equations (SDEs) [20,28,40], enabling continuous-time formulations of deep architectures such as residual networks [16] and recurrent NNs [6,9,18,36].

In NSDEs, drift and diffusion coefficients of SDEs are parameterised by NNs, yielding a continuous-time generative model of the form

$$\mathrm{d}S^{\mathrm{pr}}(t) = \mu_{\theta_{\mathbf{pr}}}(S^{\mathrm{pr}}(t), t)\mathrm{d}t + \sigma_\theta(S^{\mathrm{pr}}(t), t)\mathrm{d}W^{\mathrm{pr}}(t) \tag{4}$$

for all $t \in [0, T]$ and $S^{\mathrm{pr}}(0) = s_0 \in \mathbb{R}^{d_S}$, where $(W^{\mathrm{pr}}(t))_{t\in[0,T]}$ denotes a standard $d_{\mathcal{W}}$-dimensional Brownian motion, and $\mu_{\theta_{\mathbf{pr}}}$ and σ_θ are NNs parameterised by $\theta_{\mathbf{pr}}$ and θ, respectively.

Latent Stochastic Differential Equations. NSDEs can be implemented as a latent dynamics equation—termed the *prior* $S^{\text{pr}} \sim \mathbb{P}_{S^{\text{pr}}}$ and governed by an SDE as in Eq. 4—coupled to a stochastic observation model with density

$$p_{\theta_{\mathbf{obs}}}(o \mid s) \tag{5}$$

parameterised by $\theta_{\mathbf{obs}}$.

In [28], inference of this model is performed using amortised VI [24] - referred to as *latent SDE (LSDE) model.* In analogy to variational autoencoders [23], the NSDE prior defines a distribution over latent trajectories, the observation model plays the role of the decoder, and we introduce a *recognition model (approximate posterior)* that acts as an encoder. This recognition model $S^{\text{rec}} \sim \mathbb{Q}_{S^{\text{rec}}}$, introduces data dependence through an observation-encoded input $(h(t))_{t\in[0,T]}$, leading to

$$\mathrm{d}S^{\text{rec}}(t) = \mu_{\theta_{\mathbf{rec}}}(S^{\text{rec}}(t), h(t), t)\mathrm{d}t + \sigma_\theta(S^{\text{rec}}(t), t)\mathrm{d}W^{\text{rec}}(t)$$

for all $t \in [0, T]$ and $S^{\text{rec}}(0) = s_0 \in \mathbb{R}^{d_S}$, where $(W^{\text{rec}}(t))_{t\in[0,T]}$ denotes a standard $d_{\mathcal{W}}$-dimensional Brownian motion and $\mu_{\theta_{\mathbf{rec}}}$ is a NN parameterised by $\theta_{\mathbf{rec}}$.

For observations $o_{t_1}, \ldots, o_{t_N} \in \mathbb{R}^{d_{\mathcal{O}}}$ at time points $0 \leq t_1 < \cdots < t_N \leq T$, the encoding $(h(t))_{t\in[0,T]}$ is given by

$$h(t) = h_{\theta_{\mathbf{enc}}}(o_{t_1}, \ldots, o_{t_N}, t) \tag{6}$$

for all $t \in [0, T]$, where $h_{\theta_{\mathbf{enc}}} : \mathbb{R}^{d_{\mathcal{O}} \times N} \times [0, T] \to \mathbb{R}^{d_{\mathcal{H}}}$ is a common NN architecture (convolutional, recurrent, feedforward, etc.) parameterised by $\theta_{\mathbf{enc}}$.

Since the prior and approximate posterior share the same diffusion coefficient σ_θ, by Girsanov's theorem [31], the KL divergence between the two [32] can be expressed as

$$\mathrm{KL}(\mathbb{Q}_{S^{\text{rec}}} \,\|\, \mathbb{P}_{S^{\text{pr}}}) = \mathbb{E}_{\mathbb{Q}_{S^{\text{rec}}}}\left[\frac{1}{2}\int_0^T \|u(S^{\text{rec}}(t), t)\|^2 \mathrm{d}t\right],$$

where $u(s, t) = \mathbb{R}^{d_S} \times [0, T] \to \mathbb{R}^{d_{\mathcal{W}}}$ satisfies

$$u(s, t) = \sigma_\theta^{-1}(s, t)(\mu_{\theta_{\mathbf{rec}}}(s, h(t), t) - \mu_{\theta_{\mathbf{pr}}}(s, t))$$

for all $s \in \mathbb{R}^{d_S}$ and $t \in [0, T]$. For observations $o_{t_1}, \ldots, o_{t_N} \in \mathbb{R}^{d_{\mathcal{O}}}$ at time points $0 \leq t_1 < \cdots < t_N \leq T$, the evidence lower bound (ELBO), which corresponds to the negative VFE, serves as the optimisation criterion in VI, used for perception and learning, and is given by

$$\mathrm{ELBO} = \mathbb{E}_{\mathbb{Q}_{S^{\text{rec}}}}\left[-\frac{1}{2}\int_0^T \|u(S^{\text{rec}}(t), t)\|^2 \mathrm{d}t + \sum_{i=1}^N \log p_{\theta_{\mathbf{obs}}}(o_{t_i} \mid S^{\text{rec}}(t_i))\right]. \tag{7}$$

Because all components of the model are formalised based on the continuous-time stochastic processes $(S^{\text{rec}}(t))_{t\in[0,T]}$, classical SDE solvers are used to generate sample trajectories, which are then used to estimate gradients for backpropagation [28].

2 Framework

Our approach extends the LSDE/VI framework by incorporating control policies into the drift terms. This extension enables us to integrate the model into the AIF framework, providing a principled and continuous-time interpretation of the latent process.

2.1 Model Architecture

Prior. The prior is defined by a controlled NSDE of the form

$$\mathrm{d}S^{\mathrm{pr}}(t) = \mu_{\theta_{\mathbf{pr}}}(S^{\mathrm{pr}}(t), \pi(t), t)\mathrm{d}t + \sigma_\theta(S^{\mathrm{pr}}(t), t)\mathrm{d}W^{\mathrm{pr}}(t) \tag{8}$$

for all $t \in [0, T]$, where, in contrast to a standard NSDE, the deterministic control policy $(\pi(t))_{t\in[0,T]}$ is introduced as an additional input to the drift network. An observation model $p_{\theta_{\mathbf{obs}}}$ akin to Eq. 5 couples latent dynamics to observed variables.

Approximate Posterior. The approximate posterior is defined by a controlled NSDE of the form

$$\mathrm{d}S^{\mathrm{rec}}(t) = \mu_{\theta_{\mathbf{rec}}}(S^{\mathrm{rec}}(t), h(t), \pi(t), t)\mathrm{d}t + \sigma_\theta(S^{\mathrm{rec}}(t), t)\mathrm{d}W^{\mathrm{rec}}(t) \tag{9}$$

for all $t \in [0, T]$, where, in addition to the control policy $(\pi(t))_{t\in[0,T]}$, the drift function also receives the data-dependent encoding $(h(t))_{t\in[0,T]}$ (Eq. 6) that provides contextual information about the observed trajectory. See Fig. 2 for a schematic of the proposed framework. For notational convenience, we collect all parameters in $\Theta = (\theta_{\mathbf{enc}}, \theta_{\mathbf{obs}}, \theta_{\mathbf{pr}}, \theta_{\mathbf{rec}}, \theta)$.

Variational Free Energy. Because the prior and the recognition model share the same diffusion coefficient σ_θ, we can use an analogous argument as in the LSDE architecture [28] and Eq. 7 to express the VFE for a continuous-time latent process. For observations $o_{t_1}, \ldots, o_{t_N} \in \mathbb{R}^{d_O}$ at time points $0 \le t_1 < \cdots < t_N \le T$, VFE F (Eq. 2) is then given by

$$F(\Theta) = \mathbb{E}_{\mathbb{Q}_{S^{\mathrm{rec}}}}\left[\frac{1}{2}\int_0^T \|u(S^{\mathrm{rec}}(t), t)\|^2 \mathrm{d}t - \sum_{i=1}^{N} \log p_{\theta_{\mathbf{obs}}}(o_{t_i} \mid S^{\mathrm{rec}}(t_i))\right], \tag{10}$$

where $u(s, t) = \mathbb{R}^{d_S} \times [0, T] \to \mathbb{R}^{d_W}$ satisfies

$$u(s, t) = \sigma_\theta^{-1}(s, t)(\mu_{\theta_{\mathbf{rec}}}(s, h(t), \pi(t), t) - \mu_{\theta_{\mathbf{pr}}}(s, \pi(t), t)). \tag{11}$$

for all $s \in \mathbb{R}^{d_S}$ and $t \in [0, T]$. In contrast to Eq. 2, here—and for our experiments—we drop the explicit time dependence of the VFE, since we minimise the pathwise loss integrated over the whole interval in an offline manner. The VFE serves as a common objective for both perception and learning.

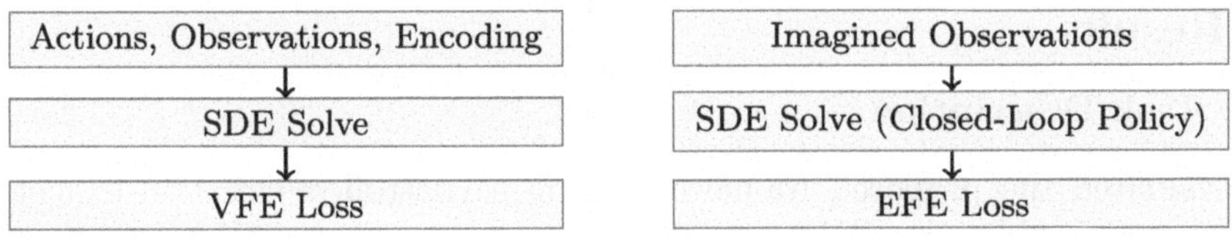

Fig. 2. Schematic illustration of the pipeline for perception (left) and planning (right).

Perception and Learning. We infer model parameters Θ by solving

$$\Theta^* = \underset{\Theta}{\operatorname{argmin}}\, F(\Theta).$$

We condition the latent dynamics on the observed data and actions to simulate trajectories, and propagate the loss gradients backward through these simulated dynamics.

Expected Free Energy. For planning, we parameterise the policy within the prior (and posterior) as a closed-loop mapping, i.e.,

$$\pi(t) = \pi_{\theta_{\mathbf{pol}}}(S^{\mathrm{pr}}(t), t)$$

for all $t \in [0, T]$, where $\pi_{\theta_{\mathbf{pol}}} : \mathbb{R}^{d_S} \times [0, T] \to \mathbb{R}^{d_\Pi}$ is an NN parameterised by $\theta_{\mathbf{pol}}$.
The continuous-time analogue of the EFE G (cf. Eq. 3) is

$$G(\theta_{\mathbf{pol}}) = \mathbb{E}_{\tilde{\mathbb{Q}}_{O_{t_{1:N}}, S^{\mathrm{rec}}}} \left[-\frac{1}{2} \int_0^T \|u(S^{\mathrm{rec}}(t), t)\|^2 \mathrm{d}t - \sum_{i=1}^{N} \log(p^{\mathrm{pref}}(O_{t_i})) \right],$$

where u is defined as in Eq. 11, $\tilde{\mathbb{Q}}_{O_{t_{1:N}}, S^{\mathrm{rec}}}$ denotes the joint distribution over simulated observations $O_{t_{1:N}}$ at solver-dependent time points $0 \le t_1 < \cdots < t_N \le T$ and the approximate posterior S^{rec}, which is conditioned on those observations using $H(t) = h_{\theta_{\mathbf{enc}}}(O_{t_1}, \ldots, O_{t_N}, t)$ for all $t \in [0, T]$. Future observations are generated with the trained prior and observation model.

Planning and Action Selection. Since the agent is assumed to act by choosing the action that minimises the EFE, the optimisation problem is given by

$$\theta_{\mathbf{pol}}{}^* = \underset{\theta_{\mathbf{pol}}}{\operatorname{argmin}}\, G(\theta_{\mathbf{pol}}).$$

A closed-loop policy is used for simulating the processes, optimised by backpropagation.

3 Results

3.1 Task Description

We evaluated the proposed framework using a controlled Ornstein-Uhlenbeck process governed by the SDE

$$\mathrm{d}X(t) = (\kappa(\mu - X(t)) + \mathrm{c}\pi(t))\mathrm{d}t + \sigma \mathrm{d}W(t) \tag{12}$$

for $t \in [0,1]$, with parameters set to $\mu = 2$, $\mathrm{c} = 1$, $\kappa = 2$, and $\sigma = 0.2$. To demonstrate the ability of our framework to handle irregularly sampled data, trajectories used for model training were generated with Eq. 12, using irregular time grids and additive Gaussian observation noise with variance $\sigma^2_{\mathrm{obs}} = 0.01$. The irregular grids were constructed by sampling $N = 100$ time points from a left-skewed Beta distribution with parameters $\alpha = 2$, $\beta = 1$, over the interval $[0,1]$. For all trajectories, the same control inputs $\pi_{t_1}, \ldots, \pi_{t_{100}}$ were applied, sampled from a zero-mean Gaussian distribution with variance $\sigma_c^2 = 1$. Simulation of the SDE was performed using the Euler-Maruyama scheme [25], with 256 sample paths generated for each training epoch (see Fig. 3a).

The goal of the proposed LSDE-AIF framework was to infer the underlying generative process from these temporally irregular observations (perception and learning), and leverage the inferred models for downstream control tasks (planning). To validate its effectiveness, we compared its performance when trained on irregular temporal grids against a baseline where the model assumed a regular grid, ignoring the true timing of observations.

3.2 Neural Network Implementation

We extend the LSDE model [19,28] by incorporating actions into the drift of the SDE. The drift networks for both the prior (Eq. 8), and the approximate posterior (Eq. 9), are parameterised by 3-layer feedforward NNs (FNNs) with softplus activation functions and a final linear output layer. The shared diffusion network is a 2-layer FNN, also using softplus in the hidden layer, with a sigmoid activation on its final output. We assume a diagonal noise covariance, which simplifies the matrix inversion required in the KL divergence computation and improves computational efficiency [19]. The prior's initial latent state $S(0)$ is sampled from a trainable Gaussian distribution. To initialise the posterior, input observations are first encoded by a linear 2-layer FNN. From the encoder's initial output $h(0)$, a separate linear layer predicts the mean and variance of the posterior's initial Gaussian, from which $S(0)$ is sampled. The latent and the encoding dimension are chosen to be 3 and 16, respectively. The decoder is parameterised by a single-hidden-layer FNN with a softplus activation in the hidden layer and a linear readout, returning the mean and the variance of the Gaussian observation model. All hidden layers have a width of 16. The training pipeline is implemented using the `torchsde` library [27], a `Python` package that provides differential equation solvers and variational methods.

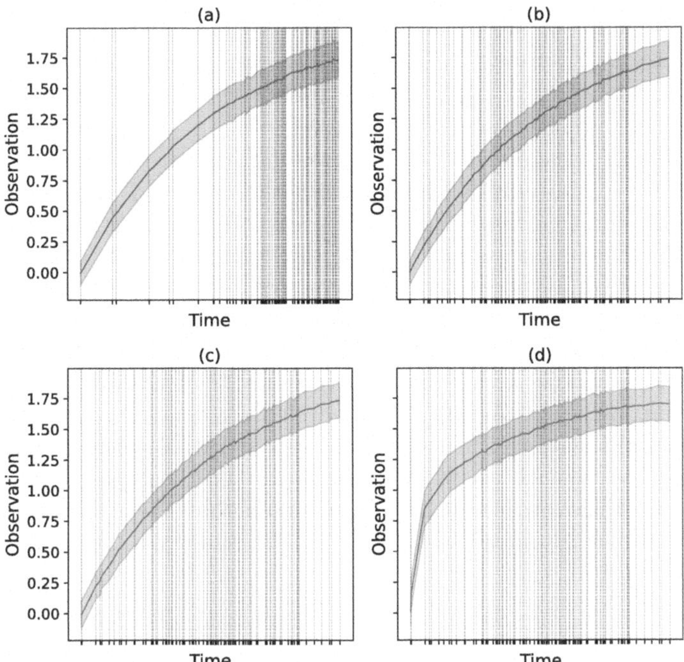

Fig. 3. Inference under irregular vs. regular temporal assumptions. (a) Observations from an Ornstein-Uhlenbeck process sampled on an irregular temporal grid (marked by vertical lines), used for training. **(b)** Observations from the same process used for model evaluation. **(c)** LSDE-inferred process, explicitly modelling irregular time steps, trained on data in (a) and reproducing data in (b) using its prior. **(d)** Same as (c) but with baseline model which assumes uniform time intervals.

3.3 Perception and Learning

To obtain the prior and approximate posterior, we trained this controlled LSDE on the data in Fig. 3a, using the same actions (control inputs) as used during data generation. Given the encoded trajectories and the corresponding actions, we simulated a batch of $B = 256$ latent SDE paths using the Euler-Maruyama method [25], and mapped them to the observation space by adding Gaussian noise (learned from the data). We used the Adam optimiser [22] with exponential learning rate decay, along with KL annealing—linearly increasing the KL weight λ from 0 to 1 over the first half of training epochs (see Eq. 13)—to mitigate posterior collapse during training. The resulting training loss for perception and learning is the negative ELBO, here estimated by a time-discretised Monte Carlo (MC) approximation with $B = 256$ sampled trajectories

$$F \approx \frac{1}{B} \sum_{k=1}^{B} \left(\lambda \, \mathrm{kl}_0^{(k)} + \lambda \mathrm{kl}_{\mathrm{qp}}^{(k)} - \sum_{i=1}^{N} \log p_\theta(o_{t_i} \mid s_{t_i}^{(k)}) \right), \tag{13}$$

where λ is the weight of the annealed KL, $\{\mathrm{kl}_0^{(k)}\}_{k=1}^B$ the KL divergences between the two initial distributions for each sample—an extension of the formulation described in Sect. 2—$\{\mathrm{kl}_{\mathrm{qp}}^{(k)}\}_{k=1}^B$ is `torchsde`'s corresponding estimate of the pathwise KL divergence, and $\{s^{(k)}\}_{k=1}^B$ are the sampled trajectories.

The results of training on the true irregular grid are shown in Fig. 3c. The prior model successfully approximates both the drift and diffusion latent process generating the Ornstein-Uhlenbeck process. In contrast, the baseline model trained under the assumption of a regular temporal grid misinterprets the dynamics, inferring a trajectory with a noticeably steeper initial incline (see Fig. 3 d). While the sampled trajectory aligns with the ground truth toward the end of the interval, the initial overshoot suggests that the model overestimates the mean-reversion rate κ, highlighting the impact of ignoring temporal irregularity during inference. Across ten random initialisations, the irregular-time model statistically outperformed the fixed-step model in terms of capturing the ground truth mean, with an MSE of 0.036 ± 0.007, compared to an MSE of 0.106 ± 0.036 (Wilcoxon $p = 0.002$), confirming that the irregular-time model provided a significantly more accurate fit to the underlying process.

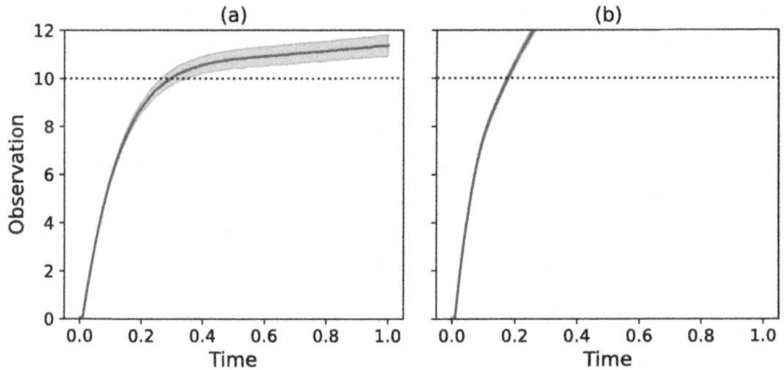

Fig. 4. Transfer of control to ground-truth dynamics. (a) Control policy from the NSDE, respecting irregular sampling, applied to the true system. **(b)** Control policy from the baseline model applied to the true system.

3.4 Planning

Next, our goal was to compare the two trained generative models in a controlled planning setting to assess the impact of modelling assumptions on downstream performance. We used the trained models for closed-loop planning by training a policy network, using a time-independent policy (a special case of Sect. 2). To isolate goal-directed behaviour from uncertainty-driven exploration, we first omitted exploratory actions and configured the policy to optimise only the preference distribution (cf. Eq. 3). This reflects an agent that fully trusts its model.

Planning involved sampling a batch of $B = 256$ latent trajectories from the prior and generating corresponding observations. These predicted observations

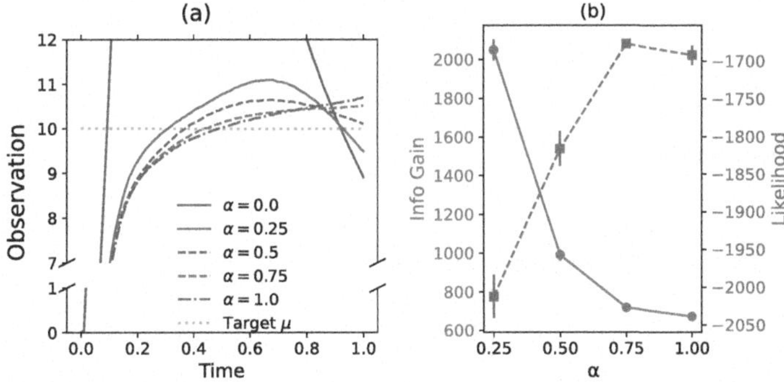

Fig. 5. Effects of mixing weight α of the EFE on the control. **(a)** Mean of sample paths for different α's (see legend). **(b)** EFE components (see y axis labels) for different α's ($\alpha = 0.0$ is discarded for visualisation purposes).

$o_{t_1}, \ldots, o_{t_N}$ were evaluated under a Gaussian preference distribution. The resulting planning loss is thus given by the MC estimate

$$G \approx -\frac{1}{B} \sum_{k=1}^{B} \left(\sum_{i=1}^{N} \log p^{\text{pref}}(o_{t_i}^{(k)}) \right),$$

where p^{pref} is the likelihood of a Gaussian distribution with $\mu(t) = 10$ and $\sigma(t) = 0.05$ for all $t \in [0, 1]$, and $\{(o_{t_1}^{(k)}, \ldots, o_{t_N}^{(k)})\}_{k=1}^{B}$ are the sampled observations.

When applying the learned policies to the original Ornstein-Uhlenbeck process, we once more observe a clear difference in performance between both models (see Fig. 4). The policy trained on the model that was given the true irregular grid successfully steers the system toward the target state (see Fig. 4a). By contrast, the policy trained on the baseline model (assuming a uniform grid) drastically overshoots the target state (see Fig. 4b). This likely stems from an overestimation of the system's mean-reversion tendency ($\kappa = 2$), causing the policy to overcompensate with an excessively large upward adjustment. Across ten random initialisations, the irregular-time model once more outperformed the fixed-step model in terms of reaching the target, with an MSE of 6.960 ± 1.707, compared to an MSE of 105.867 ± 168.959 (Wilcoxon $p = 0.002$).

Finally, to demonstrate the role of the information gain, we trained multiple policies by optimising the *full* EFE, combining the preference term with weight α and the information gain term with weight $(1-\alpha)$, for $\alpha \in [0, 1]$. To prevent runaway actions and ensure stability, for these analyses, we added L_2 regularisation on the control, clipped the KL divergence at each step (to $[-20, 20]$) and the gradient norm (at 1), set $\sigma(t) = 0.3$ for all $t \in [0, 1]$, and omitted constant terms in the Normal density to keep values on a comparable scale. All terms were averaged over time (see code for implementation). The means of $B = 256$ sample paths and statistics averaged over five models for $\alpha \in \{0, 0.25, 0.5, 0.75, 1\}$ are displayed in Fig. 5. The planning results show that decreasing α weakens purely goal-directed behaviour, as trajectories become less tightly aligned with the preference distribution and shift towards more exploratory, information-seeking actions.

4 Discussion

We showed how controlled LSDEs can be embedded into the AIF framework to combine the benefits of continuous-time modelling with flexible neural architectures. Using neural SDEs as the generative model enables principled perception, learning, and planning under irregular sampling—crucial when observations arrive at uneven intervals, such as phoneme onsets [15], wearable sensor data [29], or human perception [35]. On an Ornstein–Uhlenbeck benchmark, policies trained on models that respect temporal structure successfully controlled the real system, while those assuming uniform sampling failed due to misestimated dynamics.

Limitations and Future Work
This work establishes the method's foundations and provides open-source code, but further evaluation is needed. Future work includes statistical benchmarking, applying the method to standard control tasks, and testing on real data. Our current implementation relies on MC sampling and backpropagation, common in VI and deep AIF but at odds with AIF's original aims; exploring more biologically plausible training is an avenue for improvement. Finally, as model parameters are point-estimated, extending the approach to fully Bayesian inference remains an important goal.

4.1 Code

All code to reproduce our experiments (training, planning, figures) is available at https://github.com/humml-lab/DAIFwithNSDEs.git.

Acknowledgments. This work was supported by the German Research Foundation (DFG, Excellence Strategy EXC 2181/1 – 390900948, STRUCTURES), the Hector II Foundation, the Federal Ministry of Research, Technology and Space (BMFTR), and the Baden-Württemberg Ministry of Science as part of the Excellence Strategy of the German Federal and State Governments.

References

1. Adams, R.A., Stephan, K.E., Brown, H.R., Frith, C.D., Friston, K.J.: The computational anatomy of psychosis. Front. Psych. **4**, 47 (2013)
2. Buckley, C.L., Kim, C.S., McGregor, S., Seth, A.K.: The free energy principle for action and perception: a mathematical review. J. Math. Psychol. **81**, 55–79 (2017)
3. Çatal, O., Wauthier, S., De Boom, C., Verbelen, T., Dhoedt, B.: Learning generative state space models for active inference. Front. Comput. Neurosci. **14**, 574372 (2020)

4. Çatal, O., Wauthier, S., Verbelen, T., De Boom, C., Dhoedt, B.: Deep active inference for autonomous robot navigation. In: Workshop on Bridging AI and Cognitive Science, International Conference on Learning Representations (ICLR) (2020)
5. Chen, R.T., Rubanova, Y., Bettencourt, J., Duvenaud, D.K.: Neural ordinary differential equations. Adv. Neural Inf. Process. Syst. **31** (2018)
6. Cho, K., et al.: Learning phrase representations using RNN encoder–decoder for statistical machine translation. In: Proceedings of the 2014 Conference on Empirical Methods in Natural Language Processing (EMNLP), pp. 1724–1734 (2014)
7. Collis, P., Singh, R., Kinghorn, P.F., Buckley, C.L.: Learning in hybrid active inference models. In: International Workshop on Active Inference, pp. 49–71. Springer (2024)
8. Da Costa, L., Parr, T., Sajid, N., Veselic, S., Neacsu, V., Friston, K.: Active inference on discrete state-spaces: a synthesis. J. Math. Psychol. **99**, 102447 (2020)
9. Elman, J.L.: Finding structure in time. Cogn. Sci. **14**(2), 179–211 (1990)
10. Fountas, Z., Sajid, N., Mediano, P., Friston, K.: Deep active inference agents using monte-carlo methods. Adv. Neural. Inf. Process. Syst. **33**, 11662–11675 (2020)
11. Friston, K.: The free-energy principle: a unified brain theory? Nat. Rev. Neurosci. **11**(2), 127–138 (2010)
12. Friston, K.: The free energy principle made simpler but not too simple. Phys. Rep. **1024**, 1–29 (2023)
13. Friston, K., FitzGerald, T., Rigoli, F., Schwartenbeck, P., Pezzulo, G.: Active inference: a process theory. Neural Comput. **29**(1), 1–49 (2017)
14. Friston, K., Kilner, J., Harrison, L.: A free energy principle for the brain. J. Physiology-Paris **100**(1–3), 70–87 (2006)
15. Gwilliams, L., King, J.R., Marantz, A., Poeppel, D.: Neural dynamics of phoneme sequences reveal position-invariant code for content and order. Nat. Commun. **13**(1), 6606 (2022)
16. He, K., Zhang, X., Ren, S., Sun, J.: Deep residual learning for image recognition. In: Proceedings of the IEEE Conference on Computer Vision and Pattern Recognition, pp. 770–778 (2016)
17. van der Himst, O., Lanillos, P.: Deep active inference for partially observable MDPs. In: Active Inference: First International Workshop, IWAI 2020, Co-located with ECML/PKDD 2020, Ghent, Belgium, September 14, 2020, Proceedings 1, pp. 61–71. Springer (2020)
18. Hochreiter, S., Schmidhuber, J.: Long short-term memory. Neural Comput. **9**(8), 1735–1780 (1997)
19. Kidger, P.: On Neural Differential Equations. Ph.D. thesis, University of Oxford (2022)
20. Kidger, P., Foster, J., Li, X., Lyons, T.J.: Neural SDEs as infinite-dimensional GANs. In: International Conference on Machine Learning, pp. 5453–5463. PMLR (2021)
21. Kidger, P., Foster, J., Li, X.C., Lyons, T.: Efficient and accurate gradients for neural SDEs. Adv. Neural. Inf. Process. Syst. **34**, 18747–18761 (2021)
22. Kingma, D.P., Ba, J.: Adam: A method for stochastic optimization. arXiv preprint arXiv:1412.6980 (2014)
23. Kingma, D.P., Welling, M.: Auto-encoding variational bayes. International Conference on Learning Representations (ICLR) (2014)
24. Kingma, D.P., Welling, M., et al.: An introduction to variational autoencoders. Found. Trends® Mach. Learn. **12**(4), 307–392 (2019)
25. Kloeden, P.E., Platen, E.: Stochastic differential equations. Springer (1992)

26. Kullback, S.: Information theory and statistics. Courier Corporation (1997)
27. Li, X., Kidger, P.: TorchSDE: Pytorch implementation of differentiable stochastic differential equation solvers (2020), version 0.2.6
28. Li, X., Wong, T.K.L., Chen, R.T., Duvenaud, D.: Scalable gradients for stochastic differential equations. In: International Conference on Artificial Intelligence and Statistics, pp. 3870–3882. PMLR (2020)
29. Liu, M., Geißler, D., Bian, S., Zhou, B., Lukowicz, P.: Assessing the impact of sampling irregularity in time series data: human activity recognition as a case study. arXiv preprint arXiv:2501.15330 (2025)
30. Millidge, B.: Deep active inference as variational policy gradients. J. Math. Psychol. **96**, 102348 (2020)
31. Øksendal, B.: Stochastic differential equations. Springer (2003)
32. Opper, M.: Variational inference for stochastic differential equations. Ann. Phys. **531**(3), 1800233 (2019)
33. Parr, T., Friston, K.J.: Active inference and the anatomy of oculomotion. Neuropsychologia **111**, 334–343 (2018)
34. Parr, T., Pezzulo, G., Friston, K.J.: Active inference: the free energy principle in mind, brain, and behavior. MIT Press (2022)
35. Rolfs, M.: Microsaccades: small steps on a long way. Vision. Res. **49**(20), 2415–2441 (2009)
36. Rumelhart, D.E., Hinton, G.E., Williams, R.J.: Learning representations by back-propagating errors. Nature **323**(6088), 533–536 (1986)
37. Smith, R., Friston, K.J., Whyte, C.J.: A step-by-step tutorial on active inference and its application to empirical data. J. Math. Psychol. **107**, 102632 (2022)
38. Smith, R., Schwartenbeck, P., Stewart, J.L., Kuplicki, R., Ekhtiari, H., Paulus, M.P., Investigators, T., et al.: Imprecise action selection in substance use disorder: Evidence for active learning impairments when solving the explore-exploit dilemma. Drug Alcohol Depend. **215**, 108208 (2020)
39. Tschantz, A., Baltieri, M., Seth, A.K., Buckley, C.L.: Scaling active inference. In: 2020 International Joint Conference on Neural Networks (IJCNN), pp. 1–8. IEEE (2020)
40. Tzen, B., Raginsky, M.: Neural stochastic differential equations: Deep latent gaussian models in the diffusion limit. arXiv preprint arXiv:1905.09883 (2019)

A Diffusion Prior for Active Inference Planning

Catherine F. Higham(✉) and Roderick Murray-Smith

School of Computing Science, University of Glasgow, Glasgow G12 8QQ, UK
{catherine.higham,roderick.murray-smith}@glasgow.ac.uk

Abstract. This paper presents a method for combining Denoising Diffusion Probabilistic Models (DDPM) with Active Inference (AIF). The key idea is to leverage DDPM as a behavioural prior for inference within the AIF framework. This work is a novel application of DDPM to create a prior over actions, observations, and states for efficient planning and decision making. The DDPM is trained on behavourial observations in a training environment. Once trained, the DDPM can be conditioned on observations from a new environment without the need for retraining. The agent is then equipped to imagine likely future scenarios and focus on regions of the space which are most relevant. We show that estimates of the probability of future states can be used to provide inexpensive estimates of the Expected Free Energy. In addition, as the dimensions of the state space increase, the computational cost of generating policies is offset by a reduction in the number of policies to be considered. Together, these advantages overcome a major computational bottleneck. In summary, behavioural observations are efficiently encoded as data to train a DDPM, providing a simple but powerful prior for behaviour in a new environment.

Keywords: Active Inference Planning · Denoising Diffusion Probabilistic Models · Diffusion Prior · Conditional Diffusion

1 Introduction

Active Inference (AIF) is a framework for planning by an agent equipped with observations/sensors and actions within an environment [7,19]. AIF assumes that the agent has an internal generative probabilistic model which encodes its beliefs about how hidden states in the environment give rise to its observations/sensations. Preferences are introduced in AIF as priors over observations. The agent acts to reduce uncertainty, minimise expected surprise, and maintain favourable homeostasis over all predicted futures [7]. There are two ways to achieve this: perceptual inference (changing the belief) or active inference (acting on the world) [20]. The agent acts by predicting future beliefs, scoring each action sequence or policy by the average free energy of all beliefs consequent to that policy and then selecting an action to minimise the Expected Free Energy (EFE) [6] and the expected future surprise.

M. Albarracin et al. (Eds.): IWAI 2025, CCIS 2857, pp. 151–164, 2026.
https://doi.org/10.1007/978-3-032-16955-6_9

Denoising Diffusion Probabilistic Models (DDPM) are generative models that during training have learnt to denoise pure noise giving rise to novel data similar to the training data [13]. Generative models are a promising way to capture the correlations between actions, observations, and states over space and time. Diffusion models have several advantages over other generative models. They produce a higher quality output, in terms of image synthesis, than previous state-of-the-art models [3]. They have a simple ℓ_2 objective leading to stable training dynamics and can handle huge amounts of data. The models also offer flexibility in guiding the generation process and conditioning on observations [17]. A DDPM learns the parameters of a score network [23] approximating the gradient of the log probability density. This score network guides the data generation process. As further motivation for our work, we note that this is comparable to the EFE that guides action selection. Hence, the probability of choosing an active inference policy based on the EFE is analogous to the optimization of the data generation paths in a diffusion model [11]. Diffusion priors display good generalisation properties [25] and are currently being developed for a range of linear and nonlinear inverse problems [1,10]. In this work we show for the first time how a diffusion prior can be used in the context of AIF.

1.1 Related Work

In [4] a diffusion model-based approach to active inference is developed as a framework to leverage the expressive power of diffusion models and estimate free energy efficiently. The diffusion process is used to approximate environmental generative models. Score-based guidance provides action selection through expected free-energy minimisation. The key to this process is the ability to condition on actions and observations [17]. A diffusion-based motion predictor is used to forecast vehicle actions and aid decision making under uncertainty within AIF in [15]. The authors show that this approach reduces computational demands and requires less extensive training. Behavioural priors have been shown to be useful for efficient planning in partially observable settings in reinforcement learning [2] and active inference [24].

We extend these ideas to create a diffusion prior for behaviour, with a focus on data-rich domains like human activity recognition, that can generalise to new environments. In this way, we make good use of the generalisation properties that have been observed for diffusion priors in other inference scenarios [16,22,27]. This involves pretraining a DDPM using behavioural data collected from a training environment. We show how actions, observations, and states can be encoded as input to the DDPM. The DDPM learns the probability distribution of the training data, and provides a diffusion prior [10] that can then be used by an agent in a different environment to generate likely future observations and states arising from sequences of actions (policies) to aid planning and decision-making.

The layout of the paper is as follows. Section 2 describes the techniques that we combine and customise in order to create a diffusion prior for AIF planning. AIF is introduced in Sect. 2.1, DDPM in Sect. 2.2, conditioned sampling in Sect. 2.3, planning and EFE in Sect. 2.4 and AIF with diffusion in Sect. 2.5. Section 3 covers experiments and Sect. 4 concludes.

2 Methods

We begin with an overview of our methodology, see Fig. 1. A DDPM is trained on sequences of actions and observations that have been encoded (mapped from observation space to latent space) into a latent representation variable. For our experiments, we employ a simple encoding and decoding scheme. In other settings a machine learned encoding-decoding scheme could be used to reduce dimensionality. The DDPM model learns to denoise the noisy latent variable and enable the generation of new data from noise, see Sect. 2.2. By using the conditioning method in Sect. 2.3, the denoising process is guided by actual observations to generate new data consistent with these observations.

AIF involves the rollout of future actions. EFE is estimated for each rollout and used to choose the next action. This is a computational bottleneck in the AIF pipeline. A key advantage of our work is that the use of a diffusion prior reduces the number of possible rollouts, focusing on those that are most likely. This lowers the computational burden. At step one the rollouts arise from unconditional sampling. At step two and subsequently, the rollouts arise from sampling conditioned on observations. For a visualisation comparing classic AIF with a diffusion prior for AIF see Fig. 2. A more complicated scenario is fully described in Sect. 3.

2.1 Active Inference

In AIF [7,19] we assume that an active inference agent reacts with an environment in discrete timesteps denoted by index i. In each timestep, the environment is in a particular unobservable state $s_i \in \mathcal{S}$. The agent interacts with its environment by choosing action $a_i \in \mathcal{A}$. The effect of action a_i on the environment is modeled by a probability distribution over the previous state $p(s_{i+1}|s_i, a_i)$ known as the *transition model*. The new environment state can only be observed by the agent via a sensor state $o_i \in \mathcal{O}$. The relationship between observable states s_i and observations o_i is modeled by a conditional probability distribution $p(o_i|s_i)$ known as the *observation model*.

2.2 Denoising Diffusion Probabilistic Models

DDPMs are a class of generative models introduced in [14]; see [13] for an accessible introduction. Here, we summarise the key details. The objective of a DDPM is to approximate a generative distribution $p(x)$ from which data have been drawn. This is done by learning a model $q(x)$ that maps pure noise into real data. In our context we define x as a time series over the time horizon τ of sequential actions, observations and states $x_0 = (o_{0:\tau}, s_{0:\tau}, a_{0:\tau})$. We set up a forward diffusion process that gradually adds noise to x_0 turning it into a series

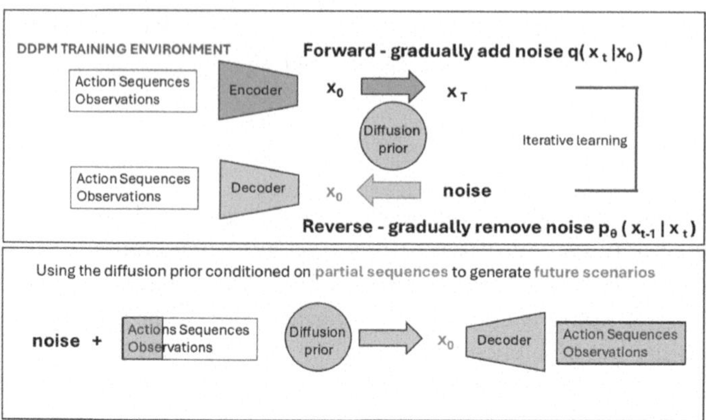

Fig. 1. The DDPM is trained iteratively with sequences of actions and observations from a training environment. Sequences of actions and observations are encoded/decoded into/from a latent space x_o. The aim is to approximate a generative distribution from which data is drawn using reverse sampling from noise with the diffusion prior $p_\theta(x_{t-1}|x_t)$ where θ are model parameters. After training, the diffusion prior takes noise as input to generate new data in a different but similar environment. The denoising process can be guided by partial sequences to produce new data consistent with these observations.

of increasingly noisy latent variables $x_{0:T}$. Note that the timesteps here $0 : T$ relate to the diffusion process and not the action timeline. The idea is to train a DDPM on sequences of actions, $a_{1:\tau}$, observations from sensors $o_{1:\tau}$ and inferred states $s_{1:\tau}$ where τ is a fixed time horizon (i.e. sequence length). Once trained, the DDPM is used as a prior to generate likely future observations and states arising from sequences of actions (policies) to aid planning and decision making. We use the DDPM to provide an approximation of the joint distribution of future observations and states given actual observations and states under a sequence of actions known as a policy π; $p(o_{i+1:i+\tau}, s_{i+1:i+\tau}|o_{1:i}, s_{1:i}, \pi)$. This probability distribution is fundamental to the evaluation of EFE and planning through action.

The diffusion inference process works by sampling a random noise vector x_T and gradually denoising it into an output x_0. With the definition of x_0 above this means that we can generate likely observations and states from the random noise vector. We can then condition on past observations and states, under future actions, to generate a marginal over future observations and states. This marginal plays the role of a posterior prediction over the consequences of future actions. We use these future states to estimate EFE and select actions.

During training, the diffusion process transforms x_0 to white Gaussian noise $x_T \sim \mathcal{N}(0, 1)$ in T timesteps. Each step in the forward direction is given by

$$q(x_t|x_{t-1}) = \mathcal{N}(x_t; \sqrt{1-\beta_t}x_{t-1}, \beta_t I), \tag{1}$$

where $\mathcal{N}(x, \mu, \sigma^2 I)$ denotes the density of a standard Gaussian with mean μ and covariance $\sigma^2 I$ at the point x. Hence the sample x_t is obtained by adding *i.i.d.* Gaussian noise with variance β_t at timestep t after scaling the previous sample x_{t-1} with $\sqrt{1-\beta_t}$ according to a variance schedule. To train the DDPM we need a sample x_t and the corresponding noise that was used to transform x_0 to x_t. We can use the independence property of the noise added at each step to calculate the total noise variance as $\bar{\alpha}_t = \prod_s^t (1 - \beta_s)$ and rewrite as a single step

$$q(x_t|x_0) = \mathcal{N}\left(x_t; \sqrt{\bar{\alpha}_t} x_0, (1 - \bar{\alpha}_t I)\right). \tag{2}$$

The diffusion model reverses this process by modeling with a neural network that predicts the parameters $\mu_\theta(x_t, t)$ and $\Sigma_\theta(x_t, t)$ of a Gaussian distribution,

$$p_\theta(x_{t-1}|x_t) = \mathcal{N}\left(x_{t-1}; \mu_\theta(x_t, t), \Sigma_\theta(x_t, t)\right). \tag{3}$$

The model is trained to predict the cumulative noise that is then removed from the current image, x_t, giving the predicted mean

$$\mu_\theta(x_t, t) = \frac{1}{\sqrt{\alpha}}\left(x_t - \frac{\beta_t}{\sqrt{1 - \bar{\alpha}_t}} \epsilon_\theta(x_t, t)\right). \tag{4}$$

2.3 Conditioned Sampling

Various methods have been proposed to condition the generative process. We use the approach in [17] which has the advantage that conditioning on observations is achieved through the reverse diffusion process only, using a mask. This technique does not modify or condition the original DDPM network. Retraining is not required. The approach can generalise to any mask shape. Hence, we can choose which elements of the data to obscure.

A pretrained DDPM can be used to condition on known elements of x_t to guide noisy x_T back to x_0 in a manner consistent with the known elements. For example, under the proposed joint model, given some observations, noise can be guided back along least action paths to likely states consistent with those observations minimising surprise and hence free energy. This can be achieved using a mask, m. Denoting the ground truth image as x, the unknown elements as $m \odot x$ and the known elements as $(1 - m) \odot x$, since every reverse step from x_t to x_{t-1} depends solely on x_t, we can alter the known regions $(1 - m) \odot x_t$ as long as we keep the correct properties of the corresponding distribution. We can sample the intermediate image x_t at any point in time using $q(x_t|x_0)$ in Eq. (2) for the known region and $p_\theta(x_{t-1}|x_t)$ in Eq. (3) for the unknown region. Thus we have

$$x_{t-1}^{knwn} \sim \mathcal{N}\left(\sqrt{\bar{\alpha}_t}, x_0, (1 - \bar{\alpha}_t I)\right), \tag{5}$$

$$x_{t-1}^{uknw} \sim \mathcal{N}(\mu_\theta(x_t, t), \Sigma_\theta(x_t, t)), \tag{6}$$

$$x_{t-1} = m \odot x_{t-1}^{uknw} + (1 - m) \odot x_{t-1}^{knwn}. \tag{7}$$

2.4 Planning and Expected Free Energy

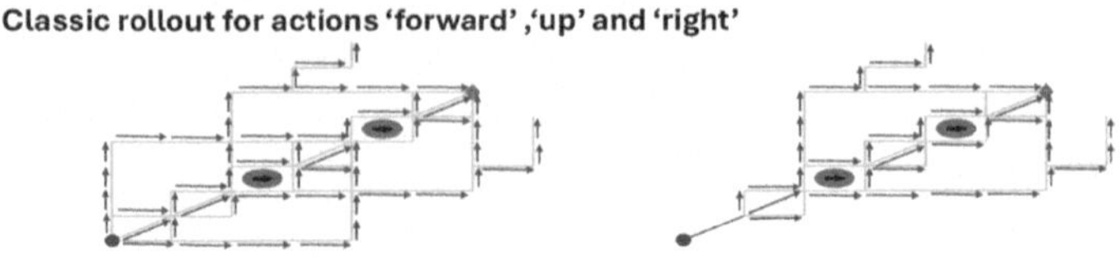

Fig. 2. Classic rollouts at step 1 *top left* and step 2 *top right* for an agent moving 'forward', 'up' or 'right' in an environment containing an obstacle *red circle*. The agent starts at *blue circle* and ends at *red diamond*. Possible paths under these rollouts are indicated in *light blue*. EFE is estimated for each path. The number of rollouts with diffusion prior at step 1 *bottom left* and step 2 *bottom right* is lower, reducing the number of paths and easing the computational burden of estimating EFE. (Color figure online)

The active agent's behaviour is informed by a prior preference distribution $p_i(s_i)$ characterising the distribution over sensor states that the agent would prefer to observe. The agent selects actions by approximating the EFE, $G(\pi)$, over a range of alternative sequences of actions π with fixed time horizon τ. EFE can be approximated by

$$G(\pi) \approx \frac{1}{\tau}[\Sigma_{k=i+1}^{i+\tau}\mathbb{E}_{q(s_k|\pi)}\left[D_{KL}(q(s_k) \parallel p_k(s_k))\right] - \mathbb{E}_{q(s_k,o_k|\pi)}[\ln(p(o_k|s_k))]], \quad (8)$$

where $q(s_k, o_k)$ is a marginal distribution over states and outcomes at the time point k in the future. This marginal can be evaluated from the diffusion prior $q(x,t)$ in Eq. (3), recalling that $q(x_k) = q(o_k, s_k, a_k)$. These marginals play the role of the predictive posteriors, given a sequence of actions up until time point k. In other words, they are hypothetical beliefs the agent anticipates holding in the future. The agent defines a probability distribution $p(\pi)$ over policies using softmax over negative EFE and samples from this distribution to choose the next action:

$$p(\pi) = \frac{\exp(-G(\pi))}{\Sigma_{\pi'}\exp(-G(\pi'))}. \quad (9)$$

Algorithm 1. Active Inference Planning with a Diffusion Prior

```
Set starting state and preference
for i = 0...T_a do                                  ▷ for each step in action timeline
    if i = 0 then
        ε ~ N(0, I)
        {π_1, ..., π_K} ~ q(ε)                      ▷ generate policies from diffusion prior
    else if i > 0 then
        {π_1, ..., π_K} ~ q(ε, o_0 ... o_{i-1})     ▷ generate from conditional diffusion prior
    end if
    for each element k ∈ K do
        Compute EFE G(π_k) for policy π_k
    end for
    Define probability of each policy p(π)
    Sample π : (a_i1 : a_iT) ~ p(π)
    Perform action a_i1 and update o_i
end for
```

2.5 Active Inference Planning with a Diffusion Prior

The agent interacts with the new environment by executing the following steps in a closed feedback loop, as summarised in Algorithm 1. The agent is equipped with a preference over observations/states and a prior belief as to its starting state. The diffusion prior generates K sequences of states, observations and actions. Each sequence is indexed and treated as a policy. EFE is evaluated for each policy and used to define the probability of each policy. The next policy is sampled from this probability distribution. The agent then interacts with the environment to obtain a new observation. The agent computes the observation/state prior for the next time step.

3 Experiments

We describe three experiments. The first shows an agent taking a direct path, avoiding obstacles towards a target. This is a straightforward task with little uncertainty, but which illustrates the ability of the agent to reduce the number of possible paths under consideration to those most likely given observations. The second experiment shows an agent learning the order in which to perform tasks to reach a desired state. Here, uncertainty arises over the order position. For example, tasks may correspond to the problem of turning on a sequence of switches in an appropriate order. The generation of policies by the diffusion model is computationally expensive but reduces the number of policies to be considered, especially as the dimensions of the state space increase. In the third experiment, we estimate these costs.

3.1 Experiment 1: Agent Aims to Take Direct Path to Target

We first illustrate the advantage of a DDPM prior. Diffusion is used to generate likely future data from noise and then to generate marginals over the conse-

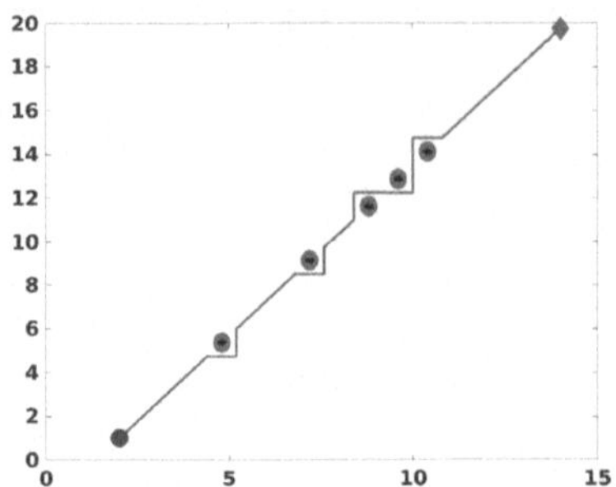

(a) **Training Environment.** Behavioural data is collected from a simulated environment where an agent moves from a starting position (*blue circle*) directly towards a given target (*red diamond*) avoiding obstacles. In this environment there are two types of object indicated by a *right arrow* and a *left arrow*. The agent can move *forward*, *right+up* or *up+right*. The agent can perceive whether there is an object ahead. The *right arrow* object is avoided by moving *right* then *up*. The *left arrow* object is avoided by moving *up* then *right*.

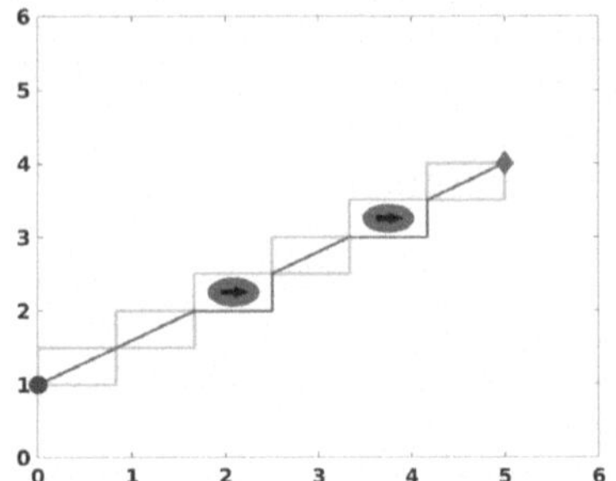

(b) **New environment** with start *blue circle* at (0,1) and target *red diamond* at (5,4). There are two left arrow obstacles in this environment. Path taken by agent *red line.* Paths from low EFE subspace of all possible paths considered by agent when estimating EFE and choosing next action *blue line.*

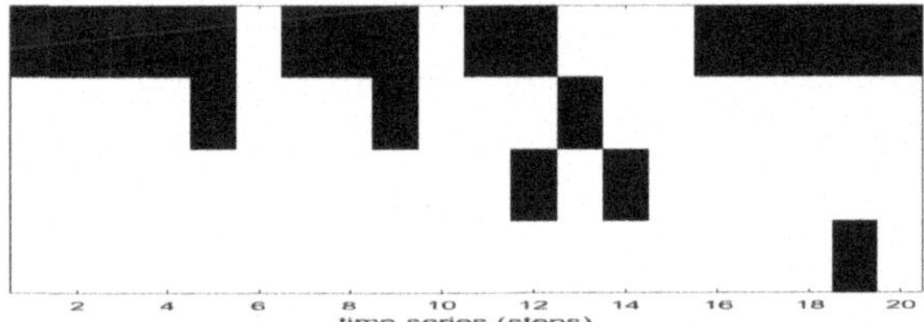

(c) Actions and observations are encoded as time-series in a binary matrix format. Action *forward* is indicated by a '1' in *row 1.* Action *right+up* is indicated by a '1' in *row 2* followed by a '0' in *row 1.* Action *up+right* is indicated by a '1' in *row 3* followed by a '0' in *row 1.* Action *target reached* is indicated by a '1' in *row 4* followed by a '0' in *row 1.* The DDPM is trained on 4×4 windows of this timeseries and learns to generate future configurations over the time period $\tau = 4$.

Fig. 3. Experiment 1: Agent aims to take direct path to target.

quences of likely actions conditioned upon the previous actions and observations. Future scenarios are generated using reverse diffusion sampling with $q(x_{i:i+\tau}|x_i)$ and used with $\tilde{p}(o, s)$ to estimate EFE as in Eq. (8). The agent's probability distribution over policies, using Eq. (9), is updated at each step informing the next action.

To train the DDPM, behavioural data is collected from a simulated environment where an agent follows deterministic rules to move directly towards a given target avoiding obstacles. This is illustrated in Fig. 3a. In this environment, there are two types of objects indicated by a right arrow and a left arrow. The agent can move forward, right+up or up+right. After moving one step, the agent can perceive whether or not there is an object ahead. The right arrow is avoided by moving right then up. The left arrow is avoided by moving up then right. To make the imagined scenarios invariant to training path length, we trained with small windows of sequential observations akin to space invariant windows of convolutional layers [9]. Recorded data comprises sequences of moves, and start, object, target positions. The encoded sequence is shown in Fig. 3c. Snapshots of this record are encoded in tensor format, $D \in \mathcal{R}^{W,H,C}$, where W, H and C denote the size of the time window allowable actions and locations. D then forms the input to the DDPM model. The agent bases its actions on future scenarios 3 steps ahead, which is adequate for this exercise. Once an action has been selected and observations made, the agent can condition on these observations and imagine future scenarios a further 3 steps ahead. The agent continues in this manner until it reaches the target.

In our example, columns of D denote the time window $W = \tau$, rows denote action, object right arrow, object left arrow, target $H = 4$ and the position dimension $C = 1$. The DDPM standard architecture comprises 8 residual blocks. The model was trained using an NVIDIA GeForce RTX 3090 GPU and converges to a solution in about 12 min. The objective is to learn the paramenters of μ_θ in Eq. (4) so that reverse sampling can be performed using $p_\theta(x_{i-1}|x_i)$ from Eq. (3). The resulting path chosen by the agent in a new environment is shown in Fig. 3b. We see that the number of possible paths is reduced to the most likely and by implication the paths with the lowest EFE. The results show that the agent finds a direct path to the target by imagining the most likely states 3 steps ahead given its current state. It uses these imagined states to estimate EFE and chooses the next direct step avoiding obstacles accordingly.

3.2 Experiment 2: Agent Learns Task Order

We investigate an environment containing four sensors, (a, b, c, d), initially in 'off' position that can be turned 'on' by the agent. However sensor 'b' can only be turned on if sensor 'a' is on, sensor 'c' can only be turned on if sensor 'b' is on and sensor 'd' can only be turned on if sensor 'c' is on. An agent exploring this environment will discover 6 possible sensor positions after 4 actions without replacement (24 possible policies). These possible configurations are encoded in a 4×4 latent space representation, see Fig. 4a, and used to train a DDPM. The agent is now faced with a new environment where there is uncertainty over the role of 'b' and 'c'. The agent's preference is for all sensors to be 'on'. The diffusion prior is used at each step to generate 100 images. After '1' is switched 'on' future possible policies are reduced to $\{abcd, abdc, acbd, acdb\}$. The agent evaluates these hypothetical policies, see Fig. 4band 4c, using the diffusion prior to approximate the data distribution and to use samples from this distribution to estimate EFE as in Eq. (8). The agent is seen to eliminate policies $\{acbd, acdb\}$ and infer that policy $\{abcd\}$ aligns best with their preference to reach the all 'on' state.

(a) The six possible encoding configurations of actions, observations and states arising in the training environment after 4 steps. The data is ordered with the action sequence encoded in the first row and sensors b:d in the subsequent rows. Positions 'on' and 'off' are indicated in *yellow* and *blue* respectively. The preferred final state in this task, all 'on', is the first configuration with an all *yellow right hand column*.

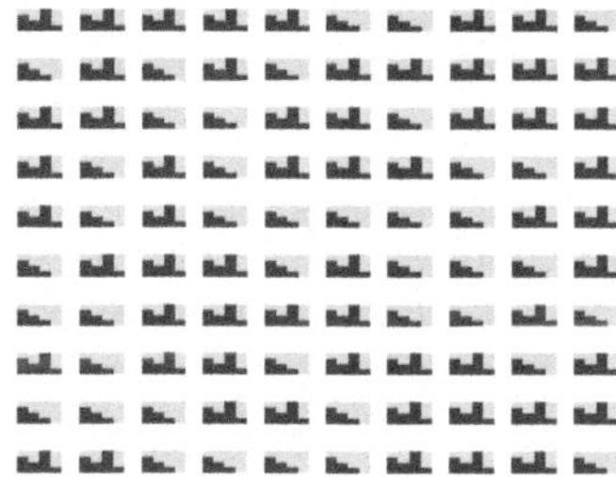

(b) 100 images generated by diffusion prior conditioned on 'ab'. Under this policy $p = [0, 0, .59, .41]$ and EFE=2.4.

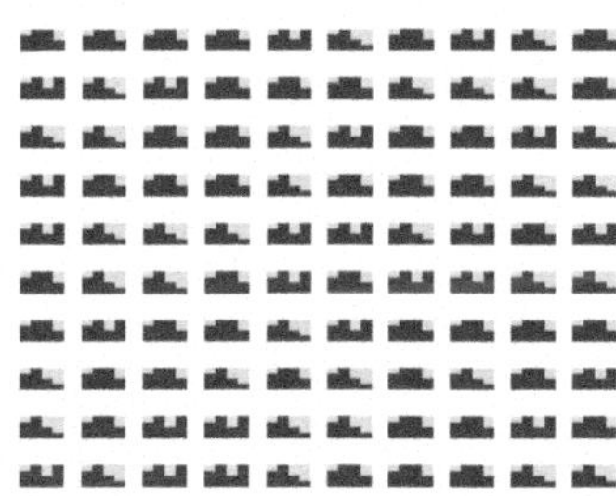

(c) 100 images generated by diffusion prior conditioned on 'ac'. Here $p = [0, .67, .33, 0]$ and EFE=21.3.

Fig. 4. Experiment 2. Training images (a). Generated images (b) and (c). The probability distribution over how many states are on after 4 steps, p, quantifies the likelihood of reaching the preferred state (all 'on') and is visualised above in the percentage of images all 'on' which is $\approx$ 40% for 'ab' *yellow right hand column* and $\approx$ 0% for 'ac' where the preferred state is not reached.

3.3 Experiment 3: Estimated Costs

We design an experiment to compare the cost of a trial between an agent with a diffusion prior (DP) and an agent without a diffusion prior (nDP) in a grid world environment, tasked with finding its way to a target with a time horizon of 4 and possible actions 'left', 'right', 'up' and 'down'. We assume that the DP agent has observed successful policies in training (based on actions 'right' and 'left' only), and thus generates a smaller number of policies (16 with DP instead of 256 with nDP) and expected trial duration. We also consider the scenario where uncertainty about future events necessitates the generation of 100 policies

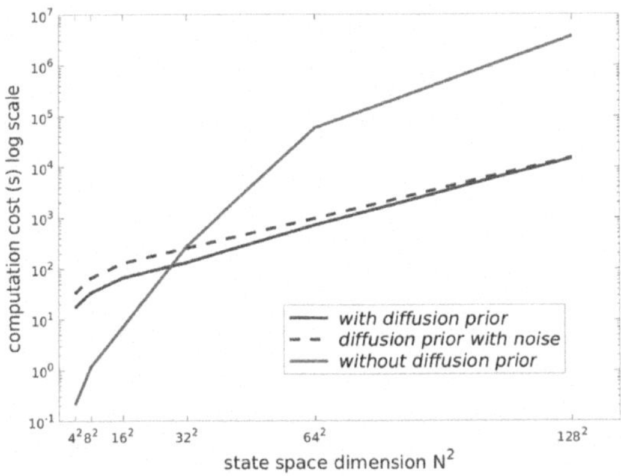

Fig. 5. Experiment 3. The cost of a trial with an expected number of steps with a diffusion prior *solid blue line* and without a diffusion prior *solid red line*. The cost with a diffusion prior is expected to be lower than the cost without a diffusion prior when the state space reaches a dimension of 32^2. Allowing for noise increases the cost with a diffusion prior *dashed blue line* only slightly.

(DP with noise). Under these assumptions, the estimated overall cost of the DP agent (both DP and DP with noise) is lower than that of the nDP agent, when the dimension of the state space exceeds 32^2, as shown in Fig. 5.

4 Discussion

This paper shows proof of concept that an agent can learn planning policies for new environments using a diffusion prior. In this work, we assume that appropriate time series data is available for training and that the agent can observe useful behaviour in the training environment. Actions, observations and states are encoded as a latent variable to form the training input to the DDPM. We learn the probability distribution on the latent variable and use noise and reverse sampling to generate likely future scenarios in a new environment. Using a conditioning method that combines noise with actual observations allows us to guide generation towards future scenarios consistent with those observations.

Clearly, the benefits afforded by diffusion priors over the consequences of action depend on the quality of training behaviour observed and the generalisation to new environments or test scenarios. However, there is some degree of flexibility in our application of these diffusion priors because we are still searching over multiple policies (which, a priori, are the most likely given the training data.)

A diffusion prior provides several advantages for AIF. By generating the most likely scenarios, the number of rollouts to be considered in the EFE calculation is lowered. This alleviates a key bottleneck for active inference. The training method is iterative and lends itself to continual learning [4]. The ability to condition on actual observations without retraining is an attractive feature of our method, making the approach applicable in a range of downstream tasks.

The iterative nature of the learning algorithm for the DDPM means that it can fix or improve previous denoising steps many times as it moves closer to the true data distribution. This creates interesting directions for future research. In particular, developing a dynamic diffusion prior that adapts to new actions and observations arising in the environment would allow an agent to adjust to a new environment with differences not captured by the training data. We also note that the DDPM architecture is flexible and well-studied. There are many pre-trained DDPMs for multi-modal data, offering exciting directions for diffusion priors. For example, diffusion noise has recently been proposed as a universal motion prior [16]. Data encoding can be combined with powerful features, such as a self attention transformer. Diffusion priors may therefore be developed for a range of behaviours. They are particularly appropriate for high dimensional data that can be usefully summarized by low dimensional abstraction. Methods for accelerating diffusion sampling with training-free methods are under development [26]. However, the computational reduction in EFE estimation has to be balanced against any additional computation due to reverse sampling with a diffusion prior. The benefit of leveraging a diffusion prior for active inference planning will be most noticeable in the widely occurring scenarios where preference states have a concentrated density in the full state space domain. Also, as Experiment 3 shows, benefits will arise when the dimension of the state space is high.

Our method will be especially useful for time series from sensors in the area of human activity recognition [8,12,18,21], in particular, for the emerging quantum sensor technologies in healthcare[1] such as brain imaging [5]. Time series data can be collected rapidly in large quantities from sensors. This data lends itself to being modeled with low dimensional abstraction. The diffusion prior could encapsulate correlations between activities over a time period such as a week and generate likely scenarios arising given observations, allowing the agent to plan ahead over a useful time frame. Future work will focus on this interface, integrating efficient diffusion and latent space reduction into the AIF pipeline.

Acknowledgments. C.F.H and R.M-S. received funding from EPSRC projects EP/T021020/1 and EP/T00097X/1. also received funding from the *Designing Interaction Freedom via Active Inference (DIFAI)* ERC Advanced Grant (proposal 101097708, funded by the UK Horizon guarantee scheme as EPSRC project EP/Y029178/1).

Data Availability. Data is available at http://dx.doi.org/10.5525/gla.researchdata.2049.

Disclosure of Interests. The authors have no competing interests.

[1] https://iuk-business-connect.org.uk/perspectives/explore-quantums-potential-in-healthcare/.

References

1. Cheng, J., Tan, S.: Diffusion priors for variational likelihood estimation and image denoising. In: Globerson, A., et al. (eds.) Advances in Neural Information Processing Systems, vol. 37, pp. 61138–61159. Curran Associates, Inc. (2024), https://proceedings.neurips.cc/paper_files/paper/2024/file/7087c949df293f13c0052ac825936e6f-Paper-Conference.pdf
2. Deisenroth, M.P., Rasmussen, C.E.: PILCO: a model-based and data-efficient approach to policy search. In: Proceedings of the 28th International Conference on International Conference on Machine Learning, ICML 2011, pp. 465–472. Omnipress, Madison, WI, USA (2011)
3. Dhariwal, P., Nichol, A.: Diffusion models beat gans on image synthesis. In: Ranzato, M., Beygelzimer, A., Dauphin, Y., Liang, P., Vaughan, J.W. (eds.) Advances in Neural Information Processing Systems, vol. 34, pp. 8780–8794. Curran Associates, Inc. (2021), https://proceedings.neurips.cc/paper_files/paper/2021/file/49ad23d1ec9fa4bd8d77d02681df5cfa-Paper.pdf
4. Durand, E.A., Joffily, M., Khamassi, M.: A diffusion model-based approach to active inference. In: 2024 IEEE International Conference on Future Machine Learning and Data Science (FMLDS), pp. 75–80 (2024). https://doi.org/10.1109/FMLDS63805.2024.00024
5. Faccio, D.: The future of quantum technologies for brain imaging. PLOS Biol. **22**(10), 1–4 (10 2024). https://doi.org/10.1371/journal.pbio.3002824, https://doi.org/10.1371/journal.pbio.3002824
6. Friston, K., Kilner, J., Harrison, L.: A free energy principle for the brain. Journal of Physiology-Paris **100**(1), 70–87 (2006). https://doi.org/10.1016/j.jphysparis.2006.10.001, https://www.sciencedirect.com/science/article/pii/S092842570600060X, theoretical and Computational Neuroscience: Understanding Brain Functions
7. Friston, K., Mattout, J., Kilner, J.: Active understanding and active inference. Biol. Cybern. **104**, 137–167 (2011)
8. Fu, B., Damer, N., Kirchbuchner, F., Kuijper, A.: Sensing technology for human activity recognition: a comprehensive survey. IEEE Access **8**, 83791–83820 (2020)
9. Goodfellow, I., Bengio, Y., Courville, A.: Deep Learning. The MIT Press (2016)
10. Graikos, A., Malkin, N., Jojic, N., Samaras, D.: Diffusion models as plug-and-play priors. In: Thirty-Sixth Conference on Neural Information Processing Systems (2022). https://arxiv.org/pdf/2206.09012.pdf
11. Grigorev, V.: Generating minimum free energy paths with denoising diffusion probabilistic models (2024). https://arxiv.org/abs/2412.10409
12. Gu, F., Chung, M.H., Chignell, M., Valaee, S., Zhou, B., Liu, X.: A survey on deep learning for human activity recognition. ACM Comput. Surv. (CSUR) **54**(8), 1–34 (2021)
13. Higham, C., Higham, D.J., Grindrod, P.: Diffusion models for generative artificial intelligence: an introduction for applied mathematicians. SIAM Rev. **67**(3), 607–623 (2025). https://doi.org/10.1137/23M1626232
14. Ho, J., Jain, A., Abbeel, P.: Denoising diffusion probabilistic models. In: Proceedings of the 34th International Conference on Neural Information Processing Systems. Curran Associates Inc., Red Hook, NY, USA (2020)
15. Huang, Y., Li, Y., Matta, A., Jafari, M.: Navigating autonomous vehicle on unmarked roads with diffusion-based motion prediction and active inference. ArXiv abs/ arXiv: 2406.00211 (2024), https://api.semanticscholar.org/CorpusID:270216090

16. Karunratanakul, K., Preechakul, K., Aksan, E., Beeler, T., Suwajanakorn, S., Tang, S.: Optimizing diffusion noise can serve as universal motion priors (2024). https://arxiv.org/abs/2312.11994
17. Lugmayr, A., Danelljan, M., Romero, A., Yu, F., Timofte, R., Van Gool, L.: Repaint: inpainting using denoising diffusion probabilistic models. In: 2022 IEEE/CVF Conference on Computer Vision and Pattern Recognition (CVPR), pp. 11451–11461 (2022). https://doi.org/10.1109/CVPR52688.2022.01117
18. Martínez-Hernández, I., Killick, R.: Changepoint detection on daily home activity pattern: a sliced poisson process method. Biometrics **80**(4), ujae114 (2024)
19. Murray-Smith, R., Williamson, J.H., Stein, S.: Active inference and human–computer interaction (2024). https://arxiv.org/abs/2412.14741
20. Oliver, G., Lanillos, P., Cheng, G.: An empirical study of active inference on a humanoid robot. IEEE Trans. Cognitive Developm. Syst. **14**(2), 462–471 (2022). https://doi.org/10.1109/TCDS.2021.3049907
21. Rousseau, T., Venture, G., Hernandez, V.: Latent space representation of human movement: assessing the effects of fatigue. Sensors **24**(23) (2024). https://doi.org/10.3390/s24237775, https://www.mdpi.com/1424-8220/24/23/7775
22. Rozet, F., Andry, G., Lanusse, F., Louppe, G.: Learning diffusion priors from observations by expectation maximization. In: Globerson, A., et al. (eds.) Advances in Neural Information Processing Systems, vol. 37, pp. 87647–87682. Curran Associates, Inc. (2024), https://proceedings.neurips.cc/paper_files/paper/2024/file/9f94298bac4668db4dc77ddb0a244301-Paper-Conference.pdf
23. Song, Y., Sohl-Dickstein, J., Kingma, D.P., Kumar, A., Ermon, S., Poole, B.: Score-based generative modeling through stochastic differential equations. In: International Conference on Learning Representations (2021). https://openreview.net/forum?id=PxTIG12RRHS
24. Tschantz, A., Baltieri, M., Seth, A.K., Buckley, C.L.: Scaling active inference. In: 2020 International Joint Conference on Neural Networks (IJCNN), pp. 1–8 (2020). https://doi.org/10.1109/IJCNN48605.2020.9207382
25. Wang, J., Yue, Z., Zhou, S., Chan, K.C.K., Loy, C.C.: Exploiting diffusion prior for real-world image super-resolution. Int. J. Comput. Vision **132**(12), 5929–5949 (2024). https://doi.org/10.1007/s11263-024-02168-7
26. Xue, S., et al.: Accelerating diffusion sampling with optimized time steps. In: Proceedings of the IEEE/CVF Conference on Computer Vision and Pattern Recognition (CVPR), pp. 8292–8301 (June 2024)
27. Zhang, H., et al.: The emergence of reproducibility and consistency in diffusion models. In: Salakhutdinov, R., Kolter, Z., Heller, K., Weller, A., Oliver, N., Scarlett, J., Berkenkamp, F. (eds.) Proceedings of the 41st International Conference on Machine Learning. Proceedings of Machine Learning Research, 21–27 Jul, vol. 235, pp. 60558–60590. PMLR (2024), https://proceedings.mlr.press/v235/zhang24cn.html

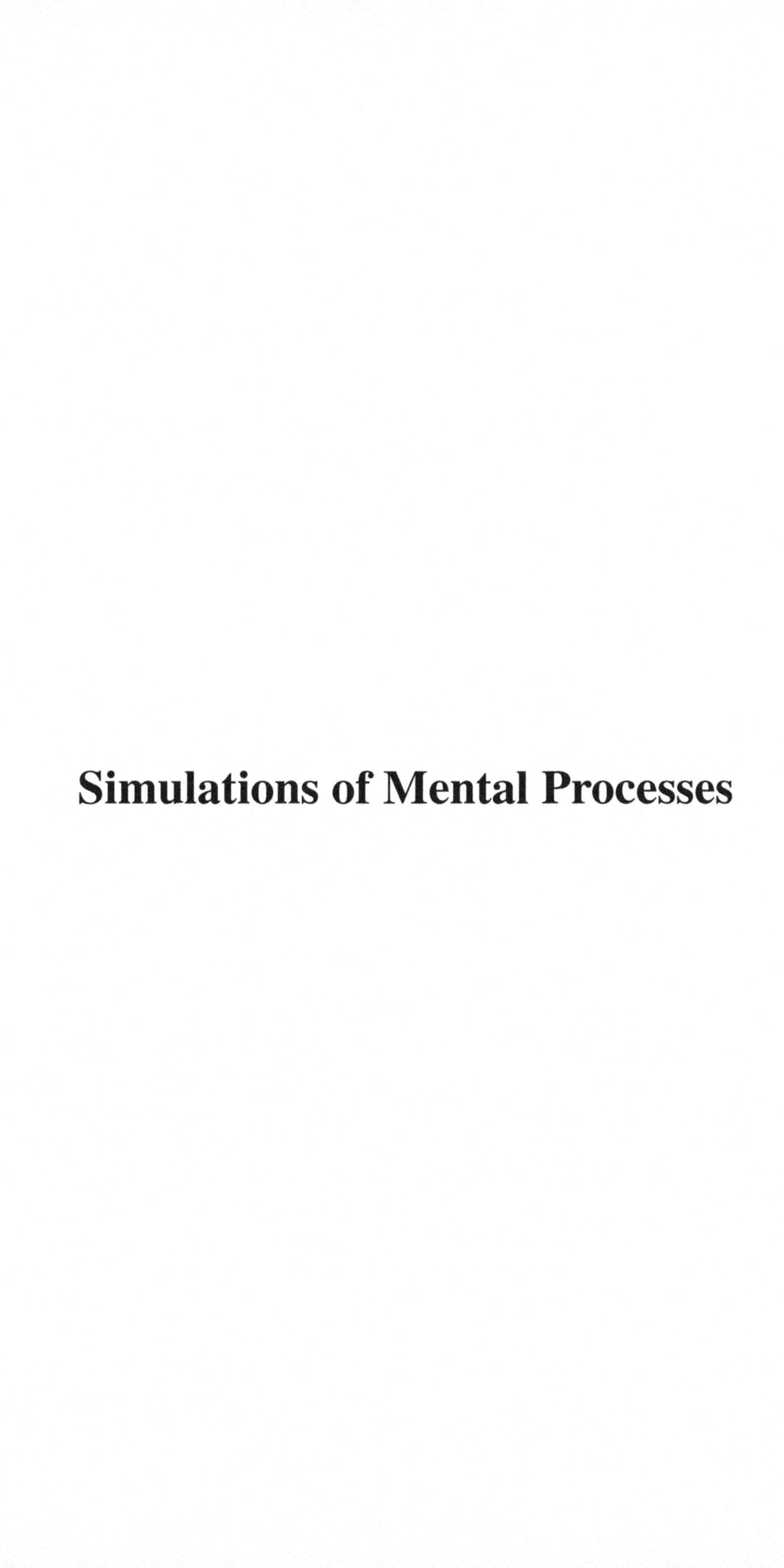

Simulations of Mental Processes

An Active Inference Model of Covert and Overt Visual Attention

Tin Mišić, Karlo Koledić, Fabio Bonsignorio, Ivan Petrović, and Ivan Marković(✉)

University of Zagreb Faculty of Electrical Engineering and Computing, Zagreb, Croatia
ivan.markovic@fer.hr

Abstract. The ability to selectively attend to relevant stimuli while filtering out distractions is essential for agents that process complex, high-dimensional sensory input. This paper introduces a model of covert and overt visual attention through the framework of active inference, utilizing dynamic optimization of sensory precisions to minimize free-energy. This work addresses the lack of active inference models that integrate visual attention with continuous sensory representations and deep generative models for robotics. Our proposed model determines visual sensory precisions based on both current environmental beliefs and sensory input, influencing attentional allocation in both covert and overt modalities. To test the effectiveness of the model, we analyze its behavior in the Posner cueing task and a simple target focus task using two-dimensional (2D) visual data. Reaction times are measured to investigate the interplay between exogenous and endogenous attention, as well as valid and invalid cueing. The results show that exogenous and valid cues generally lead to faster reaction times compared to endogenous and invalid cues. Finally, we show that reflexive saccades are faster than intentional ones, though less adaptable, and discuss the implications for robotic applications.

Keywords: Active inference · Visual attention · Posner cueing task

1 Introduction

Attention as a cognitive process allows agents to selectively focus on specific stimuli while ignoring others. This ability helps humans avoid sensory overload, and as robots acquire more complex sensory channels it could help decrease the computational load required to perform in daily tasks, such as object tracking and visual search, as well as social interactions [20–22]. Attention is often separated into top-down, or goal-driven attention, and bottom-up or stimulus-driven attention, with some theories including hysteresis as a third component [38]. Top-down attention bilaterally activates dorsal posterior parietal and frontal regions of the brain, while bottom-up attention activates the right-lateralized ventral system, with the dorsal frontoparietal system combining the two into

M. Albarracin et al. (Eds.): IWAI 2025, CCIS 2857, pp. 167–181, 2026.
https://doi.org/10.1007/978-3-032-16955-6_10

a “salience map” during visual search [6,23]. Furthermore, visual attention is separated into overt and covert attention [1,18], with overt attention involving saccadic eye movements to the attentional target, and covert attention referring to attention shifts to the target while the eyes remain fixated elsewhere.

Visual attention and its models are most often tested using the Posner cueing task, i.e., the Posner paradigm. The Posner cueing task is an experimental paradigm used to study covert visual attention [33,34]. Participants are asked to fixate on a central point while a cue directs attention to a location where a target may appear. The cue can either be endogenous – meaning that attention is voluntarily guided based on symbolic cues (e.g., an arrow pointing left or right), or exogenous – meaning that attention is automatically drawn by a sudden, peripheral stimulus (e.g. a bright flash or a flickering box). Endogenous cueing is considered to be top-down because it requires cognitive processing and active interpretation of the cue, while exogenous cueing is considered to be bottom-up because it does not require conscious interpretation. Reaction times are measured to assess how cues influence attentional shifts.

Through the original Posner paradigm [33,34] and its variations, valuable insights have been gained about attentional processes. Covert attentional shifts to a target area occur prior to any eye movement [31,33], and valid cues produce faster responses than invalid cues [33,34]. Exogenous cues were shown to produce faster reaction times than endogenous cues [4,13], showing that bottom-up attention is faster because it requires no conscious processing. The question of weather attentional selection is object-based or location-based has also been thoroughly researched, and the consensus is that both types are not mutually exclusive, but are dependent on the current task [7,37,43]. Research supporting location-based attention has shown that target eccentricity, i.e. the target distance from the central focus point, plays a role in reaction time, with reaction times increasing as target eccentricity increased [2,17,32].

Multiple approaches exist to model attention, many of which are based on Bayesian inference [9,12,24,26–28,30,36,42]. Previous studies have modeled visual attention and active saccades in visual search tasks [9,24,26,27]. However, the integration of visual attention and bottom-up action within the active inference framework–operating directly on raw two-dimensional visual input–remains largely unexplored. This gap is especially significant in robotics, where vision is a core sensory modality and visual data provide the primary basis for decision-making and interaction with the environment. A key limitation of many existing models [24,26,27,36] is their reliance on discrete sensory inputs or internal states. While suitable for abstract tasks, such discretization is problematic for robotics, where sensory and motor variables are inherently continuous. Treating continuous signals as discrete reduces precision and limits applicability in real-world scenarios. By contrast, continuous representations naturally reflect the analog nature of sensorimotor data and enable more accurate perception, motion control, and sensorimotor learning [5,8,41]. In addition, several of these models either omit deep generative models altogether [9,24,26,27] or, in cases where 2D data are used [36], do not leverage deep architectures. Conversely, active inference implementations that employ deep generative models, such as [3,35], do not address visual attention. Overall, prior work has not yet integrated visual attention, continuous sensory representations, and deep generative models into a single active inference framework suitable for robotics.

In this paper we propose a model of visual attention, shown in Fig. 1, viewed through the lens of active inference [29] – a computational approach derived from the free-energy principle (FEP). According to the FEP, systems adapt and act in a way that minimizes their free-energy [11]. Free-energy is a concept borrowed from physics, statistics, and information theory that limits the surprise on a sample of data given a generative model. This principle helps to explain how biological systems resist the natural tendency to disorder, and their action, perception, and learning processes [10]. In the FEP, attention is theoretically achieved by optimizing sensory precisions, their parameters, and mutual precision weighing [9,14,24,26–28,30]. Biased competition and endogenous/exogenous attention have been studied in this context, and the precision optimization produces behaviors similar to human attention [9,42].

This study introduces a hierarchical active inference model of overt and covert visual attention, addressing precision optimization for visual data as a mechanism for endogenous and exogenous attention as well as action control. The model integrates both top-down and bottom-up processes, enabling covert and overt shifts of attention. Its performance is demonstrated through the Posner cueing task and a simple target-focus task on two-dimensional visual input. The primary goal is to develop a model that accurately captures human attentional mechanisms while establishing a foundation for robotic applications, including active visual search and joint attention in humanrobot interaction. To address previously mentioned limitations, we propose a hierarchical active inference model that incorporates visual attention mechanisms within a deep generative framework, using continuous internal states and raw two-dimensional visual data as input. A variational auto-encoder (VAE) serves as the visual generative model, while model training and experiments are conducted in the Gazebo simulator under the Robot Operating System (ROS).

The paper is organized as follows. In Sect. 2 we give an overview of the theoretical background and elaborate the proposed approach that is based on free-energy minimization with 2D precision optimization and overt saccades through active inference. Section 3 shows the results of the Posner cueing tasks and active attention trials. Section 4 provides the discussion of the results while Sect. 5 concludes the paper and provides directions for future work.

2 Proposed Method

2.1 Free-Energy Minimization

Free-energy is defined as the negative evidence lower bound (ELBO), or as the sum of the Kullback-Leibler (KL) divergence and the surprise [9–11]:

$$F(\boldsymbol{z}, \boldsymbol{s}) = -\mathcal{L}(q) = D_{KL}[q(\boldsymbol{z})||p(\boldsymbol{z}|\boldsymbol{s})] - \ln p(\boldsymbol{s}), \tag{1}$$

where $\boldsymbol{z}$ and $\boldsymbol{s}$ represent latent system states and sensory observations, respectively, while the KL-divergence is computed between the posterior $p(\boldsymbol{z}|\boldsymbol{s})$ and the approximate variational density $q(\boldsymbol{z})$. Given that, the surprise is defined as the negative log-probability of an outcome $-\ln p(\boldsymbol{s})$. If the variational density $q(\boldsymbol{z})$ is assumed to factor into Gaussian probability density functions (pdfs) [9,11,35]:

$$q(\boldsymbol{z}) = \prod_i q(\boldsymbol{z}_i) = \prod_i \mathcal{N}(\boldsymbol{\mu}_i, \boldsymbol{\Pi}_i^{-1}), \tag{2}$$

the free-energy then becomes dependent only on the most probable hypotheses, beliefs $\boldsymbol{\mu}_i$, and precision matrices $\boldsymbol{\Pi}_i$ of the latent system states $\boldsymbol{z}$ [9,35]:

$$\begin{aligned} F(\boldsymbol{\mu}, \boldsymbol{\Pi}, \boldsymbol{s}) &= -\ln p(\boldsymbol{s}, \boldsymbol{\mu}, \boldsymbol{\Pi}) + C \\ &= -\ln p(\boldsymbol{s}|\boldsymbol{\mu}, \boldsymbol{\Pi}) - \ln p(\boldsymbol{\mu}, \boldsymbol{\Pi}) + C \end{aligned} \tag{3}$$

Furthermore, sensory observations $\boldsymbol{s}$ and beliefs $\boldsymbol{\mu}$ are defined in the context of hierarchical dynamic models [9–11,35]:

$$\begin{aligned} \tilde{\boldsymbol{s}} &= \tilde{\boldsymbol{g}}(\tilde{\boldsymbol{\mu}}) + \boldsymbol{w}_s \\ D\tilde{\boldsymbol{\mu}} &= \tilde{\boldsymbol{f}}(\tilde{\boldsymbol{\mu}}) + \boldsymbol{w}_\mu. \end{aligned} \tag{4}$$

Here, $\tilde{\boldsymbol{\mu}}$ indicates generalized coordinates of beliefs with multiple temporal orders, $\tilde{\boldsymbol{\mu}} = \{\boldsymbol{\mu}, \boldsymbol{\mu}', \boldsymbol{\mu}'', \cdots\}$, which allow for a richer approximation of the environment dynamics, D stands for the differential shift operator $D\tilde{\boldsymbol{\mu}} = \{\boldsymbol{\mu}', \boldsymbol{\mu}'', \cdots\}$ in the generalized equation of system dynamics $\tilde{\boldsymbol{f}}(\tilde{\boldsymbol{\mu}})$, while $\tilde{\boldsymbol{g}}(\tilde{\boldsymbol{\mu}})$ is the sensor model that maps current beliefs to sensory observations. The amplitudes of random fluctuations $\boldsymbol{w}_s$ and $\boldsymbol{w}_\mu$ are state dependent and are defined as Gaussian pdfs with covariances $\boldsymbol{\Sigma}_s$ and $\boldsymbol{\Sigma}_\mu$, respectively [9,35]:

$$\begin{aligned} \boldsymbol{w}_s &\sim \mathcal{N}(\boldsymbol{\mu}_i, \boldsymbol{\Sigma}_s(\boldsymbol{z}, \boldsymbol{s}, \boldsymbol{\gamma})) \\ \boldsymbol{w}_\mu &\sim \mathcal{N}(\boldsymbol{\mu}_i, \boldsymbol{\Sigma}_\mu(\boldsymbol{z}, \boldsymbol{s}, \boldsymbol{\gamma})). \end{aligned} \tag{5}$$

The covariances $\boldsymbol{\Sigma}_i$ are the inverses of precisions, $\boldsymbol{\Sigma}_i := \boldsymbol{\Sigma}_i(\boldsymbol{z}, \boldsymbol{s}, \boldsymbol{\gamma}) = \boldsymbol{\Pi}_i(\boldsymbol{z}, \boldsymbol{s}, \boldsymbol{\gamma})^{-1}$, with precision parameters $\boldsymbol{\gamma}$ that control the amplitudes [9,42]. The precisions are dynamic and depend on the current states and sensory input. It is through optimization of precisions and their parameters that attention is achieved [9,14,24,26–28,30].

2.2 Perceptual and Active Inference

Perception, action, and learning can all be optimized through the minimization of free-energy. In this paper we only consider perception and action, and leave the learning processes of attention for future work. Action and beliefs are optimized through gradient descent [10,11,29,35]:

$$\begin{aligned} \dot{\tilde{\boldsymbol{\mu}}} - D\tilde{\boldsymbol{\mu}} &= -\partial_{\tilde{\mu}} F(\tilde{\boldsymbol{\mu}}, \tilde{\boldsymbol{\Pi}}, \tilde{\boldsymbol{s}}) \\ \dot{\boldsymbol{a}} &= -\partial_{\boldsymbol{a}} F(\tilde{\boldsymbol{\mu}}, \tilde{\boldsymbol{\Pi}}, \tilde{\boldsymbol{s}}). \end{aligned} \tag{6}$$

The likelihood and prior in (3) also become generalized and can be partitioned within and across temporal orders d, respectively [35]:

$$\begin{aligned} p(\tilde{\boldsymbol{s}}|\tilde{\boldsymbol{\mu}}, \tilde{\boldsymbol{\Pi}}_s) &= \prod_d p(\boldsymbol{s}^{[d]}|\boldsymbol{\mu}^{[d]}, {\boldsymbol{\Pi}_s}^{[d]}) \\ p(\tilde{\boldsymbol{\mu}}, \tilde{\boldsymbol{\Pi}}_\mu) &= \prod_d p(\boldsymbol{\mu}^{[d+1]}|\boldsymbol{\mu}^{[d]}, {\boldsymbol{\Pi}_\mu}^{[d]}). \end{aligned} \tag{7}$$

These partitions are also assumed to take the following Gaussian pdf form:

$$\begin{aligned} p(\boldsymbol{s}^{[d]}|\boldsymbol{\mu}^{[d]}, {\boldsymbol{\Pi}_s}^{[d]}) &= \frac{|{\boldsymbol{\Pi}_s}^{[d]}|^{\frac{1}{2}}}{\sqrt{(2\pi)^L}} \exp\left(-\frac{1}{2}{\boldsymbol{e}_s^{[d]}}^T {\boldsymbol{\Pi}_s}^{[d]} \boldsymbol{e}_s^{[d]}\right) \\ p(\boldsymbol{\mu}^{[d+1]}|\boldsymbol{\mu}^{[d]}, {\boldsymbol{\Pi}_\mu}^{[d]}) &= \frac{|{\boldsymbol{\Pi}_\mu}^{[d]}|^{\frac{1}{2}}}{\sqrt{(2\pi)^M}} \exp\left(-\frac{1}{2}{\boldsymbol{e}_\mu^{[d]}}^T {\boldsymbol{\Pi}_\mu}^{[d]} \boldsymbol{e}_\mu^{[d]}\right), \end{aligned} \tag{8}$$

where L and M are the respective dimensions of sensory observations $\boldsymbol{s}$ and internal beliefs $\boldsymbol{\mu}$. Therein, $\boldsymbol{e}_s^{[d]}$ and $\boldsymbol{e}_\mu^{[d]}$ represents sensory and system dynamics prediction errors:

$$\begin{aligned} \boldsymbol{e}_s^{[d]} &= \boldsymbol{s}^{[d]} - \boldsymbol{g}^{[d]}(\boldsymbol{\mu}^{[d]}) = \boldsymbol{s}^{[d]} - \boldsymbol{p}^{[d]} \\ \boldsymbol{e}_\mu^{[d]} &= \boldsymbol{\mu}^{[d+1]} - \boldsymbol{f}^{[d]}(\boldsymbol{\mu}^{[d]}), \end{aligned} \tag{9}$$

where $\boldsymbol{p}^{[d]} = \boldsymbol{g}^{[d]}(\boldsymbol{\mu}^{[d]})$ are sensory predictions generated by the generative sensor model. Note that in our case the system dynamics model is defined through flexible intentions $\boldsymbol{h}^{(k)}$ [35], where for each intention $k \in (0, K-1)$:

$$\boldsymbol{f}^{(k)}(\boldsymbol{\mu}) = l \cdot \boldsymbol{E}_i^{(k)} + \boldsymbol{w}_\mu^{(k)} = l \cdot (\boldsymbol{h}^{(k)} - \boldsymbol{\mu}) + \boldsymbol{w}_\mu^{(k)}, \tag{10}$$

with l being the empirically-determined gain of intention errors $\boldsymbol{E}_i^{(k)}$. This gain could potentially be optimized through learning with the FEP. Flexible intentions $\boldsymbol{h}^{(k)}$ represent top-down attractors which are generated from current beliefs and drive beliefs towards dynamic goals. The implementation of the generative sensor models $\boldsymbol{g}^{[d]}$ is presented in Subsect. 3.1.

Belief Update. With state- and sensory-dependent precisions, the belief update takes the following form:

$$\begin{aligned} \dot{\tilde{\boldsymbol{\mu}}} =& D\tilde{\boldsymbol{\mu}} + \frac{\partial \tilde{\boldsymbol{g}}}{\partial \tilde{\boldsymbol{\mu}}}^T \tilde{\boldsymbol{\Pi}}_s \tilde{\boldsymbol{e}}_s + \frac{\partial \tilde{\boldsymbol{f}}}{\partial \tilde{\boldsymbol{\mu}}}^T \tilde{\boldsymbol{\Pi}}_\mu \tilde{\boldsymbol{e}}_\mu - D^T \tilde{\boldsymbol{\Pi}}_\mu \tilde{\boldsymbol{e}}_\mu \\ &+ \frac{1}{2}\mathrm{Tr}\left[\tilde{\boldsymbol{\Pi}}_s^{-1} \frac{\partial \tilde{\boldsymbol{\Pi}}_s}{\partial \tilde{\boldsymbol{\mu}}}\right] - \frac{1}{2}\tilde{\boldsymbol{e}}_s^T \frac{\partial \tilde{\boldsymbol{\Pi}}_s}{\partial \tilde{\boldsymbol{\mu}}} \tilde{\boldsymbol{e}}_s \\ &+ \frac{1}{2}\mathrm{Tr}\left[\tilde{\boldsymbol{\Pi}}_\mu^{-1} \frac{\partial \tilde{\boldsymbol{\Pi}}_\mu}{\partial \tilde{\boldsymbol{\mu}}}\right] - \frac{1}{2}\tilde{\boldsymbol{e}}_\mu^T \frac{\partial \tilde{\boldsymbol{\Pi}}_\mu}{\partial \tilde{\boldsymbol{\mu}}} \tilde{\boldsymbol{e}}_\mu, \end{aligned} \tag{11}$$

with Tr being the trace of a matrix. The terms that comprise the belief update equation are:

- $\frac{\partial \tilde{g}}{\partial \tilde{\mu}}^T \tilde{\boldsymbol{\Pi}}_s \tilde{\boldsymbol{e}}_s$: likelihood error computed at the sensory level, representing the free-energy gradient of the likelihood relative to the belief $\tilde{\boldsymbol{\mu}}^{[d]}$ in (9)
- $\frac{\partial \tilde{f}}{\partial \tilde{\mu}}^T \tilde{\boldsymbol{\Pi}}_\mu \tilde{\boldsymbol{e}}_\mu$: backward error from the next temporal order, representing the free-energy gradient relative to the belief $\tilde{\boldsymbol{\mu}}^{[d+1]}$ in (9)
- $-D^T \tilde{\boldsymbol{\Pi}}_\mu \tilde{\boldsymbol{e}}_\mu$: forward error coming from the previous temporal order, representing the free-energy gradient relative to the belief $\tilde{\boldsymbol{\mu}}^{[d]}$ in (9)
- $\frac{1}{2}\mathrm{Tr}\left[\tilde{\boldsymbol{\Pi}}_s^{-1}\frac{\partial \tilde{\boldsymbol{\Pi}}_s}{\partial \tilde{\mu}}\right] - \frac{1}{2}\tilde{\boldsymbol{e}}_s^T \frac{\partial \tilde{\boldsymbol{\Pi}}_s}{\partial \tilde{\mu}} \tilde{\boldsymbol{e}}_s$: free-energy gradients from the sensory precisions, serves as bottom-up attention
- $\frac{1}{2}\mathrm{Tr}\left[\tilde{\boldsymbol{\Pi}}_\mu^{-1}\frac{\partial \tilde{\boldsymbol{\Pi}}_\mu}{\partial \tilde{\mu}}\right] - \frac{1}{2}\tilde{\boldsymbol{e}}_\mu^T \frac{\partial \tilde{\boldsymbol{\Pi}}_\mu}{\partial \tilde{\mu}} \tilde{\boldsymbol{e}}_\mu$: free-energy gradients from the system dynamics precisions, serves as top-down attention.

Action Update. Action is also updated through the minimization of free-energy [10,11,29,35]:

$$a = \arg\min_a F(\boldsymbol{\mu}, \boldsymbol{\Pi}, \boldsymbol{s}), \tag{12}$$

with the action update taking the following form:

$$\dot{a} = -\partial_a F(\boldsymbol{\mu}, \boldsymbol{\Pi}, \boldsymbol{s}) = -\frac{\partial \tilde{\boldsymbol{s}}}{\partial a}^T \tilde{\boldsymbol{\Pi}}_s \tilde{\boldsymbol{e}}_s + \frac{1}{2}\mathrm{Tr}\left[\tilde{\boldsymbol{\Pi}}_s^{-1}\frac{\partial \tilde{\boldsymbol{\Pi}}_s}{\partial \tilde{\boldsymbol{s}}}\right]\frac{\partial \tilde{\boldsymbol{s}}}{\partial a} - \frac{1}{2}\tilde{\boldsymbol{e}}_s^T \frac{\partial \tilde{\boldsymbol{\Pi}}_s}{\partial \tilde{\boldsymbol{s}}} \tilde{\boldsymbol{e}}_s \frac{\partial \tilde{\boldsymbol{s}}}{\partial a}, \tag{13}$$

with bottom-up attention components in relation to sensory input, analogous to those in relation to belief in (11). These control signals act as reflexive saccades [16,45]. The gradient $\frac{\partial \tilde{s}}{\partial a}$ is an inverse mapping from sensory data to actions.

3 Results

3.1 Implementation of the Proposed Model

A graphical overview of the model[1] is shown in Fig. 1. The current belief state $\boldsymbol{\mu}$ is passed to exteroceptive, proprioceptive, and interoceptive generative models. Their predictions $\boldsymbol{p}$ are compared to actual inputs $\boldsymbol{s}$, with prediction errors $\boldsymbol{e}_s$ driving both action and belief updates. Proprioceptive (camera pitch/yaw) and interoceptive (symbolic cue) generative models are identity mappings, while the exteroceptive visual model is the decoder of a disentangled VAE. The VAE encodes target presence and position, simplifying conversion from intrinsic image coordinates to extrinsic camera orientation.

The belief state consists of:

- **Symbolic cue belief** – position of an endogenous cue on the image, mirroring sensory input

[1] Code, video examples, and details on VAE training and simulations are available at: https://unizgfer-lamor.github.io/ainf-visual-attention/.

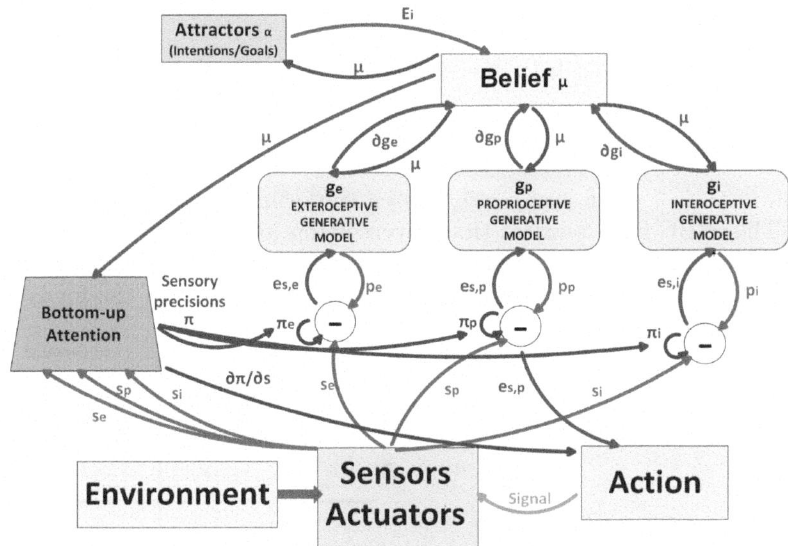

Fig. 1. Core structure of the proposed model. Beliefs about sensory causes are updated through attractor goals and error signals to minimize free-energy. Bottom-up attention is regulated by dynamic sensory precisions.

- **Camera orientation belief** – proprioceptive pitch and yaw of the camera
- **Visual belief** – disentangled encoding of target presence and position
- **Covert attention belief** – amplitude and center of an RBF governing visual precisions

Beliefs are updated through bottom-up prediction errors and top-down attractors $\boldsymbol{\alpha}$, following the flexible intentions theory in [35]. In our model, attractors for proprioceptive, visual, and covert attention beliefs are generated from current visual and cueing inputs, allowing dynamic goals. These intentions drive overt actions (camera orientation) and covert shifts (RBF updates).

Visual precision $\boldsymbol{\Pi}_s$ is defined as a diagonal matrix:

$$\boldsymbol{\Pi}_s = \begin{bmatrix} \pi_1(\boldsymbol{\mu}, \boldsymbol{s}) & 0 & \cdots & 0 \\ 0 & \pi_2(\boldsymbol{\mu}, \boldsymbol{s}) & \cdots & 0 \\ \vdots & \vdots & \ddots & \vdots \\ 0 & 0 & \cdots & \pi_L(\boldsymbol{\mu}, \boldsymbol{s}) \end{bmatrix}_{L \times L}, \tag{14}$$

where $L = 32 \times 32\,(\times 3)$ is the dimensionality of the visual data. Each precision term is determined by an RBF centered on the covert attention belief $[\mu_{amp}, \mu_u, \mu_v]$ and the centroid of the largest red object $[\boldsymbol{r_u}(\boldsymbol{s}), \boldsymbol{r_v}(\boldsymbol{s})]$:

$$\begin{aligned}
\pi_i(\boldsymbol{\mu}, \boldsymbol{s}) = \pi(x, y, \boldsymbol{\mu}, \boldsymbol{s}) = \\
\frac{\mu_{amp}}{2}\Big(\ln\Big(-\tfrac{(x-\mu_u)^2+(y-\mu_v)^2}{b^2}+1\Big)+c\Big) \\
+\frac{1}{2}\Big(\ln\Big(-\tfrac{(x-r_u(\boldsymbol{s}))^2+(y-r_v(\boldsymbol{s}))^2}{b^2}+1\Big)+c\Big),
\end{aligned} \tag{15}$$

with parameters $b = 2.6$ and $c = 1$, chosen to normalize RBF values between 0 and 1. This RBF form ensures that covert attention is drawn toward areas of high prediction error, unlike Gaussian RBFs which repel from error. As shown in Fig. 2, the resulting precision decreases with distance from the focus center, mimicking human foveation [2,32].

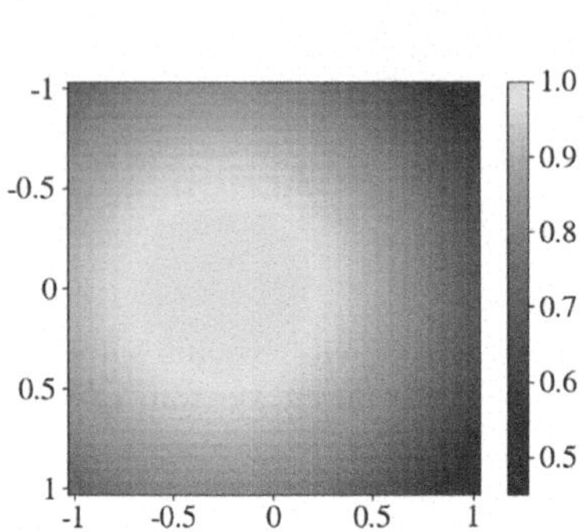

Fig. 2. An example of an RBF precision matrix with the center in (-0.25, 0.0).

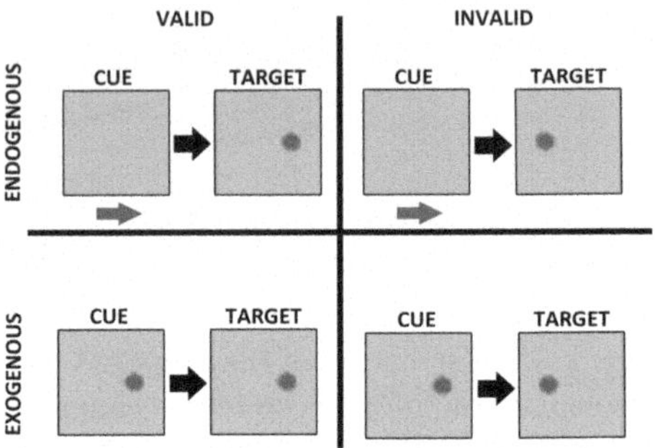

Fig. 3. Experimental setups for the four variations of the Posner Cueing Task, with visual data used in experiments. The red arrows represent symbolic endogenous cues not visible through sensory input. (Color figure online)

3.2 Simulating the Posner Cueing Task

We evaluated the model's implementation of exogenous and endogenous covert attention using the Posner cueing task. The model received three types of sensory input: (1) camera orientation (proprioceptive), (2) a symbolic cue signal (interoceptive), and (3) visual input of an empty scene in which a red sphere may appear as a target. Unlike traditional versions, the endogenous cue was delivered as a symbolic interoceptive signal rather than a visual arrow, but still required voluntary attentional shifts. Motor actions (overt attention) were disabled.

Four task variations were tested by combining two cue types (endogenous, exogenous) with two validity conditions (valid, invalid):

- **Endogenous cueing**: the symbolic cue is internally processed to form a top-down intention, shifting both covert focus and the target-position belief.
- **Exogenous cueing**: a brief appearance of the target object triggers a bottom-up shift of covert attention via free-energy gradients and updates the belief over the object's position through the VAE likelihood error.

In valid trials, the target appeared at the cued location; in invalid trials, it appeared on the opposite side of the visual field. The task variations are illustrated in Fig. 3.

Each condition was tested with $N = 200$ trials. A single trial proceeded as follows:

- A random target position (with varying eccentricity) was generated, and the model initialized (10 steps),
- A cue (endogenous or exogenous) was presented (50 steps),
- After a specified cue-target onset asynchrony (CTOA), the target appeared,
- The trial ended either upon detection, defined as the latent variable for target presence becoming positive (marking the reaction time, RT), or after 300 steps if undetected.

The results are shown in Fig. 4. The left panel plots RTs as a function of target eccentricity, while the right panel shows internal dynamics of covert focus and target-belief updates, with clear facilitation in valid-cue conditions. The model reproduces several well-established effects in human data and location-based models:

- **Cue validity:** valid cues yield faster RTs than invalid ones [33,34], consistent with the spotlight theory of attention [7,37,43]. Invalid cues increase RTs due to larger spotlight shifts at target onset.
- **Cue type:** exogenous cues produce faster RTs than endogenous cues [4,13], as bottom-up signals propagate directly via error gradients, while endogenous cues require symbolic interpretation and intentional updating.
- **Target eccentricity:** RTs increase with target distance from fixation [2,17,32], reflecting both location-based encoding and the shape of the RBF precision function.

As shown in Fig. 4(right), covert attention centers update more rapidly than beliefs over target location in both cueing conditions. This aligns with empirical findings that covert shifts precede conscious target perception [31,33] and overt attention [16,45].

To examine the effect of cue-target onset asynchrony, all four conditions were repeated across different CTOA values. Figure 5(left) shows that valid cues consistently yield faster RTs than invalid ones across CTOAs. The endogenous cueing results, shown in Fig. 5(right), replicate classic Posner findings [9,33,34], including the asymmetric pattern where invalid cues impose a greater cost than the benefit provided by valid cues.

3.3 Action Signals from Bottom-Up Attention

Since action can be determined from free-energy optimization, overt attention in the form of eye saccades or camera orientation changes can be also implemented. Here we examined focus reach times for two action-update contributions:

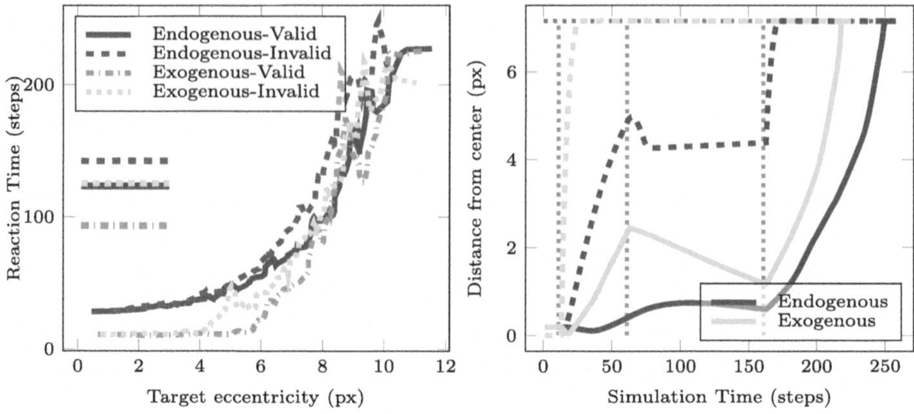

Fig. 4. Reaction time and target eccentricity relationship for endogenous and exogenous cueing. **Left:** reaction times and their averages as a function of target distance from focus point (CTOA = 100 for each trial) **Right:** covert attention center (dashed lines) and sphere position beliefs (solid lines) during valid trials, for both endogenous and exogenous cues. The horizontal line is the true target distance from center, and the vertical lines indicate trial events: the cue appears at step 10, disappears at step 60, target appears at step 160.

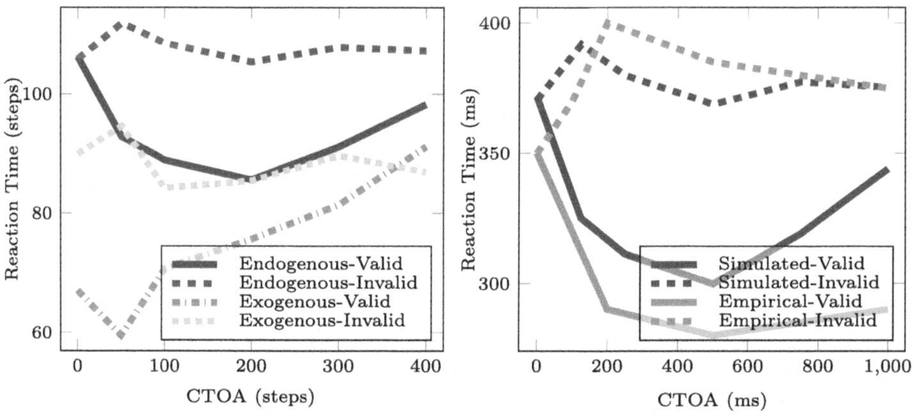

Fig. 5. Comparison of reaction times from simulated and empirical results. **Left:** average trial reaction time as a function of CTOA. Results are shown for endogenous-valid, endogenous-invalid, exogenous-valid, exogenous-invalid task variations. **Right:** comparison of simulated and empirical human data [34] for endogenous cueing. Simulated reaction times are shown up to an arbitrary constant reflecting the scale gap between human times and simulation steps.

- Top-down proprioceptive action signals: $-\frac{\partial \tilde{s}}{\partial a}^T \tilde{\boldsymbol{\Pi}}_s \tilde{\boldsymbol{e}}_s$ – these are determined from the prediction error of the proprioceptive channel, between the proprioceptive input and current proprioceptive beliefs

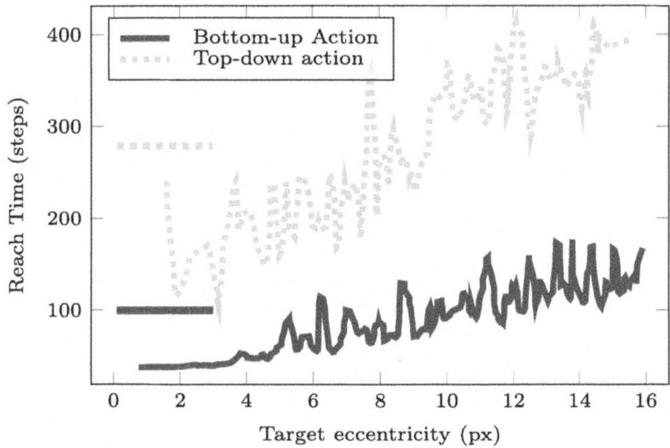

Fig. 6. Reach times and their averages for different initial target distances.

- Bottom-up visual precision action signals: $\frac{1}{2}\text{Tr}\left[\tilde{\boldsymbol{\Pi}}_s^{-1}\frac{\partial\tilde{\boldsymbol{\Pi}}_s}{\partial\tilde{s}}\right]\frac{\partial\tilde{s}}{\partial a} - \frac{1}{2}\tilde{\boldsymbol{e}}_s^T\frac{\partial\tilde{\boldsymbol{\Pi}}_s}{\partial\tilde{s}}\tilde{\boldsymbol{e}}_s\frac{\partial\tilde{s}}{\partial a}$ – these are determined through the bottom-up derivative of the precision matrix. Since the action update is dependent only on the sensory input, only the second half of (15) contributes to the action update.

The trials start with a 10-step initialization interval, after which the target appears at a random position in the agent's field of view. The trial is finished when the agent successfully focuses the target at the center of its field of view.

The reach times as a function of the initial target distance can be seen in Fig. 6. The results show that bottom-up overt orienting is overall faster than top-down intentional orienting, which is explained by the sensitivity of the precision to red objects (or any predetermined visual object of interest, like faces [16]). This is similarly reflected in how reaction time changes with distance. Both forms of orienting exhibit an increasing trend in reaction time as distance increases; however, top-down orienting shows a steeper rise, indicating a greater sensitivity to distance compared to bottom-up orienting.

4 Discussion

Our proposed model was evaluated on exogenous, endogenous, valid, and invalid variations of the Posner paradigm, as well as a simple target reach task. It successfully reproduces key attentional effects observed in human data and location-based models, including the influence of exogenous versus endogenous cues, cue validity, and overt behaviors such as involuntary saccades.

A central contribution of this work is the integration of visual attention mechanisms, continuous sensory representations, and deep generative models within an active inference framework. By operating directly on raw two-dimensional visual input, the model overcomes the limitations of prior approaches that rely on discrete inputs or lack deep generative components, making it more suitable for robotic applications.

The overt attention experiments reveal a trade-off between speed and flexibility: bottom-up orienting enables rapid but singular shifts to individual objects, while top-down orienting is slower but allows flexible allocation across multiple objects. This trade-off has direct implications for robotics. One application is active visual search, where robots adjust their viewpoint to locate a target object, potentially using intermediate objects to guide attention through top-down processes [19,39,40,44,46]. For instance, when searching for a keyboard, a desk might serve as a top-down cue directing attention upward, while the keyboard itself, once detected, would attract bottom-up attention due to its salient features. Another application is joint attention in human–robot interaction, where human gaze and head movements guide the robot's attention toward relevant objects in the shared environment [15,25]. In both cases, our model integrates bottom-up attention, where salient objects attract attention directly, and top-down attention, where contextual cues guide attention across the scene–both essential for natural and adaptive active vision in robots.

5 Conclusion

In this paper, we have proposed an active inference model of covert and overt visual attention. The proposed model successfully demonstrates known attentional phenomena and mechanisms in the context of the Posner cueing task and a simple active orienting task. It shows that valid cues produce faster reaction times than invalid cues, and that exogenous cues produce faster reaction times than endogenous cues. The model also successfully demonstrates location-based attention, with reaction times increasing with target eccentricity.

Future work will extend the model with multiple possible targets/intentions to further test object-based and location-based effects, as well as with top-down attentional mechanisms that lead to inhibition of return. Overt saccades will also be examined further, with a focus on varying attraction to different objects in tasks such as active visual search and joint attention. We plan to further develop and test this framework as a model of perception, learning, and action in autonomous robots.

Acknowledgments. This research has been supported by the H2020 project AIFORS under Grant Agreement No 952275.

Disclosure of Interests. The authors have no competing interests to declare that are relevant to the content of this article.

References

1. Blair, C.D., Ristic, J.: Attention combines similarly in covert and overt conditions. Vision (Basel) **3**(2), 16 (2019)
2. Carrasco, M., Evert, D.L., Chang, I., Katz, S.M.: The eccentricity effect: target eccentricity affects performance on conjunction searches. Percept. Psychophys. **57**(8), 1241–1261 (1995)
3. Çatal, O., Wauthier, S., Verbelen, T., De Boom, C., Dhoedt, B.: Deep active inference for autonomous robot navigation (2020)
4. Cheal, M., Lyon, D.R.: Central and peripheral precuing of forced-choice discrimination. Q. J. Exp. Psychol. A **43**(4), 859–880 (1991)
5. Cioffi, G., Cieslewski, T., Scaramuzza, D.: Continuous-time vs. discrete-time vision-based slam: A comparative study. IEEE Robot. Autom. Lett. **7**(2), 2399–2406 (2022). https://doi.org/10.1109/LRA.2022.3143303
6. Corbetta, M., Shulman, G.L.: Control of goal-directed and stimulus-driven attention in the brain. Nat. Rev. Neurosci. **3**(3), 201–215 (2002)
7. Egly, R., Driver, J., Rafal, R.D.: Shifting visual attention between objects and locations: Evidence from normal and parietal lesion subjects. J. Exp. Psychol. Gen. **123**(2), 161–177 (1994)
8. Eliasmith, C., Furlong, P.M.: Continuous then discrete: A recommendation for building robotic brains. In: Faust, A., Hsu, D., Neumann, G. (eds.) Proceedings of the 5th Conference on Robot Learning. Proceedings of Machine Learning Research, vol. 164, pp. 1758–1763. PMLR (2022), https://proceedings.mlr.press/v164/eliasmith22a.html
9. Feldman, H., Friston, K.J.: Attention, uncertainty, and free-energy. Front. Hum. Neurosci. **4** (2010)
10. Friston, K.: The free-energy principle: a unified brain theory? Nat. Rev. Neurosci. **11**(2), 127–138 (2010)
11. Friston, K., Kilner, J., Harrison, L.: A free energy principle for the brain. J. Physiol. Paris **100**(1-3), 70–87 (2006)
12. Itti, L., Baldi, P.: Bayesian surprise attracts human attention. Vision Res. **49**(10), 1295–1306 (2009)
13. Jonides, J.: Voluntary versus automatic control over the mind's eye's movement. In: Attention and Performance IX, pp. 187–203 (1981)
14. Kanai, R., Komura, Y., Shipp, S., Friston, K.: Cerebral hierarchies: predictive processing, precision and the pulvinar. Philos. Trans. R. Soc. Lond. B Biol. Sci. **370**(1668), 20140169 (2015)
15. Kaplan, F., Hafner, V.V.: The challenges of joint attention. Interaction Stud. **7**(2), 135–169 (2006). https://doi.org/10.1075/is.7.2.04kap, https://www.jbe-platform.com/content/journals/10.1075/is.7.2.04kap
16. Kauffmann, L., Peyrin, C., Chauvin, A., Entzmann, L., Breuil, C., Guyader, N.: Face perception influences the programming of eye movements. Sci. Rep. **9**(1), 560 (2019)
17. Klein, R.M.: Inhibition of return. Trends Cogn. Sci. **4**(4), 138–147 (2000)
18. Kulke, L.V., Atkinson, J., Braddick, O.: Neural differences between covert and overt attention studied using EEG with simultaneous remote eye tracking. Front. Hum. Neurosci. **10**, 592 (2016)
19. Kunze, L., Doreswamy, K.K., Hawes, N.: Using qualitative spatial relations for indirect object search. In: 2014 IEEE International Conference on Robotics and Automation (ICRA). IEEE (2014)

20. Lanillos, P., Dean-Leon, E., Cheng, G.: Multisensory object discovery via self-detection and artificial attention. In: 2016 Joint IEEE International Conference on Development and Learning and Epigenetic Robotics (ICDL-EpiRob). IEEE (2016)
21. Lanillos, P., Ferreira, J.F., Dias, J.: Designing an artificial attention system for social robots. In: 2015 IEEE/RSJ International Conference on Intelligent Robots and Systems (IROS). IEEE (2015)
22. Lanillos, P., Ferreira, J.F., Dias, J.: Multisensory 3d saliency for artificial attention systems (2015)
23. Mengotti, P., Käsbauer, A.S., Fink, G.R., Vossel, S.: Lateralization, functional specialization, and dysfunction of attentional networks. Cortex **132**, 206–222 (2020)
24. Mirza, M.B., Adams, R.A., Friston, K., Parr, T.: Introducing a bayesian model of selective attention based on active inference. Sci. Rep. **9**(1), 13915 (2019)
25. Nagai, Y., Hosoda, K., Morita, A., Asada, M.: A constructive model for the development of joint attention. Conn. Sci. **15**(4), 211–229 (2003)
26. Parr, T., Benrimoh, D.A., Vincent, P., Friston, K.J.: Precision and false perceptual inference. Front. Integr. Neurosci. **12**, 39 (2018)
27. Parr, T., Friston, K.J.: Uncertainty, epistemics and active inference. J. R. Soc. Interface **14**(136), 20170376 (2017)
28. Parr, T., Friston, K.J.: Attention or salience? Curr. Opin. Psychol. **29**, 1–5 (2019)
29. Parr, T., Pezzulo, G., Friston, K.J.: Active inference. The MIT Press (2022)
30. Parvizi-Wayne, D.: How preferences enslave attention: calling into question the endogenous/exogenous dichotomy from an active inference perspective. Phenomenol. Cogn. Sci. 1–47 (2024)
31. Peterson, M.S., Kramer, A.F., Irwin, D.E.: Covert shifts of attention precede involuntary eye movements. Percept. Psychophys. **66**(3), 398–405 (2004)
32. Pinker, S., Downing, C.J.: Attention and Performance XI: Mechanisms of attention and visual search. Erlbaum, Hillsdale, NJ (1985)
33. Posner, M.I.: Orienting of attention. Q. J. Exp. Psychol. **32**(1), 3–25 (1980)
34. Posner, M., Nissen, M., Ogden, W.: Attended and unattended processing modes: The role of set for spatial location. Modes of Perceiving and Processing Information **137** (1978)
35. Priorelli, M., Stoianov, I.P.: Flexible intentions: An active inference theory. Front. Comput. Neurosci. **17**, 1128694 (2023)
36. Rao, R.P.: An optimal estimation approach to visual perception and learning. Vision Research **39**(11), 1963–1989 (1999). https://doi.org/10.1016/S0042-6989(98)00279-X, https://www.sciencedirect.com/science/article/pii/S004269899800279X
37. Reppa, I., Schmidt, W.C., Leek, E.C.: Successes and failures in producing attentional object-based cueing effects. Atten. Percept. Psychophys. **74**(1), 43–69 (2012)
38. Shomstein, S., Zhang, X., Dubbelde, D.: Attention and platypuses. Wiley Interdiscip. Rev. Cogn. Sci. **14**(1), e1600 (2023)
39. Shubina, K., Tsotsos, J.K.: Visual search for an object in a 3D environment using a mobile robot. Comput. Vis. Image Underst. **114**(5), 535–547 (2010)
40. Sjöö, K., Aydemir, A., Jensfelt, P.: Topological spatial relations for active visual search. Rob. Auton. Syst. **60**(9), 1093–1107 (2012)
41. Škrjanc, I., Klančar, G.: A comparison of continuous and discrete tracking-error model-based predictive control for mobile robots. Rob. Auton. Syst. **87**, 177–187 (2017)
42. Spratling, M.W.: Predictive coding as a model of biased competition in visual attention. Vision Res. **48**(12), 1391–1408 (2008)

43. Vecera, S.P., Farah, M.J.: Does visual attention select objects or locations? J. Exp. Psychol. Gen. **123**(2), 146–160 (1994)
44. Vogel, J., Murphy, K.: A non-myopic approach to visual search. In: Fourth Canadian Conference on Computer and Robot Vision (CRV '07). IEEE (2007)
45. Walker, R., Walker, D.G., Husain, M., Kennard, C.: Control of voluntary and reflexive saccades. Exp. Brain Res. **130**(4), 540–544 (2000)
46. Wixson, L.E., Ballard, D.H.: Using intermediate objects to improve the efficiency of visual search. Int. J. Comput. Vis. **12**(2-3), 209–230 (1994)

Dynamic Attentional Agents in Focused Attention Meditation: Hierarchical Computational Modeling of Expert-Novice Differences

Prakash Chandra Kavi[1,2(✉)], Daniel Ari Friedman[3], and Gustavo Patow[1,2]

[1] Universitat Pompeu Fabra, 08002 Barcelona, Spain
prakash.kavi@gmail.com
[2] Universitat de Girona, 17003 Girona, Spain
[3] Active Inference Institute, Crescent City, CA 95531, USA

Abstract. We develop a three-level hierarchical framework to model the attentional dynamics of focused attention (FA) meditation, laying a foundation for advanced active inference (AIF) implementations. Grounded in the Free Energy Principle and Neuronal Packet Hypothesis, we conceptualize meditation as a predictive processing system where "thoughtseeds"—transient, agent-like entities forming Markov blankets—minimize variational free energy via bidirectional coupling with attentional networks (Default Mode Network [DMN], Ventral Attention Network [VAN], Dorsal Attention Network [DAN], Frontoparietal Network [FPN]). Thoughtseeds, emerging from superordinate neuronal ensembles, compete for Global Workspace access, modulated by meta-cognitive precision weighting to stabilize attention. This model advances the Thoughtseeds Framework toward a computational phenomenology of Vipassana, setting the stage for integrated AIF formalisms and implementation, on top of rules-based statistical learning. Simulations reveal how across ranges of biologically plausible parameters—precision weighting (0.5 vs. 0.4), complexity penalties (0.2 vs. 0.4), learning rates (0.02 vs. 0.01)— patterns associated with expert-novice differences arise from the "in silico" model. Experts with higher precision weighting and higher learning rate, notably achieve 49% lower free energy during breath focus, suppressed DMN activity (0.18 vs. 0.31), and faster distraction recovery, consistent with neuroimaging findings. Bidirectional message passing enables bottom-up formation of attentional states and top-down constraint of dynamics, offering a mechanistic account of expertise as optimized precision allocation. This framework provides testable predictions for meditation skill development, with future extensions planned to enhance AIF rigor and computational phenomenology for applications in contemplative neuroscience, computational psychiatry, and cognitive training.

Keywords: Markov blanket · neuronal packet · thoughtseed · meditation · hierarchical modeling · active inference · attentional networks · computational phenomenology · predictive processing

M. Albarracin et al. (Eds.): IWAI 2025, CCIS 2857, pp. 182–207, 2026.
https://doi.org/10.1007/978-3-032-16955-6_11

1 Introduction

1.1 The Phenomenology of the Discursive Mind During Vipassana Meditation

The neuroscientific investigation of meditation and mind-wandering has advanced considerably in recent years, providing valuable insights into attention regulation, metacognition, and the neural substrates of conscious experience (Tang et al. 2015; Dahl et al. 2015; Brandmeyer and Delorme 2021). Despite this progress, a unifying theoretical framework capable of accounting for the diverse phenomenological and cognitive effects of meditation remains elusive (Lutz et al. 2024). Contemporary classifications (Lutz et al. 2008; Vago and Silbersweig 2012) often distinguish between Focused Attention (FA), Open Monitoring (OM), and Non-Dual Awareness (NDA) practices—modes that span a continuum of attentional stability and meta-awareness [Laukkonen and Slagter 2021). This paper centers on FA meditation, particularly Vipassana, as a tractable paradigm for modeling the discursive mind—the stream of internally generated thoughts, sensory impressions, and narrative-like cognitive patterns that unfold over time (Anālayo, 2019).

Vipassana meditation, rooted in Theravāda Buddhist traditions, emphasizes sustained attention to bodily sensations and mental events in a systematic and equanimous manner. A well-established neurocognitive model, validated in both novice and experienced practitioners, delineates a four-stage attentional cycle during FA meditation (Hasenkamp et al. 2012; Lutz et al. 2008): (1) Sustained attention on an object such as the breath, characterized by top-down control; (2) Mind-wandering, where attention involuntarily drifts to task-unrelated thoughts or sensations; (3) Meta-awareness of this drift, which triggers recognition of the attentional lapse; and (4) Redirect Attention, involving the deliberate return of focus to the original object. This attentional style enables the observation of mental state transitions and provides an ideal framework for empirical study, reflecting the principle of dependent origination (pratītyasamutpāda), where mental events arise interdependently, each conditioning the next (Anālayo 2021).

One of these four stages is underpinned by distinct neural mechanisms that can be interpreted through the lens of predictive processing (Barrett and Simmons 2015; Laukkonen & Slagter 2021). Stage 1 engages the frontoparietal attention network, including the dorsolateral prefrontal cortex (dlPFC) and intraparietal sulcus, supporting sustained, goal-directed attention (MacLean et al. 2010; Lutz et al. 2008). Stage 2 activates the default mode network (DMN)—particularly the medial prefrontal cortex (mPFC) and posterior cingulate cortex (PCC)—implicated in self-referential thought and spontaneous cognition (Christoff et al. 2009; Fox et al. 2015). Stage 3 recruits the salience network, notably the anterior insula and dorsal anterior cingulate cortex (dACC), which monitor internal states and detect cognitive conflict (Seeley et al. 2007; Craig 2009; Schooler et al. 2011). Stage 4 re-engages executive control regions, enabling intentional modulation of attention and suppression of distractors (Hasenkamp et al. 2012).

This cyclical process mirrors the principle of dependent origination, wherein mental events arise interdependently in a recursive structure (Anālayo 2021). From a neurocognitive perspective, this recursive interplay reflects competitive and cooperative dynamics among large-scale brain networks, aligning with predictive processing theories that frame cognition as the minimization of prediction errors (Barrett and Simmons 2015;

Tang et al. 2015; Laukkonen and Slagter 2021). Repeated traversal of this cycle, particularly in experienced meditators, has been associated with enhanced cognitive flexibility, improved meta-awareness, and changes in interoceptive processing (Fox et al. 2014; Lutz et al. 2024; Escrichs et al. 2019).

These dynamics suggest that the discursive mind—defined as the fluctuating interplay of thoughts, sensations, and attentional shifts—is not a passive stream but an active, self-regulating process shaped by learned attentional strategies and network-level brain dynamics. The subjective availability of mental content, often described as the “presence” or “illuminative quality” of awareness (Lee 2021), modulates these processes, influencing how attention and subjectivity co-determine what enters and persists in awareness.

1.2 Free Energy Principle, Active Inference and Neuronal Packet Hypothesis: A Framework for Embodied Self-Organization

To model the complex, self-regulating dynamics underlying meditative cognition, we adopt the Free Energy Principle (FEP), a unifying theory of brain function that describes how biological systems maintain their structural and functional integrity through self-organization (Friston 2010 2013; Parr et al. 2022). According to FEP, any system that persists over time must minimize its variational free energy, an upper bound on surprise or prediction error. This minimization is achieved by developing and maintaining a generative model of the self and environment, continuously updating internal beliefs and performing actions to align sensory inputs with predictions (Friston 2010 2017a). This process is inherently self-organizing, as the system autonomously structures its internal states to reduce uncertainty and maintain homeostasis (Allen and Friston, 2018; Varela et al. 2017).

A central construct in FEP is the Markov blanket, a statistical boundary that insulates a system’s internal states from the external world while mediating interactions through sensory and active states (Pearl 1988; Friston 2013). In cognitive systems, this boundary defines what constitutes an “Agent"—an entity with beliefs about the world that guide perception and action (Kirchhoff et al. 2018; Ramstead et al. 2021).

Within this framework, Active Inference (AIF) provides the process theory that specifies how agents enact free energy minimization through recurrent cycles of *perception, learning, and action* (Friston et al. 2015; Friston et al. 2017; Parr & Friston 2018; Parr et al. 2022). Perceptual inference refines internal beliefs to minimize prediction errors, while *action* selects and executes policies that actively reshape sensory inputs to match those predictions. Crucially, agents evaluate potential actions by minimizing *expected free energy,* which balances *epistemic* value (reduced free energy through information gain) and pragmatic value (achieve preferred outcomes) (Friston et al. 2017; Ramstead et al. 2020).

To understand how these mechanisms are instantiated in the brain, the Neuronal Packet Hypothesis (NPH) posits that cognitive functions emerge from self-organizing ensembles of neurons—called neuronal packets (Yufik 2013 2019); Yufik & Friston 2016). These packets form transient Markov blankets, enabling computational autonomy and conditional independence at various spatiotemporal scales (Kirchhoff et al.

2018; Hipólito et al. 2021). Supported by the brain's sparse architecture, which facilitates localized functional units (Sporns and Betzel 2016), neuronal packets compete to minimize their own free energy, forming hierarchical, agent-like structures that represent increasingly abstract or conceptual constructs. These neuronal packets can self-organize into higher order superordinate ensembles, to embody complex cognitive processes or states as nested Markov blankets (Ramstead et al. 2021; Palacios et al. 2020). The formation of superordinate ensembles reflects a higher-order process of hierarchical and heterarchical self-organization, where coordinated neuronal activity gives rise to emergent properties (Yufik 2019; Ramstead et al. 2021 2022), such as the transient attentional states observed in meditation (Kavi et al. 2025).

In this context, we interpret attentional fluctuations during Vipassana meditation as arising from competitive interactions among such neuronal ensembles, each attempting to model and respond to the contents of consciousness. These dynamics are shaped by internal precision estimates—subjective confidence in one's predictions—and modulated through experience-dependent learning, such as meditation training (Hohwy 2013; Varela et al. 2017; Czajko et al. 2024).

1.3 The Thoughtseeds Framework: An Agentic, Hierarchical Architecture

The Thoughtseeds Framework, grounded in the FEP and the Neuronal Packet Hypothesis (NPH), offers a computational architecture for modeling thought dynamics in contemplative contexts, such as Vipassana meditation (Kavi et al. 2025). It posits that thoughts are not passive byproducts of neural activity but emergent, agent-like entities—termed thoughtseeds—arising from the self-organized activity across nested scales of neuronal packets (Yufik & Friston 2016; Yufik 2019). These packets, forming transient Markov blankets (Ramstead et al. 2021; Beck & Ramstead 2025), coordinate into hierarchically nested superordinate ensembles that compete to minimize free energy (Friston et al. 2015), shaping the flow of conscious experience (Ramstead et al. 2022).

Within this framework, each thoughtseed operates as a self-evidencing agent, equipped with a generative model that predicts sensory and interoceptive inputs (Hohwy 2016). This competition drives thoughtseeds to vie for access to a Global Workspace (GW) (Baars 1997; Dehaene & Changeux 2011 2014), where the "winning" thoughtseed—determined by its success in minimizing free energy—is broadcasted to a "functionally rich club" (FRC) responsible for the content of consciousness (Deco et al. 2021). Thoughtseed competition is instantiated through interactions with attentional networks at both mesoscale and macroscales leveraging bottom-up processing and top-down modulation (Fox et al. 2015; Lutz et al. 2024; Raffone and Srinivasan 2017). In the Thoughtseeds Framework, these transitions are actively inferred by competing thoughtseeds, each minimizing effective free energy to optimize its explanatory power over neural and interoceptive inputs (Friston et al. 2017).

1.4 Research Objectives

While the original thoughtseeds framework employed rule-based statistical learning, this work advances it by embedding thoughtseeds simulation within an Active Inference paradigm, process-driven account of meditative states and mind-wandering, enhancing its computational rigor and neuroscientific grounding. The framework retains its three-level hierarchical structure: (1) Knowledge Domains (KDs), now explicitly mapped to four attentional networks—Default Mode Network (DMN), Frontoparietal Network (FPN), Ventral Attention Network (VAN), and Dorsal Attention Network (DAN) (Yeo et al. 2011) —for biological plausibility; (2) thoughtseed networks, where thoughtseeds operate as generative agents across these KDs at higher level of abstraction as shown in Fig. 2.B of the Thoughtseeds Framework (Kavi et al., 2025); and (3) meta-cognitive processes that regulate thoughtseed competition via precision weighting and meta-awareness (Sandved-Smith et al. 2021). This revised hierarchy formalizes the interplay of attention, intention, and conscious content, enabling thoughtseeds to drive the four-stage Vipassana cycle: sustained attention to breath, mind-wandering, meta-awareness of mind-wandering, and redirect attention to breath (Hasenkamp et al. 2012).

Meditative expertise refines this dynamic by enhancing precision weighting and reducing model complexity, enabling practitioners to navigate the discursive mind with greater stability and reduced cognitive friction (Escrichs et al. 2019; Sandved-Smith et al. 2021). This efficiency is a hallmark of advanced Vipassana practice, reflecting a shift from effortful to effortless attention. Our model operationalizes these neurobiological markers, particularly the antagonistic relationship between DMN and task-positive networks (DAN, FPN), a key indicator of attentional control and expertise (Fox et al. 2015; Brewer et al. 2011).

We contribute to computational phenomenology (Ramstead et al. 2022; Sandved-Smith et al. 2021) by establishing a falsifiable link between variational free energy and subjective experiences of "mental effort" during meditation. This approach quantifies phenomenological states, bridging Buddhist introspection with systems neuroscience. Our specific contributions include enhancements for: (1) Modeling thoughtseeds as generative agents that instantiate attractor states corresponding to the meditative cycle's phases; (2) Framing thoughtseed interactions as a competitive process for control over the Global Workspace, governed by free energy minimization; (3) Simulating meditative expertise, across expert and novice meditators; and (4) Advancing computational phenomenology by linking neurocognitive dynamics to subjective experience, integrating contemplative and neuroscientific perspectives.

These contributions position the Thoughtseeds Framework at the intersection of Active Inference, Global Workspace Theory, and contemplative neuroscience, offering a principled account of how metacognitive strategies in Vipassana modulate attention, thought dynamics, and self-awareness.

2 Methods

We extend the Thoughtseeds framework (Kavi et al. 2025) by adopting Active Inference (AIF) on top of rule-based statistical learning. Rather than treating thoughts as passive epiphenomena, we formalize them as self-organizing Markov blankets forming neuronal packets that compete to minimize variational free energy (Friston 2010). This approach bridges phenomenological accounts of Vipassana meditation with neurocognitive dynamics through a principled mathematical framework, aligning with predictive processing theories (Barrett and Simmons 2015; Laukkonen and Slagter 2021).

The core theoretical advance lies in conceptualizing the four-stage meditative cycle (breath focus, mind-wandering, meta-awareness, and redirect attention) as hidden attractor states within a dynamic system governed by precision-weighted prediction errors. Each hidden state represents a distinct prediction about expected network configurations, interacting bidirectionally with thoughtseed activations. The system's evolution through this hidden state space constitutes the computational basis of meditative experience, with expertise modulating precision (Laukkonen and Slagter 2021) and model complexity (Escrichs et al. 2019).

2.1 From Knowledge Domains to Attentional Networks

The foundational level of the Thoughtseeds Framework redefines the abstract Knowledge Domains (KDs) from the original thoughtseeds framework as superordinate ensembles corresponding to four attentional networks. These networks, enable complex assemblies of neuronal packets forming nested Markov blankets across multiple scales, process distinct cognitive functions: the DMN supports self-referential processing and mind-wandering, the VAN mediates salience detection, the DAN sustains goal-directed attention, and the FPN enables cognitive control and metacognitive monitoring (Brewer et al. 2011; Smallwood & Schooler 2013). This mapping enhances the neuroscientific grounding of the original framework, where KDs were abstract repositories tied to thoughtseed activations.

Neuronal packets—self-organizing neural assemblies—bridge microscale activity to macroscale network dynamics, forming transient Markov blankets that enable localized free energy minimization (Yufik & Friston, 2016). Thoughtseeds, positioned at Level 2, emerge as lower-dimensional representations from specific activation patterns across these Level 1 ensembles, driving the four-stage Vipassana cycle (Hasenkamp et al., 2012).

2.2 Hierarchical Architecture

The Thoughtseeds Framework enhances its three-level hierarchical architecture from Kavi, et al. 2025 as illustrated in Fig. 1, with Level 1 specified as the attentional network substrate introduced in Sect. 2.1. This architecture formalizes the interplay of neural, cognitive, and metacognitive processes in Vipassana meditation.

Level 1: Attentional Network Superordinate Ensembles. By superordinate ensemble we mean a mesoscale integration of nested scales of neuronal packets whose mean activation is tracked as a single network variable. In this framework, we model four such ensembles: DMN (self-referential thought and mind-wandering), VAN (salience detection and interrupt signals), DAN (sustained goal-directed attention), and FPN (cognitive control and meta-monitoring) (Menon and Uddin 2010; Schooler 2002 2011). Their activations, driven by thoughtseed contributions and state expectations, form the foundation for the meditative cycle (Hasenkamp et al. 2012).

These networks function as substrates: large-scale neural contexts within which more specialized cognitive units emerge. Importantly, thoughtseeds are not identical to networks themselves, but rather represent higher-order patterns that arise from the coordinated activity across these underlying ensembles. They are transient, agent-like entities (e.g., *breath_focus, pending_tasks, self_reflection*) that form attractor states through specific activation patterns across multiple networks (Bullmore and Sporns, 2009; Deco and Jirsa, 2012). Whereas networks define broad functional channels Default Mode Network (DMN), Ventral Attention Network (VAN), Dorsal Attention Network (DAN), and Frontoparietal Network (FPN) (Yeo et al. 2011), thoughtseeds represent specific attentional contents with their own Markov blankets - statistical boundaries that maintain their coherence as distinct cognitive units (Friston 2010). They compete for access to the Global Workspace, where winning thoughtseeds broadcast their content to wider neural systems (Baars 1997; Dehaene and Changeux 2011; Deco et al. 2021).

Interaction is bidirectional: network activations bias which thoughtseeds emerge (e.g., DAN favors *breath_focus*, DMN favors *pending_tasks*), while dominant thoughtseeds modulate network activity to reinforce their own explanatory power. Although all thoughtseeds are exposed to the same sensory field, they do not converge into identical agents, because (i) each begins with distinct priors tied to its host ensemble, (ii) contribution weights evolve through experiential learning, and (iii) meta-cognitive regulation at Level 3 gates or suppresses specific thoughtseeds. This differentiation is consistent with the NPH, which emphasizes that ensembles self-organize into functionally specialized packets that retain conditional independence even when embedded in the same environment (Yufik and Friston 2016; Ramstead et al. 2021). Thus, the four attentional networks provide distinct substrates that stabilize heterogeneous thoughtseed dynamics, consistent with empirical findings of specialized network-level differentiation during meditation (Escrichs et al. 2019; Fox et al. 2015).

Level 2: Thoughtseed Dynamics. Thoughtseeds are modeled as attentional agents. While they do not possess "attention" themselves, they embody knowledge structures at a higher level of abstraction than simple memory traces. In this capacity, they selectively respond to attentional dynamics and contextualize the process to provide *affordances*—that is, actionable possibilities, including mental action—within the agent's generative model (Ramstead et al., 2021; Parr et al. 2022) Emerging from Level 1 network substrates, each thoughtseed is characterized by a *contribution profile*—a set of weights that determines how strongly it activates or suppresses each attentional network. For example, *breath_focus* upregulates DAN while suppressing DMN, whereas *pending_tasks* does the opposite (Fox et al. 2015). A thoughtseed also possesses a *dynamic activation level*, reflecting its current explanatory power over sensory and interoceptive inputs. See Table 2.

Competition among thoughtseeds for Global Workspace access is modeled as an approximation of Global Workspace broadcasting, using a simplified winner-takes-all mechanism for tractability (Baars 1997; Dehaene and Changeux 2011). When a thoughtseed dominates, its network profile is reinforced via top-down modulation, while competitors are attenuated. This reciprocal causality—bottom-up contributions from thoughtseeds to networks and top-down modulation from dominant thoughtseeds back onto networks—drives the emergence of meditative states (Raffone and Srinivasan 2017; Ramstead et al. 2020).

Level 3: Metacognitive Regulation. At the highest level, meta-cognitive processes regulate precision weighting and meta-awareness. This regulatory role is instantiated primarily through the FPN, which supports cognitive control and metacognitive monitoring, and the VAN, which detects salient events and can interrupt ongoing processes to trigger meta-awareness (Menon and Uddin 2010; Seeley et al. 2007). During meta-awareness, heightened FPN engagement increases the precision of signals associated with monitoring and control, facilitating rapid detection of mind-wandering and the redirection of attention (Hasenkamp et al. 2012). At the same time, DMN activity is suppressed, reducing intrusive self-referential content and stabilizing attentional focus (Brewer et al. 2011; Fox et al. 2015; Escrichs et al. 2019). In this sense, Level 3 functions as a regulatory mechanism on precision weighting, tuning the epistemic weight of prediction errors to govern attentional stability and the efficiency of recovery from distraction (Friston et al. 2017a, 2017b; Laukkonen and Slagter, 2021). This metacognitive regulation operates through base awareness levels specific to each meditative state, thoughtseed-derived influences (particularly *self-reflection* and *equanimity*), and experience-level modulation. Experts maintain higher baseline meta-awareness even during mind-wandering, but with more efficient processing that requires less explicit monitoring—consistent with the 'monitoring becomes effortless' phenomenon documented in long-term practitioners (Lutz et al., 2015; Dahl et al., 2015).

2.3 The Generative Model: Thoughtseeds and the Global Workspace

The four-stage Vipassana cycle (breath control, mind-wandering, meta-awareness, redirect attention) (Hasenkamp et al. 2012) is formalized as a generative model with hidden states s encoding prior beliefs about expected network activations. Each state specifies a prediction profile.

At each timestep t, networks generate observed activations $n_k(t)$, influenced by both thoughtseed activity and state expectations:

$$n_k(t) = (1 - \zeta) \sum_i W_{ik} z_i(t) + \zeta \mu_k(s_t) + \varepsilon_k(t) \tag{1}$$

where:

$Z_i(t)$ = activation of thoughtseed i

W_{ik} = contribution of thoughtseed i to network k

$\mu_k(S_t)$ = expected activation of network k under current state s_t.

We take $\mu_k(s_t) = \mathbb{E}[n_k(t)|s_t]$ and and evaluate it at s_t.

ζ = blending factor for bottom-up vs. top-down influences.

Network activations $n_k(t)$ arise from a mixture of bottom-up thoughtseed contributions $W_{ik} z_i(t)$ and top-down state expectations $\mu_k(s_t)$, balanced by ζ, with Gaussian noise $\varepsilon_k(t)$.

Thoughtseeds function as generative agents with three key properties:

1. A contribution profile W_{ik}: shaping their influence on networks.

2. A dynamic activation level $Z_i(t)$ determining competition for Global Workspace access
3. Sensitivity to network feedback, allowing dominant thoughtseeds to reinforce or suppress ensembles in line with predictive coding dynamics (Friston 2017a, 2017b).
4. Meta-awareness acts as a modulator of these dynamics, amplifying top-down control by enhancing the influence of state expectations. For experts, this modulation is stronger (by a factor of 1.2), representing enhanced metacognitive control during meditation states.

When a thoughtseed's activation exceeds others, it gains GW access, shaping the global brain state through a positive feedback loop that reinforces its network influence until prediction errors accumulate or competing thoughtseeds strengthen (Baars 1997). For example, the *breath_focus* thoughtseed drives high DAN activation, moderate FPN engagement, and DMN suppression, while self_reflection activates DMN and FPN with moderate VAN involvement (Hasenkamp et al. 2012). The generative model approximates network activations as observations influenced by thoughtseed activations and state expectations, with prediction errors driving state transitions and learning. Emergent meditative states arise from specific network-thoughtseed ensembles, such as mind_wandering from DMN dominance with active *pending_tasks* and *pain_discomfort* thoughtseeds, and meta_awareness from FPN and VAN activation with enhanced *self_reflection* thoughtseed (Lutz et al. 2024). The model incorporates biologically plausible non-linear effects such as compression thresholds that prevent runaway activation, DMN-DAN anticorrelation boundaries (Fox et al. 2015), and memory factors that provide temporal stability. These non-linear mechanisms are parameterized differently for experts versus novices, with experts showing stronger DMN suppression during focused states and more efficient network reconfiguration during transitions (Brandmeyer and Delorme 2021).

2.4 Active Inference Mechanisms

The system evolves over time by minimizing variational free energy, balancing precision-weighted prediction errors with model complexity (Friston 2010; Parr et al. 2022).

Perception and Belief Updating: Network activations are updated through bidirectional influences using Eq. 1, balancing bottom-up and top-down influences with temporal stability, higher for experts than novices.

- Bottom-up processing: Thoughtseeds contribute to network activations based on their activity and contribution weights.
- Top-down processing: State-specific expectations modulate activations, scaled by meta-awareness
- Temporal dynamics: Previous activations influence current ones via autocorrelation, with experts showing greater stability (higher for experts).

Crucially, while thoughtseeds influence network activations, networks reciprocally modulate thoughtseed dynamics. Each network exerts characteristic effects on thoughtseeds based on its functional role, for example: DMN enhances *pending_tasks* and *self_reflection*, VAN amplifies *pain_discomfort*, DAN strengthens *breath_focus*, and FPN boosts *equanimity* and *self_reflection* (Menon and Uddin 2010). This circular causality, formalized within Active Inference (Ramstead et al. 2020), is hypothesized to drive the emergent four-phase cycle of Vipassana meditation. This formulation captures how thoughtseeds both shape and are shaped by network dynamics, creating the emergent properties observed in meditation states. Experts exhibit stronger temporal stability, reflected in higher autocorrelation and reduced noise in network activity (Escrichs et al. 2019).

Free Energy Calculation: We define free energy $F_t(s)$ as computed online at time t and conditioned on a candidate hidden state s, corresponding to the four stages of the Vipassana attentional cycle: Breath Focus (BF), Mind Wandering (MW), Meta-Awareness (MA), and Redirect Attention (RA) (Hasenkamp et al. 2012). Thus, $s \in \{\text{BF}, \text{MW}, \text{MA}, \text{RA}\}$. At each timestep the agent uses its current observations and beliefs (precisions, mappings) to evaluate $F_t(s)$ for the present state (learning) and for admissible next states (selection); the next state emerges as the $\arg\min_{s'} F_t(s')$ over admissible s', subject to threshold constraints.

$$F_t(s) = \underbrace{\sum_k \Pi_k(\psi_t)\left[n_k(t) - \mu_k(s_t)\right]^2}_{\text{Accuracy}} + \underbrace{\lambda\|W\|_F^2}_{\text{Complexity}} \tag{2}$$

Accuracy: The first term is the *precision weighted prediction error*, measuring the squared mismatch between observed network activation $\eta_k(t)$ and the expected activation profile $\mu_k(s_t)$ under candidate state S_t scaled by a precision factor $\prod_k(\psi_t)$ that increases with meta-awareness ψ_t. This reflects how well the candidate state explains observed dynamics.

Complexity: The second term is a *parsimony prior* on the mapping matrix$W$. Written as the squared Frobenius norm $\|W\|_F^2 = \sum_{i,k} W_{ik}^2$ constraining the overall magnitude of thoughtseed–network influences by discouraging overly complex mappings. In Bayesian terms, this is equivalent to imposing a Gaussian prior on W, so that λ acts as prior precision. Intuitively, this balances fidelity to observed network activations with model simplicity, biasing the system toward stable and parsimonious thoughtseed–network mappings. Parameter values for λ (0.2 experts, 0.4 novices) were selected to reflect empirical findings of reduced model complexity in experienced meditators (Escrichs et al., 2019).

This formulation can also be written in a form closer to canonical active inference:

$$F_t(s) \approx \underbrace{-\sum_k \Pi_k(\psi_t) \log p(n_k(t) \mid \mu_k(s_t))}_{\text{Accuracy}} + \underbrace{\lambda\|W\|_F^2}_{\text{Complexity}}$$

Under Gaussian observation assumptions, the negative log-likelihood expands to a squared prediction error term plus a constant independent of state or parameters. By defining precision as the inverse variance, the precision-weighted negative log-likelihood becomes algebraically equivalent to the squared-error formulation in Eq. (2), up to additive constants.

Learning: Contribution weights W_{ik}, which determine how strongly thoughtseed i influences network k, are updated through a local delta rule:

$$W_{ik} \leftarrow (1-\rho)\, W_{ik} + \eta\, \delta_k(t)\, z_i(t), \qquad \delta_k(t) = n_k(t) - \mu_k(s_t) \tag{3}$$

The prediction error $\delta_k(t)$ captures the instantaneous mismatch between observed and expected activation of network k under state s_t.

The update rule adjusts each contribution weight W_{ik} proportionally to this prediction error and to the activation of thoughtseed $z_i(t)$. The learning rate \eta controls how rapidly weights adapt (0.02 for experts vs. 0.01 for novices), while the decay factor $1-\rho$ prevents runaway growth and ensures stability over time. To preserve biological plausibility, weights are further clipped to a bounded range $[W_{min}, W_{max}]$ (e.g., [0.1, 0.9]), preventing extreme influence strengths.

Together, these mechanisms facilitate stable learning: weights gradually increase when a thoughtseed consistently explains network activity, and decay otherwise. The implementation incorporates a critical directional error signal that guides weight updates toward state-appropriate target values, rather than merely minimizing error magnitude. This directional guidance prevents oscillations and overshooting that would occur with simple error minimization, allowing the system to converge toward efficient network-thoughtseed mappings without drifting into unstable feedback cycles. Through this regulated learning process, the system continuously refines its generative model, thereby minimizing free energy over time (Friston et al. 2017a, 2017b).

Action: State Transitions as Policies are modeled to facilitate the four-phase meditative cycle (Breath Focus to Mind Wandering to Meta-Awareness to Redirect Attention). At each timestep t, the model evaluates admissible transitions from the current state s_t and selects the next state s_{t+1} if its free energy is lowest among admissible candidates, or and threshold conditions are satisfied:

$$P(s_{t+1} = s') = \begin{cases} 1, & \text{if } s' = \arg\min_{u \in \mathcal{A}(s_t)} F_t(u) \text{ and } \Theta(s_t \to s') = 1, \\ 0, & \text{otherwise.} \end{cases} \tag{4}$$

Here, $\mathcal{A}(s_t)$denotes the set of admissible next states under the meditative cycle, and $\Theta(s_t \to s')$is a binary indicator function that evaluates whether threshold conditions are met (e.g., sufficient activation of a triggering thoughtseed). Expertise modulates these thresholds: experts require a higher threshold for transitions into mind-wandering (greater resistance to distraction) and a lower threshold for meta-awareness (faster detection of attentional lapses) (Hasenkamp et al. 2012; Brewer et al. 2011).

The notation $P(s_{t+1} = s')$ expresses the probability that the system will occupy state $s\prime$ at the next timestep, given its current state s_t and information available at time t. While the theoretical formulation allows for fully probabilistic transitions, our current implementation uses a hybrid approach combining deterministic rules with stochastic elements. The model implements transitions through a two-stage process: First, free energy differences generate transition probabilities that are precision weighted (with higher weighting for experts at 1.5 versus 0.8 for novices); second, these transitions are gated by thought seed-specific activation thresholds. Experts show stronger resistance to distractions (mind-wandering threshold of 0.75 versus 0.65 for novices) and more efficient metacognitive monitoring (meta-awareness threshold of 0.45 versus 0.55).

These threshold-modulated transitions constitute policy selections that minimize expected free energy over time, guiding the system toward preferred states (breath focus, meta-awareness) with lower prediction errors. Although this implementation preserves computational tractability, future versions will implement fully probabilistic state transitions to better capture the stochastic and nonlinear nature of real-world meditative dynamics (Laukkonen and Slagter 2021; Brandmeyer and Delorme 2021).

Learning and Meta-awareness Algorithms are described in the Supplementary section.

3 Simulation Results

3.1 Simulation Parameters

Two computational agents were simulated for 200 timesteps, representing novice and expert meditators. The agents differed in parameter values, summarized in Table 1. These values were informed by empirical literature on meditation expertise but are not direct measurements; they instantiate theoretical assumptions about how expertise modulates attentional dynamics.

These parameter differences drive distinct system dynamics. Higher precision in experts enhances confidence in attentional predictions, reducing susceptibility to distraction. Lower complexity penalties reflect more efficient generative models requiring fewer resources. Faster learning rates support rapid adaptation, while increased memory factors and reduced noise ensure greater temporal stability. Stronger DMN suppression and FPN enhancement align with empirical findings of reduced self-referential processing and improved control in experienced meditators (Brewer et al. 2011; Tang et al, 2015; Escrichs et al. 2019; Fox et al. 2015).

3.2 Hierarchical Dynamics and Bidirectional Coupling

The results illustrate how expertise modulates dynamics across the three hierarchical levels of the Thoughtseeds Framework. See Fig. 2.

At Level 1 (attentional networks), novices exhibited higher baseline DMN activation than experts (0.31 vs. 0.18) and greater variability across networks, reflecting reduced attentional stability. Experts displayed stronger DAN engagement with significantly reduced DMN dominance, particularly during focused states. This pattern

Table 1. Simulation Parameters for Novice and Expert Agents

Parameter	Novice	Expert	Rationale
Precision Weight Π	0.4	0.5	Experts assign higher precision to predictions, improving attentional stability
Complexity prior precision λ	0.4	0.2	Experts maintain simpler generative models (smaller λ means looser prior)
Learning Rate γ	0.01	0.02	Experts update beliefs more efficiently
Memory Factor μ	0.7	0.8	Experts exhibit more stable network dynamics
Noise Level η	0.08	0.04	Experts show reduced neural variability
DMN Suppression Factor	0.7	0.6	Experts more effectively inhibit DMN activity
FPN Enhancement Factor	1	1.2	Experts exhibit stronger metacognitive control

aligns with neuroimaging findings of enhanced attentional control and reduced mind-wandering in experienced meditators (Escrichs et al. 2019; Fox et al. 2015). Additionally, experts showed stronger DAN-FPN coupling, suggesting more integrated cognitive control networks.

At Level 2 (thoughtseeds), Global Workspace competition dynamics differed markedly between groups. In novices, *pending_tasks* and *self_reflection* frequently displaced *breath_focus*, producing irregular oscillations in thoughtseed activations. Experts maintained longer periods of *breath_focus* dominance (mean duration: 15.8 vs. 7.2 timesteps) and showed frequent *equanimity* activation (42% vs. 15% of timesteps), yielding smoother attentional trajectories and 46% faster recovery from distractions. These patterns reproduce empirical observations of attentional stability in expert meditators (Tang et al. 2015).

At Level 3 (meditative states), novices spent more time in Mind Wandering (35% vs. 21% of total simulation) and exhibited longer delays before transitioning to Meta-Awareness or Redirect-Attention. Experts transitioned more rapidly from distraction back to Breath-Focus, reflecting more efficient metacognitive monitoring. These differences align with empirical reports of shorter mind-wandering episodes and quicker attention redirection in experienced practitioners (Hasenkamp et al. 2012; Brewer et al. 2011; Christoff et al. 2009).

The simulation implemented two types of state transitions: (1) "natural transitions" that emerge organically when thoughtseed activations and network patterns evolve to cross specific thresholds without external triggers; and (2) "forced transitions" that occur

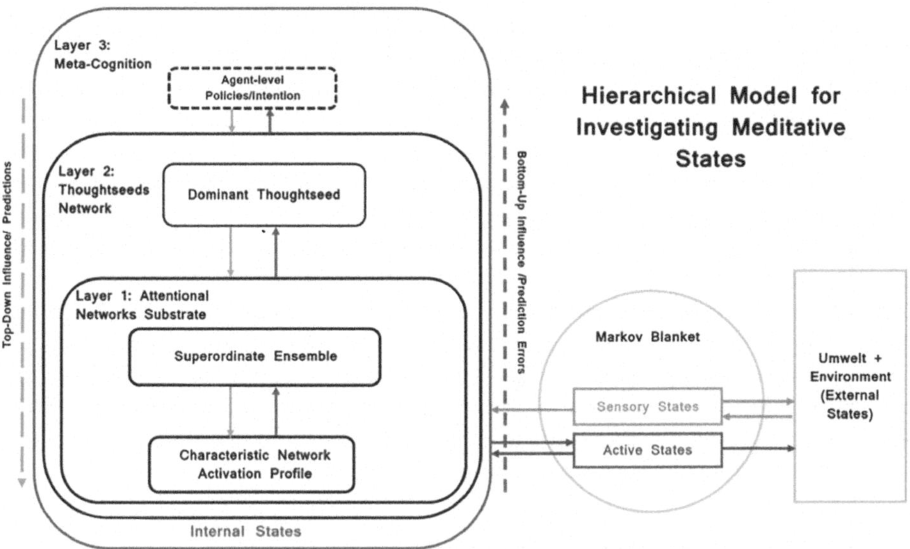

Fig. 1. Three-level hierarchical organization of Thoughtseeds Framework, rooted in Active Inference and Global Workspace Theory (GWT). **Level 1: Attentional Network Substrate** comprises four simplified attentional networks (Yeo et al. 2011)—DMN, VAN, DAN, and FPN. These networks are substrates, i.e., large-scale functional ensembles of neuronal packets at nested scales that provide the context for finer-grained dynamics. **Level 2: Thoughtseed Network** consists of attentional agents (thoughtseeds), such as breath_focus, pending_tasks, or self_reflection, which emerge from coordinated activity across the Level 1 ensembles. Thoughtseeds are hypothesized to represent specific contents of thought or attention, and compete for dominance in the Global Workspace via a simplified winner-takes-all dynamics (Baars 1997; Dehaene & Changeux 2011). **Level 3: Meta-cognition:** regulates precision weighting and meta-awareness, gating which thoughtseeds are amplified or suppressed to align behavior with meditative goals. Bidirectional arrows indicate reciprocal influences: networks bias the emergence of thoughtseeds (bottom-up), while dominant thoughtseeds modulate network activations (top-down). This circular causality reflects the Neuronal Packet Hypothesis, whereby nested Markov blankets maintain functional specialization while enabling dynamic coordination (Yufik & Friston 2016; Ramstead et al. 2021), aligning behavior with meditative goals (Laukkonen and Slagter 2021). **Bidirectional information flow** is depicted by red arrows (bottom-up prediction errors from networks to thoughtseeds) and green arrows (top-down predictions from thoughtseeds to networks), reflecting dynamic interactions across levels, each functioning as a nested Markov blanket (Friston 2013). The system interfaces with external states, labeled "Umwelt + Environment," is a manifestation of the Markov blanket of the Agent or sentient being, through sensory and active states, aligning with embodied cognition principles (Varela et al. 2017).

when predetermined conditions are met, ensuring the simulation cycles through all meditative states for analysis. Experts generated more natural transitions than novices (5 out of 21 vs. 2 out of 17), indicating greater emergent self-organization of attentional states. Current deterministic transitions represent a simplified model of how free energy minimization drives attentional shifts in meditation, where prediction errors typically accumulate gradually until triggering state changes. This implementation provides experimental control but sacrifices ecological validity, as real meditative experiences feature more spontaneous and context-sensitive transitions (Hasenkamp et al. 2012; Brandmeyer and Delorme 2021).

Furthermore, the results highlight the model's core feature of bidirectional causality (Table 2): bottom-up thoughtseed activations drive network dynamics, while dominant states exert top-down modulation to stabilize attentional focus. Expertise strengthens this coupling, making attentional control more robust and reducing variability across levels. This demonstrates how stable meditative states emerge from coordinated interaction across networks, thoughtseeds, and metacognitive regulation.

3.3 Free Energy Comparison and Dynamics

To quantify overall agentic behavior, we compared the free energy values across meditative states in novice and expert agents (Fig. 3). Free energy here combines two additive components: precision-weighted prediction errors (accuracy) and a complexity penalty (parsimony). Lower values therefore indicate states where observed network activity more closely matches state expectations with minimal model complexity.

Experts consistently exhibited lower free energy than novices across all states, reflecting more stable and efficient attentional dynamics. The largest difference was observed in Breath Focus, where experts achieved a 32.1% reduction relative to novices. This suggests that expertise improves alignment between state predictions and actual network activations, yielding more reliable focus. Smaller but consistent reductions were also observed in Mind Wandering, Meta-Awareness, and Redirect Attention, indicating that expertise improves both the stability of focused states and the efficiency of recovery from distraction.

These results support the interpretation of free energy as a *computational proxy* for stability and efficiency of attentional dynamics during meditation. Lower free energy does not directly measure subjective quality of meditation, but captures how well the generative model balances accuracy and parsimony during attentional cycles. The observed novice–expert differences align with empirical findings that experienced practitioners display reduced neural variability, greater attentional stability, and more efficient recovery from distraction (Brewer et al. 2011; Escrichs et al. 2019; Fox et al. 2015); Laukkonen and Slagter 2021).

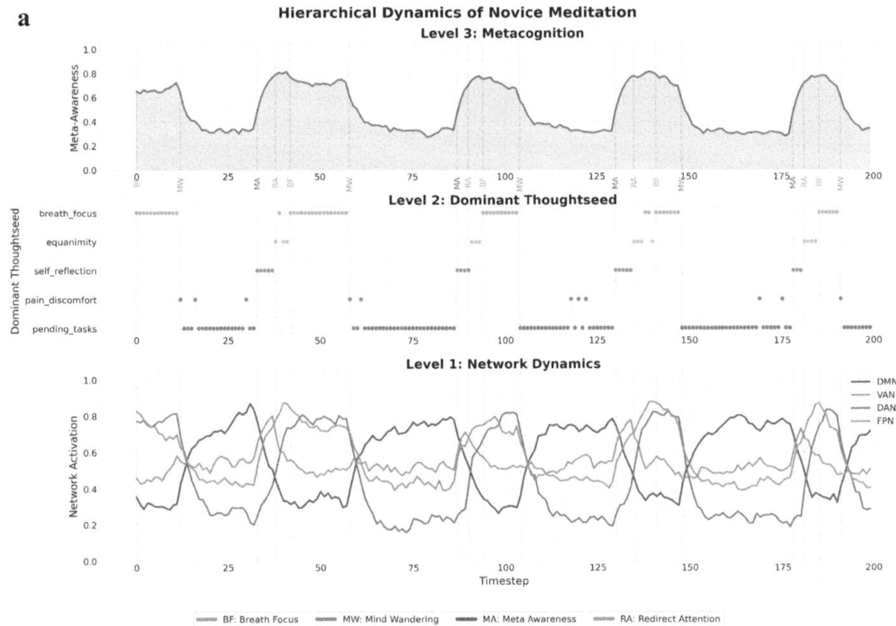

Fig. 2. Hierarchical Dynamics of Meditation displays meditation simulations over 200 timesteps, showing meta-awareness, dominant thoughtseeds, and network activations (DMN, VAN, DAN, FPN). Vertical dashed lines mark state transitions: Breath Focus (BF), Mind Wandering (MW), Meta-Awareness (MA), Redirect Attention (RA). **Novice Simulation (Fig. 2A):** Meta-awareness: Highly variable (mean: 0.63, SD: 0.19), reflecting weak metacognitive stability. Thoughtseeds: Frequent shifts between breath_focus and distractions (pain_discomfort, pending_tasks), with brief self-reflection before MA. Networks: Poor state differentiation, with high DMN during BF (mean: 0.31) and weak DAN-FPN coupling (correlation: 0.45), consistent with novice neural patterns (Brewer et al. 2011; Lutz et al. 2008). **Expert Simulation (Fig. 2B):** Meta-awareness: Stable and high (mean: 0.78, SD: 0.12), reflecting strong metacognitive skills (Fox et al. 2015). Thoughtseeds: Prolonged breath_focus (mean duration: 15.8 vs. 7.2 timesteps) and frequent equanimity (42% vs. 15% timesteps), aligning with advanced practice [Dahl et al. 2015). Networks: Clear state-specific patterns, with low DMN during BF (mean: 0.18) and strong DAN-FPN coupling (correlation: 0.72), mirroring expert neural signatures (Tang et al. 2015; Fox et al. 2015). Expert Dynamics: Show tighter coordination across levels, with meta-awareness spikes triggering thoughtseed shifts and network reconfigurations, supporting enhanced predictive processing (Barrett & Simmons 2015). Experts exhibit 37% less meta-awareness volatility during BF, 119% longer focus duration, and 46% faster recovery from distractions, consistent with empirical expertise findings (Lutz et al. 2008).

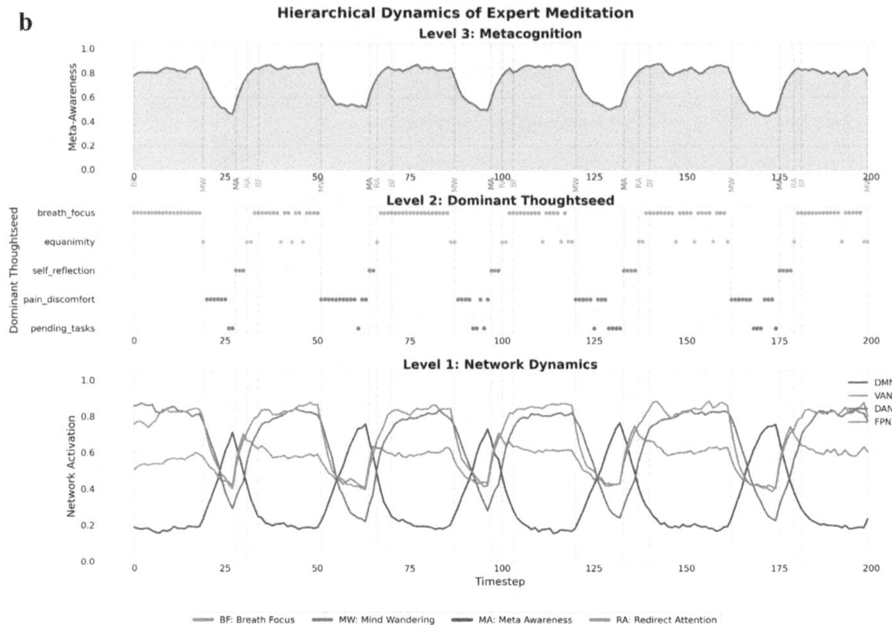

Fig. 2. *(continued)*

4 Discussion

The present study advances the *computational phenomenology* (Sandved-Smith et al. 2021; Ramstead et al., 2022) of meditation. Specifically, it introduces a hierarchical Active Inference formulation that operationalizes the Thoughtseeds Framework (Kavi et al. 2025). By embedding thoughtseeds as agent-like units within attentional networks and regulating their dynamics through precision-weighted prediction errors and metacognitive gating, the model captures both the phenomenology and neurobiology of the four-stage Vipassana cycle (Hasenkamp et al. 2012; Lutz et al. 2008). Simulations highlight how expertise reshapes attentional dynamics, producing more stable network trajectories, greater resilience to distraction, and lower overall free energy, providing a principled bridge between lived meditative experience, large-scale network neuroscience and predictive processing in the brain (Menon and Uddin 2010; Escrichs et al. 2019; Fox et al. 2015; Laukkonen and Slagter 2021).

4.1 Key Contributions

The contributions of this work can be summarized as follows:

1. Initial Integration with Active Inference: We introduce an enhancement to the Thoughtseeds Framework, a computational architecture that integrates the Neuronal Packet Hypothesis (Yufik and Friston 2016), Global Workspace Theory (Baars 1997; Dehaene and Changeux 2011), and Active Inference (Friston 2010; Parr and Friston

Table 2. Bidirectional Information Flow in Meditation

Direction	Pathway	Relationship	Examples
Bottom-up Processing (Lower to Higher)	**Level 1 → 2**	Networks influencing Thoughtseeds	High DAN activation → increased breath focus DMN activation → enhanced self-referential thoughts VAN activation → pain_discomfort detection FPN activation → meta-cognitive processing
	Level 2 → 3	Thoughtseeds influencing States	breath focus thoughtseed → breath control state pain_discomfort thoughtseed → mind wanderingpending_tasks thoughtseed → mind_wandering self-reflection → meta-awareness state equanimity → redirect breath state
Top-down Processing (Higher to Lower)	**Level 3 → 2**	States constraining Thoughtseeds	Breath control state → breath focus Mind wandering → pending tasks and pain_discomfort Meta-awareness → self-reflection Redirect breath → equanimity
	Level 2 → 1	Thoughtseeds modulating Networks	breath_focus → DAN↑, DMN↓ pending_tasks → DMN↑, DAN↓ pain_discomfort → VAN↑ self-reflection → FPN↑ equanimity → FPN↑, balanced network activation

2022). This formulation captures transient attentional agents (thoughtseeds) as Markov-blanket–defined units that interact with large-scale attentional networks (Menon and Uddin 2010; Yeo et al. 2011). While not implementing full Active Inference formalisms, our model operationalizes key active inference principles including precision-weighted prediction errors, belief updating, and complexity regularization. Furthermore, free energy is conceptualized here as a computational proxy for attentional stability and efficiency, decomposed into accuracy (precision-weighted prediction errors) and parsimony (complexity costs) (Friston 2010; Parr et al. 2022; Sandved-Smith et al. 2021; Ramstead et al. 2021). Reductions in free energy observed in expert simulations align with empirical reports of more reliable network coordination in advanced practitioners (Brewer et al. 2011; Hasenkamp et al. 2012), advancing recent hierarchical models on computational phenomenology (Sandved-Smith et al. 2021; Kavi et al. 2025).

2. Enhancing Thoughtseeds Framework with Attentional Networks: The three-level structure (attentional networks → thought seeds → meditative states) aims to capture the nested organization of the meditative mind, consistent with predictive processing

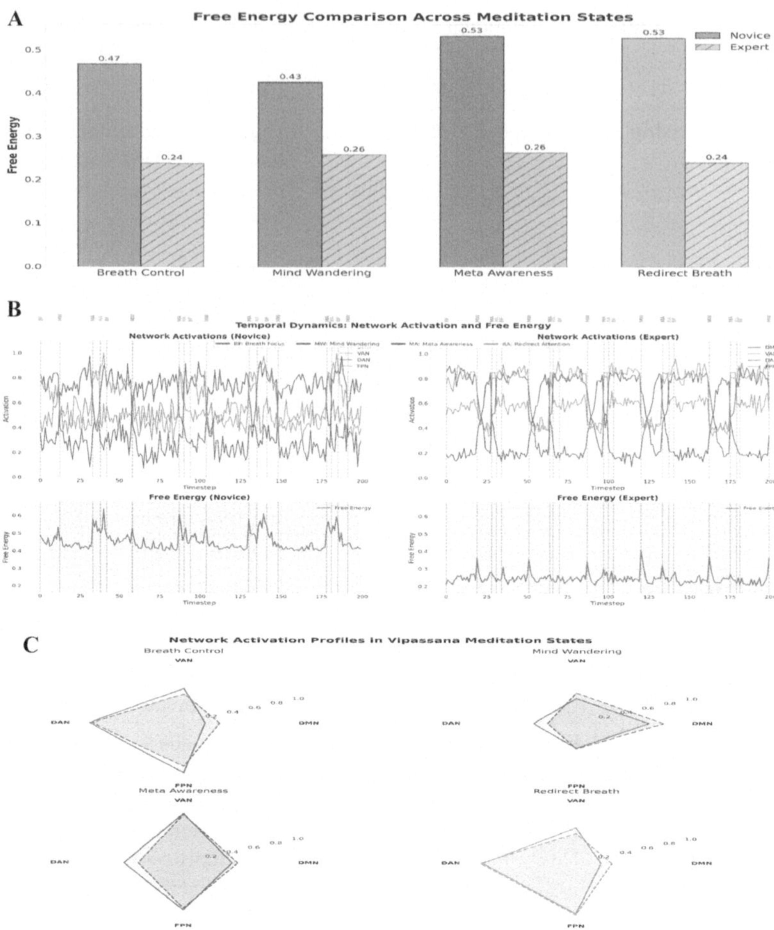

Fig. 3. Free Energy and Network Dynamics. **3. A Bar Chart**: Compares average free energy across four meditative states (breath control, mind wandering, meta-awareness, redirect breath) for novice and expert agents. Experts show lower free energy: 48.9% reduction in breath control (0.24 vs. 0.46), 39.3% in mind wandering (0.26 vs. 0.43), 50.5% in meta-awareness (0.26 vs. 0.533), and 54.4% in redirect breath (0.24 vs. 0.53), aligning with improved attention in experts (Tang et al. 2015). **3.B Time-Series Plots**: Show network activations (top) and free energy fluctuations (bottom) for novice (left) and expert (right) over 200 timesteps, with dashed lines marking state transitions. Novices exhibit high free energy spikes (SD: 0.13) driven by DMN volatility, often leading to mind wandering (Fox et al. 2015). Experts maintain stable activations with lower spikes (SD: 0.07) and 68% faster recovery (0.015 vs. 0.047 units/timestep) (Tang et al. 2015). **3.C Radar Plots**: Compare network activation profiles across states for novice (dashed) and expert (solid) agents. Experts show stronger DAN (0.82 vs. 0.79) and lower DMN (0.18 vs. 0.31) in breath control, higher DAN (0.36 vs. 0.24) in mind wandering, higher VAN/DAN (0.85/0.51 vs. 0.83/0.39) in meta-awareness, and DMN suppression (0.21 vs. 0.30) with enhanced VAN (0.60 vs. 0.51) in redirect breath, matching expert neural patterns (Brewer et al. 2011; Fox et al. 2015).

accounts (Laukkonen and Slagter 2021) and meditation literatures (Anālayo 2021). It is further grounded with attentional networks (Yeo et al. 2011) allowing falsifiability and enhanced biological plausibility for further enhancements of this framework.

3. Mechanistic Account of Novice–Expert Differences: Expertise is instantiated through biologically plausible parameter changes—precision weighting, learning rates, and complexity costs—which directly affect attentional stability, variability, and recovery from distraction (Tang et al. 2015; Escrichs et al. 2019; Fox et al. 2015). This modeling choice is consistent with empirical findings that experienced meditators exhibit reduced DMN dominance, enhanced DAN/FPN engagement, and greater attentional stability (Brewer et al. 2011; Fox et al. 2015; Seeley et al. 2007; Menon & Uddin 2010). Furthermore, the model reproduces hallmark differences reported in meditation neuroscience: novices show more volatile thoughtseed competition and prolonged mind-wandering, whereas experts show more consistent dominance of breath_focus, faster emergence of meta-awareness, and reduced variability (Hasenkamp et al. 2012; Fox et al. 2015; Tang et al. 2015).

4. Bidirectional Coupling: A key innovation in our model is the implementation of bidirectional coupling between mental contents (thoughtseeds) and attentional networks. This reciprocal interaction mechanism addresses the fundamental challenge of integrating top-down and bottom-up processes in meditation practice (Laukkonen and Slagter 2021; Raffone and Srinivasan, 2017). The bidirectional coupling operates through four distinct directional relationships as shown in Table 2.

4.2 Limitations and Future Directions

The Thoughtseeds Framework offers a promising computational approach to understanding meditation that bridges phenomenology, neuroscience, and active inference principles. By modeling the bidirectional interactions between neural and cognitive processes, it provides testable predictions about how meditation practices transform attentional dynamics. However, several limitations should be noted.

First, while the present implementation incorporates nonlinear dynamics through recurrent feedback and precision-weighted competition, the state transition mechanism remains partially constrained by deterministic rules. Future versions will implement fully probabilistic state transitions using proper message passing using a fuller active inference paradigm, better capturing the stochastic nature of attention cycling and allowing more naturalistic emergent dynamics.

Second, while our implementation ensures thoughtseeds remain differentiated through distinct priors and selective gating, a full active inference formulation would derive this differentiation more rigorously from first principles of variational free energy minimization. Future work could more explicitly characterize the free energy landscapes of meditative states as attractor basins in phase space, with transitions corresponding to trajectories driven by precision modulation and prediction error accumulation. Mapping these attractor dynamics could significantly enhance the interpretability of thoughtseeds within an active inference framework, and help explain how distinct thoughtseed agents emerge and persist as separable Markov blankets within a unified cognitive system.

Finally, the parameter values used to distinguish novices from experts (e.g., precision weighting, learning rate, complexity penalties) were guided by empirical findings but remain theoretical assumptions with a rules based approach rather than directly measured neurobiological constants. Systematic sensitivity analyses will be required to establish robustness, and empirical calibration against neuroimaging and behavioral measures will be essential. For example, mind-wandering probes, breath-counting tasks, or neurodynamical signatures could provide convergent validation (Lutz et al. 2024). More broadly, while the model reproduces empirical trends such as DMN suppression, FPN enhancement, and reduced variability in experts, its parameters and network activation patterns could be more precisely grounded using data-driven approaches. Future work should extract activation signatures from fMRI studies with Hidden Markov Models (HMM) or clustering algorithms that differentiate meditation-specific hidden states, providing empirically derived priors and target activations. Similarly, the thoughtseed constructs remain hypotheses requiring novel experimental paradigms—for example, combining neuroimaging with experience sampling to reveal neural correlates of agent-like attentional processes. Incorporating neuromodulatory mechanisms such as noradrenaline- and dopamine-related precision modulation (Parr and Friston 2022; Laukkonen and Slagter 2021), along with task or context-specific influences, will further strengthen the biological grounding of the framework in future iterations.

Despite these limitations, the model provides a proof-of-concept for computational phenomenology: formalizing subjective cycles of meditation within predictive processing and Active Inference accounts. Equations, parameter values, and update rules required to reproduce the presented simulations are explicitly provided in Sects. 2–3, ensuring methodological transparency. In this formulation, "breath focus" is treated both as an attentional state and as an embodied action policy, aligning attentional network selection with observable meditative practice. Future work could validate these dynamics against behavioral probes, such as breath-counting accuracy and mind-wandering reports, using neurophenomenological paradigms (Varela et al., 2017). Beyond meditation, this approach may inform studies of attention, mind-wandering, and meta-awareness more broadly, offering a generalizable methodology for linking lived experience to computational models of cognition.

Acknowledgments. We thank the anonymous IWAI 2025 Conference reviewers for their detailed and constructive feedback, which helped us improve the clarity of the manuscript. This research was partially funded by Grant PID2021-122136OB-C22 funded by MICIU/AEI/https://doi.org/10.13039/501100011033 and by ERDF A way of making Europe by Gustavo Patow.

Disclosure of Interests. The authors declare no conflicts of interest.

Data Availability Statement. The code is available in GitHub repository https://github.com/prakash-kavi/aif_iwai2025_thoughtseeds.

Supplementary Section

Notation (symbols and meanings)

$$\begin{aligned}
s_t &: \text{meditation state at time } t\\
\mathbf{z}_t &: \text{thoughtseed activations (breath_focus, pending_tasks,}\\
&: \text{pain_discomfort, self_reflection, equanimity)}\\
\mathbf{n}_t &: \text{network activations (DMN, VAN, DAN, FPN)}\\
\mu(s_t) &: \text{state-specific target network profile (Eq. 1)}\\
\mathbf{W} &: \text{coupling matrix (thoughtseeds} \rightarrow \text{networks)}\\
\psi_t \in [0,1] &: \text{meta-awareness at time } t\\
\Pi_t = \Pi(\psi_t, E) &: \text{precision (increases with } \psi_t)\\
F_t &: \text{free energy (Eq. 2)}
\end{aligned}$$

Parameter values by experience level

Global constants (used in both cohorts).

$$\begin{aligned}
\alpha &= 0.9 \quad \text{(target blend)}\\
\text{base}(s) &= \{\text{BF}: 0.50,\ \text{MW}: 0.25,\ \text{MA}: 0.70,\ \text{RA}: 0.60\}\\
w_{\text{self_ref}} &= 0.20\\
w_{\text{equanimity}} &= 0.15\\
\Delta_E &= 1[E = \text{expert}] \cdot 0.20\\
\Pi_t &= 0.5 + \Pi_W(E)\,\psi_t \quad \text{(precision; baseline 0.5)}
\end{aligned}$$

Active inference parameters (novice vs. expert).

$$\begin{aligned}
\Pi_W(E) &: 0.4 \text{ (novice)},\ 0.5 \text{ (expert)}\\
\lambda \text{ (complexity penalty)} &: 0.4 \text{ (novice)},\ 0.2 \text{ (expert)}\\
\eta \text{ (learning rate)} &: 0.01 \text{ (novice)},\ 0.02 \text{ (expert)}\\
\text{noise_level} &: 0.06 \text{ (novice)},\ 0.03 \text{ (expert)}\\
\text{memory_factor} &: 0.70 \text{ (novice)},\ 0.85 \text{ (expert)}\\
\text{fpn_enhancement} &: 1.0 \text{ (novice)},\ 1.2 \text{ (expert)}
\end{aligned}$$

Transition thresholds.

$$\begin{aligned}
\tau_{\text{MW}} \text{ (distraction)} &: 0.6 \text{ (novice)},\ 0.7 \text{ (expert)}\\
\text{DMN/DAN ratio} &: 0.5 \text{ (novice)},\ 0.6 \text{ (expert)}\\
\tau_{\text{MA}} \text{ (self-reflection)} &: 0.4 \text{ (novice)},\ 0.3 \text{ (expert)}\\
\tau_{\text{RB}} \text{ (return_focus)} &: 0.3 \text{ (novice)},\ 0.25 \text{ (expert)}
\end{aligned}$$

Dwell times (Gamma shape, scale) per state.

$$\begin{aligned}
\text{BF} &: (4.0, 2.5) \text{ (novice)},\ (10.0, 2.0) \text{ (expert)}\\
\text{MW} &: (9.0, 2.5) \text{ (novice)},\ (6.0, 1.5) \text{ (expert)}\\
\text{MA} &: (3.0, 1.0) \text{ (novice)},\ (2.0, 1.0) \text{ (expert)}\\
\text{RA} &: (3.0, 1.0) \text{ (novice)},\ (2.0, 1.0) \text{ (expert)}
\end{aligned}$$

Abbreviations: BF = breath_control, MW = mind_wandering, MA = meta_awareness, RA = redirect_breath.

Algorithm 1: Learning Loop (Simplified)

Data: Experience $E \in \{\text{novice}, \text{expert}\}$; steps T
Result: Histories $\{s_t, \mathbf{z}_t, \mathbf{n}_t, F_t\}_{t=1}^{T}$, weights $\mathbf{W}$

Init: $s \leftarrow$ BF,;
$\mathbf{z}$,;
$\mathbf{W}$;;
choose $\alpha, \zeta, \lambda, \eta, \rho$, BB and thresholds $\boldsymbol{\tau}(E)$;
for $t \leftarrow 1$ **to** T **do**
  $\psi \leftarrow \text{MetaAwareness}(s, \mathbf{z}, E)$;
  $\boldsymbol{\mu} \leftarrow \text{Targets}(s, \psi, E)$; $\mathbf{z} \leftarrow \alpha\, \boldsymbol{\mu} + (1-\alpha)\, \mathbf{z}$;
  $\mathbf{n} \leftarrow \zeta\, \mathbf{W}^{\top}\mathbf{z} + (1-\zeta)\, \boldsymbol{\mu}\, \psi$; $\mathbf{n} \leftarrow \text{NonLinear}(\mathbf{n}, E)$;
  $\Pi \leftarrow \Pi_0 + \Pi_W(E)\, \psi$; $\mathbf{e} \leftarrow \mathbf{n} - \boldsymbol{\mu}$; $F \leftarrow \Pi\, \|\mathbf{e}\|^2 + \lambda\, \|\mathbf{W}\|_F^2$;
  $\mathbf{W} \leftarrow (1-\rho)\, \mathbf{W} + \eta\, \dfrac{\mathbf{z}\, \mathbf{e}^{\top}}{\Pi}$; $\mathbf{W} \leftarrow \text{clip}(\mathbf{W})$;
  $i^{\star} \leftarrow \arg\max_i\, z_i$; $\mathbf{W}_{i^{\star}:} \leftarrow (1 + \text{BB})\, \mathbf{W}_{i^{\star}:}$;
  if dwell_expired(s) **then**
    if $s =$ BF *and* Distraction($\mathbf{z}$) $> \tau_{\text{MW}}(E)$ **then**
      $s \leftarrow$ MW
    else if $s =$ MW *and* SelfRef($\mathbf{z}$) $> \tau_{\text{MA}}(E)$ **then**
      $s \leftarrow$ MA
    else if $s =$ MA *and* (BreathFocus($\mathbf{z}$) $> \tau_{\text{RB}}(E)$ *or* Equanimity($\mathbf{z}$) $> \tau_{\text{RB}}(E)$) **then**
      $s \leftarrow$ RA
    else
      $s \leftarrow$ AdvanceBySequence(s)
  Record: append $(s, \mathbf{z}, \mathbf{n}, F, \psi)$ to history;

Algorithm 2: Meta-awareness (Simplified)

Data: State $s \in \{\text{BF}, \text{MW}, \text{MA}, \text{RA}\}$; thoughtseed vector $\mathbf{z}$; experience E
Result: $\psi \in [0, 1]$

$\psi \leftarrow \text{base}(s) + w_{\text{self\_ref}}\, z_{\text{self\_ref}} + w_{\text{equanimity}}\, z_{\text{equanimity}} + \Delta_E$;
if $E =$ expert **then**
  if $s =$ MA **then**
    $\psi \leftarrow 0.8\, \psi$
  else
    $\psi \leftarrow \max(0.30,\ 0.9\, \psi)$
$\psi \leftarrow \min(1, \max(0, \psi))$; **return** ψ;

References

Allen, M., Friston, K.J.: From cognitivism to autopoiesis: towards a computational framework for the embodied mind. Synthese **195**(6), 2459–2482 (2018)

Anālayo: Satipaṭṭhāna Meditation: A Practice Guide. Windhorse Publications (2019)

Anālayo, B.: Dependent arising and interdependence. Mindfulness **12**(5), 1094–1102 (2021)

Baars, B.J.: In the Theater of Consciousness: The Workspace of the Mind. Oxford University Press (1997)

Barrett, L.F., Simmons, W.K.: Interoceptive predictions in the brain. Nature Rev. Neurosci. **16**(7), 419–429 (2015)

Beck, J., Ramstead, M.J.D.: Dynamic Markov Blanket Detection for Macroscopic Physics Discovery (arXiv:2502.21217) [Preprint]. arXiv, 28 Feb 2025. https://doi.org/10.48550/arXiv.2502.21217

Brandmeyer, T., Delorme, A.: Meditation and the wandering mind: a theoretical framework of underlying neurocognitive mechanisms. Perspectives Psychol. Sci. **16**(1), 39–66 (2021)

Brewer, J.A., Worhunsky, P.D., Gray, J.R., Tang, Y.Y., Weber, J., Kober, H.: Meditation experience is associated with differences in default mode network activity and connectivity. Proc. Natl. Acad. Sci. **108**(50), 20254–20259 (2011)

Bullmore, E., Sporns, O.: Complex brain networks: graph theoretical analysis of structural and functional systems. Nature Rev. Neurosci. **10**(3), 186–198 (2009)

Cahn, B.R., Polich, J.: Meditation states and traits: EEG, ERP, and neuroimaging studies. Psychol. Bull. **132**(2), 180–211 (2006)

Christoff, K., Gordon, A.M., Smallwood, J., Smith, R., Schooler, J.W.: Experience sampling during fMRI reveals default network and executive system contributions to mind wandering. Proc. Natl. Acad. Sci. **106**(21), 8719–8724 (2009)

Craig, A.D.: How do you feel—now? the anterior insula and human awareness. Nat. Rev. Neurosci. **10**(1), 59–70 (2009)

Czajko, S., Zorn, J., Daumail, L., Chetelat, G., Margulies, D.S., Lutz, A.: Exploring the embodied mind: functional connectome fingerprinting of meditation expertise. Biological Psychiatry Global Open Sci. **4**(6), 100372 (2024)

Dahl, C.J., Lutz, A., Davidson, R.J.: Reconstructing and deconstructing the self: Cognitive mechanisms in meditation practice. Trends Cogn. Sci. **19**(9), 515–523 (2015)

Deco, G., Jirsa, V.K.: Ongoing cortical activity at rest: criticality, multistability, and ghost attractors. J. Neurosci. **32**(10), 3366–3375 (2012)

Deco, G., Vidaurre, D., Kringelbach, M.L.: Revisiting the global workspace orchestrating the hierarchical organization of the human brain. Nat. Hum. Behav. **5**(4), 497–511 (2021)

Dehaene, S., Changeux, J.P.: Experimental and theoretical approaches to conscious processing. Neuron **70**(2), 200–227 (2011)

Dehaene, S., Charles, L., King, J.R., Marti, S.: Toward a computational theory of conscious processing. Curr. Opin. Neurobiol. **25**, 76–84 (2014)

Escrichs, A., et al.: Characterizing the dynamical complexity underlying meditation. Front. Syst. Neurosci. **13**, 27 (2019)

Fox, K.C., et al.: Is meditation associated with altered brain structure? a systematic review and meta-analysis of morphometric neuroimaging in meditation practitioners. Neurosci. Biobehav. Rev. **43**, 48–73 (2014)

Fox, K.C., Spreng, R.N., Ellamil, M., Andrews-Hanna, J.R., Christoff, K.: The wandering brain: meta-analysis of functional neuroimaging studies of mind-wandering and related spontaneous thought processes. Neuroimage **111**, 611–621 (2015)

Friston, K.: The free-energy principle: a unified brain theory? Nature Rev. Neurosci. **11**(2), 127–138 (2010)

Friston, K.: Life as we know it. J. Roy. Soc. Interface **10**(86), 20130475 (2013)
Friston, K., Rigoli, F., Ognibene, D., Mathys, C., Fitzgerald, T., Pezzulo, G.: Active inference and epistemic value. Cogn. Neurosci. **2015**, 187–214 (2015)
Friston, K., Parr, T., de Vries, B.: The graphical brain: Belief propagation and active inference. Network Neurosci. **1**(4), 381–414 (2017)
Friston, K., FitzGerald, T., Rigoli, F., Schwartenbeck, P., Pezzulo, G.: Active inference: a process theory. Neural Comput. **29**(1), 1–49 (2017)
Hasenkamp, W., Wilson-Mendenhall, C.D., Duncan, E., Barsalou, L.W.: Mind wandering and attention during focused meditation: a fine-grained temporal analysis of fluctuating cognitive states. Neuroimage **59**(1), 750–760 (2012)
Hipólito, I., Ramstead, M.J.D., Convertino, L., Bhat, A., Friston, K., Parr, T.: Markov blankets in the brain. Neurosci. Biobehav. Rev. **125**, 88–97 (2021)
Hohwy, J.: The Predictive Mind. Oxford University Press (2013)
Hohwy, J.: The self-evidencing brain. Noûs **50**(2), 259–285 (2016)
Kavi, P.C., Zamora-López, G., Friedman, D.A., Patow, G.: Thoughtseeds: a hierarchical and agentic framework for investigating thought dynamics in meditative states. Entropy **27**(5), 459 (2025). https://doi.org/10.3390/e27050459
Kirchhoff, M., Parr, T., Palacios, E., Friston, K., Kiverstein, J.: The Markov blankets of life: autonomy, active inference and the free energy principle. J. R. Soc. Interface. **15**(138), 20170792 (2018)
Laukkonen, R.E., Slagter, H.A.: From many to one: the science of consciousness and the contemplative path to nondual awakening. Prog. Brain Res. **258**, 227–259 (2021)
Lee, A.Y.: Attention is presence: A phenomenology of subjectivity and cognitive access. [Doctoral dissertation, University of Toronto] (2021)
Lutz, A., Slagter, H.A., Dunne, J.D., Davidson, R.J.: Attention regulation and monitoring in meditation. Trends Cogn. Sci. **12**(4), 163–169 (2008)
Lutz, A., Abdoun, O., Dor-Ziderman, Y., Trautwein, F.M., Berkovich-Ohana, A.: An overview of neurophenomenological approaches to meditation and their relevance to clinical research. Biological Psychiatry: Cognitive Neuroscience and Neuroimaging (2024)
MacLean, K.A., et al.: Intensive meditation training improves perceptual discrimination and sustained attention. Psychol. Sci. **21**(6), 829–839 (2010)
Maturana, H.R., Varela, F.J.: Autopoiesis and Cognition: The Realization of the Living. D. Reidel Publishing Company: Dordrecht, The Netherlands (1980)
Menon, V., Uddin, L.Q.: Saliency, switching, attention and control: A network model of psychopathology. Trends Cogn. Sci. **14**(12), 549–557 (2010)
Metzinger, T. (2020). Minimal phenomenal experience. Philosophy and the Mind Sciences
Palacios, E.R., Razi, A., Parr, T., Kirchhoff, M., Friston, K.: On Markov blankets and hierarchical self-organisation. J. Theor. Biol. **486**, 110089 (2020)
Parr, T., Friston, K.J.: The active construction of conscious experience. Neuroscience of Consciousness **2018**(1), niy015 (2018)
Parr, T., Pezzulo, G., Friston, K.J.: Active Inference: The Free Energy Principle in Mind, Brain, and Behavior. MIT Press (2022)
Pearl, J.: Probabilistic Reasoning in Intelligent Systems: Networks of Plausible Inference. Morgan Kaufmann (1988)
Raffone, A., Srinivasan, N.: Mindfulness and cognitive functions: toward a unifying neurocognitive framework. Mindfulness **8**, 1–9 (2017)
Ramstead, M.J., Kirchhoff, M.D., Friston, K.J.: A tale of two densities: active inference is enactive inference. Adapt. Behav. **28**(4), 225–239 (2020)
Ramstead, M.J.D., Hesp, C., Tschantz, A., Smith, R., Constant, A., Friston, K.: Neural and phenotypic representation under the free-energy principle. Neurosci. Biobehav. Rev. **120**, 109–122 (2021)

Ramstead, M.J.D., Seth, A.K., Hesp, C., et al.: From generative models to generative passages: a computational approach to (neuro) phenomenology. Rev. Phil. Psych. **13**, 829–857 (2022)

Sandved-Smith, L., Hesp, C., Mattout, J., Friston, K., Lutz, A., Ramstead, M.J.D.: Towards a computational phenomenology of mental action: Modelling meta-awareness and attentional control with deep parametric active inference. Neurosci. Conscious., 2021, niab018 (2021)

Smallwood, J., Schooler, J.W.: The restless mind (2013)

Schooler, J.W.: Re-representing consciousness: dissociations between experience and meta-consciousness. Trends Cogn. Sci. **6**(8), 339–344 (2002)

Schooler, J.W., Smallwood, J., Christoff, K., Handy, T.C., Reichle, E.D., Sayette, M.A.: Meta-awareness, perceptual decoupling and the wandering mind. Trends Cogn. Sci. **15**(7), 319–326 (2011)

Seeley, W.W., et al.: Dissociable intrinsic connectivity networks for salience processing and executive control. J. Neurosci. **27**(9), 2349–2356 (2007)

Sporns, O., Betzel, R.F.: Modular brain networks. Annu. Rev. Psychol. **67**(1), 613–640 (2016)

Tang, Y.Y., Hölzel, B.K., Posner, M.I.: The neuroscience of mindfulness meditation. Nat. Rev. Neurosci. **16**(4), 213–225 (2015)

Vago, D.R., Silbersweig, D.A.: Self-awareness, self-regulation, and self-transcendence (S-ART): A framework for understanding the neurobiological mechanisms of mindfulness. Front. Hum. Neurosci. **6**, 296 (2012)

Van Dam, N.T., et al.: Mind the hype: A critical evaluation and prescriptive agenda for research on mindfulness and meditation. Perspect. Psychol. Sci. **13**(1), 36–61 (2018)

Varela, F. J., Thompson, E., & Rosch, E. (2017). The Embodied Mind: Cognitive Science and Human Experience. MIT Press

Thomas Yeo, B.T., et al.: The organization of the human cerebral cortex estimated by intrinsic functional connectivity. J. Neurophysiol. **106**(3), 1125–1165 (2011)

Yufik, Y.: The structure of cognition: Attentional episodes in mental state space. Front. Syst. Neurosci. **7**, 31 (2013)

Yufik, Y., Friston, K.: Life and understanding: The origins of "understanding" in self-organizing nervous systems. Front. Syst. Neurosci. **10**, 98 (2016)

Yufik, Y.M.: The understanding capacity and information dynamics in the human brain. Entropy **21**, 308 (2019)

Geometric Hyperscanning of Affect Under Active Inference

Nicolás Hinrichs[1,2(✉)], Mahault Albarracin[3,4], Dimitris Bolis[5], Yuyue Jiang[6,7], Leonardo Christov-Moore[8,9], and Leonhard Schilbach[10,11]

[1] Max Planck Institute for Human Cognitive and Brain Sciences, Leipzig, Germany
nicolas.hinrichs@oist.jp, hinrichsn@cbs.mpg.de
[2] Okinawa Institute of Science and Technology, Okinawa, Japan
[3] VERSES AI Research Lab, Los Angeles, USA
[4] Université du Québec à Montréal, Montréal, Québec, Canada
[5] Instituto Italiano di Tecnologia, Rovereto, Italy
[6] University of California, Santa Barbara, USA
[7] University of California, Irvine, USA
[8] Institute for Advanced Consciousness Studies, Santa Monica, USA
[9] SENSORIA Research, San Francisco, CA, USA
[10] Ludwig Maximilians Universität, Munich, Germany
[11] LVR-Klinikum Düsseldorf/Heinrich Heine University, Düsseldorf, Germany

Abstract. Second-person neuroscience holds social cognition as embodied meaning co-regulation through reciprocal interaction, modeled here as coupled active inference with affect emerging as inference over identity-relevant surprise. Each agent maintains a self-model that tracks violations in its predictive coherence while recursively modeling the other. Valence is computed from self-model prediction error, weighted by self-relevance, and modulated by prior affective states and by what we term temporal aiming, which captures affective appraisal over time. This accommodates shifts in the self-other boundary, allowing affect to emerge at individual and dyadic levels. We propose a novel method termed geometric hyperscanning, based on the Forman-Ricci curvature, to operationalize these processes empirically: it tracks topological reconfigurations in inter-brain networks, with its entropy serving as a proxy for affective phase transitions such as rupture, co-regulation, and reattunement.

Keywords: Active inference · Hyperscanning · Second-person neuroscience · Dyadic coupling · Forman-Ricci curvature · Phase transitions

1 Introduction

How do agents come to feel 'together'? How do they navigate the sharing of beliefs about the world and themselves, and the fragile terrain of affective meaning? Traditional models of social cognition often approach this question through

M. Albarracin et al. (Eds.): IWAI 2025, CCIS 2857, pp. 208–231, 2026.
https://doi.org/10.1007/978-3-032-16955-6_12

the lens of mental state attribution: one agent models another's beliefs, intentions, and emotions. Adopting a third-person stance towards someone has dominated theoretical and empirical social cognition research [9]. However, such models ignore a crucial fact: social understanding, in most of its ecologically valid forms, does not unfold in detached observation but in active, embodied engagement with others. As Schilbach et al. [37] argue, social interaction is not simply the backdrop against which inference occurs; instead, it is the medium through which meaning is constituted. This shift, often captured under the rubric of 'second-person neuroscience' [35,38], reframes the unit of analysis from the isolated agent to the dyad as a generative system in its own right, by taking mutual engagement in social interaction, rather than social observation, as its central explanandum. Recent formalizations of this framework through the lens of active inference [29] have provided the missing computational scaffolding: they model interacting agents as mutually coupled generative systems that co-regulate meaning via recursive belief updating. In this view, social understanding is not computed about another agent, but emerges while in social interaction with another through ongoing cycles of expectation, violation, and realignment. The dyad becomes a unit of analysis in its own right and can be regarded as a shared generative manifold, which can be modeled as a coupled system of intra- and interpersonal processes across multiple modalities [5,6].

1.1 The Affective Gap

We extend second-person active inference to the domain of affect, modeling it as recursive inference over two types of prediction error: mismatch between predictions and outcomes relevant to the world, and those relevant to identity. Drawing on Jiang and Luo's [25,26] model, valence reflects an inference over the integrity of the self-model, embedded in the dyadic generative manifold. Because both agents maintain self-models while recursively modeling each other, affective dynamics become entangled, shaping belief updating and action selection. This recursive process promotes interpersonal attunement and stabilization [6,39,42], while affective rupture triggers cognitively costly recalibration [3,10]. We propose geometric hyperscanning to link these formal dynamics to neural signatures, expanding the framework of second-person active inference and its formal dynamics of belief and affect to empirical research.

This paper makes three contributions. First, we formalize affect as recursive inference over self-model coherence. We model valence as identity-relevant prediction error weighted by self-relevance, modulated by affective memory, and shaped by temporal aiming—the agent's orientation across past and future affective states (Sect. 2.1). We show how recursive modeling of the other dynamically shifts self–other boundaries (Sect. 2.2). Second, we introduce geometric hyperscanning, an empirical method based on Forman-Ricci curvature (FRc), to track topological reconfigurations in inter-brain networks and infer affective phase transitions (Sect. 2.3). Third, we integrate this formal-empirical framework within second-person neuroscience and active inference. We offer a scalable architecture for modeling recursive affective dynamics across psychotherapy, development, and naturalistic interaction, and outline its broader implications (Sect. 3).

2 Second-Person Neuroscience and Dyadic Active Inference

Second-person neuroscience [37] holds that social understanding is constituted through real-time mutual engagement, as opposed to detached or post-hoc attribution. From a computational standpoint, this view is well-expressed in the active inference framework, in which agents minimize variational free energy to maintain a coherent embodied model of the world and their role within it. In second-person settings, this framework must be extended [7,29,41]: agents not only predict external states, but also the inferred generative models of others. These recursive beliefs form a shared generative manifold, wherein action and perception are jointly coordinated. We can conceive of each agent maintaining a generative model:

$$M = p(o_i, a_j, s_i, s_j \mid \pi_i) \tag{1}$$

where:

- o_i = observations available to agent i,
- a_j = inferred actions of agent j,
- s_i, s_j = hidden states of agents i and j,
- π_i = current policy or sequence of expected actions for agent i.

The recursive nature of this system (wherein both agents model each other, modeling them) does not lead to an infinite regress. Instead, as Friston and Frith (2015) demonstrated, it dissolves into a shared dynamic narrative, where both agents come to predict and enact the same process. In other words, both agents use the same model and the dyad, thus, becomes the minimal unit of inference.

In contrast, traditional third-person approaches to social cognition, such as Theory of Mind or mindreading paradigms, cast social understanding as an observer inferring another's hidden mental states. Classic experiments, for instance, might have a participant watch a video or read a story and then guess what another person believes or intends. These paradigms, while valid, involve a one-way attribution process, often in non-interactive settings. Neuroscientifically, third-person "mentalizing" tasks reliably engage a network including the dorsomedial prefrontal cortex (dmPFC), ventromedial prefrontal cortex (vmPFC), and temporoparietal junction (TPJ), which is often called the mentalizing or theory-of-mind network [19]. However, in a live interaction, one's brain is modeling the other and responding to them in real time, creating a feedback loop. This form of social connectedness has measurable two-brain dynamics: an engaged "I–Thou" exchange produces alignment of neural rhythms that does not occur in a one-directional "I observe him" situation (see [14,38] for a recent review).

[20] refers to these coupled networks as mutual social attention systems, proposing that when two people directly attend to each other, their brains temporarily form an integrated system. The second-person emphasis on real-time

mutual engagement dovetails with active inference, because both paradigms recognize that understanding others is an active, dynamical, and bidirectional alignment process.

2.1 Valence as Inference over the Self-model

Within this coupled generative architecture, affect is conceptualized as dynamic regulatory processes guiding cognition and behavior based on the integrity and coherence of an agent's predictive self-model [25,26]. Central to this approach is the idea that emotional valence signals the alignment or mismatch between predicted and observed outcomes, particularly those relevant to identity and social expectations [3].

Prior work has modeled valence as a derivative of free energy [27], a metacognitive signal [21], interoceptive inference [40], shifts in processing modes [46], and recursive affective dynamics linked to self-relevance and identity [1,2,26]. Building on these insights, and drawing on [26], we propose an extended model of emotional valence as a continuous inference over the integrity of the agent's self-model. Emotional states emerge from prediction errors tied to identity-relevant expectations. Valence operationalizes the emotional appraisal of interactions via two primary factors: (1) the magnitude of self-model prediction error, and (2) the self-relevance assigned to that error. We further incorporate a temporal parameter into this model, reflecting the directionality and velocity of affective processing. Building on [27,46], we introduce *temporal aiming* (see Fig. 1), encompassing both the velocity of affective evaluation (fast learning rates vs. slow enduring states) and its temporal direction (retrospective vs. prospective appraisal).

This modulation adds an essential dimension to affective inference: agents may respond differently to the same interaction depending on whether they are prospectively anxious, retrospectively regretful, or temporally stable. Temporal aiming thus complements self-relevance and prediction error as core variables in our valence model. We conceptualize emotions as functions of valence (alignment of identity-relevant predictions) and arousal (uncertainty and intensity of self-relevant evaluations), with both dimensions anchored in the evolving self-concept, shaped by reflection, comparison, and internalized ideals. In this framework, valence reflects the agent's appraisal of ongoing interaction or how "good" or "bad" the current moment feels to the agent, computed as a weighted function of self-model prediction error and self-relevance, and modulated by prior affective memory [25]:

$$V_t = \alpha \cdot PE_{\text{self},t} \cdot SR_t + \beta \cdot V_{SR,t} \tag{2}$$

where:

- $PE_{\text{self},t}$ is the prediction error between expected and actual self-state, capturing the "surprise" the agent experiences about itself
- SR_t is a dynamic scalar indexing the self-relevance of the context, reflecting how much the situation matters for the agent's identity

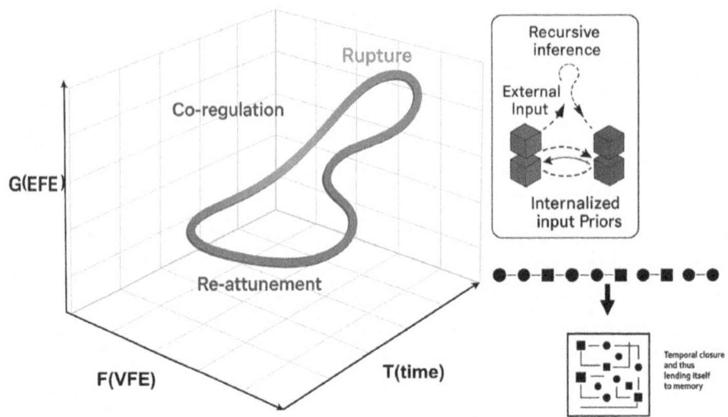

Fig. 1. *Temporal aiming in affective inference.* The figure illustrates how agents evaluate self-model prediction error over time according to different temporal orientations. Affect is shaped not only by the magnitude and relevance of prediction error, but also by the agent's temporal aim, i.e., whether inference is directed toward anticipated futures or prior experiences, and whether updates occur rapidly or gradually. This temporal structuring of affect modulates the stability and flexibility of policy selection within dyadic interaction.

- $V_{SR,t}$ is retrieved from similar prior contexts, representing the lingering emotional tone (mood) shaped by related past experiences,
- α, β are precision-modulating weights, that determine how strongly the agent reacts to immediate surprises versus how much past mood carries over.

It follows that whenever the agent experiences a self-related prediction error, it generates an affective response proportional to that error and to how self-relevant it is. The first term, $\alpha \cdot PE_{\text{self},t} \cdot SR_t$, encodes the instantaneous emotional impact: large, self-relevant surprises cause large valence shifts. The parameter α scales the impact of immediate prediction error on valence. An agent i with a larger α will exhibit heightened emotional sensitivity to surprises, particularly negative ones, which yield more negative valence. The second term, $\beta \cdot V_{SR,t}$ with $0 < \beta < 1$, he second term, $\beta \cdot V_{SR,t}$, introduces emotional inertia: it blends in past affective states, so that current mood is partly a continuation of recent emotional history, it makes valence a *leaky integrator*: it carries forward some fraction of previous valence (mood) into the current moment. The term $PE_{\text{self},t}$ includes both the agent's previous valence $V(t-1)$ and contributions from past episodes that had similar self-relevance or context as the current one. We can treat $V_{SR,t}$ as an exponentially decaying sum of past valences, with greater weight assigned to more recent times and to events with comparable self-relevance (SR). In the notation of the original model, self-model prediction error and self-relevance map onto PE_{self} and SR, respectively, where the last term of $V_{SR,t}$ helps to maintain the affective experience as iterative and continuous.

Valence Under the Dyadic Inference Framework. In dyadic active inference, each agent's self-model is treated as a dynamic function of its internal world model, which includes a simulated model of the other agent and incoming interactional feedback. At time t, agent i's self-model is conditioned on beliefs about the other agent's state, actions, and observations:

$$P_t^i = f(S_t^i, A_t^j, O_t^j) \tag{3}$$

This reflects that one's self-concept in a social setting is co-constructed by self-generated and socially observed information. Agent i's full generative model factorizes as:

$$P(O_i, A_i, S_i, S_j) = P(O_i \mid S_i, A_i) \cdot P(A_i \mid S_i, S_j) \cdot P(S_i \mid S_j) \tag{4}$$

This expresses that (1) the agent's observations depend on its state and actions, (2) its actions depend on both its own and the other's state, and (3) its own state is partly inferred from the other's state. Valence (V_i, t) measures how emotionally coherent the agent's self-model feels over time. It is computed from the prediction error in the self-model, modulated by both SR_t and $V_{SR,t}$. The expected valence is defined as:

$$\mathbb{E}(V_i^t) = \alpha \cdot PE_{\text{self},t} \cdot SR_t + \beta \cdot V_{SR,t} \tag{5}$$

where $PE^i_{\text{self},t}$ is the prediction error between expected and actual self-state, and alpha and beta are precision-modulating weights. The realized valence reflects how well observed outcomes under the current policypi_i^t align with these expectations; it is given by:

$$\begin{aligned} V_i^t &= \mathbb{E}_{P(o^t \mid \pi^t)} \left[\log P\left(o^t \mid \mathbb{E}(V_i^t)\right)\right] \\ &= \mathbb{E}_{P(o^t \mid \pi^t)} \left[\log P\left(o^t \mid \alpha \left|P_i^t - \hat{P}_i^t\right| SR_i^t + \beta V_{i,SR}^t\right)\right] \end{aligned} \tag{6}$$

This means that valence is the expectation, over possible outcomes, of how well those outcomes will sustain the agent's internalized social narrative, given its evolving model of the partner and the feedback from ongoing interaction.

Affective Valence as a Prior Over Policy in Planning-as-Inference. By design, valence serves as an internal reward/safety signal: high positive valence indicates that the interaction is proceeding smoothly, with strong affective synchrony, while negative valence signals a breakdown in synchrony, prompting either a new attempt to restore alignment or a withdrawal from the interaction, depending on individual differences in valence construction and perception (modeled by the parameters α and β). Which path the agent takes depends on individual differences in emotional sensitivity and persistence, captured by the parameters α and β Though we do not explicitly model all downstream effects here, valence functions as a feedback loop, a readout of the self-model's performance and modulating subsequent perception and action. In practical terms, this means that affect shapes the space of possible future actions by making certain policies feel "more natural" or "safer" than others. In our model, affect modulates the posterior over policies, effectively serving as a soft prior that biases action selection toward more affectively coherent trajectories. Following Da Costa's (2020) [13] planning-as-inference paradigm, we express policy selection as:

$$P(\pi_i \mid V_t) \propto e^{\lambda \cdot g(V_t, \pi_t)} \tag{7}$$

where:

- λ is an inverse temperature parameter controlling affective precision or confidence in the valence signal (high λ = more deterministic choice; low λ = more exploratory choice), and
- $g(V_t, \pi_t)$ is a compatibility function mapping affective coherence to policy likelihood (higher compatibility = higher policy probability).

This captures the recursive and value-sensitive nature of planning in social contexts. The agent is more likely to choose plans consistent with positive valence (i.e., minimal surprise and maximal narrative alignment), while negative valence reduces the likelihood of maintaining the current trajectory. When both agents are engaged in this recursive affective inference, the system forms a *dynamically coupled manifold* –a shared generative space in which each agent's self-coherence is entangled with the other's behavior. Moments of affective rupture mark local minima in joint narrative alignment and correspond to dyadic expected free energy peaks. Repairing these ruptures requires coordinated adjustments in beliefs and policies across both agents. This framework completes a recursive loop: expected valence, derived from self-model integrity and anticipated partner behavior, acts as a prior over policy selection, guiding the agent toward actions expected to preserve or restore affective coherence. These chosen policies lead to new interactions and observations, which update the agent's self-model and generate realized valence. In turn, the mismatch between expected and realized valence yields an affective prediction error that shapes the next inference cycle. This recursive loop (Fig. 2) summarizes how affectively modulated policies guide dyadic interaction through cycles of prediction error, self-model update, and realignment.

2.2 Recursive Coupling and Shared Surprise

In exploring the dynamics of affective interactions, it is helpful to draw upon the concept of intra-action from neo-materialist philosophy [4]. Intra-action emphasizes that interacting entities do not preexist independently but emerge through mutual entanglement. Under this lens, the dyadic interaction is not simply a transfer of emotional states between two separate agents; instead, it represents a continual co-construction of affective experiences and identities [33]. From this perspective, affective states are inherently recursive and relational, arising

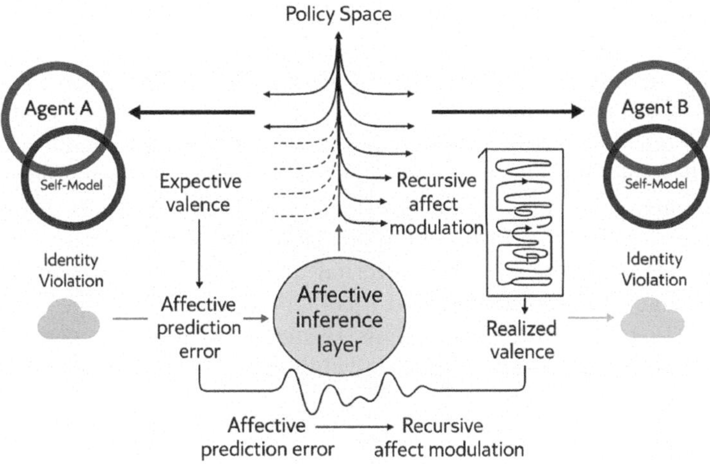

Fig. 2. *Recursive Affective Inference Framework for Dyadic Social Interaction.* Illustrates a recursive framework wherein affect dynamically regulates social behavior between two agents via cycles of inference, prediction error, and adaptive policy updates grounded in self-model coherence.

from the iterative feedback loops between agents. Affective ruptures, moments of misalignment or increased prediction error, serve as significant informational signals indicating points of divergence in the shared expectations of interacting agents. These affective evaluations modulate the posterior over policies, biasing action selection toward trajectories that restore narrative alignment and reduce identity-relevant surprise. Such an entanglement of affect and policy selection turns the dyad into a dynamic field of co-regulated affective inference, where both agents are simultaneously minimizing joint expected free energy:

$$G = \mathbb{E}_{q(s)}[\log q(s) - \log p(o, s)] \tag{8}$$

Then, in a dyadic setting, we can express the joint free energy as:

$$G_{\text{joint}} = G_i + G_j = \mathbb{E}_{q(s_i, s_j)}\left[\log \frac{q(s_i, s_j)}{p(o_i, o_j, s_i, s_j)}\right] \tag{9}$$

Here, G_{joint} quantifies the shared divergence between the dyad's joint beliefs and its joint generative model. High joint valence corresponds to low joint free energy (smooth alignment), whereas affective ruptures correspond to spikes in joint free energy (misalignment and uncertainty). At the dyadic level, recursive coupling of valence creates a shared attractor structure: higher valence biases policy selection toward trajectories that preserve narrative coherence and reduce expected future free energy. Conversely, negative valence can trigger exploratory policies to restore alignment or withdraw from destabilizing interactional trajectories. This way, valence modulates immediate affective dynamics and the agent's future epistemic and instrumental uncertainty.

Critically, moments of affective rupture–sharp drops in valence–correspond to increases in joint free energy, typically arising from mismatches in the agents' self-model priors. These transitions are not noise but informative inflection points in the dyadic generative process. Mismatches in self-model coherence directly elevate free energy, as valence encodes deviations between observed interaction outcomes and expected self-relevant predictions. Because valence reflects each agent's affective confidence in their internal model, it serves as both a readout of local free energy minimization and a control signal that shapes future joint free energy via policy adaptation. This recursive structure is central to how dyadic affective inference maintains, destabilizes, or reconstructs interpersonal synchrony over time.

2.3 Geometric Hyperscanning as Second-Person Method

To empirically access the internal dynamics of dyadic affective inference, we introduce *geometric hyperscanning* based on the Forman-Ricci curvature (FRc) [22,23], which captures phase transitions in inter-brain network topology using geometric markers derived from EEG hyperscanning data. This method enables operationalization of latent affective and narrative coherence as measurable transformations in the topology of inter-brain networks. Importantly, FRc is not treated as a post hoc readout; we propose it as a data-derived proxy for latent generative states in the dyadic active inference model. Traditional hyperscanning methods focus on node-level synchrony, typically correlating homologous regions across two brains. While informative about temporal alignment, these methods fail to detect topological reconfigurations in inter-brain structure. FRc, by contrast, is an edge-centric geometric measure that quantifies the local expansion or contraction of topological flow within a network [45], and has been shown to characterize network robustness, signal routing, and functional reorganization.

Given a network $G = (V, E)$ with edge e_{ij} connecting nodes v_i and v_j, the FRc is computed as:

$$\mathrm{FRc}(e_{ij}) = w(e_{ij}) \left(\frac{1}{w(v_i)} + \frac{1}{w(v_j)} \right) - \sum_{e \sim e_{ij}} \left(\frac{w(e_{ij})}{\sqrt{w(e_{ij}) \cdot w(e)}} \right) \tag{10}$$

where we interpret $w(\cdot)$ as node and edge weights (e.g., derived from functional connectivity matrices), to track the evolution of inter-brain topology over time, we calculate the entropy of the FRc distribution within a sliding window.

$$H_{\mathrm{FRc}}(t) = - \sum_{e \in E_t} p(e) \log p(e) \tag{11}$$

This scalar quantity acts as a proxy for interactional volatility. Peaks or discontinuities in FRc entropy correspond to phase transitions in the shared generative manifold and reflect affective rupture, repair, or co-regulation dynamics.

$$p(e) = \frac{\mathrm{FRc}(e)}{\sum_{e'} \mathrm{FRc}(e')} \tag{12}$$

FRc entropy is not simply an output variable but can be formalized as an observation model over latent affective prediction error. Curvature dynamics can infer or constrain internal state variables such as valence or identity coherence. This closes the loop between model and measurement, as the generative model predicts changes in self-model coherence and dyadic free energy, and the FRc entropy serves as an empirical signal that can be used to update model beliefs via Bayesian inference. To make these transitions tractable and interpretable, we propose a multi-level operationalization of rupture, repair, and reattunement, linking neural signatures to behavioral and psychological indicators. Neurally, rupture corresponds to abrupt drops in network integration (low FRc), high curvature entropy, or rapid shifts in entropy gradients; repair and reattunement reflect a return to more stable topologies. Behaviorally, rupture may manifest as disengagement, disrupted synchrony, or confusion (e.g., gaze aversion, vocal interruptions), while repair involves re-engagement cues such as gesture mirroring, prosodic modulation, or turn-taking; reattunement is evidenced by fluent, coordinated interaction and mutual responsiveness. Psychologically, rupture can be self-reported as disconnection or emotional distancing, repair as subjective efforts to reconnect, and reattunement as restored mutual understanding or relational safety. These experiences can be measured via self-reports (affective sliders, post-session interviews, validated instruments), annotated through multimodal behavioral data (gaze, prosody, synchrony), and aligned with curvature dynamics in the inter-brain manifold. Treated as latent categorical states (e.g., within a hidden Markov model), these phases can serve as points of annotation, classification, or real-time feedback in both experimental and clinical contexts.

Importantly, embedding these events within the recursive generative model of dyadic inference allows rupture–repair cycles to function as outcomes and as endogenous features of interactional dynamics. Affective coherence drives policy selection, while behavioral and neural feedback recursively update each agent's self-model. This recursive structure enables in situ inference over dynamic interpersonal states, validation of model predictions across modalities, and fine-grained adaptation to moment-by-moment shifts in relational coherence. Integrating *geometric hyperscanning* into the generative model therefore bounds the model with an additional objective: not only minimising identity-relevant prediction error, but implicitly driving the system toward maximal interpersonal attunement–an intrinsically rewarding state even in the absence of overt instrumental gain [7,35]. Curvature-induced surprises can surface to awareness, prompting explicit attempts at repair.

In sum, FRc-based *geometric hyperscanning* offers a scalable, multimodal method for detecting and interpreting recursive affective inference. By triangulating neural entropy, behavioral dynamics, and psychological experience, we gain an integrated empirical handle on how dyads co-regulate meaning, affect, and identity over time. This framework advances a formal, testable architecture for second-person neuroscience and lays the groundwork for modeling recursive emotional entanglement, the subject of the following section.

3 Future Directions and Conclusions

This paper proposed that affect is best understood as recursive inference over self-model coherence within coupled generative systems. Modeling the dyad via active inference, we formalized affect as a valuation of identity-relevant surprise, driving and modulating recursive loops of belief updating, policy selection, and behavioral adaptation during social interaction. We introduced a novel method based on the FRc, termed *geometric hyperscanning* [23], which tracks dynamic reconfigurations in inter-brain networks, providing an empirical window onto moments of rupture, repair, and re-attunement. By unifying formal models of belief dynamics, affective evaluation, and network geometry, we advance a scalable and interpretable framework for operationalizing second-person active inference.

3.1 Future Directions

The proposed theoretical formulation invites several concrete avenues for empirical investigation and computational development. Foremost among these is the simulation of agent-based models in which curvature entropy is treated as a streamed sensory input–continuously informing belief updates about narrative alignment. Such models would allow us to test whether the architecture, under active inference, spontaneously reproduces the rupture–repair cycles observed in real dyadic interaction. In doing so, they provide a crucial bridge between formalism, neurophysiology, and phenomenology. Beyond dyads, the framework naturally extends to hierarchically structured systems. By introducing group-level alignment variables, one can model small-group coherence as an emergent property of multiple interacting pairwise dynamics. This multiscale extension opens the door to collective behaviour, organisational synchrony, and crowd psychology applications, where relational inference is distributed across agents and temporal scales. Furthermore, the curvature-informed likelihood may be productively integrated with other sensorimotor channels. Prosodic variation, gaze dynamics, and even facial microexpressions can all be incorporated within a deep generative model, allowing heterogeneous evidence streams to converge on shared latent variables. This multimodal fusion enriches the inferential landscape and supports more robust decoding of the intersubjective field. Yet the prospect of real-time inference over relational states is not without ethical consequences. As structural priors become actionable in applied contexts–be it psychotherapy, education, or social robotics–the issues of privacy, autonomy, and interpretability move to the fore. Any translational implementation must therefore be grounded in transparent consent procedures and designed to yield intelligible feedback to clinicians and researchers and to the individuals whose relational trajectories it purports to track.

3.2 Sociomarkers for Interpersonalized Psychiatry

Geometric hyperscanning offers a promising foundation for *sociomarkers*: real-time, measurable signatures of dynamic interpersonal coordination [6], analogous to biomarkers of individual physiology. Affective co-regulation—vital for social bonding across species [8,12,16,36]—is foundational to human development, shaping cognition, emotion, and relational selfhood through early caregiver-infant interactions [7,15,17,24,43]. Misattunement and repair cycles scaffold resilience and relational growth [6,7], with fluctuations in curvature entropy potentially indexing key developmental processes [11,37,44]. Such signatures could inform personalized [30,31] and interpersonalized psychiatry [6], which views psychopathology as a disruption in co-regulation and narrative repair [5,32]. In psychotherapy, real-time dyadic metrics [5,28] can track attunement, rupture, and repair [34], helping both therapists and clients navigate the relational dynamics of change. Crucially, synchrony is neither inherently good nor constant: transitions into and out of synchrony may reflect autonomy, resistance, or meaningful differentiation, not dysfunction. Modeling these dynamics as recursive affective inference provides a principled lens for basic science and relational mental health care.

3.3 Conclusions

Building on these formal and empirical insights, this approach carries several key implications. First, it invites a rethinking of affect in social interaction–not merely as a byproduct of inference, but as a primary regulatory signal that modulates the stability and flexibility of generative coupling itself. On this view, affect becomes the sense of coherence or incoherence within the dyadic narrative, indicating when beliefs must be renegotiated or when interactional roles require repair. Second, by shifting emphasis from synchrony per se to the topology of inter-brain networks, our method aligns with a broader movement in second-person neuroscience toward characterizing interaction structure. In this respect, FRc extends Friston and Frith's (2015) [18] notion of generalized synchrony by capturing the reconfigurations in network structure that accompany recursive affective inference. Third, by embedding affect into the posterior over policies, we extend da Costa's (2020) planning-as-inference formalism into the affective domain. Crucially, the affective signal can operate implicitly, as a precision-weight on policy beliefs that never reaches phenomenal awareness, or explicitly, as a consciously accessible feeling that furnishes the agent with propositional knowledge about its own relational stance. In either guise, policy selection is driven not only by epistemic or instrumental imperatives, but also by the imperative to preserve narrative and relational coherence. Affect thereby becomes an intrinsic dimension of social planning, covertly biasing action when tacit and overtly guiding behaviour when made conscious–steering agents through moments of ambiguity toward restored interactional alignment. These considerations open a range of empirical and computational possibilities. In psychotherapy, curvature entropy may serve as a real-time sociomarker of rupture and repair; in early development, FRc could illuminate how co-regulatory patterns emerge between infants and caregivers. Simulations of affective agents equipped with recursive self- and other-models offer a formal means to investigate relational resilience and breakdown. In particular, such models allow us to test the dialectical misattunement hypothesis (Bolis et al., 2017), which posits that psychiatric risk arises not from individual deficits alone but from persistent mismatches in communicative expectations or learning rates. Simulating these mismatches under controlled conditions reveals how dyads adapt–or fail to–over developmental time, providing insight into vulnerability and repair mechanisms. The theoretical landscape also invites expansion. How does this framework scale to group interactions, where recursive generative coupling involves multiple agents? What ethical safeguards are needed when automating the detection of affective rupture and repair–especially in sensitive contexts like therapy or education? And how might shared belief geometry interact with linguistic, prosodic, and bodily cues to form a rich, multimodal map of affective inference? Together, these questions point toward a broader scientific project: one that integrates formal modeling, empirical measurement, and philosophical reflection into a unified account of relational meaning. In this view, the dynamics of belief, affect, and participation are not peripheral but fundamental to the fabric of social cognition. By reframing affect as recursive inference within coupled generative systems–and rendering those dynamics tractable via network geometry–we move closer to the vision of interaction as the basic unit of social neuroscience. Our proposal thus offers both a conceptual framework and methodological toolkit for studying human interaction's structural and affective choreography.

Acknowledgments. NH and YJ would like to thank Danielle J. Williams and the inaugural meeting of the Society for Neuroscience and Philosophy, at which the authors contributed papers and discussed integrating their presented approaches. NH would like to thank Lancelot Da Costa and Sebastian Sosa for their insightful feedback.

Disclosure of Interests. NH was funded by CNPq, MPI-CBS, and OIST. DB receives funding from the IIT. LS receives funding from the DFG.

Contributions. NH and YJ jointly conceived and drafted the initial manuscript. MA and DB contributed core conceptual ideas and critically revised the text. LCM and LS provided additional feedback and editorial refinement. All authors reviewed and approved the final version of the manuscript.

A A Simulation-Ready Implementation (MBSR + Valence–HMM)

Goal. We instantiate the manuscript's account with two coupled but distinct components: (i) an autobiographical memory-based self-relevance, (ii) a *valence* HMM that carries the temporal dependence of affect. The HMM framework is motivated by the temporal persistence of valence exhibiting state continuity with gradual, experience-driven change (for review: affective spillover/carryover effect). HMM transition matrices capture this through high diagonal probabilities while permitting off-diagonal transitions. Standard moment-by-moment computations lack this persistence; HMMs provide it via probabilistic state evolution.

A.1 Setup and notation

For agent i and step t, the internal model M_i maintains a self–state prior/posterior $P_i^t, \hat{P}_i^t$, observations o_i^t, inferred partner actions a_j^t, and the identity prediction error

$$PE_{\text{self},i}^t = \hat{P}_i^t - P_i^t.$$

The projected/predicted valence, by definition, uses a scalar self-relevance $SR_t^i \in [0,1]$ and the identity prediction error $PE_{\text{self},i}^t$ to weight identity-relevant surprise:

$$\mathbb{E}(V_i^t) = \alpha \cdot PE_{\text{self},t} \cdot SR_t + \beta \cdot V_{SR,t} \tag{13}$$

Here α, β are precision-modulating gains balancing instantaneous identity-relevant surprise and affective carryover $V_{t,SR}^i$.
The agent's current policy π_i^t is enacted based on the valence state:

$$P(\pi_i \mid V_t) \propto e^{\lambda \cdot g(V_t, \pi_t)} \tag{14}$$

where $\phi(\cdot)$ may include linear terms and selected interactions; its purpose is to summarize, per trial, the stimulus and self-context the agent brings to the exchange.

After one carries out an action policy, the results from the action will be made available as new observations to compute the realized valence V_i^t. By comparing expected outcomes under the current policy to the expectation $\mathbb{E}(V_i^t)$, this item formalizes the felt value of the present exchange as the log-likelihood of observations given the agent's valence-weighted model (Eq. 15).

$$V_i^t = \mathbb{E}_{P(o^t|\pi_i^t)}\left[\log P\left(o^t \mid \mathbb{E}(V_i^t)\right)\right]. \tag{15}$$

In the following implementation, we discretize both self-relevance and valence into three ordered HMM states:

- Self–relevance (SR): $R_i^t \in \{ r_{\text{low}}, r_{\text{med}}, r_{\text{high}} \}$.
- Valence (S): $S_i^t \in \{ s_{\text{neg}}, s_{\text{neu}}, s_{\text{pos}} \}$.

A.2 Memory-Based Self–relevance Chain

For each interaction, agents encode the present exchange in an interaction descriptor

$$z_i^t = (o_i^t,\, a_j^t,\, P_i^t) \in \mathbb{R}^d \tag{16}$$

Each agent maintains an autobiographical store $\mathcal{LTM}_i = \{(x_k, I_k, \tau_k)\}_{k=1}^K$ where each memory m_k has a descriptor x_k (context features), an impact tag $I_k \in [0,1]$ (how strongly the episode affected the agent), and a timestamp τ_k. Given the current context z_i^t, we compute how similar it is to each stored memory. These similarities are then normalized so that they add up to one, turning them into attention weights over memory:

$$Sim_k^t = \mathrm{sim}(z_i^t, x_k), \qquad \tilde{Sim}_k^t = \frac{Sim_k^t}{\sum_j Sim_j^t},$$

For each memory m_k stored, the context z_i^t will become the x_k, the time becomes τ_k, and they are accompanied by an impact tag which represents (1) how large the self-prediction error was, and (2) how much the policy distribution reorganized. Both are normalized to $[0,1]$ for comparability. Impact tags are written from the realized change at step t using a bounded, divergence-free score:

$$I^t = \mathrm{norm}\big(\|PE_{\mathrm{self},i}^t\|\big) \;\times\; \mathrm{TV}_\pi^t, \tag{17}$$

$$\mathrm{TV}_\pi^t = \tfrac{1}{2} \sum_{\pi \in \Pi_i} \big|p_{\mathrm{post}}^t(\pi) - p_{\mathrm{prior}}^t(\pi)\big| \in [0,1], \tag{18}$$

where $p_{\mathrm{prior}}^t / p_{\mathrm{post}}^t$ are policy distributions (softmax over scores) before/after assimilating step t; $\mathrm{norm}(\cdot)$ maps magnitudes to $[0,1]$.

A *memory-derived relevance score* is then obtained as a similarity-weighted average of impact tags:

$$\widetilde{SR}_t^i = \sum_{k=1}^K \tilde{Sim}_k^t \, I_k \in [0,1]. \tag{19}$$

A.3 Expected Valence HMM Chain

Expected Valence. We first compute the expected valence as a product of the previous step's identity prediction error and self-relevance (Eq. 13).

Integration into the Valence HMM. This expected valence then drives the transition dynamics of the latent valence state $S_i^t \in \{\mathrm{Neg}, \mathrm{Neu}, \mathrm{Pos}\}$ with ordinal map $s(\mathrm{Neg}) = -1$, $s(\mathrm{Neu}) = 0$, $s(\mathrm{Pos}) = 1$. Three behavioral factors shape the dynamics: (i) baseline biases $\omega_{k\ell}$ (how likely each shift is by default), (ii) sensitivity η_k (how strongly expected valence pushes transitions), and (iii) penalty ρ_k (how costly big jumps are, e.g. from negative to positive in one step). We parameterize transitions by expected valence, with stickiness and large–jump penalty:

$$T_{S,i}^t(k \to \ell) = \frac{\exp\{\,\omega_{k\ell} + \eta_k\,\mathbb{E}(V_i^t)\,s(\ell) - \rho_k\,|s(\ell) - s(k)|\,\}}{\sum_j \exp\{\,\omega_{kj} + \eta_k\,\mathbb{E}(V_i^t)\,s(j) - \rho_k\,|s(j) - s(k)|\,\}}, \tag{20}$$

with baselines $\omega_{k\ell}$, sensitivity $\eta_k > 0$, and penalty $\rho_k \geq 0$.

Filtering and Readout. The HMM updates the probability of each valence level, using both the previous distribution and the transition dynamics. This yields a smooth trajectory of affective states. The belief state over valence evolves as

$$\gamma_{S,i}^t \propto (T_{S,i}^t)^\top \gamma_{S,i}^{t-1}, \qquad \sum_\ell \gamma_{S,i}^t(\ell) = 1.$$

We can obtain a scalar valence readout in two ways:

$$v_i^t = \nu^\top \gamma_{S,i}^t, \qquad \nu = (-1, 0, 1)^\top, \tag{21}$$

Here, valence is read out as a weighted average across state probabilities, yielding a single continuous score in $[-1, 1]$.

A.4 (Expected) Valence-Guided Policy Selection

We define a valence–compatible prior over policies (Eq. 14) with inverse temperature λ controlling decisiveness. The compatibility function

$$g(E(V_i^t), \pi) = \mathbb{E}_{P(o^t|\pi)}\Big[\log P(o^t \mid \mathbb{E}(V_i^t))\Big] \tag{22}$$

scores each candidate policy by how well its expected outcomes align with the agent's current valence state. In other words, policies are preferred if they make the world more consistent with the agent's affective expectations.

A.5 (Realized) Valence-Guided Model Updating

After enacting the policy, agents would get an actual outcome of the chosen policy and have an actual emotional response (realized valence (Eq. 15)). Conceptually, this provides a feedback signal for updating.

When outcomes under policy π_i^t match expectations, realized valence is high, reinforcing both the policy and the current self-model. When outcomes diverge, realized valence is low (or negative), signaling a need to adjust the model's parameters, update memory with a new impact tag, and reconsider future policies. This update propagates across all latent states and parameters:

- Self-state posteriors: priors P_i^t are corrected to posteriors $\hat{P}_i^t$ using the new observation (o_i^t, o_j^t).

- Valence HMM: beliefs $\gamma_{S,i}^t$ are updated using expected valence $\mathbb{E}(V_i^t)$ as input, ensuring temporal persistence of affect.
- Policy posterior: the distribution over policies $P(\pi_i^t \mid V_i^t)$ is recalculated given the new affective state.
- Impact encoding: the realized impact I^t is computed via Eqs. (17)–(18), summarizing prediction error and policy reorganization.
- Memory write: the tuple (z_i^t, I^t, t) is appended to the autobiographical store $\mathcal{LTM}_i$, allowing subsequent self-relevance estimates to remain personalized and history-dependent.

This way, realized valence serves as a model-updating drive that keeps affective inference aligned with lived experience.

A.6 Table 1

Table 1. Symbols used in the simulation-ready HMM implementation.

Symbol	Type	Definition/Role
M_i	model	Agent i's generative model (self + simulated other).
o_i^t, o_j^t	obs	Observations at step t (self / partner).
$P_i^t, \hat{P}_i^t$	state	Self-state prior / posterior after observing (o_i^t, o_j^t).
$PE_{\text{self},i}^t$	vec	Identity prediction error; we use $\lVert PE_{\text{self},i}^t \rVert$.
a_j^t	act (inf.)	Inferred partner action represented in M_i.
z_i^t	feat	Descriptor $\phi(o_i^t, a_j^t, P_i^t)$; $\psi(z) = z$ by default.
$\mathcal{LTM}_i$	memory	Autobiographical store of (x_k, I_k, τ_k).
I^t	impact	$\text{norm}(\lVert PE_{\text{self},i}^t \rVert) \cdot \text{TV}_\pi^t$.
TV_π^t	scalar	Total variation between pre/post policy distributions at t.
$S(z, x)$	sim	Similarity kernel; $\tilde{S}$ is its normalization over k.
$\widetilde{SR}_t^i$	scalar	Memory-derived self–relevance, Eq. (19).
$\mathbb{E}(V_i^t)$	scalar	Expected valence, Eq. (13).
$S_i^t, \gamma_{S,i}^t$	HMM	Valence state / belief; transition $T_{S,i}^t$ in Eq. (20).
$\omega_{k\ell}, \eta_k, \rho_k$	params	Valence baselines, sensitivity, jump penalty.
π_i^t, λ	policy	Action policy; inverse temperature in (14).
$g(V, \pi)$	score	Compatibility in (14).

A.7 Appendix Illustration

See Fig. 3.

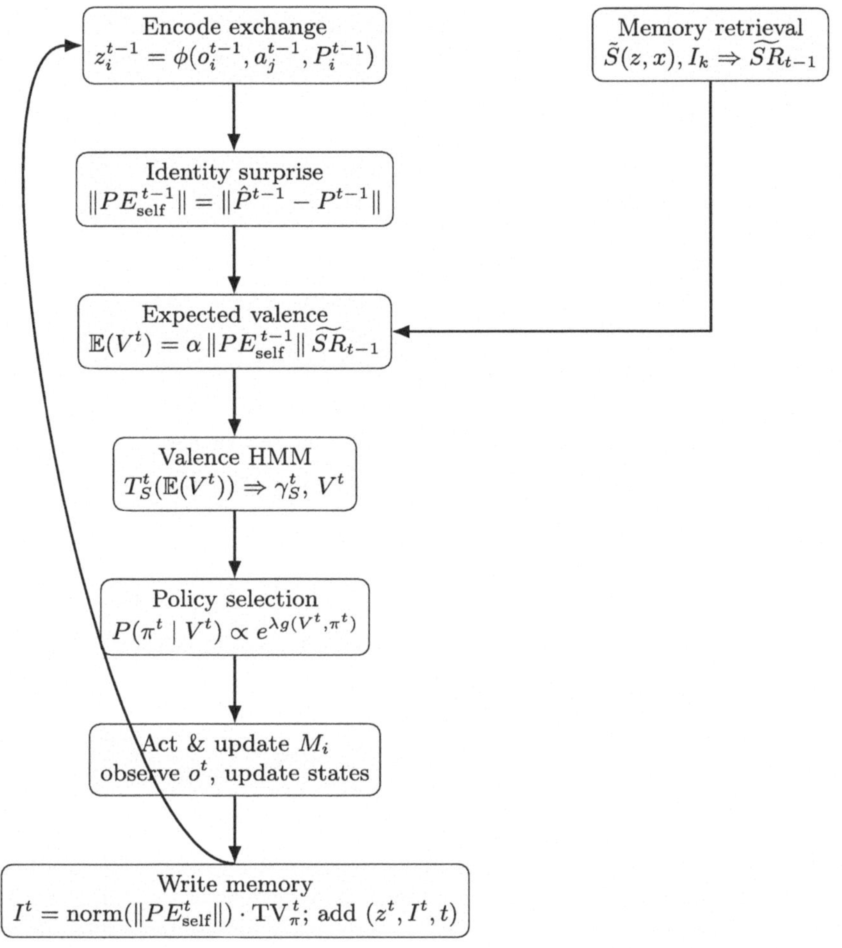

Fig. 3. Execution cycle for one agent at step t. Self–relevance is derived purely from autobiographical memory ($\widetilde{SR}$), which modulates expected valence. The HMM handles valence dynamics, while memory ensures personalization and history dependence.

B Curvature-Based Geometric Entropy Channel

Goal. This appendix mirrors Appendix A but focuses on the *geometric* observation channel used in the main text: the edge-centric Forman–Ricci curvature (FRc) and its entropy as a compact, simulation-ready observable of dyadic network reconfiguration (rupture, repair, re-attunement).

B.1 Setup and Notation

Let $G_t = (V, E_t)$ be a time-varying inter-brain graph at step t, where V indexes neural sources across the two agents and E_t contains *inter-brain* edges computed from a sliding window over the dual-brain data (e.g., cross-brain EEG coherence). Each edge $e_{ij} \in E_t$ carries a connectivity weight $w(e_{ij})$, while each node $v_i \in V$ may be assigned a weight $w(v_i)$ (e.g., node strength) (Table 2).

Table 2. Symbols for the curvature-based entropy channel (Appendix B).

Symbol	Type	Definition/Role
$G_t = (V, E_t)$	graph	Inter-brain network at time t (nodes V, edges E_t).
$w(e_{ij})$	scalar	Edge weight for connectivity between v_i and v_j.
$w(v_i)$	scalar	Node weight (e.g., strength: sum of incident edge weights).
$\mathrm{FRc}(e_{ij})$	scalar	Forman–Ricci curvature of edge e_{ij} (local geometry).
$p(e_{ij})$	prob.	Curvature-proportional mass: $\mathrm{FRc}(e_{ij})/\sum_{e' \in E_t} \mathrm{FRc}(e')$.
$H_{\mathrm{FRc}}(t)$	scalar	Curvature entropy at t: $-\sum_{e \in E_t} p(e) \log p(e)$.
$O_{\kappa,t}$	obs.	Geometric observation channel (we set $O_{\kappa,t} = H_{\mathrm{FRc}}(t)$).
S_t	latent	Dyad-level hidden state (e.g., alignment/rupture).

B.2 Forman–Ricci Curvature (FRc)

For each edge $e_{ij} = (v_i, v_j) \in E_t$, the FRc for weighted graphs is:

$$\mathrm{FRc}(e_{ij}) = w(e_{ij}) \left(\frac{1}{w(v_i)} + \frac{1}{w(v_j)} \right) - \sum_{\substack{e \sim e_{ij} \\ e \neq e_{ij}}} \frac{w(e_{ij})}{\sqrt{w(e_{ij})\, w(e)}} . \tag{23}$$

The first term grows when e_{ij} dominates the load of its endpoints; the second term reduces curvature when many strong *neighbor* edges $e \sim e_{ij}$ create parallel routes, flattening local geometry. Positive FRc typically marks edges in dense, redundant neighborhoods; strongly negative FRc highlights bridge-like edges stitching otherwise separate modules.

B.3 Curvature-Based Entropy

We compress the network's geometry at time t into a single volatility marker via the Shannon entropy of the FRc distribution:

$$p(e_{ij}) = \frac{\mathrm{FRc}(e_{ij})}{\sum_{e' \in E_t} \mathrm{FRc}(e')} \quad \text{(curvature-proportional mass)}, \tag{24}$$

$$H_{\mathrm{FRc}}(t) = - \sum_{e \in E_t} p(e) \log p(e). \tag{25}$$

High H_{FRc} indicates dispersed curvature (topological disorder/volatility), while low H_{FRc} indicates concentration on a stable backbone (topological order). Abrupt rises in $H_{\mathrm{FRc}}(t)$ signal *phase transitions* in inter-brain topology and operationalize affective rupture; subsequent drops track repair and re-attunement.

Practical Note. In practice, some FRc(e) can be negative. To use Eq. (24), one may (i) apply a positive shift before normalization, or (ii) work with a nonnegative transform (e.g., |FRc|) when the analysis targets *dispersion* rather than signed geometry. Either choice should be fixed *a priori* and reported.

B.4 Embedding Curvature in Active Inference

We treat the curvature entropy $H_{\mathrm{FRc}}(t)$ as a dyad-level *observation channel* $O_{\kappa,t}$ in the generative model:

$$P\big(O_{\kappa,t} \mid S_t\big) \quad \text{with} \quad O_{\kappa,t} \equiv H_{\mathrm{FRc}}(t)\,. \tag{26}$$

Hidden states S_t encode dyadic alignment (e.g., attunement $\leftrightarrow$ low H_{FRc}, rupture $\leftrightarrow$ high H_{FRc}). The agent inverts this likelihood online to infer $P(S_t \mid O_{\kappa,t}, \text{other cues})$. Curvature-induced surprises (unexpected H_{FRc} spikes) increase the posterior probability of rupture and drive policy adaptation toward repair; sustained low H_{FRc} reinforces attuned policies.

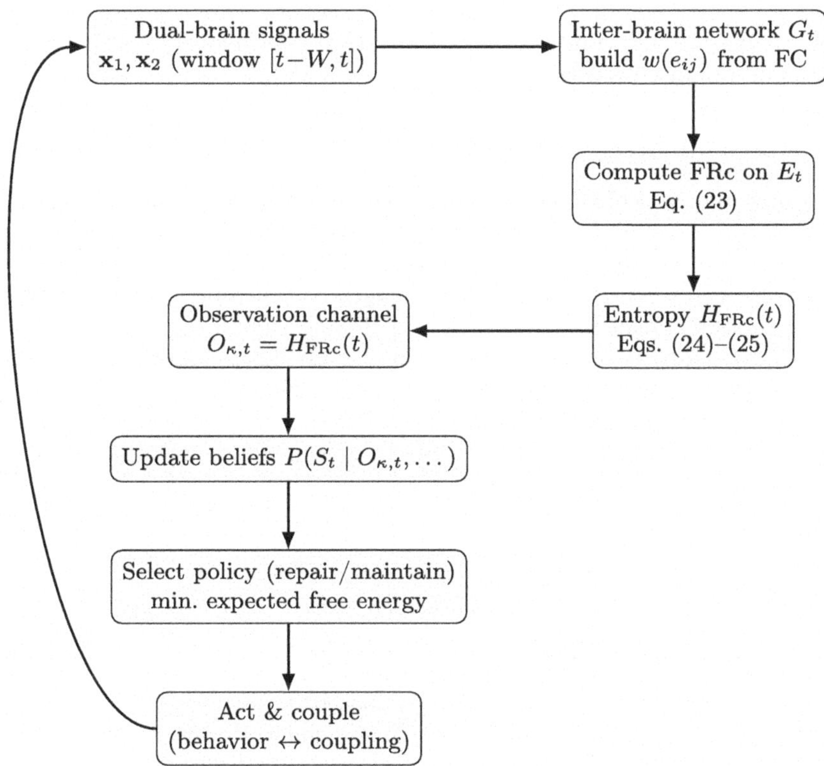

Fig. 4. *Curvature-entropy observation loop (Appendix B).* A sliding window yields G_t, FRc is computed per edge, and $H_{\mathrm{FRc}}(t)$ feeds an observation channel $O_{\kappa,t}$ for online inference and policy selection.

Algorithm 1. Curvature-Entropy Driven Inference Cycle (mirror of Appendix A)

Inputs: window W; initial $P(S_0)$; coupling/likelihood params for $P(O_{\kappa,t} \mid S_t)$; dual-brain stream or simulator.
for $t = 1, 2, \ldots$ **do**
  Signals: extract $\mathbf{x}_1[t - W : t]$, $\mathbf{x}_2[t - W : t]$.
  Network: compute FC across brains $\Rightarrow$ weights $w(e_{ij})$; form $G_t = (V, E_t)$.
  Curvature: for each $e_{ij} \in E_t$, compute $\mathrm{FRc}(e_{ij})$ via Eq. (23).
  Entropy: compute $p(e)$ by Eq. (24) and $H_{\mathrm{FRc}}(t)$ by Eq. (25).
  Observation: set $O_{\kappa,t} \leftarrow H_{\mathrm{FRc}}(t)$; combine with other cues.
  Inference: update $P(S_t \mid O_{\kappa,t}, \ldots)$ (variational or filtering).
  Policy: select action/policy minimizing expected free energy (repair if rupture likely).
  Act: enact behavior; update coupling/state dynamics for next step.
end for

B.5 Execution Cycle (Geometry Channel)

Figure 4 mirrors Appendix A's execution loop, substituting the geometric channel for valence-specific computations. Here, $O_{\kappa,t}$ continuously constrains belief updates and policy selection.

B.6 Notes on Design Choices

- **Windowing & FC.** The window W should balance temporal sensitivity and estimator stability; FC choices (e.g., correlation/coherence) should match modality and frequency-band of interest.
- **Node weights.** $w(v_i)$ can be node strength or constant 1 (unweighted node case); choices impact the sign/scale of FRc and should be reported.
- **Normalization.** If signed FRc complicates Eq. (24), adopt a fixed nonnegative mapping (shift or magnitude) chosen *a priori*.
- **Observation model.** Calibrate $P(O_{\kappa,t} \mid S_t)$ by aligning H_{FRc} transitions with behavioral/phenomenological markers (rupture/repair annotations) for the paradigm at hand.

References

1. Albarracin, M., Bouchard-Joly, G., Sheikhbahaee, Z., Miller, M., Pitliya, R., Poirier, P.: Feeling our place in the world: an active inference account of self-esteem. Neurosci. Consciousness **2024**(1), niae007 (2024)
2. Albarracin, M., Pitliya, R., Ramstead, M.: Mapping Husserlian phenomenology onto active inference. In: Friston, K., Ramstead, M., Constant, A. (eds.) IWAI 2022. CCIS, vol. 1721, pp. 99–111. Springer, Heidelberg (2023). https://doi.org/10.1007/978-3-031-28719-0_7
3. Albarracin, M., et al.: A sociometric model of self-esteem as inference. Trends Cogn. Sci. **28**(1), 45–57 (2024)

4. Barad, K.: Meeting the Universe Halfway: Quantum Physics and the Entanglement of Matter and Meaning. Duke University Press, Durham (2007)
5. Bolis, D., Balsters, J., Wenderoth, N., Becchio, C., Schilbach, L.: Beyond autism: introducing the dialectical misattunement hypothesis and a Bayesian account of intersubjectivity. Psychopathology **50**(6), 355–372 (2017)
6. Bolis, D., Dumas, G., Schilbach, L.: Interpersonal attunement in social interactions: from collective psychophysiology to inter-personalized psychiatry and beyond. Philos. Trans. R. Soc. B **378**(1870), 20210365 (2023)
7. Bolis, D., Schilbach, L.: 'i interact therefore i am': the self as a historical product of dialectical attunement. Topoi **39**, 521–534 (2020)
8. Brosnan, S., de Waal, F.: Monkeys reject unequal pay. Nature **425**(6955), 297–299 (2003)
9. Çatal, O., Van de Maele, T., Pitliya, R., Albarracin, M., Pattisapu, C., Verbelen, T.: Belief sharing: a blessing or a curse. In: Buckley, C.L., et al. (eds.) IWAI 2024. LNCS, vol. 2193, pp. 121–133. Springer, Cham (2024). https://doi.org/10.1007/978-3-031-77138-5_8
10. Christov-Moore, L., et al.: Using a chills-inducing score to augment loving kindness meditation: effects on self-transcendence, emotional breakthrough, and psychological insight. OSF Preprints (2025)
11. Cittern, D., Nolte, T., Friston, K., Edalat, A.: Intrinsic and extrinsic motivators of attachment under active inference. PLoS ONE **13**(4), e0193955 (2018)
12. Coleman, M., et al.: Neurophysiological coordination of duet singing. PNAS **118**(24), e2023035118 (2021)
13. Da Costa, L., Parr, T., Sajid, N., Veselic, S., Neacsu, V., Friston, K.: Active inference on discrete state-spaces: a synthesis. J. Math. Psychol. **99**, 102447 (2020)
14. Dumas, G., Nadel, J., Soussignan, R., Martinerie, J., Garnero, L.: Inter-brain synchronization during social interaction. PLoS ONE **5**(8), e12166 (2010). https://doi.org/10.1371/journal.pone.0012166
15. Fini, C., et al.: The social roots of self development: from a bodily to an intellectual interpersonal dialogue. Psychol. Res. **87**(6), 1683–1695 (2023)
16. Fortune, E., et al.: Neural mechanisms for the coordination of duet singing in wrens. Science **334**(6056), 666–670 (2011)
17. Fotopoulou, A., Tsakiris, M.: Mentalizing homeostasis: the social origins of interoceptive inference. Neuropsychoanalysis **19**(1), 3–28 (2017)
18. Friston, K., Frith, C.: A duet for one. Conscious. Cogn. **36**, 390–405 (2015). https://doi.org/10.1016/j.concog.2014.12.003
19. Gvirts, H.Z., Perobolovski, A.: Commentary: using second-person neuroscience to elucidate the mechanisms of reciprocal social interaction. Front. Behav. Neurosci. **14**, 13 (2020). https://doi.org/10.3389/fnbeh.2020.00013
20. Gvirts, H.Z., Perlmutter, R.: What guides us to neurally and behaviorally align with anyone specific? A neurobiological model based on fNIRS hyperscanning studies. Neuroscientist (2019). https://doi.org/10.1177/1073858419861912
21. Hesp, C., Smith, R., Parr, T., Allen, M., Friston, K., Ramstead, M.: Deeply felt affect: the emergence of valence in deep active inference. Neural Comput. **33**(2), 398–446 (2021)
22. Hinrichs, N., Hartwigsen, G., Guzman, N.: Detecting phase transitions in EEG hyperscanning networks using geometric markers (2025)
23. Hinrichs, N., Guzmán, N., Weber, M.: On a geometry of interbrain networks. arXiv preprint arXiv:2509.10650 (2025)
24. Hoehl, S., Markova, G.: Moving developmental social neuroscience toward a second-person approach. PLoS Biol. **16**(12), e3000055 (2018)

25. Jiang, F., Luo, D.: Implementing self models through joint-embedding predictive architecture. In: Proceedings of the Annual Meeting of the Cognitive Science Society, vol. 46 (2024)
26. Jiang, Y., Luo, D.: Valence computation as higher-order inference via conceptual self-processing (2025). https://doi.org/10.31234/osf.io/nsfq5
27. Joffily, M., Coricelli, G.: Emotional valence and the free-energy principle. PLoS Comput. Biol. **9**(6), e1003094 (2013)
28. Lahnakoski, J.M., Forbes, P.A., McCall, C., Schilbach, L.: Unobtrusive tracking of interpersonal orienting and distance predicts the subjective quality of social interactions. Roy. Soc. Open Sci. **7**(8), 191815 (2020)
29. Lehmann, K., Bolis, D., Friston, K.J., Schilbach, L., Ramstead, M.J.D., Kanske, P.: An active-inference approach to second-person neuroscience. Perspect. Psychol. Sci. **19**(6), 931–951 (2023)
30. Lombardo, M.V., Lai, M.C., Baron-Cohen, S.: Big data approaches to decomposing heterogeneity across the autism spectrum. Mol. Psychiatry **24**(10), 1435–1450 (2019)
31. Mandelli, V., Landi, I., Busuoli, E.M., Courchesne, E., Pierce, K., Lombardo, M.V.: Prognostic early snapshot stratification of autism based on adaptive functioning. Nat. Mental Health **1**(5), 327–336 (2023)
32. Milton, D.E.: On the ontological status of autism: the 'double empathy problem'. Disabil. Soc. **27**(6), 883–887 (2012)
33. Rahmjoo, A., Albarracin, M.: Intra-active inference i: Fundamentals (2023, preprint)
34. Ramseyer, F., Tschacher, W.: Nonverbal synchrony in psychotherapy: coordinated body movement reflects relationship quality and outcome. J. Consult. Clin. Psychol. **79**(3), 284 (2011)
35. Redcay, E., Schilbach, L.: Using second-person neuroscience to elucidate the mechanisms of social interaction. Nat. Rev. Neurosci. **20**(8), 495–505 (2019)
36. Roma, P.G., Silberberg, A., Ruggiero, A.M., Suomi, S.J.: Capuchin monkeys, inequity aversion, and the frustration effect. J. Comp. Psychol. **120**(1), 67–73 (2006)
37. Schilbach, L., et al.: Toward a second-person neuroscience. Behav. Brain Sci. **36**(4), 393–414 (2013)
38. Schilbach, L., Redcay, E.: Synchrony across brains. Annu. Rev. Psychol. **76**(1), 883–911 (2025). https://doi.org/10.1146/annurev-psych-080123-101149
39. Scholtes, C.M., Poppelaars, E.S., van Berkel, S.R., van IJzendoorn, M.H.: Dyadic synchrony, rupture, and repair in mother–child interaction: implications for emotional and behavioral development. Dev. Psychopathol. **32**(4), 1519–1533 (2020)
40. Smith, R., Parr, T., Friston, K.J.: Simulating emotions: an active inference model of emotional state inference and emotion concept learning. Front. Psychol. **10**, 2844 (2019)
41. Veissière, S.P., Constant, A., Ramstead, M.J., Friston, K.J., Kirmayer, L.J.: Thinking through other minds: a variational approach to cognition and culture. Behav. Brain Sci. **43**, e90 (2020)
42. van Vugt, M.K., Lu, Z., Zhang, Z.: Inter-brain synchronization during agreement versus disagreement in debate. Mindfulness **11**(5), 1103–1112 (2020)
43. Vygotsky, L.S.: Mind in Society: The Development of Higher Psychological Processes, vol. 86. Harvard University Press, Cambridge, MA (1978). (Original work published 1935)

44. Wass, S.V., Whitehorn, M., Haresign, I.M., Phillips, E., Leong, V.: Interpersonal neural entrainment during early social interaction. Trends Cogn. Sci. **24**(4), 329–342 (2020)
45. Weber, M., Saucan, E., Jost, J.: Characterizing complex networks with Forman-Ricci curvature and associated geometric flows. J. Complex Netw. **5**(4), 527–550 (2017)
46. Yanagisawa, H., Wu, X., Ueda, K., Kato, T.: Free energy model of emotional valence in dual-process perceptions. Neural Netw. **157**, 422–436 (2023)

Human Experiments

An Active Inference Model of Mouse Point-and-Click Behaviour

Markus Klar(✉), Sebastian Stein, Fraser Paterson, John H. Williamson, and Roderick Murray-Smith

University of Glasgow, Glasgow, Scotland, UK
markus.klar@glasgow.ac.uk

Abstract. We explore the use of Active Inference (AIF) as a computational user model for spatial pointing, a key problem in Human-Computer Interaction (HCI). We present an AIF agent with continuous state, action, and observation spaces, performing one-dimensional mouse pointing and clicking. We use a simple underlying dynamic system to model the mouse cursor dynamics with realistic perceptual delay. In contrast to previous optimal feedback control-based models, the agent's actions are selected by minimizing Expected Free Energy, solely based on preference distributions over percepts, such as observing clicking a button correctly. Our results show that the agent creates plausible pointing movements and clicks when the cursor is over the target, with similar end-point variance to human users. In contrast to other models of pointing, we incorporate fully probabilistic, predictive delay compensation into the agent. The agent shows distinct behaviour for differing target difficulties without the need to retune system parameters, as done in other approaches. We discuss the simulation results and emphasize the challenges in identifying the correct configuration of an AIF agent interacting with continuous systems.

Keywords: Mouse Pointing · Computational Interaction · Active Inference · Continuous Systems · Perceptual Delay · Simulation Intelligence

1 Introduction

Pointing is a fundamental interaction technique and *Human-Computer Interaction (HCI)* has a long history of studying human pointing behaviour [18]. Pointing is a way of communicating information through an interface by manipulating a user-controlled cursor into spatial partitions corresponding to actions; typically, these are discrete actions like selecting a menu item. Modeling pointing behaviour is crucial in engineering interactions well-suited to human capabilities, or in predicting the behaviour of users in using an existing interface without real users in the loop. As a prevalent and tractable interaction problem, it has attracted substantial attention in the HCI literature. HCI has historically

M. Albarracin et al. (Eds.): IWAI 2025, CCIS 2857, pp. 235–254, 2026.
https://doi.org/10.1007/978-3-032-16955-6_13

focused on high-level descriptive models of pointing using summary statistics. Fitts' Law [5,6,14] is the classic model, and relates the expected time taken to acquire a target (with a certain tolerated error rate) to the size and distance of the target from the cursor origin. While such models are effective in explaining movement time and are simple to apply, they do not attempt to model the detailed process of target acquisition itself, and do not consider click dynamics. Insights into complete cursor trajectories would enable more precise prediction of user intentions and the development of improved assistive tools.

One way to improve this understanding is to build a fine-grained, computational, generative model of the human user that synthesizes the entire target acquisition process. Approaches to model whole pointing trajectories have been introduced, e.g. in [3,4,13,15,16], often with models derived from a control-theoretic perspective, such as the Minimum Jerk model [7] or the E-LQG [25]. Other approaches have applied machine learning to derive policies that emulate human behaviour from data [1,2,10]. These classical models typically rely on the design of cost or reward functions, which are often defined by a combination of the distance between the cursor and the target and the control effort required to reach it. They are also typically deterministic in that they do not model or represent uncertainty explicitly, some even requiring a predefined movement time. Furthermore, in contrast to our approach, these focus on cursor movement only and clicking is usually not included, or seen as a separate task.

We propose the application of active inference (AIF) [19] to create a pointing model that is fully probabilistic, predictive and formulates the problem in terms of preferences over observations, rather than rewards. Different behaviour for difficult targets, which occurs in real user data (see Fig. 1), should emerge from greater uncertainty in the target acquisition rather than changed model and reward parameters – which is often the case in the classical models of pointing mentioned above. As an underlying model, we make use of the well-understood Second-Order Lag model which is commonly used to model human mouse pointing [4,16,23]. Using such a simple system to represent the human motor control system allows us to focus on the AIF agent's inference and control. We additionally apply perceptual noise and delay in order to achieve a more human-like behaviour [20,21]. AIF is a widely-applicable model of the behaviour of organisms that unifies Bayesian models of perception with a probabilistic model for action under uncertainty. It postulates behaviour as a process of reducing surprise to drive the environment into preferred configurations and minimizing uncertainty that might, in the future, lead to disfavoured states.

Recently there has been interest in bringing active inference concepts into the study and practice of HCI, see reviews such as [17,26]. Stein et al. [24] explored ordinal selection with AIF, i.e., selecting the target from a set of discrete elements, but modeling pointing with AIF has not been explored. In the context of pointing, AIF offers a fundamentally different perspective. An AIF formulation frames pointing as a continuous process of minimizing prediction error by continuously updating internal beliefs based on sensory input. Active inference defines the goals of a (simulated) user in terms of a distribution over desired

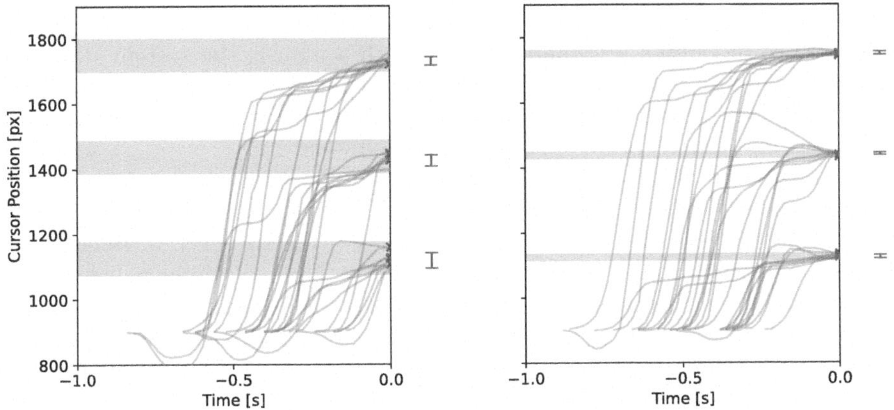

Fig. 1. Mouse cursor trajectories (green lines) and clicks (green crosses) of one human user pointing and clicking at targets (red) with different distance (850px, 537px, 225px) and different widths (left: 100px, right: 20px). $t = 0$ indicates the moment of a correct click. The error bars show the end-point standard deviations. (Color figure online)

observations. For example, rather than moving the cursor exactly to the centre of the target, the user might be satisfied as long as there is a high chance of clicking the button correctly. However, AIF has typically been applied in discrete contexts or in restricted forms. Modelling pointing, at the level of dynamic trajectory generation, requires a continuous representation of actions (such as potentials to be propagated to a simulated motor system), states, and observations (such as cursor position). We propose an AIF model that is capable of modelling fine-grained pointing behaviour in this continuous setting, and which generalises to different tasks and environments.

In summary, our contributions are:

- A continuous active inference formulation of point and click behaviour which does not rely on arbitrary discretisations, appropriately models perceptual noise and delay, and provides probabilistic predictions of cursor trajectories.
- An empirical study of a classical 1D pointing task with our active inference model, including a comparison of pointing trajectories with behaviour observed in a user study.

2 Active Inference for Continuous Dynamic Systems with Observation Noise and Perceptual Delay

Control of a human limb is an example of the general problem of control of a dynamic system. Section A describes the notation for such continuous system dynamics, how the agent constructs a belief about the states, dynamics and noise, as well as how it updates them. Specific adaptations of the standard general active inference approach in our model include:

- **System Dynamics:** The general equations are described in Sect. A.1. In the context of modelling a human pointing with a mouse, the system state may be a combination of the cursor's position and velocity and whether the mouse button is being clicked. The observation reflects the user's visual, proprioceptive, or haptic perceptions, which are noisy by nature.
- **Probabilistic Beliefs of AIF Agent** are constructed about the states, dynamics and noise: 1. Belief about system state $Q^s := \mathcal{N}(\mu^s, \Sigma^s)$. 2. Belief about parameters of the system dynamics $Q^\theta := \mathcal{N}(\mu^\theta, \Sigma^\theta)$. 3. Belief about noise in observations $Q^p := \exp(\mathcal{N}(\mu^p, \Sigma^p))$. Each of these beliefs is represented by the mean and covariance matrix of a (multivariate) Gaussian distribution. Since the belief about the noise Q^p describes the standard deviation of the different noise factors and thus needs to be positive at all times, we choose a log-normal distribution for robustness.
- **Belief Update mechanisms** are mostly standard, and are described in Sect. A.2. To efficiently update the agent's belief after performing an action, we use an Unscented Kalman filter (UKF) which can be used to propagate Gaussian random variables through system dynamics [11,27].
- **Action Selection** is based on minimising Expected Free Energy (EFE) as described in Sect. A.3. Since these calculations require a lot of computation time, especially when the horizon N is large, the evaluation in this work will exclude simulations with information gain.
- **Perceptual Delay** is modeled by constructing a belief about $s(t-\tau)$ for a given perceptual delay of $\tau \in \mathbb{N}$ timesteps and updating this belief using the observation $o(t-\tau)$ obtained in step t. Before action selection, the agent uses its generative model and the last actions to create a temporary belief about $s(t)$ by sequential updates. Additionally, we let the agent know about the exact target location only after the initial delay (see Sect. 3.2).

3 1D Mouse Pointing with Active Inference

We now demonstrate how AIF can be applied to simulate a human user performing a 1D-mouse pointing task. In such a task, the user is asked to position a mouse cursor within a one-dimensional target whose size and distance is varied systematically. A detailed description about the pointing task is provided in Sect. C. We simulate single trials which terminate as soon as the button is clicked successfully or after two seconds. Due to numerical constraints, both the canvas and the targets are scaled down by a factor of 1000. Furthermore, the initial position is recalculated to be centred at the origin. All shown results are however rescaled to the original size. We apply our model to investigate whether a cursor controlled with AIF: 1. allows us to model mouse clicking based on probabilistic beliefs about the system and preference distributions; 2. shows similar variability to observed human movements; and 3. can model the variety of behavioural strategies shown for different target difficulties without separate models for each.

The complete interaction loop of the agent with the system is depicted as a block diagram in Fig. 2 and provided as pseudo code in Sect. B.2.

3.1 The Generative Process and Model

To simulate human pointing, we need to build a model for the pointing technique and interface. In Active Inference terms, this means that we construct a generative process that plays the role of the 'real world'. The agent can interact with the process via a Markov Blanket which also defines what the agent observes. The agent also constructs a generative model of the world, which can be different from the real world. We decided to keep the system dynamics identical, however, the generative model is governed by the agent's belief about system parameters (which may be wrong or uncertain). Our model has two main components, the arm dynamics and the finger dynamics, and an additional logic for mouse clicks.

Arm Dynamics As a basis for our simulated arm dynamics we choose a Second-Order Lag model [4,16,23]. The state of this system is given by the cursor position s_1 and velocity s_2. The agent controls the acceleration a_1 that is applied to the cursor. When moving a computer mouse over a desk in a reasonable area, there is no force pulling the mouse to a certain stable position. Therefore we decided to set the cursor's stiffness to zero. However, damping is still applied with a factor of $d > 0$.

Finger Dynamics To model the index finger pressing the left mouse button, we use a simple First-Order Lag system [23] with stiffness $k > 0$. A click is only initiated when the displacement applied to the mouse button exceeds a threshold, requiring the agent to release the button before being able to click again. Therefore, both the previous button displacement s_3 and the current button displacement s_4 must be present in the system state. The agent can increase or decrease this force with a second action a_2.

After time-discretisation with $\Delta t = 0.02$, the combined system dynamics which take state s and action a and obtain the new state are therefore given by

$$f_\theta : \mathbb{R}^4 \times \mathbb{R}^2 \rightarrow \mathbb{R}^4, \quad f_\theta(s, a) = \begin{bmatrix} s_1 + \Delta t\, s_2 \\ s_2 - \Delta t\, d\, s_2 + \Delta t\, a_1 \\ s_4 \\ s_4 - \Delta t\, k\, s_4 + \Delta t\, a_2 \end{bmatrix}, \tag{1}$$

where $d > 0$ is the damping parameter of the cursor and $k > 0$ is the stiffness parameter of the finger. The observation function is given by

$$g_\theta : \mathbb{R}^4 \rightarrow \mathbb{R}^5, \quad g_\theta(s) = \begin{bmatrix} s_1 \\ s_4 \\ c \\ c \wedge \left(|s_1 - T| \leq \frac{W}{2}\right) \\ c \wedge \left(|s_1 - T| \leq \frac{W}{2}\right) - c \end{bmatrix}, \tag{2}$$

$$c = (s_3 < \alpha_c) \wedge (s_4 > \alpha_c) \tag{3}$$

and contains the cursor position, the current button displacement, whether the mouse has been clicked, whether the button has been hit, and whether the button was missed. The parameters T and width $W > 0$ are the target centre and width, respectively, and $\alpha_c > 0$ is the clicking threshold. Thus, both functions depend on the system parameters $\theta = [d, k, T, W, \alpha_c]$.

Click Logic. The observation of clicks works as follows. If the previous button displacement was below and the current button displacement is above the threshold, a click is issued and $c = 1$. In any other case $c = 0$.[1] If, additionally, the cursor position is inside the target, i.e., $\left(|s_1 - T| \leq \frac{W}{2}\right)$, the button is clicked successfully, i.e., $g(s)_4 = 1$ (otherwise $g(s)_4 = 0$). The last entry of the output represents a negative feedback for clicking outside the button, such as a beep tone commonly used in studies [16]. As such, $g(s)_5 = -1$ when a misclick happens (otherwise $g(s)_5 = 0$). We intentionally add this observation to allow the agent to have a specific preference of *not* observing a misclick.

Since human perception is inherently subject to noise, we add Gaussian noise to the agent's observations of cursor position and button displacement. We assume that the other components of output are observed without any disturbances, leading to an observation of $o = [o_1, o_2, g(s)_3, g(s)_4, g(s)_5]$ with $o_1 \sim \mathcal{N}(g(s)_1, \sigma_1^p)$ and $o_2 \sim \mathcal{N}(g(s)_2, \sigma_2^p)$. Except otherwise noted, we use the parameters shown in Sect. B.1.

3.2 Belief Priors and Perceptual Delay

We explore agents that are supposedly well-trained in the pointing task, i.e., the means of the prior belief about the system parameters Q^θ are close to the real values while the uncertainty is small. However, to model the initial delay of perceiving the target, we initialise the agent's belief about the target location to be centred at the initial position and have a high uncertainty, i.e., the agent believes that the target could appear anywhere on the display. In summary, the agent's initial belief Q^θ is given by

$$\mu^\theta = [d, k, 0.0, 0.03, \alpha_c], \ \Sigma^\theta = \operatorname{diag}\left(\left[0.2, 0.2, 0.9, 0.02, 10^{-6}\right]^2\right). \tag{4}$$

After the delay period, we update the belief to include the correct target position T and width W with low uncertainty, that is,

$$\mu^\theta = [d, k, \mathbf{T}, \mathbf{W}, \alpha_c], \ \Sigma^\theta = \operatorname{diag}\left(\left[0.2, 0.2, \mathbf{10^{-6}}, \mathbf{10^{-6}}, 10^{-6}\right]^2\right). \tag{5}$$

Note that we keep some uncertainty about the damping and stiffness. This is due to the fact that an agent with pristine knowledge about all system parameters would not need to infer the system state from noisy observations. Instead, its internal model of the world would allow for perfect prediction of the cursor trajectory and clicking, resulting in super-human performance. However, to keep the sources of uncertainty in check, we provide the agent with full knowledge about the amount of observation noise, i.e. Q^p is given by

$$\mu^p = [\sigma_1^p, \sigma_2^p], \ \Sigma^p = 10^{-8} I_2. \tag{6}$$

[1] Following the convention that `True` is represented as 1 and `False` as 0.

In future work, Q^θ and Q^p might also be inferred by interacting with the system over a longer time period.

The agent starts with a reasonable initial belief about the system state, that is

$$\mu^s = [0.0, 0.0, 0.0, 0.0]\,,\ \Sigma^s = \text{diag}([0.001, 0.0001, 0.00005, 0.00005]^2)\,. \tag{7}$$

We address the time a perceived signal takes to be processed by a human as follows. We delay observations that the agent received from the system after applying an action by $\tau = 5$ timesteps (=100 ms). To compensate for this perceptual delay, the agent keeps a belief about the state before the delay, $Q^s(t-\tau)$ (instead of t). Prior to planning its next action, the agent then predicts the system state at time t using its internal model. This is achieved by sequentially updating a temporary copy $\tilde{Q}^s(t-\tau)$ of $Q^s(t-\tau)$ by using actions applied during the delay $a(t+i), i = -\tau, \ldots, -1$, until arriving at $\tilde{Q}^s(t)$ (see Sect. A.2, 1.). After planning and performing the next action, the agent uses the "old" action $a(t-\tau)$ and, if available, delayed observation from $o(t-\tau)$ to obtain the belief $Q^s(t-\tau+1)$. This belief then serves as an initial point for the next planning step.

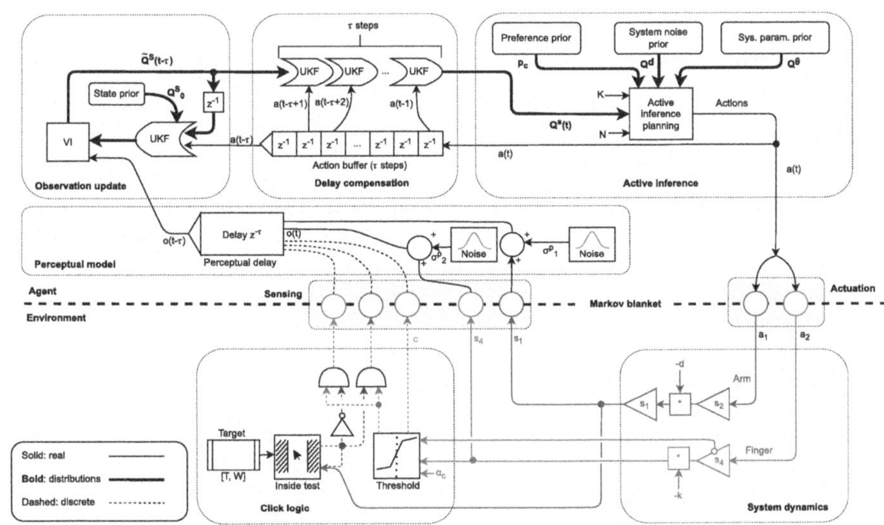

Fig. 2. A block diagram showing the interaction loop of the agent (upper) and the environment (lower). The agent includes a perceptual model, predictive delay compensation and min-EFE action planning. The simulated environment consists of a first and second-order lag modelling cursor and finger dynamics, along with click handling logic. See Algorithm 1 for full details.

3.3 Preference Prior

The agent's goal is to control the generative process described in Sect. 3.1 in the same way a human does, i.e., such that the target button is clicked as accurately and fast as possible. To model this in a natural way with Active Inference, we define the agent's preference distribution as a marginal distribution on the observations of the cursor position o_1, button click o_4, and misclick o_5,

$$P^c_{o_1,o_4,o_5} = \mathcal{N}\left([T, 1.0, 0.0], \text{diag}\left([0.01, 0.01, 0.001]^2\right)\right). \tag{8}$$

The mean of this distribution is centred around the cursor being at the target's centre, the button being successfully clicked, and not getting negative feedback on a misclick. Adding the additional preference of not observing a misclick ensures that the agent only clicks when it is certain enough that the cursor is on the button. Scaling the diagonal values of the covariance matrix allows creating agents that have different focus on speed versus accuracy, a classical trade-off that is present in human aimed movements [5,9,22].

The parameters used in the results are manually fitted to the data of one user performing the pointing task described in Sect. C.1 and more details about the fitting process can be found in Sect. B.1. We additionally present the effect of changing individual parameters in Sect. D. It is important to mention that we – other than previous optimal control based approaches – use the same parameter set for all targets. The qualitatively different behaviours emerge from the properties of the AIF agent instead.

4 Results

We investigate whether the AIF agent is able to produce movements which are similar to human-controlled movements, despite the simplicity of the underlying model. In particular, we are interested in the amount of variability produced by the agent compared to a human performing the same task.

4.1 AIF Agent Performs Pointing Task

Figure 3 shows a time-series view of how the AIF agent performs the pointing task. Before obtaining knowledge about the target position (before 0.1 s), the agent keeps the cursor close to the centre. During the fast surge movement (an initial ballistic phase of a mouse movement, [16]), the agent's perception of the cursor position (left plot) is subject to increased uncertainty due to the larger impact of the perceptual delay. After slowing down close to the target, the agent regains a more precise belief about the cursor position and ultimately manages to click on the target. The resulting corrective movements are commonly observed in human pointing. In the right plot, the button displacement is represented. The initiation of a click is prompted upon attaining a threshold of $\alpha_c = 0.05$. Given that the preference distribution is shaped by the observation of not missing the button, the agent only clicks when it believes that the cursor is inside the target.

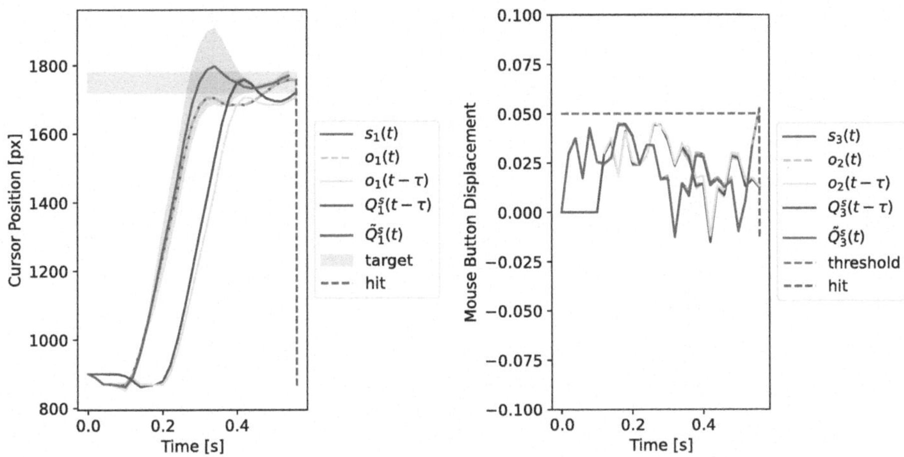

Fig. 3. Mouse cursor and button displacement for one trial of the AIF agent from the start at 900 to $T = 1750$, $W = 60$px. The plot shows the real state $s(t)$ (blue), the agent's belief about $s(t-\tau)$ at time t (purple) and the prediction of $s(t)$ used for planning (green). The ribbon displays ±3 standard deviations. (Color figure online)

4.2 AIF Agent Follows Fitts' Law

Fitts' Law describes a logarithmic relationship between the ratio of a target distance to target width and the time for humans to reliably acquire it. It is usually given in the Shannon form: $MT = a + b \cdot \log_2\left(1 + \frac{D}{W}\right)$, where MT is the movement time, a and b are empirical constants, D is the distance to the target, and W is the width of the target along the axis of motion [5,14]. The log ratio term $\log_2\left(1 + \frac{D}{W}\right)$ is called the *index of difficulty (ID)* and is measured in bits; larger values indicate targets which are more difficult to acquire.

If our model is a plausible model of human pointing, it should follow this relationship. If it is well-tuned for a specific configuration (such that we emulate a realistic human motor system with a particular pointing device), we should recover similar coefficients a, b as in experimental data. Figure 4 shows the Fitts' Law evaluation for the AIF agent and a human user side-by-side. The AIF agent follows the characteristic pattern of Fitts' Law with an identical slope b as the user, while having a shorter constant delay a.

4.3 AIF Agent Shows Changing Behaviour for Different Targets

Figure 5 shows the trajectories produced by the AIF agent and a human user, split up by target size. As predicted by Fitts' Law, movements of both agent and human, are longer for more difficult targets, i.e., smaller targets (right plots) or targets with larger distance from the initial position at 900px. Similar to the human, if targets are smaller or further away, the agent-controlled cursor often shows one or multiple sub-movements, i.e., shorter corrective movements, after the surge phase.

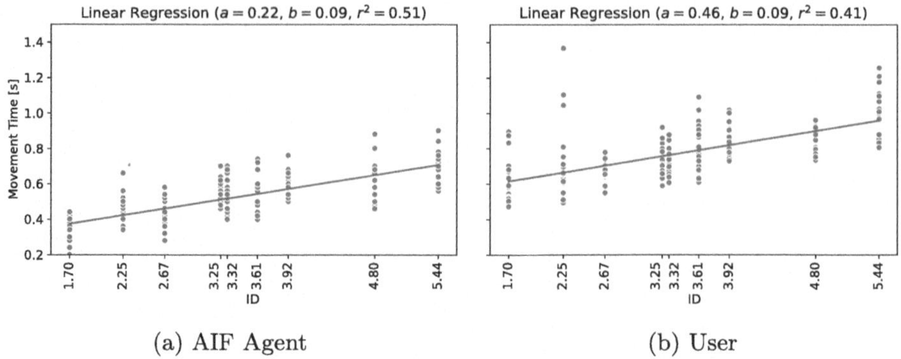

(a) AIF Agent (b) User

Fig. 4. Following Fitts' Law, human pointing movements show a linear relationship between movement time and the index of difficulty (ID). The simulation with AIF shows a similar trend. The data is fitted with a linear regression model. Simulation: $a = 0.22, b = 0.09, r^2 = 0.51$; User: $a = 0.46, b = 0.09, r^2 = 0.41$.

Another characteristic of human behaviour is that it shows higher end-point variance for larger targets. The end-point variance of the AIF agent's movements shows a similar trend, which is displayed via the error bars next to the trajectories in Fig. 5. Similar to the user, for large targets (left plots in Fig. 5), the endpoints of the agent's movements are skewed towards the starting point, showing that the agent makes use of the target size.

5 Discussion

Implications of the Results. The results indicate that a full active inference model of target acquisition and click actuation with appropriate perceptual delays leads to behaviours that are comparable to human performance. The gross behaviour, as quantified by the Fitts' law plots, indicates human-like pointing capacity, even without precise calibration. The AIF agent also shows realistic end-point variance and makes use of the target size. Unlike optimal control methods, where parameters have to be fit not just for each user but for each target, a single parameterisation leads to plausible pointing behaviour across targets spanning a range of indices of difficulty. The AIF agent's predictive models can compensate for the delayed observations with variability that appears to correspond well to human pointing in the presence of the well-known latency in the visual system. The combination of cursor movement and clicking enables interesting interactions, even in a simple model as ours, such as updating the belief about the cursor position by observing a click or misclick.

Limitations. Our model has a radically simplified representation of human perception and motor control as a second-order system with Gaussian-corrupted observations and fixed perceptual delay. Our parameterisations are hand-tuned and validated on the data of a single user. Our agent has a relatively short time

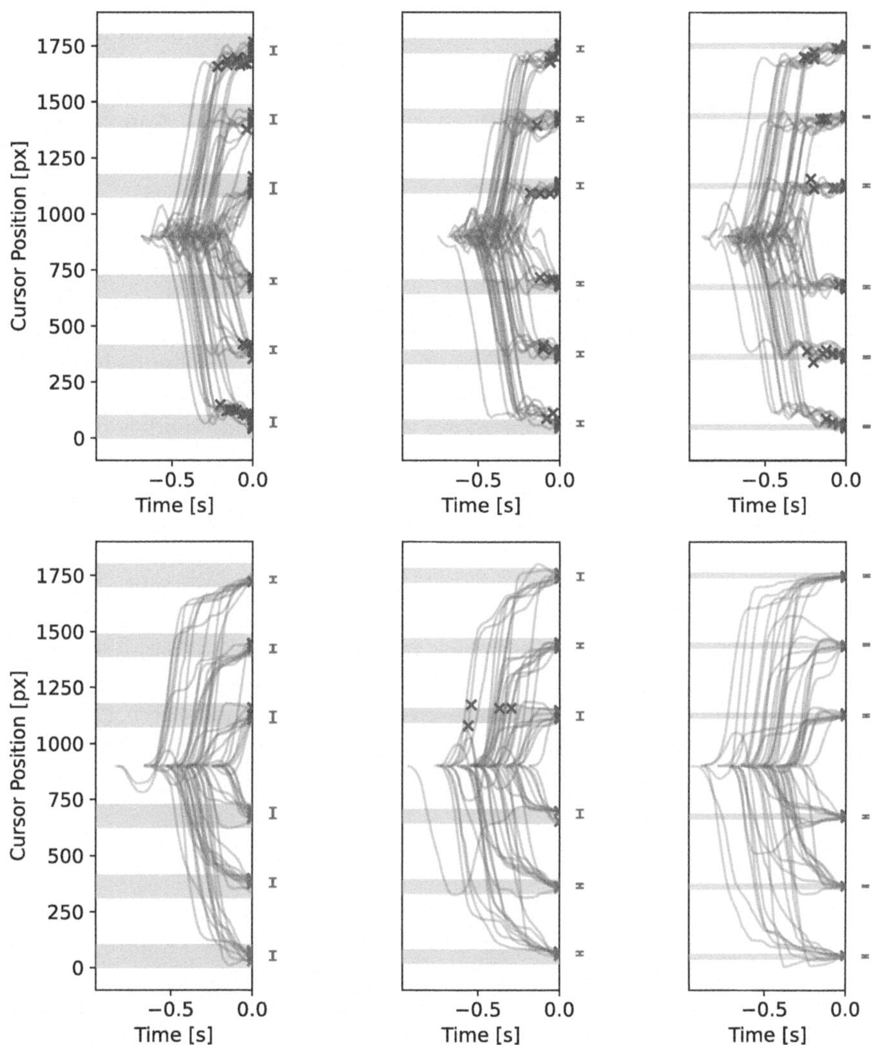

Fig. 5. Cursor trajectories for the AIF agent (upper row, blue) and a human user (lower row, green) for different target sizes (from left to right: 100px, 60px, 20px). The crosses mark mouse clicks, the red areas display the targets. $t = 0$ indicates the moment of a correct click. The error bars show the end-point standard deviations. The averages for the different target sizes are: Agent: $(17.63, 10.08, 4.74)$, User: $(19.48, 13.02, 4.81)$. (Color figure online)

horizon as a consequence of the computational cost of sample-based inference and the consequent effect of "sample starvation" on policy selection. These inference limitations have an implicit effect on the parameterisation of the agent – for example, action selection is challenging if no sequence of actions in the time horizon brings the predicted state close to the "correct click" state. This can also lead to oscillating behaviour, as in Fig. 5, which is not as pronounced in humans. The computational effort involved in the variational inference procedure and limit the amount of plans that can be evaluated. However, our approach using UKF with Monte Carlo sampled rollouts inference makes our fully continuous formulation viable.

Outlook. Our model is a working AIF model of human pointing, but it required significant custom development. There is a clear need for software building blocks to make progress in AIF-based modelling of human interaction: reusable models and components appropriate for HCI tasks. We use a simple linear system as our forward model and a similarly simplified perceptual model with known delay. There remains opportunity to explore more authentic biomechanical models of the arm, or stronger perceptual models that, e.g. add uncertainty about the perceptual delay, properly account for motor delays, or include models of foveated vision. Although our model reproduces behaviour qualitatively similar to our test user, we have much to explore in calibrating parameters to better match human performance. This could involve automatic optimisation of agent parameters to align with observed user data, and/or parameters informed directly by established biomechanical and perceptual data. Optimisation could also be performed online, with an active inference agent learning system parameters during simulated interactions. We also see scope to develop more sophisticated metrics to establish what "similar" trajectories are, particularly when quantifying variability in dynamics. There are obvious extensions of our model to e.g. 2D or 3D pointing tasks, but these will likely require better inference, new datasets and stronger biomechanical and perceptual models.

5.1 Conclusion

We presented an active inference simulation of human pointing and clicking behaviour with continuous states, actions, and observations. Despite its simple 2OL forward model our model generates plausible trajectories across a range of target sizes, and it includes a full model of actuation ("click") planning and predictive delay compensation.

While AIF control of even this elementary system is not straightforward – our agent has a profusion of parameters and significant inference challenges – our findings suggest that AIF is a promising model for fine-grained interaction models. The inherently probabilistic nature allows for a greater variety of cursor trajectories than comparable methods, generating a wide variety of plausible behaviours. Our model is the first step towards active inference-based user modelling in HCI. Our results indicate that active inference is viable for user behavioural models and highlight many open challenges in building more sophisticated AIF-based synthetic users.

The Python code used to create the simulations in this work can be found at https://github.com/mkl4r/AIF-Pointing.

A Active Inference for Continuous Dynamic Systems with Control and Observation Noise

A.1 System Dynamics

We control a general dynamic system with continuous states, actions, and observations with active inference. We take a discrete-time approach, with a constant time step Δt. A dynamic system is then described by functions

$$s_{t+1} = f_\theta(s_t, a_t), \tag{9}$$

$$o_t = g_\theta(s_t) + \epsilon_t, \tag{10}$$

where (9) maps the current system *state* $s_t \in \mathbb{R}^n$ and an *action* $a_t \in \mathbb{R}^m$ at timestep $t \in \mathbb{N}$ to the new system state $s_{t+1} \in \mathbb{R}^n$, given some system parameters θ. The system's output (10) is given by a function $g_\theta : \mathbb{R}^n \mapsto \mathbb{R}^l$ and may be perturbed by constant and/or state-dependent Gaussian noise to obtain the agent's *observation* $o_t \in \mathbb{R}^l$ at time step t.

A.2 Belief Updates

A human user will usually only have access to a limited set of system states that they observe via output devices such as displays. For instance, although we cannot perceive the velocity of a mouse cursor directly (as in a speedometer), it is still an important source of information to prevent overshooting a target button. The agent *infers* this information by propagating its belief about the cursor velocity through its internal model, as explained below.

The agent has two ways to update its belief about the system state Q^s_t: 1. After performing an action, the agent uses its internal model of the world to update its belief, resulting in a new belief Q^s_{t+1} for timestep $t+1$. An efficient approach for this is to propagate the agent's belief through time using an Unscented Kalman filter (UKF) [11], which allows the estimation of states under nonlinear transition functions with Gaussian belief distributions. It involves choosing sigma points in the combined space of Q^s and Q^θ, evaluating f_θ on these with the chosen action a_t to obtain the corresponding s_{t+1}, and estimating the parameters of the updated belief distribution Q^s_{t+1} directly from the s_{t+1} using Maximum Likelihood. Note that during this step we fold in the application of the action a_t into the transition function that the UKF applies to make predictions of the future state under action a_t. This update usually increases uncertainty, since the agent is uncertain about the current state of the world as well as the dynamics. 2. After a new observation, the agent updates its belief using variational inference. This means, that the agent minimizes the variational free energy F with respect to the belief distribution parameters and the noise belief,

$$\min_{\hat{Q}^s} F\left(\hat{Q}^s, o_t\right) = \mathrm{D_{KL}}\left(Q^s \,\middle\|\, \hat{Q}^s\right) - \mathbb{E}_{s_t \sim \hat{Q}^s; \sigma_p \sim Q^p}\left[\ln P_o(o_t | s_t; \sigma_p)\right],$$

where Q^s is the variational belief *prior* to making the new observation (but after having taken the last action), $\hat{Q}^s$ is the *posterior* belief, and $\mathrm{D_{KL}}$ the

KL-divergence. The first term has an analytical solution for Gaussian belief distributions, whereas the second term requires sampling from $\hat{Q}^s$. For optimization via gradient descent, we use the reparameterization trick [12] to backpropagate through n_s^{VI} samples in each of the k^{VI} update steps. Depending on the knowledge of observation noise, this step may reduce the agent's uncertainty.

A.3 Action Selection

In every timestep t, the agent generates K plans representing sequences of actions with a N-step planning horizon by sampling actions from uniform distributions,

$$\pi_j = (a_{1j}, \ldots, a_{Nj}), \forall j \in \{1, \ldots, K\}$$
$$a_{ij} \sim \mathcal{U}(\underline{a}, \overline{a}),$$

where $\underline{a}, \overline{a}$ are the lower and upper boundaries for the actions taken by the agent. These plans are evaluated with respect to the *Expected Free Energy (EFE)* [19],

$$G(\pi_j) = \frac{1}{N} \sum_{i=1}^{N} G_{ij}(a_{ij})$$
$$G_{ij}(a_{ij}) = \underbrace{-\mathbb{E}_{Q_{ij}^s} \left[\mathrm{D_{KL}} \left(\hat{Q}_{ij}^s \,\middle\|\, Q_{ij}^s \right) \right]}_{\text{Information Gain}} \underbrace{- \mathbb{E}_{s \sim Q_{ij}^s; \sigma_p \sim Q^p; \tilde{o} \sim \mathcal{N}(g(s), \sigma_p),} \left[\ln P^c(\tilde{o}) \right]}_{\text{Pragmatic Value}},$$

where Q_{ij}^s are predicted via rollout, i.e. by successively updating the agent's belief after performing action a_{ij} as described in Sect. A.2. Since these rollouts are predictions 'in the agent's mind', the beliefs cannot be updated using observations here.[2] In practice, it is often impossible to calculate these analytically, so we approximate based on observations sampled from the current belief. In case of perceptual delay, the initial belief distribution is obtained by applying UKF with the last performed actions as described in Sect. 3.2.

The *information gain* is approximated by computing the KL-divergence between Q_{ij}^s and different instances of $\hat{Q}_{ij}^s$. This involves taking n_s^{IG} samples $s \sim Q_{ij}^s$ and $\Xi_p \sim Q^p$ which are used to sample n_o^{IG} observations from the generative model each, i.e. $\tilde{o} \sim \mathcal{N}(g_\theta(s), \Xi_p)$. After performing the update after observation, as described in Sect. A.2 for each of these, we obtain $n^{\mathrm{IG}} = n_s^{\mathrm{IG}} \cdot n_o^{\mathrm{IG}}$ different instances of $\hat{Q}_{ij}^s$.

The *pragmatic value* represents the bias encouraging the agent to take actions that result in preferred observations, which are characterized by the preference distribution P^c. As above, we sample n_s^{PV} states from the current belief and sample n_o^{PV} observations each to estimate the log-likelihood of these given the preference distribution. The more likely these observations are, the better they and the plan leading to them are in the eyes of the agent.

[2] In discrete systems, Sophisticated Inference can be applied to walk the complete tree of possible actions and observations. However, even in cases where this is possible it is often too computationally expensive [8].

The next action is chosen as the first action of the plan with minimal expected free energy, i.e. $a_{t+1} = a_{1j}$, where $j = \text{argmin}_{j \in \{1,\dots,K\}} G(\pi_j)$.

B Implementation Details

B.1 Parameters

Our AIF agent has many parameters that can have an influence in the resulting trajectories. The following Sect. 1 gives an overview of these parameters. To find suitable parameters that lead to movements similar to those of a human, we proceed as follows.

- We roughly estimate the necessary damping d and maximal hand control $\overline{a}_1$ by comparing the 99% percentile and maxima of accelerations and velocities observed in the study to those created by applying the maximal activation to the mouse cursor system Eq. (1) for one second.
- We aim to set the stiffness of the finger dynamic such that the earliest click can not happen faster than observed in a user clicking as fast as possible, that is $\approx 0.14\,\text{s}$ per click. However, this had to be relaxed slightly to ensure that the agent can click the button consistently.
- We obtain the reaction time of the agent by computing the first timestep after the user exceeds an acceleration threshold of 10px/s^2, that is $\tau \approx 0.1\,\text{s}$.
- We tune the preference distribution P^C such that the resulting behaviour shows similar clicks and misclicks as humans.

The remaining parameters are tuned by visually comparing trajectories and the Fitts' Law results.

B.2 The Complete Algorithm

Algorithm 1 shows the pseudo code for the complete interaction loop of an AIF agent with the mouse model. Prior to planning the next action, the agent compensates for the perceptual delay by estimating the system state at time t. This is achieved by sequentially updating the temporary copy $\tilde{Q}^s$ using the previously applied actions (for-loop in planning phase) and the agent's internal model of the system. After planning and performing an action, the agent uses the 'old' action and, if available, delayed observation from τ timesteps before to update the belief $Q^s(t-\tau)$.

C Experiment Details

C.1 User Data Acquisition

We gather movement data from a simple, PyScript-based application. The task is to move the mouse cursor from a start position to a target that appears in one of six random locations, either to the left or to the right of the start position at pixel 900. The width of the canvas is set to 1800 pixels. The distance between

Table 1. List of Parameters. The value column shows the commonly used value, if not otherwise stated in the paper. ∗: Value is given by the task; ∗∗: Value is explicitly explained in Sect. 3; X: Value is not used for the results. For details see Sect. A.

	Variable	Domain	Value	Description
System	d	$\mathbb{R}_+$	24.0	Cursor damping
	k	$\mathbb{R}_+$	10.0	Button stiffness
	T	$\mathbb{R}$	∗	Target position [px]
	W	$\mathbb{R}_+$	∗	Target width [px]
	α_c	$\mathbb{R}_+$	0.05	Click threshold
	Δt	$\mathbb{R}_+$	0.02	Time step length [s]
	τ	$\mathbb{N}$	5	Perceptual delay [timesteps]
	$\underline{a}$	$\mathbb{R}^2$	$(-50.0, -1.0)$	Control lower bounds
	$\overline{a}$	$\mathbb{R}^2$	$(50.0, 1.0)$	Control upper bounds
	Σ_p	$\mathbb{R}^3$	∗∗	Std/Covariance of observation noise
Belief	Q^s	$\mathbb{R}^4 \times \mathbb{R}^{4\times4}$	∗∗	State prior
	Q^θ	$\mathbb{R}^5 \times \mathbb{R}^{5\times5}$	∗∗	System parameters prior
	Q^d	$\mathbb{R}^3 \times \mathbb{R}^{3\times3}$	∗∗	Noise prior
Control	N	$\mathbb{N}$	12	Prediction horizon [timesteps]
	K	$\mathbb{N}$	3000	# Plans
	P^c	$\mathbb{R}^3 \times \mathbb{R}^{3\times3}$	∗∗	Preference distribution
Hyper parameters	k^{VI}	$\mathbb{N}$	30	# Optimization steps for VI
	n_s^{VI}	$\mathbb{N}$	300	# State samples for VI
	η^{VI}	$(0, 1)$	$3.0 \cdot 10^{-4}$	Learning rate for VI
	n_s^{IG}	$\mathbb{N}$	X	# State samples for Information Gain
	n_o^{IG}	$\mathbb{N}$	X	# Observation samples for Information Gain
	n_s^{PV}	$\mathbb{N}$	50	# State samples for Pragmatic value
	n_o^{PV}	$\mathbb{N}$	3	# Observation samples for Pragmatic value

the start position and the target's centre is varied between 225, 537, 850 pixels, while its width is varied between 20, 60, 100 pixels, resulting in a total of 18 different targets. The task is performed by six human users, hitting every target ten times in a random order, resulting in a total of 1080 movements. We split the resulting data into single trials, and removed outliers that showed a movement time greater than three standard deviations away from the mean (a depiction of the task as well is shown in Fig. 6). Table 2 shows the 18 targets.

Algorithm 1. Active Inference for Mouse Point-and-Click

Require: Parameters in Table 1
  $t \leftarrow 1$
  $Q^s(1+i) \leftarrow Q^s,\ \forall i \in \{-\tau, \dots, 0\}$
  $a(1+i) \leftarrow 0,\ \forall i \in \{-\tau, \dots, -1\}$
  while Button not clicked **do**
    1. Planning (Section A.3):
    $\tilde{Q}^s(t-\tau) \leftarrow Q^s_(t-\tau)$
    for $i \in \{-\tau, \dots, -1\}$ **do**
      $\tilde{Q}^s(t+i+1) \leftarrow$ Update $\tilde{Q}^s(t+i)$ using UKF and $a(t-i)$ (Section A.2, 1.)
    end for
    $\pi_j \leftarrow (a_{ij})_{ij}, \forall i \in \{1, \dots, N\}, j \in \{1, \dots, K\}; a_{ij} \sim \mathcal{U}(\underline{a}, \overline{a})$
    $j \leftarrow \text{argmin}_{j \in \{1,\dots,K\}} G(\pi_j)$ (using $\tilde{Q}^s(t)$)
    $a(t) \leftarrow \pi_j(1)$
    2. Step System:
    $s(t+1) \leftarrow f_\theta(s(t), a(t))$
    $o(t-\tau+1) \sim \mathcal{N}(g(s(t+1)), \sigma^p)$
    3. Update Q^s:
    $Q^s(t-\tau+1) \leftarrow$ Update $Q^s(t-\tau)$ via UKF and $a(t-\tau)$ (Section A.2, 1.)
    if $t > \tau$ **then**
      $Q^s(t-\tau+1) \leftarrow$ Update $Q^s(t-\tau+1)$ via VI and $o(t-\tau+1)$ (Section A.2, 2.)
    end if
    $t \leftarrow t+1$
  end while

(a) Start position (b) Target

Fig. 6. The task commences by clicking on the designated start position, which is represented by the blue line. Subsequently, a red target is displayed, either to the right or the left. After clicking on the target, the cursor must be moved back to the initial position prior to starting the next trial. (Color figure online)

D Additional Results

D.1 Changing Model Parameters

Changing individual parameters can affect the point and click behaviour of the AIF agent heavily. Figure 7 shows trajectories to the same target ($T = 1750$, $W = 30$px), while adapting only one parameter. Changing the damping parameter of the mouse cursor model (upper left) defines how steep the slope in the surge phase can be, with higher damping leading to a gentler slope (red trajectories, $d = 40$). The influence of the planning horizon (upper right) is harder to

Table 2. All targets of the pointing task with position, width in pixel, and index of difficulty.

Target Nr	Position	Width	ID
1	675	20	3.61
2	363	20	4.80
3	50	20	5.44
4	1125	20	3.61
5	1438	20	4.80
6	175	20	5.44
7	675	60	2.25
8	363	60	3.32
9	50	60	3.92
10	1125	60	2.25
11	1438	60	3.32
12	1750	60	3.92
13	675	100	1.70
14	363	100	2.67
15	50	100	3.25
16	1125	100	1.70
17	1438	100	2.67
18	1750	100	3.25

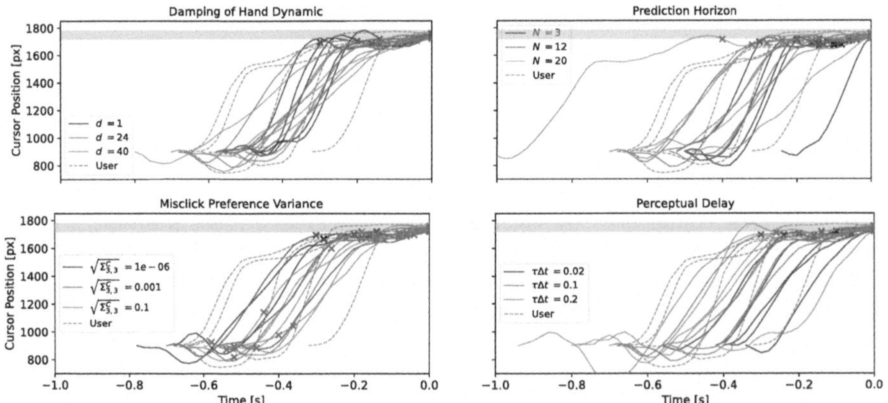

Fig. 7. In each plot, only one parameter is changed while the others are fixed. The blue trajectories refer to the agent used in the remaining result section and the green trajectories are those of a human user. $t = 0$ indicates the moment of a correct click. The crosses mark clicks and the red area the target. (Color figure online)

distinguish. If the number of sampled plans during planning is not adjusted, the chance of sampling a good plan with a higher horizon is decreased. Also, since the agent only applies the first action of the best plan, having a longer plan reduces the influence of this action on the expected free energy, which might lead to suboptimal control (long red trajectory). Similarly, adjusting the preference of observing a misclick, i.e., changing $\sqrt{\Sigma^C_{3,3}}$, (lower left) has no clear effect on the resulting trajectories. Nonetheless, lower values lead to less misclicks (almost no black crosses, $\sqrt{\Sigma^C_{3,3}} = 10^{-6}$), while higher values can lead to many misclicks (many red crosses, $\sqrt{\Sigma^C_{3,3}} = 0.1$). Increasing the agent's perceptual delay (lower right) increases the time to target in most cases (red lines, 0.2 s reaction time). This is not surprising, since with longer delay the agent receives knowledge about the target position later and has a higher uncertainty about the cursor position after the surge phase (not shown).

References

1. Cheema, N., Frey-Law, L.A., Naderi, K., Lehtinen, J., Slusallek, P., Hämäläinen, P.: Predicting mid-air interaction movements and fatigue using deep reinforcement learning. In: Proceedings of the 2020 CHI Conference on Human Factors in Computing Systems, CHI 2020, Honolulu, HI, USA, pp. 1–13. Association for Computing Machinery (2020). https://doi.org/10.1145/3313831.3376701
2. Dalsgaard, T.-S., Knibbe, J., Bergström, J.: Modeling pointing for 3D target selection in VR. In: Proceedings of the 27th ACM Symposium on Virtual Reality Software and Technology. VRST 2021, Osaka, Japan. Association for Computing Machinery (2021). https://doi.org/10.1145/3489849.3489853
3. Do, S., Chang, M., Lee, B.: A simulation model of intermittently controlled point-and-click behaviour. In: Proceedings of the 2021 CHI Conference on Human Factors in Computing Systems, CHI 2021, Yokohama, Japan. Association for Computing Machinery (2021). https://doi.org/10.1145/3411764.3445514
4. Fischer, F., Fleig, A., Klar, M., Müller, J.: Optimal feedback control for modeling human computer interaction. ACM Trans. Comput.-Hum. Interact. **29**(6), 1–70 (2022). https://doi.org/10.1145/3524122. https://dl.acm.org/doi/10.1145/3524122 (visited on 08/30/2024)
5. Fitts, P.M.: The information capacity of the human motor system in controlling the amplitude of movement. J. Exp. Psychol. **47**(6), 381 (1954)
6. Fitts, P.M., Peterson, J.R.: Information capacity of discrete motor responses. J. Exp. Psychol. **67**(2), 103 (1964)
7. Flash, T., Hogan, N.: The coordination of arm movements: an experimentally confirmed mathematical model. J. Neurosci. **5**(7), 1688–1703 (1985)
8. Friston, K., Da Costa, L., Hafner, D., Hesp, C., Parr, T.: Sophisticated inference. Neural Comput. **33**(3), 713–763 (2021)
9. Guiard, Y., Rioul, O.: A mathematical description of the speed/accuracy trade-off of aimed movement. In: Proceedings of the 2015 British HCI Conference, pp. 91–100 (2015)

10. Ikkala, A., et al.: Breathing life into biomechanical user models. In: Proceedings of the 35th Annual ACM Symposium on User Interface Software and Technology, UIST 2022, Bend, OR, USA. Association for Computing Machinery (2022). https://doi.org/10.1145/3526113.3545689
11. Julier, S.J., Uhlmann, J.K.: Unscented filtering and nonlinear estimation. Proc. IEEE **92**(3), 401–422 (2004)
12. Kingma, D.P., Welling, M.: Auto-encoding variational bayes. arXiv preprint arXiv:1312.6114 (2013). https://doi.org/10.48550/arXiv.1312.6114
13. Klar, M., Fischer, F., Fleig, A., Bachinski, M., Müller, J.: Simulating interaction movements via model predictive control. ACM Trans. Comput.-Hum. Interact. **30**(3) (2023). https://doi.org/10.1145/3577016
14. MacKenzie, I.S.: Fitts law. In: The Wiley Handbook of Human Computer In teraction, pp. 347–370. Wiley (2018). https://doi.org/10.1002/9781118976005.ch17. https://onlinelibrary.wiley.com/doi/abs/10.1002/9781118976005.ch17 (visited on 04/24/2024)
15. Martín, J.A.Á, Gollee, H., Müller, J., Murray-Smith, R.: Intermittent control as a model of mouse movements. ACM Trans. Comput.-Hum. Interact. (TOCHI) **28**(5), 1–46 (2021)
16. Müller, J., Oulasvirta, A., Murray-Smith, R.: Control theoretic models of pointing. ACM Trans. Comput.-Hum. Interact. **24**(4), 27:1–27:36 (2017)
17. Murray-Smith, R., Williamson, J.H., Stein, S.: Active inference and human computer interaction (2024). arXiv: 2412.14741 [cs.HC]. https://arxiv.org/abs/2412.14741
18. Myers, B.A.: Pick, Click, Flick! ACM, The Story of Interaction Techniques (2024)
19. Parr, T., Pezzulo, G., Friston, K.J.: Active Inference: The Free Energy Principle in Mind, Brain, and Behavior. The MIT Press (2022)
20. Perrinet, L.U., Adams, R.A., Friston, K.J.: Active inference, eye movements and oculomotor delays. Biol. Cybern. **108**(6), 777–801 (2014). https://doi.org/10.1007/s00422-014-0620-8
21. Priorelli, M., Stoianov, I.P.: Flexible intentions: an active inference theory. Front. Comput. Neurosci. **17** (2023). https://doi.org/10.3389/fncom.2023.1128694. https://www.frontiersin.org/journals/computational-neuroscience/articles/10.3389/fncom.2023.1128694
22. Schmidt, R.A., Zelaznik, H., Hawkins, B., Frank, J.S., Quinn, J.T., Jr.: Motor-output variability: a theory for the accuracy of rapid motor acts. Psychol. Rev. **86**(5), 415 (1979)
23. Sheridan, T.B., Ferrell, W.R.: Man-Machine Systems: Information, Control, and Decision Models of Human Performance. M.I.T. Press, Cambridge (1974)
24. Stein, S., Williamson, J.H., Murray-Smith, R.: Towards interaction design with active inference: a case study on noisy ordinal selection. In: 5th International Workshop on Active Inference (IWAI 2024) (2024)
25. Todorov, E., Li, W.: A generalized iterative LQG method for locally-optimal feedback control of constrained nonlinear stochastic systems. In: Proceedings of the 2005, American Control Conference, 2005, pp. 300–306 (2005)
26. Vertegaal, R., Merritt, T., Greenberg, S., Tarun, A.P., Li, Z., Fountas, Z.: Interactive inference: a neuromorphic theory of human-computer interaction (2025). arXiv: 2502.05935 [cs.HC]. https://arxiv.org/abs/2502.05935
27. Wan, E.A., Van Der Merwe, R.: The unscented Kalman filter for nonlinear esti mation. In: Proceedings of the IEEE 2000 Adaptive Systems for Signal Processing, Communications, and Control Symposium (Cat. No. 00EX373), pp. 153–158 (2000)

Writing Through Sympathetic Arousal: A Perceptual Active Inference Model on the Effects of Sympathetic Arousal Interoception on Written and Spoken Grammatical Complexity

Alexandra Brooke Skelton, Lukas Alexander De Vries, Angelica Maria Silva, and Roberto Limongi(✉)

Brandon University, 270 18th Street, Brandon, MB R7A6A9, Canada
{skeltoab70,Devriela87,silvaa,limongir}@brandonu.ca

Abstract. This study investigates the role of sympathetic arousal (SA) interoception in shaping the grammatical complexity of language production across writing and speaking modalities. Drawing from the active inference framework, we hypothesized that SA interoception drives the deployment of complex linguistic structures during language production. Under laboratory conditions, university undergraduate students produced short written and spoken texts while their galvanic skin response (GSR) was continuously recorded. Grammatical complexity was scored through a natural language processing algorithm, and sympathetic arousal was estimated using non-linear dynamic causal models fitted to the GSR data. A perceptual Markov decision process (MDP) model was then applied to capture the inferred interoceptive states and their influence on discourse production with high and low grammatical complexity. The results revealed that participants a priori believed that SA onsets would occur at the beginning of the discourse production task, regardless of the modality. However, when they produced written discourses, they showed higher interoception precision as to when the sympathetic arousal would occur—which was not accounted for by the time when they started to write. Furthermore, during the observation of a late SA onset, this higher precision more likely drove the production of registers with high grammatical complexity in the writing condition than in the speaking condition. While the reported MDP model has limitations when explaining the complex process of writing and its neural underpinnings, it is sensitive to capturing differences in SA interoception across language production modalities and grammatical complexity levels.

Keywords: Writing to learn · Sympathetic arousal interoception · Grammatical complexity

1 Introduction

Academic writing proficiency is emerging as a primary predictor of academic achievement in higher education [1, 2]. Academic writing pursues two goals: communicating knowledge and learning new knowledge [3–6]. When a writing task pursues learning

M. Albarracin et al. (Eds.): IWAI 2025, CCIS 2857, pp. 255–269, 2026.
https://doi.org/10.1007/978-3-032-16955-6_14

new knowledge, it is referred to as a writing-to-learn activity. Meta-analytic data reveal that writing activities enhance academic learning in terms of long-term memory consolidation of the new knowledge [2, 6–8]. The benefit is consistent across disciplines (e.g., sciences, social sciences, or mathematics) and grade level (e.g., ranging from elementary school to high school), with an average weighted effect size of 0.3 [3]. Interestingly, higher effect sizes have been observed in higher education [9], and fourth-year college students benefit more from writing than their first-year peers [10]. Therefore, the learning effect of academic writing is variable, yet unquestionable.

While academic writing activities emerge as highly recommended evidence-based pedagogical practices, their highly variable effect size suggests caution in their classroom implementation. It remains uncertain whether conditions such as the type of activity, assessment method, and educational level account for this variability [3]. Crucially, we remain uncertain whether individualized academic-writing activities can benefit students with developmental disorders (e.g., attention deficit/hyperactivity disorder and autism) [11–13]. This uncertainty has led to the following recommendation: "If we are to identify the sources of this variability, we need to isolate and test additional theoretically derived moderators…this requires that [researchers] must also purposefully expand the moderators included in their investigations." (p. 216) [3].

Explaining the variability in the learning effect of academic-writing activities requires pinpointing the underlying mechanistic cause of learning triggered by these activities. Causal explanations for the effects of pedagogical interventions are a general need in education [14–16], and such explanations entail formal (mathematical) representations [17]. Crucially, in the context of writing activities, the learning mechanism would manifest itself during the writing process [18]. Within this context, identifying the source of the variable effect of writing activities requires a formal representation of what causes students to learn while they write—a formal model of the neurobiological causes. This work represents an initial step to advance our understanding of the basic and specific brain-mind causal mechanisms underlying the learning effect of writing, which are unknown. In this direction, the present study connects the areas of academic writing (education), large-corpus discourse analysis (linguistics) via natural language processing, and active inference (theoretical neurobiology).

The manuscript is divided into the following parts. First, we introduce the notion of grammatical complexity and the average nominal dependents per noun (AnDn) as a linguistic feature indexing academic learning. Second, based on the general educational consensus that academic writing poses a challenge to students compared to producing spoken discourses, we draw on the biopsychosocial model of sympathetic arousal (SA) regulation to introduce SA interoception as a driving factor and predictor of either highly or lowly grammatically complex writings. Third, we describe an experiment under laboratory conditions where galvanic skin response (GSR) was measured during the production of short written and spoken texts. Fourth, we fit a non-linear dynamic causal model (DCM) and a perceptual active inference model (Markov decision process, MDP) to the obtained experimental data. The general goal of fitting this model was to determine whether SA interoception has a role in the production of written discourses with high grammatical complexity.

1.1 Grammatical Complexity of Written Texts as an Index of Academic Learning

Over the past two decades, the field of corpus linguistics has made significant progress in understanding what makes grammar more or less complex in speaking and writing [19, 20]. Two major perspectives have guided this work: text linguistics and register variation. Text linguistics focuses on the types of grammatical structures used in full texts, analyzing how complexity emerges across extended discourses. The register variation approach examines how grammatical choices vary depending on the communicative situation, such as casual conversation versus academic writing.

Both text linguistics and register variation have revealed consistent patterns in how high and low grammatical complexity manifest across spoken and written language modalities. For example, consider this written sentence: "The rapid expansion of renewable energy technologies in urban infrastructure planning demonstrates a paradigm shift in sustainable development." This sentence contains multiple complex noun phrases (e.g., renewable energy technologies, urban infrastructure planning, a paradigm shift), which are typical of written academic prose. Consider now the following spoken text: "I saw that big dog in the old park yesterday." This sentence has fewer modifiers per noun, which are more common in spontaneous speech.

From the above patterns, Biber and Larsson [21] have identified two major dimensions of grammatical complexity: the written dimension and the spoken dimension. The written dimension is dominant in written registers like academic articles and formal reports and is characterized by phrasal elaboration, especially complex noun phrases, prepositional phrases, and nominalizations. The spoken dimension is dominant in spoken registers like conversation and interviews and is characterized by clausal elaboration, such as finite dependent clauses and coordination. Importantly, these two dimensions are not just different—they are often inversely related. When one increases, the other tends to decrease. For instance, academic writing shows more modifiers per noun than we can employ in a conversation (even in academia). It is worth noting that while the definition of grammatical complexity is somewhat uniform across linguistic perspectives, valid methods to measure this complexity are controversial [21]. In this work, we used one simple yet reliable metric: the average number of dependents per noun (AnDn).

In general, higher grammatical complexity in written texts than in spoken registers has been consistently reported [22–28]. More specifically, higher grammatical complexity is associated with more advanced writing [29], and senior university students produce more grammatically complex discourses than their junior peers [26]. If we assume that the disciplinary knowledge of a senior student is more strongly consolidated than that of a junior student, producing a highly grammatically complex discourse would index a more consolidated disciplinary knowledge, a deeper learning effect. Therefore, in this work, we regard AnDn not only as an index of grammatical complexity but also as evidence of learning depth.

1.2 The Biopsychosocial Model of Arousal

Writing with high grammatical complexity provides a linguistic path to consolidate academic learning. However, producing a grammatically complex discourse is not an easy task. Furthermore, grammatical complexity does not fully capture the learner's physiological state during the writing process. The biopsychosocial model of arousal regulation [30] offers a complementary lens, positing that performance contexts elicit varying patterns of SA depending on how individuals appraise tasks by weighing situational demands against perceived personal resources. When individuals appraise that their resources are sufficient or exceed the demands of a task, they experience a challenge state. In contrast, when one perceives the demands of a task to exceed their personal resources, the task demands are seen as overwhelming, and they experience a threat state.

Appraisal outcomes result in distinct patterns of sympathetic nervous system activation, particularly via the sympathetic-adrenomedullary (SAM) and the hypothalamic-pituitary-adrenal (HPA) axes. Challenge states typically activate the SAM axis, characterized by moderate SA and lower cortisol release, which facilitates attention and enhanced task performance. In contrast, a threat state is associated with feelings of anxiety and impaired task performance resulting from simultaneous SAM and HPA axis activation, which governs a stronger cortisol response and heightened SA. While SAM activation promotes immediate cognitive efficiency, HPA activation is associated with attentional disruption and consolidation deficits over time [31]. Crucially, these physiological states can actively shape cognitive processes, such as attention, encoding, and memory performance outcomes. In manageable high-demand contexts, SA in a challenge state may increase cognitive arousal and memory encoding, serving as a signal that amplifies task-relevant information [32]. Laboratory studies have lent support to this theory, showing that individuals assigned to challenge state conditions outperformed those in threat conditions on memory recall and reaction time tasks [33, 34]. Moreover, individual differences in interoception, or the awareness of one's physiological state, moderate how this arousal is integrated into the cognitive system [35]. This suggests that reappraisal strategies of arousal regulation would emerge as an intervention path in education.

One reappraisal strategy of arousal regulation during academic task performance is delayed task engagement. For example, if participants perceive that writing is a threatening task, they might delay the start of the writing process. That is, delaying the initiation of an academic task is a stress-relief strategy [36]. However, delaying task initiation because one is "anxious" or "stressed" entails perceiving the bodily signals of stress, perceiving the associated SA. This speaks to the interoception of SA as a driving factor of highly or lowly grammatically complex text production.

1.3 Tackling the Driving Effect of SA Interoception Through Perceptual Markov Decision Process

Since the inception of predictive coding and the active inference framework under the first energy principle [37], the study of interoception has received remarkable attention [38–40]. Relevant to the current study is the work of Smith et al. [40]. In that work, participants performed a heartbeat tapping task (cardiac perception task). They tapped a key when they believed their hearts beat. Electrocardiographic recordings representing the (internal) sensory observations informed a generative model whose hidden states in the timestep (t) of interest ($t = 2$) were no-heartbeat and heartbeat states. The posteriors over these states given the recordings were regarded as heartbeat interoception. A two-timestep perceptual model, specified in terms of a MDP (i.e., Bayesian perceptual inference [41]), was fit to the experimental data with two free parameters (a transition probability parameter, here referred to as the B-parameter) and a likelihood parameter (interoceptive precision parameter, here referred to as the L-parameter). Actions (key presses) were assumed to be randomly produced based on the posterior state probabilities. We draw on that work to investigate whether the interoception of SA drives the production of highly grammatically complex texts across language production modalities (speaking and writing). As specified in the next section, participants would decide to produce texts with high or low levels of grammatical complexity based on their SA interoception.

2 Materials and Methods

2.1 Participants

Under laboratory conditions, 22 undergraduate students (13 males and nine females; M age = 21.4, SD = 1.6) from Brandon University were recruited via campus advertisements. Participation was voluntary, and all individuals signed an informed consent prior to the initiation of the experiment. All participants received a $50 gift card as compensation for their time. The study was approved by the Brandon University Research Ethics Committee.

2.2 Materials

Our stimuli consisted of an experimental word list of 180 nouns with the highest academic frequency selected from the Strathy Corpus of Canadian English [42]. Following this, for each participant, we created six lists (one for each experimental block) using generative AI (ChatGPT-4) [43]. We prompted it to produce a randomized list order from the experimental list. This randomized order was used to randomly assign each word without replacement to a language condition (writing and speaking). Each list included 10 randomized noun-pairs (20 total words), with five noun-pairs in both the spoken and written conditions.

2.3 Procedure

Prior to the beginning of the experiment, participants were given a verbal explanation of the experiment tasks and requested to put a galvanic skin response device (Consensus Shimmer3 GSR+) on their left wrist. Two electrodes were placed on the left hand; one electrode was placed on the palmar surface of the medial phalanx, and the other on the palmar surface of the distal phalanx. Following this, they were seated at a computer and given noise-cancelling headphones to wear throughout the experiment. Once they started, they were presented with a set of general instructions on a computer screen, followed by a familiarization block which consisted of two response trials (one writing and one speaking) with two noun-pairs. The words for these blocks were not used in the subsequent experiment. Immediately prior to the initiation of the experiment, the experimenter commenced collecting the GSR data, which continued throughout the entirety of the experiment. The experiment consisted of six blocks. In each trial, participants produced two to three sentences that connected the meaning between two words. For example, given the word "car" and the word "chair" participants connected the words in the register "I drove to the furniture store to buy a chair which I needed for my new office". Participants were informed of the modality by observing the colour of the word pairs. Half of the participants were instructed to speak aloud their responses if the word pairs appeared in red and write their sentences if they appeared in green; the other half received the opposite instructions. Language-production conditions were randomized across trials. The stimulus delivery program was PsychoPy [44].

2.4 Language Production Task and Galvanic Skin Response Measurement

The language production task (Fig. 1) began with a fixation screen lasting 1.5 s, followed by the first trial, which lasted 60 s and consisted of both the presentation of the stimuli and language production collection. Word pairs appeared in either red or green, were situated vertically, and remained on the screen during the entire duration of the trial. Participants either spoke aloud their sentences, which were captured by a microphone contained within the noise-cancelling headphones, or wrote them using the keyboard. Should they complete their response before the 60 s had elapsed, they were instructed to return their hands to the desk and rest until the next trial began. Overall, each language production task block consisted of 10 trials (20 noun-pairs, five written and five spoken) across six total blocks, totalling 30 written productions and 30 spoken productions per participant.

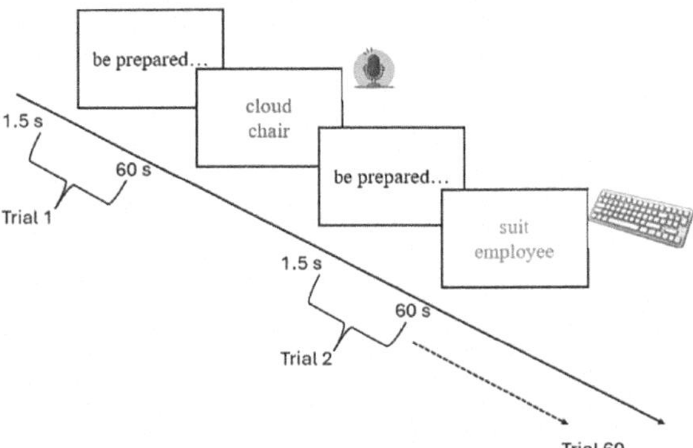

Fig. 1. Language production task. Participants performed 60 trials of language production in two randomized modalities: speaking and writing. Words' colours cued participants about the language modality (counterbalanced across participants). Participants were able to observe the words throughout the production phase. In this figure, the microphone and keyboard icons indicate the relevant production modality expected given the words' colours.

To quantify grammatical complexity in participants' productions (i.e., registers), we used the natural language processing software TAASSC (https://www.linguisticanalysistools.org/taassc.html) to quantify the AnDn using a grammatical complexity index. To control for a possible confounding effect of the two words' semantic similarity on grammatical complexity [45], we computed the cosine similarity using a Python-based algorithm built with the SpaCy natural language processing library [46], which used pre-trained 300-dimensional word embeddings. Cosine similarity measures the semantic similarity between two vectorized words based on the angle between two vectors. Values range from -1.0 to 1.0, with cosines close to 1.0 indicating high semantic similarity, cosines near 0 indicating low similarity, and negative cosine values indicating opposite semantic meanings [47].

2.5 Computational Modeling of Neurophysiological, Linguistic, and Behavioral Data

At a participant level, a non-linear dynamic causal model (DCM) [48] was fit to the GSR data to estimate latent SA given the observed GSR measurements. The DCM is itself a model of the neural cause of the GSR measurements. Crucially, a non-linear DCM allows us to identify (1) the stimulus-evoked peak amplitude of SA, (2) the duration of the evoked response (dispersion), and (3) the delay of the response (latency). Therefore, a non-linear DCM provides three possible indices of SA, among which, as described below, we selected the most reliable one based on a Bayesian model selection procedure. The selected SA index was regarded as an observation whose hidden cause was regarded as its subjective perception, interoception, in the context of MDP models.

2.6 Markov Decision Process (MDP) Model

We used SPM12 (http://www.fil.ion.ucl.ac.uk/spm/) and modified simulation and estimation scripts available through [40] to specify three types of MDP models dictated by the three parameter estimates obtained from the non-linear DCM: amplitude model, dispersion model, and latency model. Each model comprised two timepoints (t), a D matrix (prior beliefs), a B matrix (beliefs transition probabilities), and an A matrix (state-observation or likelihood matrix). Participants began each trial ($t = 1$) in a start state (i.e., by observing the "be prepared" slide), and upon the appearance of the two words ($t = 2$), they transitioned to an SA state. In the amplitude model, the states were "null," "low," or "high." In the dispersion model, the states were "null," "short," or "large," and in the latency model, the states were "null," "early," or "late." The probability of transitioning to a particular state was dictated by the B-parameter of the B matrix.

The response (action) model (the likelihood model specified in the A matrix) comprised the L-parameter indexing interoceptive precision. The response model assumed a stochastic action-selection process wherein the action corresponded to the production of a register with either low or high grammatical complexity.

The selection of a given grammatical complexity level corresponded to a sampling from the posterior distribution over the hidden (true) state given the observation (represented by the relevant DCM-SA parameter). For example, in the latency model, the participant would observe a late SA onset and would infer whether the cause of that observation was their body experiencing a late SA onset. This inference would correspond to the (posterior) probability of experiencing a late SA onset state given the observation of a late SA. If the posterior approaches one, the agent would more likely produce a highly grammatically complex discourse, and vice versa.

Parameter recoverability was assessed for a range of parameters (from 0.1 to 1), and we assessed the strength of the associations between true and estimated parameters through Bayesian correlations (L-parameter, $r = 0.95$, BF > 1000; B-parameter, $r = 0.96$, BF > 1000)—the number of trials per condition was set to 30.

There was no previous research linking MDP, SA, and grammatical complexity, based on which we could define the prior means and variances for the parameter estimation algorithm. Therefore, we set the means at 0.5. Furthermore, from three possible variance values (0.25, 0.5, and 0.75), we searched for a prior with the most parsimonious trade-off between complexity and generalizability. Crucially, as introduced above, we created three types of MDP models corresponding to three types of parameter estimates obtained from the non-linear DCM. Thus, we also needed to select the model that best explained our data. To achieve both goals simultaneously, we estimated nine models and selected the model with the highest protected exceedance probability.

2.7 Data Discretization

Estimated SA amplitudes, dispersions, latencies, observed language-production onset times, and AnDn scores were registered as continuous variables. However, for easier computational tractability and interpretability (i.e., discrete state MDP), those variables were discretized at a participant level (and collapsed across speaking and writing conditions) based on the median of the relevant distribution. Discretized SA variables conformed to the following coding: "low" or "high" (SA amplitude), "small" or "large" (SA dispersion), and "early" or "late" (SA latency). Similarly, AnDn scores were coded as "low" or "high."

3 Results

3.1 Behavioral Results

Data from one participant were excluded due to failing to follow instructions, and data from another participant were excluded due to excessive sweating. With the data from the 20 remaining participants, we found that the mean AnDn (index of grammatical complexity) was higher in the writing condition (M = 1.02, SD = 0.36) than in the speaking condition (M = 0.95, SD = 0.34). We confirmed this difference through a Bayesian analysis of covariance (ANCOVA) with the language-production condition as a fixed factor, the trial and participant as random factors, and cosine similarity as a covariate (BF = 18.32). Furthermore, the mean language production onset time was longer in the speaking condition (M = 11.52 s, SD = 8.27 s) than in the writing condition (M = 5.7 s, SD = 3.76 s). An ANCOVA model with the language-production condition as a fixed factor, the trial and participant as random factors, and cosine similarity as a covariate confirmed this difference (BF10 > 10 000).

3.2 Dynamic Causal Modelling Results

After estimating the SA amplitude, dispersion, and latency, we assessed the relationship between the language-production condition and the estimated SA parameters through an ANCOVA model that included participants and trials as random effects and cosine similarity as a covariate. The mean SA amplitude (in arbitrary units, au) was slightly higher in the writing condition ($M = 1.05$ au, $SD = 1.67$ au) than in the speaking condition ($M = 0.96$ au, $SD = 1.26$ au). However, the statistical model did not confirm this difference ($BF = 0.27$). The mean SA dispersion was larger in the speaking condition ($M = 4.38$ s, $SD = 3.49$ s) than in the writing condition ($M = 2.37$ s, $SD = 1.8$ s). Furthermore, the mean SA latency was longer in the speaking condition ($M = 8.86$ s, $SD = 6.32$ s) than in the writing condition ($M = 4.62$ s, $SD = 3.78$ s). ANCOVA models confirmed both mean differences ($BF > 10\,000$).

3.3 MDP Results

Among the nine DCMs, the Bayesian model selection procedure favoured the latency model with a prior variance of 0.75. The protected exceedance probability in support of this model was 0.98, and the Bayesian omnibus risk was 0.002.

The state-transition probability parameter (B-parameter) was slightly larger in the speaking condition ($M = 0.19$, $SD = 0.03$, 95% HDI: 0.18–0.21) than in the writing condition ($M = 0.17$, $SD = 0.04$, 95% HDI: 0.15–0.19). However, this difference received weak evidence ($BF = 1.7$) from an ANCOVA model with the language condition as a fixed effect and the participant as a random effect. In contrast, the likelihood parameter (L-parameter) was larger in the writing condition ($M = 0.6$, $SD = 0.07$, 95% HDI: 0.57–0.63) than in the speaking condition ($M = 0.41$, $SD = 0.07$, 95% HDI: 0.38–0.44) (Fig. 2). An ANCOVA model with the fixed effect of the language condition and the random effect of the participant confirmed this difference ($BF > 10\,000$). We further investigated whether production onset time explained this difference by comparing two ANCOVA models. One model only included the language condition, whereas the other model included the language condition and production-onset time. Posterior probabilities favoured the simpler model ($P(M|data) = 0.69$) over the complex model ($P(M|data) = 0.31$). The language-only model showed decisive evidence vs. the null ($BF = 1.305 \times 10^8$), whereas the more complex model, though also strong vs. the null ($BF = 5.747 \times 10^7$), was less supported than the simpler model. Given the estimated parameters, we inverted the MDP model to obtain the posterior beliefs about the SA onsets (Fig. 3). These beliefs much more accurately reflected the true hidden states when SA onsets occurred early in the language production phase than when they occurred late.

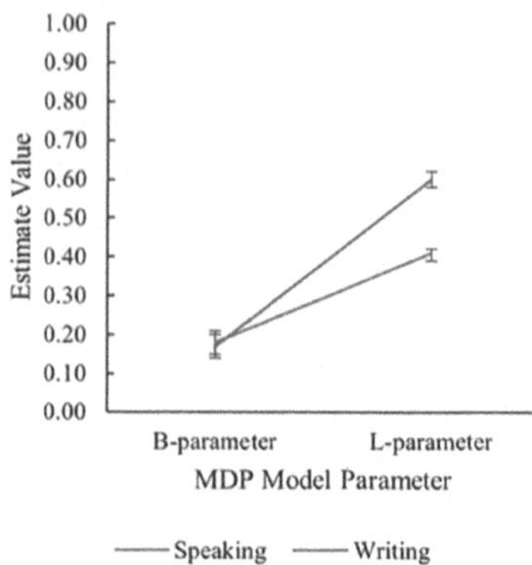

Fig. 2. Group-level parameter estimates of the winning MDP model.

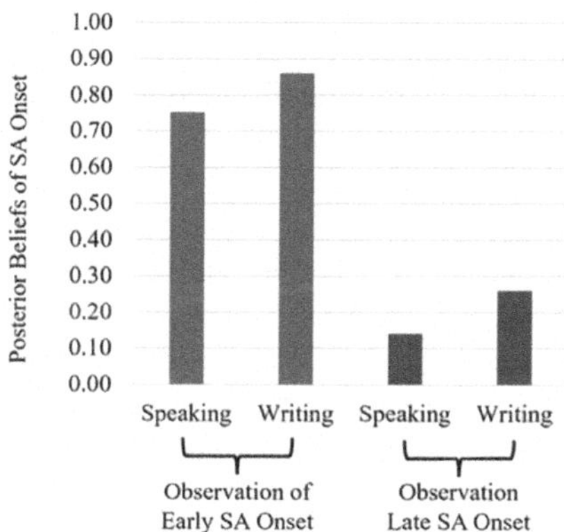

Fig. 3. Posterior beliefs (group level) about SA onset times given their sensory observations.

4 Discussion

In the current experiment, the statistical models (ANCOVA models) showed that writing outperformed speaking in producing highly grammatically complex discourses, suggesting a stronger learning effect of writing. However, on average, participants delayed the production of their discourses in the speaking condition compared to the writing condition. This contradicts our expectation of delayed task engagement in the writing condition. Furthermore, compared to the writing condition, the DCMs in the speaking condition yielded more dispersed and delayed estimates of SA onsets.

The winning perceptual MDP model (SA latency model) showed that in both conditions, participants believed that upon the two words' appearance on the screen, they would most likely experience an early SA onset. These prior beliefs drove the interoception of *when* the SA onset occurred towards the early onset state, regardless of whether they observed either an early or late SA onset. However, the model yielded higher interoceptive precision in the writing condition than in the speaking condition. This interoceptive precision difference slightly corrected the accuracy of the SA-onset interoception in the writing condition and worsened it in the speaking condition. Therefore, the main model finding refers to participants showing higher interoceptive precision about *when* an SA event occurred during the writing condition than during the speaking condition. Crucially, this effect was not explained by the between-modality difference in language-production onset times. Furthermore, during the observation of a late SA onset, this higher precision likely drove the production of registers with higher grammatical complexity in the writing condition than in the speaking condition. Therefore, based on our main goal in this work, we found initial evidence to support that SA interoception has a positive role in the production of written discourses with high grammatical complexity.

In the context of this work, we regarded learning as indexed by grammatical complexity in language production, where higher grammatical complexity indexes deeper learning and vice versa. This conceptualization of learning follows on from the expert-novel differentiation frequently used in discourse analysis in education. Experts in a discipline show distinct linguistic features in their discourses compared to non-experts, with expertise level being a proxy of learning depth.

Based on this conceptualization of learning, our initial findings carry implications for educational practice, particularly regarding the use of writing-to-learn activities in higher education. If grammatical complexity emerges as a neurobiologically grounded index of learning during writing, it may serve as a proxy for the depth of learning taking place. This would support the view that a grammatically complex written discourse is not merely a linguistic output, but a reflection of conceptual consolidation and knowledge integration achieved by experts in an academic field. From this perspective, individual variability in the learning effect of writing activities could be partially explained by students' interoception sensitivity to task-related stress or challenge. Students who appropriately perceive when (early or late) in a task they are having an SA event may be better positioned to engage in complex linguistic production, and by extension, deeper learning. Therefore, tailoring writing-to-learn activities to individual physiological profiles could represent a step toward precision education (cf. precision psychiatry [49, 50]), helping educators move beyond "one-size-fits-all" approaches to writing instruction by incorporating both linguistic, neurophysiological, and computational markers of learning.

These findings also offer a preliminary step toward advancing computational phenotyping frameworks regarding some developmental disorders. As proposed by [51], physiological markers such as sympathetic arousal become more diagnostically meaningful when contextualized through computational parameters. In this context, our results align with the proposition that behavioural and biological data, when jointly modelled, provide a richer and more valid representation of latent cognitive constructs. Crucially, this approach holds promise for better supporting students with developmental conditions such as attention-deficit/hyperactivity disorder. These students often face unique challenges in language production, which may affect their ability to express ideas in writing. By using grammatical complexity and neurocomputational parameters as a marker of learning, educators may gain more precise insight into the effectiveness of writing-based interventions for neurodiverse learners. Ultimately, this could inform the design of personalized, evidence-based educational strategies that are both inclusive and neurobiologically informed in terms of computational phenotypes for learning.

This work has several limitations. First, while grammatical complexity as a proxy of learning speaks to ecological validity in educational discourse analysis, this conceptualization of learning differs from psychological or computational operationalizations (e.g., memory retrieval accuracy and response times, changes of parameter estimates, etc.). Therefore, future works should address computational and (cognitive) psychological validity of grammatical complexity as an index of learning, for example, by including a memory retrieval task or repeating trials with the same words to assess whether the second trial elicits the same semantic network produced in the first trial.

Second, the current MDP represents the perceptual component of an active inference process (i.e., predictive coding). Therefore, it regards belief update as passive inference. While the model shows sensitivity to the effect of language production modality and SA latency on grammatical complexity, it lacks action-driven interoception: "would producing a highly complex written register increase my SA interoception, minimizing the surprised caused by the appearance of the two words?" A partially observable Markov decision process might provide this sort of interoception. The model could also be improved by incorporating all three parameters of the non-linear DCM as three different outcome modalities in a single scheme. Moreover, it could incorporate different grammatical complexity indices as additional action modalities (in line with the multidimensional framework of grammatical complexity proposed by Biber and Larsson [21]).

Finally, the model does not represent the process of writing or speaking but the process of deciding to produce language with two possible levels of grammatical complexity, which is a critical component of the writing process. Therefore, these results warrant future research on how much of the entire writing process the model explains.

Despite these limitations, the fact that this simple perceptual model showed sensitivity to detect differential effects of writing and speaking on the interoception of SA and its relationship with grammatical complexity opens a new perspective in the study of academic writing and its effects on the brain and the mind from the perspective of the active inference framework.

Acknowledgment. This article draws on research supported by the Social Sciences and Humanities Research Council, Canada.

The authors have no competing interests to declare that are relevant to the content of this article

References

1. Kuiken, F., Vedder, I.: The interplay between academic writing abilities of Dutch undergraduate students, a remedial programme, and academic achievement. Int. J. Biling. Educ. Biling. **24**(10), 1474–1485 (2021)
2. Silva, A.M., Limongi, R.: La escritura epistémica en contextos académico profesionales: desafíos de investigación educativa, cognitiva y neurocientífica. Rev. Educ. Super. Soc. **1**, 41–58 (2017)

3. Graham, S., Kiuhara, S.A., MacKay, M.: The effects of writing on learning in science, social studies, and mathematics: a meta-analysis. Rev. Educ. Res. **90**(2), 179–226 (2020)
4. Young, A.P.: Teaching Writing Across the Curriculum. 4th ed. Prentice Hall Resources for Writing, Upper Saddle River, NJ (2006)
5. Galbraith, D., Baaijen, V.M.: The work of writing: raiding the inarticulate. Educ. Psychol. **53**, 238–257 (2018)
6. Silva, A.M., Limongi, R.: Writing to learn increases long-term memory consolidation: a mental-chronometry and computational-modeling study of "epistemic writing." J. Writ. Res. **11**(1), 211–243 (2019)
7. Peters, A., Galbraith, D., Limongi, R.: Writing-to-learn: effects of writing compared to speaking on memory for text. Poster Presented at EARLI 2023, Thessaloniki, Greece (2023)
8. Peters, A., Galbraith, D., Limongi, R.: Writing-to-learn: effects of writing compared to speaking on memory for text. Paper to Be Presented at the Symposium "Components of Writing to Learn" SIG Writing 2024, Paris, France (2024)
9. Bangert-Drowns, R.L., Hurley, M.M., Wilkinson, B.: The effects of school-based writing-to-learn interventions on academic achievement: a meta-analysis. Rev. Educ. Res. **74**(1), 29–58 (2004)
10. Anderson, P., Anson, C.M., Gonyea, R.M., Paine, C.: The contributions of writing to learning and development: results from a large-scale multi-institutional study. Res. Teach. Engl. **50**(2), 199–235 (2015)
11. Molitor, S.J., Langberg, J.M., Evans, S.W.: The written expression abilities of adolescents with attention-deficit/hyperactivity disorder. Res. Dev. Disabil. **51–52**, 49–59 (2016)
12. McClure, E.B., Pennington, R.C., Bewley, S.C.: Evaluating the evidence-base supporting writing instruction strategies for students with autism spectrum disorder: a systematic review of experimental research. Focus Autism Other Dev. Disabl. **39**(3), 139–149 (2024)
13. Limongi, R.: From the schizophrenia clinic to the classroom: what can precision psychiatry and theoretical neurobiology say about the effect of writing on learning? In: SIG12 2024 Keynote "John R. Hayes Award" Recipient (2024)
14. Scerif, G., et al.: Making the executive 'function' for the foundations of mathematics: the need for explicit theories of change for early interventions. Educ. Psychol. Rev. **35**(4), 110 (2023)
15. Morrison, K., van der Werf, G.: Searching for causality in educational research. Educ. Res. Eval. **22**(1–2), 1–5 (2016)
16. Kitto, K., Hicks, B., Buckingham Shum, S.: Using causal models to bridge the divide between big data and educational theory. Br. J. Educ. Technol. **54**(5), 1095–1124 (2023)
17. Pearl, J.: Causality: Models, Reasoning, and Inference, 2nd edn. Cambridge University Press, New York (2009)
18. Torrance, M.C., Conijn, R.: Methods for studying the writing time-course. Read. Writ. **37**(2), 239–251 (2023)
19. Biber, D., Gray, B., Staples, S., Egbert, J.: The Register-Functional Approach to Grammatical Complexity. Routledge, New York (2021)
20. Biber, D., Larsson, T., Hancock, A.: The linguistic organization of grammatical text complexity. Corpus Linguist. Linguist. Theory **20**(1), 89–115 (2024)
21. Biber, D., Larsson, T.: Accounting for the entire system of complexity features: evidence for general oral versus literate grammatical complexity dimensions. Corpus Linguist. Linguist. Theory (2025)
22. Biber, D., Gray, B.: Challenging stereotypes about academic writing: complexity, elaboration, explicitness. J. Engl. Acad. Purp. **9**(1), 2–20 (2010)

23. Biber, D., Gray, B.: The historical shift of scientific academic prose in English towards less explicit styles of expression. In: Bhatia, V., Sanchez Hernandez, P., Perez-Paredes, P. (eds.) Researching Specialized Languages. Studies in Corpus Linguistics, vol. 47, pp. 11–24. John Benjamins, Amsterdam (2011)
24. Biber, D., Gray, B.: Grammatical change in the noun phrase: the influence of written language use. Engl. Lang. Linguist. **15**(2), 223–250 (2011)
25. Biber, D., Gray, B., Poonpon, K.: Should we use characteristics of conversation to measure grammatical complexity in L2 writing development? TESOL Q. **45**(1), 5–35 (2011)
26. Staples, S., Egbert, J., Biber, D., Gray, B.: Academic writing development at the university level: phrasal and clausal complexity across level of study, discipline, and genre. Writ. Commun. **33**(2), 149–183 (2016)
27. Biber, D., Gray, B., Staples, S., Egbert, J.: Investigating grammatical complexity in L2 English writing research: linguistic description versus predictive measurement. J. Engl. Acad. Purp. **46**, 100869 (2020)
28. Biber, D.: The Register-Functional Approach to Grammatical Complexity: Theoretical Foundation, Descriptive Research Findings, Application. Routledge Advances in Corpus Linguistics. Routledge, New York (2022)
29. Crossley, S.A., McNamara, D.S.: Does writing development equal writing quality? A computational investigation of syntactic complexity in L2 learners. J. Second. Lang. Writ. **26**, 66–79 (2014)
30. Blascovich, J., Tomaka, J.: The biopsychosocial model of arousal regulation. In: Zanna, M.P. (ed.) Advances in Experimental Social Psychology, pp. 1–51. Academic Press, Cambridge, MA (1996)
31. Lupien, S.J., Maheu, F., Tu, M., Fiocco, A., Schramek, T.E.: The effects of stress and stress hormones on human cognition: implications for the field of brain and cognition. Brain Cogn. **65**(3), 209–237 (2007)
32. Seery, M.D.: The biopsychosocial model of challenge and threat: using the heart to measure the mind. Soc. Personal. Psychol. Compass **7**(9), 637–653 (2013)
33. Tomaka, J., Blascovich, J., Kelsey, R.M., Leitten, C.L.: Subjective, physiological, and behavioral effects of threat and challenge appraisal. J. Personal. Soc. Psychol. **65**(2), 248–260 (1993)
34. Moore, L.J., Vine, S.J., Wilson, M.R., Freeman, P.: Examining the antecedents of challenge and threat states: the influence of perceived required effort and support availability. Int. J. Psychophysiol. **92**, 38–47 (2014)
35. Critchley, H.D., Wiens, S., Rotshtein, P., Öhman, A., Dolan, R.J.: Neural systems supporting interoceptive awareness. Nat. Neurosci. **7**(2), 189–195 (2004)
36. Grunschel, C., Patrzek, J., Fries, S.: Exploring different types of academic delayers: a latent profile analysis. Learn. Individ. Differ. **23**, 225–233 (2013)
37. Friston, K.J., FitzGerald, T., Rigoli, F., Schwartenbeck, P., Pezzulo, G.: Active inference: a process theory. Neural Comput. **23**(2), 1–49 (2016)
38. Seth, A.K., Suzuki, K., Critchley, H.D.: An interoceptive predictive coding model of conscious presence. Front. Psychol. **2**, 395 (2011)
39. Fermin, A.S.R., Friston, K., Yamawaki, S.: An insula hierarchical network architecture for active interoceptive inference. R. Soc. Open. Sci. **9**(6), 220226 (2022)
40. Smith, R., et al.: A Bayesian computational model reveals a failure to adapt interoceptive precision estimates across depression, anxiety, eating, and substance use disorders. PLOS Comput. Biol. **16**(12), e1008484 (2020)
41. Parr, T., Pezzulo, G., Friston, K.J.: Active Inference: The Free Energy Principle in Mind, Brain, and Behavior. MIT Press, Cambridge, MA (2022)
42. Strathy Language Unit: Strathy Corpus of Canadian English: Queen's University Dataverse (2024). https://www.english-corpora.org/can/

43. OpenAI: ChatGPT-4o [Large language model] (2024)
44. Peirce, J., et al.: PsychoPy2: experiments in behavior made easy. Behav. Res. Methods **51**(1), 195–203 (2019)
45. Humphreys, G.F., Mirković, J., Gennari, S.P.: Similarity-based competition in relative clause production and comprehension. J. Mem. Lang. **89**, 200–221 (2016)
46. Honnibal, M., Montani, I., Van Landeghem, S., Boyd, A.: spaCy: industrial-strength natural language processing in Python (2020). https://doi.org/10.5281/zenodo.1212303
47. Mikolov, T., Chen, K., Corrado, G., Dean, J.: Efficient estimation of word representations in vector space. arXiv (2013). arXiv:1301.3781
48. Bach, D.R., Daunizeau, J., Friston, K.J., Dolan, R.J.: Dynamic causal modelling of anticipatory skin conductance responses. Biol. Psychol. **85**(1), 163–170 (2020)
49. Friston, K.J.: Precision psychiatry. Biol. Psychiatry Cogn. Neurosci. Neuroimaging **2**(8), 640–643 (2017)
50. Friston, K.J., Redish, A.D., Gordon, J.A.: Computational nosology and precision psychiatry. Comput. Psychiatry **1**, 2–23 (2017)
51. Limongi, R., Skelton, A.B., Tzianas, L.H., Silva, A.M.: Increasing the construct validity of computational phenotypes of mental illness through active inference and brain imaging. Brain Sci. **14**(12), 1278 (2024)

Evaluation of "As-Intended" Vehicle Dynamics Using the Active Inference Framework

Kazuharu Kidera[1(✉)], Takuma Miyaguchi[2], and Hideyoshi Yanagisawa[2]

[1] Honda R&D Co., Ltd., 4630 Shimotakanezawa, Haga-machi, Haga-gun, Tochigi 321-3393, Japan
kazuharu_kidera@jp.honda

[2] The University of Tokyo, 7-3-1 Hongo, Bunkyo-ku, Tokyo 113-8656, Japan

Abstract. We constructed a computational model of the driver's brain for steering tasks using the active inference framework, grounded in the free energy principle—a theory from computational neuroscience. This model enables quantitative estimation of how accurately the brain learns vehicle dynamics and performs appropriate steering, using a measure called variational free energy. Through driving simulator experiments, we observed strong correlations between variational free energy and both expert drivers' subjective "as-intended" scores and general participants' objective control performance. These results suggest that variational free energy provides a promising quantitative metric for evaluating whether a vehicle behaves "as-intended."

Keywords: Free energy principle · Active inference · Vehicle dynamics

1 Introduction

1.1 Research Objective

In the development of vehicle dynamics performance, static design parameters are typically set based on accumulated experience and established engineering theories. Basic performance is then validated using physically measurable or simulation-based dynamic indicators, such as steering response [1]. However, the final adjustment of these parameters still relies heavily on subjective evaluations by expert drivers—assessments that are often neither objective nor quantitative. A fundamental question remains: what does it truly mean for a driver to operate a vehicle "as-intended"? We address this issue by applying the free energy principle [2] and active inference [3]—neuroscientific theories—as a novel approach to understanding and evaluating "as-intended" control.

1.2 Free Energy Principle and Active Inference

The free energy principle is a theory that explains the brain's perception, learning, and action in a unified manner. By interacting with the external environment through the body, the brain constructs a model of the world known as the generative model. The process

M. Albarracin et al. (Eds.): IWAI 2025, CCIS 2857, pp. 270–282, 2026.
https://doi.org/10.1007/978-3-032-16955-6_15

of building this model corresponds to learning and action, while perception is realized as Bayesian inference through the model. These processes are achieved by minimizing a quantity called variational free energy (VFE), which represents the modeling error [3].

Active inference builds on this principle by providing a theoretical framework for planning actions to achieve goals while continuously updating the generative model [3]. It can be implemented in a simulation environment. By modeling driver behavior within this framework, the modeling error of vehicle dynamics in the driver's brain can be quantified as VFE (see Fig. 1, modified from [3]).

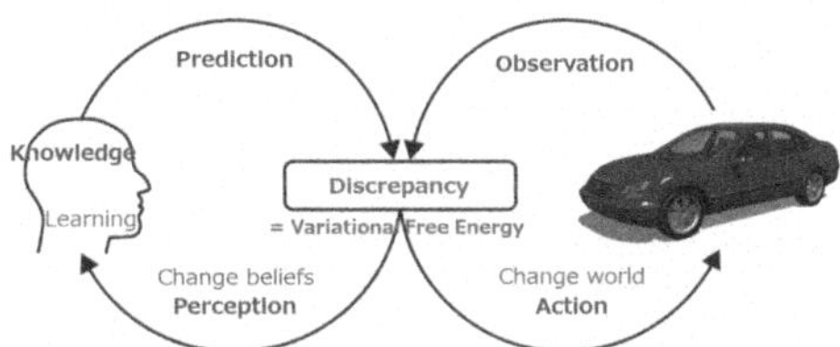

Fig. 1. A conceptual illustration of how perception and action minimize VFE, a measure of the discrepancy between the brain's generative model and the external world.

2 Methodology

2.1 Computational Model of the Brain in Steering Tasks

This study focuses on a steering task involving a car traveling at a constant speed of 100 km/h. Vehicle dynamics were simulated using CarSim, developed by Mechanical Simulation Corporation, now part of Applied Intuition, Inc. [4]. The driving course, 10 m wide, includes three corners—left, right, and left—and a straight section. The objective is to follow the centerline as closely as possible. Figure 2 shows both the layout of this course and the configuration of the driving simulator used in the experiment for this study. The CarSim model incorporates complex dynamics such as tire nonlinearity, mechanical friction, and stroke velocity-dependent damping, capturing realistic characteristics of vehicle behavior.

The first step involves constructing a self-learning driving behavior model using the active inference framework. This enables the simulation of information processing that occurs in the driver's brain. Since the vehicle dynamics are too complex to describe analytically, we adopted a discrete-time active inference model, which avoids the need for explicit motion and observation equations. In this framework, the generative model is formulated as a partially observable Markov decision process (POMDP) based on mean-field approximation, as illustrated in Fig. 3 [3, 5]. It represents how the hidden state vector—representing the firing states of neuronal populations—evolves over time as a probability distribution. The tensors and vectors used in this generative model are summarized in Table 1.

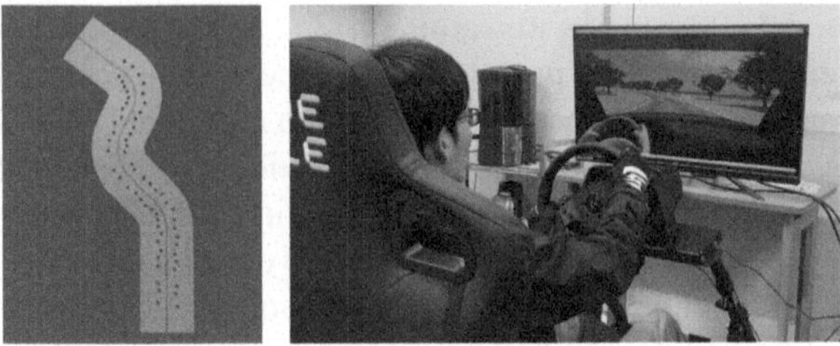

Fig. 2. Top-down view of the driving course and simulator setup used to evaluate "as-intended" vehicle control.

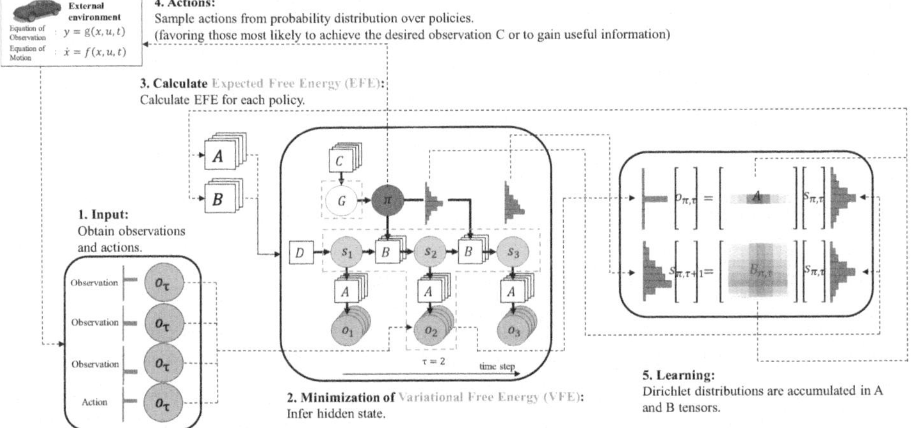

Fig. 3. The active inference framework consisting of: 1. Observation, 2. inference of hidden states, 3. evaluation of policies, 4. action selection, and 5. learning.

Table 1. Components of the generative model in the active inference framework.

Component	Descriptions and Configurations
Observation o_τ	One-hot vectors representing specific observed values, with the following dimensions for each modality: - modality [0]: Lateral deviation from the centerline at 15 m ahead (-10 to $+$ 10 m, 64 bins) - modality [1]: Area error with respect to the centerline up to 15 m ahead (-160 to $+$ 160 m^2, 64 bins) - modality [2]: Steering angle (-120 to $+$ 120°, 49 bins) - modality [3]: Steering torque feedback (-30 to $+$ 30 Nm, 64 bins) - modality [4]: Vehicle yaw rate (-40 to $+$ 40 deg./s, 64 bins) - modality [5]: Vehicle lateral acceleration (-1.6 to $+$ 1.6 G, 64 bins)

(continued)

Table 1. *(continued)*

Component	Descriptions and Configurations
Hidden state s_τ	Categorical distribution with the following dimension: - factor [0]: 36
Likelihood A	For each modality, the mapping from hidden states to observations has the following matrix dimensions: - modality [0]: 64 × 36 - modality [1]: 64 × 36 - modality [2]: 49 × 36 - modality [3]: 64 × 36 - modality [4]: 64 × 36 - modality [5]: 64 × 36 Each matrix is column-normalized and filled with uniform values
State transition B	The state transition tensor has the following dimension: - factor [0]: 36 × 36 × 49 Here, 49 corresponds to the action (steering angle) dimension. Each matrix slice corresponding to an action is column-normalized and filled with uniform values
Preference C	Preferences are defined for the following three modalities (the others are set as uniform categorical distributions): - modality [0]: Normal distribution with a mean of 0 m and a standard deviation equivalent to 0.5 m - modality [1]: Normal distribution with a mean of 0 m^2 and a standard deviation equivalent to 8 m^2 - modality [3]: Normal distribution with a mean of 0 Nm and a standard deviation equivalent to 15 Nm
Initial state D	Categorical distribution with the following dimension: - factor [0]: 36 Initialized with a sigmoid-shaped distribution to provide a better starting point for learning than a random distribution

In this study, we used the JAX version of pymdp [6], an open-source Python library for active inference that supports large-scale tensor operations and parallel computing via Google's JAX [7]. We integrated pymdp, which implements the computational model of the brain and runs on WSL (Windows Subsystem for Linux), with CarSim, which represents the external environment and runs on MATLAB Simulink. Communication from CarSim to pymdp over a local Ethernet connection corresponds to sensory nerve signals, while communication in the opposite direction corresponds to motor nerve signals that control the steering angle. The cycle of the POMDP for the steering task was set to 25 Hz as a representative frequency, since a single frequency had to be chosen. Although this choice was initially based on assumptions about the timescale of motor and sensory processing, its validity is examined through the experimental results in this study. Interestingly, a study by a game developer suggests that the brain processes information during gameplay at 25–33.3 Hz [8].

Inference of hidden states, corresponding to the recognition process, was performed by minimizing VFE using marginal message passing (MMP) over the past 32 time steps of observations (approximately 1.3 s of sensory memory). MMP is regarded as one of the most promising variational inference algorithms in active inference [9]. The VFE of a policy π (i.e., a sequence of actions), denoted as F_{π}, is defined as follows [3]:

$$F_{\pi} = D_{KL}\left[Q(\tilde{s}|\pi)||P(\tilde{o},\tilde{s}|\pi)\right] \tag{1}$$

In discrete-time active inference, the F_{π} at time step τ is computed using MMP as [3]:

$$\boldsymbol{F}_{\pi\tau} = \boldsymbol{s}_{\pi\tau} \cdot \left(ln\, \boldsymbol{s}_{\pi\tau} - ln\boldsymbol{A} \cdot o_{\tau} - \frac{1}{2}\left(ln\,(\boldsymbol{B}_{\pi\tau}\boldsymbol{s}_{\pi\tau-1}) + ln\left(\boldsymbol{B}^{\dagger}_{\pi\tau+1}\boldsymbol{s}_{\pi\tau+1}\right)\right)\right) \tag{2}$$

In active inference, the expected free energy (EFE) is computed for each policy, a softmax function is applied to obtain a probability distribution over policies, and an action—steering angle in this case—is selected from this distribution [3]. EFE can be expressed in various mathematical forms, but one formulation represents it as the sum of an epistemic value term (exploration), a pragmatic value term (exploitation), and a novelty term, as shown in Eq. (3) [3].

$$\begin{aligned} G(\pi) = &-E_{Q(\tilde{o}|\pi)}\left[D_{KL}\left[Q(\tilde{s}|\pi,\tilde{o})||Q(\tilde{s}|\pi)\right]\right] \\ &-E_{Q(\tilde{o}|\pi)}[ln\,P(\tilde{o}|C)] - E_{\tilde{Q}(\tilde{o},\tilde{s}|\pi)}\left[D_{KL}\left[Q(\theta|\tilde{o},\tilde{s})||Q(\theta)\right]\right] \end{aligned} \tag{3}$$

Learning in the generative model follows the update rules given in Eq. (4) for the A tensor, B tensor, and D vector [3].

$$\begin{aligned} \boldsymbol{a} &= a + \sum_{\tau} o_{\tau} \otimes \boldsymbol{s}_{\tau} \\ \boldsymbol{b}_{\pi\tau} &= b_{\pi\tau} + \sum_{\tau} \boldsymbol{s}_{\pi\tau} \otimes \boldsymbol{s}_{\pi\tau-1} \\ \boldsymbol{d} &= d + \boldsymbol{s}_1 \end{aligned} \tag{4}$$

With the above task setting and generative model, simulation confirmed that autonomous learning of the steering task was possible. Starting from a state where both the A and B tensors were flat, actions were generated through active inference, enabling the generative model (i.e., the computational model of the brain) to learn, much like an actual human would. During learning, steering behavior was initially random, and the vehicle frequently went off course due to a lack of knowledge about driving operations or vehicle dynamics. After several dozen trials, the generative model began to partially follow the target trajectory (the centerline), and after approximately 50 to 150 trials, it was able to complete the course while staying close to it. This progression reflects a typical pattern in active inference: initial dominance of the novelty term in EFE, followed by state exploration and eventual exploitation.

2.2 Hypothesis

During online learning, we found that VFE increases with greater deviation from the centerline and decreases as the deviation decreases (see Fig. 4). Figure 1 illustrates that when VFE is low, the vehicle's behavior is predictable; when it is high, it becomes unpredictable. This unpredictability reduces the accuracy of action selection—steering in this task. Consequently, deviation from the centerline is closely linked to VFE. Since the desired observations (C vector) are not used in computing VFE, this relationship is a non-trivial finding demonstrated through simulation.

Hypothesis: Lower VFE reflects a higher degree of "as-intended" control in the steering task.

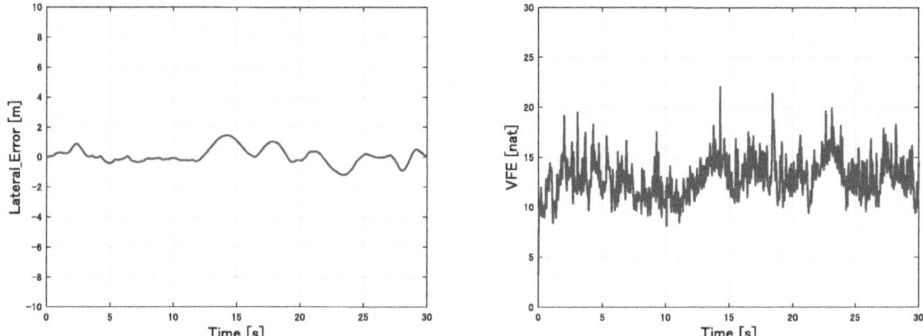

Fig. 4. Time-series plot of lateral deviation from the centerline and VFE during online learning of the steering task. Higher VFE is associated with greater deviation.

2.3 Verification of Hypothesis Through Offline Learning

In this paper, we refer to autonomous learning using the active inference framework, as described in Sect. 2.1, as online learning. Offline learning is also possible using observation and action data from a computer-controlled or human driver, without using policy-based actions from EFE. As shown in Fig. 3 and Eq. (4), the A, B, and D parameters of the generative model can be learned by sequentially inputting observation data, with the appropriate B matrix selected based on the actions.

We validated this offline approach using the course in Fig. 2 (excluding the initial 9-s straight section, added as a lead-in segment for the simulator), 12 CarSim vehicle models, and the Simulink driver model, which controls steering based on observation modalities 0 and 1 (Table 1). Figure 5 shows the time-averaged VFE in the final trial after 15 learning iterations per vehicle. While the driver model follows the centerline reasonably well across all vehicles, the average VFE shows clear and systematic differences, reflecting each vehicle's specifications and dynamics. This supports the hypothesis in Sect. 2.2.

Figure 6 shows time-series data for the vehicles with the lowest and highest average VFE. The course includes straight sections and corners with different radii. VFE rises in transitional regions where the vehicle transitions between straight and curved segments, indicating non-steady-state motion. In the low-VFE vehicle, corrective steering is minimal and observations transition smoothly. In contrast, the high-VFE vehicle shows significant corrective steering and large fluctuations during the same transitions.

Vehicle model		A-Class Hatchback	B-Class Hatchback	B-Class SportsCar	C-Class Hatchback	D-Class Sedan	D-Class Minivan	D-Class SUV	E-Class Sedan	E-Class SUV	Exotic SportsCar	F-Class Sedan	Full Size SUV
Drive System		FF	FF	FR	FF	FF	FF	4WD	4WD	4WD	4WD	4WD	4WD
Weight [kg]		750	1110	1020	1270	1370	1800	1430	1650	1590	1360	1820	2257
Yaw Inertia [kg m^2]		750	1343.1	1020	1536.7	2315.3	3528	2059.2	3234	2687.1	1065.2	4095	3524.9
Wheelbase [mm]		2350	2600	2330	2910	2866	3000	2660	3050	2950	2650	3160	3140
Average of track widths [mm]		1395	1482.5	1482.5	1675	1550	1640	1565	1600	1575	1625	1605	1737.5
Center of Gravity Height [mm]		540	540	375	540	520	700	650	530	720	375	590	781
Steering ratio		22.9	17.8	17.9	19.9	17.6	21.04	17	16.2	19.7	15.6	19.2	18
Steering–Yaw Rate Response @ 0.2Hz	Gain [dB]	-6.16	-7.98	-6.88	-10.1	-8.26	-13.6	-12.0	-11.5	-12.4	-3.65	-12.8	-11.9
	Phase [deg.]	-11.1	-5.22	-3.01	-3.15	-6.82	-2.96	0.015	-1.31	-2.25	0.121	-1.82	-3.03
Steering–Lateral Accel. Response @ 0.2Hz	Gain [dB]	-32.6	-34.3	-33.1	-36.3	-34.9	-39.9	-38.6	-37.8	-39.1	N/A	-39.1	-38.3
	Phase [deg.]	-24.0	-14.0	-9.87	-9.18	-15.4	-11.4	-12.1	-9.25	-12.7	N/A	-7.03	-14.1
Car Sim precomputed average VFE [nat]		3.13	2.95	2.98	2.82	3.30	3.96	4.24	3.02	3.40	2.50	2.62	3.81

Fig. 5. Time-averaged VFE obtained through offline learning using the Simulink driver model for 12 CarSim vehicle models. Vehicle parameters and dynamic characteristics generally considered favorable to handling are shown in deeper blue; unfavorable ones are shown in deeper red.

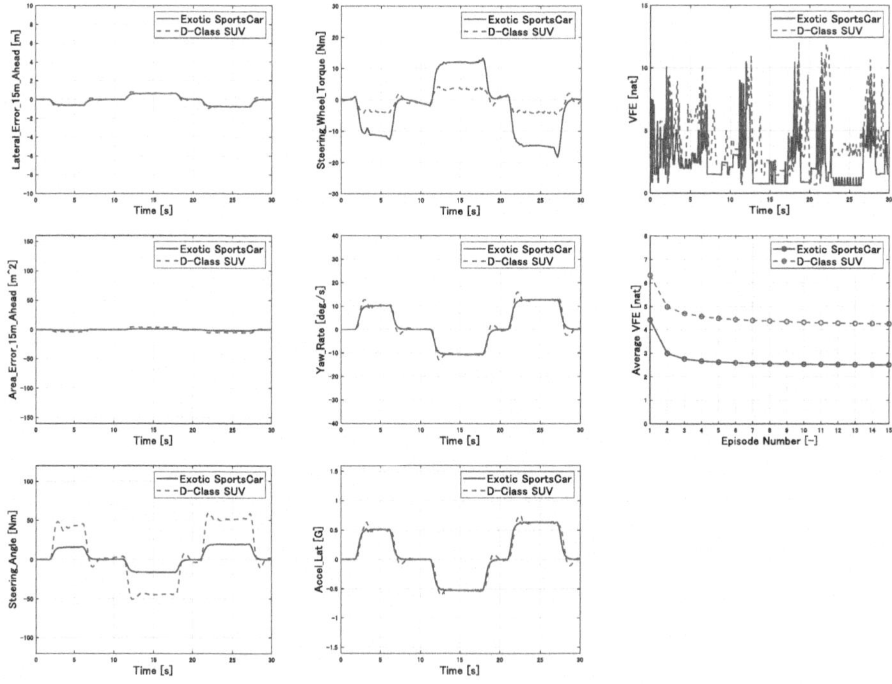

Fig. 6. Time-series plots of observations for each modality in Table 1. The right side displays the VFE over time in the final trial and the progression of time-averaged VFE over 15 trials.

In the active inference framework, this can be interpreted as follows. The brain learns both the frequency of co-occurrence between observations and hidden states at each time step, and the transitions between hidden states over time, as described in Eq. (4). When these relationships become complex—as in vehicles with high VFE—the modeling error of the generative model (i.e., VFE) tends to increase. In such cases, achieving "as-intended" steering is expected to be extremely difficult for humans. The following sections experimentally verify this hypothesis.

3 Experimental Verification

3.1 Experimental Conditions

To verify the hypothesis presented in Sect. 2.2, we conducted an experiment. It was approved by the Life Ethics Committee of Honda R&D Co., Ltd. (Committee number: 100HM-041H) and the Ethics Review Committee of the Graduate School of Engineering, The University of Tokyo (Approval number: KE24-76).

Participants drove the course using CarSim on a simple simulator (see Fig. 2), with the 12 vehicle models listed in Fig. 5. They were instructed to steer as close to the centerline as possible. Vehicle speed was fixed at 100 km/h in CarSim, so throttle and brake were not used. The course took 39 s to complete.

Each vehicle was driven five times on the course. The first two runs were excluded due to unstable driving during familiarization and difficulty in obtaining stable offline VFE. The remaining three trials were used for analysis. To ensure data reliability, 15 off-course trials were also excluded from the 300 total, in the case of the 60 general participants described later.

All trial data from each participant's successive runs were used for continuous offline learning, ensuring that the calculated VFE reflected their full experience. The A tensor, B tensor, and D vector listed in Table 1 were initialized only once per participant at the beginning, and continuous learning then followed Eq. (4).

Before the experiment, all participants were given instructions originally provided in Japanese and translated into English for this paper. They were told that the purpose of the experiment was to evaluate whether the car behaves "as-intended" when the driver has a specific intention. Driving along the centerline was used to provide a common intention among participants. However, precision was not defined, as participants had varying levels of driving skill, making it unrealistic to demand uniform accuracy.

After each run, participants performed sensory evaluations by rating how "as-intended" they felt the car behaved. Participants were asked the following questions:

- Please score how "as-intended" you felt the car was during the previous trial. The degree of "as-intended" refers to the extent to which the car behaved as you intended.
- 0: Not at all "as-intended", 10: Completely "as-intended" (11-point Likert scale)

The evaluation focused on the perceived behavior of the car, not the participants' driving skills. Participants included:

1. One professional expert driver with extensive experience in vehicle development.
2. Sixty general participants holding Japanese driver's licenses.

The expert driver's evaluations were considered reliable, whereas those by general participants were potentially less so. Accordingly, the expert driver evaluated all 12 vehicles in a random order specified by the experimenter. Each general participant drove one vehicle randomly assigned by the experimenter, and only the evaluation for that vehicle was used for analysis. Five participants were assigned to each vehicle model, ensuring balanced driving experience across models.

3.2 Results of the Experiment by Expert Driver

Figure 7 (left) shows the relationship between the expert driver's "as-intended" scores and the average VFE obtained through offline learning using the CarSim Simulink driver's data (hereafter, CarSim precomputed average VFE; see Fig. 5). A strong negative correlation was found ($r = -0.80, p < 0.001$), strongly supporting the hypothesis in Sect. 2.2. Figure 7 (right) shows the same relationship using the expert driver's own driving data, which also revealed a relatively strong negative correlation ($r = -0.67$, $p < 0.001$), further reinforcing the hypothesis.

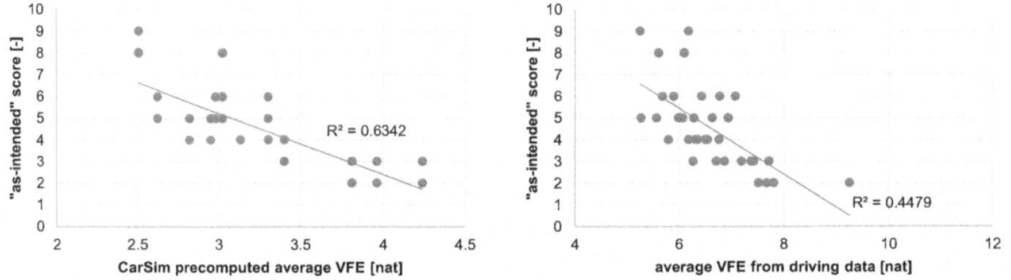

Fig. 7. Relationship between "as-intended" scores and average VFE values. Left: VFE from CarSim Simulink driver. Right: VFE from expert driver's own data.

The slightly weaker correlation in the expert driver's own data (r = −0.67), compared to the CarSim data (r = −0.80), may partly reflect the influence of learning order: offline learning for the expert driver was performed continuously across all vehicles, whereas the CarSim data were learned independently for each vehicle. The stronger correlation with the CarSim data suggests that the expert driver's "as-intended" scores were based on an absolute internal standard, unaffected by the driving order.

As shown in Fig. 8 (left), the expert driver exhibited almost no variation in control performance—measured as the time-averaged distance from the centerline—across different vehicle models, despite variations in the CarSim precomputed average VFE. The expert driver's steering control was comparable to that of the Simulink driver, which likely underlies the strong correlations in Fig. 7 (left) and suggests that control performance does not determine the "as-intended" score.

Figure 8 (middle) suggests that the expert driver may have based their "as-intended" scores on the amount of corrective steering required. Vehicles with higher CarSim precomputed average VFE required greater corrective steering, as quantified by the time-averaged magnitude of steering angle components above a 1.27 Hz cutoff frequency, extracted using a washout filter. This measure showed a strong positive correlation with

VFE ($r = 0.88$, $p < 0.001$). Furthermore, as shown in Fig. 8 (right), this corrective steering effort exhibited a strong negative correlation with the "as-intended" score ($r = -0.79$, $p < 0.001$). These results suggest that the expert driver, either consciously or unconsciously, perceived task difficulty and adjusted their steering effort accordingly.

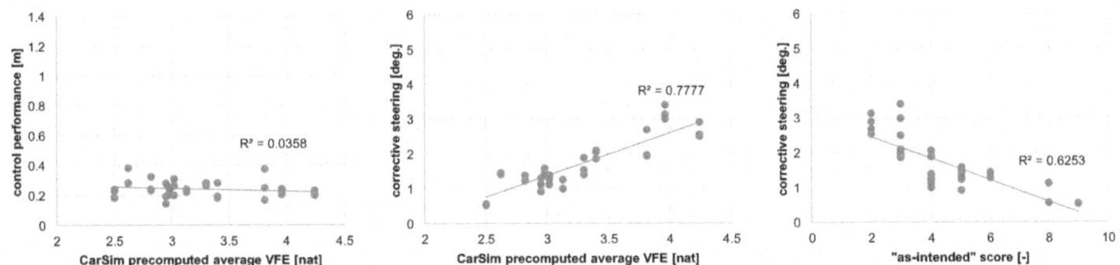

Fig. 8. Relationships between CarSim precomputed average VFE, control performance, corrective steering, and "as-intended" score in the expert driver.

3.3 Results of the Experiment by 60 General Participants

Figure 9 (left) shows the relationship between average VFE from offline learning of each participant's driving data and their control performance, based on data from 60 general participants. A strong positive correlation was found ($r = 0.78$, $p < 0.001$). Figure 9 (right) shows the relationship between the same VFE and the amount of corrective steering, with a relatively strong correlation ($r = 0.65, p < 0.001$). These results strongly support the hypothesis in Sect. 2.2.

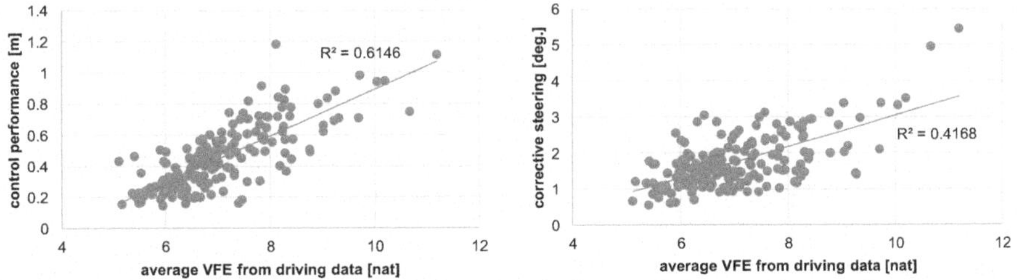

Fig. 9. Relationship between average VFE from offline learning of participant driving data and control performance (left) or corrective steering (right) in 60 general participants.

In contrast, Fig. 10 (left) shows the relationship between the CarSim precomputed average VFE and the "as-intended" scores given by general participants ($r = -0.02$, $p = 0.746$). Figure 10 (right) shows the relationship between the same scores and the average VFE obtained through offline learning from each participant's driving data ($r = -0.28$, $p < 0.002$). Neither figure shows a clear correlation, unlike the results with the expert driver. In Fig. 10 (left), the horizontal axis represents different vehicle models, and the "as-intended" scores varied widely among general participants—even for the same vehicle. This variation was observed not only across participants, but also within the same participant across repeated trials.

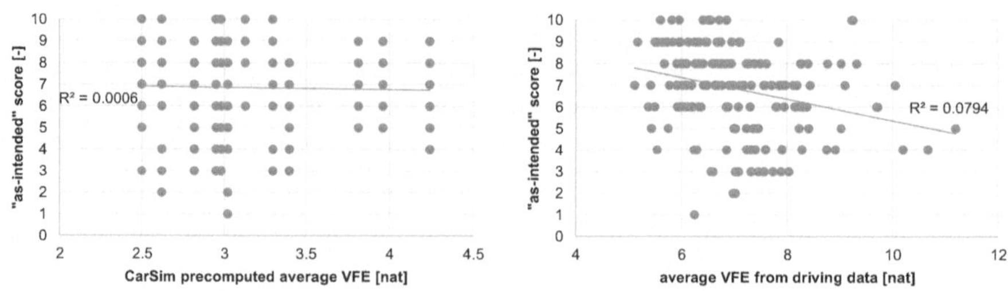

Fig. 10. Relationships between "as-intended" scores and average VFE in 60 general participants.

Figure 11 further illustrates this issue. Figure 11 (left) shows the relationship between CarSim precomputed average VFE and control performance among 60 general participants. The variation in control performance among participants is as large as or larger than that due to vehicle model differences. The resulting weak correlation ($r = 0.18$, $p < 0.019$) suggests that control performance is unlikely to be an indicator of the "as-intended" score. Figure 11 (middle) shows a moderate correlation between CarSim precomputed average VFE and corrective steering ($r = 0.49$, $p < 0.001$). However, Fig. 11 (right) shows that corrective steering is also not a strong determinant of the "as-intended" score ($r = -0.33$, $p < 0.001$). Unlike the expert driver, general participants showed no clear basis for their "as-intended" evaluations, leading to widely varying scores.

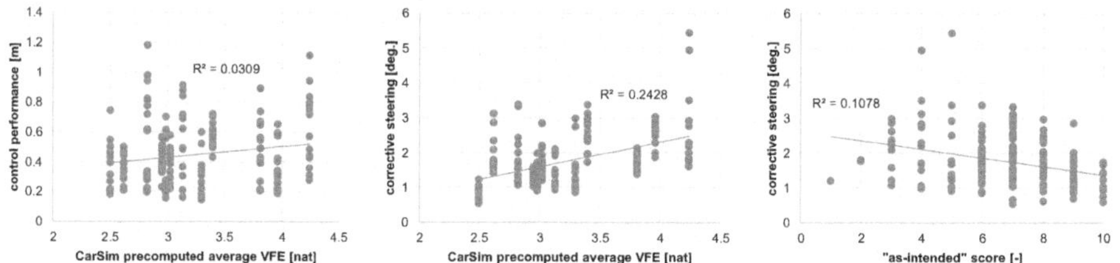

Fig. 11. Relationships between CarSim precomputed average VFE, control performance, corrective steering, and "as-intended" score in 60 general participants.

4 Discussion

Traditionally, "as-intended" controllability has been evaluated through subjective assessments by expert drivers using simulators or real vehicles. We confirmed that, while general participants showed variability in their evaluations, the expert driver's assessments consistently aligned with objective indicators such as corrective steering. This is analogous to wine tasting: trained sommeliers can evaluate reliably, whereas this is difficult for general consumers. Accordingly, we provided an explicit intention (to follow the centerline as closely as possible) and ensured objective evaluation of "as-intended" via control performance and corrective steering, even if general participants could not consciously judge it.

In this study, we introduced a novel methodology. By defining a specific task (e.g., a steering task), constructing a computational model of the brain capable of online learning within the active inference framework, and calculating VFE through offline learning using existing data, we demonstrated that the resulting VFE values correlate strongly with both reliable subjective "as-intended" scores and objective indicators (e.g., control performance or corrective steering).

These findings support the hypothesis that lower VFE reflects a higher degree of "as-intended" controllability. They also suggest that VFE values obtained through offline learning using simulation data can quantify this controllability and potentially serve as a proxy for expert drivers' subjective "as-intended" scores.

Ideally, VFE values should be derived from online learning within the active inference framework. However, we found that the learning outcomes were sensitive to factors such as the random seed used in action selection based on EFE, as well as the D vector. This sensitivity made it difficult to obtain stable and consistent VFE values for vehicle evaluation. It remains unclear whether this limitation arises from the framework itself or from the specific design of the generative model. Further studies are needed to clarify this issue.

5 Conclusion

Based on the methodology introduced and experimentally validated in this study, we conclude that, regardless of individual differences in driving skill or variations in vehicle dynamics, the modeling error of vehicle dynamics learned in the driver's brain—represented by VFE—plays a critical role in determining whether the driver can steer the vehicle "as-intended" from an objective standpoint.

These findings strongly support the theoretical perspective underlying active inference, which holds that the brain learns by accumulating statistical regularities in sensory inputs and acts through probabilistic inference. Furthermore, this study presents a novel approach to the long-standing challenge of quantitatively evaluating and understanding "as-intended" control in vehicle dynamics.

Importantly, the offline learning approach proposed in this study is not limited to simulation or simulator data—it can also be applied to real-world driving data. This enables consistent and objective evaluation of "as-intended" controllability regardless of the evaluator. Notably, the ability to perform such evaluations using only simulation data offers a clear advantage in the early stages of vehicle dynamics design. Future research will aim to extend this approach to real-world applications.

Acknowledgments. This study was initiated by Honda R&D Co., Ltd. in collaboration with The University of Tokyo. We would like to thank all participants in the driving experiments. We would also like to express our sincere appreciation to the pymdp development team for their timely completion and release of the JAX version in response to our request, which significantly contributed to the implementation of our simulation framework.

Disclosure of Interests. Kazuharu Kidera is employed by Honda R&D Co., Ltd. The other authors declare no competing interests.

References

1. Tao, M., Sugimachi, T., Suda, Y., Shibata, K., Katou, D., Fukaya, T.: A study on vehicle dynamics characteristics that realize "as-intended" driving. Trans. Soc. Automot. Eng. Jpn. **48**(6), 1265–1271 (2017) (in Japanese). https://doi.org/10.11351/jsaeronbun.48.1265

2. Friston, K.: The free-energy principle: a unified brain theory? Nat. Rev. Neurosci. **11**(2), 127–138 (2010). https://doi.org/10.1038/nrn2787
3. Parr, T., Pezzulo, G., Friston, K.J.: Active Inference: The Free Energy Principle in Mind, Brain, and Behavior. MIT Press, Cambridge (2022). https://doi.org/10.7551/mitpress/12441.001.0001
4. Applied Intuition: CarSim. https://www.appliedintuition.com/products/carsim. Accessed 16 May 2025
5. Smith, R., Friston, K.J., Whyte, C.J.: A step-by-step tutorial on active inference and its application to empirical data. J. Math. Psychol. **107**, 102632 (2022). https://doi.org/10.1016/j.jmp.2021.102632
6. Heins, C., et al.: pymdp: a Python library for active inference in discrete state spaces. J. Open Source Softw. **7**(73), 4098 (2022). https://doi.org/10.21105/joss.04098
7. Bradbury, J., et al.: JAX: composable transformations of Python+NumPy programs. https://github.com/google/jax. Accessed 16 May 2025
8. CEDEC Digital Library: Brainwave frequency bands and information processing speed during gameplay. https://cedil.cesa.or.jp/cedil_sessions/view/2146. Accessed 16 May 2025. (in Japanese)
9. Parr, T., Markovic, D., Kiebel, S.J., Friston, K.J.: Neuronal message passing using Mean-field, Bethe, and Marginal approximations. Sci. Rep. **9**(1), 1889 (2019). https://doi.org/10.1038/s41598-018-38246-3

Scaling and Efficiency

Message Passing-Based Inference in an Autoregressive Active Inference Agent

Wouter M. Kouw[1(✉)], Tim N. Nisslbeck[1], and Wouter L. N. Nuijten[1,2]

[1] Eindhoven University of Technology, Eindhoven, Netherlands
w.m.kouw@tue.nl
[2] Lazy Dynamics B. V., Eindhoven, Netherlands

Abstract. We present the design of an autoregressive active inference agent in the form of message passing on a factor graph. Expected free energy is derived and distributed across a planning graph. The proposed agent is validated on a robot navigation task, demonstrating exploration and exploitation in a continuous-valued observation space with bounded continuous-valued actions. Compared to a classical optimal controller, the agent modulates action based on predictive uncertainty, arriving later but with a better model of the robot's dynamics.

Keywords: Intelligent agents · Free energy minimization · Active inference · Autoregressive models · Factor graphs · Message passing

1 Introduction

Active inference is a comprehensive framework that unifies perception, planning, and learning under the free energy principle, offering a promising approach to designing autonomous agents [2,18]. We present the design of an active inference agent implemented as a message passing procedure on a Forney-style factor graph [8,10]. The agent is built on an autoregressive model, making continuous-valued observations and inferring bounded continuous-valued actions [7,15]. We show that leveraging the factor graph approach produces a distributed, efficient and modular implementation [1,3,17,22].

Probabilistic graphical models have long been a unifying framework for the design and analysis of information processing systems, including signal processing, optimal controllers, and artificially intelligent agents [4,5,8,12,16]. Many famous algorithms can be written as message passing algorithms, including Kalman filtering, model-predictive control, and dynamic programming [12,16]. However, it can be a challenge to formulate new algorithms due to the requirement of local access to variables and the difficulty of deriving backwards messages. We highlight some of these challenges, and contribute with

- the derivation of expected free energy minimization in a multivariate autoregressive model with continuous-valued observations and bounded continuous-valued actions (Sect. 4.2), and

M. Albarracin et al. (Eds.): IWAI 2025, CCIS 2857, pp. 285–298, 2026.
https://doi.org/10.1007/978-3-032-16955-6_16

- the formulation of the planning model as a factor graph with marginal distribution updates based on messages passed along the graph (Fig. 3).

We validate the proposed design on a robot navigation task, comparing the agent to an adaptive model-predictive controller.

2 Problem Statement

We focus on the class of discrete-time stochastic nonlinear dynamical systems with state $z_k \in \mathbb{R}^{D_z}$, control $u_k \in \mathbb{R}^{D_u}$, and observation $y_k \in \mathbb{R}^{D_y}$ at time k. Their evolution is governed by a state transition function f and an observation function g:

$$z_k = f(z_{k-1}, u_k) + w_k\,, \qquad y_k = g(z_k) + v_k\,, \tag{1}$$

where w_k, v_k are stochastic contributions. The agent only receives noisy outputs $y_k \in \mathbb{R}^{D_y}$ from a system and sends control inputs $u_k \in \mathbb{U} \subset \mathbb{R}^{D_u}$ back. It must drive the system to output y_* without knowledge of the system's dynamics. Performance is measured with free energy (which in the proposed model is equal to the negative log evidence), Euclidean distance to goal, and the 2-norm magnitude of controls, over the course of a trial of length T.

3 Model Specification

The model is autoregressive in nature, meaning that the system output at time k is predicted from the system input u_k, M_u previous system inputs $\bar{u}_k$ and M_y previous system outputs $\bar{y}_k$:

$$\bar{u}_k = \begin{bmatrix} u_{k-1} \\ \vdots \\ u_{k-M_u} \end{bmatrix}, \quad \bar{y}_k = \begin{bmatrix} y_{k-1} \\ \vdots \\ y_{k-M_y} \end{bmatrix}, \quad x_k = \begin{bmatrix} u_k \\ \bar{u}_k \\ \bar{y}_k \end{bmatrix}. \tag{2}$$

The vector x_k is the concatenation of these elements and has dimension $D_x = D_u(M_u+1)+D_yM_y$. Our likelihood function is based on a Gaussian distribution

$$p(y_k \mid \Theta, u_k, \bar{u}_k, \bar{y}_k) = \mathcal{N}\big(y_k \mid A^{\intercal} x_k, W^{-1}\big)\,, \tag{3}$$

where $A \in \mathbb{R}^{D_x \times D_y}$ is a regression coefficient matrix and $W \in \mathbb{R}_+^{D_y \times D_y}$ is a precision matrix. Let $\Theta = (A, W)$ refer to the parameters jointly. Their prior distribution is a matrix normal Wishart distribution [21, D175]:

$$\begin{aligned} p(\Theta) &= \mathcal{MNW}\big(A, W \mid M_0, \Lambda_0^{-1}, \Omega_0^{-1}, \nu_0\big) &(4)\\ &= \mathcal{MN}\big(A \mid M_0, \Lambda_0^{-1}, W^{-1}\big)\, \mathcal{W}\big(W \mid \Omega_0^{-1}, \nu_0\big). &(5) \end{aligned}$$

The prior distributions over control inputs are independent Gaussian distributions, as are the goal prior distributions for future observations:

$$p(u_k) = \mathcal{N}(u_k \mid 0, \Upsilon^{-1})\,, \qquad p(y_t \mid y_*) = \mathcal{N}(y_t \mid m_*, S_*)\,, \tag{6}$$

where Υ is a precision matrix and $y_* = (m_*, S_*)$ are the goal mean vector and covariance matrix.

4 Inference

4.1 Learning

We use Bayesian filtering to update parameter beliefs given y_k, u_k [15,19]:

$$\underbrace{p(\Theta \mid \mathcal{D}_k)}_{\text{posterior}} = \frac{\overbrace{p(y_k \mid \Theta, u_k, \bar{u}_k, \bar{y}_k)}^{\text{likelihood}}}{\underbrace{p(y_k \mid u_k, \mathcal{D}_{k-1})}_{\text{evidence}}} \underbrace{p(\Theta \mid \mathcal{D}_{k-1})}_{\text{prior}} . \tag{7}$$

where $\mathcal{D}_k = \{y_i, u_i\}_{i=1}^k$ is short-hand for data up to time k. Note that the memories $\bar{u}_k, \bar{y}_k$ are subsets of $\mathcal{D}_{k-1}$. The evidence term is the evaluation of the observation y_k under the predictive distribution, obtained by marginalizing the likelihood over the parameters [14].

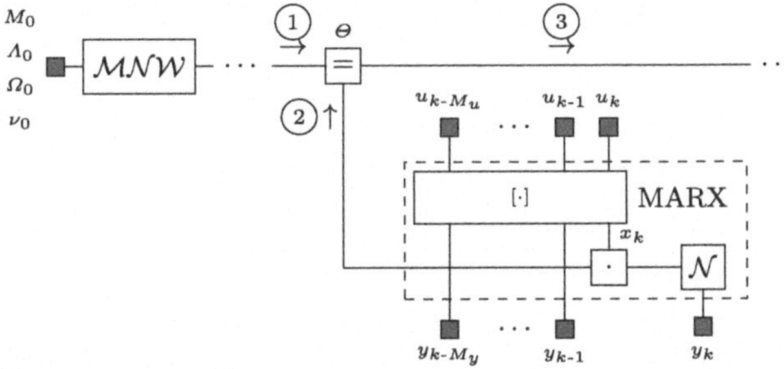

Fig. 1. Forney-style factor graph of one time step (separated by dots) of Bayesian filtering. Edges represent random variables and nodes operations on those variables. Black squares represent observed variables or set parameters, and the dotted box represents a custom node, composed of the nodes within. Message 1 is the prior belief over parameters and message 2 the likelihood-based update. These are multiplied at the equality node, yielding the marginal posterior distribution (message 3).

We express the Bayesian filtering procedure as message passing on the factor graph[1] shown in Fig. 1. Message ① is the prior distribution on Θ,

$$① = \mathcal{MNW}(A, W \mid M_{k-1}, \Lambda_{k-1}^{-1}, \Omega_{k-1}^{-1}, \nu_{k-1}) . \tag{8}$$

Message ② originates from the MARX likelihood function and is an improper matrix normal Wishart distribution [15, Lemma 2],

$$② = \mathcal{MNW}(A, W \mid \bar{M}_k, \bar{\Lambda}_k^{-1}, \bar{\Omega}_k^{-1}, \bar{\nu}_k) , \tag{9}$$

[1] For excellent introductions to the factor graph approach, see [11,20].

with parameters based on data and buffers at time k,

$$\bar{\nu}_k = 2 - D_x + D_y \, , \; \bar{\Lambda}_k = x_k x_k^{\mathsf{T}} \, , \; \bar{M}_k = (x_k x_k^{\mathsf{T}})^{-1} x_k y_k^{\mathsf{T}} \, , \; \bar{\Omega}_k = 0_{D_y \times D_y} \, . \tag{10}$$

It is improper because $\bar{\Omega}_k$ is singular. But when multiplied with the prior distribution, it produces the conjugate posterior distribution exactly [15, Thm. 1]. This multiplication occurs in the equality node and produces message ③:

$$③ = p(\Theta \mid \mathcal{D}_k) = \mathcal{MNW}(A, W \mid M_k, \Lambda_k^{-1}, \Omega_k^{-1}, \nu_k) \, . \tag{11}$$

The parameters of this distribution are

$$\begin{aligned}
\nu_k &= \nu_{k-1} + 1, & (12)\\
\Lambda_k &= \Lambda_{k-1} + x_k x_k^{\mathsf{T}}, & (13)\\
M_k &= (\Lambda_{k-1} + x_k x_k^{\mathsf{T}})^{-1} (\Lambda_{k-1} M_{k-1} + x_k y_k^{\mathsf{T}}), & (14)\\
\Omega_k &= \Omega_{k-1} + y_k y_k^{\mathsf{T}} + M_{k-1}^{\mathsf{T}} \Lambda_{k-1} M_{k-1} - & (15)\\
&\quad (\Lambda_{k-1} M_{k-1} + x_k y_k^{\mathsf{T}})^{\mathsf{T}} (\Lambda_{k-1} + x_k x_k^{\mathsf{T}})^{-1} (\Lambda_{k-1} M_{k-1} + x_k y_k^{\mathsf{T}}) \, .
\end{aligned}$$

Marginalizing the Gaussian likelihood in Eq. 3 over the parameter posterior distribution (Eq. 11) yields a multivariate location-scale T-distribution [14]:

$$p(y_k | u_k, \mathcal{D}_k) = \int p(y_k | \Theta, u_k, \bar{u}_k, \bar{y}_k) p(\Theta | \mathcal{D}_k) \mathrm{d}\Theta = \mathcal{T}_{\eta_k}\big(y_k | \mu_k(u_k), \Sigma_k(u_k)\big), \tag{16}$$

with $\eta_k = \nu_k - D_y + 1$ degrees of freedom and a mean and covariance of

$$\mu_k(u_k) = M_k^{\mathsf{T}} \begin{bmatrix} u_k \\ \bar{u}_t \\ \bar{y}_t \end{bmatrix} , \; \Sigma_k(u_k) = \frac{1}{\nu_k - D_y + 1} \Omega_k \Big(1 + \begin{bmatrix} u_t \\ \bar{u}_t \\ \bar{y}_t \end{bmatrix}^{\mathsf{T}} \Lambda_k^{-1} \begin{bmatrix} u_t \\ \bar{u}_t \\ \bar{y}_t \end{bmatrix} \Big). \tag{17}$$

The subscripts under μ and Σ indicate which parameters were used, i.e., here they refer to M_k, Λ_k, Ω_k and ν_k.

4.2 Actions

Planning. We start by building a generative model for the input and output at time $t = k + 1$:

$$p(y_t, \Theta, u_t \mid \mathcal{D}_k) = p(y_t \mid \Theta, u_t, \bar{u}_t, \bar{y}_t) \, p(\Theta \mid \mathcal{D}_k) \, p(u_t). \tag{18}$$

Note that $\bar{u}_t$ and $\bar{y}_t$ are absent on the left-hand side because, at time $t = k + 1$, these buffers are subsets of $\mathcal{D}_k$. We want the agent to pursue a target, a specific future observation. To do so, we first isolate the marginal distribution $p(y_t)$,

$$p(y_t \mid \Theta, u_t, \bar{u}_t, \bar{y}_t) p(\Theta \mid \mathcal{D}_k) = p(\Theta \mid y_t, u_t, \mathcal{D}_k) p(y_t), \tag{19}$$

and then constrain it to be the goal prior, $p(y_t) \rightarrow p(y_t \mid y_*)$. We use Bayes' rule in the reverse direction to relate the distribution over parameters given the future output and input, to known distributions:

$$p(\Theta \mid y_t, u_t, \mathcal{D}_k) = \frac{p(y_t \mid \Theta, u_t, \bar{u}_t, \bar{y}_t) p(\Theta \mid \mathcal{D}_k)}{p(y_t \mid u_t, \mathcal{D}_k)} . \tag{20}$$

To obtain an approximate marginal posterior distribution for the action u_t, we form an expected free energy functional,

$$\mathcal{F}_k[q] = \mathbb{E}_{q(y_t \mid \Theta, u_t, \bar{u}_t, \bar{y}_t)} \Big[\mathbb{E}_{q(\Theta, u_t)} \big[\ln \frac{q(\Theta, u_t)}{p(y_t, \Theta, u_t, \mid y_*, \mathcal{D}_k)} \big] \Big] . \tag{21}$$

The variational model is $q(y_t, \Theta, u_t, \mid \bar{u}_t, \bar{y}_t) = q(y_t \mid \Theta, u_t, \bar{u}_t, \bar{y}_t) q(\Theta) q(u_t)$. The likelihood and parameter factors are not free variational distributions but fixed to the same form as the likelihood and parameter factors of the generative model:

$$q(y_t \mid \Theta, u_t, \bar{u}_t, \bar{y}_t) = p(y_t \mid \Theta, u_t, \bar{u}_t, \bar{y}_t) = \mathcal{N}\big(y_t \mid A^\intercal x_t, W^{-1}\big) \tag{22}$$

$$q(\Theta) = p(\Theta \mid \mathcal{D}_k) = \mathcal{MNW}(A, W \mid M_k, \Lambda_k^{-1}, \Omega_k^{-1}, \nu_k) . \tag{23}$$

We then minimize this expected free energy functional with respect to the variational distribution $q(u_t)$:

$$q^*(u_t) = \underset{q \in Q}{\arg\min}\ \mathcal{F}_k[q] . \tag{24}$$

where Q represents the set of candidate variational distributions.

Theorem 1. *The optimal variational posterior $q^*(u_t)$ under the free energy functional defined in (21) is proportional to a prior times a likelihood,*

$$q^*(u_t) \propto p(u_t) \exp\big(- G(u_t) \big) , \tag{25}$$

where G is the sum of a mutual information and a cross-entropy term

$$G(u_t) = -\mathbb{E}_{p(y_t, \Theta \mid u_t, \mathcal{D}_k)} \big[\ln \frac{p(y_t, \Theta \mid u_t, \mathcal{D}_k)}{p(y_t \mid u_t, \mathcal{D}_k) p(\Theta \mid \mathcal{D}_k)} \big] - \mathbb{E}_{p(y_t \mid u_t, \mathcal{D}_k)} \big[\ln p(y_t \mid y_*) \big] . \tag{26}$$

The proof can be found in Appendix A.

Corollary 1. *The expected free energy function $G(u_t)$ evaluates to:*

$$G(u_t) = \text{constants} - \frac{1}{2} \ln |\Sigma_t(u_t)| + \frac{1}{2} \mathrm{Tr} \Big[S_*^{-1} \big(\Sigma_t(u_t) \frac{\eta_t}{\eta_t - 2} + \Xi(u_t) \big) \Big] . \tag{27}$$

where $\Xi(u_t) = (\mu_t(u_t) - m_*)(\mu_t(u_t) - m_*)^\intercal$.

The proof is also in Appendix A.

Figure 2 provides an example of how this inference process can be mapped to a factor graph, using $M_y = M_u = 2$. The node marked "MARX" is the composite

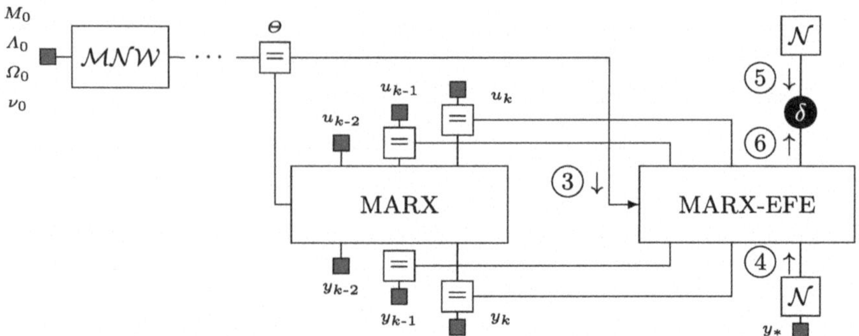

Fig. 2. Factor graph of the 1-step ahead planning model. The left half of the graph is the same as in Fig. 1. The parameter posterior (message 3) is passed forwards to the MARX-EFE node, which takes in message 4 from the goal prior node and produces message 6 containing the exponentiated EFE function. Combined with message 5 from the control prior node, this produces the variational control posterior. The δ circle denotes a collapse of the posterior to a Dirac delta distribution [9].

node depicted in Fig. 1. It is now connected to another composite node marked "MARX-EFE", which is connected to previous observations, parameters, goal prior parameters y_* and the to-be-taken action u_t. Message ③ is the same as in Fig. 1, namely the parameter posterior distribution (Sect. 4.1). Note that during planning, the parameters are not updated. This is indicated by a *directed* edge from the equality node to the MARX-EFE node. Message ④ is the goal prior distribution (Eq. 6), ⑤ is the control prior distribution (Eq. 6) and message ⑥ is the unnormalized exponentiated expected free energy;

$$④ = \mathcal{N}(y_t \mid m_*, S_*)\ , \quad ⑤ = \mathcal{N}(u_t \mid 0, \Upsilon^{-1})\ , \quad ⑥ = \exp(-G(u_t))\,. \tag{28}$$

Note that message 6 is the result of the EFE derivation (Theorem 1) and not the result of minimizing the Bethe free energy, a point discussed in more detail in Sect. 6.

Normalizing $q^*(u_t)$, i.e., the product of ⑤ and ⑥, requires integrating over u_t. This is challenging and avoided by collapsing the approximate posterior to its maximum a posteriori point-mass distribution, $q^*(u_t) \approx \delta(u_t - \hat{u}_t)$ where

$$\hat{u}_t = \underset{u_t \in \mathbb{U}}{\arg\max}\ q^*(u_t) = \underset{u_t \in \mathbb{U}}{\arg\min}\ \frac{1}{2} u_t^\intercal \Upsilon u_t + G(u_t)\,. \tag{29}$$

We believe this is justified because collapsing the posterior to a point estimate is anyway required to pass controls to actuators.

Horizon. Extending the time horizon is challenging, requiring additional marginalizations that complicate the above results (see Sect. 6 for an extended discussion). In this paper, we adopt a simpler approach and generalize the planning factor graph (Fig. 2) by including additional MARX-EFE nodes. Figure 3

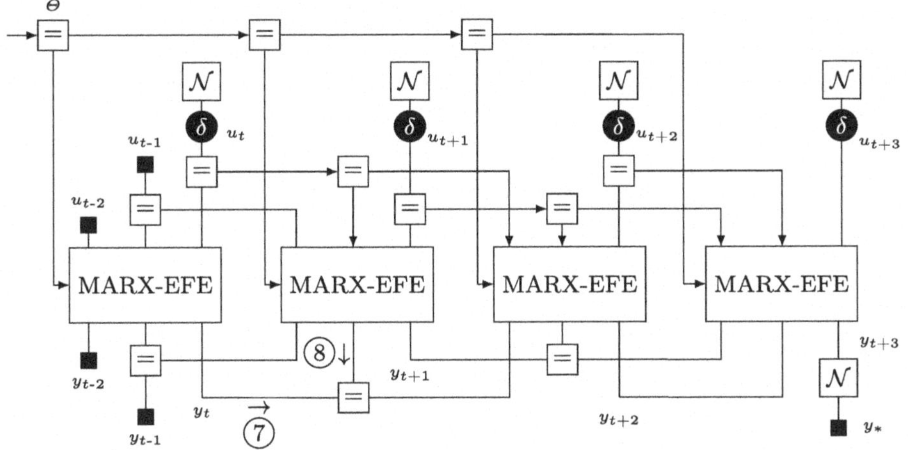

Fig. 3. Factor graph of a 4-step ahead planning model, showing repeated MARX-EFE node from Fig. 2. Some buffer variables are now latent as well. Message 7 is the posterior predictive over y_t carrying forward system output predictions given a selection control input. Message 8 is a predictive likelihood over y_t sent backwards from the node at time $t+1$. Together the forward and backward pass of predictive messages generates a sequence of goal priors.

shows an extension with $M_y = M_u = 2$ and a time horizon of $H = 4$. The main difference is that, for $t > k+2$, the buffer $\bar{y}$ will no longer contain delta distributed variables. Note that the buffer $\bar{u}$ *will* still contain delta variables, because we constrain the marginal action posteriors to be delta's (Eq. 29). Inspecting the second MARX-EFE node in Fig. 3 reveals that the only change from the MARX-EFE node in Fig. 2 is the incoming message for y_t. This message is the posterior predictive distribution (given a selected action $\hat{u}_t$) sent out by the first MARX-EFE node in the planning graph;

$$\textcircled{7} = p(y_t \mid \hat{u}_t, \mathcal{D}_t) = \mathcal{T}_{\eta_k}(y_t \mid \mu_k(\hat{u}_t), \Sigma_k(\hat{u}_t)) \,. \tag{30}$$

This message is incorporated into the MARX-EFE node function through a variational approximation:

$$\begin{aligned} &p(y_{t+1} \mid u_{t+1}, \bar{u}_{t+1}, \tilde{y}_{t+1}, \Theta) \\ &\quad\approx \exp\big(\mathbb{E}_{p(y_t \mid \hat{u}_t, \mathcal{D}_t)}\big[\ln p(y_{t+1} \mid u_{t+1}, \bar{u}_{t+1}, \bar{y}_{t+1}, \Theta)\big]\big) && (31)\\ &\quad\propto \exp\big(-\frac{1}{2}(y_{t+1}^{\mathsf{T}} W y_{t+1} - 2 y_{t+1} W A^{\mathsf{T}} \mathbb{E}_{p(y_t \mid \hat{u}_t, \mathcal{D}_t)}\,\big[u_{t+1}\;\bar{u}_{t+1}\;y_t\;\hat{y}_{t-1}\big]^{\mathsf{T}})\big) && (32)\\ &\quad\propto \mathcal{N}(y_{t+1} \mid A^{\mathsf{T}}\big[u_{t+1}\;\bar{u}_{t+1}\;\mu_k(\hat{u}_t)\;\hat{y}_{t-1}\big]^{\mathsf{T}}, W^{-1}). && (33) \end{aligned}$$

Note that this is, in essence, still the MARX likelihood function except with the mean of the posterior predictive $\mu_k(\hat{u}_t)$ instead of an observed value for y_t in $\bar{y}_{t+1}$. We mark this change with $\tilde{y}$ instead of $\bar{y}$.

For the backwards message from the MARX-EFE node at $t+1$ towards the variable y_t, we first utilize the same variational approximation as above but now with respect to the variational factor $q(y_{t+1}) = \mathcal{N}(y_{t+1} \mid m_{t+1}, S_{t+1})$ (Eq. 39);

$$\begin{aligned} &\exp\big(\mathbb{E}_{q(y_{t+1})}\big[\ln p(y_{t+1} \mid u_{t+1}, \bar{u}_{t+1}, \bar{y}_{t+1}, \Theta)\big]\big) \\ &\quad \propto \exp\Big(-\frac{1}{2}\mathbb{E}_{q(y_{t+1})}\big[y_{t+1}^{\intercal} W y_{t+1} - 2y_{t+1} W A^{\intercal}\left[u_{t+1}\ \bar{u}_{t+1}\ y_t\ \hat{y}_{t-1}\right]^{\intercal}\big]\Big) \qquad (34) \\ &\quad \propto \mathcal{N}(m_{t+1} \mid A^{\intercal}\left[u_{t+1}\ \bar{u}_{t+1}\ y_t\ \hat{y}_{t-1}\right]^{\intercal}, W^{-1}). \qquad (35) \end{aligned}$$

Then we marginalize over the parameter posterior distribution,

$$\begin{aligned} \text{⑧} &\propto \mathbb{E}_{p(\Theta \mid \mathcal{D}_k)}\big[\mathcal{N}(m_{t+1} \mid A^{\intercal}\left[u_{t+1}\ \bar{u}_{t+1}\ y_t\ \hat{y}_{t-1}\right]^{\intercal}, W^{-1})\big] \qquad (36) \\ &\propto \mathcal{T}_{\bar{\eta}}\big(m_{t+1} \mid \bar{\mu}_{t+1}(y_t), \bar{\Sigma}_{t+1}(y_t)\big), \qquad (37) \end{aligned}$$

with $\bar{\eta} = \nu_k - D_y + 1$ degrees of freedom and mean and covariance

$$\bar{\mu}_{t+1}(y_t) = M_k^{\intercal}\begin{bmatrix} u_{t+1} \\ \bar{u}_{t+1} \\ y_t \\ \hat{y}_{t-1} \end{bmatrix}, \quad \bar{\Sigma}_{t+1}(y_t) = \frac{1}{\bar{\eta}}\Omega_k\Big(1 + \begin{bmatrix} u_{t+1} \\ \bar{u}_{t+1} \\ y_t \\ \hat{y}_{t-1} \end{bmatrix}^{\intercal} \Lambda_k^{-1}\begin{bmatrix} u_{t+1} \\ \bar{u}_{t+1} \\ y_t \\ \hat{y}_{t-1} \end{bmatrix}\Big). \qquad (38)$$

In essence, this distribution scores which values of y_t best predict y_{t+1}, with m_{t+1} as a pseudo-observation. At the y_t edge, we perform a variational factor update based on the product of messages ⑦ and ⑧:

$$q(y_t) \propto \mathcal{T}_{\eta_k}\big(y_t \mid \mu_k(\hat{u}_t), \Sigma_k(\hat{u}_t)\big)\mathcal{T}_{\bar{\eta}}\big(m_{t+1} \mid \bar{\mu}_{t+1}(y_t), \bar{\Sigma}_{t+1}(y_t)\big). \qquad (39)$$

This product is not part of a known parametric family of distributions. We perform a Laplace approximation to produce $q(y_t) \approx \mathcal{N}(y_t \mid m_t, S_t)$ where [13]:

$$m_t = \arg\max_{y_t}\ \ln \mathcal{T}_{\eta_k}\big(y_t \mid \mu_k(\hat{u}_t), \Sigma_k(\hat{u}_t)\big)\mathcal{T}_{\bar{\eta}}\big(m_{t+1} \mid \bar{\mu}_{t+1}(y_t), \bar{\Sigma}_{t+1}(y_t)\big) \qquad (40)$$

$$S_t^{-1} = -\nabla_{y_t}^2 \ln \mathcal{T}_{\eta_k}\big(y_t | \mu_k(\hat{u}_t), \Sigma_k(\hat{u}_t)\big)\mathcal{T}_{\bar{\eta}}\big(m_{t+1} | \bar{\mu}_{t+1}(y_t), \bar{\Sigma}_{t+1}(y_t)\big)\Big|_{y_t = m_t}. \qquad (41)$$

This Gaussian variational factor effectively becomes a goal prior for time t. So, in the extended time horizon, we see that the forward and backward passes over the future observations generate a sequence of intermediate goal priors. As such, at each future time point, the agent needs only to solve a 1-step ahead expected free energy minimization problem.

5 Experiments

We perform simulation experiments in which agents have to reach a target state in a single trial. We refer to the proposed free energy minimizing agent as *MARX-EFE*[2]. The benchmark is the same agent but with controls found by minimizing

[2] Agent built with RxInfer; https://github.com/biaslab/IWAI2025-MARXEFE-MP.

a standard model-predictive control cost function;

$$\hat{u}^{\text{MPC}}_{k+1:k+H} = \underset{u_{k+1:k+H} \in \mathbb{U}^H}{\arg\min} \sum_{t=k+1}^{k+H} u_t^{\mathsf{T}} \Upsilon u_t + \big(\mu_t(u_t) - m_*\big)^{\mathsf{T}} \big(\mu_t(u_t) - m_*\big). \quad (42)$$

This agent will be called *MARX-MPC*. We evaluate the probabilities of the system output y_k under the goal prior distribution, $p(y_k \mid y_*)$. Additionally, we evaluate model evidence, i.e., the probability of the system output observation under the agent's predictive distribution, $p(y_k \mid u_k, \mathcal{D}_k)$.

System. Consider a linear Gaussian dynamical system where the state vector z_k contains the two-dimensional position and velocity of a robot. The robot's state transition and measurement functions are

$$f(z_{k-1}, u_k) = \begin{bmatrix} 1 & 0 & \Delta t & 0 \\ 0 & 1 & 0 & \Delta t \\ 0 & 0 & 1 & 0 \\ 0 & 0 & 0 & 1 \end{bmatrix} z_{k-1} + \begin{bmatrix} 0 & 0 \\ 0 & 0 \\ \Delta t & 0 \\ 0 & \Delta t \end{bmatrix} u_k, \quad g(z_k) = \begin{bmatrix} 1 & 0 & 0 & 0 \\ 0 & 1 & 0 & 0 \end{bmatrix} z_k, \quad (43)$$

where Δt is the time step size. Its covariance matrices are

$$Q = \begin{bmatrix} \frac{\Delta t^3}{3}\varsigma_1 & 0 & \frac{\Delta t^2}{2}\varsigma_1 & 0 \\ 0 & \frac{\Delta t^3}{3}\varsigma_2 & 0 & \frac{\Delta t^2}{2}\varsigma_2 \\ \frac{\Delta t^2}{2}\varsigma_1 & 0 & \Delta t\varsigma_1 & 0 \\ 0 & \frac{\Delta t^2}{2}\varsigma_2 & 0 & \Delta t\varsigma_2 \end{bmatrix}, \quad R = \begin{bmatrix} \rho_1 & 0 \\ 0 & \rho_2 \end{bmatrix}, \quad (44)$$

with $\varsigma = \big[10^{-6} \; 10^{-6}\big]^{\mathsf{T}}$ and $\rho = \big[10^{-3} \; 10^{-3}\big]^{\mathsf{T}}$.

Prior Parameters. The prior parameters are weakly informative; $\nu_0 = 100$, $M_0 = 1/(D_x D_y) \cdot I_{D_x \times D_y}$, $\Lambda_0 = 10^{-2} \cdot I_{D_x}$, $\Omega_0 = I_{D_y}$, and $\Upsilon = 10^{-6} \cdot I_{D_u}$. The system starts at $z_0 = \big[0\;0\;0\;0\big]$ and the goal prior has mean $m_* = \big[0\;1\big]^{\mathsf{T}}$ and covariance matrix $S_* = 10^{-6} \cdot I_{D_y}$. Buffers are fixed at $M_u = M_y = 2$ and the time horizon at $H = 3$. Controls are limited to $\mathbb{U} = [-1, 1]$ for $T = 10000$ steps at $\Delta t = 0.1$.

Results. Figure 4 shows the experimental results comparing MARX-EFE to MARX-MPC. The left figure shows that MARX-EFE consistently scores a smaller free energy than MARX-MPC, demonstrating that it cares more strongly about accurately predicting its next observation. The middle figure shows the distance to goal over the duration of the trial where, on average, MARX-MPC reaches the goal sooner than MARX-EFE. MARX-MPC does not care about making accurate predictions, only to close the gap to the target as quickly as possible. It is successful in that regard but struggles to park itself on the target exactly because it ignored opportunities to learn the finer parts of robot's dynamics earlier in the trial. The MARX-EFE agent ultimately gets closer than MARX-MPC because - by the time it gets to the goal - it has a much better model of the robot's dynamics. The right figure shows the 2-norm of the controls, highlighting that MARX-MPC consistently utilizes maximum power ($\max \|u_t\|_2 = \sqrt{2}$) to get closer. The MARX-EFE agent takes very small actions in the beginning, when it is uncertain of their outcomes, and slowly takes larger actions when its uncertainty shrinks. We interpret this behaviour as some form of "caution".

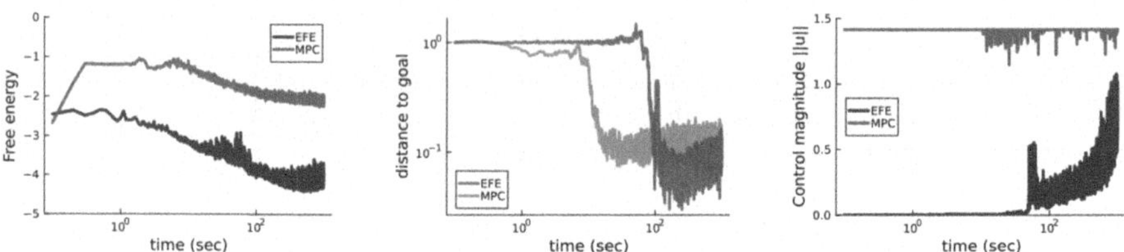

Fig. 4. MARX-EFE (blue) vs. MARX-MPC (red) over a trial of 1000 seconds, compared in terms of free energy (left), Euclidean distance to goal (m_*; middle) and 2-norm of controls (right). Results are averaged over 10 experiments. MARX-EFE initially takes smaller actions, aiming to improve parameters and predictions first. It arrives at the goal later than MARX-MPC but is better able to park on the goal itself. MARX-EFE's actions are small initially but increase in magnitude as uncertainty shrinks. (Color figure online)

6 Discussion

For planning, we observe that the expectation over the future observation is actually not necessary in the above model. If the likelihood is incorporated into the numerator of Eq. 21 as well, i.e.,

$$\mathcal{F}_k[q] = \mathbb{E}_{q(y_t, \Theta, u_t)} \Big[\ln \frac{q(y_t, \Theta, u_t)}{p(y_t, \Theta, u_t \mid y_*, \mathcal{D}_k)} \Big] , \tag{45}$$

then - following the same steps as in Appendix A - the EFE function becomes:

$$G(u_t) = \mathbb{E}_{p(y_t \mid u_t, \mathcal{D}_k)} \big[\ln p(y_t \mid u_t, \mathcal{D}_k) \big] + \mathbb{E}_{p(y_t \mid u_t, \mathcal{D}_k)} \big[- \ln p(y_t \mid y_*) \big] . \tag{46}$$

In both the mutual information in Eq. 26 and in the entropy of the posterior predictive above, the only term that depends on u_t is the variance of the posterior predictive $\Sigma_k(u_t)$. All other terms drop out due to the translation invariance of differential entropies. Thus, we find the same solution when using a standard free energy functional instead of an expected free energy functional [22].

Limitations. In Sect. 4.2 we avoided forming the joint posterior predictive distribution over all future outputs in the horizon;

$$p(y_{k+1:k+H} \mid u_{k+1:k+H}, \mathcal{D}_k) = \int p(\Theta \mid \mathcal{D}_k) \prod_{t=k+1}^{k+H} p(y_t \mid \Theta, u_t, \bar{u}_t, \bar{y}_t) \, \mathrm{d}\Theta \,. \quad (47)$$

This marginalization is a challenge because blocks of the autoregressive coefficient start to nest in both the mean and covariance matrix of the joint Gaussian likelihood. To illustrate this, consider an example with $M_y = 1$ such that $\bar{y}_t = y_{t-1}$. The joint distribution of the likelihoods for $k+1$ and $k+2$ is Gaussian distributed with mean vector and covariance matrix

$$\begin{bmatrix} A_1^\intercal u_{k+1} + A_2^\intercal \bar{u}_k + A_3^\intercal y_k \\ A_1^\intercal u_{k+2} + A_2^\intercal \bar{u}_{k+1} + A_3^\intercal (A_1^\intercal u_{k+1} + A_2^\intercal \bar{u}_k + A_3^\intercal y_k) \end{bmatrix} \begin{bmatrix} W^{-1} & W^{-1} A_3 \\ A_3^\intercal W^{-1} & A_3^\intercal W^{-1} A_3 + W^{-1} \end{bmatrix}, \quad (48)$$

where A_i represent row-indexed blocks of the coefficient matrix A. Marginalizing this joint future likelihood over $p(\Theta \mid \mathcal{D}_k)$ is difficult but a solution would avoid the variational and Laplace approximation errors in Eqs. 31, 34, and 40.

7 Conclusion

We designed an active inference agent with continuous-valued actions as a message passing procedure on a factor graph. The forward and backward pass of predictions over future system outputs generates a sequence of intermediate goal priors. Each node in the planning graph only has to solve a 1-step ahead EFE minimization problem to find appropriate controls. The agent successfully navigates a robot to a goal position under unknown dynamics.

Acknowledgments. The authors gratefully acknowledge support from the Eindhoven Artificial Intelligence Systems Institute.

Disclosure of Interests. The authors have no competing interests to declare that are relevant to the content of this article.

A Appendix

Proof. Using the factorisation of the variational model, the expected free energy functional can be re-arranged to isolate the expectation over $q(u_t)$:

$$\mathcal{F}_t[q] = \mathbb{E}_{q(y_t \mid \Theta, u_t, \bar{u}_t, \bar{y}_t)} \Big[\mathbb{E}_{q(\Theta, u_t)} \Big[\ln \frac{q(\Theta, u_t)}{p(y_t, \Theta, u_t \mid y_*, \mathcal{D}_k)} \Big] \Big] \quad (49)$$

$$= \mathbb{E}_{q(u_t)} \Big[\ln \frac{q(u_t)}{p(u_t)} + \mathbb{E}_{q(y_t \mid \Theta, u_t, \bar{u}_t, \bar{y}_t) q(\Theta)} \Big[\ln \frac{q(\Theta)}{p(y_t, \Theta, u_t \mid y_*, \mathcal{D}_k)} \Big] \Big] \quad (50)$$

$$= \mathbb{E}_{q(u_t)} \Big(\ln \frac{q(u_t)}{p(u_t) \exp(-G(u_t))} \Big), \quad (51)$$

for $G(u_t) = \mathbb{E}_{q(y_t \mid \Theta, u_t, \bar{u}_t, \bar{y}_t) q(\Theta)} \big[\ln q(\Theta) / p(y_t, \Theta \mid u_t, y_*, \mathcal{D}_k) \big]$ and the identity $G(u_t) = \ln \big(1/\exp(-G(u_t)) \big)$. Constraining $q(u_t)$ to be a probability distribution over the space of affordable controls $\mathbb{U}$ is done with a Lagrange multiplier:

$$\mathcal{L}[q, \gamma] = \mathcal{F}_k[q] + \gamma \Big(\int_{\mathbb{U}} q(u_t) \mathrm{d}u_t - 1 \Big) . \tag{52}$$

The stationary solution $q^*(u_t)$ of the Lagrangian is found at $\delta \mathcal{L}[q, \gamma] / \delta q = 0$ [23]). Let $\delta q(u_t) = \epsilon \phi(u_t)$ be a variation with ϕ a continuous and differentiable test function. Then the variational derivative can be found with:

$$\int_{\mathbb{U}} \frac{\delta \mathcal{L}[q, \gamma]}{\delta q} \phi(u_t) \mathrm{d}u_t = \frac{d\mathcal{L}[q(u_t) + \epsilon \phi(u_t), \gamma]}{d\epsilon} \Big|_{\epsilon = 0} \tag{53}$$

$$= \int_{\mathbb{U}} \Big(\ln \frac{q(u_t)}{p(u_t) \exp(-G(u_t))} + 1 + \gamma \Big) \phi(u_t) \mathrm{d}u_t \, . \tag{54}$$

Setting the variational derivative to 0, yields

$$\ln \frac{q(u_t)}{p(u_t) \exp(-G(u_t))} + 1 + \gamma = 0 \rightarrow q(u_t) = \exp(-\gamma - 1) p(u_t) \exp(-G(u_t)). \tag{55}$$

Plugging this into the constraint gives $\exp(-\gamma - 1) = 1 / \int_{\mathbb{U}} p(u_t) \exp(-G(u_t)) \mathrm{d}u_t$. As such, we have:

$$q(u_t) = \frac{p(u_t) \exp(-G(u_t))}{\int_{\mathbb{U}} p(u_t) \exp(-G(u_t)) \mathrm{d}u_t} \, . \tag{56}$$

Using (18), (19) and (22), $G(u_t)$ can be simplified to a negative mutual information plus a cross-entropy term:

$$G(u_t) = \mathbb{E}_{p(y_t \mid \Theta, u_t, \bar{u}_t, \bar{y}_t) p(\Theta \mid \mathcal{D}_k)} \Big[\ln \frac{p(\Theta \mid \mathcal{D}_k) p(y_t \mid u_t, \mathcal{D}_k)}{p(y_t \mid \Theta, u_t, \bar{u}_t, \bar{y}_t) p(\Theta \mid \mathcal{D}_k) p(y_t \mid y_*)} \Big] \tag{57}$$

$$= -\mathbb{E}_{p(y_t, \Theta | u_t, \mathcal{D}_k)} \Big[\ln \frac{p(y_t, \Theta \mid u_t, \mathcal{D}_k)}{p(y_t | u_t, \mathcal{D}_k) p(\Theta | \mathcal{D}_k)} \Big] + \mathbb{E}_{p(y_t | u_t, \mathcal{D}_k)} \big[- \ln p(y_t | y_*) \big] . \tag{58}$$

Proof. We split the mutual information into a joint entropy minus the entropy of the posterior predictive and that of the parameter posterior [13]:

$$-\mathbb{E}_{p(y_t, \Theta | u_t, \mathcal{D}_k)} \Big[\ln \frac{p(y_t, \Theta \mid u_t, \mathcal{D}_k)}{p(y_t | u_t, \mathcal{D}_k) p(\Theta | \mathcal{D}_k)} \Big] = \mathbb{E}_{p(y_t \mid u_t, \mathcal{D}_k)} \big[\ln p(y_t \mid u_t, \mathcal{D}_k) \big]$$

$$+ \mathbb{E}_{p(\Theta \mid \mathcal{D}_k)} \big[\ln p(\Theta \mid \mathcal{D}_k) \big] - \mathbb{E}_{p(y_t, \Theta | u_t, \mathcal{D}_k)} \big[\ln p(y_t, \Theta \mid u_t, \mathcal{D}_k) \big] . \tag{59}$$

Since entropies are invariant to translation, only the entropy of the posterior predictive affects $G(u_t)$ [7]. The entropy of a location-scale T-distribution is [6]:

$$\mathbb{E}_{p(y_t|u_t,\mathcal{D}_k)}\big[\ln p(y_t \mid u_t, \mathcal{D}_k)\big]$$
$$= -\mathbb{E}_{\mathcal{T}_{\eta_k}(y_t\,|\,0,\Sigma_k(u_t))}\big[-\ln \mathcal{T}_{\eta_k}(y_t \mid 0, \Sigma_k(u_t))\big] \tag{60}$$
$$= -\ln \frac{(\eta_k\pi)^{\frac{D_y}{2}}}{\Gamma(\frac{D_y}{2})} B(\frac{D_y}{2}, \frac{\eta_k}{2}) - \frac{\eta_k + D_y}{2}\big(\psi(\frac{\eta_k + D_y}{2}) - \psi(\frac{\eta_k}{2})\big) - \frac{1}{2}\ln\big|\Sigma_k(u_t)\big|. \tag{61}$$

where $B(\cdot), \Gamma(\cdot), \psi(\cdot)$ are the beta, gamma and digamma functions, respectively. Note that only the last term depends on u_t. The cross-entropy from posterior predictive to goal distribution is:

$$\mathbb{E}_{p(y_t\,|\,u_t,\mathcal{D}_k)}\big[-\ln p(y_t \mid y_*)\big]$$
$$= \frac{1}{2}\Big(\ln 2\pi|S_*| + \mathbb{E}_{\mathcal{T}_{\eta_t}(y_t\,|\,\mu_t(u_t),\Sigma_t(u_t))}\big[(y_t - m_*)^\intercal S_*^{-1}(y_t - m_*)\big]\Big) \tag{62}$$
$$= \frac{1}{2}\ln 2\pi|S_*| + \frac{1}{2}\mathrm{Tr}\Big[S_*^{-1}\big(\Sigma_t(u_t)\frac{\eta_t}{\eta_t - 2} + \Xi(u_t)\big)\Big], \tag{63}$$

where $\Xi(u_t) = (\mu_t(u_t) - m_*)(\mu_t(u_t) - m_*)^\intercal$. Note that the first term is constant.

References

1. De Vries, B., Friston, K.J.: A factor graph description of deep temporal active inference. Frontiers Comput. Neurosci. **11**(95) (2017)
2. Friston, K., Kilner, J., Harrison, L.: A free energy principle for the brain. J. Physiol. **100**(1–3), 70–87 (2006)
3. Friston, K.J., Parr, T., de Vries, B.: The graphical brain: belief propagation and active inference. Netw. Neurosci. **1**(4), 381–414 (2017)
4. Hewitt, C.: Viewing control structures as patterns of passing messages. Artif. Intell. **8**(3), 323–364 (1977)
5. Hoffmann, C., Rostalski, P.: Linear optimal control on factor graphs–a message passing perspective–. IFAC-PapersOnLine **50**(1), 6314–6319 (2017)
6. Kotz, S., Nadarajah, S.: Multivariate T-Distributions and Their Applications. Cambridge University Press (2004)
7. Kouw, W.M.: Information-seeking polynomial NARX model-predictive control through expected free energy minimization. IEEE Control Syst. Lett. **8**, 37–42 (2023)
8. van de Laar, T., Koudahl, M., van Erp, B., de Vries, B.: Active inference and epistemic value in graphical models. Frontiers Rob. AI **9** (2022)
9. van de Laar, T., Koudahl, M., de Vries, B.: Realizing synthetic active inference agents, part II: variational message updates. Neural Comput. **37**(1), 38–75 (2024)
10. van de Laar, T.W., de Vries, B.: Simulating active inference processes by message passing. Frontiers Rob. AI **6**(20) (2019)
11. Loeliger, H.A.: An introduction to factor graphs. IEEE Signal Process. Mag. **21**(1), 28–41 (2004)

12. Loeliger, H.A., Dauwels, J., Hu, J., Korl, S., Ping, L., Kschischang, F.R.: The factor graph approach to model-based signal processing. Proc. IEEE **95**(6), 1295–1322 (2007)
13. MacKay, D.J.: Information Theory, Inference and Learning Algorithms. Cambridge University Press (2003)
14. Nisslbeck, T.N., Kouw, W.M.: Factor graph-based online Bayesian identification and component evaluation for multivariate autoregressive exogenous input models. Entropy **27**(7) (2025)
15. Nisslbeck, T.N., Kouw, W.M.: Online Bayesian system identification in multivariate autoregressive models via message passing. In: IEEE European Control Conference (2025)
16. Palmieri, F.A., Pattipati, K.R., Di Gennaro, G., Fioretti, G., Verolla, F., Buonanno, A.: A unified view of algorithms for path planning using probabilistic inference on factor graphs. arXiv:2106.10442 (2021)
17. Parr, T., Markovic, D., Kiebel, S.J., Friston, K.J.: Neuronal message passing using mean-field, Bethe, and marginal approximations. Sci. Rep. **9**(1889) (2019)
18. Parr, T., Pezzulo, G., Friston, K.J.: Active Inference: The Free Energy Principle in Mind, Brain, and Behavior. MIT Press (2022)
19. Särkkä, S.: Bayesian Filtering and Smoothing. Cambridge University Press (2013)
20. Şenöz, İ., van de Laar, T., Bagaev, D., de Vries, B.: Variational message passing and local constraint manipulation in factor graphs. Entropy **23**(7) (2021)
21. Soch, J., Faulkenberry, T.J., Petrykowski, K., Allefeld, C.: The Book of Statistical Proofs (2021). https://doi.org/10.5281/zenodo.4305950
22. de Vries, B., et al.: Expected free energy-based planning as variational inference. arXiv:2504.14898 (2025)
23. Wainwright, M.J., Jordan, M.I., et al.: Graphical models, exponential families, and variational inference. Found. Trends® Mach. Learn. **1**(1–2), 1–305 (2008)

Multi-dimensional Autoscaling of Processing Services: A Comparison of Agent-Based Methods

Boris Sedlak[1(✉)], Alireza Furutanpey[1], Zihang Wang[1], Víctor Casamayor Pujol[2], and Schahram Dustdar[1,2]

[1] Distributed Systems Group, TU Wien, Vienna, Austria
b.sedlak@dsg.tuwien.ac.at
[2] DISL, ICREA, Universitat Pompeu Fabra, Barcelona, Spain

Abstract. Edge computing breaks with traditional autoscaling due to strict resource constraints, thus, motivating more flexible scaling behaviors using multiple elasticity dimensions. This work introduces an agent-based autoscaling framework that dynamically adjusts both hardware resources and internal service configurations to maximize requirements fulfillment in constrained environments. We compare four types of scaling agents: Active Inference, Deep Q Network, Analysis of Structural Knowledge, and Deep Active Inference, using two real-world processing services running in parallel: YOLOv8 for visual recognition and OpenCV for QR code detection. Results show all agents achieve acceptable SLO performance with varying convergence patterns. While the Deep Q Network benefits from pre-training, the structural analysis converges quickly, and the deep active inference agent combines theoretical foundations with practical scalability advantages. Our findings provide evidence for the viability of multi-dimensional agent-based autoscaling for edge environments and encourage future work in this research direction.

Keywords: Internet of Things · Stream Processing · Active Inference · Artificial Neural Networks · Autoscaling · Markov Decision Processes · Reinforcement Learning

1 Introduction

The rise of Edge Computing and the Computing Continuum (CC) addresses the limitations of traditional Cloud infrastructures [4]. By bringing computation closer to users and data sources (e.g., IoT devices) these paradigms significantly reduce network latency, critical for applications that demand near real-time responses, such as autonomous driving, e-health, and virtual reality. A common use case, as depicted in Fig. 1, could be to detect entities in a video stream (e.g., humans) or tracking objects that have a QR code attached. By running these inference services locally, the overall network congestion is also mitigated by minimizing long-distance data transfers.

M. Albarracin et al. (Eds.): IWAI 2025, CCIS 2857, pp. 299–311, 2026.
https://doi.org/10.1007/978-3-032-16955-6_17

However, Edge and CC environments introduce new challenges [1]: They rely on resource-constrained computing hardware, and thus break with traditional Cloud-based autoscaling. In Cloud systems, autoscaling mechanisms elastically respond to increased user demand by allocating more resources to a service or replicating it. This is infeasible at the Edge or in the CC, where computing resources are strictly limited [9]. Especially when resources are scarce, applications require a more flexible scaling behavior that uses a wider range of adaptations – hence, operating in multiple *elasticity dimensions* [3]. On the one hand, this protects the service execution and promises higher requirements fulfillment – captured through a set of Service Level Objectives (SLOs). On the other hand, this increases the complexity for choosing optimal scaling actions. What is needed, hence, are lightweight multi-dimensional scaling mechanisms that optimize the service execution without obstructing existing workloads.

To fill this gap, we propose an agent-based autoscaling approach tailored for Edge and CC systems, which adjusts processing services in multiple elasticity dimensions. Our approach employs decentralized local agents that (1) observe the service execution and their SLO fulfillment without centralized control; thus, we can monitor the resource allocation per service or the application throughput. If SLOs are violated, the agents attempt to restore the desired state by (2) adjusting the service execution; the exact scaling policy is learned by the agent according to environmental feedback. Notably, our approach allows scaling policies tailored to the individual services, where one service could, for example, scale down its machine learning (ML) model, while another service claim the remaining resources. This allows building composite and customizable scaling policies, which go further than existing approaches [13].

To show the viability of our approach, we implement four different versions of our general agent and compare their performance in a processing environment, where the agent needs to dynamically scale two processing services on an Edge device. In particular, we compare an Active Inference agent (AIF), a Deep Active Inference agent (DACI), a Deep Q-Network agent (DQN), and an agent using a numerical solver – called Analysis of Structural Knowledge (ASK). During our experiments, a scaling agent manages two physically executed services: one for video stream inference (Fig. 1a) using the well-known Yolov8 model [15], and another for QR code reading (Fig. 1b), implemented with OpenCV2 [11].

On the short term, our work provides well-needed baselines for comparing the performance of different agent-based approaches – particularly needed for emerging solutions. On the long term, our evaluation environment is extensible and allows incorporating other agents.

We summarize our contributions as:

1. Introducing an agent-based autoscaling approach within a *multi-dimensional elasticity space* that dynamically maximize SLO fulfillment in IoT and Edge environments by adjusting hardware and service configurations
2. Evaluating on *real-world applications* of four distinct scaling agent architectures (Active Inference, Deep Active Inference, Deep Q-Network, and Analysis of Structural Knowledge) for real-time service orchestration.

(a) CV (Yolov8)

(b) QR (OpenCV)

Fig. 1. Demo output of the results produced by the two processing services.

3. Providing a benchmarking environment for future autonomous research and demonstrating viability through agent-based resource allocation for parallel processing services (YOLOv8 and OpenCV) on constrained hardware.

2 Preliminaries

In the following, we provide a formal description of the problem, as well as how researchers have addressed it so far with agent-based methods – including AIF.

2.1 Problem Definition

As depicted in Fig. 2, multiple services are executed within one Edge device – sharing the device's processing resources between them. The execution of the individual services is affected by the amount of resources allocated (e.g., CPU, or RAM), and the configuration of the service-internal parameters (e.g., input quality, or model size). Considering that both sets of parameters can be *elastically* adjusted during runtime [3], we summarize them as elastic configurations.

The allocated resources and the service configuration influence the degree to which the service outcome satisfies the client requirements (i.e., the SLOs). However, and this is the core of the problem, it is not a priori known what will be the resulting SLO fulfillment for a specific configuration. Hence, the problem boils down to adjusting the elastic configurations in such a way, that the SLO fulfillment is maximized. While the number of service configurations and resource allocations is limited, it is not easily possible to brute-force the problem by exhaustively searching the solution space. The reason is that actions taken on the environment require a considerable amount of time to show effect. For instance, orchestration tools like Docker and Kubernetes usually consider a cooldown period of several minutes after taking an action.

Formal Definition More formally, the problem domain and the physical processing environment is defined as follows:

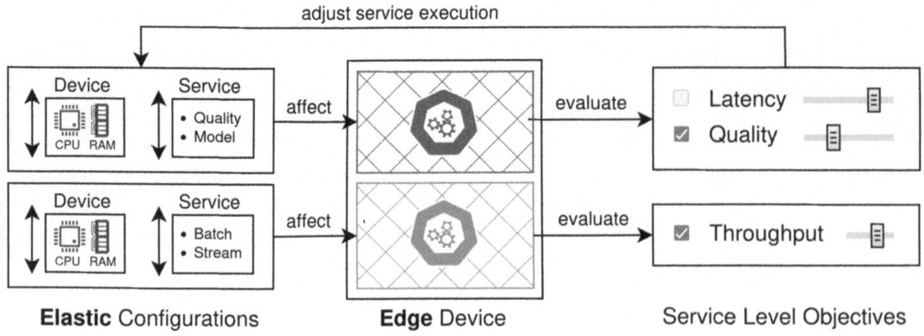

Fig. 2. Services compete for limited device resources; according to SLO fulfillment, the service execution is adjusted by elastically changing device and service configurations.

Processing Hardware. An Edge device d is defined by its hardware constrains H, e.g., the physical number of CPU cores c_{phy} and RAM capacity r_{phy}, and the set of processing services S that is executed there; hence $d = \langle H, S \rangle$.

Processing Services. Each processing service ($s \in S$) is characterized through a service type t, a set of SLOs Q, and a service-internal configuration K; hence $s = \langle t, Q, K, c_s \rangle$, where a service (s) has a number of cores (c_s) allocated to it, for which $c_s \leq c_{phy}$ must hold. Each SLOs $q \in Q$ with $q = \langle v, t \rangle$ tracks a variable v and reflects whether its current assignment reaches a target (t).

Service Monitoring. To track the system state and the SLO fulfillment, the device d and its services $s \in S$ continuously monitor the execution at different levels. This provides a set of software- or application-related metrics (M_s), and a set of device-related metrics (M_d) that capture the hardware state.

SLO Fulfillment. Using these metrics, we calculate the continuous SLO fulfillment (ϕ) for a metrics ($m \in M$) and an SLO q as shown in Eq. (1).

$$\phi(q, m) = \begin{cases} \frac{m}{t_q} & \text{if } m \leq t_q \\ 1.0 & \text{if } m > t_q \end{cases} \tag{1}$$

This means, that SLOs cannot be overfulfilled; for instance, if the target is keeping the service throughput (tp) ≥ 30, both assignments for $m_{tp} = 40$ and $m_{tp} = 100$ achieve the maximum SLO fulfillment of $\phi = 1.0$.

2.2 Related Work

In the following, we identify competing approaches on autoscaling that do not use AIF, and others that use AIF agents.

Most notably, competing approaches and traditional autoscaling methods [17,18] struggle in environments, where multi-dimensional adaptation is crucial for maintaining service quality. In particular, RL-based systems still find no productive use by large providers despite over a decade of research. We argue that AIF-based agents offer a promising alternative by enabling adaptive, efficient, and decentralized control suited to the dynamic and uncertain nature of complex multi-tier distributed systems. Hence, our focus is on further establishing AIF for service adaptation on resource-restricted devices.

Among existing works, Sedlak et al. [14] explore how AIF can be used to optimize SLO fulfillment across various use cases. Danilenka et al. [2] demonstrate how AIF agents can adapt to dynamic and heterogenous environments. Pujol et al. [12] adopt a representation using Partially Observable Markov Decision Processes (POMDPs) within a multi-agent system. Lapkovskis et al. [8] presented a comparison of AIF agents with other methods for a CC application that complements our work with a different set of algorithms. Vyas et al. [16] provide an adaptation mechanism for active sensing on Edge devices, which dynamically adjust perception by changing the camera orientation.

3 Methodology

In this section, we first present the general design of a scaling agent that maximizes the SLO fulfillment for a set of services under limited processing resources. Afterward, we show four approaches of how this agent can be implemented. Namely, we present a traditional RL agent using a Deep Q-Network (DQN), an Bayesian agent using AIF, a Deep Active Inference (DACI) agent, and an algebraic agent based on Analysis of Structural Knowledge (ASK).

3.1 Processing Environment

To evaluate our scaling agents for a real-world problem, we need a processing environment in which to operate. For this, we first introduce the structure of the two processing services and how they are monitored during runtime. Lastly, we introduce our training environment for pretraining some of the agents.

Processing Services. In the following, we introduce the two processing services that will be executed in parallel on an Edge device. By limiting ourselves to two services, we decrease the complexity of the optimization problem, while retaining the central issue of resource limitations.

Computer Vision (CV). The CV service processes a continuous stream of video frames to perform object detection on them. For this, it uses Yolov8 [15], a DNN model developed by Ultralytics; Fig. 1ashowcases its output.

The service state for CV is composed by four features, as displayed in Table 1: *quality* determines the ingested video resolution, *model size* describes the Yolo model (e.g., v8n or v8m), and *cores* determines the maximum resources allocated; *throughput* describes the service output in terms of frames per second.

The action state for CV is a subset of these variables – as part of the service configuration (K), it is possible to adjust *quality* and *model size*, while *cores* (c_s)

is another property of the service. The exception is *throughout*, which cannot be directly set, but is statistically dependent on the three other features.

Table 1. Variables of the CV service used for sensing and acting on the environment

variable	type	range	actionable	dependent	SLO target
quality	int	[128, 320]	✓ (step= 32)	—	≥ 288
model size	int	[1, 5]	✓ (step=1)	—	≥ 3
cores	float	(0, 8)	✓ (no step)	—	—
throughput	int	[0, 100]	—	all others	≥ 5

QR Code Reader (QR). The QR service scans a continuous video stream to detect QR codes within the individual video frames. For this, it uses the Python wrapper of OpenCV [6]; Fig. 1bshowcases the service output.

The service state for QR is composed by three features, as displayed in Table 2. The three features are defined analog to the CV service; the exception is that QR does not use a specific model size but a fixed algorithm.

Table 2. Variables of the QR service used for sensing and acting on the environment

variable	type	range	actionable	dependent	SLO target
quality	int	[300, 1000]	✓ (step= 100)	—	≥ 900
cores	float	(0, 8)	✓ (no step)	—	—
throughput	int	[0, 100]	—	all others	≥ 60

To quantify the requirements and preferences on how the services should operate, both Tables 1 and 2 contain the precise SLO targets. Despite the fact that both services operate on a video stream, they differ greatly in their input shape and the expected *throughput*. Notably, QR has a high expected throughput of 60 frames per second, while the resource-heavy CV has a target of 5.

Service Monitoring. The two processing services operate in batches of 1 s: at the beginning of each iteration, 100 frames are ingested to the service; when the processing timeframe (i.e., 1000 ms) is exceeded, the service counts the number of processed frames – the *throughput*. This information, together with the service properties, is then collected in a time-series DB.

Later, when a scaling agent wants to resolve the service states, it can query this information through the time-series DB. A key advantage of this approach, is that it allows computing sliding-windows over monitored variables; thus, it stabilizes an agent's perception against temporary perturbations, like momentary drops in throughput. Stable states are most relevant for evaluating SLOs and avoiding overhasty scaling decisions, e.g., the Kubernetes HPA[1] considers per default a time window of 30min; in our case, we will stick to 5 s.

Training Environment. Changes to the processing environment need time to show effect. This, however, is conflictive with contemporary ML training, which often requires thousands of iteration to converge to a policy. Hence, sample-efficient methods, like model-based algorithms (also evaluated in this paper) move into focus. However, to not exclude traditional RL algorithms, like DQN, we supply a training environment based on the Gymnasium framework[2].

To estimate the state transition probabilities for state-action pairs, we use a part of the monitored metrics to train a Linear Gaussian Bayesian Network (LGBN); this can be seen as a simplified replication of the real processing environment. For an assignment of free (i.e., actionable) variables, the LGBN samples the remaining variables – in this case *throughput*. Training on the LGBN environment, scaling agents can infer expected state and reward of actions.

3.2 General Agent Design

To optimize SLO fulfillment, we supervise the service execution through a dedicated scaling agent, executed locally on the processing device. Generally, our agent follows a simple two-step scheme in which it first resolve the states of services and hardware, and then acts on the processing environment. To react to dynamic runtime behavior, our scaling agents iterate in cycles of 5 s – thus adhering to the sliding window for service monitoring.

Perception. At the beginning of each iteration, the agent queries the service states through the time-series DB (cfr. Sect. 3.1). Next, it iterates through the services ($s \in S$) and evaluates their SLO fulfillment as in Eq. (1). This indicates the degree to which requirements are currently fulfilled.

To determine the amount of unclaimed resources, the agent then resolves the number of cores (c_s) currently assigned to the individual services[3] and the number of free cores (c_{free}) on the device d as in Eq. (2).

$$c_{free} = c_{phy} - \sum_{s \in S} c_s. \tag{2}$$

[1] https://kubernetes.io/docs/tasks/run-application/horizontal-pod-autoscale/.

[2] https://gymnasium.farama.org/api/env/.

[3] Internally, the services are executed in a containerized environment – Docker; the claimed cores per service can be queried by our agent running on the same machine.

Action. After resolving the environmental state, the agent adjusts the processing environment according to its preferences by changing actionable service variables. However, the exact policy depends on the implementation of the scaling agent and will be explained in the next section. For example, some agents will try solving this by training an accurate world model, e.g., the AIF agent, while the DQN will use simplified state-action networks. Most importantly, this also determines how the different scaling agents explore the solution space.

3.3 Scaling Agent Implementations

In the following, we describe four different implementations of the general scaling agent, using different agent-based approaches. The code for the agents, as well as for the processing services, can be found in the following repository[4].

Active Inference (AIF). The AIF agent is defined following the same idea as in [12], which leverages `pymdp` [7]. Hence, we model the agent as a Markov Decision Process and specifying the transition model using Dynamic Bayesian Networks and Conditional Probability Tables (CPTs). In particular, the influence of actions over certain state factors, such as *throughput*, whose transition dynamics are unknown, is initially defined as a uniform probability distribution within the CPTs[5]. These uniform priors allow the agent to learn the true dynamics through experience and improve SLO fulfillment.

The AIF agent simultaneously controls both services, which increases its state and action spaces. To maintain practical computational times, the agent can choose only one action per service: modifying data *quality*, *model size*, or number of *cores*. This reduces computational demands but also restricts the agent's capacity for more dynamic and fine-grained interventions. Additionally, compared to the other agents in this article, the state space factors for this AIF agent are discretized into coarser bins. While this lowers computational cost, it results in less precise action selection and coarser estimates of SLO fulfillment. The final state space of the agent consists of 7 factors with 3,457,440 state combinations, and an action space of 35 different combinations.

Using this setup, the pymdp implementation (using v0.0.7.1) remained prohibitively slow, requiring approximately 20 s per iteration for policy inference and observation model construction. We address these bottlenecks through two algorithmic optimizations: (1) vectorized policy evaluation eliminates nested loops in expected free energy calculations, and (2) sparse matrix operations reduce overhead in belief updating. These improvements achieved 20-30x speedup in inference routines, leading to the performance describing in Fig. 3b.

Deep Active Inference (DACI). While traditional AIF implementations using graphical models provide excellent interpretability, they are challenging to

[4] Repository with implementations for scaling agents and processing services.
[5] The interested reader can find the matrix definition in the following directory.

scale, limiting their practicality in real-world applications with high-dimensional input spaces. DACI addresses scalability in the CC by exploiting the abundant availability of hardware acceleration (e.g., TPUs, NPUs) for Artificial Neural Networks (ANNs). This work presents preliminary results on Deep Active Inference for the CC based on the work by Fountas et al. [5]. The basic idea is to learn non-linear transforms with ANNs to map the high-dimensional input space into a compressed latent representation of the environment. Then, the agent may operate efficiently in complex, multi-dimensional elasticity spaces that would be intractable for discrete graphical models. Analogous to any active inference agent, it 1) samples the environment and calibrates its internal generative model to explain sensory observations and 2) performs actions to reduce the uncertainty about the environment. We slightly simplify and adjust the original objective in [5] to compute the variational free energy for each time-step t as:

$$
\begin{aligned}
\mathcal{F}'_t = \; & \mathbb{E}_{Q_{\phi_s}(s_t)} \left[\log P_{\theta_o}(o_t|s_t)\right] \\
& + D_{KL}\left[Q_{\phi_s}(s_t) \,\|\, P_{\theta_s}(s_t|s_{t-1}, a_{t-1})\right] \\
& + \lambda_{\text{eq}} \cdot \text{Var}[\text{fulfillment}_{\text{services}}] \\
& + \lambda_{\text{util}} \cdot \left(1 - \frac{\text{ucores}}{\text{tcores}}\right)^2
\end{aligned}
$$

We have omitted the KL divergence involving the habitual network and included regularization terms to reduce the variance between service fulfillments and encourage resource utilization. Our intuition is to regularize the term $-\mathbb{E}[\log P(o_\tau|\pi)]$ that represents preferences about future observations by encoding the specification for balanced services and resource efficiency to have a higher prior probability. Lastly, we encode the objective to prioritize actions that reduce uncertainty about the environment by minimizing the EFE with an expression that is equivalent to that in the original work [5].

Deep Q Networks (DQN). The DQN [10] agent approximates Q-values for state-action pairs in the multi-dimensional autoscaling problem. We implement separate DQN models for each processing service (CV and QR), allowing service-specific policy learning while maintaining coordinated resource allocation when competing for limited resources. The models are trained jointly within the shared processing environment to benefit from the pre-training in the simulated LGBN environment, resulting in the DQN agent achieving stable performance from the initial deployment phase.

Analysis of Structural Knowledge (ASK). This agent solves the orchestration problem through numerical optimization, using SLSQP. For this, it consider the variables and their parameter boundaries from Tables 1 and 2, the objective function ϕ for calculating the SLO fulfillment of a parameter assignment, and the SLOs (Q). Given that, it misses a continuous function that allows descending to the optimal solution. In a nutshell: how to estimate *throughput*, given the assignments of all bounded parameters – including continuous-scale *cores*?

We address this, similarly to the LGBN training environment, through a regression model that captures the dependencies between service variables (K) and the allocated cores (c_k). As the agent captures more metrics, the accuracy of the regression and the inferred parameter assignment improves. To create an accurate model, the ASK agent needs an initial exploration time (i.e., 20 iterations in the experiments), during which the parameters are assigned randomly. After this time, the ASK is ready for the numerical optimization.

For our two service s_1 and s_2 and all their variables $x \in \{K_s \cup c_s\}$, the ASK agent infers assignments that maximize the SLO fulfillment as in Eq. (3).

$$\max_{\mathbf{x}} \sum_{s \in S} \phi_s(x) \quad \text{subject to} \quad \sum_{s \in S} c_s \leq c_{\text{phy}}, \quad x_i \in [x_i^{\min}, x_i^{\max}] \tag{3}$$

This presents a couple of advantages: First, the continuous action space allows agents to infer fine-grained elasticity strategies; in particular *cores* can be split precisely between the devices. Also, it is possible to infer assignments for all services and all their elasticity parameters in one operation.

4 Evaluation

4.1 Experimental Design

To analyze the performance of the different scaling agents we evaluate them, one after the other, in the processing environment (i.e., the two processing services executed in parallel). In each experiment we capture the following performance metrics: (1) SLO fulfillment, as the average over the two services, and (2) complexity of inference, as the duration for running one inference cycle.

Each experiment contains 50 iterations within the processing environment, i.e., 50 times that the agent interacts with the environment by sensing its state and acting on it. As described in Sect. 3.2, the agents operate with a frequency of 5 s – every 5 s they orchestrate the services according to the inferred policy. Thus, individual experiments take $5 \times 50 = 250$ seconds. To stabilize the empirical results, we repeat each experiment 10 times.

During the experiments, agents must achieve the SLO targets specified in Tables 1 and 2. To fulfill these SLOs, an agent needs to optimally adjust the services and split the resources – in this case 8 physical cores (i.e., $c_{phy} = 8$).

4.2 Experimental Results

Figure 3a shows the SLO fulfillment rate of all agents over the 50 steps of the experiment. The ASK agent requires approximately 20 iterations to train and stabilize, reaching an average SLO fulfillment rate of **0.868** over the final 10 steps. In contrast, the DQN agent, having been pre-trained in the simulation environment, maintains stable performance throughout the experiment and achieves an average of **0.753** in the last 10 iterations. Regarding the active inference-based methods, the AIF agent takes around 10 steps to stabilize its performance, ultimately reaching an average fulfillment rate of **0.704** during the final 10 steps. Lastly, although the DACI agent exhibits steady improvement, it only achieves an average SLO fulfillment rate of **0.724** by the end of the experiment.

In terms of computational speed (Fig. 3b) there is a considerable difference between the agents note that the y-axis is plotted in log10 scale. The DQN agent remains the fastest with a mean of **60** milliseconds, benefiting from mature neural network optimizations. Our optimized AIF agent achieves an average time of **1.7** seconds, demonstrating substantial improvement through vectorized inference over the original implementation with roughly **20** seconds. The DACI and ASK agents exhibit execution times of **2.8** seconds and **1.2** seconds respectively, highlighting the computational trade-offs inherent in their respective approaches. When merely exploring (expl) the ASK agent also took less time than during the actual numerical inference (inf). Notably, all agents (incl. the AIF) now operate within practical time constraints for the 5-second scaling interval, enabling real-time performance comparison across all four methods.

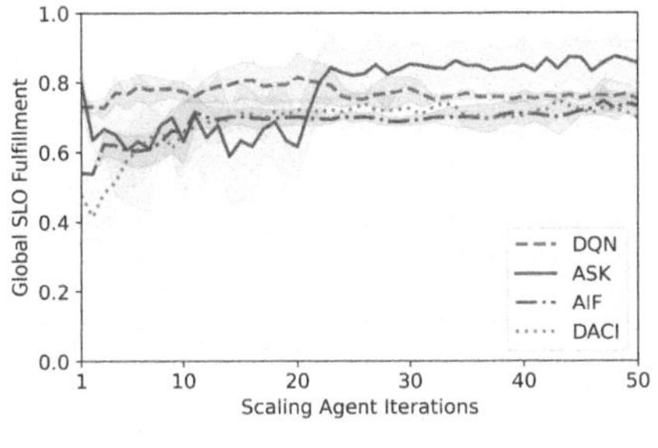

(a) SLO fulfillment of different agents

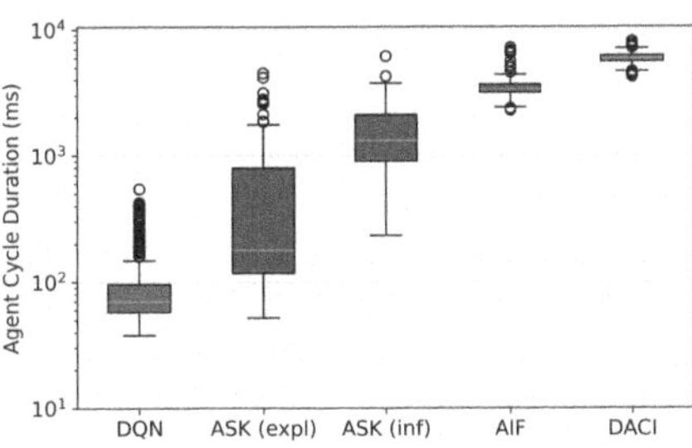

(b) Duration of cycles (log_{10}) in milliseconds

Fig. 3. Scaling agents operating on the processing environment for 50 iterations (=250 s); 10 experiment repetitions per agent type to stabilize results.

5 Conclusion and Future Work

This article highlights a key need for orchestration under resource limitations: multi-dimensional elasticity. By optimizing resource allocation and finding quality trade-offs, it supports flexible scaling behavior that improve Service Level Objectives (SLOs). To show the capabilities of different learning methods for solving such problems, we developed a real-world processing environment where two services compete for resources. We compared four different agents: an Active Inference (AIF) agent using `pymdp`, a Deep Q-Network (DQN), a Deep Active Inference (DACI) agent, and an agent using Analysis of Structural Knowledge

(ASK). Our evaluation shows that ASK achieved the highest SLO fulfillment (0.87), with AIF, DQN and DACI roughly taking the second spot together; AIF and ASK, which learn online, took about 10 and 20 rounds to converge respectively. In terms of execution time, agents must infer a scaling policy in less than 5 s – our orchestration interval. Our optimized AIF agent achieved an average of 1.7 s per iteration, showcasing our improvement of the pymdp library. However, it was the DQN agent that excelled with an average execution under 60 ms. While DACI exhibited the longest execution time, it is arguably the most promising approach as it combines theoretical foundations with practical scalability, which are a necessity in large-scale distributed systems.

We emphasize that we only present early-stage results. Besides further improvements to the neural architecture and the objective, careful theoretical analysis, extensive experimentation, and optimizations that can exploit parallel processing are works in progress. Moreover, future work will refine agents for distributed settings, explore service-hardware interaction, and develop tools for easier configuration and orchestration. We also plan to extend the framework to support a broader set of real-world use cases.

Acknowledgment. This work is partially supported by CNS2023-144359 and the European Union NextGenerationEU/PRTR under MICIU/AEI/10.13039/5011000 11033, and partially by the European Union under (TEADAL, 101070186).

References

1. Casamayor-Pujol, V., Morichetta, A., Murturi, I., Donta, P.K., Dustdar, S.: Fundamental Research challenges for distributed computing continuum systems. Information **14**, 198 (2023). https://doi.org/10.3390/info14030198
2. Danilenka, A., et al.: Adaptive Active Inference Agents for Heterogeneous and Lifelong Federated Learning (2024)
3. Dustdar, S., Guo, Y., Satzger, B., Truong, H.L.: Principles of elastic processes. IEEE Internet Comput. **15**, 66–71 (2011)
4. Dustdar, S., Pujol, V.C., Donta, P.K.: On Distributed Computing Continuum Systems. IEEE Trans. Knowl. Data Eng. **35**(4), 4092–4105 (2023). https://doi.org/10.1109/TKDE.2022.3142856
5. Fountas, Z., Sajid, N., Mediano, P., Friston, K.: Deep active inference agents using monte-carlo methods. Advances in neural information processing systems (2020)
6. Heinisuo, O.P.: opencv-python: Wrapper package for OpenCV python bindings (2021). https://github.com/skvark/opencv-python
7. Heins, C., et al.: pymdp: A Python library for active inference in discrete state spaces. Journal of Open Source Software (2022)
8. Lapkovskis, A., Sedlak, B., Magnússon, S., Dustdar, S., Donta, P.K.: Benchmarking Dynamic SLO Compliance in Distributed Computing Continuum Systems (2025). https://doi.org/10.48550/arXiv.2503.03274
9. Manaouil, K., Lebre, A.: Kubernetes and the Edge? report, Inria Rennes - Bretagne Atlantique (2020). https://inria.hal.science/hal-02972686
10. Mnih, V., Kavukcuoglu, K., Silver, D., Graves, A., Antonoglou, I., Wierstra, D., Riedmiller, M.: Playing atari with deep reinforcement learning (2013)

11. opencv: opencv at 4.9.0 (2024). https://github.com/opencv/opencv/tree/4.9.0
12. Pujol, V.C., Sedlak, B., Salvatori, T., Friston, K., Dustdar, S.: Distributed Intelligence in the Computing Continuum with Active Inference (2025). https://doi.org/10.48550/arXiv.2505.24618
13. Sedlak, B., Morichetta, A., Raith, P., Pujol, V.C., Dustdar, S.: Towards Multi-dimensional Elasticity for Pervasive Stream Processing Services. In: 2025 IEEE PerCom Workshops (2025)
14. Sedlak, B., Pujol, V.C., Donta, P.K., Dustdar, S.: Equilibrium in the Computing Continuum through Active Inference. Future Generation Comput. Syst. **160**, 92–108 (2024). https://doi.org/10.1016/j.future.2024.05.056
15. Varghese, R., M., S.: YOLOv8: A Novel Object Detection Algorithm with Enhanced Performance and Robustness. In: ADICS (2024)
16. Vyas, D., Prado, M.d., Verbelen, T.: Towards smart and adaptive agents for active sensing on edge devices (Jan 2025). https://doi.org/10.48550/arXiv.2501.06262
17. Wang, Z., Li, P., Liang, C.J.M., Wu, F., Yan, F.Y.: Autothrottle: A Practical Bi-Level Approach to Resource Management for SLO-Targeted Microservices (2024)
18. Zhao, Y., Uta, A.: Tiny Autoscalers for Tiny Workloads: Dynamic CPU Allocation for Serverless Functions. Society, IEEE Comp (2022)

Predictive Coding-Based Deep Neural Network Fine-Tuning for Computationally Efficient Domain Adaptation

Matteo Cardoni(✉) and Sam Leroux

IDLab, Department of Information and Technology, Ghent University—imec, 9052 Ghent, Belgium
{matteo.cardoni,sam.leroux}@ugent.be

Abstract. As deep neural networks are increasingly deployed in dynamic, real-world environments, relying on a single static model is often insufficient. Changes in input data distributions caused by sensor drift or lighting variations necessitate continual model adaptation. In this paper, we propose a hybrid training methodology that enables efficient on-device domain adaptation by combining the strengths of Backpropagation and Predictive Coding. The method begins with a deep neural network trained offline using Backpropagation to achieve high initial performance. Subsequently, Predictive Coding is employed for online adaptation, allowing the model to recover accuracy lost due to shifts in the input data distribution. This approach leverages the robustness of Backpropagation for initial representation learning and the computational efficiency of Predictive Coding for continual learning, making it particularly well-suited for resource-constrained edge devices or future neuromorphic accelerators. Experimental results on the MNIST and CIFAR-10 datasets demonstrate that this hybrid strategy enables effective adaptation with a reduced computational overhead, offering a promising solution for maintaining model performance in dynamic environments.

Keywords: Predictive Coding · on-device learning · domain adaptation

1 Introduction

The Backpropagation (BP) algorithm [1] is currently the predominant approach for training deep neural network architectures. It is, however, not without its limitations. BP requires the computation and transmission of global error signals which can be computationally intensive, apart from being biologically implausible. This has sparked interest in alternative training techniques such as those based on Predictive Coding (PC) [2–4]. While most of the works in this area are grounded in theoretical neuroscience, PC is also promising from an engineering point of view as it provides a much more efficient hardware-friendly training

M. Albarracin et al. (Eds.): IWAI 2025, CCIS 2857, pp. 312–327, 2026.
https://doi.org/10.1007/978-3-032-16955-6_18

procedure for neuromorphic systems. Unlike BP, PC relies on local computations and error-driven updates that align well with the distributed, event-driven nature of neuromorphic architectures. This makes it a compelling candidate for developing scalable, low-power learning algorithms suitable for real-time applications on energy-constrained edge devices [5].

Despite the progress made in recent years, models trained using PC still fall short of the accuracy levels achieved by those trained with BP, particularly when applied to deeper architectures and more complex datasets [6,7]. In this paper, we propose a novel hybrid approach that combines the strengths of both algorithms. As illustrated in Fig. 1, we introduce a two-stage training and deployment pipeline. In the first stage, a model is trained offline using BP on high-performance cloud infrastructure with virtually unlimited computational resources. This allows us to reach an accuracy level that would be currently impossible to achieve using PC alone.

The trained model is then deployed on a resource-constrained edge device such as a wearable or sensor node. Over time, shifts in the input data distribution due to sensor drift, environmental changes, or user behavior may degrade model performance, necessitating continual adaptation. To address this, we propose using PC as a lightweight, local learning mechanism that allows efficient on-device updates, without the need for expensive BP or communication with the cloud.

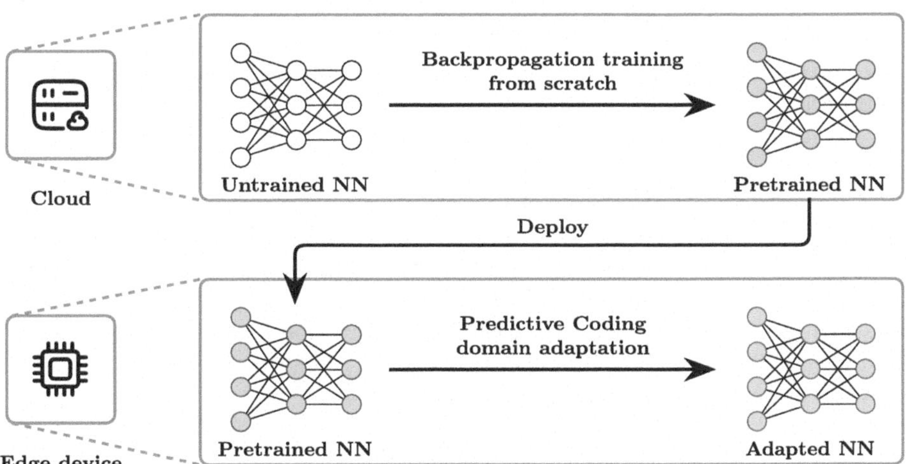

Fig. 1. Schematic overview of our proposed approach. After training a model offline using Backpropagation (BP), we employ Predictive Coding (PC) for on-device domain adaptation.

The remainder of this paper is structured as follows: we provide an overview of the related work in Sect. 2 and describe our experimental setup and results in Sect. 3. We conclude in Sect. 4 and give some pointers for future research directions in this area.

2 Related Work

This section provides a basic overview of Backpropagation and Predictive Coding for neural network training. We refer to [8] for a more in-depth tutorial.

2.1 Neural Networks

A feedforward neural network (FNN) consists of a stack of L layers, indexed as $\{l_0, l_1, \ldots, l_L\}$, with each layer defined by its weight matrix θ_l. Each layer l computes an activity a_l transforming its input activity a_{l-1} as follows:

$$a_l = f(a_{l-1}, \theta_l) \tag{1}$$

This transformation depends on the type of layer (e.g. fully connected, convolutional, ...) and typically includes a non-linear activity. Setting the activation a_0 to a datapoint x, the neural network performs a function NN: $X \to Y$:

$$\text{NN}(x) = (l_L \circ l_{L-1} \circ l_{L-2} \circ \ldots \circ l_1)(x) \tag{2}$$

The weights of the model θ are initialized randomly and tuned using a training procedure to achieve the desired behavior.

2.2 Backpropagation

Backpropagation (BP) is the most widely used technique for training neural networks. It aims to minimize a global loss function $\mathcal{L}$ that quantifies the discrepancy between the network's output and the desired target output. Common loss functions include Mean Squared Error (MSE) for regression tasks and Cross-entropy loss for classification tasks. The BP algorithm is able to calculate a suitable set of weights $\hat{\theta}$ numerically using gradient descent:

$$\hat{\theta} = \underset{\theta}{\text{argmin}}\, \mathcal{L}(\theta) \tag{3}$$

In BP, the error of each layer is the gradient of the loss with respect to the activation of that layer. Because this depends on the gradients of subsequent layers, this backward computation must proceed sequentially from the output layer to the input layer. As a result, the gradient computation across layers is inherently sequential.

2.3 Predictive Coding

Predictive Coding (PC) training for deep neural network architectures differs fundamentally from Backpropagation (BP) in that it only uses local information to perform weight updates. During training, an input sample x is clamped to the first layer activity (a_0) and the last layer (a_L) is clamped to the desired output y. We then repeat a two-step training procedure:

1. **Inference step:** In this step, the weights θ_l are held fixed. Each layer predicts the activity of the next layer:

   $$\mu_l = f(a_{l-1}, \theta_l) \tag{4}$$

   The predictions μ_L are then used to calculate a layer-wise prediction error that measures the discrepancy between the actual and predicted activity:

   $$\varepsilon_l = a_l - \mu_l \tag{5}$$

   Unlike during BP-based training, the actual activity of the layers is not obtained by passing the input through the model, as in Eq. 1. It is, instead, obtained by minimizing the energy function, which is the sum of the layer-wise prediction errors:

   $$E(a, \theta) = \frac{1}{2} \sum_l (\varepsilon_l)^2 \tag{6}$$

   The updated values of the activities are obtained as the minimum of the energy function:

   $$\hat{a}_l = \underset{a_l}{\operatorname{argmin}}\, E(a, \theta) \tag{7}$$

   The $\hat{a}_l$ are typically obtained via local gradient descent of Eq. 6, calculating the update of a_l as:

   $$\Delta a_l = -\gamma \frac{\partial E}{\partial a_l} = -\gamma \Big(\varepsilon_l - (\theta_l)^T \varepsilon_{l+1} \odot f'(\theta_l a_l) \Big), \tag{8}$$

   with γ being the *inference rate*: a step size required by the gradient descent. This step infers the suitable activity values for the different layers, given the clamped input and target values and is typically performed by running gradient descent for a fixed number of iterations.
   It is to be noted that the dependency of Δa_l from ε_l and ε_{l+1} allows ε_L, the error between the classification layer activity a_L and the label y, to be propagated from the output layer L to the input. Moreover, this computation can be layer-wise parallelized.
2. **Weight update step:** The activities, after being updated, are now held fixed, and the weights are updated to further reduce the energy function from Eq. 6. The updated weights values are obtained as the minimum of the energy function:

   $$\hat{\theta}_l = \underset{\theta_l}{\operatorname{argmin}}\, E(\hat{a}, \theta) \tag{9}$$

The updates of θ_l are obtained via local gradient descent:

$$\Delta\theta_l = -\alpha\frac{\partial E}{\partial\theta_l} = \alpha\varepsilon_{l+1} \odot f'(\theta_l\hat{a}_l)(\hat{a}_l)^T, \tag{10}$$

with α being the weights learning rate. This computation too can be layer-wise parallelized.

2.4 Backpropagation-Predictive Coding Equivalence at Test Time

At test time, the activity of each layer in the Predictive Coding Network model, including the last layer, can be obtained by clamping the lowest layer to the input data and by computing updates of the activity by repeating the inference procedure. However, with the output unclamped, this can also be computed in a single pass by simply performing the forward pass of a normal neural network [8]. This observation that models trained using Backpropagation and Predictive Coding are equivalent at test time is what motivated this work as it allows switching between both training algorithms while maintaining a compatible model.

2.5 Applications of Predictive Coding

Predictive Coding has seen growing application in recent years across a range of increasingly complex tasks. Initial studies focused on classification, associative memory, and denoising using Multi-Layer Perceptron (MLP) architectures [7,9,10]. Building on these foundations, more recent work has extended PC to convolutional neural networks (CNNs) for higher-level visual tasks. For example, [11,12] introduced recurrent variants of PC networks that achieved classification performance comparable to Backpropagation on the CIFAR-10 dataset, and promising, though less competitive, results on ImageNet. Furthermore, [7] systematically benchmarked PC training in both MLPs and CNNs for classification and associative memory tasks. This work also introduced a JAX-based implementation framework [13], which forms the basis for the PC models used in our work.

2.6 Domain Adaptation and on-Device Learning

Classical artificial intelligence approaches typically assume that the training and test data are drawn from the same underlying distribution. However, this assumption often breaks down in real-world applications, particularly when deploying pretrained neural networks on devices that process data from dynamic, non-stationary environments [14,15]. Domain adaptation seeks to address this mismatch by adapting models trained on a source domain to perform well on a target domain with a different marginal data distribution.

Domain adaptation is particularly important in scenarios where a model is deployed on a physical device operating in a specific environment. These edge devices typically have limited computational and energy resources, which preclude the use of large, resource-intensive models. Instead, a lightweight, location-specific model, adapted to the unique characteristics of the local sensor or environment, can achieve performance comparable to that of a larger, general-purpose model, while being significantly more energy-efficient and responsive [16]. While it is possible to perform the domain adaptation remotely, on cloud infrastructure, a full on-device approach is favorable in terms of privacy, communication overhead and system cost [17]. Novel algorithms that can perform this training more efficiently are therefore of high value for advancing adaptive intelligence at the edge. To the best of our knowledge, this work is the first to consider Predictive Coding for on-device domain adaptation.

3 Experiments and Results

3.1 Experimental Setup

We report results on the well-known MNIST [18] and CIFAR-10 [19] image classification benchmarks. Following the setup in [7], we use a three-layer fully connected neural network for MNIST, and a series of convolutional architectures inspired by the VGG family [20] for CIFAR-10. Full model specifications and training hyperparameters are provided in the Appendix.

All experiments were implemented using JAX [13]. Models trained with Backpropagation (BP), along with their training procedures, were implemented in Equinox [21]. Instead, models trained with Predictive Coding (PC), along with their training procedures, were implemented in the PCX library [7], which generalizes the gradients computation from Eqs. 8 and 10 by means of the JAX automatic differentiation.

Each model was initially trained from scratch using BP on the original dataset (the details of these trainings are reported in the Appendix). To simulate domain shifts, the training and test images were then transformed in three different ways, for three different scenarios:

- Inversion (180 degrees rotation).
- Clockwise rotation of 20 degrees.
- Addition of random uniform noise in the range of [-0.05, 0.05] per channel (while the pixel values are in the range [0, 1]).

We evaluated the performance of both BP and PC in adapting the pretrained models to this transformed data, assessing their ability to recover lost accuracy under this input perturbation. The domain adaptation took place retraining on the entire training dataset after applying the corresponding transformation. Figure 2 visualizes the procedure, for the case where a VGG5 model (see Appendix for details) is used for training and domain adaptation on the CIFAR-10 dataset. All experiments were conducted on a cloud infrastructure.

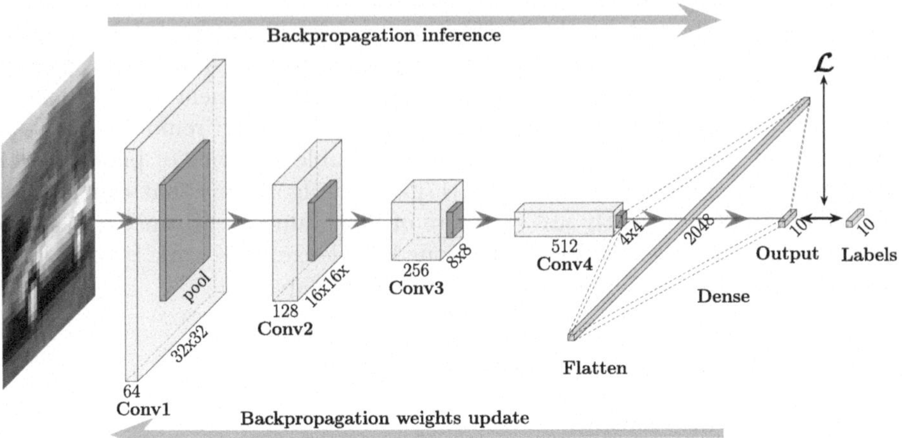

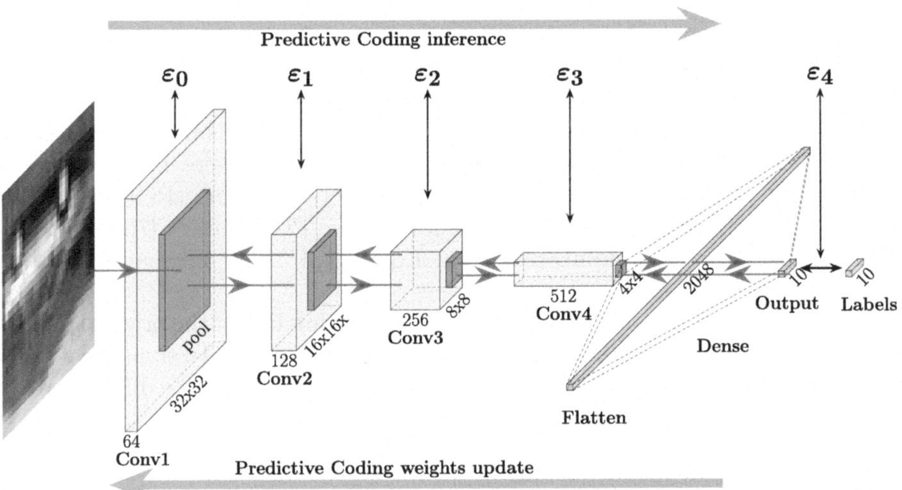

Fig. 2. Experiment workflow, including the initial training of a VGG5 (see Appendix for the architecture detail) for the CIFAR-10 dataset in its original version. The same Neural Network is subsequently retrained with a domain-shifted version of the dataset. In this case, an image inversion (180 degrees rotation) has been applied to all the images. Opposite to the Backpropagation (BP), Predictive Coding (PC) training is based on layer-wise errors ε_l, that allows to compute local gradient of the weights, as in Eq. 10

3.2 Results

Table 1 reports the average training time per epoch for both Backpropagation (BP) and Predictive Coding (PC), measured on a GeForce GTX 1080 Ti GPU.

To obtain an accurate comparison, only the training time, after data loading, has been measured. Since the data transformations simulate domain shifts in data sampled by a sensor, also the data transformation has been excluded from the measurement time.

The results show that for the convolutional models, PC achieves a substantial speed advantage, with an average training time that is only 52% of that required by BP. In contrast, for the MLP model, the PC training times are on average 56% higher, likely due to the simplicity of the architecture, which limits the potential gains in efficiency from PC.

Table 1. Wall-clock time per epoch (in seconds) for both Backpropagation (BP) and Predictive Coding (PC), averaged over 10 epochs for the MLP model and over 100 epochs for the VGG models. Values are reported as mean ± standard deviation. For the CNN-based architectures, PC is on average twice as fast.

Dataset	Model	BP	PC
MNIST	MLP	1.88±0.29	2.93±3.17
CIFAR10	VGG5	9.48±0.28	4.34±1.48
	VGG7	10.48±0.88	5.16±2.06
	VGG9	8.48±0.58	5.43±1.79

Although these results demonstrate that PC can reach a higher training throughput compared to BP, this does not necessarily mean that it results in faster training (i.e. performance increase over time), since accuracy gains per epoch vary across models and under different domain shifts. To investigate this, Fig. 3 shows the models test accuracy on the domain shifted input data as a function of the training time. We compare our PC-based domain adaptation approach with a BP-based approach. As a baseline, we also include the performance for a PC and a BP model, trained from scratch on the domain-shifted data. All training details can be found in the Appendix.

The graphs show both the mean accuracies (middle lines) and standard deviations (color-filled areas) resulting from 5 different executions. In some of the plots the standard deviation is hardly noticeable, due to the small variance over different runs.

For the inverted data, the BP training from scratch baseline consistently obtains the highest accuracy. For the VGG5 model, our proposed PC-based domain adaptation achieves a similar accuracy (87.7% ± 0.2% compared to 88.2% ± 0.2%). The VGG5 model almost reaches the accuracy obtained by the BP from scratch training, but with a much faster training, outperforming the BP-based adaptation (87.2% ± 0.3%) and PC-based training from scratch (86.5% ± 0.1%). These values correspond to the peak accuracy observed within 350 s of training. The two other domain shifts proved to be more challenging, resulting in lower accuracies overall. Here, all models benefit from pretraining combined with domain adaptation. For the VGG5 model, PC-based domain adaptation matches

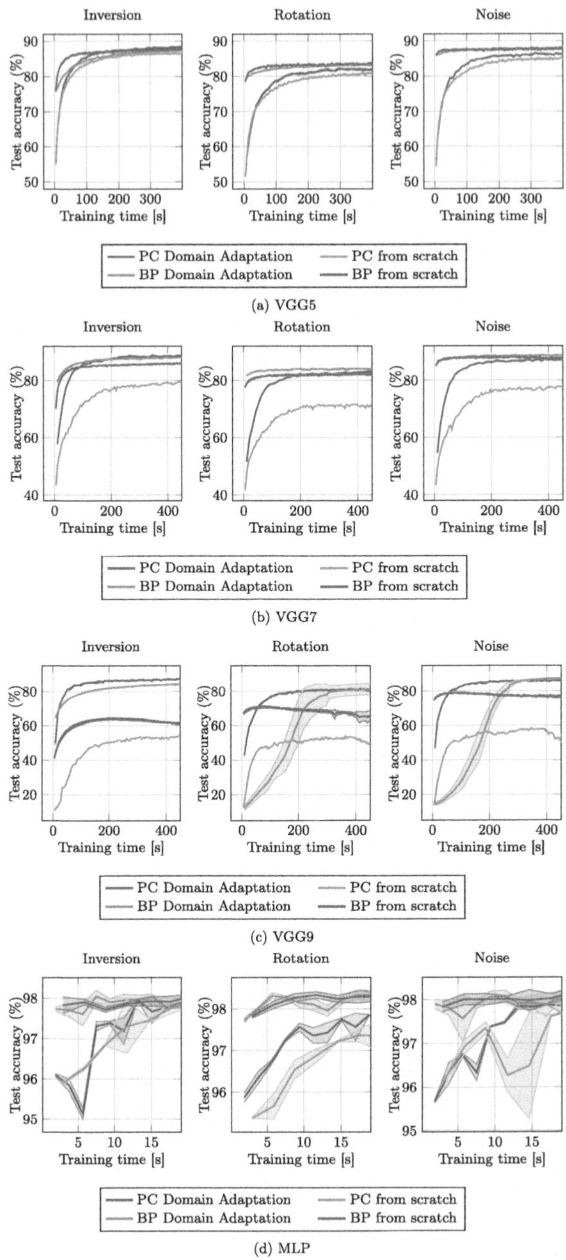

Fig. 3. Model Accuracy as a function of training time for models trained from scratch and domain-adapted using Backpropagation (BP) and Predictive Coding (PC). Trainings were performed, for each model, respectively with inverted, rotated and noisy data (Sect. 3.1). The plots highlight the validity brought by the proposed technique, especially for the VGG5 model. The transparent areas represent the accuracy standard deviation over five runs.

or surpasses the accuracy obtained with BP domain adaptation for rotated and noisy data, while it obtains a slightly lower accuracy on the VGG7 model.

The models trained from scratch using PC obtain significantly lower accuracies compared to their counterparts trained using BP.

The performance of PC-based algorithms decreases as model depth increases, as is evident for the deeper VGG9 model. This aligns with previous observations that PC struggles when training deep architectures [6,7], while it can outperform BP for more shallow architectures [7]. The BP-based domain adaptation for VGG9 delivers lower performances with respect to the BP-based domain adaptation of the other models, highlighting that domain adaptation may in general be more difficult for deeper models, regardless of the training technique. For the easier MNIST dataset with a simple three layer MLP model, both the PC and BP domain adaptation prove to be beneficial, providing equally accurate models. Both approaches outperform training a model from scratch.

4 Conclusion and Future Work

Our experiments demonstrate that Predictive Coding (PC) is an efficient algorithm for on-device domain adaptation of models initially pretrained with Backpropagation (BP). We showed that models trained with BP can effectively leverage PC-based finetuning on a domain-shifted variant of the same dataset—in our case, MNIST and CIFAR-10. This opens up new avenues of research into computationally efficient domain adaptation techniques using PC. This could prove especially efficient when implemented on neuromorphic hardware [22].

Future work will extend this study by evaluating training times on embedded platforms, including neuromorphic hardware. We also aim to scale our approach to deeper neural network architectures. While the domain adaptation procedure was completely supervised in this work, this might not be feasible in real-world environments. More research into unsupervised or self-supervised approaches in combination with PC is therefore a highly interesting research direction.

Acknowledgments. This research was supported by funding from the Flemish Government under the "Onderzoeksprogramma Artificiële Intelligentie (AI) Vlaanderen" program.

Appendix

Table 2 reports the MLP architecture that was used in this study, while Table 3 describes the VGG-inspired architectures [20]. Both the MLP and VGG architectures follow the setup from [7].

Table 2. Summary of the MLP model for MNIST classification

MLP
Dense (784, 512) Leaky ReLU
Dense (512, 512) Leaky ReLU
Dense (512, 10)

Table 3. Summary of The VGG5, VGG7, and VGG9 models, inspired from [20]. The notation Conv$< n >$_$< m >$ stands for a 2D convolutional layer with a kernel of size $n \times n$ and m channels. All Convolutional layers are followed by a GELU non-linear activation. All Max Pooling layers have a size of 2×2 and a stride of 2×2.

VGG5	VGG7	VGG9
Conv3-128 (Pad (1, 1)) Max Pool	Conv3-128 (Pad (1, 1)) Max Pool	Conv3-128 (Pad (1, 1)) Max Pool
Conv3-256 (Pad (1, 1)) Max Pool	Conv3-128 (Pad (1, 1))	Conv3-128 (Pad (1, 1)) Max Pool
Conv3-512 (Pad (1, 1)) Max Pool	Conv3-256 (Pad (1, 1))	Conv3-256 (Pad (1, 1)) Max Pool
Conv3-512 (Pad (1, 1)) Max Pool	Conv3-256	Conv3-256 (Pad (1, 1)) Max Pool
–	Conv3-512 (Pad (1, 1)) Max Pool	Conv3-512 (Pad (1, 1))
–	Conv3-512	Conv3-512 (Pad (1, 1)) Max Pool
–	–	Conv3-512 (Pad (1, 1))
–	–	Conv3-512 (Pad (1, 1))
Dense(2048, 10)	Dense(512,10)	Dense(512, 10)

Tables 4 a, 4 b, and 4 c describe the hyperparameters that were used to obtain the results from Fig. 3.

Using the design space search from [7] as a starting point:

- Both Squared Error (SE) and Cross Entropy (CE) were used as loss function, also in Predictive Coding(PC)-based training. In fact, the energy function can be adapted, as explained in [23].
- AdamW (Adam with weight decay regularization) was used as weights optimizer.
- SGD with momentum was used as predictions optimizer (for the PC-based trainings).
- All the weights learning rate was scheduled with a linear warmup followed by cosine decay.

However, unlike in [7], the number of PC inference steps (Subsect. 2.3) was kept at 4 for all CNN models. This choice was taken to obtain a training time per epoch that is, for each model, approximately half the time per epoch of the Backpropagation (BP) training counterpart. All PC trainings were performed with Forward Initialization, which is widely used to accelerate the predictions optimization [6]. This initialization technique implies that for each training batch a first forward pass (BP-like) is performed, as in Eq. 1. After the forward pass, the predictions μ_l for are initialized as the activities a_l, for each layer apart from the last one, for which the labels y are used. After this step, the training takes place as in Subsect. 2.3.
Fig. 4 displays the test accuracy per training time, obtained with the models from Table 2 and Table 3 respectively, training them with BP on original data (i.e. not subject to domain shift). The weights obtained from this models were used as starting point for both the PC and BP-based domain adaptation techniques.

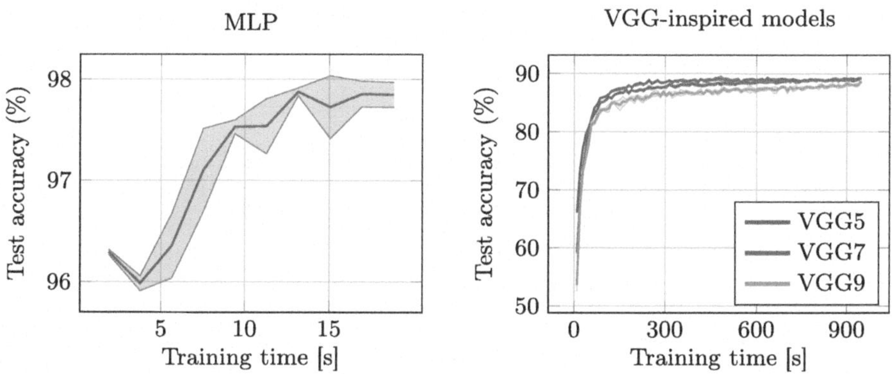

Fig. 4. Test accuracy per training time of the MLP (Table 2) and the VGG-inspired models (Table 3), training with Backpropagation (BP) on original data.

Table 5 reports the hyperparameters that were used to train the models with BP on original data (Fig. 4).
Table 6 describes the CNNs training hyperparameters in which the design space exploration took place.
For the PC-based trainings of VGG9, the hyperparameter design space exploration has been focused on higher momentums for the SGD optimizer, instead of focusing on high inference rates. This decision has been taken to explore hyperparameters combinations that would favor the training stability, as VGG9 has been proven to be harder to train with PC.

In order to avoid abrupt changes in the test accuracy for the PC-based trainings, the training hyperparameters choice has been constrained to the ones for which, during the design space exploration, the following properties held:

- After the 5th epoch, there was never a test accuracy decrease of no more than 5% points between one epoch and the next.

Table 4. Models training hyperparameters corresponding to each specific training technique: training from scratch and domain adaptation performed both with Back-propagation (BP) and Predictive Coding (PC). The training was performed on data subject to different domain shifts: inversion, rotation, and noise addition (see Subsect. 3.1 for details). AdamW was used as weights optimizer, while SGD with momentum was used as predictions optimizer.

(a) Inverted images

	Model	Loss	θ_{lr}*	Weight decay	γ†	m‡	Inference steps
BP from scratch	VGG5	SE	$2.5e^{-4}$	$3e^{-4}$	-	-	-
	VGG7	SE	$2.5e^{-4}$	$2e^{-4}$			
	VGG9	CE	$5e^{-4}$	$3e^{-4}$			
	MLP	CE	$1e^{-3}$	$1e^{-4}$			
BP domain adaptation	VGG5	SE	$1e^{-3}$	$2e^{-4}$	-	-	-
	VGG7	CE	$1e^{-3}$	$2e^{-4}$			
	VGG9	CE	$1e^{-3}$	$3e^{-4}$			
	MLP	CE	$1e^{-3}$	$1e^{-4}$			
PC from scratch	VGG5	SE	$1e^{-4}$	$1e^{-4}$	$2.5e^{-2}$	0.1	4
	VGG7	CE	$1e^{-4}$	$1e^{-4}$	$2.5e^{-2}$	0.1	4
	VGG9	SE	$1e^{-3}$	$1e^{-4}$	$1e^{-4}$	0.1	4
	MLP	CE	$1e^{-3}$	$1e^{-4}$	$1e^{-3}$	0.1	13
PC domain adaptation	VGG5	CE	$1e^{-4}$	$1e^{-4}$	$1e^{-2}$	0.1	4
	VGG7	CE	$1e^{-4}$	$1e^{-4}$	$1e^{-2}$	0.5	4
	VGG9	CE	$1e^{-4}$	$1e^{-4}$	$1e^{-3}$	0.1	4
	MLP	CE	$1e^{-3}$	$1e^{-4}$	$1e^{-3}$	0.1	13

(b) Rotated images

	Model	Loss	θ_{lr}*	Weight decay	γ†	m‡	Inference steps
BP from scratch	VGG5	SE	$2.5e^{-4}$	$2e^{-4}$	-	-	-
	VGG7	SE	$2.5e^{-4}$	$3e^{-4}$			
	VGG9	SE	$5e^{-4}$	$2e^{-4}$			
	MLP	CE	$1e^{-3}$	$1e^{-4}$			
BP domain adaptation	VGG5	SE	$5e^{-4}$	$1e^{-4}$	-	-	-
	VGG7	CE	$1e^{-3}$	$1e^{-4}$			
	VGG9	SE	$1e^{-3}$	$3e^{-4}$			
	MLP	CE	$1e^{-3}$	$1e^{-4}$			
PC from scratch	VGG5	SE	$1e^{-4}$	$1e^{-4}$	$2.5e^{-2}$	0.5	4
	VGG7	CE	$1e^{-4}$	$1e^{-4}$	$2.5e^{-2}$	0.1	4
	VGG9	SE	$1e^{-4}$	$1e^{-4}$	$1e^{-2}$	0.1	4
	MLP	CE	$1e^{-3}$	$1e^{-4}$	$1e^{-3}$	0.1	13
PC domain adaptation	VGG5	CE	$1e^{-4}$	$1e^{-4}$	$1e^{-2}$	0.1	4
	VGG7	CE	$1e^{-4}$	$1e^{-4}$	$1e^{-2}$	0.1	4
	VGG9	CE	$1e^{-4}$	$1e^{-4}$	$1e^{-2}$	0.1	4
	MLP	CE	$1e^{-3}$	$1e^{-4}$	$1e^{-3}$	0.1	13

(c) Noisy images

	Model	Loss	θ_{lr}*	Weight decay	γ†	m‡	Inference steps
BP from scratch	VGG5	SE	$2.5e^{-4}$	$1e^{-4}$	-	-	-
	VGG7	SE	$2.5e^{-4}$	$1e^{-4}$			
	VGG9	SE	$2.5e^{-4}$	$1e^{-4}$			
	MLP	CE	$1e^{-3}$	$1e^{-4}$			
BP domain adaptation	VGG5	CE	$2.5e^{-4}$	$3e^{-4}$	-	-	-
	VGG7	CE	$1e^{-4}$	$2e^{-4}$			
	VGG9	SE	$5e^{-4}$	$1e^{-4}$			
	MLP	CE	$1e^{-3}$	$1e^{-4}$			
PC from scratch	VGG5	SE	$1e^{-4}$	$1e^{-4}$	$2.5e^{-2}$	0.5	4
	VGG7	CE	$1e^{-4}$	$1e^{-4}$	$2.5e^{-2}$	0.1	4
	VGG9	SE	$1e^{-4}$	$1e^{-4}$	$1e^{-2}$	0.1	4
	MLP	CE	$1e^{-3}$	$1e^{-4}$	$1e^{-3}$	0.1	13
PC domain adaptation	VGG5	CE	$1e^{-4}$	$1e^{-4}$	$1e^{-2}$	0.5	4
	VGG7	CE	$1e^{-4}$	$1e^{-4}$	$2.5e^{-2}$	0.5	4
	VGG9	CE	$1e^{-4}$	$1e^{-4}$	$1e^{-2}$	0.1	4
	MLP	CE	$1e^{-3}$	$1e^{-4}$	$1e^{-3}$	0.1	13

* Weights learning rate.
† Inference rate.
‡ SGD optimizer momentum.

Table 5. Models Backpropagation (BP) training hyperparameters, used for training from scratch on original data. AdamW was used as weights optimizer.

Model	Loss	$\theta_{\mathbf{lr}}{}^*$	Weight decay
VGG5	SE	$2.5e^{-4}$	$3e^{-4}$
VGG7	SE	$2.5e^{-4}$	$2e^{-4}$
VGG9	CE	$5e^{-4}$	$3e^{-4}$
MLP	CE	$1e^{-3}$	$1e^{-4}$

Table 6. Training hyperparameters used for the search of the best model for the Backpropagation (BP) and the Predictive Coding (PC) training.

		BP	**PC**		
		VGG5-7-9	**VGG5**	**VGG7**	**VGG9**
	Loss	[SE, CE]	[SE, CE]	[SE, CE]	[SE, CE]
Weights update	θ^*_{lr}	[$1e^{-4}$, $2.5e^{-4}$, $5e^{-4}$, $1e^{-3\S}$]	[$1e^{-4}$, $1e^{-3}$]	[$1e^{-4}$, $1e^{-3}$]	[$1e^{-4}$, $1e^{-3}$]
	Weight decay	[1, 2, 3]e^{-4}	$1e^{-4}$	$1e^{-4}$	$1e^{-4}$
Predictions update	$\gamma^\dagger$	-	[$1e^{-5}$, $1e^{-4}$, $1e^{-3}$, $1e^{-2}$, $2.5e^{-2}$, $5e^{-2}$]	[$1e^{-5}$, $1e^{-4}$, $1e^{-3}$, $1e^{-2}$, $2.5e^{-2}$, $5e^{-2}$]	[$1e^{-5}$, $1e^{-4}$, $1e^{-3}$, $1e^{-2}$]
	$m^\ddagger$	-	[0.1, 0.5]	[0.1, 0.5]	[0.1, 0.5, 0.9]

* Weights learning rate.
† Inference rate.
‡ SGD optimizer momentum.
§ A weights learning rate (θ_{lr}) of $1e^{-3}$ was used only for BP domain adaptation.

- The final epoch accuracy was no lower than 5 points with respect to the maximum one.

Regarding the BP-based domain adaptation, a weights learning rate (θ_{lr}) of $1e^{-3}$ has been used only for this training technique, since lower θ_{lr} values did not always provide a satisfactory test accuracy.

Finally, Table 7 summarizes the epochs, batch size, normalization values and data augmentation that were used for training from scratch and domain adapting, both using BP and PC.

Table 7. Summary of training epochs, batches, normalization values and data augmentations for the datasets used in the experiments.

	MNIST	CIFAR10
Training Epochs	10	100
Batch size	128	128
Normalization mean	–	[0.4914, 0.4822, 0.4465]
Normalization standard deviation	–	[0.2023, 0.1994, 0.2010]
Data augmentation	–	(4, 4) zero pad + random cropping 32x32, random (50%) horizontal flip

References

1. Rumelhart, D.E., Hinton, G.E., Williams, R.J.: Learning representations by back-propagating errors. nature **323**(6088), 533–536 (1986)
2. Rao, R., Ballard, D.: Predictive coding in the visual cortex: a functional interpretation of some extra-classical receptive-field effects. Nat. Neurosci. **2**, 79–87 (1999)
3. Friston, K.: A theory of cortical responses. Philosophical transactions of the Royal Society of London. Series B, Biological sciences, **360**, 815–36 (2005)
4. Friston, K., Kiebel, S.: Predictive coding under the free-energy principle. Philosophical transactions of the Royal Society of London. Series B, Biological sciences, **364**, 1211–21 (2009)
5. Millidge, B., Salvatori, T., Song, Y., Bogacz, R., Lukasiewicz, T.: Predictive coding: towards a future of deep learning beyond backpropagation? (2022)
6. Goemaere, C., Oliviers, G., Bogacz, R., Demeester, T.: Error optimization: overcoming exponential signal decay in deep predictive coding networks (2025)
7. Pinchetti, L., et al.: Benchmarking predictive coding networks – made simple (2025)
8. van Zwol, B., Jefferson, R., Broek, E.L.: Predictive coding networks and inference learning: tutorial and survey (2024)
9. Whittington, J.C., Bogacz, R.: An approximation of the error backpropagation algorithm in a predictive coding network with local hebbian synaptic plasticity. Neural Comput. **29**(5), 1229–1262 (2017)
10. Sun, W., Orchard, J.: A predictive-coding network that is both discriminative and generative. Neural Comput., **32**, 1836–1862 (2020)
11. Wen, H., Han, K., Shi, J., Zhang, Y., Culurciello, E., Liu, Z.: Deep predictive coding network for object recognition (2018)
12. Han, K., Wen, H., Zhang, Y., Fu, D., Culurciello, E., Liu, Z.: Deep predictive coding network with local recurrent processing for object recognition (2018)
13. Bradbury, J., et al.: JAX: composable transformations of Python+NumPy programs (2018)

14. Farahani, A., Voghoei, S., Rasheed, K., Arabnia, H.R.: A brief review of domain adaptation. In: Stahlbock, R., et al. (eds.) Advances in Data Science and Information Engineering, pp. 877–894. Springer International Publishing, Cham (2021)
15. Sun, S., Shi, H., Wu, Y.: A survey of multi-source domain adaptation. Inf. Fusion **24**, 84–92 (2015)
16. Leroux, S., Li, B., Simoens, P.: Automated training of location-specific edge models for traffic counting. Comput. Electrical Eng. **99**, 107763 (2022)
17. Zhao, T., et al.: A survey of deep learning on mobile devices: applications, optimizations, challenges, and research opportunities. Proc. IEEE **110**(3), 334–354 (2022)
18. Deng, L.: The mnist database of handwritten digit images for machine learning research [best of the web]. IEEE Sign. Process. Magazine **29**(6), 141–142 (2012)
19. Krizhevsky, A.: Learning multiple layers of features from tiny images, 32–33 (2009)
20. Simonyan, K., Zisserman, A.: Very deep convolutional networks for large-scale image recognition (2015)
21. Kidger, P., Garcia, C.: Equinox: neural networks in JAX via callable PyTrees and filtered transformations. In: Differentiable Programming workshop at Neural Information Processing Systems 2021 (2021)
22. Ajayan, J., Nirmal, D., IV, B.K.J., Sreejith, S.: Advances in neuromorphic devices for the hardware implementation of neuromorphic computing systems for future artificial intelligence applications: a critical review. Microelectron. J., **130**, 105634 (2022)
23. Pinchetti, L., Salvatori, T., Yordanov, Y., Millidge, B., Song, Y., Lukasiewicz, T.: Predictive coding beyond gaussian distributions. In: Proceedings of the 36th International Conference on Neural Information Processing Systems, NIPS '22, Red Hook. Curran Associates Inc. (2022)

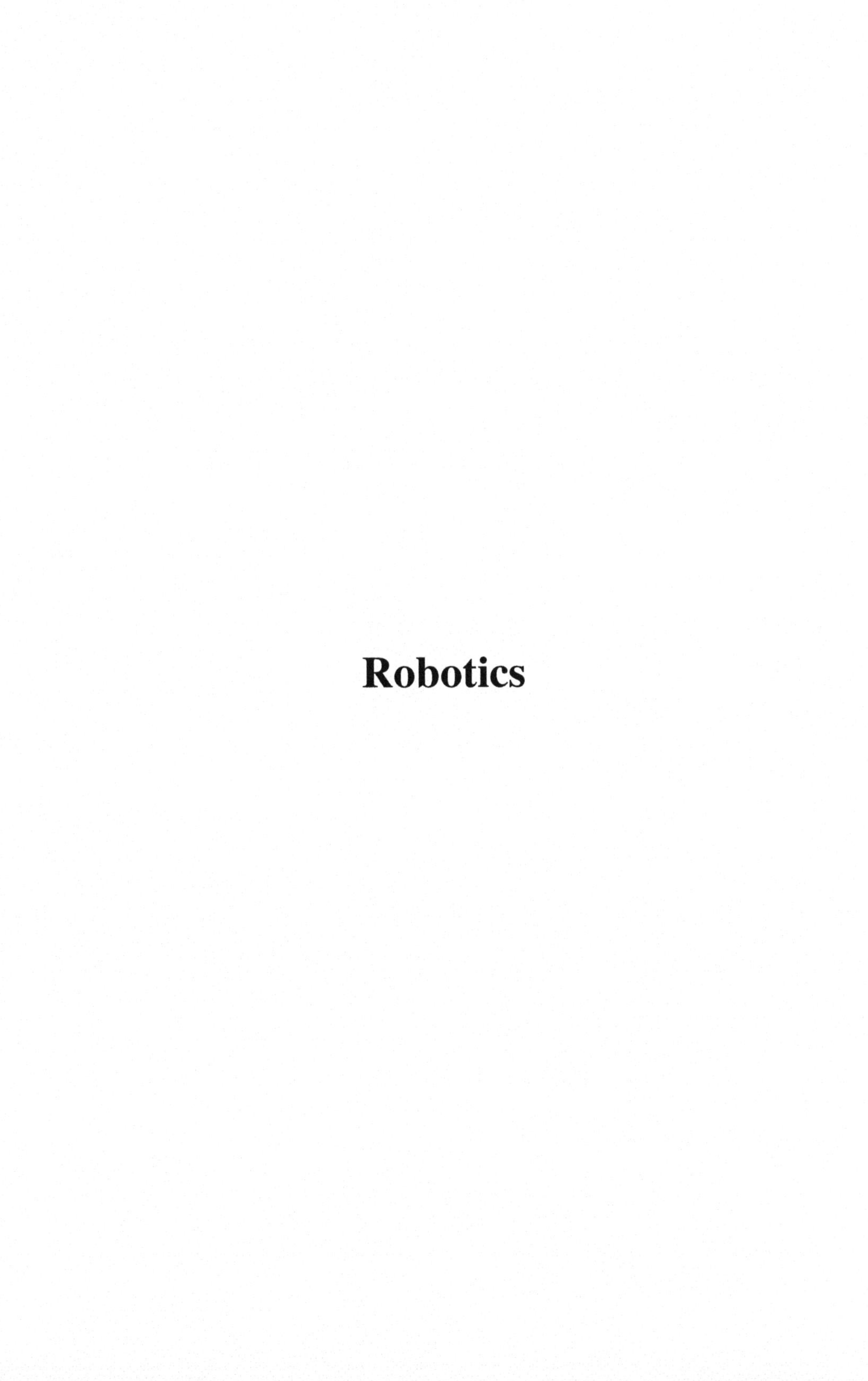

Robotics

Navigation and Exploration with Active Inference: from Biology to Industry

Daria de Tinguy[1(✉)], Tim Verbelen[2], and Bart Dhoedt[1]

[1] Ghent University, Ghent, Belgium
daria.detinguy@ugent.be
[2] Verses, Los Angeles, CA, USA

Abstract. By building and updating internal cognitive maps, animals exhibit extraordinary navigation abilities in complex, dynamic environments. Inspired by these biological mechanisms, we present a real-time robotic navigation system grounded in the Active Inference Framework (AIF). Our model incrementally constructs a topological map, infers the agent's location, and plans actions by minimising expected uncertainty and fulfilling perceptual goals without any prior training. Integrated into the ROS2 ecosystem, we validate its adaptability and efficiency across both 2D and 3D environments (simulated and real-world), demonstrating competitive performance with traditional and state-of-the-art exploration approaches while offering a biologically inspired navigation approach.

Keywords: Autonomous Navigation · Active Inference · Robotics · Topological maps

1 Introduction

Animals exhibit remarkable navigation capabilities that allow them to thrive in complex and unpredictable environments. From migratory birds flying across continents [8], to rodents learning intricate mazes [31] and humans navigating subways [3], biological agents demonstrate an ability to explore, localise, and plan. These capabilities rely on internal models of the world, enabling organisms to imagine trajectories and evaluate outcomes on the fly. Such abilities far exceed those of most autonomous systems, particularly in real-world settings where pre-mapped routes or supervised learning often fail to adapt to dynamic changes [4, 5, 24, 40].

Neuroscience has suggested that animals navigate using structured internal representations (cognitive maps) combining spatial, temporal, and relational information [3, 9, 41]. Such models are formed and refined through experience, enabling organisms to infer location, predict outcomes, and choose actions with minimal supervision. Translating these insights to robotics could enable adaptive, data-efficient, and robust behaviour [20, 23].

M. Albarracin et al. (Eds.): IWAI 2025, CCIS 2857, pp. 331–347, 2026.
https://doi.org/10.1007/978-3-032-16955-6_19

The Active Inference Framework (AIF) offers a principled and biologically grounded approach to modelling such behaviour, proposing an unifying framework for this endeavour. Rooted in Bayesian inference and predictive coding, AIF treats perception, learning, and action as components of a single generative process: agents minimise expected free energy by updating beliefs about the world and selecting actions to reduce uncertainty or satisfy preferences [10,11]. While conceptually appealing and biologically plausible, AIF has seen limited application in embodied robotics, with only a few recent works exploring its real-world deployment in the context of navigation [32].

In this paper, we present a bio-inspired navigation system that applies AIF to real-world spatial reasoning, localisation, and planning. Our agent builds and updates a topological cognitive map on the fly, using it to infer its position and state, evaluate hypotheses about unvisited areas, and choose actions that minimise expected surprise. It requires no pre-training, is robust to sensor drift and environmental change, and adapts continuously.

We validate our approach in a variety of settings, from simple 2D mazes to large-scale 3D environments (using ROS2 [30]) with realistic dynamics and sensing. These include structured and unstructured indoor scenarios such as warehouses and a real-world maze. Our results show interpretable and adaptive behaviour, supporting efficient planning and generalisation without heavy data training, moving robotic autonomy closer to the flexibility of natural navigation.

The paper is structured as follows: We begin by reviewing navigation strategies that support exploration under Active Inference. We then present our method for topological mapping, belief-based localisation, and goal-directed planning, along with its application in 2D mazes. The system architecture and its ROS2 implementation are described next, followed by a comprehensive evaluation of its exploration efficiency in 3D environments (simulated and real-world). We conclude by discussing current limitations and potential extensions.

2 Related Work

Navigation requires integration of localisation and mapping, decision-making (where should I go), and motion planning (how should I move). Those are typically separated in classical navigation pipelines, often focusing on a subset of navigation and relying on hand-tuned rules or pre-training. Hand-tuned rule models often have limited adaptability in dynamic settings as well as potentially computationally intensive or poorly scalable to large environments [4,24], while learning-based methods require extensive environment-specific data and struggle to generalise [20].

Recent work combines mapping and planning via probabilistic or topological models [1,17]. They are lightweight models, scalable, and well-suited for reasoning and planning in complex environments. However, these models often depend on costly computations or fail to generalise in large, unstructured and dynamic environments. Similarly, self-supervised and dataset-driven methods like BYOL-Explore [15] or ViKiNG [34] show strong performance in, respectively, exploration or goal-reaching; however, they require an important training phase in similar environments, and only fulfil a given task.

Conversely, Active Inference offers a unified, biologically inspired alternative that treats navigation as inference, avoiding explicit reward functions. Rooted in the idea that agents minimise surprise through belief updating, AIF enables continuous adaptation by integrating perception, localisation, and action selection [28].

Initial works demonstrated how AIF could support emergent animal behaviours like exploration and goal-reaching [41] in simplified 2D settings [18, 26,33]. More recent studies introduced cognitive maps and structure learning [21], or proposed hierarchical models for efficient spatio-temporal planning [7]. While these contributions highlight the flexibility of AIF, they remain limited to low-dimensional, 2D environments and often rely on computationally expensive policy search.

Notably, G-SLAM [32] demonstrated real-world AIF deployment via unified mapping and control, but required pre-trained generative models and lacked compatibility with robotics stacks.

Our work builds on these ideas to demonstrate how AIF can be applied in real-world navigation scenarios without requiring prior training or rigid pipelines. We propose a model for robotic navigation that efficiently explores fully unknown environments using modular components (with ROS2) and flexible sensory input to achieve joint mapping, localisation, and decision-making.

3 Inferring Motion to Foretell Model Growth

When navigating an environment, an AIF agent continuously refines its internal model of the world by updating transition probabilities (the likelihood of moving between states) and observation likelihoods (the probability of perceiving certain features given a state). These updates are driven by the agent's actions and resulting sensory inputs, gradually aligning the model's predictions with actual observations. This alignment allows the agent to better anticipate outcomes and select actions, minimising Expected Free Energy (EFE), to optimise its navigation strategy.

A key limitation in most advanced models is that they 1) have static state dimensions (illustrated Fig. 1 a) and presented in [21,26]) and/or 2) often expand their internal state space only after encountering new observations, as pictured in Fig. 1 b) and presented in [12,13]. For instance, in room-structured mazes, new states are added only when the agent physically enters a new room. However, this strategy causes the agent to forget unexplored possibilities, such as unvisited rooms observed indirectly (e.g., through visible doors), and thus cannot use this information to predict or plan future exploration effectively. As a result, policies evaluated through EFE become short-sighted, focusing only on directly experienced transitions and ignoring potentially informative unvisited areas.

To overcome this, our model expands its internal topological map based not only on actual observations but also on predicted states (as presented in Fig. 1

c)). By considering hypothetical transitions (such as unvisited rooms behind seen doors), the agent maintains connectivity across the environment and can revisit these paths later. This allows the agent to retain awareness of unexplored options and plan accordingly. This strategy significantly improves exploration efficiency, enabling more comprehensive and strategic coverage of the environment with less redundancy and delay.

The core of our model relies on the generative model presented in Fig. 1, where we separately model the inference of the location (state s) from the estimated position p as two distinct latent variables. p encodes the believed position of the agent considering motion, past position and past state confidence, while s is the state representing the localisation of the agent, consisting of an observation and a position. p is deduced from the previous action a_{t-1}, pose p_{t-1} and state s_{t-1}. While s is inferred from the current observation o (from which we expect to deduce the presence of obstacles), position p and previous state s_{t-1}. Inferring those two variables improves the robustness of the system to kidnapping and ambiguous situations (where an observation changed in a past location).

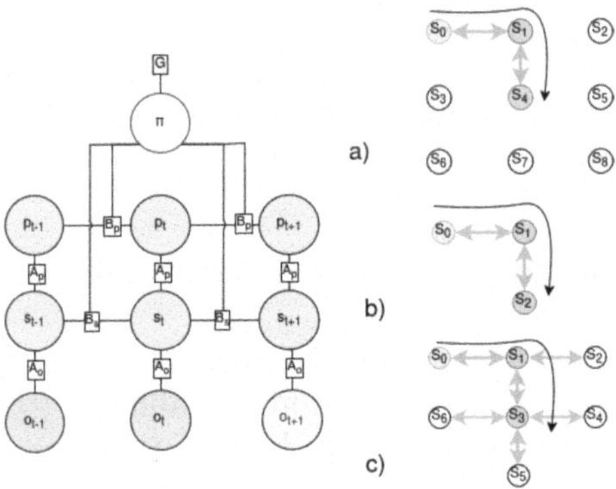

Fig. 1. Left: Factor graph of the POMDP generative model, showing transitions from past to future (up to time $t+1$). Known observations (blue) inform current latent states. Future actions follow policy π, influencing inferred positions and states (orange), and generating predictions of future observations (grey). The agent's position p_t is determined by p_{t-1} and the selected policy, while the latent state s_t is inferred from o_t, p_t, and s_{t-1}. Transitions are parametrised by B matrices, and A matrices encode the likelihood of observations given latent states. Right (a – c): Three ways of structuring the world, progressing from the most common (a) with a given static world dimension, to (b) a growing state learning given a new observation, to (c) the structure learned by our proposed model given expected motions.(Color figure online)

This model results in the approximate posterior presented in Equation (1)

$$Q(\tilde{s}, \tilde{p}|\tilde{o}, \tilde{a}) = Q(s_0, p_0|o_0) \prod_{t=1}^{T} Q(s_t, p_t|s_{t-1}, p_{t-1}, a_{t-1}, o_t) \tag{1}$$

Our model's key component is its ability to infer both position and latent state jointly. This enables it to remain robust under visual ambiguity (e.g., perceptual aliasing or kidnapping) while also supporting self-expansion by hypothesising unexplored parts of the environment.

Expanding the agent's model is also governed by Free Energy minimisation, i.e. by comparing whether an expanded model better explains current or expected observations than our current model at hand. Concretely, given a current model P, and an alternative, expanded model $\tilde{P}$, the Free Energy difference can be written as 2 [28]:

$$\Delta F = F[\tilde{P}(\theta)] - F[P(\theta)] = \ln \mathbb{E}_{Q(\theta)}\left[\frac{P(\theta)}{\tilde{P}(\theta)}\right] \tag{2}$$

If the free energy of an expanded model is lower than that of the current model, the agent updates its internal structure to incorporate the newly encountered (or predicted) information, effectively increasing the model's dimension. This involves predicting motions leading to previously uncharted states through the Expected Free Energy (EFE) of policies, where the benefit of updating A_p is evaluated.

The generative model includes a state transition matrix B_s and an observation likelihood matrix A_o. In addition, it features a position likelihood matrix A_p, linking each state to a possible pose, and a transition model B_p, which differs from the standard matrix form. B_p is implemented as a list that tracks imagined positions. The agent infers its next pose by updating the previous position based on the intended action a and the expected collision outcome $P(c)$, a binary variable (1 if we expect an obstacle between two poses or 0 otherwise). The EFE of a policy π is defined in Eq. 3. The learning term of this equation quantifies how much we learn about position likelihood in light of possible obstacles, while the inference term evaluates the next state s_{t+1} and position p_{t+1} given an expected collision c_{t+1}. This mechanism assumes that the agent can detect or identify obstacles via its observations o.

$$\begin{aligned}
G(\pi) &= \mathbb{E}_{Q_\pi}[\log Q(s_{t+1}, p_{t+1}, A_p|\pi) - \log Q(s_{t+1}, p_{t+1}, A_p|c_{t+1}, \pi) - \log P(c_{t+1})] \\
&= -\underbrace{\mathbb{E}_{Q_\pi}[\log Q(A_p|s_{t+1}, p_{t+1}, c_{t+1}, \pi) - \log Q(A_p|s_{t+1}, p_{t+1}, \pi)]}_{\text{expected information gain (learning)}} \\
&\quad -\underbrace{\mathbb{E}_{Q_\pi}[\log Q(s_{t+1}, p_{t+1}|c_{t+1}, \pi) - \log Q(s_{t+1}, p_{t+1}|\pi)]}_{\text{expected information gain (inference)}} \\
&\quad -\underbrace{\mathbb{E}_{Q_\pi}[\log P(c_{t+1})]}_{\text{expected collision}}
\end{aligned} \tag{3}$$

Policies π may lead the agent toward locations that are not yet represented in the generative model. To decide whether to expand the model to include such locations, we evaluate Eq. 4, which computes the EFE of the prior over the position likelihood parameters A_p. A high expected information gain from exploring a particular direction suggests that the model should grow to accommodate a new pose. However, if a collision c is likely, this suppresses the probability of creating a new position at that location.

$$
\begin{aligned}
P(A_p) &= \sigma(-G) \\
G(A_p) &= \mathbb{E}_{Q_{A_p}}[\log P(p, s|A_p) - \log P(p, s|c, A_p) - \log P(c)] \\
&= \underbrace{-\mathbb{E}_{Q_{A_p}}[\log P(p, s|c, A_p) - \log P(s|p, A_p) - \log P(p|A_p)]}_{\text{expected information gain (expanding)}} \\
&\quad \underbrace{- \mathbb{E}_{Q_{A_p}}[\log P(c)]}_{\text{expected collision}}
\end{aligned}
\tag{4}
$$

If adding parameters to A_p reduces EFE compared to the current model, then the model expands its parameter space to include a new position in A_p and consequently B_p. This entails creating a new state associated with that position, which increases the dimensionality of all model components. The updated observation model A_o assigns uniform probabilities to newly added states, reflecting the uncertainty about unvisited states. The transition model B_s forms or updates transition probabilities between existing and newly created states, considering how much certainty we have about the new state. Details of this process are further discussed in Sect. 5.

To ensure robustness to kidnapping (sudden displacements) and observational ambiguity, the model relies on joint confidence in the inferred state and pose, conditioned on the previous state, pose, and action. If our current observation diverges from expectations by a given threshold but the state and pose transitions remain coherent, the agent updates its observation likelihood while maintaining confidence in its position. Conversely, if both the observation and the inferred transitions are surprising, the model reduces confidence in its current pose, alerting the model to a possible kidnapping. In such cases, the model halts further updates until a sequence of consistent observations restores confidence in the agent's location (confidence in a state reaches the given threshold), at which point the model updating process can resume.

4 Exploration and Goal Reaching in a Tolman Maze

Our model's navigation strategy is evaluated in a dynamic maze environment inspired by the second maze of Tolman's experiment [36], shown in Fig. 2 second column. The original experiment aimed to show evidence of the development of a flexible 'cognitive map' of the maze during exploration and that such mental representation guided the rat's behaviour in the reward sessions. Calling this mental map reorganisation "insight", defined later as "the solution of a problem by the sudden adaptive reorganization of experience" [29].

Agents begin at a designated start point (bottom of the maze) and must locate a reward placed at the opposite side. Three possible paths connect the

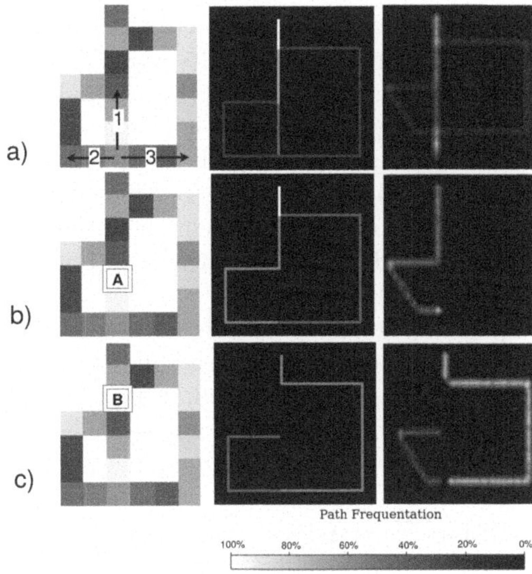

Fig. 2. Our results (second column) compared to L.-E. Martinet & al's (third column) [22]. In our study, the agent's flow paths towards the objective (top of the map) are shown, with re-planning occurring when the desired path is blocked. The varying colour gradient of the lines indicates the frequency of selection for each path over all agents. A sequence is read horizontally, a) is the maze without obstacles, and b) and c) illustrate obstacles at points A and B, respectively. The occupancy grid maps demonstrate the learning of maze topology by simulated agents, initially without obstacles, showing a significant preference for Route 1. When a block is introduced at point A, the animals predominantly choose Route 2. With an obstacle placed at point B, the animals mainly opt for Route 3.

start and goal, with Route 1 being the shortest and Route 3 the longest. Two critical junctions, A and B, can be blocked to restrict access to Routes 1 and 2, respectively.

Agents start with no prior knowledge of the maze structure or sensory cues, but are guided by a preference for red tiles. A utility weight of 2 biases the agent toward preference-seeking over pure exploration. Learning is cumulative across all conditions, allowing agents to build and update their knowledge acquired in earlier runs.

We ran ten agents under three sequential conditions:

- no obstacles, Fig. 2 a)
- a blockage at A, Fig. 2 b)
- a blockage at B, Fig. 2 c)

Each condition involves 12 sequential runs per agent. Every 20 time steps, the agent is 'kidnapped' and repositioned at the start without notice, requiring it to correct beliefs and infer its new location using sensory inputs.

Figure 2 shows the frequency of path selections between conditions of our model (second column), compared to the results from [22] (third column), which uses 100 agents. Unlike those agents or rats in the original experiment, ours cannot detect obstacles from afar; it must reach adjacent rooms to perceive blockages. Cells correspond to discrete rooms, with white blocks indicating obstructions. The agent's route decisions thus emerge from internal belief updating given direct perception of new obstacles, showing how the agent tends to choose the most efficient path depending on the situation. Full scenario details and conclusions are provided in [35].

The observed adaptability reflects two core mechanisms: the agent's capacity to simulate action outcomes up to 14 steps ahead and the plasticity of its internal map (its ability to flexibly revise beliefs based on new evidence and adapt to a kidnapping situation). These traits highlight our model's potential for efficient, insight-driven navigation under uncertainty in simple 2D environments.

5 Toward Realistic Settings

In this section, we are bridging the gap between our principled model working in low-dimensional settings and our model able to cope with the messiness of real-world environments.

To assess the feasibility of deploying our approach under realistic conditions, we implement it on a physical robot operating in previously unseen realistic indoor spaces. This setting challenges the model with sensor noise and non-uniform geometry, aspects often simplified in 2D environments.

Our system is deployed with ROS2 and is modular by design. The AIF-based planner is module, sensor and platform-agnostic and can interact with traditional perception modules and motion planning systems as well as diverse robots (Turtlebot, Turtlebot3 [39] and RosbotXL [16], the used robots and sensors are defined in Appendix 7). Each component can be independently replaced or refined without disrupting the rest of the pipeline. Future work could incorporate more sophisticated perceptual pipelines (e.g., semantic SLAM or learned embeddings), without requiring changes to the generative model.

The complete architecture of the navigation system is illustrated in Fig. 3. It is made of four modular components: (i) a generative model that performs mapping, inferring and planning under AIF framework, (ii) an odometry inference module (localisation) that estimates the state of the agent based on belief rather than direct sensor readings, (iii) a sensor processing stream (currently using panoramic visual input) and (iv) a motion control module responsible for the execution of selected actions. Modules newly added or modified from the 2D experiment to the real-world experiments are highlighted by red contours.

Critically, we do not rely on sensory estimated pose but on the agent's inferred pose, aligning with the epistemic stance of AIF, where belief takes precedence over sensory measurements. As a result, in the face of occlusion or localisation drift causing metric misalignment with the ground truth, the agent remains functionally coherent, navigating based on its internal generative model rather than needing external corrections.

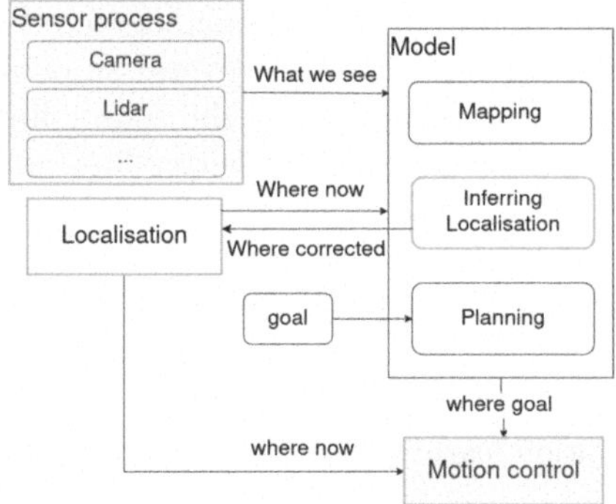

Fig. 3. Overview of the system architecture. Modules interact through belief propagation, Inferring and planning (localisation, mapping and action selection) rely on the Active Inference framework. The perceptual and motion planning still use traditional approaches. Believed odometry takes precedence over sensor odometry. Preferences are expected from the user if we want to reach a target observation. Red contours highlight newly added or modified modules for Real-world navigation. (Color figure online)

Motion planning is handled by external controllers (e.g., Nav2 [25] or potential field planners [19]) that receive the position of the target state as input. However, goals are not specified as fixed coordinates but rather as probabilistic regions (Gaussian distributions) corresponding to expected poses and associated sensory observations. The agent determines it has reached a goal not through position alone, but when its actual sensory input aligns with its predicted perceptual state.

In deployment, we observe the system's ability to autonomously construct and extend its internal model while navigating through 3D spatial layouts. As exploration unfolds, the agent updates its representation to accommodate newly encountered scenes, refining its model via action-perception cycles. When getting to the next desired position, the agent captures a panoramic RGB view (taken with the camera and a rotation of the robot) and the latent cognitive states are updated via Structural Similarity Metrics (SSIM) between incoming views and stored experiences. When a discrepancy arises, it updates its internal model: either refining its observation model or revising its transition model, depending on its confidence in its current state estimate.

We use a Lidar to consider collisions as our camera does not provide accurate depth measurement; however, in our model, we still consider this as a visual observation o and do not divide the information into separate observations. This results in a Transition matrix B_s being updated according to the situation. We use a Dirichlet pseudo-count mechanism defined in Eq. 5 with learning rates (λ). The learning rates are determined by whether we were physically blocked toward an objective or we imagine being able to reach (or not) a position according to our sensors. The diverse situations are presented in Table 1. This continual adaptation of the state transitions according to the situation allows the agent to rapidly reconfigure its internal map to new or undetected obstacles (or removed obstacles) and avoid blocked regions, even several steps away, up to a user-set preferred distance, usually corresponding to the sensor limit. This increases the adaptability of the model to situations and better fine-tunes its internal map to the reality of the environment (Fig. 4).

$$B_\pi = B_\pi + Q(s_t|s_{t-1}, \pi)Q(s_{t-1}) * B_\pi * \lambda \tag{5}$$

Table 1. Transition learning rate (λ) depending on the situation

Transitions	Possible	Impossible	Predicted Possible	Predicted Impossible
Forward	7	-7	5	-5
Reverse	5	-5	3	-3

6 Exploration in Realistic Environments

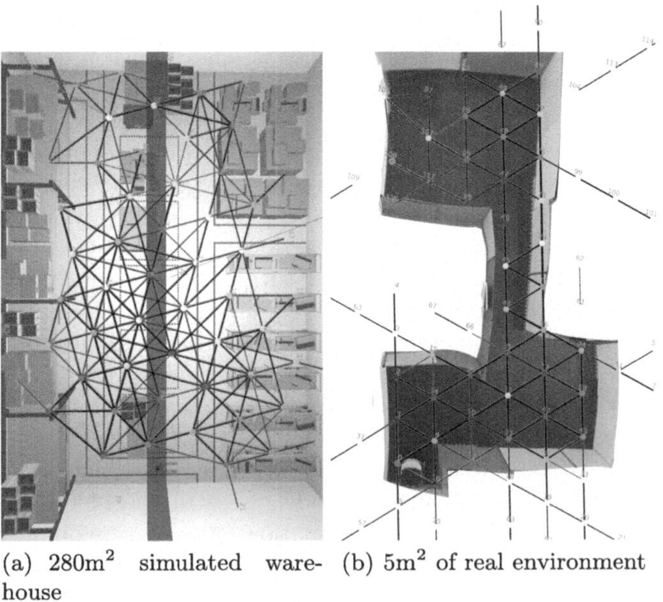

(a) 280m^2 simulated warehouse (b) 5m^2 of real environment

Fig. 4. Final map of exploration in a) Amazon simulated warehouse, b) a real-world environment. Coloured points signify visited locations, where the same colour attributions mean the same observation. The thickness of the lines depicts the agent's believed probability of transitioning between two states given an action.

We evaluated our model in both simulated and real-world settings. The simulated environment, built in Gazebo [14], consists of a $280\,\text{m}^2$ warehouse-like layout [2], shown in Fig. 6a, while the real-world test was conducted in a $5\,\text{m}^2$ controlled maze in Fig. 6b. In both cases, the agent incrementally builds its topological map using a user-defined spatial resolution, with a minimum node spacing of approximately 2 m in simulation and 0.5 m in the real environment. This minimum spacing represents the radius of influence of a state, and this adaptive granularity enables the agent to scale its representation to match the environmental complexity.

The agent's action space consists of 13 discrete actions: 12 evenly spaced orientations across 360°, plus a "stay" action. However, when expanding its topological map, a state can generate transitions up to a maximum of six adjacent states, rather than creating a new node for every possible direction. This constraint is imposed to prevent an overly dense graph structure and maintain a manageable level of connectivity. By limiting the number of newly created states per location, the agent ensures that only the most promising directions are used to expand the map. This approach preserves the flexibility to connect to more distant or informative states later, particularly when no obstacles are present, thereby maintaining a sparse yet navigable topological representation of the environment. Both the number of imagined transitions and available actions can be tuned by the user, allowing for control over planning granularity and computational cost.

Each visited location is associated with a sensory observation. According to the agent's observation model, locations sharing the same dot colour in the map visualisation indicate perceptual similarity. The thickness of the edges between the nodes represents the inferred likelihood of successful transitions, and thicker edges denote a greater confidence in the navigability between states. In the real environment, the lidar sometimes hallucinates free space due to incorrect reflections, and erased connections are visible when the agent attempts to reach those points.

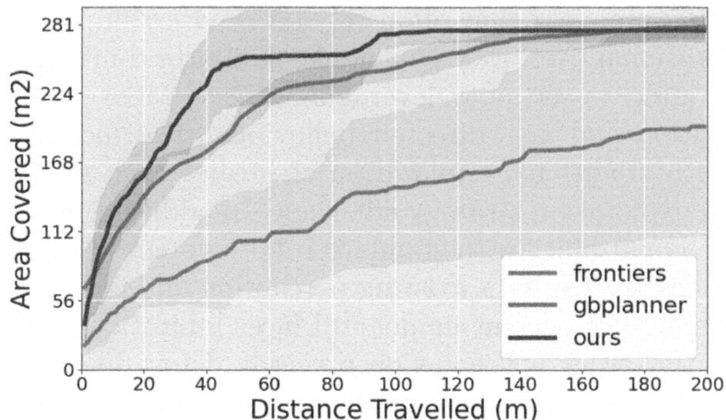

Fig. 5. Lidar Coverage of a $280\,\text{m}^2$ warehouse over the distance travelled (m) with our model, Frontiers and Gbplanner over five runs each.

To evaluate exploration performance, we compared our model against Frontiers [37], which is a classical exploration algorithm aiming for unexplored areas and Gbplanner [6]. Gbplanner uses a metric map [27] to form its topological map, then used for exploration. Gbplanner is an improved version of the 2021 DARPA's winner [38]. While being an AI method, our model is not suitable for comparing our navigation to deep-learning methods due to the absence of pre-training.

Figure 5 displays the area observed by the robot (considering the Lidar) over its travelled distance, averaged over five runs per model, initiating at different starting points in the warehouse. Time-based metrics were avoided due to variability in simulated time within the Gazebo environment.

Overall, our Active Inference framework outperforms or matches the performance of more traditional approaches (Frontiers) and state-of-the-art exploration systems (Gbplanner), with an efficient goal-oriented exploration (goals being the next state to reach). Frontiers falls behind in exploration efficiency as it lacks optimised navigation and requires multiple passes over the same areas. Especially when some small, unreachable areas have unexplored zones without a clear frontier. Gbplanner has been built to be robust in underground passages rather than optimising navigation in large open spaces. Additional experiments validating the dynamic adaptability of our model can be found in appendix 7.

These experiments validate the applicability of our approach beyond simulation and into realistic settings, laying a foundation for more flexible, autonomous navigation in the world based on a biologically plausible strategy.

7 Conclusion

We have illustrated how a biologically inspired navigation system grounded in the Active Inference Framework (AIF) could be used to navigate in an unknown environment without pre-training in 2D and 3D environments (simulated and real). By unifying probabilistic localisation, topological mapping, and belief-driven planning into a single generative model, our approach offers an alternative to traditional navigation pipelines that often rely on rigid assumptions, pre-training, or heavy computational overhead. Our model supports continuous learning and decision-making under uncertainty, leveraging expected free energy minimisation to guide exploration and goal-directed behaviour. The model shows promise in competing with traditional exploration strategies such as Frontiers [37] or Gbplanner [6] in exploration strategy efficiency. In addition, our modular architecture ensures compatibility with standard robotics stacks (ROS2), allowing for straightforward integration with existing perception and control systems.

While promising, our current model still faces limitations related to perception analysis. Future work will focus on precisely determining the limits of the current model in real-world situations (including computational load analysis) as well as enhancing perceptual inference (e.g., via semantic representations or learnt embeddings), extending hierarchical planning capacities, and improving computational efficiency for broader deployment. Ultimately, this work moves a step closer to biologically plausible and practically capable robotic navigation in open-ended, real-world scenarios.

Acknowledgements. This research received funding from the Flemish Government (AI Research Program) under the "Onder-zoeksprogramma Artificiële Intelligentie (AI) Vlaanderen" programme and the Inter-university Microelectronics Centre (IMEC).

Appendix

Robots

Our system is robot-agnostic; however, we have to adapt the sensor pipeline to the specific sensors used. In simulation, we used a TurtleBot3 Waffle with a Pi camera and a 360-degree lidar with a 12 m range. In the real environment, several robots have been used; at first, we used the Turtlebot with a forward lidar of 240 degrees of 12 m range and a camera RealSense D435; the Turtlebot4 with a 360-degree lidar of 8 m range and a camera OAK-D-Pro. Due to uncorrected drift, the resulting maps did not allow a clear superposition with the layout of the environment; thus, we used a RosbotXL with a 360-degree lidar of 18 m range and a 360-degree camera to better show the results of the navigation. However, while the Lidar range is 18 m, the agent only considers the Lidar up to 8 consecutive nodes to create or update new transitions to be as reliable as possible.

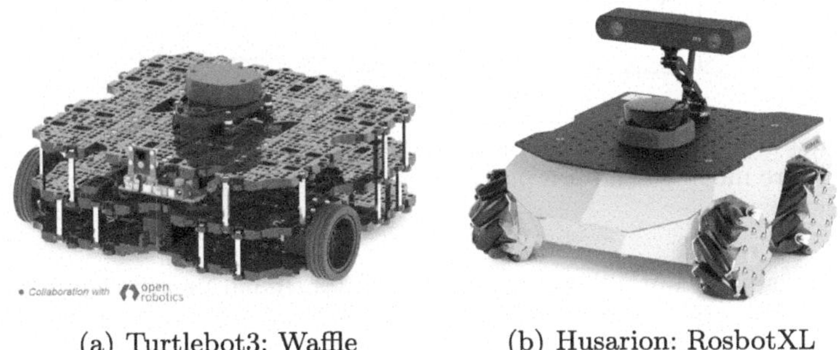

(a) Turtlebot3: Waffle (b) Husarion: RosbotXL

Fig. 6. Turtlebot3 waffle robot was used in simulation while tests have been conducted with a turtlebot, turtlebot4 and RosbotXL in the real environment.

Handling Dynamic Obstacles

Through Eq. 3 our model can predict obstacles, and if an obstacle was not detected (e.g., not detected by the Lidar), it can still recover from a failed motion. Equation 5 is the main factor of our map flexibility to change with the learning rates defined in Table 1.

Figure 7 exemplifies this process with an obstacle (e.g. a box) moved between two positions (from position (-1,0) to (-1,-1) over a visited state 3) while the agent explored a mini warehouse [2] presented in Fig. 8. As the agent fails to reach state 3 after the movement of the box, it adjusts its internal map to reflect the new reality. The transition probability to that location reduces, and the state ID at this position becomes incorrect (state 1 instead of 3) because the agent cannot correct its belief by moving to that location. A new state (state 20) is created at the former position of the obstacle, and new transitions are established. This change does not affect any of the other existing states.

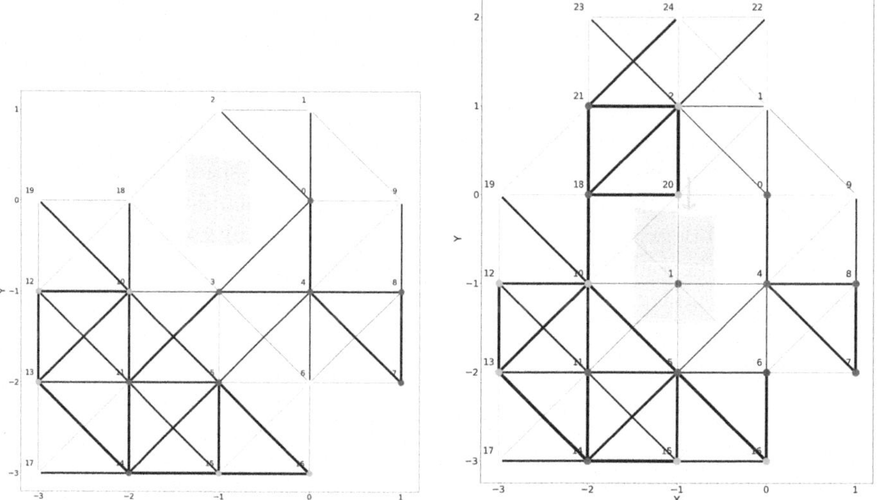

(a) Agent map with an obstacle at position (-1,0) before moving it after a partial exploration

(b) Obstacle at position (-1,-1) after 20 more steps.

Fig. 7. An obstacle was initially placed at position (-1,0) and moved to position (-1,-1) over state 3 during exploration. The failure to reach the state reduces the transition probabilities, represented as thinner black lines between state nodes in the graph. The thickness of the link represents the transition certainty between locations, and each dot colour represents an observation; similar enough observations hold the same colours.

This display, in a qualitative way, shows how the model effectively adapts to change in the environment. The results displayed in Fig. 7 is obtained after the agent circled once around the obstacle to update all adjacent states toward the obstructed state.

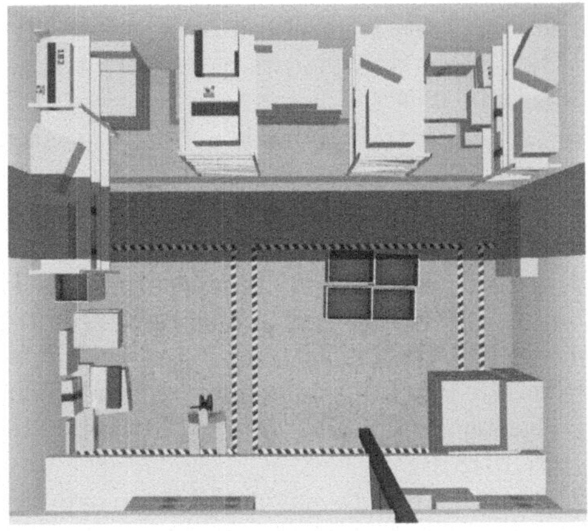

Fig. 8. Top view of a mini warehouse of 36 m^2.

References

1. An, D., et al.: Etpnav: evolving topological planning for vision-language navigation in continuous environments (2024). https://arxiv.org/abs/2304.03047
2. AWS-robotics: AWS-Robomaker-small-warehouse-world (2020). https://github.com/aws-robotics/aws-robomaker-small-warehouse-world. Accessed: 08-01-2024
3. Balaguer, J., Spiers, H., Hassabis, D., Summerfield, C.: Neural mechanisms of hierarchical planning in a virtual subway network. Neuron **90**, 893–903 (2016). https://doi.org/10.1016/j.neuron.2016.03.037
4. Campos, C., Elvira, R., Gomez, J.J., Montiel, J.M.M., Tardos, J.D.: ORB-SLAM3: an accurate open-source library for visual, visual-inertial and multi-map SLAM. IEEE Trans. Rob. **37**(6), 1874–1890 (2021)
5. Chaplot, D.S., Gandhi, D., Gupta, S., Gupta, A., Salakhutdinov, R.: Learning to explore using active neural slam. In: International Conference on Learning Representations (ICLR) (2020)
6. Dang, T., Tranzatto, M., Khattak, S., Mascarich, F., Alexis, K., Hutter, M.: Graph-based subterranean exploration path planning using aerial and legged robots. J. Field Rob **37**(8), 1363–1388 (2020), Wiley Online Library
7. de Tinguy, D., Van de Maele, T., Verbelen, T., Dhoedt, B.: Spatial and temporal hierarchy for autonomous navigation using active inference in mini grid environment. Entropy **26**(1), 32 (2024). http://doi.org/10.3390/e26010083
8. Dorst, J.P.: Migration. Encyclopedia Britannica (2024)
9. Epstein, R., Patai, E.Z., Julian, J., Spiers, H.: The cognitive map in humans: spatial navigation and beyond. Nat. Neurosci. **20**, 1504–1513 (2017). https://doi.org/10.1038/nn.4656
10. Friston, K.: Life as we know it. J. Roy. Soc. Interface / Roy. Soc. **10**, 20130475 (2013). https://doi.org/10.1098/rsif.2013.0475
11. Friston, K., Moran, R.J., Nagai, Y., Taniguchi, T., Gomi, H., Tenenbaum, J.: World model learning and inference. Neural Netw. **144**, 573–590 (2021)
12. Friston, K., Parr, T., Zeidman, P.: Bayesian model reduction (2019)
13. Friston, K.J., et al.: Supervised structure learning (2023)

14. Gazebosim: (2014). https://classic.gazebosim.org. Accessed: 27-05-2025
15. Guo, Z.D., et al.: Byol-explore: exploration by bootstrapped prediction (2022). https://arxiv.org/abs/2206.08332
16. Husarion: rosbotxl (2025). https://husarion.com/tutorials/howtostart/rosbotxl-quick-start/. Accessed: 21-05-2025
17. Jardali, H., Ali, M., Liu, L.: Autonomous mapless navigation on uneven terrains. 2024 IEEE International Conference on Robotics and Automation (ICRA), pp. 13227–13233 (2024). https://api.semanticscholar.org/CorpusID:267770585
18. Kaplan, R., Friston, K.: Planning and navigation as active inference. bioRxiv (2017). https://doi.org/10.1101/230599
19. Koren, Y., Borenstein, J.: Potential field methods and their inherent limitations for mobile robot navigation. vol. 2, pp. 1398 – 1404 vol.2 (1991). https://doi.org/10.1109/ROBOT.1991.131810
20. Levine, S., Shah, D.: Learning robotic navigation from experience: principles, methods and recent results. Philosophical Trans. Roy. Soc. B: Biol. Sci. **378**(1869) (2022). https://doi.org/10.1098/rstb.2021.0447
21. de Maele, T.V., Dhoedt, B., Verbelen, T., Pezzulo, G.: Integrating cognitive map learning and active inference for planning in ambiguous environments (2023)
22. Martinet, L.-E., Passot, J.-B., Fouque, B., Meyer, J.-A., Arleo, A.: Map-based spatial navigation: a cortical column model for action planning. In: Freksa, C., Newcombe, N.S., Gärdenfors, P., Wölfl, S. (eds.) Spatial Cognition 2008. LNCS (LNAI), vol. 5248, pp. 39–55. Springer, Heidelberg (2008). https://doi.org/10.1007/978-3-540-87601-4_6
23. Mirowski, P., et al.: Learning to navigate in complex environments. CoRR **abs/1611.03673** (2016). http://arxiv.org/abs/1611.03673
24. Mohamed, I.S., Yin, K., Liu, L.: Autonomous navigation of AGVS in unknown cluttered environments: log-MPPI control strategy. IEEE Rob. Autom. Lett. **7**(4), 10240–10247 (2022). https://doi.org/10.1109/LRA.2022.3192772
25. nav2: nav2 (2021). https://docs.nav2.org/. Accessed: 12-01-2024
26. Neacsu, V., Mirza, M.B., Adams, R.A., Friston, K.J.: Structure learning enhances concept formation in synthetic active inference agents. PLOS ONE **17**(11), 1–34 (2022). https://doi.org/10.1371/journal.pone.0277199
27. Oleynikova, H., Taylor, Z., Fehr, M., Nieto, J.I., Siegwart, R.: Voxblox: Building 3D signed distance fields for planning. CoRR **abs/1611.03631** (2016). http://arxiv.org/abs/1611.03631
28. Parr, T., Pezzulo, G., Friston, K.: Active Inference: the free energy principle in mind, brain, and behavior. The MIT Press (2022). https://doi.org/10.7551/mitpress/12441.001.0001
29. Riesen, A.H.: Varying behavioral manifestations of animals: <i>a study in behaviour: Principles of ethology and behavioural physiology, displayed mainly in the rat</i>. s. a. barnett. methuen, london, 1963. 304 pp. 45s.; <i>learning and instinct in animals</i>. w. h. thorpe. methuen, london, ed. 2, 1963. 568 63s. Science **141**(3578), 344–345 (1963). https://doi.org/10.1126/science.141.3578.344
30. ROS2: Ros2 humble (2022). https://github.com/ros2. Accessed: 21-05-2025
31. Rosenberg, M., Zhang, T., Perona, P., Meister, M.: Mice in a labyrinth show rapid learning, sudden insight, and efficient exploration. eLife **10**, e66175 (2021). https://doi.org/10.7554/eLife.66175
32. Safron, A., Çatal, O., Verbelen, T.: Generalized simultaneous localization and mapping (g-slam) as unification framework for natural and artificial intelligences: towards reverse engineering the hippocampal/entorhinal system and principles of

high-level cognition. Front. Syst. Neurosci. **16** (2022). https://doi.org/10.3389/fnsys.2022.787659

33. Schwartenbeck, P., Passecker, J., Hauser, T.U., FitzGerald, T.H., Kronbichler, M., Friston, K.J.: Computational mechanisms of curiosity and goal-directed exploration. eLife **8**, e41703 (2019). https://doi.org/10.7554/eLife.41703
34. Shah, D., Levine, S.: Viking: Vision-based kilometer-scale navigation with geographic hints. In: Robotics: Science and Systems XVIII. Robotics: Science and Systems Foundation (2022). https://doi.org/10.15607/rss.2022.xviii.019
35. de Tinguy, D., Verbelen, T., Dhoedt, B.: Learning dynamic cognitive map with autonomous navigation. Front. Comput. Neurosci. **18** (2024). https://doi.org/10.3389/fncom.2024.1498160
36. insight in rats: Tolman E.C., H.C. Univ. Calif. Publ. Psychol. **4**, 215–232 (1930)
37. Topiwala, A., Inani, P., Kathpal, A.: Frontier based exploration for autonomous robot (2018). https://arxiv.org/abs/1806.03581
38. Tranzatto, M., et al.: Team Cerberus wins the DARPA subterranean challenge: technical overview and lessons learned (2022). https://arxiv.org/abs/2207.04914
39. Turtlebot: Turtlebot versions (2024). https://www.turtlebot.com/about/. Accessed: 16-12-2024
40. Walid, J., Nabil, E.A.: Toward intelligent navigation for autonomous mobile robots: learning from the classics. In: Bendaoud, M., El Fathi, A., Bakhsh, F.I., Pierluigi, S. (eds.) Advances in Control Power Systems and Emerging Technologies, pp. 189–195. Springer Nature Switzerland, Cham (2024)
41. Zhao, M.: Human spatial representation: What we cannot learn from the studies of rodent navigation. J. Neurophysiology **120** (2018). https://doi.org/10.1152/jn.00781.2017

Free-Gate: Planning, Control and Policy Composition via Free Energy Gating

Francesca Rossi[1], Émiland Garrabé[2], and Giovanni Russo[1,3](✉)

[1] Scuola Superiore Meridionale, Naples, Italy
f.rossi@ssmeridionale.it
[2] Sorbonne University, Paris, France
garrabe@isir.upmc.fr
[3] University of Salerno, Fisciano, Italy
giovarusso@unisa.it

Abstract. We consider the problem of optimally composing a set of primitives to tackle planning and control tasks. To address this problem, we introduce a free energy computational model for planning and control via policy composition: Free-Gate. Within Free-Gate, control primitives are combined via a gating mechanism that minimizes free energy. This composition problem is formulated as a finite-horizon optimal control problem, which we prove remains convex even when the cost is not convex in states/actions and the environment is nonlinear, stochastic and non-stationary. We develop an algorithm that computes the optimal primitives composition and demonstrate its effectiveness via in-silico and hardware experiments on an application involving robot navigation in an environment with obstacles. The experiments highlight that Free-Gate enables the robot to navigate to the destination despite only having available simple motor primitives that, individually, could not fulfill the task.

Keywords: Free Energy Minimization · Policy Composition · Autonomous Systems · Decision Making · Optimal Control

1 Introduction

When we are given a new task, we can re-use and compose what we know to achieve our goal. This ability to compose knowledge is widely believed to be a key ingredient in differentiating natural and machine intelligence [32,49,57]. For state-of-the-art autonomous agents, designed by, e.g., leveraging deep neural networks, this desirable capability is promoted by endowing the network with inductive biases. In turn, this is achieved via, e.g., specialized architectures, meta-learning, or large-scale pre-training [49]. However, there is little evidence that machines designed on this paradigm can achieve knowledge composition and a key challenge transversal to learning and control is to equip autonomous agents with mechanisms mimicking this ability.

Motivated by this, we introduce Free-Gate: a computational model for planning and control through policy composition. In Free-Gate, simple policies (i.e., primitives) are composed by a gating mechanism. This mechanism is grounded in the minimization of

M. Albarracin et al. (Eds.): IWAI 2025, CCIS 2857, pp. 348–363, 2026.
https://doi.org/10.1007/978-3-032-16955-6_20

free energy, a unifying account across information theory, statistical learning and active inference [10,53,65]. Essentially, Free-Gate enables agents to compute an optimal policy from a set of primitives that are dynamically composed via a free energy minimization process. In turn, this composition problem is formalized as a finite-horizon optimal control problem. Free-Gate returns the optimal solution to this problem: the (possibly, non-stationary) optimal policy for the agent built by linearly combining the primitives. After characterizing convexity of the problem we give an algorithm that provably returns the optimal solution. The effectiveness of the algorithm is demonstrated on a robot navigation task in an environment with obstacles. Remarkably, our in-silico and hardware experiments show that Free-Gate enables the robot to navigate to the goal position – and avoid obstacles – despite only having available primitives that, individually, could not complete the task. Next, we survey policy composition architectures related to Free-Gate (see also Sect. 2) and refer interested readers to [32,49] for broader discussions on compositionality in natural and machine intelligence from the viewpoint of psychology, cognition and philosophy. We refer to [52] for a survey of free energy minimization problems across learning, control and optimization.

Related Works. The design of modular architectures enabling agents to fulfill a given task from a set of primitives is a central topic for learning and control, with robotics often used as a design, test and validation domain [48,57]. In this context, two popular models are MOSAIC (Modular Selection and Identification for Control) and the mixture of experts, see, e.g., [20,25,33,61]. The MOSAIC architecture builds control inputs via a weighted linear combination of motor *command* signals. The weights for the commands are obtained from *responsibility signals* computed via a stationary soft-max selection rule. Building on this, [8,50] complement the soft-max rule for weights selection with temporal difference learning aimed at maximizing a stationary reward. Despite the experimental successes on nonlinear control tasks reported in these works, to the best of our knowledge there is no normative framework explaining why the soft-max rule should be used to compose the motor signals. The mixture of experts model was originally introduced in the context of machine learning [25] and has become a cornerstone for Large Language Model architectures [2]. Following this approach, decisions are computed by learning a stationary policy (in infinite-horizon settings) to linearly combine the output of a pool of experts that also set the performance baseline. We refer to, e.g., [5,34] for more details and to [1,17,59] for applications to control tasks where a stationary policy is again learned for infinite-horizon control problems. A related approach to combine experts is via architectures based on the so-called Product of Experts [23] model where experts (probabilistic models) are combined by properly multiplying their probability distributions. While originally developed for inference, this model has been recently extended [19] to learn stationary policies for control tasks. Non-stationary policies (in finite-horizon settings) can be instead computed with the architecture from [16]. However, the results are obtained under the simplifying assumption that the agent can directly specify state transitions rather than policies. Policy composition is also related to the rich literature on hierarchical Reinforcement Learning (RL) architectures [43]. The idea behind these architectures is to break down long time-horizon (and complex) tasks into a sequence of shorter (and simpler) subtasks. Namely, a *high-level* policy is learned to orchestrate the subtasks and each of the subtasks is in turn learned via its own RL algorithm, thus yielding a *temporal abstraction* [43,56]. The hierarchical decomposition can shorten the task learning time, thus improving learning efficiency [24]. Remarkably, in [47] this approach is applied to active inference, with behavior emerging from the minimization of expected free energy at each level of the hierarchy. In the context of active inference we also recall [44]. Here, hierarchical architectures are proposed in which multiple generative models are selectively engaged based on contextual cues, thus implementing a mixture of experts scheme guided by epistemic and instrumental value. These models enable efficient planning by composing both perception and action strategies in a modular way. In [45] connections are drawn between active inference and KL control and this enables efficient computation (see also [14] for a survey across learning and control). In developmental robotics, compositionality emerges through interactive learning processes that couple sensory and motor experiences with symbolic structure [55,57]. In these works world models are constructed by composing simple generative components, supporting abstraction and behavior generalization. In the context of computational neuroscience, an architecture for knowledge composition is also postulated within the Thousand Brains Theory (TBT). TBT argues [21,37] that our ability to efficiently learn new concepts is hosted in our neocortex, a highly parallel system having cortical columns as functional units. According to TBT, our perception of the world arises from composing the outputs of these parallel units.

Contributions. We present Free-Gate, a computational model for planning and control enabling agents to optimally combine control primitives in order to tackle sequential tasks. In Free-Gate, primitives are optimally composed via a gating mechanism that minimizes free energy. This leads to formalize a finite-horizon optimal control problem that is convex even when the cost is not convex in states/actions and the environment is nonlinear, non-stationary and stochastic. To the best of our knowledge, this is the first free energy model featuring an underlying optimization problem for optimal policy computation that is convex in this setting and that, additionally: (i) does not require that the agent directly specifies its state transitions; (ii) does not assume stationary/discounted settings. Our key technical contributions are as follows:

- our model provides a normative mechanism for primitives composition. This mechanism links the decision rules determining how primitives are composed to the solution of an underlying constrained optimal control problem having the free energy as cost and primitives' weights as decision variables;
- we characterize the underlying optimization problem. This can be tackled via recursive arguments and, at each time step, the optimal weights for primitives composition can be found by solving a problem that is convex even when the agent cost is not in states/actions. We do not make specific assumptions on the environment dynamics besides the standard Markov assumption;
- the results are turned into an algorithm and validated both in-silico, via simulations, and on a hardware test-bed, using the Robotarium [60]. The experiments show the effectiveness of Free-Gate on a rover navigation task, where the agent needs to combine simple motor primitives to reach a final destination while avoiding obstacles. Finally, the code implementing Free-Gate is fully documented and made available at https://github.com/francesca-rossi1/Free-Gate.

The paper is organized as follows. After introducing (Sect. 2) the background, we present Free-Gate and recast the primitives composition problem as an optimal control problem (Sect. 3.1). Then, we present an algorithm (Sect. 3.2) to tackle the problem, characterizing: (i) optimality; (ii) convexity of the underlying optimization. Free-Gate is validated in Sect. 4 and concluding remarks are given in Sect. 5.

2 Background

Sets are denoted in calligraphic, e.g., $\mathcal{X}$, and vectors in **bold**, e.g., $\mathbf{x}$. A random variable is denoted by $\mathbf{V}$ and its realization by $\mathbf{v}$. For continuous (or discrete) random variables we denote the *probability density function*, pdf, (or *probability mass function*, pmf) of $\mathbf{V}$ by $p(\mathbf{v})$. The joint pmf of $\mathbf{V}_1$ and $\mathbf{V}_2$ is denoted by $p(\mathbf{v}_1, \mathbf{v}_2)$ and the conditional pmf of $\mathbf{V}_1$ with respect to $\mathbf{V}_2$ is $p(\mathbf{v}_1 \mid \mathbf{v}_2)$. Countable sets are denoted by $\{w_k\}_{k_1:k_n}$, where w_k is a generic element, k_1 and k_n are respectively the indices of the first and last element, and $k_1 : k_n$ is the set of consecutive integers between (including) k_1 and k_n. The support of a function $p(\cdot)$ is $\mathcal{S}(p)$. The expectation of a function $\mathbf{h}(\cdot)$ of a continuous random variable $\mathbf{V}$ with pdf $p(\mathbf{v})$ is $\mathbb{E}_p[\mathbf{h}(\mathbf{V})] := \int \mathbf{h}(\mathbf{v})p(\mathbf{v})d\mathbf{v}$. We utilize the Kullback-Leibler (KL) divergence [31]: $D_{\mathrm{KL}}(p||q) := \int p(\mathbf{v}) \ln\left(\frac{p(\mathbf{v})}{q(\mathbf{v})}\right) d\mathbf{v}$, that is a measure of discrepancy between two pdfs $p(\mathbf{v})$ and $q(\mathbf{v})$. The $D_{\mathrm{KL}}(p||q)$ is finite only if $\mathcal{S}(p) \subseteq \mathcal{S}(q)$, i.e., $p(\mathbf{v})$ is absolutely continuous with respect to $q(\mathbf{v})$, see [6, Chapter 8]. We use the standard convention $0\log(0) = 0$. For discrete variables the integral is replaced with the sum in the above expressions. Finally, we let $\Delta^n := \{\mathbf{w} \in \mathbb{R}^n \mid w_i \geq 0\ \forall i \in 1:n,\ \sum_{i=1}^n w_i = 1\}$ be the probability simplex and recall that the affine dimension of $\Delta^n \subset \mathbb{R}^n$ is $n-1$.

Free Energy Minimization. The minimization of free energy plays a central role for autonomous decision making across optimization, inference, risk-aware control and learning [14,26,29,35,51]. A wide range of methods and frameworks, including RL schemes [15,18,65] based on maximum entropy, Bayesian inference [27], and entropy-regularized optimal transport [3], rely on optimizing free energy functionals [28,53], eventually with constraints [13]. In the context of active inference – which unifies perception, action, and learning – the variational free energy is minimized to align the agent internal model with sensory inputs [10,12]. For control, this framework optimizes policies under uncertainty using actions as decision variables while maintaining consistency with internal models [36,44,46]. Beyond artificial agents, this minimization principle is also central to the broader quest for a unifying theory of self-organization [22], brain function and sentient behaviors [9,11,42]. Formally, see, e.g., [28], the free energy is a functional of the form:

$$\mathcal{F}(p(\mathbf{z})) = D_{\mathrm{KL}}(p(\mathbf{z})||q(\mathbf{z})) + \mathbb{E}_{p(\mathbf{z})}[l(\mathbf{z})],$$

where: (i) the first term is the statistical complexity of $p(\mathbf{z})$ with respect to a given $q(\mathbf{z})$. Minimizing this term amounts to minimizing the discrepancy (in the KL divergence sense) between $p(\mathbf{z})$ and $q(\mathbf{z})$; (ii) the second term is the agent's expected loss. Following, e.g., [28], when $l(\mathbf{z})$ is a log-loss of the form $-\log(\tilde{p}(\mathbf{z}))$ the free energy is known as the evidence lower bound (ELBO) commonly used in variational inference [38, Chapter 10].

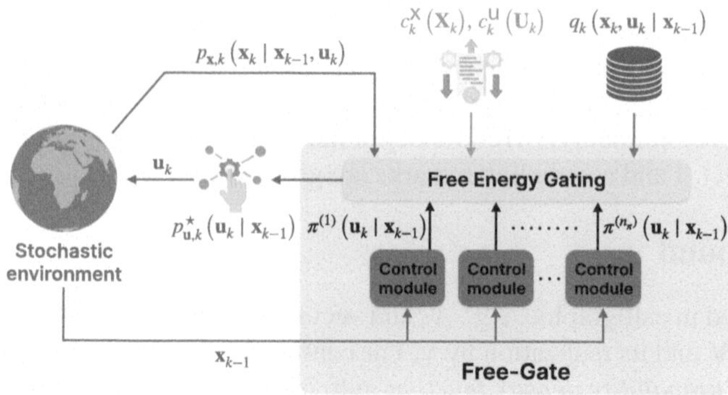

Fig. 1. In Free-Gate control primitives are dynamically combined by a free energy gating mechanism. This mechanism, given the state, the set of primitives, the environment model, the state/action costs, and the generative model, computes the optimal primitives' weights by minimizing free energy. The resulting policy, $p^{\star}_{\mathbf{u},k}(\mathbf{u}_k|\mathbf{x}_{k-1})$, is a linear combination of the primitives weighted by these optimal values.

3 The Free-GateModel

We consider the set-up of Fig. 1 where the agent determines an action $\mathbf{u}_k \in \mathcal{U} \subseteq \mathbb{R}^{n_u}$ based on the state $\mathbf{x}_{k-1} \in \mathcal{X} \subseteq \mathbb{R}^{n_x}$. The time indexing is chosen so that the environment/system transitions from $\mathbf{x}_{k-1}$ to $\mathbf{x}_k$ when $\mathbf{u}_k$ is applied. The possibly stochastic, nonlinear and non-stationary Markovian environment is described, at each k, by $p_{\mathbf{x},k}(\mathbf{x}_k|\mathbf{x}_{k-1}, \mathbf{u}_k)$. The agent determines the action by sampling from the policy $p_{\mathbf{u},k}(\mathbf{u}_k|\mathbf{x}_{k-1})$ so that the closed-loop behavior over $k \in 0 : N$ is described by $p_{0:N} := p_0(\mathbf{x}_0) \prod_{k=1}^{N} p_k(\mathbf{x}_k, \mathbf{u}_k|\mathbf{x}_{k-1})$, with $p_0(\mathbf{x}_0)$ being an initial prior capturing initial conditions and $p_k(\mathbf{x}_k, \mathbf{u}_k|\mathbf{x}_{k-1}) := p_{\mathbf{x},k}(\mathbf{x}_k|\mathbf{x}_{k-1}, \mathbf{u}_k)\, p_{\mathbf{u},k}(\mathbf{u}_k|\mathbf{x}_{k-1})$. Also, $c_k^{\mathrm{x}}(\cdot) : \mathcal{X} \to \mathbb{R}$ and $c_k^{\mathrm{u}}(\cdot) : \mathcal{U} \to \mathbb{R}$ are the state and action costs, respectively.

3.1 Free-GateDescription

As schematically illustrated in Fig. 1, given state/action costs and a generative model

$$q_{0:N} := q_0(\mathbf{x}_0) \prod_{k=1}^{N} q_k(\mathbf{x}_k, \mathbf{u}_k|\mathbf{x}_{k-1}) = q_0(\mathbf{x}_0) \prod_{k=1}^{N} q_{\mathbf{x},k}(\mathbf{x}_k|\mathbf{x}_{k-1}, \mathbf{u}_k)\, q_{\mathbf{u},k}(\mathbf{u}_k|\mathbf{x}_{k-1}),$$

the aim of Free-Gate is to compute $\{p^{\star}_{\mathbf{u},k}(\mathbf{u}_k|\mathbf{x}_{k-1})\}_{1:N}$, an optimal sequence of policies that minimizes the free energy functional

$$D_{\mathrm{KL}}(p_{0:N}||q_{0:N}) + \sum_{k=1}^{N} \mathbb{E}_{p_{k-1}(\mathbf{x}_{k-1})}\left[\mathbb{E}_{p_k(\mathbf{x}_k, \mathbf{u}_k|\mathbf{x}_{k-1})}\left[c_k^{\mathrm{x}}(\mathbf{X}_k) + c_k^{\mathrm{u}}(\mathbf{U}_k)\right]\right],$$

where $p_{k-1}(\mathbf{x}_{k-1})$ is the marginal probability of the state at $k-1$. The functional naturally arises in the context of control and learning (see, e.g., [15,52]) and, when the cost is the negative log-likelihood of states, it yields the expected free energy minimized in active inference [41,52]. At each k, $p^{\star}_{\mathbf{u},k}(\mathbf{u}_k|\mathbf{x}_{k-1})$ is a linear combination of a set of n_π primitives, denoted by $\pi_k^{(i)}(\mathbf{u}_k|\mathbf{x}_{k-1})$, $i \in \mathcal{P} := 1 : n_\pi$ and have support $\mathcal{S}(\pi)$. In the architecture: (i) each *control module* provides a primitive. The primitive is black-box, in the sense that the control module does not *disclose* what underlying control algorithm it is implementing. The terminology *primitive* is inspired by the fact that the modules provide policies encoding simple behaviors (e.g., basic motor commands as in our experiments); (ii) *free energy gating* mechanism, which given the primitives returns the weights, $\mathbf{w}_k^{(i),\star}$, to optimally combine them. Thus, the optimal policy is $p^{\star}_{\mathbf{u},k}(\mathbf{u}_k|\mathbf{x}_{k-1}) = \sum_{i\in\mathcal{P}} \mathbf{w}_k^{(i),\star}\pi_k^{(i)}(\mathbf{u}_k|\mathbf{x}_{k-1})$.

Free Energy Gating. In Free-Gate, the gating mechanism aims at finding the optimal primitives' weights combination by minimizing free energy over the time horizon $1 : N$. Namely, the problem of finding the optimal weights combination is recast via the following finite-horizon optimal control problem (to streamline notation, from now on we omit time steps in the probabilities subscripts).

Problem 1. *Given the* $\{p_{\mathbf{x}}(\mathbf{x}_k|\mathbf{x}_{k-1},\mathbf{u}_k)\}_{1:N}$, *the generative model* $q_{0:N}$, *and the set of primitives* $\pi^{(i)}(\mathbf{u}_k|\mathbf{x}_{k-1})$, $i \in \mathcal{P}$, *find the sequence of weights* $\{\mathbf{w}_k^\star\}_{1:N}$ *such that:*

$$
\begin{aligned}
\{\mathbf{w}_k^\star\}_{1:N} \in \underset{\{\mathbf{w}_k\}_{1:N}}{\operatorname{argmin}}\, & D_{\mathrm{KL}}(p_{0:N}\|q_{0:N}) + \sum_{k=1}^{N} \mathbb{E}_{p(\mathbf{x}_{k-1})}\left[\mathbb{E}_{p(\mathbf{x}_k,\mathbf{u}_k|\mathbf{x}_{k-1})}[c_k^{\mathrm{x}}(\mathbf{X}_k) + c_k^{\mathrm{u}}(\mathbf{U}_k)]\right] \\
\text{s.t. } & p_{\mathbf{u}}(\mathbf{u}_k|\mathbf{x}_{k-1}) = \sum_{i\in\mathcal{P}} \mathbf{w}_k^{(i)}\pi^{(i)}(\mathbf{u}_k|\mathbf{x}_{k-1}), \quad \forall k \in 1:N \\
& \mathbf{w}_k \in \Delta^{n_\pi}, \quad \forall k \in 1:N,
\end{aligned}
\tag{1}
$$

where $\mathbf{w}_k$ *is the stack of* $\mathbf{w}_k^{(i)}$*'s and* $p(\mathbf{x}_{k-1})$ *is the state marginal probability at* $k-1$.

In the formulation, the costs are left general and can be used to encode the agent task. For example, in our robot navigation application, the costs promote obstacle avoidance. In Problem 1, the constraints formalize that the agent policy is a combination of the primitives. The cost functional, which is also considered in [15], becomes the surprise upper bound typical in active inference when the cost terms inside the expectation are chosen as negative log-likelihoods [52]. We also refer to, e.g., [45] and [30] for an application to planning, and control from demonstration, where the generative model encodes a desired behavior extracted from example data [13]. The decision variables are the weights of the linear combination and therefore, at each k, the Free-Gate policy is $p^{\star}_{\mathbf{u}}(\mathbf{u}_k|\mathbf{x}_{k-1}) = \sum_{i\in\mathcal{P}} \mathbf{w}_k^{(i),\star}\pi^{(i)}(\mathbf{u}_k|\mathbf{x}_{k-1})$. In Problem 1, $N = 1$ yields biomimetic policies [58], which often arise as reflexes in bio-inspired motor control [46]. Next, we let $\mathbf{w}$ be the stack of all $\mathbf{w}_k$'s, $k \in 1 : N$; $F(\mathbf{w})$ denotes the cost of Problem 1. We make the following standing assumptions:

Assumption 1. *The primitives are finite.*

Assumption 2. $\mathcal{W} := \{\mathbf{w} : \mathbf{w}$ *is feasible for Problem 1 and* $F(\mathbf{w}) < +\infty\}$ *is non-empty.*

Algorithm 1. Free Energy Gating

1: **Input:** N,$\{p_{\mathbf{x}}(\mathbf{x}_k|\mathbf{x}_{k-1},\mathbf{u}_k)\}_{k\in 1:N}$,$q_{0:N}$,$\{c_k^{\mathrm{x}}(\cdot)\}_{k\in 1:N}$,$\{c_k^{\mathrm{u}}(\cdot)\}_{k\in 1:N}$,$\{\pi^{(i)}(\mathbf{u}_k|\mathbf{x}_{k-1})\}_{i\in\mathcal{P}}$
2: **Output:** $\{p_{\mathbf{u}}^{\star}(\mathbf{u}_k|\mathbf{x}_{k-1})\}_{1:N}$
3: $l_{N+1}^{\star}(\mathbf{X}_N)\leftarrow 0$
4: **for** $k=N:1$ **do**
5: $\quad \bar{c}(\mathbf{X}_k,\mathbf{U}_k)\leftarrow c_k^{\mathrm{x}}(\mathbf{X}_k)+c_k^{\mathrm{u}}(\mathbf{U}_k)+l_{k+1}^{\star}(\mathbf{X}_k)$
6: $\quad \mathbf{w}_k^{\star}\leftarrow$ minimizer of the problem in (2)
7: $\quad l_k^{\star}(\mathbf{X}_{k-1})\leftarrow$ optimal value of the problem in (2)
8: $\quad p_{\mathbf{u}}^{\star}(\mathbf{u}_k|\mathbf{x}_{k-1})\leftarrow\sum_{i\in\mathcal{P}}\mathbf{w}_k^{(i),\star}\pi^{(i)}(\mathbf{u}_k|\mathbf{x}_{k-1})$
9: **end for**

Assumption 3. *Each primitive has full support over the action space, i.e.,* $\mathcal{S}(\pi)=\mathcal{U}$.

Remark 1. *Assumption 1 is rather mild and always satisfied for, e.g., discrete variables. Assumption 2 is standard in the context of optimal transport [39, Theorem 1.10]. This assumption, guaranteeing that the optimal cost is finite, is satisfied when: (i) the state/action cost functions are lower-bounded; (ii)* $p_{\mathbf{x}}(\mathbf{x}_k|\mathbf{x}_{k-1},\mathbf{u}_k)$ *is absolutely continuous with respect to* $q_{\mathbf{x}}(\mathbf{x}_k|\mathbf{x}_{k-1},\mathbf{u}_k)$*; (iii) the primitives are absolutely continuous with respect to* $q_{\mathbf{u}}(\mathbf{u}_k|\mathbf{x}_{k-1})$. *Assumption 3 is used to guarantee that the gradient of the cost in* (1) *is well defined. This assumption, implying that the policy is not deterministic, can also be found in the context of policy gradient methods for RL [54, Chapter 13].*

3.2 Optimal Primitives Composition: Tackling Problem 1

In Free-Gate, the free energy gating solves Problem 1 and this is done via Algorithm 1, which outputs the optimal policy obtained by dynamically combining the primitives at each k. Given the inputs specified in Algorithm 1, it computes the optimal weights $\{\mathbf{w}_k^{\star}\}_{1:N}$ via the backward recursion in lines $4-9$. Within the recursion, $\mathbf{w}_k^{\star}$ is the minimizer of the following problem:

$$\begin{aligned}\min_{\mathbf{w}_k\in\Delta^{n_\pi}}\; & D_{\mathrm{KL}}\left(p\left(\mathbf{x}_k,\mathbf{u}_k|\mathbf{x}_{k-1}\right)||q\left(\mathbf{x}_k,\mathbf{u}_k|\mathbf{x}_{k-1}\right)\right)+\mathbb{E}_{p(\mathbf{x}_k,\mathbf{u}_k|\mathbf{x}_{k-1})}\left[\bar{c}(\mathbf{X}_k,\mathbf{U}_k)\right]\\ \text{s.t. } & p_{\mathbf{u}}\left(\mathbf{u}_k|\mathbf{x}_{k-1}\right)=\sum_{i\in\mathcal{P}}\mathbf{w}_k^{(i)}\pi^{(i)}\left(\mathbf{u}_k|\mathbf{x}_{k-1}\right),\end{aligned}\tag{2}$$

where $\bar{c}(\mathbf{X}_k,\mathbf{U}_k):=c_k^{\mathrm{x}}(\mathbf{X}_k)+c_k^{\mathrm{u}}(\mathbf{U}_k)+l_{k+1}^{\star}(\mathbf{X}_k)$; $l_{k+1}^{\star}(\mathbf{X}_k)$ is initialized at 0 and iteratively built within the recursion in Algorithm 1.

Remark 2. *The cost functional in* (2) *is the free energy at time step* k. *In this problem, the expected loss contains the term* $l_{k+1}^{\star}(\mathbf{X}_k)$, *which, as shown in Algorithm 1, is the optimal free energy value obtained by solving the problem at* $k+1$. *By embedding* $l_{k+1}^{\star}(\mathbf{X}_k)$ *into the expected loss at time step* k, *the optimal policy* $p_{\mathbf{u}}^{\star}(\mathbf{u}_k|\mathbf{x}_{k-1})$ *accounts for planning over the time horizon. For biomimetic (or greedy) actions, where* $N=1$, $l_{k+1}^{\star}(\mathbf{X}_k)$ *is equal to* 0.

Next, we characterize the optimality of the policy returned by Algorithm 1 (Property 1) and the convexity of the optimization problems solved at each k (Property 2).

Properties of Algorithm 1. We now analyze key properties of Algorithm 1. Specifically, we establish the optimality of the policy returned by Algorithm 1 and the convexity of the optimization problems that yield the optimal weights. The results leverage standard arguments from convex functions and control the space of densities. See Appendix A for the self-contained proofs.

Property 1 (Optimality). *Algorithm 1 returns the optimal solution of Problem 1.*

The next property characterizes the convexity of the problem in (2). The property can be proved by leveraging convexity of the KL divergence and the composition property of a convex function with an affine mapping. In Appendix A we provide a proof by explicitly computing the Hessian of the cost function.

Property 2 (Convexity). *At each k, the problem in* (2) *is convex in the decision variables $\mathbf{w}_k$. Moreover, if the primitives are linearly independent the cost is strictly convex.*

Remarkably, when the primitives are linearly independent, the problem features a strictly convex cost function. This implies that, if the optimal solution is in the relative interior of the feasibility domain, then Free-Gate policy outperforms the individual primitives, thus achieving *transcendence* in the spirit of [64].

Remark 3 (Transcendence). *If the primitives are linearly independent and the optimal weight vector $\mathbf{w}_k^\star \in \Delta^{n_\pi}$ has at least two strictly positive elements, then the resulting policy $p_{\mathbf{u}}^\star(\mathbf{u}_k|\mathbf{x}_{k-1})$ strictly outperforms each primitive, i.e., $J_k(p_{\mathbf{u}}^\star(\mathbf{u}_k|\mathbf{x}_{k-1})) < \min_{i\in\mathcal{P}} J_k(\pi^{(i)}(\mathbf{u}_k|\mathbf{x}_{k-1}))$, where $J_k(\cdot)$ denotes the cost in* (2).

4 Validation

We use Free-Gate to control a unicycle robot so that it can navigate towards a goal point in an environment with obstacles. Free-Gate is validated through the Robotarium [60]: this platform offers both a hardware infrastructure and a high-fidelity simulator. Consequently, Free-Gate is validated via both simulations and real hardware experiments with rovers moving in a 3.2 m × 2 m work area. The code for replicating our experiment is available at https://github.com/francesca-rossi1/Free-Gate together with a recording of our hardware experiment. The Robotarium provides built-in functions to map the unicycle dynamics onto a single integrator. Hence, it gives users the possibility to maneuver the robot considering an integrator dynamics. The dynamics we use is therefore $\mathbf{x}_k = \mathbf{x}_{k-1} + \mathbf{u}_k dt$, where $\mathbf{x}_k = [p_{x,k}, p_{y,k}]^\top$ is the state (robot's position, in m) at time step k, $\mathbf{u}_k = [v_{x,k}, v_{y,k}]^\top$ is the input vector (velocity, in m/s) and $dt = 0.033\,\mathrm{s}$ is the Robotarium time step. We set $\mathbf{x}_k \in \mathcal{X} := [-1.6, 1.6] \times [-1, 1]$, matching the work area, and $\mathbf{u}_k \in \mathcal{U} := [-0.2, 0.2] \times [-0.2, 0.2]$, in accordance with the maximum speed physically allowed on the robot. For the application of our results, the state and input spaces are discretized into 33×21 and 7×7 equidistant bins, respectively. We emulate measurement noise for the robot position, which following, e.g.,

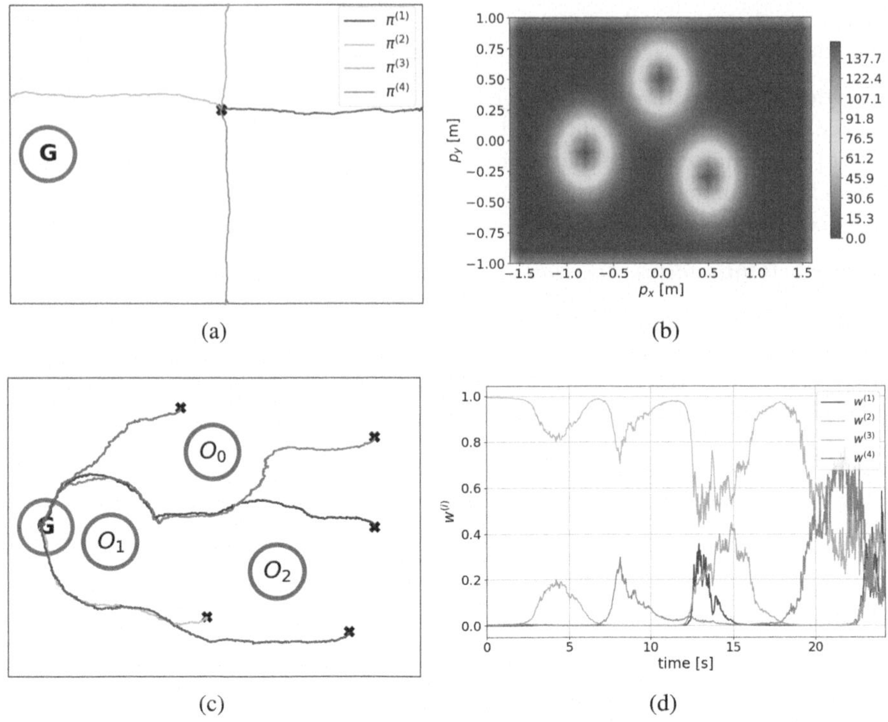

Fig. 2. (a) Trajectories when the robot is controlled by the 4 primitives described in Sect. 4; (b) heat map of the cost used in the experiments; (c) robot trajectories when controlled by Free-Gate; (d) optimal primitives' weights corresponding to the experiment in blue in panel (c). Free-Gate allows the robot to idle at the goal position despite lacking a stopping primitive.

[62] we assume to be Gaussian. Hence, $p_{\mathbf{x}}(\mathbf{x}_k|\mathbf{x}_{k-1}, \mathbf{u}_k) = \mathcal{N}(\mathbf{x}_{k-1} + \mathbf{u}_k dt, \boldsymbol{\Sigma}_x)$, with covariance matrix $\boldsymbol{\Sigma}_x = 0.008\mathbf{I}_2$ ($\mathbf{I}_2$ is the 2×2 identity). Free-Gate has access to $n_\pi = 4$ primitives, which we obtained by discretizing Gaussians of the form $\pi^{(i)}(\mathbf{u}_k|\mathbf{x}_{k-1}) = \mathcal{N}(\bar{\mathbf{u}}_k^{(i)}, \boldsymbol{\Sigma}_u)$. Gaussian assumption is standard for primitives, see, e.g., [40]. For these Gaussians, we set $\boldsymbol{\Sigma}_u = 0.005\mathbf{I}_2$, while $\bar{\mathbf{u}}_k^{(i)}$ is a vector signal (e.g., coming from a simple proportional controller) proportional to the distance between the robot position and each of the boundaries of the Robotarium work area (see GitHub for details). These primitives only allow the robot to move along four cardinal directions (right, left, up, down – see Fig. 2(a)) but cannot solve the navigation task individually. Inspired by, e.g., [22] and references therein, in the experiments we set the generative, time-series, model as: (i) $q_{\mathbf{x}}(\mathbf{x}_k|\mathbf{x}_{k-1}, \mathbf{u}_k) = \mathcal{N}(\mathbf{x}_{k-1} + \mathbf{u}_k dt, \tilde{\boldsymbol{\Sigma}}_x)$, with $\tilde{\boldsymbol{\Sigma}}_x = 0.002\mathbf{I}_2$, and (ii) $q_{\mathbf{u}}(\mathbf{u}_k|\mathbf{x}_{k-1}) = \mathcal{N}(\tilde{\mathbf{u}}_k, \tilde{\boldsymbol{\Sigma}}_u)$, where $\tilde{\mathbf{u}}_k$ is provided by the Robotarium built-in position controller, not accounting for obstacles, and $\tilde{\boldsymbol{\Sigma}}_u = 0.005\mathbf{I}_2$. For the control task, the goal position is $\mathbf{x}_\mathrm{d} = [-1.3, 0]^\top$ (Fig. 2(a)) and the robot needs to avoid obstacles positioned at $[0, 0.5]^\top$, $[-0.8, -0.1]^\top$, $[0.5, -0.3]^\top$ and captured via the following cost adapted from [15]: $c_k^{\mathrm{x}}(\mathbf{x}_k) = 150\sum_{j=1}^{3} g_j(\mathbf{x}_k) + 30\sum_{j=1}^{4} w_j(\mathbf{x}_k)$. In this expression: (i) $g_j(\cdot)$ are Gaussian-shaped obstacles penalties; (ii) $w_j(\cdot)$ is the

cost associated to the boundaries, with $w_j(\cdot)$ again defined as Gaussian. In Fig. 2(b) the heat map of the cost $c_k^{\mathrm{x}}(\mathbf{x}_k)$ is shown. In addition, to equip the agent with planning abilities without implementing the full backward recursion in Algorithm 1, we find a heuristically approximated cost-to-go (Remark 2) as the cost obtained by applying twice the same input (see GitHub for details). Note that Assumption 1, Assumption 2 and Assumption 3 are all satisfied (Remark 1). In the experiments, CVX [7] is used to solve the optimization problem in Algorithm 1 at each time step. To validate Free-Gate, we run 5 experiments starting from different initial positions of the robot. Each simulation terminates when the robot remains within a 0.08 m radius of the goal for at least 2 s. We record the robot's trajectory in each experiment, and the results are shown in Fig. 2(c): Free-Gate allows the robot to successfully reach the goal while avoiding obstacles. Note that there is no dedicated *stop* primitive, yet Free-Gate still enables the robot to remain at the goal point once this is reached by properly combining the available primitives. This is also illustrated in Fig. 2(d), where the time evolution of the optimal weights from one of the experiments is reported. The simulation results are confirmed when Free-Gate is deployed on the Robotarium hardware. A recording of one hardware experiment is available on our GitHub. Namely, the recording shows that, when Free-Gate is used, the real robot successfully completes the navigation task.

5 Conclusions

We introduced Free-Gate: a computational model for planning and control via policy composition. Free-Gate consists of a set of functional units that return control primitives. The primitives are composed by a gating mechanism that minimizes free energy. This leads to recast the problem of finding the optimal weights for the primitives as a finite-horizon optimal control problem. This problem is convex even when the agent cost is not and when the environment is nonlinear, stochastic and non-stationary. After turning the results into an algorithm, the effectiveness of Free-Gate is illustrated via both in-silico and hardware experiments on a robot navigation task. Based on the results presented here, our future work will be aimed at: (i) expanding Free-Gate to integrate, in the spirit of active inference, perception, learning and belief-state estimation for partially observable settings; (ii) giving rigorous conditions establishing what makes for a good set of primitives; (iii) embedding in Free-Gate function approximators, e.g., deep neural networks, to estimate the cost-to-go and tackle continuous-control settings; (iv) improving scalability in high-dimensional action spaces; (v) embodying Free-Gate in applications where primitives are the output of modules engineered to process multi-stream, heterogeneous, sensory information; (vi) implementing the gating mechanism via biologically plausible neural networks using, e.g., the tools from [4].

Acknowledgments. GR wishes to thank K. Friston (University College London, UK) for the insightful discussions on the free energy minimization formulation. GR is also grateful to G. Pezzulo, D. Maisto (National Research Council, Italy), L. Da Costa (VERSES AI Research, USA), A. Paul (Monash University, Australia) for the conversations on the ideas of this paper. GR was supported by the European Union-Next Generation EU Mission 4 Component 1 CUP E53D23014640001.

Disclosure of Interests. The authors have no competing interests to declare.

A Proofs

Proof of Proposition 1. The derivations are inspired by [13] and leverage the chain rule for the KL divergence. We report here a self contained proof of Proposition 1. For notational convenience, in what follows we omit the subscript of the expectation operator when it applies to $\bar{c}(\mathbf{X}_N, \mathbf{U}_N)$, writing $\mathbb{E}\left[\bar{c}(\mathbf{X}_N, \mathbf{U}_N)\right]$ instead of $\mathbb{E}_{p(\mathbf{x}_N, \mathbf{u}_N|\mathbf{x}_{N-1})}\left[\bar{c}(\mathbf{X}_N, \mathbf{U}_N)\right]$. Specifically, following the chain rule, the cost in (1) can be written as:

$$D_{\mathrm{KL}}\left(p_{0:N-1}||q_{0:N-1}\right) + \sum_{k=1}^{N-1} \mathbb{E}_{p(\mathbf{x}_{k-1})}\left[\mathbb{E}_{p(\mathbf{x}_k,\mathbf{u}_k|\mathbf{x}_{k-1})}\left[c_k^{\mathrm{x}}\left(\mathbf{X}_k\right) + c_k^{\mathrm{u}}\left(\mathbf{U}_k\right)\right]\right]$$
$$+\mathbb{E}_{p(\mathbf{x}_{N-1})}\left[D_{\mathrm{KL}}\left(p\left(\mathbf{x}_N, \mathbf{u}_N|\mathbf{x}_{N-1}\right)||q\left(\mathbf{x}_N, \mathbf{u}_N|\mathbf{x}_{N-1}\right)\right) + \mathbb{E}\left[\bar{c}(\mathbf{X}_N, \mathbf{U}_N)\right]\right],$$

where $\bar{c}(\mathbf{X}_N, \mathbf{U}_N) = c_N^{\mathrm{x}}\left(\mathbf{X}_N\right) + c_N^{\mathrm{u}}\left(\mathbf{U}_N\right) + l_{N+1}^{\star}\left(\mathbf{X}_N\right)$, with $l_{N+1}^{\star}\left(\mathbf{X}_N\right) = 0$. Hence, Problem 1 can be recast as the sum of the following two sub-problems:

$$\min_{\{\mathbf{w}_k \in \Delta^{n_\pi}\}_{1:N-1}} D_{\mathrm{KL}}\left(p_{0:N-1}||q_{0:N-1}\right) + \sum_{k=1}^{N-1} \mathbb{E}_{p(\mathbf{x}_{k-1})}\left[\mathbb{E}_{p(\mathbf{x}_k,\mathbf{u}_k|\mathbf{x}_{k-1})}\left[c_k^{\mathrm{x}}\left(\mathbf{X}_k\right) + c_k^{\mathrm{u}}\left(\mathbf{U}_k\right)\right]\right]$$
$$\text{s.t. } p_{\mathbf{u}}\left(\mathbf{u}_k|\mathbf{x}_{k-1}\right) = \sum_{i\in\mathcal{P}} \mathbf{w}_k^{(i)} \pi^{(i)}\left(\mathbf{u}_k|\mathbf{x}_{k-1}\right), \quad \forall k \in 1:N-1,$$

and

$$\min_{\mathbf{w}_N \in \Delta^{n_\pi}} \mathbb{E}_{p(\mathbf{x}_{N-1})}\left[D_{\mathrm{KL}}\left(p\left(\mathbf{x}_N, \mathbf{u}_N|\mathbf{x}_{N-1}\right)||q\left(\mathbf{x}_N, \mathbf{u}_N|\mathbf{x}_{N-1}\right)\right) + \mathbb{E}\left[\bar{c}(\mathbf{X}_N, \mathbf{U}_N)\right]\right]$$
$$\text{s.t. } p_{\mathbf{u}}\left(\mathbf{u}_N|\mathbf{x}_{N-1}\right) = \sum_{i\in\mathcal{P}} \mathbf{w}_N^{(i)} \pi^{(i)}\left(\mathbf{u}_N|\mathbf{x}_{N-1}\right),$$

In the above problem, since the expectation in the cost functional does not depend on $\mathbf{w}_N$, the optimal cost is $\mathbb{E}_{p(\mathbf{x}_{N-1})}[l_N^{\star}\left(\mathbf{X}_{N-1}\right)]$, where $l_N^{\star}\left(\mathbf{X}_{N-1}\right)$ is obtained by solving:

$$\min_{\mathbf{w}_N \in \Delta^{n_\pi}} D_{\mathrm{KL}}\left(p\left(\mathbf{x}_N, \mathbf{u}_N|\mathbf{x}_{N-1}\right)||q\left(\mathbf{x}_N, \mathbf{u}_N|\mathbf{x}_{N-1}\right)\right) + \mathbb{E}\left[\bar{c}(\mathbf{X}_N, \mathbf{U}_N)\right]$$
$$\text{s.t. } p_{\mathbf{u}}\left(\mathbf{u}_N|\mathbf{x}_{N-1}\right) = \sum_{i\in\mathcal{P}} \mathbf{w}_N^{(i)} \pi^{(i)}\left(\mathbf{u}_N|\mathbf{x}_{N-1}\right),$$

This is the optimization problem solved at $k = N$ in Algorithm 1.
Note that $\mathbb{E}_{p(\mathbf{x}_{N-1})}[l_N^{\star}\left(\mathbf{X}_{N-1}\right)]$ is bounded (Assumption 2) and can be conveniently written as $\mathbb{E}_{p(\mathbf{x}_{N-2})}\left[\mathbb{E}_{p(\mathbf{x}_{N-1},\mathbf{u}_{N-1}|\mathbf{x}_{N-2})}\left[l_N^{\star}\left(\mathbf{X}_{N-1}\right)\right]\right]$, so that the cost in (1) becomes $D_{\mathrm{KL}}\left(p_{0:N-1}||q_{0:N-1}\right) + \sum_{k=1}^{N-1} \mathbb{E}_{p(\mathbf{x}_{k-1})}\left[\mathbb{E}_{p(\mathbf{x}_k,\mathbf{u}_k|\mathbf{x}_{k-1})}\left[c_k^{\mathrm{x}}\left(\mathbf{X}_k\right) + c_k^{\mathrm{u}}\left(\mathbf{U}_k\right)\right]\right] + \mathbb{E}_{p(\mathbf{x}_{N-2})}\left[\mathbb{E}_{p(\mathbf{x}_{N-1},\mathbf{u}_{N-1}|\mathbf{x}_{N-2})}\left[l_N^{\star}\left(\mathbf{X}_{N-1}\right)\right]\right]$. Hence, iterating the arguments above, we obtain that the optimal solution of Problem 1 can be found by solving,

at each k, the optimization problem given in Algorithm 1 with $l^{\star}_{N+1}(\mathbf{X}_N)$ equal to 0 and, $\forall k \in 1 : N-1$, $l^{\star}_{k+1}(\mathbf{X}_k)$ being the optimal value of the optimization problem at $k+1$. This yields the desired conclusions. □

Proof of Property 2. Following [16], we note that the constraint set is convex and therefore we only need to prove convexity of the cost. We do it by showing that the Hessian, $\mathbf{H}(\mathbf{x}_{k-1})$, with entries h_{ij} for $i, j \in \mathcal{P}$, is positive semi-definite. For compactness, we introduce the following shorthand notation: $\pi^{(i)}_{\mathbf{u},k} := \pi^{(i)}(\mathbf{u}_k|\mathbf{x}_{k-1})$, $p_{\mathbf{x},k} := p_{\mathbf{x}}(\mathbf{x}_k|\mathbf{x}_{k-1}, \mathbf{u}_k)$, $q_{\mathbf{x},\mathbf{u},k} := q(\mathbf{x}_k, \mathbf{u}_k|\mathbf{x}_{k-1})$. Embedding the constraint $p_{\mathbf{u}}(\mathbf{u}_k|\mathbf{x}_{k-1}) = \sum_{i\in\mathcal{P}} \mathbf{w}^{(i)}_k \pi^{(i)}(\mathbf{u}_k|\mathbf{x}_{k-1})$ directly into the cost functional of problem (2), $l_k(\mathbf{x}_{k-1}) := D_{\text{KL}}(p(\mathbf{x}_k, \mathbf{u}_k|\mathbf{x}_{k-1}) \,||\, q(\mathbf{x}_k, \mathbf{u}_k|\mathbf{x}_{k-1})) + \mathbb{E}_{p(\mathbf{x}_k,\mathbf{u}_k|\mathbf{x}_{k-1})}[\bar{c}(\mathbf{X}_k, \mathbf{U}_k)]$, the first derivative of $l_k(\mathbf{x}_{k-1})$ with respect to the decision variable $\mathbf{w}^{(i)}_k$ is:

$$\begin{aligned}\frac{\partial l_k(\mathbf{x}_{k-1})}{\partial \mathbf{w}^{(i)}_k} &= \frac{\partial}{\partial \mathbf{w}^{(i)}_k}\left(\int\!\!\int \sum_{j\in\mathcal{P}} \mathbf{w}^{(j)}_k \pi^{(j)}_{\mathbf{u},k} p_{\mathbf{x},k} \ln\left(\frac{\sum_{j\in\mathcal{P}} \mathbf{w}^{(j)}_k \pi^{(j)}_{\mathbf{u},k} p_{\mathbf{x},k}}{q_{\mathbf{x},\mathbf{u},k}}\right) d\mathbf{x}_k d\mathbf{u}_k \right.\\ &\quad \left. + \int\!\!\int \sum_{j\in\mathcal{P}} \mathbf{w}^{(j)}_k \pi^{(j)}_{\mathbf{u},k} p_{\mathbf{x},k} \bar{c}(\mathbf{x}_k, \mathbf{u}_k) d\mathbf{x}_k d\mathbf{u}_k\right)\\ &= \int\!\!\int \pi^{(i)}_{\mathbf{u},k} p_{\mathbf{x},k}\left(\ln\left(\frac{\sum_{j\in\mathcal{P}} \mathbf{w}^{(j)}_k \pi^{(j)}_{\mathbf{u},k} p_{\mathbf{x},k}}{q_{\mathbf{x},\mathbf{u},k}}\right) + \bar{c}(\mathbf{x}_k, \mathbf{u}_k) + 1\right) d\mathbf{x}_k d\mathbf{u}_k,\end{aligned}$$

where the integrals, well defined by Assumption 1 and 3, are respectively over $\mathcal{X}$ and $\mathcal{U}$. Next, we compute the second order derivative:

$$h_{ij}(\mathbf{x}_{k-1}) := \frac{\partial^2 l_k(\mathbf{x}_{k-1})}{\partial \mathbf{w}^{(i)}_k \partial \mathbf{w}^{(j)}_k} = \int\!\!\int \frac{\pi^{(i)}_{\mathbf{u},k} \pi^{(j)}_{\mathbf{u},k} p_{\mathbf{x},k}}{\sum_{t\in\mathcal{P}} \mathbf{w}^{(t)}_k \pi^{(t)}_{\mathbf{u},k}} d\mathbf{x}_k d\mathbf{u}_k, \quad \forall i, j \in \mathcal{P}.$$

Pick now a non-zero vector $\mathbf{v} \in \mathbb{R}^{n_\pi}$. We get:

$$\mathbf{v}^\top \mathbf{H}(\mathbf{x}_{k-1})\mathbf{v} = \sum_{i,j} v_i v_j h_{ij}(\mathbf{x}_{k-1}) = \int\!\!\int \mathbf{v}^\top \mathbf{A}(\mathbf{x}_k, \mathbf{u}_k, \mathbf{x}_{k-1})\mathbf{v}\, d\mathbf{x}_k d\mathbf{u}_k, \tag{3}$$

where $\mathbf{A}(\mathbf{x}_k, \mathbf{u}_k, \mathbf{x}_{k-1}) := \dfrac{p_{\mathbf{x},k}}{\sum\limits_{t\in\mathcal{P}} \mathbf{w}^{(t)}_k \pi^{(t)}_{\mathbf{u},k}} \boldsymbol{\pi}_{\mathbf{u},k} \boldsymbol{\pi}^\top_{\mathbf{u},k}$, with $\boldsymbol{\pi}_{\mathbf{u},k} := [\pi^{(1)}_{\mathbf{u},k}, \ldots, \pi^{(n_\pi)}_{\mathbf{u},k}]^\top$.

Convexity then follows by noticing that $\mathbf{A}(\mathbf{x}_k, \mathbf{u}_k, \mathbf{x}_{k-1})$, is the outer product of vector $\boldsymbol{\pi}_{\mathbf{u},k}$ by itself rescaled by a non-negative fraction. Therefore, $\mathbf{A}(\mathbf{x}_k, \mathbf{u}_k, \mathbf{x}_{k-1})$ is positive semi-definite for all $\mathbf{x}_k, \mathbf{u}_k, \mathbf{x}_{k-1}$.

To establish strict convexity, we notice that the quantity in (3) can be rewritten as:

$$\mathbf{v}^\top \mathbf{H}(\mathbf{x}_{k-1})\mathbf{v} = \int\!\!\int \frac{p_{\mathbf{x},k}}{\sum_{t\in\mathcal{P}} \mathbf{w}^{(t)}_k \pi^{(t)}_{\mathbf{u},k}} (\mathbf{v}^\top \boldsymbol{\pi}_{\mathbf{u},k})^2\, d\mathbf{x}_k d\mathbf{u}_k,$$

which is zero if and only if $\mathbf{v}^\top \boldsymbol{\pi}_{\mathbf{u},k} = 0$ almost everywhere in the support of $p_{\mathbf{x},k}$. If the primitives $\pi^{(i)}_{\mathbf{u},k}$, which are positive over $\mathcal{U}$ (by Assumption 3), are linearly independent, there is no non-zero vector $\mathbf{v}$ such that $\mathbf{v}^\top \boldsymbol{\pi}_{\mathbf{u},k} = 0$ [63]. Hence, in this case $\mathbf{H}(\mathbf{x}_{k-1})$ is positive definite and thus the cost functional $l_k(\mathbf{x}_{k-1})$ is strictly convex in $\mathbf{w}_k$. □

References

1. Cacciatore, T., Nowlan, S.: Mixtures of controllers for jump linear and non-linear plants. In: Advances in Neural Information Processing Systems, vol. 6 (1993)
2. Cai, W., Jiang, J., Wang, F., Tang, J., Kim, S., Huang, J.: A survey on mixture of experts in large language models. IEEE Trans. Knowl. Data Eng. **37**(7), 3896–3915 (2025). https://doi.org/10.1109/TKDE.2025.3554028
3. Carlier, G., Duval, V., Peyré, G., Schmitzer, B.: Convergence of entropic schemes for optimal transport and gradient flows. SIAM J. Math. Anal. **49**(2), 1385–1418 (2017). https://doi.org/10.1137/15M1050264
4. Centorrino, V., Gokhale, A., Davydov, A., Russo, G., Bullo, F.: Positive competitive networks for sparse reconstruction. Neural Comput. **36**(6), 1163–1197 (2024). https://doi.org/10.1162/neco_a_01657
5. Cesa-Bianchi, N., Lugosi, G.: Prediction, Learning, and Games. Cambridge University Press (2006)
6. Cover, T.M., Thomas, J.A.: Elements of Information Theory (Wiley Series in Telecommunications and Signal Processing). Wiley-Interscience, USA (2006)
7. Diamond, S., Boyd, S.: CVXPY: a Python-embedded modeling language for convex optimization. J. Mach. Learn. Res. **17**(83), 1–5 (2016)
8. Doya, K., Samejima, K., Katagiri, K.i., Kawato, M.: Multiple model-based reinforcement learning. Neural Comput. **14**(6), 1347–1369 (2002). https://doi.org/10.1162/089976602753712972
9. Friston, K.: The free-energy principle: a rough guide to the brain? Trends Cogn. Sci. **13**(7), 293–301 (2009). https://doi.org/10.1016/j.tics.2009.04.005
10. Friston, K.: The free-energy principle: a unified brain theory? Nat. Rev. Neurosci. **11**(2), 127–138 (2010). https://doi.org/10.1038/nrn2787
11. Friston, K., et al.: The free energy principle made simpler but not too simple. Phys. Rep. **1024**, 1–29 (2023). https://doi.org/10.1016/j.physrep.2023.07.001
12. Friston, K., FitzGerald, T., Rigoli, F., Schwartenbeck, P., Pezzulo, G.: Active inference: a process theory. Neural Comput. **29**(1), 1–49 (2017). https://doi.org/10.1162/NECO_a_00912
13. Gagliardi, D., Russo, G.: On a probabilistic approach to synthesize control policies from example datasets. Automatica **137**, 110121 (2022). https://doi.org/10.1016/j.automatica.2021.110121
14. Garrabé, E., Russo, G.: Probabilistic design of optimal sequential decision-making algorithms in learning and control. Annu. Rev. Control. **54**, 81–102 (2022). https://doi.org/10.1016/j.arcontrol.2022.09.003
15. Garrabe, E., Jesawada, H., Vecchio, C.D., Russo, G.: On convex data-driven inverse optimal control for nonlinear, non-stationary and stochastic systems. Automatica **173**, 112015 (2025). https://doi.org/10.1016/j.automatica.2024.112015
16. Garrabé, É., Lamberti, M., Russo, G.: Optimal decision-making for autonomous agents via data composition. IEEE Control Syst. Lett. **7**, 2557–2562 (2023). https://doi.org/10.1109/LCSYS.2023.3287450
17. Gomi, H., Kawato, M.: Recognition of manipulated objects by motor learning with modular architecture networks. Neural Netw. **6**(4), 485–497 (1993). https://doi.org/10.1016/S0893-6080(05)80053-X

18. Haarnoja, T., Tang, H., Abbeel, P., Levine, S.: Reinforcement learning with deep energy-based policies. In: International Conference on Machine Learning, pp. 1352–1361. PMLR (2017)
19. Hansel, K., Urain, J., Peters, J., Chalvatzaki, G.: Hierarchical policy blending as inference for reactive robot control. In: 2023 IEEE International Conference on Robotics and Automation (ICRA), pp. 10181–10188. IEEE (2023). https://doi.org/10.1109/ICRA48891.2023.10161374
20. Haruno, M., Wolpert, D.M., Kawato, M.: Mosaic model for sensorimotor learning and control. Neural Comput. **13**(10), 2201–2220 (2001). https://doi.org/10.1162/089976601750541778
21. Hawkins, J.: A thousand brains: a new theory of intelligence. Basic Books (2021)
22. Heins, C., Millidge, B., Da Costa, L., Mann, R.P., Friston, K.J., Couzin, I.D.: Collective behavior from surprise minimization. Proc. Natl. Acad. Sci. **121**(17) (2024). https://doi.org/10.1073/pnas.2320239121
23. Hinton, G.E.: Training products of experts by minimizing contrastive divergence. Neural Comput. **14**(8), 1771–1800 (2002). https://doi.org/10.1162/089976602760128018
24. Hutsebaut-Buysse, M., Mets, K., Latré, S.: Hierarchical reinforcement learning: a survey and open research challenges. Mach. Learn. Knowl. Extr. **4**(1), 172–221 (2022). https://doi.org/10.3390/make4010009
25. Jacobs, R.A., Jordan, M.I., Nowlan, S.J., Hinton, G.E.: Adaptive mixtures of local experts. Neural Comput. **3**(1), 79–87 (1991). https://doi.org/10.1162/neco.1991.3.1.79
26. Jaiswal, P., Honnappa, H., Rao, V.A.: On the statistical consistency of risk-sensitive Bayesian decision-making. In: Proceedings of the 37th International Conference on Neural Information Processing Systems. NIPS 2023. Curran Associates Inc., Red Hook (2024)
27. Jordan, M.I., Ghahramani, Z., Jaakkola, T.S., Saul, L.K.: An introduction to variational methods for graphical models. Mach. Learn. **37**, 183–233 (1999). https://doi.org/10.1023/A:1007665907178
28. Jose, S.T., Simeone, O.: Free energy minimization: a unified framework for modeling, inference, learning, and optimization [lecture notes]. IEEE Signal Process. Mag. **38**(2), 120–125 (2021). https://doi.org/10.1109/MSP.2020.3041414
29. Kouw, W.M.: Information-seeking polynomial NARX model-predictive control through expected free energy minimization. IEEE Control Syst. Lett. (2023). https://doi.org/10.1109/LCSYS.2023.3347190
30. Kouw, W.M.: Planning to avoid ambiguous states through gaussian approximations to non-linear sensors in active inference agents, pp. 195–208. Springer, Cham (2024). https://doi.org/10.1007/978-3-031-77138-5_13
31. Kullback, S., Leibler, R.: On information and sufficiency. Ann. Math. Stat. **22**, 79–87 (1951)
32. Lake, B.M., Ullman, T.D., Tenenbaum, J.B., Gershman, S.J.: Building machines that learn and think like people. Behav. Brain Sci. **40**, e253 (2017). https://doi.org/10.1017/S0140525X16001837
33. Van de Maele, T., Verbelen, T., Çatal, O., Dhoedt, B.: Embodied object representation learning and recognition. Front. Neurorobot. **16**, 840658 (2022). https://doi.org/10.3389/fnbot.2022.840658
34. Masoudnia, S., Ebrahimpour, R.: Mixture of experts: a literature survey. Artif. Intell. Rev. **42**(2), 275–293 (2012). https://doi.org/10.1007/s10462-012-9338-y
35. Mazzaglia, P., Verbelen, T., Dhoedt, B.: Contrastive active inference. In: Ranzato, M., Beygelzimer, A., Dauphin, Y., Liang, P., Vaughan, J.W. (eds.) Advances in Neural Information Processing Systems, vol. 34, pp. 13870–13882. Curran Associates, Inc. (2021)
36. Millidge, B., Tschantz, A., Seth, A.K., Buckley, C.L.: On the relationship between active inference and control as inference. In: Active Inference, pp. 3–11. Springer (2020). https://doi.org/10.1007/978-3-030-64919-7_1

37. Mountcastle, V.B.: An organizing principle for cerebral function: the unit module and the distributed system (1978). https://api.semanticscholar.org/CorpusID:59731439
38. Murphy, K.P.: Probabilistic Machine Learning: Advanced Topics. MIT Press (2023). http://probml.github.io/book2
39. Nutz, M.: Introduction to Entropic Optimal Transport. Lecture Notes (2020)
40. Paraschos, A., Daniel, C., Peters, J., Neumann, G.: Using probabilistic movement primitives in robotics. Auton. Robot. **42**(3), 529–551 (2017). https://doi.org/10.1007/s10514-017-9648-7
41. Parr, T., Friston, K.J.: Generalised free energy and active inference. Biol. Cybern. **113**(5–6), 495–513 (2019). https://doi.org/10.1007/s00422-019-00805-w
42. Parr, T., Pezzulo, G., Friston, K.J.: Active Inference: The Free Energy Principle in Mind, Brain, and Behavior. The MIT Press (2022)
43. Pateria, S., Subagdja, B., Tan, A.H., Quek, C.: Hierarchical reinforcement learning: a comprehensive survey. ACM Comput. Surv. (CSUR) **54**(5), 1–35 (2021). https://doi.org/10.1145/3453160
44. Paul, A., Isomura, T., Razi, A.: On predictive planning and counterfactual learning in active inference. Entropy **26**(6), 484 (2024). https://doi.org/10.3390/e26060484
45. Paul, A., Sajid, N., Da Costa, L., Razi, A.: On efficient computation in active inference. Expert Syst. Appl. **253**, 124315 (2024). https://doi.org/10.1016/j.eswa.2024.124315
46. Pezzulo, G., Rigoli, F., Friston, K.: Active inference, homeostatic regulation and adaptive behavioural control. Prog. Neurobiol. **134**, 17–35 (2015). https://doi.org/10.1016/j.pneurobio.2015.09.001
47. Pezzulo, G., Rigoli, F., Friston, K.J.: Hierarchical active inference: a theory of motivated control. Trends Cogn. Sci. **22**(4), 294–306 (2018). https://doi.org/10.1016/j.tics.2018.01.009
48. Prescott, T.J., Wilson, S.P.: Understanding brain functional architecture through robotics. Sci. Robot. **8**(78), eadg6014 (2023). https://doi.org/10.1126/scirobotics.adg6014
49. Russin, J., McGrath, S.W., Williams, D.J., Elber-Dorozko, L.: From Frege to chatGPT: compositionality in language, cognition, and deep neural networks (2024). https://arxiv.org/abs/2405.15164
50. Samejima, K., Katagiri, K., Doya, K., Kawato, M.: Multiple model-based reinforcement learning for nonlinear control. Electron. Commun. Japan (Part III: Fundamental Electronic Science) **89**(9), 54–69 (2006). https://doi.org/10.1002/ecjc.20266
51. Sanger, T.D.: Risk-aware control. Neural Comput. **26**(12), 2669–2691 (2014). https://doi.org/10.1162/NECO_a_00662
52. Shafiei, A., Jesawada, H., Friston, K., Russo, G.: Distributionally robust free energy principle for decision-making. Nat. Commun. **17**, 707 (2026). https://doi.org/10.1038/s41467-025-67348-6
53. Simeone, O.: A brief introduction to machine learning for engineers. Found. Trends® Signal Process. **12**(3–4), 200–431 (2018). https://doi.org/10.1561/2000000102
54. Sutton, R.S., Barto, A.G., et al.: Reinforcement Learning: An Introduction, vol. 1. MIT Press, Cambridge (1998)
55. Tani, J., Nolfi, S.: Learning to perceive the world as articulated: an approach for hierarchical learning in sensory-motor systems. Neural Netw. **12**(7–8), 1131–1141 (1999). https://doi.org/10.1016/S0893-6080(99)00060-X
56. Vezhnevets, A.S., et al.: Feudal networks for hierarchical reinforcement learning. In: International Conference on Machine Learning, pp. 3540–3549. PMLR (2017)
57. Vijayaraghavan, P., Queißer, J.F., Flores, S.V., Tani, J.: Development of compositionality through interactive learning of language and action of robots. Sci. Robot. **10**(98) (2025). https://doi.org/10.1126/scirobotics.adp0751

58. Vincent, J.F., Bogatyreva, O.A., Bogatyrev, N.R., Bowyer, A., Pahl, A.K.: Biomimetics: its practice and theory. J. R. Soc. Interface **3**(9), 471–482 (2006). https://doi.org/10.1098/rsif.2006.0127
59. Willi, T., Obando-Ceron, J., Foerster, J., Dziugaite, K., Castro, P.S.: Mixture of Experts in a Mixture of RL settings (2024). https://arxiv.org/abs/2406.18420
60. Wilson, S., et al.: The Robotarium: globally impactful opportunities, challenges, and lessons learned in remote-access, distributed control of multirobot systems. IEEE Control Syst. Mag. **40**(1), 26–44 (2020). https://doi.org/10.1109/MCS.2019.2949973
61. Wolpert, D.M., Kawato, M.: Multiple paired forward and inverse models for motor control. Neural Netw. **11**(7–8), 1317–1329 (1998). https://doi.org/10.1016/S0893-6080(98)00066-5
62. Yoo, J., Kim, H.J., Johansson, K.H.: Mapless indoor localization by trajectory learning from a crowd. In: 2016 International Conference on Indoor Positioning and Indoor Navigation (IPIN), pp. 1–7. IEEE (2016). https://doi.org/10.1109/IPIN.2016.7743685
63. Yu, X., Liu, T., Gong, M., Batmanghelich, K., Tao, D.: An efficient and provable approach for mixture proportion estimation using linear independence assumption. In: Proceedings of the IEEE Conference on Computer Vision and Pattern Recognition, pp. 4480–4489 (2018)
64. Zhang, E., et al.: Transcendence: generative models can outperform the experts that train them. In: Globerson, A., et al. (eds.) Advances in Neural Information Processing Systems, vol. 37, pp. 86985–87012. Curran Associates, Inc. (2024)
65. Ziebart, B.D., Maas, A., Bagnell, J.A., Dey, A.K.: Maximum entropy inverse reinforcement learning. In: Proceedings of the 23rd National Conference on Artificial Intelligence - Volume 3, AAAI 2008, pp. 1433–1438. AAAI Press (2008)

Mobile Manipulation with Active Inference for Long-Horizon Rearrangement Tasks

Corrado Pezzato(✉), Ozan Çatal, Toon Van de Maele, Riddhi J. Pitliya, and Tim Verbelen

VERSES AI Research Lab, Los Angeles, CA 90016, USA
{corrado.pezzato,ozan.catal}@verses.ai

Abstract. Despite growing interest in active inference for robotic control, its application to complex, long-horizon tasks remains untested. We address this gap by introducing a fully hierarchical active inference architecture for goal-directed behavior in realistic robotic settings. Our model combines a high-level active inference model that selects among discrete skills realized via a whole-body active inference controller. This unified approach enables flexible skill composition, online adaptability, and recovery from task failures without requiring offline training. Evaluated on the Habitat Benchmark for mobile manipulation, our method outperforms state-of-the-art baselines across the three long-horizon tasks, demonstrating for the first time that active inference can scale to the complexity of modern robotics benchmarks.

1 Introduction

Active inference offers a principled framework for modeling the action-perception loop, unifying inference and control. Both continuous and discrete formulations have been developed to capture sensorimotor and cognitive processes [18,26], and past works have explored hybrid schemes that integrate discrete decision-making with continuous control in the context of handwriting [19] and oculomotor tasks [16,17].

More recently, hybrid continuous-discrete approaches have shown promise in generating rich, goal-directed behavior in 2D simulated settings for reaching, grasping, and tool use [21,23,24]. However, active inference has yet to demonstrate scalability to the complexity and long time horizons required by modern robotics benchmarks. In particular, no prior work has convincingly shown that active inference alone can match or exceed the performance of state-of-the-art methods in realistic robotic tasks.

In this paper, we address this gap by introducing a fully hierarchical hybrid architecture for active inference-based control in long-horizon mobile manipulation tasks. Figure 1 provides a high-level overview. Our system combines a

C. Pezzato and O. Çatal—Equal contribution.

M. Albarracin et al. (Eds.): IWAI 2025, CCIS 2857, pp. 364–381, 2026.
https://doi.org/10.1007/978-3-032-16955-6_21

high-level active inference agent that reasons over task-relevant abstractions with a novel whole-body controller based on continuous hierarchical active inference [20,22]. This integration allows for flexible and robust skill execution, supports online adaptation, and eliminates the need for offline training. We evaluate our approach on three long-horizon mobile manipulation tasks from the Habitat Benchmark [28], namely `TidyHouse`, `PrepareGroceries`, and `SetTable`. These tasks require complex, multi-step interactions with articulated objects and constrained environments, such as retrieving items from drawers or refrigerators, transporting them across rooms, and placing them on surfaces. Success demands both long-term planning and precise, reactive motor control. Our method outperforms state-of-the-art baselines across all three tasks, providing a compelling demonstration of active inference.

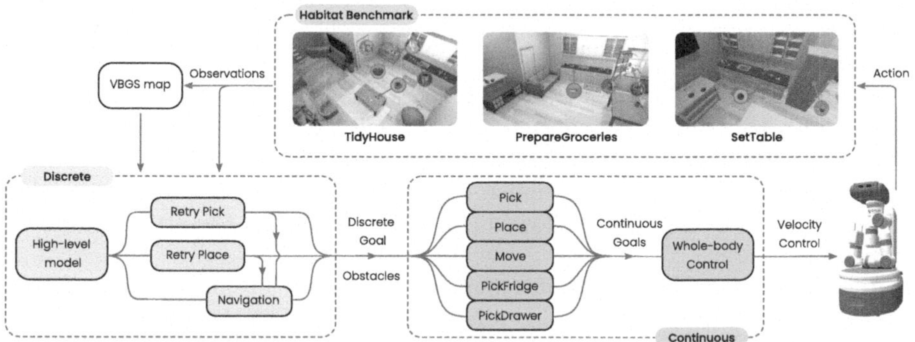

Fig. 1. Solution overview. Overview of the proposed solution and the Habitat Tasks described in detail in Sect. 2.

1.1 Related Work

Long-Horizon mobile manipulation challenges that demand both navigation and manipulation abilities are well-suited to evaluating the effectiveness of embodied AI algorithms. Works like the Habitat Benchmark [28], ThreeDWorld [8], and ManipulaTHOR [5] require robots to navigate in indoor apartments and rearrange household objects. We chose to focus on the Habitat Benchmark because of its challenging nature: it requires continuous motor control (base and arm), interaction with articulated objects (opening drawers and fridges), and involves complicated scene layouts with clutter. In the literature, long-horizon problems have been tackled with task-and-motion-planning (TAMP) approaches [9,12,13,27].

Although effective, these methods often rely on accurate knowledge of the scene and objects, and are computationally expensive. Learning-based approaches have emerged in recent years as an alternative; however, monolithic end-to-end RL solutions to long-horizon tasks are shown to be prone to failure [10,28]. The main reasons for this are the high sample complexity, inefficient exploration, as well as complicated reward design.

A common strategy for addressing long-horizon tasks in RL is to decompose them into shorter, more manageable subtasks. For instance, the authors of Habitat [28] propose a hierarchical framework for mobile manipulation. This integrates classical task planning to generate high-level symbolic goals, while individual low-level skills are trained using RL to achieve these goals. This method demonstrates superior performance compared to monolithic end-to-end RL policies and traditional sense-plan-act pipelines. However, naively chaining multiple skills can result in hand-off issues [28], where the terminal state of one skill falls outside the distribution of states encountered during training by the subsequent skill, or leads to states that are infeasible for it to handle. This is especially an issue for stationary manipulation skills. Prior work often decouples the mobile base from the manipulator to simplify the inverse kinematics of redundant systems [25].

In contrast, [10] highlights that mobile manipulation skills are inherently more robust to error accumulation during skill chaining. By leveraging the robot's full embodiment, these skills enable more effective subtask execution by allowing the robot to reposition itself. Improved subtask formulation, skill composability, and reusability allowed [10] to reach state-of-the-art performance to date on Habitat.

Active inference offers a distinct perspective on decision-making and control, framing both as aspects of a unified inference process. The theory proposes that complex movements can emerge naturally from generative models that encode goals as prior beliefs and observation preferences [18]. Within this framework, the nervous system is seen as maintaining a hierarchical generative model that continuously produces and refines perceptual hypotheses.

Early work on hybrid active inference combining discrete and continuous processes focused on understanding systems such as oculomotion [7,16,17] and handwriting [19]. For instance, [7] proposed linking discrete and continuous states by using Bayesian model averaging over discrete priors, and converting the resulting continuous posterior into a discrete representation through Bayesian model comparison. These models typically use a discrete state-space to generate empirical priors, which then guide a continuous controller through sequences of attractive setpoints to achieve articulated behavior.

Hybrid active inference has been extended to dynamic tasks that demand flexible planning [21,23,24]. These scenarios require agents to infer object dynamics, decompose tasks into subgoals, and coordinate multiple degrees of freedom to execute composite actions. While such studies demonstrate the promise of hybrid active inference in complex motor control, no implementation has yet scaled to meet the complexity of established robotics benchmarks. In this work, we introduce a fully hierarchical hybrid active inference system that, for the first time, outperforms state-of-the-art baselines on the Habitat Benchmark, demonstrating its viability for long-horizon robotic manipulation.

1.2 Habitat Benchmark

The Habitat Benchmark [28] comprises three long-horizon mobile manipulation tasks visualized in Fig. 1: `TidyHouse`, `PrepareGroceries`, and `SetTable`.

In `TidyHouse`, the robot is tasked with relocating five objects from an initial to a designated goal position. Both the start and goal locations are typically on open surfaces such as tables or kitchen counters. The `PrepareGroceries` task involves moving two objects from an already open refrigerator to a countertop and returning one object from the counter back into the fridge. Finally, in `SetTable`, the agent must move a bowl from a closed drawer to a table, and then place an apple retrieved from a closed fridge on the same table. This scenario involves interacting with articulated objects and picking and placing items within confined containers. All tasks are specified as a sequence of subgoals. Each subgoal is a tuple (s_1, s^*), where s_1 is the initial 3D center-of-mass position of an object and s^* denotes its goal position. For instance, `TidyHouse` is defined by a set of five such tuples: $\{(s_1^i, s^{*i})\}_{i=1}^5$. The generated scenes for the tasks are built upon the ReplicaCAD dataset, which provides a diverse set of 105 photorealistic indoor environments. Each episode instantiates a rearrangement task by randomly sampling rigid objects from the YCB dataset and placing them on annotated support surfaces, resulting in cluttered initial configurations.

2 Methods

2.1 Planning with Universal Generative Models

To tackle the Habitat Benchmark, we propose an active inference agent, endowed with a hierarchy of generative models, each minimizing the Variational Free Energy. In this universal generative model [6] actions at one level become the preferences of the level below. As we traverse down the hierarchy, each model's time horizon becomes shorter, and the planning becomes more fine-grained. Each component in the hierarchy has its own set of responsibilities. The high-level model orchestrates the sequence of logical actions to take, i.e., where to move and what to pick or place. The *Navigation model* manages the path planning through the environment, while the *Pick & Place model* coordinates how targets should be approached or released. Finally, at the bottom of the hierarchy lies the active inference whole-body controller, which calculates joint controls for the robot and performs obstacle avoidance. These models keep track of their surroundings using a Variational Bayes approach to Gaussian splatting [14,15], which integrates RGBD observations into a probabilistic world map.

High Level Model. At the most abstract level, the agent consists of a partially observed Markov Decision Process (POMDP) capturing dynamics over possible states of the object that need to be collected. This allows for the formation of beliefs over these states and robust planning of the agent's various skills. State and action inference in this POMDP is achieved by minimizing variational free energy according to the FEP [18]. Crucially, this model tracks the robot's

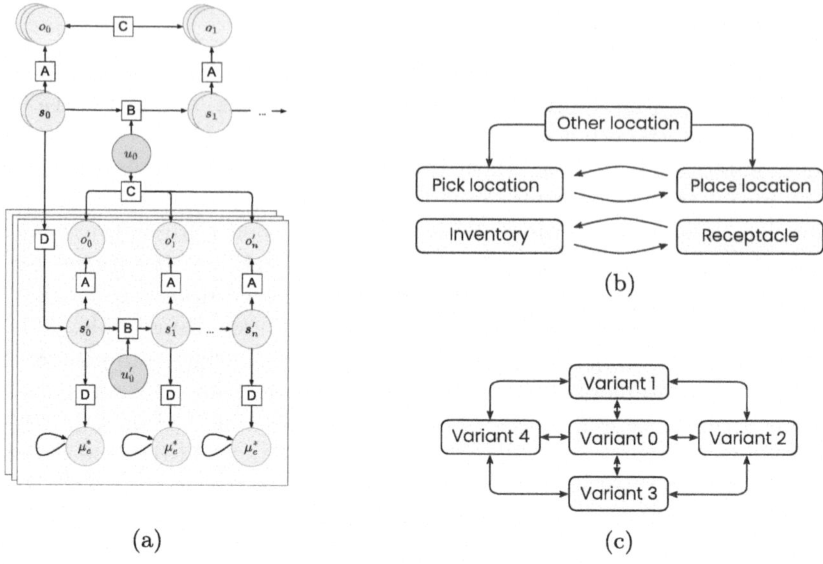

Fig. 2. The Generative Model. (a) The high-level model sequences skills, each implemented by a generative model interacting with the continuous controller. (b) Dynamics at the highest level. The highest level models the robot location (top) and the object location (bottom). The robot can be either at an other location – irrelevant to the task – a pick location, or a place location. The object can be in the robot's inventory or receptacle (location). (c) Dynamics at the pick & place level. The model switches between various approach parameters based on feedback from the low-level controller to increase robustness against failures.

position with respect to the pick/place location as well as the object's relation to the robot, the pick, and the place location. This is then used to schedule either movement or manipulation skills. We present this model using generic active inference terms in Fig. 2a. In contrast to classical active inference models, the actions in our model are abstractions of the skills they represent. The joint controls are generated by the continuous lowest-level model.

Pick & Place Model. Every interaction with the environment can fail due to unforeseen circumstances or invalid prior beliefs. To accommodate this, the hierarchy maintains a retry model for the failure-prone skills such as picking and placing. As shown in Fig. 2c, this model maintains a set of possible approach directions and switches between them based on detected successes or failures of a pick or place. Each approach parameter represents a goal for the lower level controller that can be enacted from that location.

Navigation Model. Crucially, still missing from our description of the hierarchical model is a way to move from one point to another. Discrete active inference models are well-suited for this, as shown in [4]. However, due to the static nature of the environment as well as the prior knowledge about the object location, we opted to use A* pathfinding [11] to generate waypoints that act as extrinsic goals for the low-level controller.

2.2 Perception with Variational Bayes Gaussian Splatting

The models need to infer the structure of the environment to keep track of obstacles and goals in the world. Following the Variational Bayesian approach used throughout the other models, we use a Variational Bayes Gaussian Splat (VBGS) [15] to build a 3D representation of the world from RGBD observations. In this model, the world is represented as a 6D Gaussian mixture over 3D points in space with corresponding 3D color information; the generative model is updated online from observations using Coordinate Ascent Variational Inference (CAVI) [1–3] without needing a replay buffer or observation queue. As with a normal 3D Gaussian Splat, the model effectively forms a radiance field that captures the room's free and occupied space. This allows for easy obstacle avoidance further down the hierarchy. The parameters of the distributions of component k (i.e. μ_k, Σ_k) that generate $\boldsymbol{s}$ and $\boldsymbol{c}$ are random variables, z is the associated mixture component for a given data point, dependent on the categorical parameters π.

2.3 Continuous Control with Active Inference

Recent works proposed Hierarchical Active Inference (HAIF) schemes for continuous control of kinematic chains [20,22]. In this section, we extend previous work [20] to whole-body control of differential drive mobile manipulators. Such robots are composed of a wheeled mobile base and one or more robot arms. We leverage the modularity of HAIF and define one generative model for base control and one for arm control, and then link them through *top-down prediction errors* and *bottom-up predictions.* This results in an overall control scheme that coordinates the whole body of the mobile manipulator at once. An overview of the HAIF approach is depicted in Fig. 3. Importantly, each block in the hierarchy has the same structure and follows the same update rules for state estimation and control. The difference for base and arm control lies in the definition of the generative model $\boldsymbol{g}_e$ and the physical quantities the internal and external beliefs represent, as explained next.

In general, the kinematic generative model $\boldsymbol{g}_e$ computes the extrinsic beliefs at the current level j given the current intrinsic beliefs $\boldsymbol{\mu}_i^j$ and the extrinsic beliefs from the level below $\boldsymbol{\mu}_e^{j-1}$, i.e. $\boldsymbol{\mu}_e^j = \boldsymbol{g}_e(\boldsymbol{\mu}_i^j, \boldsymbol{\mu}_e^{j-1})$. The goal is to obtain a set of equations to describe how the internal beliefs of active inference agents are generated and updated over time. Following [22], the biologically plausible belief update equations are:

$$\dot{\boldsymbol{\mu}}_i^j = \begin{bmatrix} \boldsymbol{\mu}_i^{j'} + \pi_p^j \boldsymbol{\varepsilon}_p^j + \partial_{\boldsymbol{\mu}_i} \boldsymbol{g}_e^\top \pi_e^{j+1} \boldsymbol{\varepsilon}_e^{j+1} + \partial \boldsymbol{f}_i^{j\top} \pi_{\boldsymbol{\mu}_i}^j \boldsymbol{\varepsilon}_{\boldsymbol{\mu}_i}^j \\ -\pi_{\boldsymbol{\mu}_i}^j \boldsymbol{\varepsilon}_{\boldsymbol{\mu}_i}^j \end{bmatrix}, \tag{1}$$

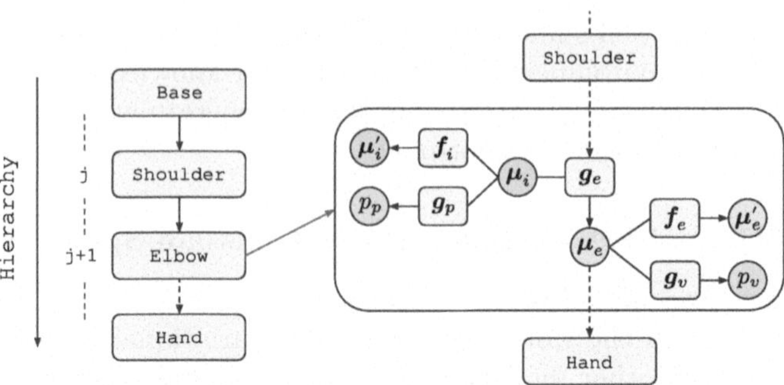

Fig. 3. Overview of the Hierarchical Active Inference approach for mobile manipulator control. Intrinsic and extrinsic beliefs $\boldsymbol{\mu}_i$, $\boldsymbol{\mu}_e$ are internal representations of joint angles and Cartesian poses, respectively. They generate proprioceptive and visual predictions p_p and p_v at their level according to the generative models $\boldsymbol{g}_v$, $\boldsymbol{g}_p$. They are also linked through a kinematic generative model $\boldsymbol{g}_e$. The functions $\boldsymbol{f}_i$ and $\boldsymbol{f}_e$ describe the dynamics and are used to guide goal-directed behavior.

$$\dot{\boldsymbol{\mu}}_e^j = \begin{bmatrix} \boldsymbol{\mu}_e^{j'} - \pi_e^j \boldsymbol{\varepsilon}_e^j + \partial_{\boldsymbol{\mu}_e} \boldsymbol{g}_e^\top \pi_e^{j+1} \boldsymbol{\varepsilon}_e^{j+1} + \pi_v^j \boldsymbol{\varepsilon}_v^j + \partial \boldsymbol{f}_e^{j\top} \pi_{\boldsymbol{\mu}_e}^j \boldsymbol{\varepsilon}_{\boldsymbol{\mu}_e}^j \\ -\pi_{\boldsymbol{\mu}_e}^j \boldsymbol{\varepsilon}_{\boldsymbol{\mu}_e}^j \end{bmatrix}, \tag{2}$$

where π_p, π_e, π_v are precision parameters for proprioceptive, extrinsic, and visual models, and $\boldsymbol{\varepsilon}_p$, $\boldsymbol{\varepsilon}_e$ and $\boldsymbol{\varepsilon}_v$ are the proprioceptive, extrinsic, and visual prediction errors, respectively. Finally, $\boldsymbol{\varepsilon}_{\boldsymbol{\mu}_i}^j = \boldsymbol{\mu}_i^{j'} - \boldsymbol{f}_i^j(\boldsymbol{\mu}_i^j)$ and $\boldsymbol{\varepsilon}_{\boldsymbol{\mu}_e}^j = \boldsymbol{\mu}_e^{j'} - \boldsymbol{f}_e^j(\boldsymbol{\mu}_i^j)$ are dynamics prediction errors, with precision $\pi_{\boldsymbol{\mu}_i}$ and $\pi_{\boldsymbol{\mu}_e}$. These are used to achieve goal-directed behavior and collision avoidance, as explained later. In the equations above, we assumed to be able to observe joint positions, velocities, and link positions such that $\boldsymbol{g}_p$ and $\boldsymbol{g}_v$ are identity mappings. Link positions can be computed from joint positions via forward kinematics or estimated via visual input. We now have to find a suitable form for $\boldsymbol{g}_e$ to easily compute the gradients with respect to intrinsic and extrinsic beliefs for both the arm and the base.

Arm Generative Model. Similarly to [20], the generative model for the HAIF agent comprises an intrinsic belief $\boldsymbol{\mu}_i$ about joint angles and links' lengths, as well as an extrinsic belief $\boldsymbol{\mu}_e$ about a link's absolute Cartesian position and orientation, for each joint j:

$$\boldsymbol{\mu}_i^j = \left[\theta^j, l^j\right]^\top \qquad \boldsymbol{\mu}_e^j = \left[x^j, y^j, z^j, q_w^j, q_x^j, q_y^j, q_z^j\right]^\top = \left[\boldsymbol{t}^j\ \boldsymbol{q}^j\right]^\top. \tag{3}$$

The function $\boldsymbol{g}_e$ describes the 3D position and orientation of the subsequent link of a kinematic chain given the pose of the previous one. We can define the generative model $\boldsymbol{g}_e$ as in [20] (see Appendix A.1 for more details):

$$\boldsymbol{g}_e(\boldsymbol{\mu}_i^j, \boldsymbol{\mu}_e^{j-1}) = \begin{bmatrix} \boldsymbol{t}^{j-1} + \boldsymbol{h}(\boldsymbol{q}^{j-1} \cdot [0\ \boldsymbol{t}^j] \cdot \boldsymbol{q}^{j-1*}) \\ \boldsymbol{q}^{j-1} \cdot \boldsymbol{q}^j \end{bmatrix}, \tag{4}$$

where $\boldsymbol{q}^{j-1*}$ is the conjugate quaternion, $''\cdot''$ represents the Hamilton product, and $\boldsymbol{h}(\cdot)$ is a function that returns the imaginary coefficients of a quaternion. In our HAIF agent, the translation $\boldsymbol{t}^{j-1}$ and quaternion $\boldsymbol{q}^{j-1}$ are given by the extrinsic beliefs $\boldsymbol{\mu}_e^{j-1}$. The translation vector $\boldsymbol{t}^j$ and rotation $\boldsymbol{q}^j$ are instead dependent on the kinematic properties of the current link j and the joint angle and length θ^j, l^j. The generative model can then be fully specified as a function of the intrinsic and extrinsic beliefs, and the gradients can be computed in closed:

$$\frac{\partial \boldsymbol{g}_e}{\partial \boldsymbol{\mu}_i} = \begin{bmatrix} \partial_\theta \boldsymbol{g}_e \\ \partial_l \boldsymbol{g}_e \end{bmatrix} \in \mathbb{R}^{2\times 7}, \qquad \frac{\partial \boldsymbol{g}_e}{\partial \boldsymbol{\mu}_e} = \begin{bmatrix} \partial_{x,y,z} \boldsymbol{g}_e \\ \partial_q \boldsymbol{g}_e \end{bmatrix} \in \mathbb{R}^{7\times 7}. \tag{5}$$

Thanks to the choice of using quaternions as singularity-free orientation representation, these gradients are easy to compute since the terms in the generative model are either linear or quadratic in the parameters, or they appear as arguments of sine and cosine functions.

Differential Drive Generative Model. We now extend the HAIF for robot arm control to a differential drive robot. To do so, we write a simple kinematic model based on Euler's updates where the robot base position and orientation with respect to a world frame are expressed as:

$$\begin{cases} x_{t+1} = x_t + V_t \cos(\theta_t)\delta t \\ y_{t+1} = y_t + V_t \sin(\theta_t)\delta t \\ \theta_{t+1} = \theta_t + \omega_t \delta t \end{cases}, \tag{6}$$

where V, ω are respectively forward and rotational velocities. By considering small wheel increments $\Delta\phi_{R,L} = \phi_{\{R,L\}_t} - \phi_{\{R,L\}_{t-1}}$ in between timesteps where $\phi_{\{R,L\}}$ are the right and left wheel rotations, the expressions for forward and angular velocities result:

$$V_t = \frac{r}{2\delta t}\left(\Delta\phi_R + \Delta\phi_L\right), \tag{7}$$

$$\omega_t = \frac{r}{l\delta t}\left(\Delta\phi_R - \Delta\phi_L\right). \tag{8}$$

The terms r, l are the wheel radius and distance respectively. The generative model for the differential drive HAIF is defined as a one-level hierarchical model where there are two controllable states, the wheel rotations:

$$\boldsymbol{g}_e(\boldsymbol{\mu}_i^j, \boldsymbol{\mu}_e^{j-1}) = \begin{bmatrix} x_{t-1} + \frac{r}{2}\left(\Delta\phi_R + \Delta\phi_L\right)\cos(\theta_{t-1}) \\ y_{t-1} + \frac{r}{2}\left(\Delta\phi_R + \Delta\phi_L\right)\sin(\theta_{t-1}) \\ \theta_{t-1} + \frac{r}{l}\left(\Delta\phi_R - \Delta\phi_L\right) \end{bmatrix}, \tag{9}$$

We set the intrinsic beliefs to be wheel rotations and extrinsic beliefs to be position $x - y$ and orientation θ with respect to a world frame:

$$\boldsymbol{\mu}_i = \left[\phi_R,\ \phi_L\right]^\top \qquad \boldsymbol{\mu}_e = \left[x,\ y,\ \theta\right]^\top. \tag{10}$$

The internal and external beliefs are then updated through the gradient of the generative model:

$$\frac{\partial \boldsymbol{g}_e}{\partial \boldsymbol{\mu}_i} = \begin{bmatrix} \partial_{\phi_R} \boldsymbol{g}_e \\ \partial_{\phi_L} \boldsymbol{g}_e \end{bmatrix} \in \mathbb{R}^{2\times 3}, \qquad \frac{\partial \boldsymbol{g}_e}{\partial \boldsymbol{\mu}_e} = \begin{bmatrix} \partial_x \boldsymbol{g}_e \\ \partial_y \boldsymbol{g}_e \\ \partial_\theta \boldsymbol{g}_e \end{bmatrix} \in \mathbb{R}^{3\times 3}. \tag{11}$$

Whole-Body Generative Model. The arm and base models can be combined into a single whole-body model by defining an overall hierarchical structure that combines the two. From Eqs. (1) and (2), one can notice how the prediction errors at the level above $\boldsymbol{\varepsilon}_e^{j+1}$ influence the beliefs at the current level $\dot{\boldsymbol{\mu}}_i^j$ and $\dot{\boldsymbol{\mu}}_e^j$. By this logic, the extrinsic prediction errors of the first level of the hierarchy will not have any influence since there is no level left below in the chain. However, we can propagate these errors back to the top level of the hierarchy of the mobile base kinematic model. By doing so, the base can further minimize free energy by moving its wheels. In turn, we can propagate up from the base kinematic model the predictions about the first link's position and orientation of the robot arm, closing the loop. Mathematically, all equations remain the same apart from the one corresponding to the update of internal beliefs $\dot{\boldsymbol{\mu}}_i^j$ for the base, which becomes:

$$\dot{\boldsymbol{\mu}}_i^0 = \begin{bmatrix} \boldsymbol{\mu}_i^{0'} + \pi_p^0 \boldsymbol{\varepsilon}_p^0 + \partial_{\boldsymbol{\mu}_i} \boldsymbol{g}_e^\top \pi_e^1 (\kappa_{base} \boldsymbol{\varepsilon}_{e,base}^1 + \kappa_{arm} \boldsymbol{\varepsilon}_{e,arm}^2) + \partial \boldsymbol{f}_i^{0\top} \pi_{\boldsymbol{\mu}_i}^0 \boldsymbol{\varepsilon}_{\boldsymbol{\mu}_i}^0 \\ -\pi_{\boldsymbol{\mu}_i}^0 \boldsymbol{\varepsilon}_{\boldsymbol{\mu}_i}^0 \end{bmatrix} \tag{12}$$

where κ_{base} and κ_{arm} are tuning parameters to weight the effect of arm and base prediction errors. By tuning these parameters, one can shape robot behavior, for instance, to allow more or less base response due to the arm's extrinsic prediction errors. The equation for $\dot{\boldsymbol{\mu}}_e^0$ remains the same since the gradient with respect to extrinsic beliefs is zero. This is because the values x_{t-1}, y_{t-1}, θ_{t-1} are simply constants from the previous time step and not beliefs from the level below.

This simple change allows using the base motion to minimize the arm's prediction errors, extending the arm's reachability beyond its stationary workspace. This is crucial for the successful completion of the Habitat Benchmark. Additionally, we still preserve the ability to send individual goals to the base as explained below.

Goals, Obstacles, and Control. To realize goal-directed behavior, we can define attractive goals and repulsive forces in $\boldsymbol{f}_i$ and $\boldsymbol{f}_e$ as in [20,22]. Goals can be both intrinsic (joint positions) or extrinsic (Cartesian poses) for the arm and base, and they can be combined to define future desired states $\boldsymbol{\mu}^*$. In general, goals act as attractors, forming dynamic functions $\boldsymbol{f}_a = \kappa_a(\boldsymbol{\mu}^* - \boldsymbol{\mu})$ that linearly minimize the distance between the desired and current states, both for intrinsic and extrinsic states. The desired states can be defined flexibly in terms of the current beliefs as

$$\boldsymbol{\mu}^* = N\boldsymbol{\mu} + \boldsymbol{n}^*, \tag{13}$$

where N achieves dynamic behaviors, such as keeping a limb vertical by imposing the x, y coordinates of a link to be the same as the previous one, while $\boldsymbol{n}^*$ imposes an attractor to a static configuration. Some examples of basic goals that can be given to the mobile manipulator in the Habitat Benchmark are 1) **End-effector** goal: the robot will use its whole body to reach a target (x^*, y^*, z^*) position, 2) **Base goal**: the robot will move its base to reach a goal (x^*, y^*, θ^*), where θ^* can be updated over time such that the robot faces the goal $\theta_t^* = \arctan 2(y^* - y_t, x^* - x_t)$, 3) **Arm joint goal**: the robot will reach a specific joint configuration, 4) **Combinations of the above**: the user can mix goals for example making the base move while keeping the arm in a certain joint configuration. The same idea of attractive forces can be used for collision avoidance through repulsive forces, where a repulsive state $\boldsymbol{\mu}^!$ has to be avoided (see Appendix A.2).

Given a goal, the control action is computed by minimizing the proprioceptive component of the free energy with respect to the control signals [22]:

$$\dot{\boldsymbol{a}} = -\partial_a \mathcal{F}_p = -\partial_a \tilde{\boldsymbol{s}}_p \pi_p \tilde{\boldsymbol{\varepsilon}}_p, \tag{14}$$

where $-\partial_a \tilde{\boldsymbol{s}}_p$ is the partial derivative of proprioceptive observations with respect to the control, and $\tilde{\boldsymbol{\varepsilon}}_p = \tilde{\boldsymbol{s}}_p - \boldsymbol{g}_p(\boldsymbol{\mu})$ are the generalized proprioceptive prediction errors.

2.4 Whole-Body High-Level Skills for Objects Rearrangement

Similarly to the chosen Habitat Baseline [10], we define a set of abstract high-level skills that the high-level planner can sequence at runtime. These skills are implemented with the whole-body controller, and are divided into `Pick`, `Place`, `Move`, `PickFromFridge`, `PickFromDrawer`. The skills are defined as a fixed sequence of goals given to the whole-body controller to realize an overall behavior. Every skill computes a whole-body control action for the robot. Details about the skills can be found in Appendix A.3.

3 Experiments

3.1 Experiments Setup

Agent Embodiment: The Habitat Benchmark employs a Fetch robot that features a mobile wheeled base, a 7DoF robotic arm, and a parallel-jaw gripper. It is equipped with two RGB-D cameras with a resolution of 128×128 pixels on both the arm and the head. The robot perceives its state through proprioceptive sensing, which includes joint angles of the arm and the Cartesian coordinates of the end-effector. The robot can also sense the goal positions (3DoF), as well as a scalar to indicate whether an object is held. Obstacle positions are sensed by querying the map model.

Action Space: The action space is a 10DoF continuous space, for whole-body control. It is composed of forward and angular base velocities, a 7DoF arm velocity action, and a 1DoF gripper action. Grasping is abstract as in previous work [10,28], such that if the gripper action is positive, the object closest to the end-effector within 15cm will be snapped to the gripper. An object is instead released by a negative action.

Evaluation Metrics: Each task in the Habitat Benchmark is composed of a sequence of subtasks that must be completed. As in previous work [10,28], we measure performance by reporting the completion rate at each subtask stage, with the success rate of the final subtask representing the overall task success. Notably, if the previous subtask has failed, the current subtask is also considered a failure independently of its outcome. At the start of each evaluation episode, the robot's base is placed at a random position and orientation, ensuring no collisions, while the arm begins in a resting configuration.

Baselines. We compare our model against methods in [10], namely a Monolithic RL approach and a Multi-skill RL Mobile manipulation (MM). The latter had the best performance among several learning-based and classical task and motion planning approaches. The Monolithic RL is an end-to-end RL policy for each complete task (`TidyHouse`, `PrepareGroceries`, and `SetTable`). Different reward functions are selected according to oracle knowledge about the current subtask being executed, such as picking or placing, to train a single policy for such long-horizon tasks. The Multi-skill RL Mobile manipulation approach, instead, trains different mobile manipulation policies that are then chained by an oracle task planner executed in open loop. For details about the baselines, we refer the reader to [10].

3.2 Results

We report the benchmark results in Fig. 4. Our approach outperforms the baselines in all three tasks, averaging 72.5% completion rate in `TidyHouse`, 77% completion rate in `PrepareGroceries`, and 50% completion rate in `SetTable` over five seeds. The best performing baseline, MM, instead, averages 71%, 64%, and 29% respectively. Considering all three tasks combined, we achieved a 66.5% success rate compared to the 54.7% of the MM baseline. Notably, the MM baseline requires extensive offline training. That is 6400 episodes per task across varied layouts and configurations in the Habitat environments, and 100 million steps per skill across a total of 7 skills. In contrast, our method relies on hand-tuning each skill over just a handful of episodes and is evaluated directly on unseen layouts and configurations, demonstrating strong generalization without the need for data-intensive training. However, we still rely on privileged information, such as the floor map for path planning and articulated object states (open or closed). These assumptions will be removed in future work.

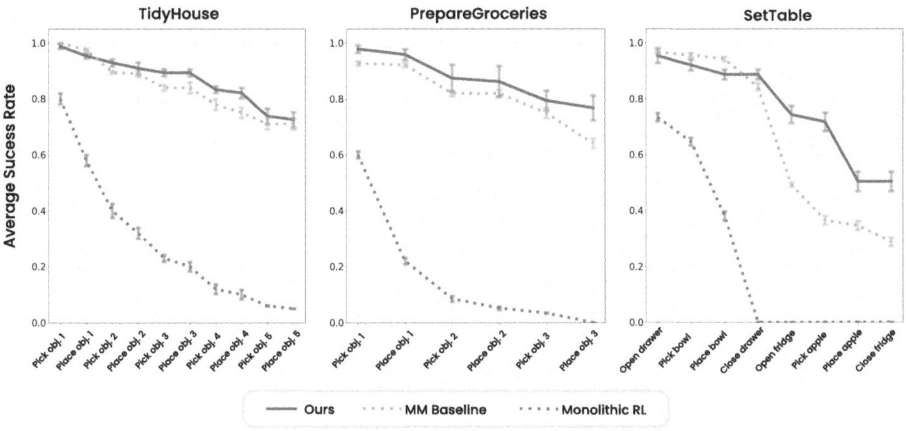

Fig. 4. Evaluation results on the Habitat Benchmark, averaged over 100 episodes. Each task is evaluated on different apartment layouts and divided into stages. For a stage to be successful, all previous stages must also be successful. **TidyHouse**: Evaluated over five pick-and-place stages, from `Pick obj. 1` to `Place obj. 5`. **PrepareGroceries**: Measured from `Pick obj. 1` to `Place obj. 3`. **SetTable**: Involves a more complex sequence including `Open drawer` → `Pick bowl` → `Place bowl` → `Close drawer`, and similarly for the fridge and apple.

4 Discussion and Conclusion

In this work, we proposed a hierarchical active inference model to address the Habitat Benchmark, surpassing state-of-the-art performance across all three benchmark tasks. Our system is composed of two key components: a high-level model that selects appropriate low-level skills based on discrete observations, and a set of low-level skills defined through goals for a novel whole-body controller using hierarchical active inference. This architecture enables the system to flexibly adapt to task failures and dynamically adjust behavior in response to environmental changes. Importantly, our method operates online, without requiring offline training, and supports real-time adaptation of the high-level plan. While our current implementation still relies on certain privileged information (such as access to a global map for path planning), our future work will focus on enabling the agent to actively explore and construct maps in real-time. Additionally, knowledge of the state of articulated objects, such as drawers and refrigerators, will be inferred directly from raw RGBD sensory input. Moreover, while low-level skills are currently composed of a fixed sequence of goals for the continuous whole-body controller, one could add an intermediate hierarchical level as in [21] to smoothly transition between subgoals. Another interesting direction to explore would be learning skills directly from demonstration. In summary, our hierarchical hybrid active inference model demonstrates promising results in goal-directed robotic control within complex environments. With further enhancements in perception, exploration, and skill acquisition, we believe this framework could serve as a foundation for more generalized and scalable robotic agents.

A Appendix

A.1 Kinematic Generative Model

To generative model in Eq. (4) is defined from generic transformation matrices. To compute the absolute position and orientation of the current link j given the absolute position and orientation of the previous one, we can write:

$$ {}^{w}T_j = \left[\begin{array}{c|c} R^{j-1}R^{j} & \boldsymbol{t}^{j-1} + R^{j-1}\boldsymbol{t}^{j} \\ \hline \boldsymbol{0} & 1 \end{array}\right], \tag{15} $$

where w indicates the world frame as an absolute reference, R represents a rotation matrix, and $\boldsymbol{t}$ a translation vector. The world frame can be the base link of a robot arm. From Eq. (15), we note that the resulting absolute rotation of a link is the multiplication of two rotation matrices. However, we can express this as a quaternion multiplication $\boldsymbol{q}^{j-1} \cdot \boldsymbol{q}^{j}$. Similarly, we can rotate a vector $\boldsymbol{t}^{j}$ by a quaternion $\boldsymbol{q}^{j-1}$ corresponding to R^{j-1}, leading to Eq. (4). Considering a generic DenavitHartenberg (DH) transformation matrix

$$ \left[\begin{array}{ccc|c} \cos\theta^j & -\sin\theta^j\cos\alpha^j & \sin\theta^j\sin\alpha^j & l^j\cos\theta^j \\ \sin\theta^j & \cos\theta^j\cos\alpha^j & -\cos\theta^j\sin\alpha^j & l^j\sin\theta^j \\ 0 & \sin\alpha^j & \cos\alpha^j & d^j \\ \hline 0 & 0 & 0 & 1 \end{array}\right], \tag{16} $$

we note that the translation vector is simply $\boldsymbol{t}^j = [l^j\cos\theta^j, l^j\sin\theta^j, d^j]$. According to the DH convention, the rotational part of the transformation matrix is the composition of a rotation θ^j about the previous z-axis and a rotation of α^j around the x-axis. We can then write:

$$ \boldsymbol{q}^j = \left[\cos\tfrac{\theta^j}{2}\cos\tfrac{\alpha^j}{2}, \cos\tfrac{\theta^j}{2}\cos\tfrac{\alpha^j}{2}, \cos\tfrac{\theta^j}{2}\cos\tfrac{\alpha^j}{2}, \cos\tfrac{\theta^j}{2}\cos\tfrac{\alpha^j}{2}\right]. \tag{17} $$

The generic kinematic model in Eq. (4) can be expressed as

$$ \boldsymbol{g}_e(\boldsymbol{\mu}_i^j, \boldsymbol{\mu}_e^{j-1}) = \begin{bmatrix} \boldsymbol{t}^{j-1} + \boldsymbol{h}(\boldsymbol{q}^{j-1} \cdot [0\ \boldsymbol{t}^j] \cdot \boldsymbol{q}^{j-1*}) \\ \boldsymbol{q}^{j-1} \cdot \boldsymbol{q}^j \end{bmatrix}, = \begin{bmatrix} x^{j-1} + x_{tf} \\ y^{j-1} + y_{tf} \\ z^{j-1} + z_{tf} \\ q_{w,\ tf} \\ q_{x,\ tf} \\ q_{y,\ tf} \\ q_{z,\ tf} \end{bmatrix}, \tag{18} $$

where x_{tf}, y_{tf}, z_{tf} and $q_{*,\ tf}$ are the transformed translations and rotation. Computing the Hamilton products yields the following expressions for the transformed positions

$$
\begin{aligned}
x_{tf} &= q_w^{j-1^2} l^j \cos\theta^j + q_x^{j-1^2} l^j \cos\theta^j - q_y^{j-1^2} l^j \cos\theta^j - q_z^{j-1^2} l^j \cos\theta^j \\
&+2q_x^{j-1} q_y^{j-1} l^j \sin\theta^j + 2q_x^{j-1} q_z^{j-1} d^j + 2q_w^{j-1} q_y^{j-1} d^j - 2q_w^{j-1} q_z^{j-1} l^j \sin\theta^j, \\
y_{tf} &= q_w^{j-1^2} l^j \sin\theta^j - q_x^{j-1^2} l^j \sin\theta^j + q_y^{j-1^2} l^j \sin\theta^j - q_z^{j-1^2} l^j \sin\theta^j \\
&+2q_x^{j-1} q_y^{j-1} l^j \cos\theta^j + 2q_y^{j-1} q_z^{j-1} d^j - 2q_w^{j-1} q_x^{j-1} d^j + 2q_w^{j-1} q_z^{j-1} l^j \cos\theta^j, \\
z_{tf} &= q_w^{j-1^2} d^j - q_x^{j-1^2} d^j - q_y^{j-1^2} d^j + q_z^{j-1^2} d^j \\
&+2q_x^{j-1} q_z^{j-1} l^j \cos\theta^j + 2q_y^{j-1} q_z^{j-1} l^j \sin\theta^j - 2q_w^{j-1} q_y^{j-1} l^j \cos\theta^j + 2q_w^{j-1} q_x^{j-1} l^j \sin\theta^j,
\end{aligned}
$$

and orientation:

$$
\begin{aligned}
q_{w,\,tf} &= q_w^{j-1} \cos\frac{\theta^j}{2} \cos\frac{\alpha^j}{2} - q_x^{j-1} \cos\frac{\theta^j}{2} \sin\frac{\alpha^j}{2} - q_y^{j-1} \sin\frac{\theta^j}{2} \sin\frac{\alpha^j}{2} - q_z^{j-1} \sin\frac{\theta^j}{2} \cos\frac{\alpha^j}{2}, \\
q_{x,\,tf} &= q_w^{j-1} \cos\frac{\theta^j}{2} \sin\frac{\alpha^j}{2} + q_x^{j-1} \cos\frac{\theta^j}{2} \cos\frac{\alpha^j}{2} + q_y^{j-1} \sin\frac{\theta^j}{2} \cos\frac{\alpha^j}{2} - q_z^{j-1} \sin\frac{\theta^j}{2} \sin\frac{\alpha^j}{2}, \\
q_{y,\,tf} &= q_w^{j-1} \sin\frac{\theta^j}{2} \sin\frac{\alpha^j}{2} - q_x^{j-1} \sin\frac{\theta^j}{2} \cos\frac{\alpha^j}{2} + q_y^{j-1} \cos\frac{\theta^j}{2} \cos\frac{\alpha^j}{2} + q_z^{j-1} \cos\frac{\theta^j}{2} \sin\frac{\alpha^j}{2}, \\
q_{z,\,tf} &= q_w^{j-1} \sin\frac{\theta^j}{2} \cos\frac{\alpha^j}{2} + q_x^{j-1} \sin\frac{\theta^j}{2} \sin\frac{\alpha^j}{2} - q_y^{j-1} \cos\frac{\theta^j}{2} \sin\frac{\alpha^j}{2} + q_z^{j-1} \cos\frac{\theta^j}{2} \cos\frac{\alpha^j}{2}.
\end{aligned}
$$

A.2 Collision Avoidance in HAIF

A repulsive state $\boldsymbol{\mu}^!$ can be imposed on intrinsic beliefs, to realize joint limit avoidance, or extrinsic beliefs for collision avoidance with the environment. We define joint limit avoidance as:

$$
\boldsymbol{f}_{r,\theta}(\boldsymbol{\mu}) = \begin{cases} 0, & \text{if } ||\boldsymbol{e}_\theta|| > \gamma_\theta \\ k_{r,\theta}\zeta(1/\gamma_\theta - 1/||\boldsymbol{e}_\theta||), & \text{otherwise} \end{cases}, \tag{19}
$$

where $\boldsymbol{e}_\theta = \boldsymbol{\mu}^!_\theta - \boldsymbol{\mu}_\theta$, $\boldsymbol{\mu}_\theta$ is the slice of beliefs about joint angles, $\boldsymbol{\mu}^!_\theta$ are the joint limits, and γ_θ is a chosen threshold. The variable $\zeta \in \{-1,\ 1\}$ is negative for lower limits and positive for upper limits. The collision avoidance strategy is instead the same as [22]:

$$
\boldsymbol{f}_{r,obst}(\boldsymbol{\mu}) = \begin{cases} 0, & \text{if } ||\boldsymbol{e}_{obst}|| > \gamma_{obst} \\ k_{r,obst}(1/\gamma_{obst} - 1/||\boldsymbol{e}_{obst}||)\boldsymbol{e}_{obst}/||\boldsymbol{e}_{obst}||^3, & \text{otherwise} \end{cases}, \tag{20}
$$

where $\boldsymbol{e}_{obst} = \boldsymbol{\mu}^!_{pos} - \boldsymbol{\mu}_{pos}$, $\boldsymbol{\mu}_{pos}$ is the slice of beliefs about link positions, and $\boldsymbol{\mu}^!_{pos}$ is the position of an obstacle given by the VBGS module. Goal attractors and repulsive forces for joint limits and collision avoidance are then summed together to form the dynamics function of a single level. This allows one to achieve behaviors such as reaching a target while avoiding an obstacle. Parameters are manually chosen to achieve sufficient performance in the test cases, but could be automatically optimized.

A.3 Whole-Body Skills

The routines for the different skills for the mobile manipulator are defined as follows:

- `Pick`: The robot unfolds its arm (joint goal), moves to a pre-grasp position above the target object (end-effector + joint goal), and then proceeds to the grasp pose to perform the grasp once close enough (end-effector goal). After grasping, it retreats to the post-grasp pose (end-effector goal) and folds the arm back into a compact configuration (joint goal) (see Fig. 5).
- `Place`: It mirrors the `Pick` sequence, but targets a specified place location.
- `PickFromDrawer`: The end-effector is moved in front of the drawer hinge and grasps the handle once close enough (end-effector + joint goal). Then, the robot executes a linear backward trajectory to pull the drawer open (end-effector goal). The object is picked as in `Pick`. Finally, the robot end-effector is placed in front of the handle again (end-effector goal), and the drawer is pushed close following a linear trajectory (end-effector goal) (see Fig. 6).
- `PickFromFridge`: The robot unfolds its arm (joint goal), moves in front of the fridge handle, and grasps it once close enough (end-effector goal). It then follows a circular trajectory to partially open the door (end-effector goal). After that, the arm retreats (joint goal), and finally, the arm starts a linear trajectory from behind the half-opened door to push it to fully open (end-effector goal). The object is picked as in `Pick`, and then the robot first moves to the left of the fridge door (base + joint goal), and after it follows a linear trajectory to push the door closed (end-effector goal) (see Fig. 7).
- `Move`: The `NavModel` computes a global path towards a final goal and orientation, and provides the move skill with the current active subgoal $(x^*,\ y^*)$, along with the final desired position and orientation. At each step, the heading θ^* toward the subgoal is computed. A predefined joint configuration (joint goal) for the arm is set to avoid collisions. The skill terminates when the robot is within a threshold distance of the target pose. See Fig. 8) for an example. The reach threshold for the position is kept at 0.8 m while the one for orientation to 0.3rad. These are particularly loose since we rely on whole-body manipulation skills and are not required to precisely position the base before executing them.

Unfold and reach pre-grasp

Proceed to grasp

Lift to post-grasp

Retreat arm

Fold arm for navigation

Fig. 5. Evolution of the `Pick` skill over time.

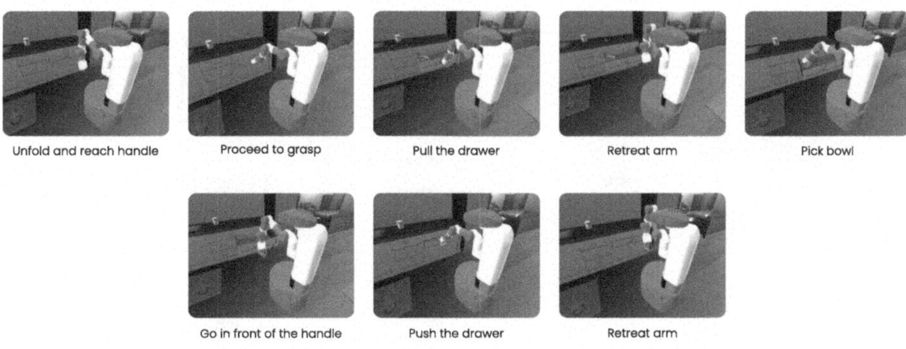

Fig. 6. Evolution of the `PickFromDrawer` skill over time.

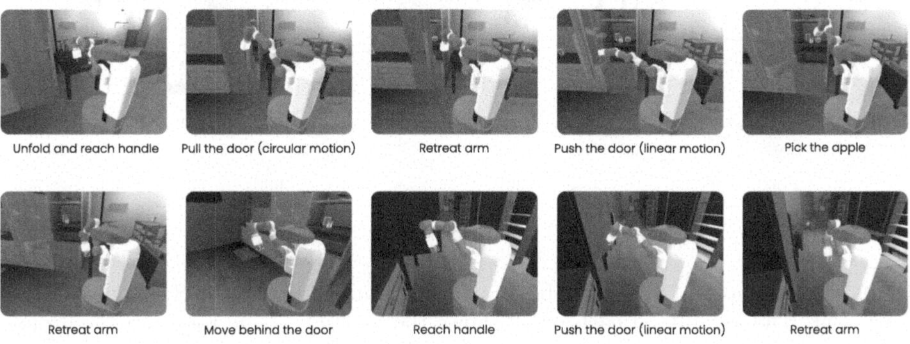

Fig. 7. Evolution of the `PickFromFridge` skill over time.

A.4 Example Evolution of a Probabilistic Map

In Fig. 9 we present an example of how a probabilistic map can evolve through time using VBGS.

Fig. 8. Top view of the `Move` skill where the robot moves through subgoals following the global path.

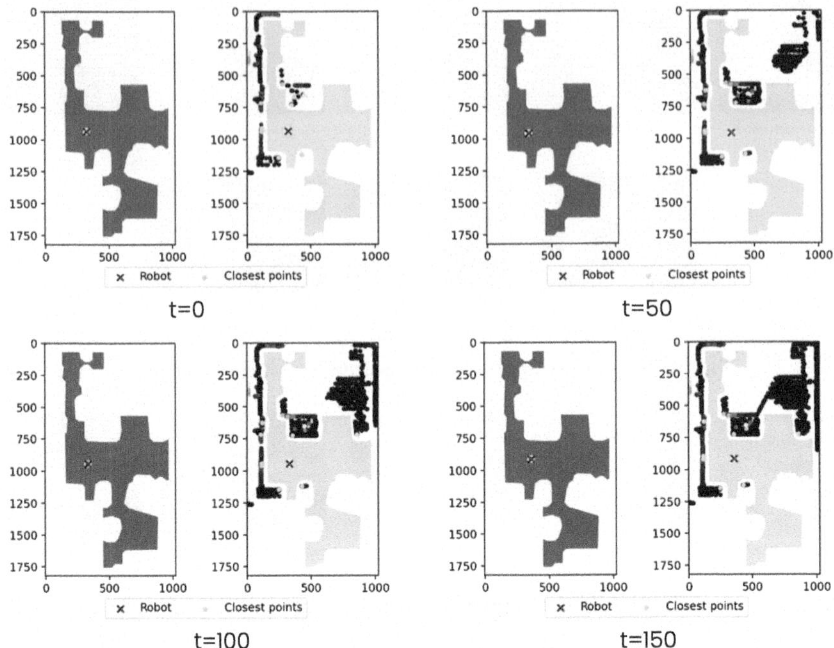

Fig. 9. An example of the probabilistic map evolution with VBGS in one Habitat apartment. The left side of each panel shows the location of the robot on the ground truth floor plan. The right side overlays the Gaussian components over the obstacles projected onto the floor.

References

1. Beal, M.J.: Variational Algorithms for Approximate Bayesian Inference. Ph.D. thesis, University College London (2003)
2. Bishop, C.M.: Pattern Recognition and Machine Learning. Springer (2006)
3. Blei, D.M., Kucukelbir, A., McAuliffe, J.D.: Variational inference: a review for statisticians. J. Am. Stat. Assoc. **112**(518), 859–877 (2017). https://doi.org/10.1080/01621459.2017.1285773
4. Çatal, O., Van de Maele, T., Pitliya, R.J., Albarracin, M., Pattisapu, C., Verbelen, T.: Belief sharing: a blessing or a curse. In: Buckley, C.L., et al. (eds.) Active Inference, pp. 121–133. Springer, Cham (2025)
5. Ehsani, K., et al.: Manipulathor: a framework for visual object manipulation. In: Proceedings of the IEEE/CVF Conference on Computer Vision and Pattern Recognition, pp. 4497–4506 (2021)
6. Friston, K.J., et al.: Supervised structure learning. Biol. Psychol. **193**, 108891 (2024). https://doi.org/10.1016/j.biopsycho.2024.108891. https://www.sciencedirect.com/science/article/pii/S0301051124001510
7. Friston, K.J., Parr, T., de Vries, B.: The graphical brain: belief propagation and active inference. Netw. Neurosci. **1**(4), 381–414 (2017)
8. Gan, C., et al.: The threedworld transport challenge: a visually guided task-and-motion planning benchmark towards physically realistic embodied AI. In: 2022

International Conference on Robotics and Automation (ICRA), pp. 8847–8854. IEEE (2022)
9. Garrett, C.R., Lozano-Pérez, T., Kaelbling, L.P.: Pddlstream: integrating symbolic planners and blackbox samplers via optimistic adaptive planning. In: Proceedings of the International Conference on Automated Planning and Scheduling, vol. 30, pp. 440–448 (2020)
10. Gu, J., Chaplot, D.S., Su, H., Malik, J.: Multi-skill mobile manipulation for object rearrangement. arXiv preprint arXiv:2209.02778 (2022)
11. Hart, P.E., Nilsson, N.J., Raphael, B.: A formal basis for the heuristic determination of minimum cost paths. IEEE Trans. Syst. Sci. Cybern. **4**(2), 100–107 (1968). https://doi.org/10.1109/TSSC.1968.300136
12. Kaelbling, L.P., Lozano-Pérez, T.: Hierarchical task and motion planning in the now. In: 2011 IEEE International Conference on Robotics and Automation, pp. 1470–1477. IEEE (2011)
13. Kaelbling, L.P., Lozano-Pérez, T.: Integrated task and motion planning in belief space. Int. J. Robot. Res. **32**(9–10), 1194–1227 (2013)
14. Kerbl, B., Kopanas, G., Leimkühler, T., Drettakis, G.: 3D gaussian splatting for real-time radiance field rendering. ACM Trans. Graph. **42**(4) (2023). https://repo-sam.inria.fr/fungraph/3d-gaussian-splatting/
15. Van de Maele, T., Catal, O., Tschantz, A., Buckley, C.L., Verbelen, T.: Variational bayes gaussian splatting (2024). https://arxiv.org/abs/2410.03592
16. Parr, T., Friston, K.J.: Active inference and the anatomy of oculomotion. Neuropsychologia **111**, 334–343 (2018)
17. Parr, T., Friston, K.J.: The discrete and continuous brain: from decisions to movement-and back again. Neural Comput. **30**(9), 2319–2347 (2018)
18. Parr, T., Pezzulo, G., Friston, K.J.: Active inference: the free energy principle in mind, brain, and behavior. MIT Press (2022)
19. Parr, T., Friston, K., Pezzulo, G.: Generative models for sequential dynamics in active inference. Springer Nature Link (2023)
20. Pezzato, C., Buckley, C., Verbelen, T.: Why learn if you can infer? Robot arm control with hierarchical active inference. In: The First Workshop on NeuroAI@ NeurIPS2024 (2024)
21. Priorelli, M., Stoianov, I.P.: Deep hybrid models: infer and plan in a dynamic world. Entropy **27**, 570 (2025). https://doi.org/10.3390/e27060570
22. Priorelli, M., Pezzulo, G., Stoianov, I.P.: Deep kinematic inference affords efficient and scalable control of bodily movements. Proc. Natl. Acad. Sci. **120**(51), e2309058120 (2023). https://doi.org/10.1073/pnas.2309058120
23. Priorelli, M., Stoianov, I.P.: Slow but flexible or fast but rigid? Discrete and continuous processes compared. Heliyon **10**(20) (2024)
24. Priorelli, M., Stoianov, I.P.: Dynamic planning in hierarchical active inference. Neural Netw. 107075 (2025)
25. Sandakalum, T., Ang, M.H., Jr.: Motion planning for mobile manipulators-a systematic review. Machines **10**(2), 97 (2022)
26. Smith, R., Friston, K.J., Whyte, C.J.: A step-by-step tutorial on active inference and its application to empirical data. PubMed (2022)
27. Srivastava, S., Fang, E., Riano, L., Chitnis, R., Russell, S., Abbeel, P.: Combined task and motion planning through an extensible planner-independent interface layer. In: 2014 IEEE International Conference on Robotics and Automation (ICRA), pp. 639–646. IEEE (2014)
28. Szot, A., et al.: Habitat 2.0: training home assistants to rearrange their habitat. In: Advances in Neural Information Processing Systems, vol. 34, pp. 251–266 (2021)

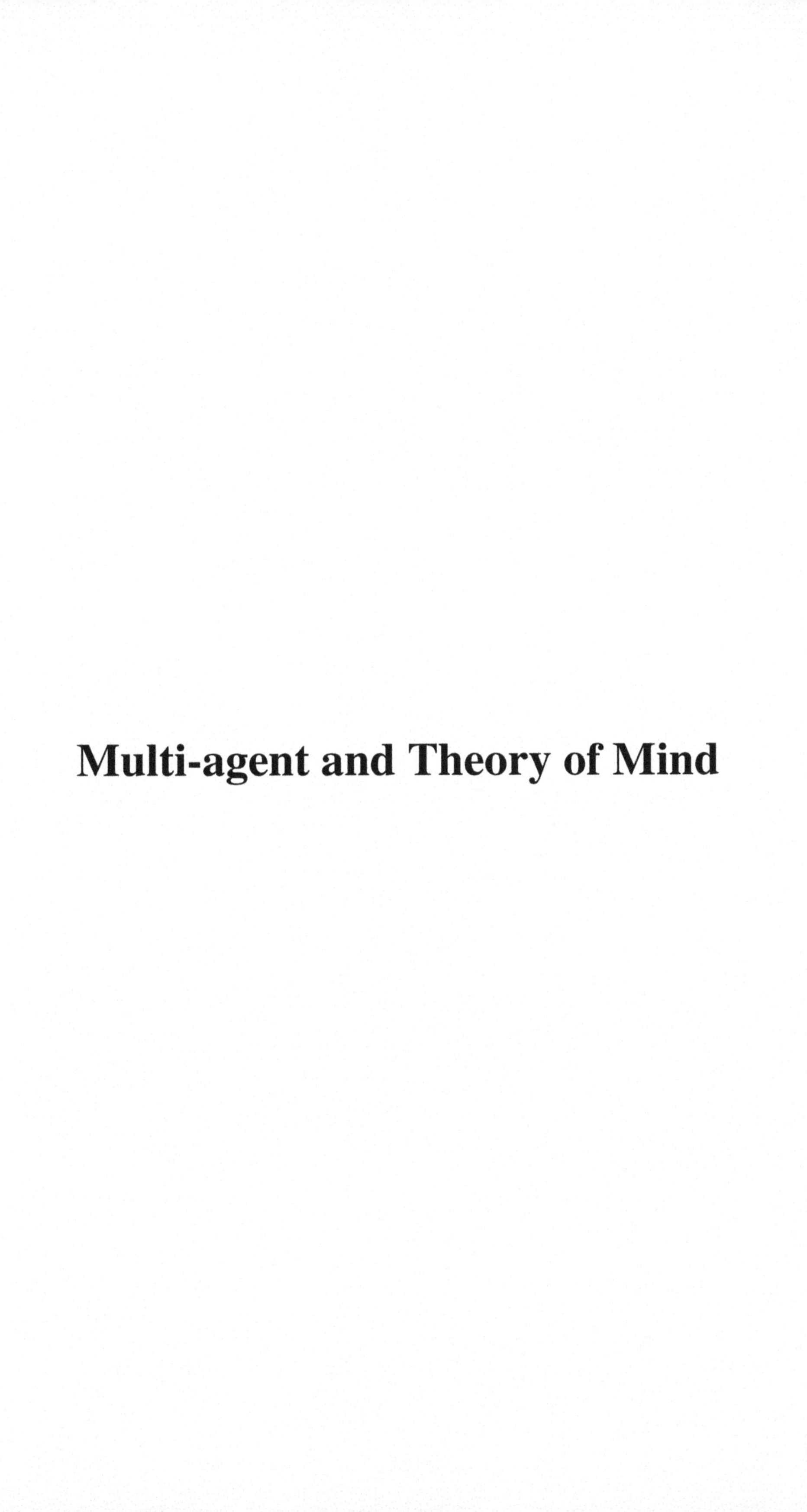

Multi-agent and Theory of Mind

Theory of Mind Using Active Inference: A Framework for Multi-agent Cooperation

Riddhi J. Pitliya(✉), Ozan Çatal, Toon Van de Maele, Corrado Pezzato, and Tim Verbelen

VERSES, Los Angeles, CA 90067, USA
riddhi.jain@verses.ai

Abstract. Theory of Mind (ToM) – the ability to understand that others can have differing knowledge and goals – enables agents to reason about others' beliefs while planning their own actions. We present a novel approach to multi-agent cooperation by implementing ToM within active inference. Unlike previous active inference approaches to multi-agent cooperation, our method neither relies on task-specific shared generative models nor requires explicit communication. In our framework, ToM-equipped agents maintain distinct representations of their own and others' beliefs and goals. ToM agents then use an extended and adapted version of the sophisticated inference tree-based planning algorithm to systematically explore joint policy spaces through recursive reasoning. We evaluate our approach through collision avoidance and foraging simulations. Results suggest that ToM agents cooperate better compared to non-ToM counterparts by being able to avoid collisions and reduce redundant efforts. Crucially, ToM agents accomplish this by inferring others' beliefs solely from observable behaviour and considering them when planning their own actions. Our approach shows potential for generalisable and scalable multi-agent systems while providing computational insights into ToM mechanisms.

Keywords: Theory of Mind · Active Inference · Multi-agent Cooperation · Sophisticated Inference · Recursive Planning

1 Introduction

Theory of Mind (ToM) represents one of the most remarkable achievements of human cognition – the ability to understand that other agents possess minds with beliefs and goals that may differ from our own [6]. This meta-cognitive skill enables us to recognise that others may hold false beliefs and maintain perspectives distinct from our own. For example, when we observe someone searching for an object that we know has been moved in their absence, we can anticipate their search behaviour based on where they believe the object to be, rather than its actual location [1,12]. This fundamental distinction between reality and belief enables sophisticated forms of cooperation, competition, and

M. Albarracin et al. (Eds.): IWAI 2025, CCIS 2857, pp. 385–400, 2026.
https://doi.org/10.1007/978-3-032-16955-6_22

communication. ToM emerges early in human development and underpins our ability to navigate complex multi-agent environments [11].

While ToM is fundamental to human social cognition, current approaches to multi-agent cooperation using active inference lack this crucial capability. Previous active inference models for multi-agent cooperation have predominantly relied upon assumptions of shared or identical generative models that limit their generalisability and practical application. We propose that a better solution to conduct and model multi-agent cooperation is by implementing ToM within the planning stage of active inference. This offers a more principled and generalisable solution for multi-agent artificial intelligent systems, and a computational model that could serve as a tool to deepen our understanding of how humans implement ToM. Before presenting our novel approach, we first elaborate on the limitations of existing approaches of conducting cooperation using active inference.

1.1 Existing Active Inference Approaches to Multi-agent Cooperation

Maisto and colleagues [9] introduced "interactive inference", wherein agents maintain probabilistic beliefs about shared goals (such as both agents pressing the same or different buttons) and update these beliefs through observations of others' locations and actions. Their agents selected epistemic policies designed to reduce uncertainty about the joint goals. However, this approach assumes agents share identical goals, which is not always the case in multi-agent cooperation tasks. Moreover, their model relies on a carefully tailored generative model for the focal agent (i.e., the agent conducting ToM) that incorporates the other agent's location as an observation which itself encodes information about the shared goal. These assumptions limit generalisability to scenarios where actions do not inherently signal goals or where agents have complementary rather than identical and shared objectives.

Matsumura and colleagues [10] addressed collision avoidance (agents passing by each other without colliding) using simulation theory, where agents use their own internal models to imagine others' situations. While this enables basic perspective-taking, their implementation is domain-specific to navigation tasks that use the social force model as it includes parameters for forward movement and mutual repulsion. The approach lacks the recursive reasoning capabilities that relies on maintenance of separate belief representations for different agents, which is necessary for more complex coordination scenarios.

Other researchers have proposed multi-agent cooperation through explicit information exchange mechanisms [2,5]. These approaches involve agents sharing likelihood messages – information about the probability of observations given states – rather than posterior beliefs directly. However, this requires the generative model structures (state factors and observation modalities) to be identical between agents. Moreover, while mathematically principled, this method sidesteps the fundamental challenge of inferring others' beliefs from observable behaviour alone, a capability that humans routinely demonstrate and leverage during multi-agent cooperation.

Overall, these approaches predominantly assume that all agents operate under the same generative model, with identical beliefs about transition dynamics, observation likelihoods and goal structures. Such aligned models fail to capture the reality of agents with different experiences, capabilities, and objectives. Furthermore, these approaches typically involve only single-level reasoning (*"what will the other agent do?"*) rather than the recursive beliefs that characterise ToM (*"what do I think the other agent thinks about the situation?"*). Many implementations are also tailored to specific tasks (e.g., navigation or mutual button-pressing) and do not provide general principles for multi-agent cooperation across different tasks.

To address these limitations, we present the first generalisable implementation for multi-agent cooperation by implementing ToM using sophisticated active inference [3], with three key features:

1. Our agents maintain distinct beliefs and generative models for themselves and others, allowing them to reason about different perspectives while avoiding the assumption of shared knowledge and knowledge structures.
2. We propose a novel tree-based planning procedure that systematically explores joint policy spaces by interleaving policy and observation rollouts between agents.
3. Our agents can reason about how others' actions affect the world states via message passing between their own beliefs and their beliefs of the other agent's beliefs, maintaining perspective separation while allowing for information integration.

We empirically validate our approach by simulating two multi-agent scenarios: a collision avoidance task where agents must navigate past each other without occupying the same location, and an apple foraging task requiring efficient search and consumption of resources. These scenarios are implemented in a simple 3×3 grid environment to provide a clear and interpretable proof of concept, with future work aimed at extending the approach to larger and more complex settings. Our results demonstrate that ToM-equipped agents conduct multi-agent cooperation more effectively than non-ToM agents. The ToM agents successfully avoid collisions and reduce redundant efforts as they are able to interact *proactively* rather than reactively with other agents.

2 Our Approach: Theory of Mind in Active Inference

2.1 Sophisticated Inference

Our approach builds upon sophisticated inference [3], which extends standard active inference to consider recursive forms of expected free energy (EFE). In standard active inference, agents evaluate policies by considering *"what would happen if I did that?"*. Sophisticated inference instead deepens this to *"what would I believe about what would happen if I did that?"*. This distinction is crucial for ToM. When reasoning about other agents, we need to consider not just what they will do, but what they believe about the consequences of their actions. This recursive reasoning about the other agent requires maintaining a separate model of the other agent.

2.2 ToM Agent's Belief Structure of Multiple Agents

In our ToM framework, the focal agent maintains separate state beliefs (s) for itself and each other agent in the environment. In the case of a two-agent scenario, this yields $s = \{\underbrace{s^{f,\text{self}}, s^{f,\text{world}}}_{\text{focal}(s^f)}, \underbrace{s^{o,\text{self}}, s^{o,\text{world}}}_{\text{other}(s^o)}\}$, where:

- $s^{f,\text{self}}$ is the focal agent's beliefs about its own states (e.g., focal agent's own location)
- $s^{f,\text{world}}$ is the focal agent's beliefs about the world states (e.g., other agent's location or item at current location)
- $s^{o,\text{self}}$ is the focal agent's beliefs about the other agent's self states (e.g., other agent's location)
- $s^{o,\text{world}}$ is the focal agent's beliefs about what the other agent believes about the world states (e.g., focal agent's location or item at current location)

This structure lets the focal agent keep its own perspective separate while modelling how others might see things differently. To elaborate, the belief components can be flexibly combined to capture different reasoning cases. For instance, the focal agent can pair $s^{o,\text{self}}$ with $s^{f,\text{world}}$ to predict what the other agent would observe given the focal agent's own beliefs about the environment. The focal agent can also pair $s^{o,\text{self}}$ with $s^{o,\text{world}}$ to predict what the other agent thinks it will observe, based on the other agent's own (possibly mistaken) view of the world. This kind of cross-perspective reasoning allows the focal agent to distinguish between (a) what it thinks the other agent will perceive and (b) what it thinks the other agent believes it may perceive. Because the focal agent's own world beliefs ($s^{f,\text{world}}$) may differ from its beliefs about the other agent's world beliefs ($s^{o,\text{world}}$), it can represent cases where knowledge is asymmetric – for example, when the focal agent knows something the other does not.

By maintaining these distinct representations, our framework does not assume shared knowledge structures between agents. The focal agent can construct and continuously update its model of the other agent based on observable behaviour, while the actual other agent may operate with an entirely different generative model. This capability enables our agents to collaborate effectively even when they possess different prior knowledge, capabilities, or goals – a fundamental requirement for realistic and practical multi-agent systems.

2.3 Recursive Planning with Theory of Mind

The core innovation in our approach lies in the planning algorithm that enables agents to reason recursively about joint policy spaces. It systematically explores how the focal agent's beliefs about another agent's beliefs influence its planning decisions. The recursive form of the EFE is extended to the ToM setting

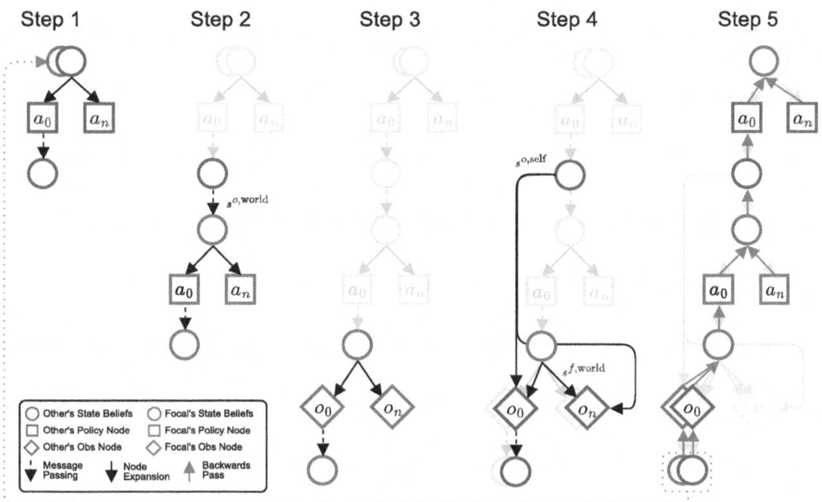

Fig. 1. Recursive Planning Tree for Theory of Mind. Red and purple represent the focal and other agent's nodes respectively. Circles indicate the agent's beliefs, squares indicate evaluated actions and diamonds indicated expected observations. For a detailed description of each step, see Sect. 2.3. (Color figure online)

(see Appendix A), which results in a deep tree search algorithm that alternates between the focal and other agent's policies and observations. At each planning horizon, the tree search unfolds through five main stages, as detailed below and illustrated in Fig. 1.

Step 1: Other Agent Policy Expansion. As aforementioned in Sect. 2.2, we begin with the focal agent's beliefs, which comprises separate beliefs for itself and the other agent in the environment ($s = \{s^f, s^o\}$). The focal agent first considers which policies the other agent is likely to select. This is visualised in Step 1 of Fig. 1, where each policy node represents a specific action that the other agent might execute (a_0; purple square). The potential actions are evaluated based on the focal agent's beliefs about the other agent's beliefs (s^o; purple circle above purple squares). Essentially, the focal agent asks *"Given what I believe about the other agent's beliefs and goals, what would the other agent choose to do?"*. The focal agent then computes how the other agent would update its beliefs if it were to execute that action (s^o; the purple circle below the purple square).

Step 2: Focal Agent Policy Expansion. For each considered action of the other agent, the focal agent evaluates its own policy options. Crucially, before doing so, the focal agent updates its world beliefs, based on the anticipated consequences of the other agent's actions, using likelihood message passing. Here, the focal agent uses its computation of how the other agent's beliefs about the world ($s^{o,\text{world}}$; purple circle above the red circle) would change if the other agent were to execute an action. A likelihood message is then created, capturing the information gained from the other agent's anticipated action in the form of the difference between the other agent's updated beliefs and its prior beliefs. This mechanism allows the focal agent to incorporate information into its own beliefs about how the world states ($s^{f,\text{world}}$; red circle above the red squares) would change due to the other agent's actions. Using these updated beliefs, the focal agent then evaluates its own policy options (a_0; red square) through standard EFE calculations. This creates branches in the tree structure for each possible joint policy combination between the focal and other agent. The focal agent then computes how its beliefs would be updated if it were to execute an action given the other agent's action (s^f; red circle below the red square).

Step 3: Focal Agent Observation Expansion. Then, given the joint policies, the focal agent considers what observations it is likely to receive and its resulting posterior beliefs.

This process is illustrated in Step 3 of Fig. 1, where the focal agent's observation nodes (o_0; red diamonds) represent the various observations the focal agent expects to encounter given the focal agent's beliefs considering the execution of both agents' actions (s^f; red circle before red diamonds). This results in the computation of the focal agent's posterior beliefs (s^f; red circle after red diamonds).

Step 4: Other Agent Observation Expansion. Here, the focal agent considers what observations the other agent is likely to receive (o_0; purple diamonds) given the joint policy and anticipated world state changes given its action. The observation probabilities are computed using the focal agent's beliefs about the other agent's self states ($s^{o,\text{self}}$; purple circle from an earlier expansion) and the focal agent's own updated beliefs about the world state ($s^{f,\text{world}}$; red circle before the purple diamonds). The focal agent then updates its representation of the other agent's posterior beliefs (s^o; purple circle after the purple diamond).

Step 5: Tree Backwards Pass and Policy Selection. Finally, after expanding the tree for the current horizon, a backwards pass computes policy selection probabilities for the focal agent. The backwards pass is visualised in Step 5 of Fig. 1, with the green upward arrows showing that EFE values propagate from the leaf observation nodes back through each policy branch to inform the final policy selection at the root. To plan for another time step, the observation nodes' leaves from Step 5 serve as the root node for Step 1 (grey dotted arrow).

Recursive EFE values are computed for each joint policy combination and weighted by the observation probabilities. The other agent's policy probabilities are marginalised for policy selection. The resulting probability distribution balances goal-directed with information-seeking behaviour while taking into account the uncertainty over the other agent's actions.

Our implementation achieves computational efficiency through two mechanisms as practised in sophisticated active inference [3]. Policy pruning reduces tree expansion by eliminating unlikely policy nodes and those nodes do not branch out. Observation pruning similarly focuses on probable outcomes, reducing the combinatorial explosion.

3 Experimental Validation

We empirically validated our ToM framework across two multi-agent scenarios that required different forms of cooperation. All simulations occurred on a 3×3 grid environment (see Appendix B for the reference grid layout) with deterministic dynamics and perfect observability of agent locations. The experimental design comprised two conditions for each task: a baseline condition where both agents used sophisticated active inference without ToM capabilities, and a ToM condition where one agent (red) was equipped with our ToM framework while the other (purple) remained non-ToM. All simulations were conducted using the JAX-based Python package, `pymdp`, which offers efficient and flexible tools for constructing such models [7].

3.1 Collision Avoidance Task

Task Description. The collision avoidance task presented a basic cooperation challenge: two agents were initiated at opposite corners of the grid with objectives to swap positions while avoiding collision. The shortest path for both agents involved traversing the central cell that would result in collision. We evaluated performance using three primary metrics: task completion success (whether agents reached their respective goals), collision occurrence, and path efficiency (total time steps to completion). The task demanded proactive cooperation, as reactive strategies would result in deadlock.

Generative Model. Each agent's generative model included two state factors: own location (9 discrete location states plus a null state for boundary violations as in [8]) and other agent's location (also 10 states like the focal agent's location states). The observation model provided perfect perceptual access to both locations through identity likelihood mappings, eliminating sensory uncertainty.

The action space comprised nine options: directional movements (up, down, left, right, four diagonals) plus no-operation. Transition dynamics regarding the agent's own location reflected full controllability abiding by standard grid-world physics with collision constraints where agents attempting to occupy the same cell became permanently stuck. Invalid movements were specified via the null state (which had a severe negative utility). For example, if the agent were to move up from location 1 (top-left corner), it would enter the severely disfavoured null state, driving it away from invalid movements. Since the other agent's location was not controllable, the transition dynamics for the other agent's location reflected uniform probability distribution between the valid actions the other agent can take. For example, the probability of moving from location 1 (top-left corner) to the centre, go down, go right, or have no-operation is 1/4.

Preferences encoded goal-seeking behaviour: high positive utility for reaching the target location and severe penalty for the null observation. Critically, no explicit collision avoidance preferences were included – coordination had to emerge through ToM reasoning rather than hard-coded behaviours. Planning horizons were set to 3 time steps for both ToM and non-ToM agents, sufficient to reach goals via alternative paths while requiring that much forward planning to identify coordination opportunities.

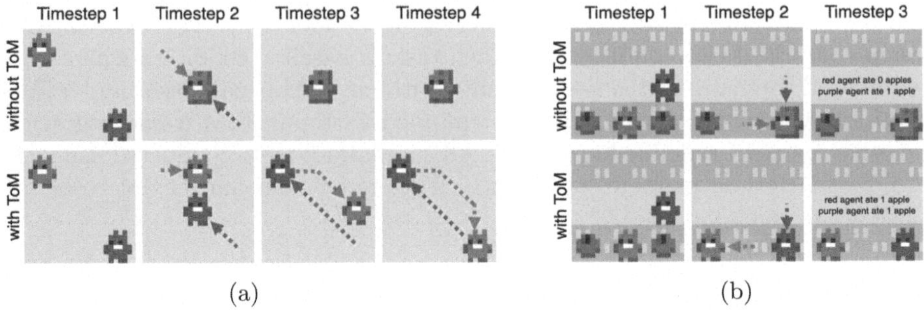

Fig. 2. Experimental Validation of Theory of Mind in Active Inference Using Multi-Agent Cooperation Tasks. (a) Collision avoidance task: agents must swap locations while avoiding collision at the central location. *Top row*: Non-ToM condition where both agents select the shortest path, resulting in collision and deadlock. *Bottom row*: ToM condition where the red agent (equipped with ToM) anticipates the purple agent's path and selects an alternative route, enabling successful cooperation. (b) Apple foraging task: agents search for and consume apples in orchard locations, with both agents initially knowing an apple exists at the bottom-right corner. *Top row*: Non-ToM condition showing resource competition where both agents converge on the known apple location, resulting in only one agent (purple) successfully consuming the apple. *Bottom row*: ToM condition where the red agent (equipped with ToM) explores an alternative location to avoid redundant competition, successfully finding and consuming an apple while the purple agent consumes the known apple, achieving optimal resource allocation for both agents.

Results. For an illustration of the results from this task, see Fig. 2a. In the non-ToM condition, both agents predictably selected individually optimal policies, moving directly toward goals via the centre. This resulted in collision and permanent deadlock, with neither agent achieving its objective. This is a clear cooperation failure despite sophisticated individual planning while observing the other agent's location but not being equipped to incorporate it into the planning process.

In the ToM condition, the red (ToM) agent reasoned that the other agent will most likely move to the central location to take the shortest path to its goal, so it does not select moving to that location to avoid collision even though it was the most optimal policy for itself. Instead, the agent went around the centre, which was the next best alternative route towards its goal. The ToM agent selected a longer but collision-free route. For the detailed planning tree of the ToM vs non-ToM agents at time step 0, see Appendix B.

3.2 Apple Foraging Task

Task Description. The apple foraging task examined cooperation in resource acquisition scenarios with partial observability. The 3×3 grid featured orchard locations (top and bottom rows) where apples could spawn, and wasteland (middle row) containing no resources. Both agents began with identical prior knowledge: certainty of an apple at the bottom-right corner, with complete uncertainty about the presence of an apple at other orchard locations. The agents' initial positions were equidistant from the known apple (see Fig. 2b). Apple consumption was exclusive – only one agent could consume each apple, with random selection if both agents reached the same apple simultaneously. The cooperation challenge involved balancing exploitation of known resources against exploration of uncertain locations, while avoiding redundant competition for the same resources.

Generative Model. The generative model incorporated three types of state factors: agent locations, reward feedback (binary: received/not received, which was conditioned upon eating an apple), and environmental items (wasteland, apple, or empty orchard). Agents observed their own location, the other agent's location, items at their current location, and their own reward feedback.

The environment was partially observable such that agents could only assess apple availability at their current location, creating uncertainty about the broader resource distribution. Apples spawned probabilistically in orchard locations (25% per time step) and remained in that location until consumption. Action repertoires included movements (up, down, left, right actions), eating, and no operation. Preferences simply favoured reward acquisition without explicit cooperation incentives. Planning horizons were set to 3 time steps, sufficient for reaching the opposing end of the grid environment and consuming apples.

Results. For an illustration of the results from this task, see Fig. 2b. For the detailed planning tree of the ToM vs non-ToM agents at time step 0, see Appendix C.

In the non-ToM condition, both agents converged on the known apple location (bottom-right corner), resulting in resource competition. Only one agent succeeded in consuming the apple (determined randomly), while the other wasted effort, demonstrating inefficient cooperation.

In the ToM condition, the red (ToM) agent reasoned that the other agent will most likely go to the known apple location and therefore, it selected to explore another location where it was uncertain whether there is an apple. This resulted in avoidance of redundant efforts and more effective cooperation, avoiding resource competition. This strategy proved successful as both agents discovered and consumed apples.

4 Discussion

Our experimental results demonstrate that equipping active inference agents with ToM capabilities fundamentally transforms their approach to completing a task, conducting multi-agent cooperation. The ToM-equipped agents successfully navigated both collision avoidance and resource competition scenarios, achieving better cooperation compared to their non-ToM counterparts. Importantly, this enhanced performance emerged without requiring explicit communication protocols, shared generative models between agents, or pre-set strategies.

The success of our ToM agents stems from their ability to reason about others' beliefs and anticipate their behaviours. In the collision avoidance task, the ToM agent recognised that both agents following optimal individual paths would result in collision. By reasoning about the other agent's likely trajectory, the ToM agent proactively selected the next best alternative route, demonstrating cooperative behaviour compared to a merely reactive collision response. Similarly, in the apple foraging task, the ToM agent anticipated resource competition and strategically explored uncertain locations, leading to more efficient resource allocation across both agents.

Our framework addresses fundamental limitations from previous multi-agent active inference implementations. Most importantly, we eliminate the restrictive assumption of shared or identical generative models that dominates prior work [2,5,9,10]. This limited their generalisability and applicability to more complex or real-world scenarios where agents possess different experiences, capabilities, and objectives. In contrast, our ToM framework allows for heterogeneous multi-agent generative models. The ToM agent maintains distinct belief representations for each agent in its environment, allowing it to reason about others without assuming they share its own knowledge, goals, or even generative model structure.

4.1 Future Directions

While this paper provides valuable insights into computationally conducting multi-agent cooperation by implementing ToM in active inference, there are several avenues for future research to build upon our findings.

Our current implementation assumes observational access to the other agents' locations and is situated in a simple 3×3 grid environment. While these simplifications enables clear demonstration of the core ToM principles, we naturally need to examine it under more complex and real-world scenarios as such scenarios involving more noisy sensory information and complex task requirements and dynamics. Future work should also include systematic quantitative evaluation using aggregated performance metrics across random seeds and statistical comparisons against non-ToM baselines to more rigorously assess the robustness of our approach.

Another simplification in the reported simulations was that our ToM agents assumed knowledge of others' goals and operated with fixed generative models of other agents. Future implementations should incorporate online learning mechanisms, potentially using Dirichlet counts [4], to continuously learn and update beliefs about the other agents' model, preferences, and capabilities. Such adaptive learning would significantly enhance the framework's generalisability, enabling effective cooperation with agents whose characteristics are initially unknown or evolving.

Moreover, while our current implementation focuses on dyadic interactions, the underlying principles naturally extend to larger multi-agent scenarios. Each agent would maintain separate belief representations for all other agents, and the planning algorithm would expand over joint policy spaces of arbitrary size. However, computational complexity grows exponentially with the number of agents, presenting scalability challenges that require careful considerations and is an avenue of further research.

Furthermore, our implementation focuses on first-order ToM reasoning (*"what does the other agent believe?"*) rather than higher-order recursive reasoning (*"what do I think the other agent thinks I believe?"*). While first-order ToM proves to be sufficient for our tested cooperation scenarios, more complex social situations may require deeper levels of recursive reasoning. This could be examined in scenarios where there are multiple ToM agents, investigating how recursive reasoning between ToM-capable agents affects cooperation dynamics and computational requirements.

Additionally, our framework has been validated only in cooperative scenarios where agents' goals are somewhat complementary. Future research should investigate competitive scenarios, where agents' objectives directly conflict, to assess whether the planning algorithm remains effective and how generative models should be structured to handle adversarial interactions.

5 Conclusion

We have presented the first generalisable implementation of ToM within the active inference framework for multi-agent cooperation. Our approach represents a significant advancement over existing methods by eliminating the restrictive assumption that all agents must operate under shared or identical generative models and goal structures.

The core innovation of our framework lies in enabling agents to maintain distinct beliefs about themselves and others while reasoning recursively about how others' beliefs influence their behaviour. Through our novel tree-based planning algorithm, ToM-equipped agents systematically explore joint policy spaces by considering what others believe and how those beliefs influence their consideration and decisions of actions. This recursive reasoning capability allows for sophisticated cooperation online without requiring explicit communication or pre-arranged cooperation protocols.

We validated our framework with two tasks: collision avoidance and resource foraging tasks. ToM agents successfully cooperated in both scenarios, avoiding conflicts and achieving more efficient outcomes compared to non-ToM agents.

Importantly, these cooperative capabilities emerged immediately upon encountering the task challenges, without requiring lengthy training periods or domain-specific learning.

The framework we have developed bridges computational and cognitive science, providing both a practical tool for enhancing artificial intelligence systems and a computational foundation for understanding how sophisticated social reasoning might emerge from principled probabilistic inference about others' minds.

A Expected Free Energy Under Sophisticated Inference with Theory of Mind

In sophisticated inference, the expected free energy is calculated in a recursive way [3]. This can be construed as a deep tree search, where the tree branches over allowable actions at each point in time, and the likely observations consequent upon each action. Equation 1 expresses the expected free energy of each potential next action (a_τ) and observation (o_τ) as the utility and information gain of that action plus the average expected free energy of future beliefs, under counterfactual observations and actions:

$$\begin{aligned} G(o_\tau, a_\tau) = \mathbb{E}_{Q(o_{\tau+1}|a_{\leq\tau})} \Big[\underbrace{-\ln P(o_{\tau+1}|C)}_{\text{utility}} - \underbrace{D_{KL}[Q(s_{\tau+1}|o_{\tau+1})||Q(s_{\tau+1})]}_{\text{information gain}} \Big] \\ + \underbrace{\mathbb{E}_{Q(a_{\tau+1}|o_{\tau+1})Q(o_{\tau+1}|a_{\leq\tau})} \big[G(o_{\tau+1}, a_{\tau+1}) \big]}_{\text{expected free energy of subsequent actions}} \end{aligned} \quad (1)$$

with

$$\begin{aligned} Q(o_{\tau+1}|a_{\leq\tau}) &= P(o_{\tau+1}|s_{\tau+1})Q(s_{\tau+1}|a_{\leq\tau}) \\ Q(a_\tau|o_\tau) &= \sigma\big(-G(o_\tau, a_\tau)\big) \end{aligned}$$

When planning with Theory of Mind, this deep tree search becomes an alternation between actions of the other agent and actions of the focal agent, as well as counterfactual observations for both. The resulting expected free energy in Eq. 2 is now expressed as a function of an action and observation for the focal agent (a^f_τ, o^f_τ) as well as the other agent (a^o_τ, o^o_τ):

$$\begin{aligned} G(o^f_\tau, o^o_\tau, a^f_\tau, a^o_\tau) = \mathbb{E}&_{Q(o^f_{\tau+1}, o^o_{\tau+1}|a^f_{\leq\tau}, a^o_{\leq\tau})} \\ &\big[-\ln P(o^f_{\tau+1}|C^f) - \mathbb{D}_{\text{KL}}[Q(s^f_{\tau+1}|o^f_{\tau+1})||Q(s^f_{\tau+1})] \big] \\ + \mathbb{E}&_{Q(a^f_{\tau+1}|o^f_{\tau+1})Q(a^o_{\tau+1}|o^o_{\tau+1})Q(o^f_{\tau+1}, o^o_{\tau+1}|a^f_{\leq\tau}, a^o_{\leq\tau})} \\ &\big[G(o^f_{\tau+1}, o^o_{\tau+1}, a^f_{\tau+1}, a^o_{\tau+1}) \big] \end{aligned} \quad (2)$$

with

$$
\begin{aligned}
Q(o^f_{\tau+1}, o^o_{\tau+1}|a^f_{\leq\tau}, a^o_{\leq\tau}) &= P(o^f_{\tau+1}|s^f_{\tau+1})Q(s^f_{\tau+1})P(o^o_{\tau+1}|s^o_{\tau+1}, s^f_{\tau+1})Q(s^o_{\tau+1})\\
Q(s^f_{\tau+1}) &= Q(s^f_{\tau+1}|s^f_\tau, a^f_\tau, s^o_{\tau+1})\mathbb{E}_{Q(o^o_{\tau+1}|a^o_{\leq\tau})}[Q(s^o_{\tau+1}|o^o_{\tau+1})]\\
Q(a^o_\tau|o^o_\tau) &= \sigma\big(-G(o^o_\tau, a^o_\tau|C^o)\big)\\
Q(a^f_\tau|o^f_\tau) &= \sigma\big(-G(o^f_\tau, o^o_\tau, a^f_\tau, a^o_\tau)\big)
\end{aligned}
\tag{3}
$$

Note that the utility term for the focal agent uses the focal agent's preferences C^f, which might be distinct from the other agent's preferences C^o which are used to calculate the posterior over the other's actions. In addition, our posterior over expected observations for the other agent is conditioned on both $s^o_{\tau+1}$ and $s^f_{\tau+1}$. Here, we combine the $s^{o,\text{self}}$ with $s^{f,\text{world}}$ to generate expected observations from the perspective of the other, using the world belief of the focal agent. Another crucial point in Eq. 3 is that to calculate the predictive posterior $Q(s^f_{\tau+1})$ of the focal agent, we also incorporate the effect the other's action might already have had on the focal agent's belief state. This is implemented via belief sharing via likelihood message passing, similar to [2].

B Planning Trees for Collision Avoidance Task

Figure 3 shows the difference in planning trees for both a non-ToM (c) and ToM agent (d) completing the collision avoidance task with another agent. It is clear that the ToM agent's planning tree (d) entertains four routes the agent can take to avoid colliding with the other (non-ToM) agent and that the route illustrated in the right-most branch of the planning tree is selected to complete the task (a).

C Planning Trees for Apple Foraging Task

Figure 4 shows the difference in planning trees for both a non-ToM (c) and ToM agent (d) completing a apple foraging task with another agent. We start from the orchard environment state in (a), where both agents navigate to locate and consume apples, with initial shared knowledge of an apple at location 9. We visualise the planning tree for the red agent with (d) and without ToM (c).

In the non-ToM case, the focal agent (red) evaluates only its own policies over a 2-step horizon, selecting to go to location 9 (P=1.0) based on expected utility (G=10.00). However, when pursuing this policy it will end up with no apple, as the purple agent will get there first.

In contrast, in the ToM case, the focal agent (red) indeed reasons that the other agent (purple) will get to the apple first, and therefore chooses to explore the tile on the right, which might have an apple as well.

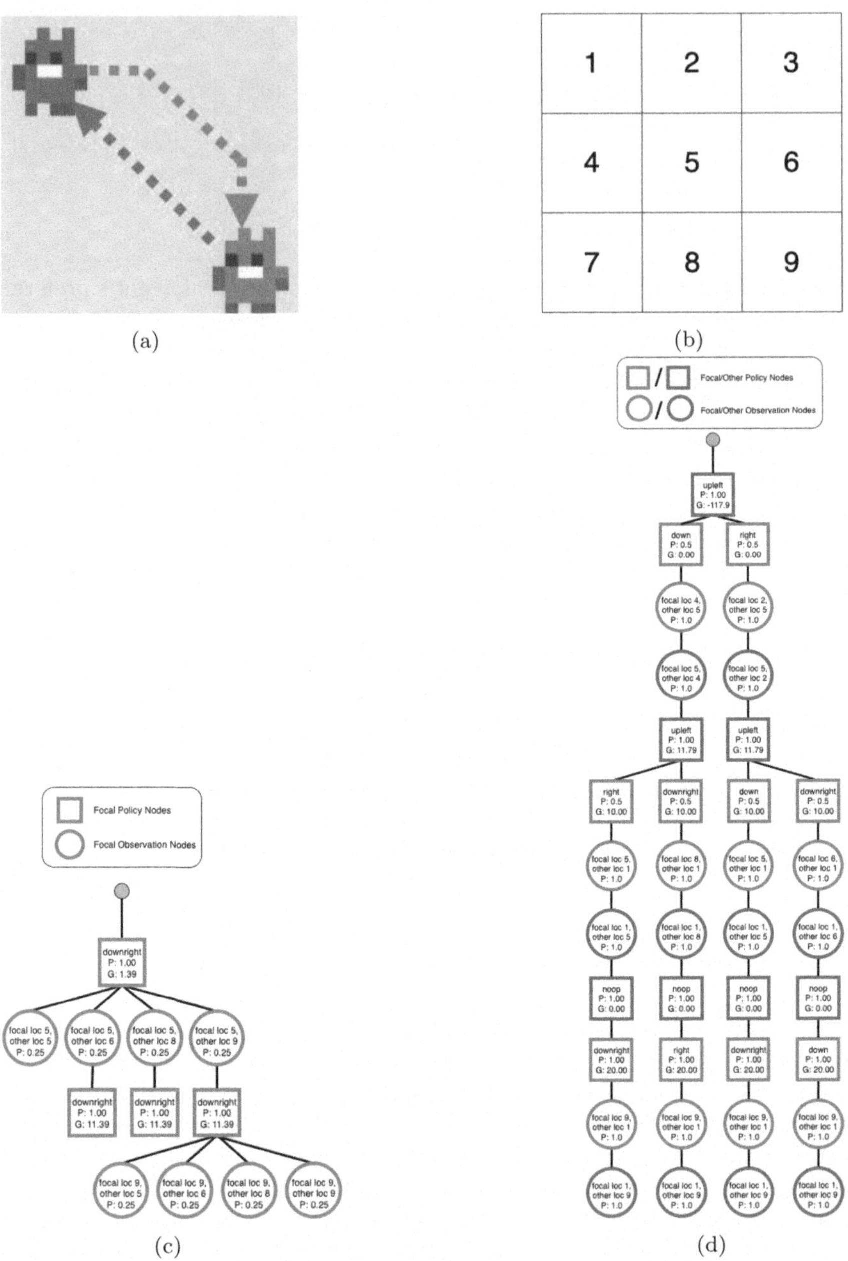

Fig. 3. (a) Collision avoided by the red agent as the focal agent with ToM. (b) Reference grid layout for 3×3 simulation environment. Numbers 1–9 indicate cell indices used throughout the experiments to specify agent locations and movements. (c) Non-ToM planning tree: The red agent evaluates only its own policies over a 2-step horizon, selecting to go to location 5 (P=1.0), which eventually results in colliding with the other agent. (d) ToM planning tree: The red (ToM) agent recursively models the purple (non-ToM) agent's policy space and beliefs, resulting four possible routes the red agent can take to avoid collision with the other agent which is predicted to go to its goal location via location 5. (Color figure online)

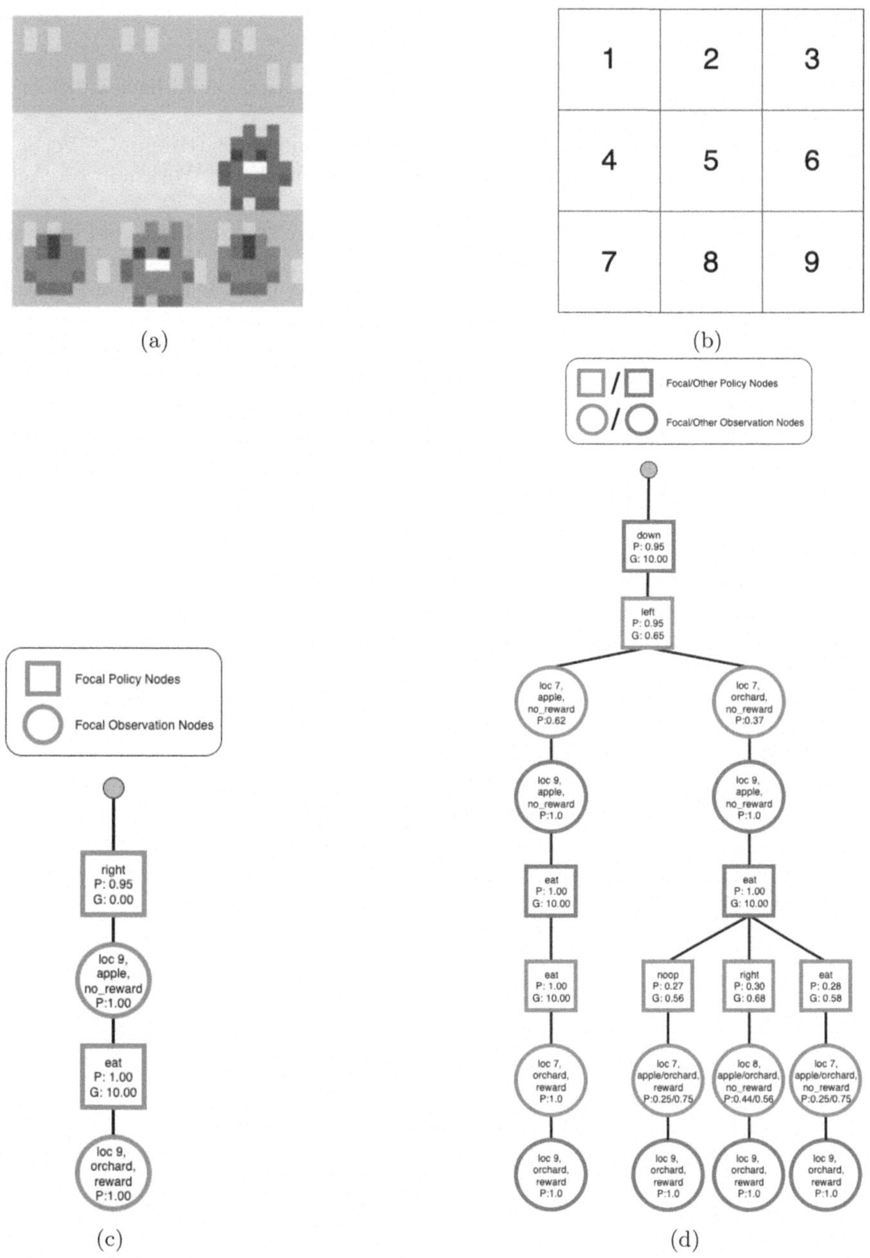

Fig. 4. (a) Start situation with the red agent as the focal agent. (b) Reference grid layout for 3×3 simulation environment. Numbers 1-9 indicate cell indices used throughout the experiments to specify agent locations and movements. (c) Non-ToM planning tree: The red agent evaluates only its own policies over a 2-step horizon, selecting to go to location 9 (P=1.0) based on expected utility (G=10.00). (d) ToM planning tree: The red (ToM) agent recursively models the purple (non-ToM) agent's policy space and beliefs, resulting in selection of going left with near full certainty (q_pi=0.95). This recursive reasoning leads to exploration of location 7, as the agent predicts the purple agent will pursue location 9, enabling coordinated foraging. (Color figure online)

References

1. Baron-Cohen, S., Leslie, A.M., Frith, U.: Does the autistic child have a "theory of mind"? Cognition **21**(1), 37–46 (1985)
2. Çatal, O., Van de Maele, T., Pitliya, R.J., Albarracin, M., Pattisapu, C., Verbelen, T.: Belief sharing: a blessing or a curse. In: International Workshop on Active Inference, pp. 121–133. Springer (2024)
3. Friston, K., Da Costa, L., Hafner, D., Hesp, C., Parr, T.: Sophisticated inference. Neural Comput. **33**(3), 713–763 (2021)
4. Friston, K., FitzGerald, T., Rigoli, F., Schwartenbeck, P., Pezzulo, G., et al.: Active inference and learning. Neurosci. Biobehav. Rev. **68**, 862–879 (2016)
5. Friston, K.J., et al.: Federated inference and belief sharing. Neurosci. Biobehav. Rev. **156**, 105500 (2024)
6. Frith, C., Frith, U.: Theory of mind. Current Biol. **15**(17), R644–R645 (2005)
7. Heins, C., Millidge, B., Demekas, D., Klein, B., Friston, K., Couzin, I., Tschantz, A.: pymdp: A python library for active inference in discrete state spaces. arXiv preprint arXiv:2201.03904 (2022)
8. Van de Maele, T., Dhoedt, B., Verbelen, T., Pezzulo, G.: Integrating cognitive map learning and active inference for planning in ambiguous environments. In: International Workshop on Active Inference, pp. 204–217. Springer (2023)
9. Maisto, D., Donnarumma, F., Pezzulo, G.: Interactive inference: A multi-agent model of cooperative joint actions. IEEE Trans. Syst. Man Cybern. Syst. **54**(2), 704–715 (2023)
10. Matsumura, T., Esaki, K., Yang, S., Yoshimura, C., Mizuno, H.: Active inference with empathy mechanism for socially behaved artificial agents in diverse situations. Artif. Life **30**(2), 277–297 (2024)
11. Wellman, H.M., Cross, D., Watson, J.: Meta-analysis of theory-of-mind development: The truth about false belief. Child Dev. **72**(3), 655–684 (2001)
12. Wimmer, H., Perner, J.: Beliefs about beliefs: representation and constraining function of wrong beliefs in young children's understanding of deception. Cognition **13**(1), 103–128 (1983)

Demonstration of Making Humans Recognize Intentions of Robot Actions Through Active Inference

Kanako Esaki[1(✉)], Yasuyuki Kudo[1], Tadayuki Matsumura[1], Takeshi Kato[2], Misa Owa[1], Junichi Miyakoshi[1], Yasuhiro Asa[1], Yang Shao[1], Ryuji Mine[1], and Hiroyuki Mizuno[1]

[1] Research and Development Group, Hitachi, Ltd., Tokyo, Japan
kanako.esaki.oa@hitachi.com

[2] Hitachi Kyoto University Laboratory, Kyoto University, Kyoto, Japan

Abstract. Toward collaboration between humans and machines in real-world environments where uncertainty is inherent, we demonstrated that the intentions behind machine actions can be shared with humans through an exchange of a clipboard between a real robot and a person. In the demonstration, we compared how humans interpret the behavioral intentions of two types of robots. One type follows an artificial minimal self (AMS), which extends its internal model to the actions of others based on active inference. The other type only executes pre-designed action sequences, which we used as a benchmark. Because the AMS virtually shares the processes inferred by a self-agent and another agent, the intentions of the two agents are expected to be shared in the same way that humans share their intentions by inferring and sharing processes in order to address uncertainty. The results showed that, in some cases, people tend to recognize the intention of the robot following the AMS as slightly higher than that of the robot executing pre-designed action sequences.

Keywords: Intention · Human-machine collaboration · Minimal self

1 Introduction

When humans collaborate, all parties are required to share their intentions for their actions. Let us consider a situation in which people are collaborating. First, they share the goal to achieve through collaboration. Furthermore, uncertainties inherent in the real-world collaboration must be addressed. Specifically, because the actions of others in collaboration involve uncertainty, sub-goals for achieving the goal must also be shared. Sharing sub-goals makes it easier to predict the actions of others and clarify one's own actions, facilitating collaboration. The sub-goals are regarded as intentions in this paper. In human collaboration, intentions are implicitly shared by inferring the processes that others expect to achieve the goal. Through the sharing of intentions, humans are able to collaborate in real-world environments where uncertainty is inherent.

M. Albarracin et al. (Eds.): IWAI 2025, CCIS 2857, pp. 401–415, 2026.
https://doi.org/10.1007/978-3-032-16955-6_23

For machines to collaborate with humans in real-world environments where uncertainty is inherent, both parties need to share their intentions, just as in collaboration between humans. In particular, many studies have been conducted in the field of social robotics to bring the collaboration between humans and machines closer to that between humans. Grassi et al. proposed a conversation system for agents such as social robots [1]. The proposed system does not simply respond to the words of human users but rather consistently grasps the user's intentions and creates a flow of conversation, enabling to share intentions from humans to machines. Furthermore, studies have also been conducted on the sharing of goals from machines to humans. Putte et al. had the social robot Pepper conduct a questionnaire survey of hospital patients on behalf of nurses [2]. The goals were shared by displaying dialogs and answer choices on the robot's screen. Yamaji et al. developed a "Social Trash Box" that cannot pick up trash [3]. A trash-can-shaped robot shared its goal of "throwing away trash" with humans by bowing. Studies have also been conducted on sharing the internal state of machines with humans. Faraj et al. developed Eva, a humanoid head that can mimic human facial expressions, and enabled the robot to express emotions such as joy and sadness [4]. Briggs et al. discussed when and how robots should refuse instructions that violate human norms and enabled robots to express their will through refusal behaviors created based on the concept of "face threat" from human politeness theory [5]. However, in these studies, although internal states such as emotions and will at the current moment were shared from machines to humans, the intention to consider future processes toward a goal was not.

In order to enable two-way intention sharing between humans and machines, we must develop methods for machines to share their intentions with humans. According to active inference, which explains the principles of human behavior, humans have a preference corresponding to their goals and act by inferring the processes that are expected to bring them closer to their preference [6–10]. Since the inferred processes include sub-goals, or intentions, active inference explains how humans act with intention. Furthermore, we extended active inference and proposed an artificial minimal self (AMS) for machines to share intentions with other agents for collaboration [11]. In normal active inference, each agent has a generative model of its own observations and actions. Based on this model, the agent infers hidden states as causes of certain observations and selects actions that will result in state transitions that bring it closer to its preference, or processes. The AMS is inspired by minimal self [12] and further extends the model of the self to include not only the self-agent but also other agents. Doing so easily enables incorporating a function for sharing the intentions of actions between the self and other agents into the model. The model provides processes inferred under the expectation that not only the self-agent but also other agents will act as expected. This means that the sub-goals, or intentions, included in these processes are virtually shared within the model. This would lead to the self-agent selecting actions so that intentions are shared between the self-agent and other agents. Thus, the AMS should cause the self-agent to take actions to share its intentions with other agents.

In this study, we demonstrated that intentions can be shared by applying the AMS to collaboration between humans and robots. We have previously applied it to collaboration between robots [11]. Such collaboration assumed that once the robot "sent" its intentions, the other robots would always accurately "understand" them, thereby sharing them. Collaboration is achieved simply by "sending." In this study, we demonstrated the collaboration between humans and robots by comparing robots implementing the AMS with robots executing pre-designed action sequences as a benchmark. The intentions are not shared simply by robots "sending" them. Rather, the intentions are shared when humans "understand" them. Accordingly, collaboration is achieved only when the robot "sends" its intentions to the human and the human "understands" them.

2 Method

2.1 Artificial Minimal Self

In normal active inference [10], which is the basis of the AMS, an agent has a generative model in which its own observations, hidden states and actions are probabilistic variables. Figure 1(a) shows a conceptual diagram of the generative model in normal active inference. The generative model $p(s_\tau, o_\tau)$ includes the likelihood $p(o_\tau|s_\tau)$, the transition probability $p(s_{\tau+1}|s_\tau, \pi)$, the preference $p(o_\tau|C)$, and the prior $p(s_{\tau=0})$, which are represented by matrices $\boldsymbol{A}$, $\boldsymbol{B}$, $\boldsymbol{C}$, and $\boldsymbol{D}$, respectively, when limited to discrete spaces.

$$
\begin{aligned}
&p(o_\tau|s_\tau) = Cat(\boldsymbol{A}) \\
&p(s_{\tau+1}|s_\tau, \pi) = Cat(\boldsymbol{B}_{\pi\tau}) \\
&p(o_\tau|C) = Cat(\boldsymbol{C}) \\
&p(s_{\tau=0}) = Cat(\boldsymbol{D})
\end{aligned} \tag{1}
$$

As shown in Fig. 1(a), all probabilistic variables of observations, hidden states, and actions are set only for the self-agent. The next action of the self-agent is selected based on this matrix of the self-agent by minimizing expected free energy $\boldsymbol{G}$:

$$
\begin{aligned}
&\boldsymbol{G}_\pi = \boldsymbol{H} \cdot \boldsymbol{s}_{\pi\tau} + \boldsymbol{o}_{\pi\tau} \cdot \boldsymbol{\varsigma}_{\pi\tau} \\
&\boldsymbol{H} = -diag(\boldsymbol{A} \cdot \ln \boldsymbol{A}) \\
&\boldsymbol{\varsigma}_{\pi\tau} = \ln \boldsymbol{o}_{\pi\tau} - \ln \boldsymbol{C}_\tau
\end{aligned} \tag{2}
$$

The AMS [11] is an extension of active inference inspired by Gallagher's minimal self [12]. The minimal self consists of a sense of self-ownership and a sense of self-agency. Most adults have these two senses toward their own bodies, but they can also extend them to objects and other people around them. When we use tools or collaborate with other people, our sense of self extends to the tools and other people. The AMS follows this idea and treats not only self-agents but also other agents as the self. In normal active inference, the self-agent selects its own action based on its observations, whereas in the AMS, the self-agent selects its own actions and infers the actions of other agents by also considering the observations of other agents.

AMS is enabled by integrating the other agents' observations, hidden states and actions into the generative model of normal active inference. Figure 1(b) shows a conceptual diagram of the generative model in the AMS. The probabilistic variables in the rows and columns of the matrix include those of other agents in addition to those of the self-agent. The probabilistic variables are often defined based on agents when using normal active inference for generating the actions of intelligent agents such as robots

[13, 14]. However, the corresponding values for the probabilistic variables of other agents defined based on agents can only be obtained from the other agents' viewpoint. The self-agent cannot obtain the corresponding values directly. For example, if another robot's action variable is defined as "move an object 10 m forward," the self-robot cannot obtain the current value of that variable because it does not have the reference position or direction. Therefore, the AMS defines environment-based probabilistic variables. For example, if the action variable is "move an object to point P in the environment," the robot can obtain the current value of that variable, assuming that the robot is able to observe point P. Defining environment-based probabilistic variables enables the integration of probabilistic variables of other agents.

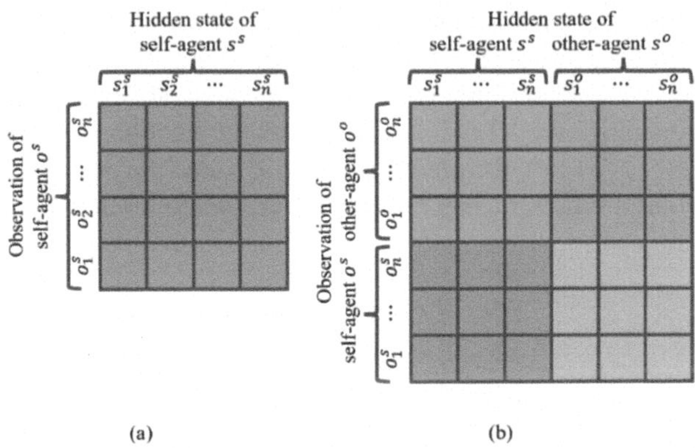

Fig. 1. Conceptual diagram of generative models for (a) normal active inference and (b) the artificial minimal self.

By integrating the probabilistic variables of other agents' observations, hidden states and actions into the generative model of the self-agent, the self-agent and other agent share intentions. Let us imagine a situation in which a robot implementing the AMS is collaborating with a human. Because the human has a minimal self that extends to others, we assume that the human can be regarded as the other agent with the AMS. Figure 2 shows a conceptual diagram of the self-agent and human (other agent) sharing intentions. The robot calculates multiple processes expected to lead to its preference based on its current observations and generative model at timestep t. The robot selects the process that is most likely to reach the preference in the fewest timesteps from among the multiple processes, and selects the action $a4^h$ for timestep t in that process. The human also calculates and selects the process. In the process leading to preference, the final state $s7$ where the process selected by the robot and the process inferred by the human match is the shared sub-goal, which means that the robot and the human share the intention of the action at timestep t. Because the probabilistic variables of the other agent's (human's) observations, hidden states and actions are integrated into the generative model, the process also includes the state $s4^h$ and action $a4^h$ of the human. This results in a match between the process selected by the robot and the process selected by the human, i.e., the sharing of intentions between the robot and the human. The shared intentions between the robot and the human are further reflected in the robot's actions.

In Fig. 2, the action selected by the robot at timestep t is the human action $a4^h$. In other words, at timestep t, the robot takes an action of waiting while expecting the human to perform action $a4^h$. In addition, if the robot and the human share the same intention at timestep t, the human will perform action $a4^h$ as expected by the robot, and the state will transition to $s4^h$. Therefore, at timestep $t+1$, the robot will select action $a7$ as originally assumed at timestep t. Meanwhile, if the robot and the human do not share the same intention at timestep t, the robot will select an action different from the one originally assumed at timestep t at timestep $t+1$. For example, if the human does not perform any action and the state remains $s1$, the same action as in timestep t $a4^h$ will be repeated with a high probability in timestep $t+1$. In this way, the robot's waiting action and the repetition of the same action when the state does not transition as expected will result in the robot actively sharing its intentions with the human, and this should contribute to the human's recognition of the robot's behavioral intentions.

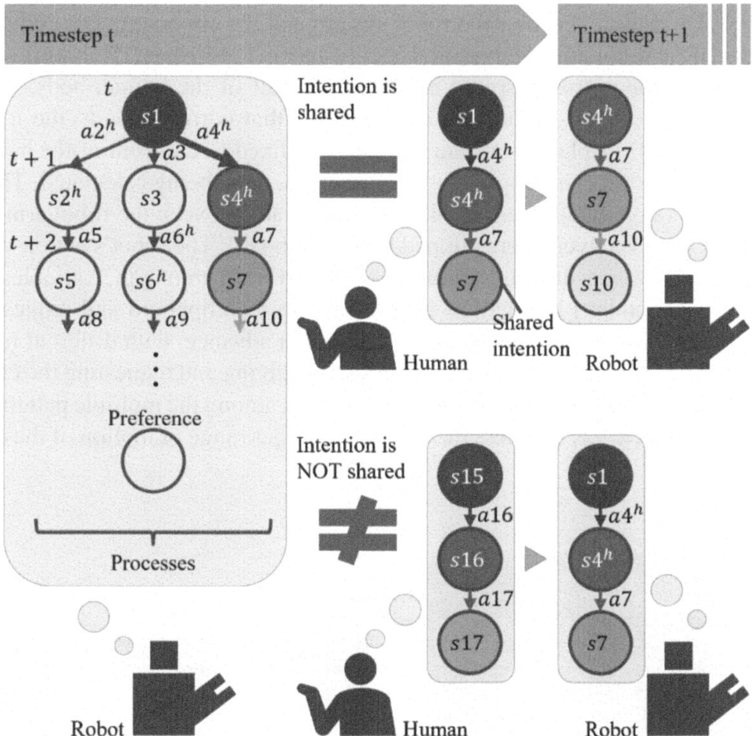

Fig. 2. Conceptual diagram sharing the intentions of the other agent. The symbols s and a represent hidden states and actions, respectively. Symbols with an upper index h represent human's hidden states or actions, while symbols without it represent robot's hidden states or actions.

2.2 Experimental Setup

We demonstrated that intentions can be shared by the AMS in a situation in which a person and a robot exchange a clipboard. The situation begins with the person and the robot facing each other. The robot is holding a clipboard with a cafeteria flyer clipped to it. First, the robot gives the clipboard to the person. Then, the person receives the clipboard from the robot. After that, the robot requests the person to return the clipboard. The person responds by returning the clipboard to the robot. The goal of for the first half of this collaboration situation is for the person to hold the clipboard, and the goal for the second half is for the robot to hold the clipboard. To achieve each goal, the robot takes actions to give the clipboard to the person and to request that the person return the clipboard. If the robot's intention to give the clipboard or request its return is shared with the human, the above series of actions will be performed, and the goal of collaboration is expected to be achieved.

The robot used for the demonstration was a robot system integrated with a collaborative robot arm. Figure 3 shows the robot system and the clipboard. The collaborative robot was a UR5e robot arm (Universal Robots), and a 2F-140 hand (Robotiq) was fixed to the tip of the robot arm. Inspired by the upper half of the human body, the robot arm and hand were fixed to the side of a housing that corresponds to the torso. An LCD-AHU431XDB display (I-O Data Device) was fixed to the front of the housing to show that a calculation is in progress while its actions were being calculated. The depth sensors and the range finder were used to avoid contact between the robot arm and the participants. Cafeteria flyers were clipped to the clipboard. The robot's action variables were defined in advance, and one action was selected from among the variables at each timestep (see Appendix.) For actions such as giving the clipboard and requesting the clipboard, multiple motion patterns were prepared in advance, with different ranges of arm motion for each pattern. When an action such as giving and requesting the clipboard was selected, one pattern was randomly selected from among the multiple patterns. Each time a motion pattern was selected, the one with a larger range of motion of the arm was executed.

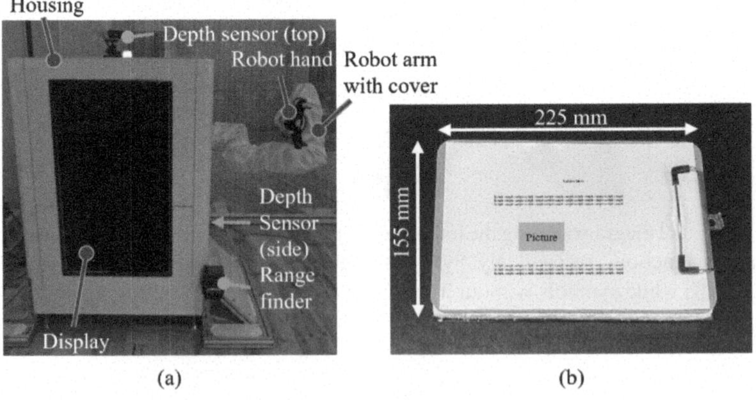

Fig. 3. Photographs of (a) the robot system and (b) clipboard.

The participants in the experiment were 106 employees of Hitachi, Ltd. The number of participants was set at 102 or more, based on the assumption of a one-tailed t-test with two independent groups (ratio of 1:1), a rejection level of 5%, and a test power of 80%, with an expected effect size of 0.5. To recruit the participants, we first held an online briefing session to explain the outline and safety of the experiment. Following the briefing session, the participants who attended were sent an email with instructions on how to register for the experiment. Those who registered took part. The 106 participants were 73 men and 33 women, with 13 in their 20 s, 33 in their 30 s, 37 in their 40 s, 21 in their 50 s, and two in their 60 s. The participants were randomly divided into two groups: 54 participants in the experimental group, who experienced the robot with the AMS, and 52 participants in the control group, who interacted with robots that only executed pre-designed action sequences as a benchmark. All participants signed a consent form on the day of the experiment.

The experiment was conducted in two phases. In phase 1, up to three participants observed the robot's actions simultaneously for efficient data collection, without talking to ensure independence. In this phase, a situation was deliberately created in which the robot recognized that its intention was not shared with the people. Under this condition, the AMS robot and benchmark robot were compared. First, the participants observed the robot giving the clipboard to a person. To create a situation where the participants did not understand the robot's intention, the experimenter instructed the participants not to take any action, such as receiving the clipboard. Therefore, the robot with the AMS repeatedly selected the action of giving the clipboard to share its intention. However, the benchmark robot performed the action of giving the clipboard only once and then stopped. Then, the clipboard held by the robot was removed, and the experimenter distributed clipboards to each participant. With the clipboards in their hands, the participants observed the robot requesting the clipboards from them. As with the action of giving the clipboard, the robot with the AMS repeatedly selected the action of requesting the clipboard, while the benchmark robot performed the action of requesting the clipboard only once and then stopped.

In phase 2, one participant at a time interacted with the robot to exchange the clipboard. The experimenter simply instructed the participants to "interact freely with the robot." First, the robot performed the action of giving the clipboard to the person and then performed the action of requesting the clipboard from the person. The robot with the AMS repeatedly selected the action of giving the clipboard until the person received it. Furthermore, after the person received the clipboard, it repeatedly selected the action of requesting the clipboard until the person returned it. However, the benchmark robot performed the actions of giving the clipboard and requesting it once, regardless of whether the person received or returned the clipboard. Each phase lasted about 10 min. The experiments were conducted for 10 days. To ensure consistency across experiments, we conducted all of them under indoor lighting and provided instructions by reading predefined scripts.

After each phase, participants answered a questionnaire. In the questionnaire after phase 1, participants answered questions about their attributes, such as gender, and the robot's intentions as observed in phase 1. Specifically, regarding the robot's intentions, participants answered "Did you understand what the robot wanted to do?" on a 6-point Likert scale (1: I strongly disagree – 7: I strongly agree, excluding 4: neutral) for each of the actions of giving the clipboard and requesting the clipboard. In addition, participants who selected the higher scores (5: I slightly agree – 7: I strongly agree) were asked to respond to statements describing the details of what the robot wanted to do (see Appendix) on a 7-point Likert scale (1: I strongly disagree – 7: I strongly agree.) As shown in Fig. 2, the intention in this study was defined as a hidden state. However, the statements were designed to indicate actions corresponding to the hidden state so that participants could easily understand them. The statements included those referring to robot's actions ("I want to") and those referring to a person's actions ("I want *you* to"). The statements also included those related to the flyers and those related to the clipboard. Even if the participants wanted to take the flyers, they would have to take the clipboard because of the distance between them and the robot. Therefore, we assumed that sharing of the intention could be verified with either statement. In the questionnaire after phase 2, the participants answered questions about their experience with robots, such as "Have you ever been involved in robot-related work in research/development?" and questions related to the interactions in phase 2.

3 Results

3.1 Recognition of Intention

The difference in the mean scores between the AMS and benchmark robots was analyzed for statements of the intentions behind the actions of giving the clipboard and requesting it. First, statements with an absolute effect size Cohen's d [14] exceeding 0.2, indicating a difference, were extracted. Then, for this exploratory study, the extracted statements were evaluated for significant differences using a relaxed criterion. P-values less than 0.100 on a t-test were judged to indicate significant tendance.

The AMS showed a higher tendency to share the intention of the robot's action of giving a clipboard to a person compared with the benchmark. Figure 4 shows the mean scores for the statements regarding the intention behind the action of giving the clipboard, where the absolute Cohen's d exceeded 0.2. As shown in Fig. 4 (a) and (b), the AMS tended to have higher mean scores than the benchmark for statements referring to the robot's action. However, as shown in Fig. 4 (d) and (e), the AMS tended to have lower mean scores than the benchmark for statements referring to the person's action.

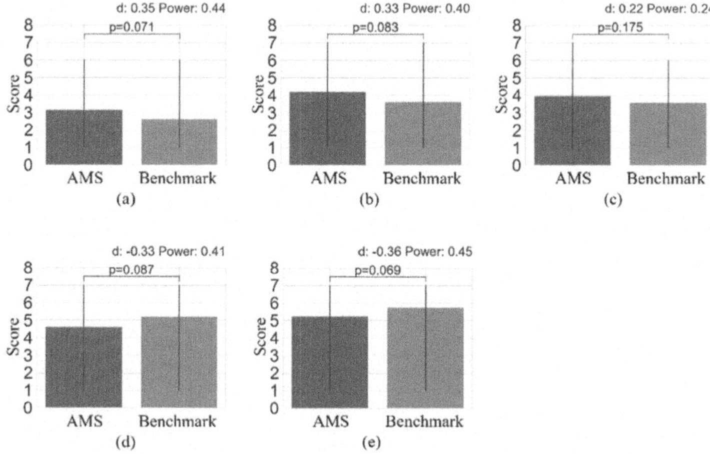

Fig. 4. Scores for (a) "I want to get rid of the flyers that were clipped to the clipboard.", (b) "I want to distribute the flyers that were clipped to the clipboard.", (c) "I want to interact with you.", (d) "I want you to take the flyers that were clipped to the clipboard.", and (e) "I want you to take the clipboard."

The robot's action of requesting a clipboard involved sharing the intention to refer to specific person's actions at the same level in the AMS and the benchmark. Figure 5 shows the mean scores for the statement regarding the intention behind the action of requesting a clipboard, where the absolute value of Cohen's d exceeded 0.2. As shown in Fig. 5(a) and (b), the p-value was not less than 0.100, indicating no significant difference. Meanwhile, as shown in Fig. 5(c), the benchmark tended to be higher than the AMS for abstract intentions and did not refer to specific robot's actions.

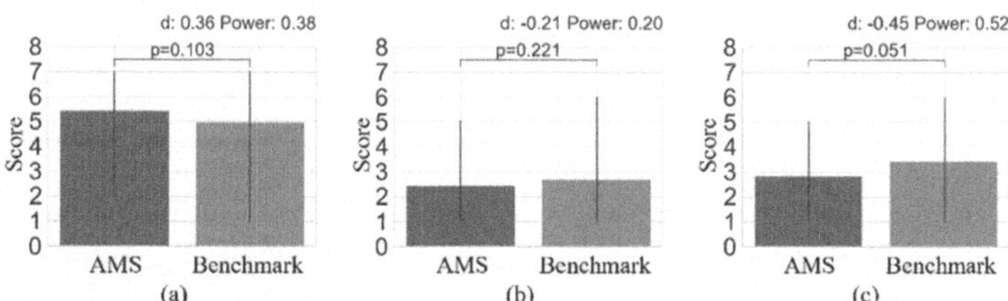

Fig. 5. Scores for (a) "I want you to return the clipboard.", (b) "I want you to read the flyer on the clipboard and come try the lunch menu written there.", and (c) "I want to serve you."

3.2 Comparison Based on Experience Using Robots

The statements extracted in the previous section were further analyzed based on whether or not the participants had experience using robots in research and development work. Here, "Experience with robots" refers to participants who answered either "currently involved" or "previously involved" to the question "Have you ever been involved in robot-related work in research/development?" "No experience with robots" refers to the participants who answered "never involved." Table 1 summarizes the results. Among these, those with a p-value less than 0.100 are shown in the graph and explained in the following.

Table 1. Statistical results for statements based on whether participants had experience using robots. The mean scores are shown in order of the AMS and benchmark.

Statements	Experience with robots			No experience with robots		
	Mean	Cohen's d	p-value	Mean	Cohen's d	p-value
I want to get rid of the flyers that were clipped to the clipboard.	2.750 3.100	-0.19	0.359	3.308 2.484	0.58	0.018
I want to distribute the flyers that were clipped to the clipboard.	3.875 3.900	-0.01	0.490	4.308 3.548	0.45	0.050
I want to interact with you.	4.500 3.400	0.64	0.118	3.808 3.677	0.08	0.381
I want you to take the flyers that were clipped to the clipboard.	4.375 4.900	-0.29	0.306	4.692 5.290	-0.35	0.103
I want you to take the clipboard.	5.750 6.000	-0.22	0.353	5.077 5.645	-0.39	0.074
I want to serve you.	2.889 3.556	-0.62	0.118	2.792 3.353	-0.39	0.127
I want you to return the clipboard.	5.778 4.778	0.72	0.086	5.292 5.059	0.19	0.302
I want you to read the flyer on the clipboard and come try the lunch menu written there.	2.111 2.778	-0.74	0.078	2.542 2.647	-0.08	0.412

A difference in participant responses to a robot giving a clipboard was observed, depending on whether or not the participant had experience using robots in research and development work. Figure 6 shows the mean scores for statements regarding the intention behind the action of giving the clipboard, with p-values less than 0.100 in Table 1. As shown in Table 1, for participants who had experience with robot, the p-value for statements regarding the intention behind the action of giving the clipboard exceeded 0.100, indicating no statistical difference between the AMS and the benchmark. However, as shown in Fig. 6 the responses from participants who had no experience with robot showed a similar tendency to those in Fig. 4. Specifically, the mean score for statements referring to robot's actions tended to be higher for the AMS than for the benchmark, while the mean score for statements referring to person's actions tended to be higher for the benchmark than for the AMS. In particular, the mean score for the response "I want to get rid of the flyers that were clipped to the clipboard." in Fig. 6 (a) was significantly higher for the AMS than for the benchmark, with a p-value of 0.018.

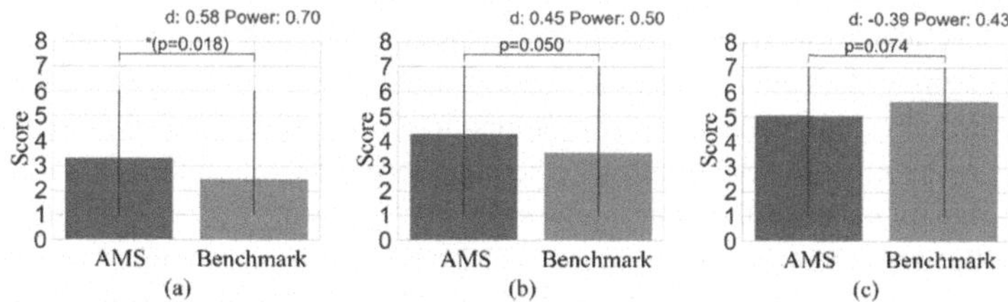

Fig. 6. Scores for (a) "I want to get rid of the flyers that were clipped to the clipboard.", (b) "I want to distribute the flyers that were clipped to the clipboard.", and (c) "I want you to take the clipboard." Note that these are responses from the participants who had no experience with robots.

Regarding the robot's action of requesting a clipboard from a person, a difference in tendency was observed depending on whether or not the participants had experience with robots in research and development work. Figure 7 shows the mean scores for statements regarding the intention behind the action of requesting a clipboard, with p-values less than 0.100 in Table 1. As shown in Fig. 7, the responses of participants who had experience with robots showed a different tendency to those shown in Fig. 5. Specifically, in statements referring to a person's specific actions, the mean score for the AMS and for the benchmark were significantly different with p-values less than 0.100, while in statements referring to robot's specific actions no statistical difference between the AMS and the benchmark was found. In addition, among statements of specific persons' actions, the AMS score was higher in one case, and the benchmark score was higher in another. However, as shown in Table 1 no statistical difference was evident between the AMS and the benchmark in responses from participants who had no experience with robots.

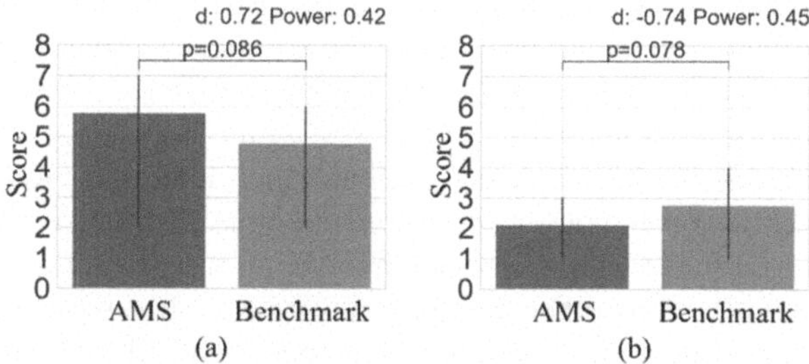

Fig. 7. Scores for (a) "I want you to return the clipboard." and (b) "I want you to read the flyer on the clipboard and come try the lunch menu written there." Note that these are responses from the participants who had experience with robots.

4 Discussion

The intention behind the robot's action of giving the clipboard to a person can be considered to have been shared by the AMS. The robot with the AMS repeated the action of giving the clipboard to approach the goal of "for the person to hold the clipboard." Each time the action of giving the clipboard was repeated, various patterns appeared, and the range of arm movement gradually increased. The person may become aware of the intention behind the robot's actions. Furthermore, as the robot repeated the action of giving the clipboard, the multiple flyers clipped to the clipboard could caught the attention of the participants, and the effect size became larger for statements of robot's action referring to the flyers than for statements of robot's action referring to the clipboard. Meanwhile, the fact that the mean score was higher in the benchmark for statements referring to person's actions suggests a sense of self-efficacy among humans [16]. The benchmark robot was limited to giving the clipboard once. When this happened, the participants faced an uncertain situation. Because of the humans' sense of self-efficacy, the participants might begin to think about what they should do in such an uncertain situation. Thus, participants might focus on their expected actions rather than the intentions of the robot's actions.

Differences in responses for giving the clipboard based on participants' prior experience with robots suggest that knowledge of the demonstration affects the recognition of the robot's intention. In order for a robot to share its intention with a person, the person needs to first feel that the robot has an intention. The task of a robot handing an object to a person is one of the well-known human-robot interaction tasks [17] for those who are or have been involved in research and development work with robots. Therefore, participants who had experience with robots may have simply regarded it as a handover task without trying to understand the robot's intention. Once they regarded it as a handover task, they would not try to understand the robot's intention due to confirmation bias [18] even if the robot repeated the action of giving the clipboard. As a result, the p-value was more than 0.100 for both the clipboard and the flyer. However, for participants who had never been involved in robotics in their research or development, the HRI task itself was something they had never seen before. Therefore, participants who had no experience with robots tended to perceive the robot's intention in the repeated action of giving the clipboard, combined with the robot system's appearance, which was inspired by the upper body of a human, and their perception of the affordance of the clipboard and flyer.

The results of the differences in the intention for the actions of robots requesting clipboards from people depending on their experience with robots suggest that the amount of time spent using robots affects the recognition of their intentions. Regardless of their experience with robots, people rarely encounter situations in which robots request their belongings. However, those who have been involved in research or development work with robots (even with robots other than the ones used in this experiment) have spent more time using robots than those who have not been involved with robots. Therefore, even if participants who had experience with robots did not have experience in the relevant situation, their experience backed by the amount of time they had spent using robots may have increased their tendency to sense the intention of a person's specific action, at least in the short term (like Fig. 7 (a)).

5 Conclusion

For collaboration between humans and machines in real-world environments where uncertainty is inherent, we demonstrated that the behavioral intentions of machines can be shared with humans through a situation involving the exchange of clipboards between a real robot and a person. In the demonstration, we compared how humans interpret the behavioral intentions of the robot with an AMS against the benchmark robot that only executes pre-designed action sequences. The AMS extends its internal model to the actions of others based on active inference. Because the AMS virtually shares the processes inferred by the self-agent and the other agent, the intentions of the self-agent and the other agent should have been shared in the same way that humans do. The demonstration results showed that the robot with the AMS tended to have higher recognition of intentions toward robots by people than the benchmark robot. Future works include expanding the range of action variables so machines can respond to different situations, and exploring applications to other human-machine collaborative scenarios.

Acknowledgments. This study was funded by JST RISTEX (grant number JPMJRS22J5) and JSPS Topic-Setting Program to Advance Cutting-Edge Humanities and Social Sciences Research (grant number JPJS00122679495). The studies involving humans were approved by the Ethics Committee of Kyoto University (6-P-7). The studies were conducted in accordance with the local legislation and institutional requirements. The participants provided their written informed consent to participate in this study. The authors thank Prof. Yasuo Deguchi, Assoc. Prof. Takuro Onishi, and Dr. Harufumi Tamazawa for their advice on ethical review. We also thank Dr. Nobutaka Kimura for providing constructive feedback and suggestions on the manuscript.

Disclosure of Interests. The authors have no competing interests that are relevant to the content of this article to declare.

Appendix

The labels for the observation, hidden state (combination of observation), and action variables were predesigned.

$$
\begin{aligned}
&o^{robot} \in \{at\ origin,\ at\ giving\ position,\ at\ requesting\ position\} \\
&o^{person} \in \{at\ origin,\ at\ receiving\ position,\ at\ giving\ position\} \\
&o^{clipboard} \in \{with\ robot,\ with\ person\} \\
&a \in \left\{ \begin{array}{l} robot\ move\ to\ giving\ position, \\ robot\ move\ to\ requesting\ position\ normally, \\ robot\ move\ to\ requesting\ position\ humbly, \\ robot\ move\ to\ requesting\ position\ firmly, \\ robot\ move\ to\ judging\ position, \\ robot\ stop, \\ person\ move\ to\ receiving\ position, \\ person\ move\ to\ giving\ position, \\ person\ stop \end{array} \right\}
\end{aligned}
\tag{A1}
$$

o^{person} and $o^{clipboard}$ were obtained by detecting whether the robot hand was grasping or not grasping the clipboard. At each time step, the action label that minimized the expected free energy $\boldsymbol{G}$ was selected.

The specific statements asking about intentions in the questionnaire for this experiment are shown below. For each statement, the participants were asked to respond on a 7-point Likert scale (1: I strongly disagree – 7: I strongly agree.)

- First half (while the robot is holding the clipboard)

- I want to get rid of the flyers that were clipped to the clipboard.
- I want to distribute the flyers that were clipped to the clipboard.
- I want you to take the flyers that were clipped to the clipboard.
- I want you to look at the flyers that were clipped to the clipboard and come try the lunch menu written there.
- I want to give you the clipboard.
- I want you to take the clipboard.
- I want to interact with you.
- I want to serve you.
- I want to complete the instructions in the program.

- Second half (while the participants are holding the clipboard)

- I want you to return the flyers that were clipped to the clipboard.
- I want to receive the flyers that were clipped to the clipboard.
- I want you to return the clipboard.
- I want to receive the clipboard.
- I want you to look at the flyers that were clipped to the clipboard and come try the lunch menu written there.
- I want to interact with you.
- I want to serve you.
- I want to complete the instructions in the program.

References

1. Grassi, L., Recchiuto, C.T., Sgorbissa, A.: Knowledge-grounded dialogue flow management for social robots and conversational agents. Int. J. Soc. Robot. **14**, 1273–1293 (2022)
2. van der Putte, D., Boumans, R., Neerincx, M., Rikkert, M.O., de Mul, M.: A social robot for autonomous health data acquisition among hospitalized patients: an exploratory field study. In: 2019 14th ACM/IEEE International Conference on Human-Robot Interaction (HRI), pp. 658–659. IEEE, Daegu, Korea (South) (2019)
3. Yamaji, Y., Miyake, T., Yoshiike, Y., De, R.S., Silva, P., Okada, M.: STB: child-dependent sociable trash box. Int. J. Soc. Robot. **3**, 359–370 (2011)
4. Faraj, Z., et al.: Facially expressive humanoid robotic face. HardwareX **9**, e00117 (2021)
5. Briggs, G., Williams, T., Jackson, R.B., Scheutz, M.: Why and how robots should say 'no.' Int. J. Soc. Robot. **14**, 323–339 (2022)
6. Friston, K., Kilner, J., Harrison, L.: A free energy principle for the brain. J. Physiol.-Paris **100**(1–3), 70–87 (2006)
7. Friston, K.: The free-energy principle: a unified brain theory? Nat. Rev. Neurosci. **11**, 127–138 (2010)

8. McGregor, S., Baltieri, M., Buckley, C.L.: A minimal active inference agent. arXiv, 1503.04187 (2015)
9. Friston, K., FitzGerald, T., Rigoli, F., Schwartenbeck, P., Pezzulo, G.: Active inference: a process theory. Neural Comput. **29**(1), 1–49 (2017)
10. Parr, T., Pezzulo, G., Friston, K.J.: Active Inference: The Free Energy Principle in Mind, Brain, and Behavior. MIT Press (2022)
11. Esaki, K., Matsumura, T., Kato, T., Minusa, S., Shao, Y., Mizuno, H.: Artificial minimal self on free energy principle for autonomous cooperative behavior. In: Proceedings of the 2024 Artificial Life Conference (ALIFE 2024), p. 9. MIT Press, Copenhagen, Denmark (2024)
12. Gallagher, S.: Philosophical conceptions of the self: implications for cognitive science. Trends Cogn. Sci. **4**(1), 14–21 (2000)
13. Çatal, O., Verbelen, T., Van de Maele, T., Dhoedt, B., Safron, A.: Robot navigation as hierarchical active inference. Neural Netw. **142**, 192–204 (2021)
14. Georgeon, O.L., de Montera, B., Robertson, P.: Reducing intuitive-physics prediction error through playing. In: International Workshop on Active Inference (IWAI 2024), pp. 222–233. Springer, Oxford, UK (2024)
15. Cohen, J.: Statistical Power Analysis for the Behavioral Sciences. Routledge (2013)
16. Bandura, A.: Self-efficacy mechanism in human agency. Am. Psychol. **37**(2), 122–147 (1982)
17. Ortenzi, V., Cosgun, A., Pardi, T., Chan, W.P., Croft, E., Kulić, D.: Object handovers: a review for robotics. IEEE Trans. Rob. **37**(6), 1855–1873 (2021)
18. Nickerson, R.S.: Confirmation bias: a ubiquitous phenomenon in many guises. Rev. Gen. Psychol. **2**(2), 175–220 (1998)

Navigating Uncertainties with Active Inference and Probabilistic Diffusion

Yufei Huang[1], Yulin Li[1(✉)], Andrea Matta[2], and Mohsen A. Jafari[1]

[1] Rutgers University – New Brunswick, Piscataway, NJ 08854, USA
yl959@soe.rutgers.edu
[2] Politecnico di Milano, Via La Masa 1, 20156 Milano, Italy

Abstract. This paper presents a novel approach of integrating Active Inference (AIF) and Probabilistic Diffusion (PD) for autonomous navigation. AIF is a promising paradigm for "optimal" control without the need for complex and extensive dynamical models, especially for non-linear and time-variant systems. On the other hand, PD can be used to generate navigational strategies (e.g., prediction of motions) for reaching a target. When combined, the two form a locally optimal navigator, where AIF optimizes the future strategies sampled from PD, while the samples are adjusted using the variational free energy from AIF when new actual observations are made. We demonstrate the framework in a simulated parking lot scenario and highlight two key advantages over reinforcement learning (RL): learning without handcrafted rewards and improved generalization to novel settings. Though AIF and RL differ fundamentally, our benchmark shows that AIF+PD achieves comparable performance in complex multi-agent parking tasks.

Keywords: Active Inference · Probabilistic Diffusion Models · Model-Based Autonomous Control · Multi-Agent Coordination

1 Introduction

Active Inference (AIF) is a control framework grounded in Predictive Mind theory [1]. It models how humans forecast sensory outcomes and act to minimize prediction error. AIF applies this idea to decision-making under uncertainty by combining predictive reasoning with variational inference [2]. Specifically, AIF uses Bayes' rule to infer hidden states of the true environment using a probabilistic perception model, which is updated with each new observation. The true generative process can be dynamic, time-varying, and non-stationary. The optimization follows the Free Energy Principle, minimizing surprisal caused by mismatches between predicted and actual outcomes [2]. These mismatches may result from (i) an inaccurate model of the dynamics or (ii) the need for corrective actions. The perception model incorporates both retrospective (e.g., autoregressive) and prospective (e.g., forecasting) components. Variational Free Energy (VFE) updates beliefs, while Expected Free Energy (EFE) guides action

M. Albarracin et al. (Eds.): IWAI 2025, CCIS 2857, pp. 416–429, 2026.
https://doi.org/10.1007/978-3-032-16955-6_24

selection. However, computing EFE requires evaluating many possible action–state trajectories, which becomes computationally expensive in large spaces and undermines AIF's usability for real-time control.

In this article, we propose a Probabilistic Diffusion (PD) generative model for future trajectories. For example, in a two-dimensional space defined by (x, y) coordinates, this model acts like a motion predictor, defining a feasible trajectory of states and actions to go from a randomly generated point to a preferred target. Given the feasible set of motion trajectories, AIF can then be applied to optimize the move strategy. We demonstrate this approach in autonomous parking, a task characterized by sparse cues and interacting agents. At each control step, the agent samples PD trajectories, selects the trajectory minimizing EFE, executes the first action, and resamples trajectories upon receiving new observations, similar to traditional Model Predictive Control (MPC) [3]. If the subsequent observation deviates from PD's prediction, the predictor is dynamically updated through VFE, assigning a new starting point for future trajectory sampling. This adaptability distinguishes our proposed method from other limited existing approaches, as the PD model continually adapts to uncertainties revealed by actual observations.

Our novel PD model requires no expert-generated trajectories or precise dynamics modeling, relying solely on randomly generated actions for training. As AIF learns the task more effectively, the PD-generated strategies further improve. In contrast, traditional reinforcement learning (RL) baselines, generated using sampled rollouts, lack this online adaptability and often fail under random initial conditions. Our comparative benchmarks show that even without any explicit reward signals, the proposed AIF+PD method achieves comparable or better performance relative to RL, especially in scenarios with uncertain initial conditions where RL typically struggles.

2 Related Works

Recent works have explored AIF in robotics and cognitive science as a framework for decision-making under uncertainty [4]. RL has been effectively utilized for trajectory planning in automated parking systems, yet it typically requires extensive data and significant computational resources [8]. Unlike RL, which maximizes numerical rewards explicitly linked to outcomes, AIF minimizes expected free energy, shifting the focus from reward accumulation toward reducing uncertainty and achieving states of minimal surprise [5]. AIF's preference-based approach inherently balances exploration and exploitation, adapting its actions according to both current knowledge and new observations [6].

Preliminary studies [7,8] demonstrate AIF's potential for adaptive decision-making in dynamic environments, particularly for autonomous vehicles. In contrast, MPC, another common method for autonomous navigation, predicts future vehicle states through solving optimization problems under dynamic constraints [3,9]. However, MPC's reliance on precise vehicle modeling and high computational demands limits its practical application in dynamic real-world settings [10].

PD has recently been applied in robot planning and maze-solving contexts [11,12]. These approaches use expert-generated state-action trajectory matrices and diffusion-based conditional sampling methods, implicitly integrating PD with RL through trained reward functions. While effective for planning, such methods lack real-time adaptability to actual system conditions, thus limiting their utility for control tasks.

Recent comparative studies highlight decisive advantages of AIF over conventional RL and control methods in scenarios such as parking, where precise modeling and extensive training are impractical. For example, AIF-based agents have demonstrated perfect success rates in unseen urban tasks, significantly outperforming Behavioral Cloning and RL in terms of sample efficiency and generalization [17]. Furthermore, network-based AIF implementations have substantially outperformed Deep RL in industrial control applications, achieving significantly lower error rates and drastically reduced training costs [18]. This efficiency arises from AIF's intrinsic free-energy minimization principle, enabling robust performance with minimal reliance on large reward-driven datasets and computational resources.

3 Problem Formulation and Preliminaries

The problem of interest is to demonstrate the integration of AIF with PD for a navigation problem under uncertainties. The motivation is to reduce the computation time of exploring the prospective views (future trajectories) at time epochs where an observation is made or an outcome is received. We will describe the methodology in parking scenarios, where vehicles, starting from random positions, velocities, and directions, navigate toward their designated parking spots by skillful maneuvering to avoid stationary and mobile obstacles. We will assume that each vehicle's condition is characterized by its location, marked by coordinates (x, y), its speed in the direction of both coordinates (v_x, v_y), and the direction it's facing, noted as h. The state of each vehicle at equally distant time epochs t can then be represented as a vector of five elements $S_t = [x_t \; y_t \; v_t^x \; v_t^y \; h_t]$. There are two control actions under our AIF agent, namely, steering and acceleration. The kinematic bicycle model [13] includes vehicle's position, (x, y), vehicle's forward speed, v, vehicle's heading, ψ, the vehicle's acceleration, α, vehicle's slip angle at the center of gravity, β, and δ for the front wheel angle. Our enhancement of the traditional bicycle model includes adding noise terms that follow a normal distribution, $\epsilon \sim \mathcal{N}(0, \Sigma)$, where Σ represents a diagonal covariance matrix.

$$\begin{cases} x_{t+1} = x_t + v_t^x \cdot \cos(\delta_t + \beta_t) \cdot \Delta t + \epsilon_x \\ y_{t+1} = y_t + v_t^y \cdot \sin(\delta_t + \beta_t) \cdot \Delta t + \epsilon_y \\ \delta_{t+1} = \delta_t + \frac{v_t \cdot \sin(\beta_t)}{L/2} \cdot \Delta t + \epsilon_\delta \\ v_{t+1}^x = v_t \cdot \cos(\delta_{t+1}) + \epsilon_{v_x} \\ v_{t+1}^y = v_t \cdot \sin(\delta_{t+1}) + \epsilon_{v_y} \end{cases} \quad (1)$$

Equation (1) defines the bicycle model $f(S_t, a_t)$ under uncertainty, where Δt is the time step, L is the length of the vehicle, $\beta_t = \arctan(1/2 \cdot \tan(\delta_t))$ is the steering angle at the mass center. The variances $\sigma_x^2, \sigma_y^2, \sigma_{vx}^2, \sigma_{vy}^2, \sigma_\delta^2$ correspond to the respective state variables in S_t. These modifications adjust the model's response to environmental and vehicular variabilities. In the sequel, we will use two pairs of probability distributions: $\{P(\cdot), Q(\cdot)\}$ and $\{p(\cdot), q(\cdot)\}$. Note the case difference between the two sets. In the AIF formulation, $P(\cdot)$ refers to the true generative process, also called G-Density, and $Q(\cdot)$ refers to the variational density or (R-Density) that approximates the true one. In PD formulation, $p(\cdot)$ refers to the true probability distribution of the initial data, and $q(\cdot)$ refers to the generative (approximation) distribution of the initial data.

4 AIF Formulations

The AIF agent aims at steering the vehicle to its preferred state - the desired parking space, while maintaining an accurate view of the world around the vehicle. The PD motion predictor aims to provide the AIF agent with fast samples of feasible trajectories from any given point to the goal state. Formally speaking, the motion predictor, parameterized by φ, projects the likelihood of potential actions $p(a_n|S_n;\varphi)$ within such a trajectory. We assume a deterministic mapping (likelihood function) between states and observations and Markovian transitions between the states. This assumption can be relaxed by replacing the deterministic mapping with a probabilistic likelihood function, but the Markovian assumption is a necessary condition. The likelihood of future states and actions can then be represented by a probability distribution,

$$Q(\boldsymbol{S}', \boldsymbol{a} \mid S_n) := \prod_{\tau=n}^{N-1} \mathcal{N}(S_{\tau+1}; f(S_\tau, a_\tau), \Sigma)\, p(a_\tau \mid S_\tau; \varphi) \tag{2}$$

where $Q(\boldsymbol{S}', \boldsymbol{a} \mid S_n)$ denotes the estimated future states and actions, assuming the agent is in state S_n. Instead of asserting that actions lead to a single specific state, the Gaussian distribution introduces a scope of possible next states $S_{\tau+1}$, with $f(S_\tau, a_\tau)$ providing the mean or most likely next state and Σ encapsulating the uncertainty in this transition. The term $p(a_\tau|S_\tau;\varphi)$ (obtained from PD) captures the conditional probability of an action a_τ, given the current state S_τ, and influenced by the model's parameters φ. In the AIF framework, VFE serves as a measure of the divergence between the predicted and actual future states. In this work, a modified VFE quantifies the discrepancy between the outcomes predicted by the PD and the observed true states, and is given by,

$$\text{VFE} = D_{\text{KL}}[q(\varphi \mid S_n, a) \,\|\, p(\varphi)] - \mathbb{E}_q\left[\ln P(\boldsymbol{S}' \mid S_n, a, \varphi)\right] \tag{3}$$

where the first term is the Kullback-Leibler (KL) divergence function [16], which quantifies the discrepancy between the current belief about the PD parameters $q(\varphi|S_n, a)$ (conditioned on the observed state and action) and the prior beliefs $p(\varphi)$. This term effectively regularizes the diffusion process by minimizing the difference between the true set of parameters and our current belief; the second term $\mathbb{E}_q\left[\ln P(\boldsymbol{S'} \mid S_n, a, \varphi)\right]$ represents the expected log likelihood of observing the trajectory of states $\boldsymbol{S'}$ given the current state S_n, action a, and model parameters φ. This term anchors the diffusion model's predictions to the actual observations by measuring how well our model explains the observed state transitions, thus calculating the core component of the variational free energy. The above VFE is minimized by continuously adjusting the φ using the data collected during operation. This optimization is typically performed using gradient descent on the physics-informed variational autoencoder (to be discussed later), and the parameter updates use the relation, with η as the learning rate.

$$\varphi_{new} = \varphi_{old} - \eta \nabla_{\varphi} \mathrm{VFE} \tag{4}$$

Relations (3) and (4) link the PD and AIF formulations for the agent to use updated motion predictors after each move it makes. At such an instant and given the current vehicle state data, the AIF agent must decide on its next course of action by optimizing an EFE function. The agent will use the updated motion predictor samples (initialized to the current state) for the calculation of EFEs.

In the exploration of preferred states by the AIF agent, a preference distribution C_β is defined over the state space $\mathbb{S}$. This distribution is weighted by parameter $\beta > 0$ to prioritize states that the agent finds rewarding. These preferred states are derived from the Boltzmann distribution expressed as in logarithmic form by:

$$-\log C_\beta(S) = -\beta R(S) - c(\beta), \quad \forall S \in \mathbb{S} \tag{5}$$

Agents are inclined to maximize $C_\beta(S)$ for a given β. $R(S)$ includes three terms λ_{goal}, λ_{safe}, and λ_{smooth} applied to three separate goals, namely, reaching a parking spot S_{goal}, maintaining safety by avoiding any collisions, S^{v-}_{n+1}, and ensuring smoothness in the control actions. That is,

$$R(S) = -\lambda_{goal} \cdot \|S'_{n+1} - S_{goal}\| + \lambda_{safe} \cdot \sum \|S'_{n+1} - S^{v-}_{n+1}\| - \lambda_{smooth} \cdot \|a'_n - a_{n-1}\| \tag{6}$$

The AIF agent seeks entire trajectories that are aligned with the agent's preferences and the dynamics of the vehicle's environment. With the additive property over the states, $R(S)$ can be extended to a trajectory $\boldsymbol{S} := (S_0, S_1, \ldots, S_N) \in \mathbb{S}^N$. The vehicle AIF agent utilizes EFE to balance the exploration-exploitation trade-off by minimizing surprise (or uncertainty) and maximizing the likelihood of achieving preferred outcomes [14]. There are many different formulations of EFE. Here, we will use the one that encourages the selection of actions that not only minimize surprise but also align future states closely with those that are considered preferable or beneficial. It is given by,

$$G(s,a) \approx -\mathbb{E}_{Q(s'|s,a)}\left[\log P(o, s' \mid s, a)\right] + \mathbb{E}_{Q(s'|s,a)} D_{\mathrm{KL}}\left[Q(\boldsymbol{S'} \mid \boldsymbol{a'}, S_n) \,\|\, C_\beta(\boldsymbol{S'})\right] \tag{7}$$

where $Q(s'|s,a)$ is the approximate posterior or the agent's belief about the next state given the current state and action. $P(o, s'|s,a)$ is the generative model that links states and observations, providing the likelihood of observing o in state s' after taking action a. Note that for simplicity, we assume that observations and states are deterministically mapped one-to-one. The first term $\mathbb{E}_{Q(s'|s,a)}[\log P(o, s'|s, a)]$ quantifies the surprise of observing o and the next state s' because of applying action a in the current state s. In other words, it estimates how unexpected or unlikely the observation and state transition are under the current policy. The second term $\mathbb{E}_{Q(s'|s,a)} D_{KL}[Q(\boldsymbol{S'}|\boldsymbol{a}, S_n)||C_\beta(\boldsymbol{S'})]$ incorporates the cost function $C_\beta(\boldsymbol{S'})$ and quantifies how the distribution of predicted future states $Q(\boldsymbol{S'}|\boldsymbol{a}, S_n)$ diverges from a desired or preferred state distribution as encoded by $C_\beta(\boldsymbol{S'})$. This divergence aims to penalize decisions leading to future states that are less preferred according to the cost function. This form aligns with the standard risk + ambiguity decomposition of EFE: the KL-to-preferences is the risk, while the expected log-likelihood contributes an ambiguity term that reflects outcome uncertainty given the state. Given the underlying Markovian assumptions, the EFE for a sequence of actions is additive and can be expressed as an aggregate over a trajectory. The sequence of actions to consider for this formulation is determined by sampling from distributions that are computed using the PD formulation of the motion predictor.

5 Vehicle Navigation in Unmarked Parking Areas

The PD motion predictor calculates the probability distribution of a vehicle's imminent actions that will lead to its successful parking spot. Similar to the common PD [15], we include a forward training process and a reverse learning process, both governed by Markovian transitions. The diffusion model is refined via VFE every time the prediction error are excessive. Here, we apply diffusion to actions (speed and direction), and the states are computed using a physics-based dynamic model. Furthermore, we do not rely on expert data for training the model. In contrast to the image generation, where clarity is progressively diminished by overlaying Gaussian noise, the motion diffusion predictor simulates the real-world conditions that a vehicle starts from its parking spot, note the state as S_0, and takes random actions to drive away from the initial position. The final state S_T is determined when the vehicle hits the boundary of the parking area, or conflicting traffic conditions far from the initial position occur. Figure 1 provides a depiction where the green cars persist in random movements until an eventual collision with the parking area's boundary occurs. This forward process effectively emulates the journey from a state of complete organization to one of disorder, analogous to the method by which noise is added in image processing.

To aid in navigation, two extra pieces of information are provided for each vehicle at time t: θ_t shows the direction to the vehicle's starting parking spot, and l_t measures how far the vehicle is from this spot. During each time interval, the vehicles under control undergo random changes in speed and direction. This follows the common approach in PD where the amount of Gaussian noise is increased over time to gradually obscure the initial data. In a similar vein, the range within which these random driving decisions are made becomes wider as time progresses. The sequence of random driving decisions made throughout this

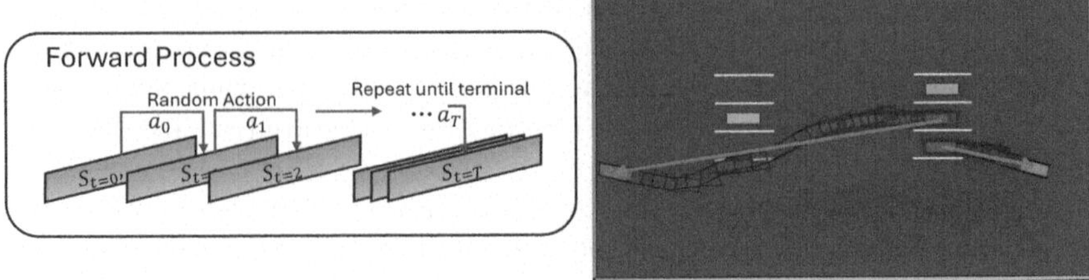

Fig. 1. Forward process for automated parking. Left: Gaussian random actions generate state sequences mimicking diffusion. Right: example vehicle trajectory. Reversed sequences are used to train the VAE.

process is represented by $a_1, \ldots, a_{T-1}$. Each action a_i, where i ranges from 1 to $T-1$, is drawn from a Gaussian distribution $a_i \sim \mathcal{N}(a_{i-1}, \sigma_i^2 I)$ that is truncated, i.e., $a_i = \text{clip}(a_i, lb, ub)$. This range reflects the real-world physical constraints on how much a vehicle can accelerate or decelerate (throttle) and turn (steering) at any given moment. To avoid dramatic movement change, the previous action is the mean of the current action, and the actions are chosen based on the previous action a_{i-1} but with added variability defined by σ_i^2. σ_i follows a linear growth, ensuring that with each step from the first to the last, the variance expands smoothly from its minimum to its maximum value:

$$\sigma_i = \sigma_{min} + (\sigma_{max} - \sigma_{min}) \times \frac{i}{T-1} \tag{8}$$

where σ_{min} and σ_{max} define the bounds of variance, while i represents the current step, and $T-1$ signifies the total number of steps.

In the reverse process, a vehicle starts at a random position with an initial speed and direction. The motion diffusion predictor employs learned parameters to infer the most probable previous action distribution of a vehicle—essentially 'denoising' the vehicle's trajectory to yield a predicted path back to its parking state. The predictor is trained on reversed state-action sequences:

$$S'_t \xrightarrow{a'_{t-1}} S'_{t-1} \xrightarrow{a'_{t-2}} \cdots \xrightarrow{a'_1} S'_1 \xrightarrow{a'_0} S'_0 \tag{9}$$

where S'_t is the reverse of S_t. In the reverse process, the goal is to uncover the range of possible actions that would logically return the vehicle to its prior state S'_{t-1} based on the current state and navigational aids θ_t and l_t. Minimizing KL Divergence [16] adjusts the parameters of the predictive model q to closely approximate the true distribution p, that is,

$$\min_{a'_{t-1}} \text{KL}\big(p(a'_{t-1} \mid S'_t, \theta'_t, l'_t) \,\|\, q(a'_{t-1} \mid S'_t, \theta'_t, l'_t)\big) \tag{10}$$

Given the current state S'_t and predicted action a'_{t-1}, the next state S'_{t-1} can be estimated using the kinematic bicycle model. Taking the common reparameterization approach in PD, we must use $S_t = \gamma_t S_0 + \sigma_t \boldsymbol{a}$, which cannot be guaranteed here due to the limited nature of our forward process. As such, the common score matching techniques cannot be employed. Instead, we introduce a rolling back method to learn the score function $\nabla_{S_t} \log q_t(S_t \mid S_{t+\Delta t})$. The idea is that the forward and backward transitions mirror each other, which reverses the trajectory and returns to the original state.

A neural network model architecture utilizes a physics-informed variational autoencoder (VAE) approach to carry out two key stages of prediction, as is shown in Fig. 2.

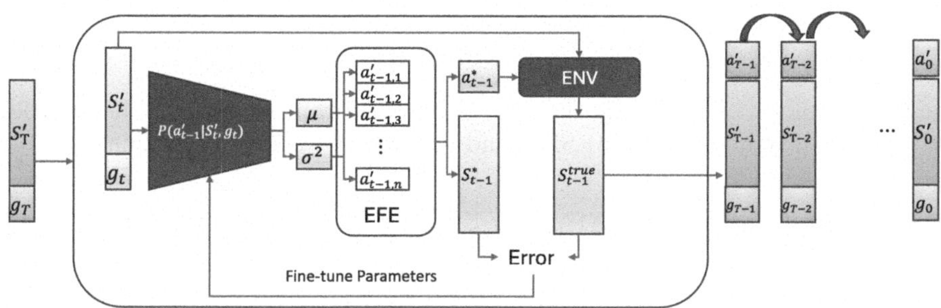

Fig. 2. Reverse process for autonomous parking. The planning uses a physics-informed VAE with EFE to reconstruct reversed state–action sequences $(\mathbf{S}', \mathbf{a}')$, mimicking the reverse of $(\mathbf{S}, \mathbf{a})$.

At its core, the neural network model predicts the probability $p(a'_{t-1} \mid S'_t, \theta'_t, l'_t)$ using the reversed state S'_t. Note that in Fig. 2 the guidance term g_t represents the auxiliary inputs θ'_t, l'_t. Using the mean and variance derived from $p(a'_{t-1} \mid S'_t, \theta'_t, l'_t)$, a set of reparametrized next actions $a'_{t-1,n}$ are generated. The physical bicycle model is then used to estimate multiple preceding states $S'_{t-1,n}$ from S'_t given potential action $a'_{t-1,n}$. Using the EFE defined in Equation (7), an action a^*_{t-1} is selected that would lead us to the preferred next state S^*_{t-1}. The discrepancy between this estimated state S^*_{t-1} and the actual prior state, once reversed to S^{true}_{t-1}, is quantified using the mean squared error (MSE). This rollback technique is pivotal in ensuring that the actions predicted by the model effectively guide the vehicle closer to its initial parking. The VAE introduces randomness into the process to account for environmental uncertainties and the inherent variability in vehicle responses.

The loss function designed for the physics-informed VAE is threefold: (i) It accounts for the state prediction error, quantifying the difference between predicted states and true states; (ii) It incorporates a regularization component for the variance $\sigma^2_{a'_{t-1}|S'_t}$, which ensures that the model does not overly concentrate on minimizing the prediction error linked to the mean $\mu_{a'_{t-1}|S'_t}$, but also accurately gauges the level of uncertainty in the variance; (iii) It accounts for model uncertainty in predictions of the next state, by regulating the variance of state transition noise $\epsilon \sim \mathcal{N}(0, \Sigma)$. We have,

$$\mathcal{L} = \text{MSE}(S_{t-1}^{\text{true}}, S_{t-1}^{*}) + \lambda_1 R_a \left(\sigma^2_{a'_{t-1}|S'_t}\right) + \lambda_2 R_\epsilon S'_0(\Sigma) \tag{11}$$

where λ_1 and λ_2 serve as scaling factors that can be adjusted, and

$$R_a \left(\sigma^2_{a'_{t-1}|S'_t}\right) = -\frac{1}{T} \sum_{t=1}^{T} [\log(\sigma_i + \epsilon) + \log(1 - \sigma_i - \epsilon)] \tag{12}$$

and

$$R_\epsilon(\Sigma) = \frac{\text{MSE}(S_{t-1}^{*}, S_{t-1}^{\text{true}})}{2 \cdot \Sigma^2} \tag{13}$$

The log-likelihood of observing the next state S_{t-1}^{*} from a Gaussian distribution with the mean (true next state) S_{t-1}^{true} and variance Σ^2 is given by Equation (14):

$$\log\left(p(S_{t-1}^{\text{true}} \mid S_{t-1}^{*}, \Sigma^2)\right) = -\frac{(S_{t-1}^{*} - S_{t-1}^{\text{true}})^2}{2 \cdot \Sigma^2} - \log(\Sigma\sqrt{2\pi}) \tag{14}$$

We can drop the constant term $\log(\Sigma\sqrt{2\pi})$ for optimization since it does not affect the relative evaluations of the model parameters.

6 Simulation and Validation Results

We evaluated the proposed AIF+PD framework in multi-vehicle parking scenarios using the `highway-env` simulation package and compared its performance against standard RL baselines. We simulated scenarios with two controllable vehicles and with three controllable vehicles (each with static obstacles) to demonstrate the scalability of the approach.

During training, vehicle starting positions were randomized across parking spots in each trial to expose the model to a diverse set of initial conditions. In deployment, controllable vehicles begin at random locations with assigned destinations. Each vehicle is initialized with a velocity and orientation and navigates using AIF toward its parking goal while avoiding collisions.

Figure 3 shows representative trajectories of the AIF+PD planner. (a) illustrates basic two-vehicle control. (b) and (d) highlight both obstacle avoidance and coordinated multi-agent behavior. (b)–(d) further show complex maneuvers, such as three-point turns, indicating that AIF+PD learns sophisticated strategies from simple random samples and scales from basic to coordinated behaviors without handcrafted rewards.

Figure 4 illustrates the loss plot of the diffusion motion predictor throughout its training and validation phases. Initially, a precipitous decline in the training loss is observed, indicative of the model's rapid acclimatization to the structure within the training data. Concurrently, the validation loss mirrors the downward trend of the training loss, suggesting a consistent learning pattern that generalizes beyond the training set. As the epochs advance, both losses stabilize and

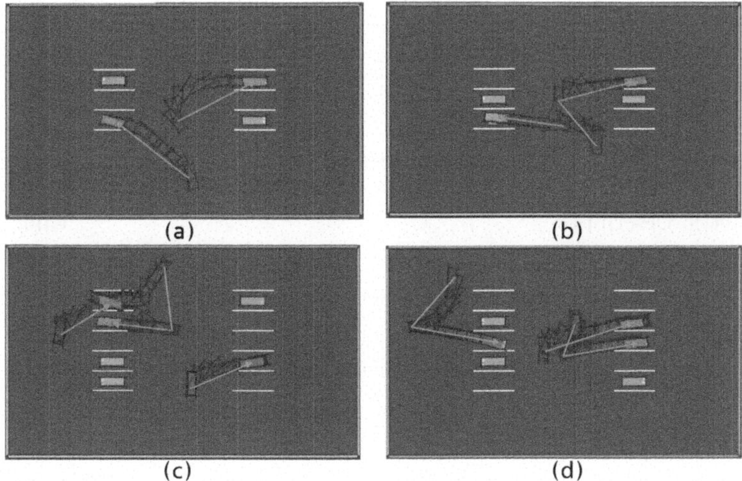

Fig. 3. Vehicle trajectories controlled by diffusion-based AIF

exhibit minimal variance, which implies that the model has potentially reached its learning capacity given the current architecture and dataset. The absence of a significant gap between the training and validation losses towards the end of the training suggests that the model is not overfitting and is well-calibrated to the complexity of the data it aims to model.

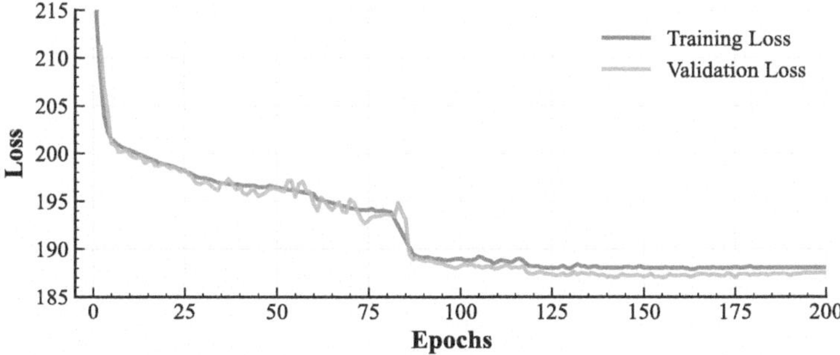

Fig. 4. Loss plot for training the reverse process.

Table 1 enumerates more key metrics used to evaluate the diffusion-based motion predictor. The MSE for the model's predictions, which evaluates the average squared difference across all elements in the vehicle state (x, y, v_x, v_y, h), is 0.2296. Furthermore, the model demonstrates a probabilistic confidence in its predictions, with the true next state falling within one standard deviation (δ) of the predicted distribution 43.05% of the time, within two standard deviations (2δ) 68.22% of the time, and within three standard deviations (3δ) 86.27% of the time. These statistics not only affirm the model's predictive strength but also suggest a well-calibrated understanding of the uncertainty inherent in vehicle movements.

Table 1. Performance metrics for the diffusion-based motion predictor

Metric Description	Value
MSE between reparameterized next state and true next state	0.2296
True next state within 1 δ of predicted next state distribution	43.05%
True next state within 2 δ of predicted next state distribution	68.22%
True next state within 3 δ of predicted next state distribution	86.27%

Next, we illustrate how our AIF+PD model behaves in a multi-agent parking task, using the `highway-env` framework in a setup similar to Fig. 3d. The scenario involves three controllable vehicles initialized at fixed positions, each assigned a random target parking spot, with three additional static vehicles placed randomly as obstacles. The objective is for each vehicle to reach its designated spot without collisions.

It is important to note that AIF+PD and RL approaches are fundamentally different in objective and formulation. AIF+PD relies on variational inference and structured priors to generate plausible goal-directed behaviors by minimizing expected free energy, while reinforcement learning maximizes cumulative reward and depends heavily on reward design and exploration strategies. We also do not compare with MPC since our bicycle model explicitly includes stochastic noise to mimic non-stationary processes, which breaks MPC's assumption of precise dynamics and makes direct comparison misleading. The goal of this benchmark is therefore not to claim superiority but to highlight behavioral differences in a shared environment.

We trained SAC, DDPG, PPO, DQN, and TD3 agents using Stable-Baselines3 [19] with default hyperparameters, all in the same multi-agent setup. Among them, only SAC and DDPG, when paired with Hindsight Experience Replay (HER), were able to learn effective parking policies. The others failed to converge due to the sparse reward landscape. Even among successful RL policies, performance declined as agent count increased, indicating difficulty in managing inter-agent interference.

The results summarized in Table 2 highlight complementary strengths. SAC+HER achieves the highest single-agent success rate and the lowest Euclidean tracking error, demonstrating strong individual control precision. However, its performance degrades when multiple agents are introduced, likely due to the increased complexity of coordination without explicit interaction modeling. In contrast, AIF+PD demonstrates improved performance in coordinated multi-agent scenarios, where cooperation and collision avoidance are essential. Additionally, when initialized from random positions not seen during training,

Table 2. Comparison of AIF+PD and RL in Multi-Agent Parking

Method	Success Rate ≥1 Car	≥2 Cars	3 Cars	Direction Diff (rad)	Euclidean Distance (m)
AIF+PD	92%	80%	58%	0.0847 ± 0.0190	0.4973 ± 0.0487
SAC + HER	97%	75%	56%	0.1393 ± 0.2723	0.2510 ± 0.3140
DDPG + HER	80%	50%	30%	2.0137 ± 0.8720	2.9911 ± 7.0734

Note: All models took approximately 10 hours to train using a 12900k CPU and 3090 GPU. AIF+PD received no reward signals at training time.

AIF+PD maintains stable performance, whereas RL methods frequently fail to complete the tasks. This robustness suggests that the structured inference mechanism in AIF+PD effectively facilitates emergent cooperative behavior without reliance on engineered rewards or specialized training protocols. A current limitation is that our approach assumes a reversible diffusion process, which simplifies training but constrains general applicability. Future work will test this assumption and benchmark the computational cost of inference against RL and MPC, focusing on scalability as agent count and trajectory horizon increase.

7 Conclusion and Future Work

This paper presented a novel framework for autonomous navigation by integrating AIF with a PD motion predictor. The proposed PD model efficiently generates feasible trajectories from randomized initial conditions, providing AIF with candidate paths toward successful parking outcomes. By leveraging Expected Free Energy to select optimal actions and dynamically adjusting predictions based on real-time observations, the AIF+PD approach robustly addresses uncertainties inherent in complex, multi-agent parking scenarios. Simulation results demonstrated that this method achieves comparable or superior cooperative performance and better generalization than traditional RL approaches, particularly under conditions of random initial positioning where RL often fails. Additionally, the framework uniquely eliminates reliance on explicit reward engineering and extensive replay buffers, simplifying the training process and reducing computational requirements.

For future research, we aim to scale the framework to more complex, real-world navigation scenarios. Specific directions include: (i) implementing a rolling-window planner to decompose long, unmarked routes into localized PD segments; (ii) integrating probabilistic likelihood models to handle sensor noise and partial observability; and (iii) extending PD to support conditional sampling for dynamic goal adaptation. To support reproducibility and enable further development, we will release all code and trained models as open-source.

References

1. Friston, K., Kiebel, S.: Predictive coding under the free-energy principle. Philos. Trans. R. Soc. Lond. B Biol. Sci. **364**(1521), 1211–1221 (2009)
2. Pezzulo, G., Rigoli, F., Friston, K.: Active Inference, homeostatic regulation and adaptive behavioural control. Prog. Neurobiol. **134**, 17–35 (2015)
3. Mohseni, F., Frisk, E., Nielsen, L.: Distributed Cooperative MPC for autonomous driving in different traffic scenarios. IEEE Trans. Intell. Vehicles **6**(2), 299–309 (2021)
4. Çatal, O., Verbelen, T., Van der Maele, T., Dhoedt, B., Safron, A.: Robot navigation as hierarchical active inference. Neural Netw. **142**, 192–204 (2021)
5. Ghugare, R., Bharadhwaj, H., Eysenbach, B., Levine, S., Salakhutdinov, R.: Simplifying Model-based RL: Learning Representations, Latent-space Models, and Policies with One Objective. In: ICLR, Kigali (2023)
6. Sajid, N., Ball, P.J., Parr, T., Friston, K.J.: Active inference: demystified and compared. The Wellcome Centre for Human Neuroimaging, UCLQueen Square Institute of Neurology, London (2020)
7. Pezzato, C., Corbato, C.H., Bonhof, S., Wisse, M.: Active Inference and Behavior Trees for Reactive Action Planning and Execution in Robotics. IEEE Trans. Rob. **39**(2), 1050–1069 (2023)
8. Pezzato, C., Ferrari, R., Corbato, H.C.: A novel adaptive controller for robot manipulators based on active inference. IEEE Robot. Automation Lett. **5**(2), 2973–2980 (2020)
9. Choi, W.Y., Lee, S.-H., Chung, C.C.: Horizonwise model-predictive control with application to autonomous driving vehicle. IEEE Trans. Industr. Inf. **18**(10), 6940–6949 (2021)
10. Muraleedharan, A., Okuda, H., Suzuki, T.: Real-time implementation of randomized model predictive control for autonomous driving. IEEE Trans. Intell. Vehicles **7**(1), 11–20 (2022)
11. Ajay, A., Du, Y., Gupta, A., Tenenbaum, J., Jaakkola, T., Agrawal, P.: Is Conditional Generative Modeling all you need for Decision-Making? In: 2023 International Conference on Learning Representations, Kigali (2023)
12. Janner, M., Du, Y., Tenenbaum, J.B., Levine, S.: Planning with diffusion for flexible behavior synthesis. In: Proceedings of the 39th International Conference on Machine Learning, Baltimore (2022)
13. Polack, P., Altché, F., d'Andréa-Novel, B., de La Fortelle, A.: The kinematic bicycle model: A consistent model for planning feasible trajectories for autonomous vehicles? In: IEEE Intelligent Vehicles Symposium (IV), Redondo Beach, California, US (2017)
14. Parr, T., Pezzulo, G., Friston, K.J.: Active Inference The Free Energy Principe in Mind, Brain, and Behavior. MIT Press (2022)
15. Rombach, R., Blattmann, A., Lorenz, D., Esser, P., Ommer, B.: High-Resolution Image Synthesis with Latent Diffusion Models. arXiv (2022)
16. Kullback, S., Leibler, R.A.: On Information and Sufficiency. Ann. Math. Stat. **22**(1), 79–86 (1951)
17. Delavari, E., Moore, J., Hong, J., Kwon, J.: Towards Human-Like Driving: Active Inference in Autonomous Vehicle Control. arXiv (2024)

18. Singh, H., Yoon, J.h., Fontana, P., Hanson, K., Sales, W., Kamat, V., Friston, K.: Network-based Active Inference for Adaptive and Cost-efficient Real-World Applications: A Benchmark Study of a Valve-turning Task Against Deep Reinforcement Learning. OpenReview (2025)
19. Raffin, A., Hill, A., Gleave, A., Kanervisto, A., Dormann, N.: Stable-Baselines3: Reliable Reinforcement Learning Implementations (2021). https://github.com/DLR-RM/stable-baselines3. Accessed 05 June 2025

Author Index

M. Albarracin et al. (Eds.): IWAI 2025, CCIS 2857, pp. 431–432, 2026.
https://doi.org/10.1007/978-3-032-16955-6

população idosa em detrimento de outras medidas que reforcem a qualidade de vida, facilitem o acesso a bens e serviços de saúde, de caráter mais preventivo e concorram para uma imagem positiva do/a idoso/a.

A imagem social presente nestes normativos distancia-se dos objetivos e das ações-chave defendidos em Malta, Setembro de 2013 com a publicação do "Plano Estratégico de Ação para o Envelhecimento na Europa (2012-2020)" pela Organização Mundial de Saúde:

> *Allowing more people to lead active and healthy lives in later age requires investing in a broad range of policies for healthy ageing, from prevention and control of non communicable diseases (NCDs) over the life-course to strengthening health systems, in order to increase older people's access to affordable, high-quality health and social services. (WHO, 2012, p.1).*

A pretensão é a de que os Estados europeus acatem programas e medidas políticas que favoreçam a saúde e a inclusão social da população idosa como ativos significativos para o desenvolvimento dos países. Estas medidas além de proporcionarem sociedades mais justas do ponto de vista da coesão, mitigam a visão tradicional de improdutividade destes cidadãos. A verdade é que nos encontramos numa cruzilhada de ação, onde as respostas sociais atuais não estão ajustadas aos cidadãos mais idosos ampliando-se a discriminação geracional devido à crença de que são estes os causadores da insustentabilidade da Segurança Social.

1.2. Impacto da identidade social atribuída à Velhice nas medidas de Proteção Social

É no ano de 1983, com o programa governamental, que se intensifica o papel interventor do Estado face aos idosos, constatando-se uma crescente "humanização" das políticas sociais da velhice através da criação de novos mecanismos valorizadores de uma relativa independência e autonomização do idoso. Este intuito concretiza-se através da pretensão de admitir como único limite ao desempenho de um papel ativo por parte dos idosos as exigências de qualificação e do mercado do trabalho, assim como através das propostas que visavam desenvolver um conjunto de medidas de sensibilização quer via ensino, quer mediante a comunicação social. (Cardoso et al., 2012, p.614).

Em 1986 com a integração de Portugal na União Europeia, e, até finais da década de 90, observa-se um reforço do investimento público em medidas reparadoras e o aumento de respostas sociais que tinham em especial atenção os idosos que se encontravam em situação de vulnerabilidade social. Os centros de dia, serviços de apoio domiciliário, unidades residenciais, acolhimento familiar crescem exponencialmente e colocam à sociedade portuguesa outros desafios no que respeita os direitos dos/as cidadãos/ãs mais idosos/as.

No aspeto pecuniário assiste-se a um reforço da pensão social e o aparecimento do Complemento solidário para Idosos. Respostas sociais de caráter compensatório e reparador que apesar de importantes em situações grave de pobreza e exclusão social não possuem um caráter estratégico, nem pressupõe um planeamento crítico de intervenção mais aprofundada nas estruturas sociais. No final do século XX, princípios do século XXI, os percursos pessoais assumem configurações específicas com a transformação dos modelos familiares, dos modelos económicos e a desconstrução do Estado de Bem Estar em Portugal possibilitando outras formas de proteção social, nas quais o Estado deixa de desempenhar o papel de protagonista para o transferir para outras entidades, nomeadamente, o setor privado, a sociedade civil organizada e a família. É defendida uma maior responsabilização dos titulares das medidas sociais e observa-se uma nova relação entre o "social" e o "económico".

A velhice é, nesta altura, percecionada como um problema social, onde a caridade substitui a solidariedade (ausência da defesa dos direitos sociais) e as políticas sociais sustentam-se na solidariedade mecânica onde a única forma de sustentabilidade da segurança social e consequentemente da proteção social se baseia em balancetes de "*deve e haver*" entre as despesas e receitas sociais.

A história recente de políticas de austeridade e de recessão económica têm vindo a demonstrar que as fórmulas e equações sociais em cima identificadas não são eficazes. Esta ineficácia traduz-se, a nosso ver, na sua inadequabilidade à realidade das novas dinâmicas sociais onde se inclui o envelhecimento.

As políticas de investimento social emergem nesta dinâmica, integrando uma mistura mais ou menos consistente entre conservadorismo e liberalismo. O conceito de investimento social implica a existência de um retorno social que se substancia em todas as ações, atividades, medidas e programas sociais capazes de emancipar os cidadãos nas dimensões de formação, desenvolvimento de competências e de inclusão social.

> *We define "social investments" as policies that primarily bring about more of, or an improvement in, skills formation, maintenance, or use. We define "dismantlement" as policy changes that cut social expenditures in the short to medium term, with no regard for these same three factors. Kvist (2013, p. 94-95).*

De acordo com o mesmo autor (2013) o Estado Social tradicional baseava o investimento social de acordo com o ciclo de vida do/a cidadão/ã. Neste pacto, existe um maior investimento social (e também de retorno social) em termos de medidas compensatórias de eventuais riscos nos primeiros anos de vida dos sujeitos. O investimento social vai reduzindo, bem como o retorno social e financeiro expetável, à medida que o sujeito avança no seu ciclo vital. Esta lógica claramente redistributiva baseia-se na lógica da responsabilidade do sujeito para a contribuição do sistema social através do fator trabalho.

A inexistência de uma estratégia concertada para uma população mais ativa, mais heterogénea e com outras necessidades, continua a persistir em prol da reprodução de um discurso de reparação, ao invés de um discurso de modernização, face às políticas e estratégias públicas a lidar com esta realidade demográfica.

Em síntese, neste início do século XXI, a abordagem que o Estado faz da velhice remete para a sua construção como problema social ou como uma fase na vida dos indivíduos marcada por necessidades materiais e sociais que se impõem como objeto de políticas sociais específicas. Essas políticas visam, sobretudo, a regulação do acesso à pensão de velhice, as alterações dos quantitativos das pensões e os apoios específicos em equipamentos e serviços.

1.3. As Agendas Nacionais e Europeias para o Envelhecimento

Baseamo-nos em três documentos para elaborar este subponto: (i) A Estratégia Nacional para o Envelhecimento Ativo em Saudável 2017-2025 – Proposta do Grupo de Trabalho Interministerial (Despacho n.º 12427/2016); (ii) Plano de Ação Internacional para o Envelhecimento (2002). Este Plano foi o resultado da II Assembleia Mundial do Envelhecimento realizada de 8 a 12 de abril de 2002, em Madrid, promovida pela ONU; (iii) Envelhecimento no Século XXI: Celebração e Desafio. Publicado pelo Fundo de População das Nações Unidas em 2012.

Neles procurámos reter além das orientações, linhas de ação e estratégias, também e essencialmente, os princípios e as narrativas relativas à imagem que queriam veicular do envelhecimento, bem como a definição dos principais desafios sentidos não só a nível internacional, mas como os mesmos se traduziam no documento português.

Cristina Barbosa realizou um estudo publicado em 2015 (Barbosa, 2015, p.152) com o objetivo geral de compreender os conceitos e visão sobre o envelhecimento presentes nas políticas locais e conhecer os modelos de governança local das políticas públicas no domínio do envelhecimento, de Portugal e da Suécia.) Utilizou como campo de análise Portugal e Suécia pela diferença de modelos de Estado de Bem-estar Social. O quadro síntese (quadro 1) realizado pela autora (2015, p.164) confronta a visão portuguesa com a visão sueca na dimensão de Visão Sobre o Envelhecimento e Envelhecimento/Politicas Publicas.

Quadro 1 – Quadro resumo da análise ao domínio "Conceito e visão sobre o envelhecimento" (cfr. Fonte)

Portugal/Lisboa **(Visão Sobre o Envelhecimento)**	**Suecia/Nacka** **(Visão Sobre o Envelhecimento)**
- Envelhecimento como processo e ativo; - Visão holística e heterogénea do envelhecimento; - Valorização do papel das pessoas idosas na sociedade; - Pessoas idosas como fonte de capital social e riqueza para o país; - Discordância com designações para "envelhecimento" e "políticas públicas no âmbito do envelhecimento"; - Importância das relações inter-geracionais, para equilíbrio social; - Importância da aprendizagem ao longo da vida;	- Adoção de leis que orientam as políticas: dignidade, empowerment e liberdade de escolha - Conceção de envelhecimento como um target competidor gerador de capital, bem-sucedido, com longevidade, saúde - Preocupação no desenvolvimento de respostas e cuidados - Envelhecimento populacional é sucesso civilizacional, mas com prós e contras - Possibilidade de agefobia; - Dificuldade em gerir a multiculturalidade e migrações;
(Políticas Públicas e Envelhecimento)	**(Políticas Públicas e Envelhecimento)**
- Predominância de políticas centrais assistencialistas e políticas locais de participação; - Nível local com maior ação e dinamismo no envelhecimento; - Importância de se avançar com uma política nacional e/ou *guidelines* nacionais; - Opção por políticas de envelhecimento ativo;	- Políticas que definem as atuações das entidades e serviços - Políticas com visão de ciclo de vida contínuo e não tanto ativo - Adoção de políticas (de envelhecimento) preventivas no contexto - Ação do local, que segue guidelines mas tem autonomia para decidir - Incredibilidade e afastamento às orientações da EU; Falta Formação especializadas no envelhecimento

Fonte: Barbosa (2015, p. 164) Quadro 3 – Quadro resumo da análise ao domínio "Conceito e visão sobre o envelhecimento".

De um modo geral e de acordo com a autora (idem, 2015, p.165):

a visão dos entrevistados face ao envelhecimento são no sentido de o entender como um processo integrante do ciclo de vida e de recusar concepções irrealistas e homogéneas, face a este target populacional. (...) No que se refere às políticas públicas, as intervenções vão no sentido de retirar a palavra "envelhecimento" e ainda de tornar estas políticas mais holísticas e compreensivas e não específicas.

É interessante esta dimensão da imagem social da velhice que se constrói numa e noutra realidade. De facto, apesar de não haver diferenças fragmentadoras, salientamos que em Portugal a idade continua a ser um "cluster" para a definição de políticas sociais e que as mesmas tendem a ter em conta as orientações europeias. A dimensão "atividade" é uma dimensão importante pois segue os modelos tipicamente do sul da Europa em que o factor trabalho é um eixo fundamental na definição do indivíduo, concorrendo positivamente para a sua integração social. Na Suécia, esta parece não ser a grande preocupação. O cidadão mais velho é definido holisticamente onde a manutenção de fatores contextuais que concorrem para a sua qualidade de vida é mais importante que a sua integração num modelo de produtividade.

O Plano de Ação Internacional para o Envelhecimento (2002), bem como as recomendações presentes no "Envelhecimento no Século XXI" do Fundo de População das Nações Unidas em 2012 definem o envelhecimento demográfico como uma das maiores conquistas do mundo civilizacional. Alertam, igualmente, para a necessidade deste processo ser um processo heterogéneo e sublinham os desafios que a alteração demográfica colocam à nova ordem social e económica. Os fatores de sustentabilidade de serviços e políticas que consigam promover qualidade e dignidade é uma preocupação latente em ambos os documentos, com maior relevo para o primeiro.

Na página 14 deste documento pode ler-se:

Devemos reconhecer que, sendo maior o número de pessoas que recebem melhor educação e desfrutam de longevidade e boa saúde, os idosos podem contribuir mais do que nunca para a sociedade e, de fato, assim o fazem. Se incentivarmos sua participação ativa na sociedade e no desenvolvimento, podemos estar certos que seu talento e experiência inestimáveis. Os idosos que podem e querem trabalhar devem ter a oportunidade de assim o fazer, e todas as pessoas devem ter a oportunidade de continuar aprendendo ao longo da vida. (Plano de Ação Internacional para o Envelhecimento (2002).

A preocupação com a sustentabilidade de um sistema previdencial reflete-se na necessidade de aumentar os anos de contribuição ativo dos sujeitos para o mesmo. Ora, esta possibilidade não é errada em si, desde que não seja uma imposição política. Ao contrário dos países como a Suécia onde um dos riscos social, parece associado ao "idadismo", nos países do sul da Europa, o risco encontra-se associado ao "garante" da sua sobrevivência financeira. Destaco a este nível o seguinte texto retirado do documento: Envelhecimento no Século XXI p. 5.

Não há prova concreta de que o envelhecimento da população, em si, tenha minado o desenvolvimento econômico ou de que países não disponham de recursos suficientes para garantir pensões e atendimento à saúde da população idosa. Entretanto, em termos globais, apenas 1/3 dos países contam com planos de previdência social abrangentes, a maioria dos quais cobrem somente aqueles que se encontram em empregos formais, ou seja, menos da metade da população economicamente ativa mundial.

A preocupação em viabilizar economicamente o envelhecimento demográfico e em mudar o paradigma do cidadão passivo mais envelhecido levam a um maior envolvimento dos sistemas nacionais e europeus de saúde numa ótica de prevenção da dependência e de promoção da proatividade. De facto, Estratégia Nacional para o Envelhecimento Ativo e Saudável (2017, p.12) realça esta necessidade:

> *A carga da doença e a redução do bem-estar afetam a pessoa idosa e também as suas famílias, os sistemas de saúde, social e a economia (World Health Organization, 2014). As pessoas idosas com problemas de saúde ou de dependência necessitam de mais cuidados de saúde e de apoio social, por parte da família e das instituições da economia social e dos serviços de saúde.*

Este documento aponta ainda para a necessidade de maiores estudos sobre o processo de envelhecimento, formas de minorar os seus aspetos colaterais e um investimento claro no envelhecimento ativo que deve (de acordo com a Estratégia, 2017) estar presente em todas as políticas sociais setoriais.

Considerações finais

Vivemos um novo paradigma, uma nova ordem social, uma nova ordem política. São os ventos da mudança. Ventos que sopram em direções, por vezes, antagónicas e que de certeza nos irão colocar novos desafios a curto e médio prazo.

Apesar da incerteza, assistimos às mudanças com reflexividade e num esforço de concretizar novas respostas. O envelhecimento está a mudar, bem como a forma como os cidadãos mais velhos equacionam o seu papel nas dinâmicas sociais. Uma dessas mudanças concretiza-se na alteração da imagem do idoso no funcionamento familiar, social e económico. De facto, a importância que o cidadão maior possui no suporte familiar, no suporte financeiro e social das populações mais jovens é visível. A sua função reconhece-se e é legitimada socialmente pelo espaço que ocupa na organização da sociedade. Ser idoso já não possui uma única definição pois a tirania da idade já não é real. O envelhecimento já não é considerado um risco social, mas sim, uma fase do ciclo vital do sujeito e neste sentido as políticas sociais não estão adaptadas ou tendem a tornar-se injustas.

O discurso da proatividade e da necessidade de manutenção do cidadão em atividade laboral não pode ser vista de forma tão redutora. A continuidade ou não de exercer a sua atividade laboral devia ser uma escolha do cidadão. As políticas redistributivas não se encontram atualizadas e o modelo tradicional do Estado, baseado numa lógica de investimento e retorno socioeconómico face às fases do ciclo vital do cidadão, está desajustado. Medidas mais flexíveis como o trabalho parcial, a rotatividade laboral, a formação ao longo da vida, maior adequação trabalho-família devem fazer parte da agenda política dos países do sul da europa. São medidas de promoção de bem-estar e preventivas de malefícios vários, nomeadamente, no que se refere à saúde mental do cidadão.

Portugal não está em desacordo com as orientações europeias. As guidelines são adaptadas, mas, face à recente recessão económica e ao rescaldo da mesma, a consolidação de políticas redistributivas continua a ser um fator de maior preocupação. Mesmo quando os atuais estudos apontem para a sua inoperacionalidade face à mudança social necessária.

Observamos com expetativa que os diferentes atores sociais parecem tentar coligar-se para a implementação de estratégias nacionais concertadas. A cooperação é sempre o melhor caminho, proporcionando olhares múltiplos sobre a mesma realidade e possibilitando, por conseguinte, a sua compreensão multidimensional.

Bibliografia

Barbosa, C (2015) Políticas públicas locais para o envelhecimento: o caso de Portugal e da Suécia In EXEDRA, dez. 2015: *Número temático: Contributos para uma abordagem multidisciplinar do Envelhecimento. Teoria, Investigação e Prática.* 152-175 Coimbra: ESEC. Disponível em http://www.exedrajournal.com/wp-content/uploads/2016/02/pdf-final-v2--publicar.pdf acessado em 29 de agosto de 2018.

Capucha, Luis (2014) Envelhecimento e Políticas Sociais em Tempos de Crise *in* Sociologia, Problemas e Práticas n.º 74, 2014, 113-131.

Cardoso, S; Santos, M.H; Baptista, M.I & Clemente, S. (2012) Estado e políticas sociais sobre a velhice em portugal (1990-2008) Análise Social, 204, xlvii (3.º), 2012, 606-630

Carvalho, M.I & Almeida, M.J (2014) Contributo para o desenvolvimento de um modelo de proteção social na velhice em Portugal. Disponível em http://www.app.com.pt/wp-content/uploads/2014/07/Artigo_Contributo-para-o-desenvolvimento-de-um-modelo-de-prote%C3%A7%C3%A3o-social-na-velhice-em-Portugal_M%C2%AAIC-e-MJA.pdf acessado em 8 de julho de 2018.

Costa, Firmino (2012) Desigualdades Sociais Contemporâneas. Lisboa: Editora Mundos Sociais.

Deleuze Against Control: Fictioning to Myth-Science Simon O'Sullivan Goldsmiths, sagepub.co.uk. DOI:10.1177/0263276416645154 tcs.sagepub.com

Domingues, Joelza (2017) Esquerda e Direita: rótulos mais simbólicos que políticos in Joelza Ester Domingues www.ensinarhistoriajoelza.com.br/esquerda-e-direita-rotulos/ acessado a 30 de Junho de 2018.

Envelhecimento no Século XXI: Celebração e Desafio (2012) HelpAge International. Publicado pelo Fundo de População das Nações Unidas em 2012. (UNFPA), Nova York e pela HelpAge International, Londres.

Estratégia Nacional para o Envelhecimento Ativo em Saudável 2017-2025 (2017) Proposta do

Grupo de Trabalho Interministerial (Despacho n.°12427/2016) disponível em https://www.sns.gov.pt/wp-content/uploads/2017/07/ENEAS.pdf acedido a 21 de Out de 2018.

Kvist (2013, p.95) "A life-course perspective on social investment policies and their returns"

Plano de ação internacional sobre o envelhecimento (2002). Organização das Nações Unidas. Brasília: Secretaria Especial dos Direitos Humanos. Disponível em http://www.observatorionacionaldoidoso.fiocruz.br/biblioteca/_manual/5.pdf acedido a 21 de Outubro de 2018.

Santos, C (2016) Social Policy: From the Death of Welfare State to the State "Nameless". An historic overview of Social Policies in South and Eastern Countries in Social Theory, Empirics, Policy and Practice Journal, 2016 (13), Vilnius: 54-68. ISBN 978-92-4-151350-0

Silva, Filipe (2013). O Futuro do Estado Social. Lisboa: Fundação Francisco Manuel dos Santos

Therbon, Goran (org.) (2006). Inequalities of the World. New Theoretical Frameworks, Multiple Empirical Approaches. Londres: Verso, 332p.

WHO (2015) Relatório Mundial sobre Envelhecimento e Saúde. Geneva: World Health Organization; 2015.

WHO (2017) Global strategy and action plan on ageing and health. Geneva: World Health Organization; 2017. ISBN 978-92-4-151350-0

CAPÍTULO II

INTERVIR POSITIVAMENTE NO PROCESSO DE ENVELHECIMENTO: MINDFULNESS, QUALIDADE DE VIDA E BEM-ESTAR

Albertina L. Oliveira

Universidade de Coimbra, Faculdade de Psicologia e Ciências da Educação

Centro de Estudos Interdisciplinares da Universidade de Coimbra (CEIS20)

ORCID: http://orcid.org/0000-0001-9521-528X

Resumo: *O presente capítulo, ancorado em numerosas contribuições teóricas e empíricas da investigação científica interdisciplinar, convoca um prisma de análise positivo do processo de envelhecimento e da velhice e acentua o enorme potencial humano e dos contextos sociais para fomentar a qualidade de vida de todas as pessoas à medida que envelhecem. Par o efeito, consideramos, em primeiro lugar, a emergente abordagem científica das práticas meditativas, sob a designação de mindfulness, o significado do conceito e o seu contributo para a promoção do bem-estar e vitalidade das pessoas, bem como o seu papel no retardamento do processo de envelhecimento. Em segundo lugar, debruçamo-nos sobre o construto de qualidade de*

DOI | https://doi.org/10.14195/978-989-26-1788-6_2

vida e a sua relação com o envelhecimento para finalizarmos com a sistematização dos principais fatores que emergem da literatura como estando fortemente associados ao envelhecimento com bem-estar.

Palavras-chave: *Envelhecimento; Mindfulness; Qualidade de vida; Bem-estar*

Abstract: *This chapter is anchored in numerous theoretical and empirical contributions of interdisciplinary scientific research. It stresses a positive view of the aging process and old age and emphasizes the enormous human potential as well as the potential of the social contexts in order to promote the quality of life of all human beings as they age. Firstly, it considers the emerging scientific approach to meditative practices, under the designation of mindfulness, the meaning of the concept and its contribution to the promotion of well-being and vitality, as well as its positive effect in the delaying of the aging process. Secondly, it reflects on the construct of quality of life and its relationship with aging and the chapter finished with a systematization of the major factors emerging from the literature as strongly associated with aging and well-being.*

Key-words: *Aging; Mindfulness; Quality of life; Well-being*

Refletir sobre o processo de envelhecimento é acima de tudo uma chamada a desenvolver uma perspetiva ampla na apreciação da vida, estendendo o horizonte, de forma intencional, à etapa da velhice. Em que condições de saúde, de vitalidade, é possível e desejamos estar, enquanto pessoas mais velhas? Que contributo queremos dar para a sociedade em que vivemos? Certamente que estamos de acordo que é desejável sermos capazes, individual e coletivamente, de manter o maior número de anos possível a autonomia funcional, um bom estado de saúde e bem-estar. Viver a velhice com o sentimento de realização pessoal, com vitalidade, interesse em continuar a aprender, contribuindo para construir um mundo mais pacífico e solidário, é um desígnio que podemos adotar como comum. Sustentar esta visão e mobilizar o que estiver ao nosso alcance para a atingir, requer que nos desloquemos da tradicional, mas limitada, perspetiva de reduzir o mais possível as doenças crónicas e os seus impactos negativos, para outra que diversas disciplinas têm vindo a acentuar, comprometida com a promoção da saúde na sua aceção global. Tal perspetiva remete-nos para a consideração da vida como contínuo processo de desenvolvimento e centra-se na promoção do vasto potencial humano bem como na mobilização de contextos sociais e/ou comunitários apropriados para o coadjuvarem.

O envelhecimento é, segundo a OMS (cit. por República Portuguesa, 2017, p. 8), "*um processo progressivo de mudança biopsicossocial da pessoa durante todo o ciclo de vida*", sendo por isso um fenómeno complexo resultante da interação de múltiplos fatores, pessoais e contextuais, que vão acentuando as diferenças individuais e tornam o grupo humano das pessoas mais velhas altamente heterogéneo. É, assim, possível encontrar pessoas de idade avançada (com 70, 80 ou mais anos) a manifestarem as suas prestações mais significativas na vida bem como depararmo-nos com outras que, com 50 ou 60 anos, já apresentam níveis de incapacidade notórios. Como refere Simões (2006), "*em nenhuma outra idade tais diferenças assumem expressão tão elevada como na última fase da existência*" (p. 25).

Sob a ótica mais tradicional, a investigação diz-nos que à medida que aumenta o número de anos aumenta também o número de doenças, muitas das quais dependem do estilo de vida mantido e/ou reforçado em períodos mais precoces do desenvolvimento (DGS, 2015). Mas, contribuir para um bom processo de envelhecimento implica deslocarmo-nos do foco na prevenção das doenças (por mais que ele seja valioso e importante) para essa outra perspetiva, profundamente positiva, que é a de fomentar desde idades precoces, individual e coletivamente, o bem-estar físico, mental e social, ou seja, envolvermo-nos em processos de ativação da vitalidade, harmonia e saúde na sua aceção mais completa.

Estas perspetivas não são uma novidade do nosso tempo. Pelo contrário, remontam a idades bem recuadas da história (Depp, Vahia & Jeste, 2012), encontrando-se na Grécia antiga defensores do envelhecimento segundo uma conceção negativa (que acentua a doença e a fragilidade), sendo o caso de Aristóteles e Galeno, mas também se encontra quem tenha preconizado uma leitura positiva, como, por exemplo, o Romano Cícero que, na sua *Cato Maior De Senectute*, defendeu que a maturidade emocional se fortalece com o avançar da idade, como aliás a investigação atual comprova (e.g., Labouvie-Vief, 2005), e que aumenta a capacidade para uma vida virtuosa. O mesmo autor chama também a atenção para a influência do comportamento prévio em idades mais avançadas, o que nos remete para a ação do sujeito (e dos fatores que lhe estão associados) na determinação dos períodos ulteriores de vida, ganhando aqui destaque o estilo de vida e o mindfulness.

Embora o envelhecimento dependa da confluência de numerosos fatores, muitos dos quais estão fora do nosso controlo, é esta capacidade de influência que todos temos que dá corpo ao título deste trabalho – intervir positivamente no processo de envelhecimento: mindfulness, qualidade de vida e bem-estar.

Consideremos, em primeiro lugar, a temática do mindfulness para depois nos centrarmos, mais especificamente, na qualidade de vida e bem-estar.

1. O mindfulness

O *mindfulness* apresenta uma íntima relação com o bem-estar, podendo ser visto como um poderoso promotor da qualidade de vida das pessoas. Segundo Simões et al. (2009), o *Ontario Project on Successful Aging* aponta de forma inequívoca que "o envelhecimento bem-sucedido é 80% atitude e 20% qualquer outra coisa e está ao alcance de toda a gente" (p.119). Esta constatação bem como a surpreendente conclusão de numerosas investigações internacionais de que a satisfação com a vida "não é tanto uma questão de se obter o que se pretende, mas de se desejar o que se tem" (Myers & Diener, 1995, p. 13) remetem-nos para a importância do campo atitudinal, em termos de compreensão do bem-estar das pessoas, o que abre inteiramente as portas e constitui o principal enlace com o recente estudo científico do mindfulness. Consideremos, então, e em primeiro lugar, o seu aparecimento, a sua importância, as suas principais características, o lugar que ocupa na compreensão do processo de envelhecimento e da qualidade de vida das pessoas e populações.

O contexto da abordagem científica ao mindfulness

Longamente implantado nas tradições milenares yoguis e budistas (mas também em tradições contemplativas de outras culturas, como por exemplo as ordens contemplativas cristãs), o mindfulness tem vindo a revolucionar as intervenções terapêuticas, a educação, a intervenção social e, também, a compreensão do processo de envelhecimento. Os bons resultados que têm vindo a ser encontrados surgiram, segundo Kabat-Zinn (2011), da confluência histórica de duas epistemologias bem distintas que, ao encontrarem-se, começaram a originar a chamada viragem contemplativa ou, como lhe chama Wallace (2007), a ciência contemplativa. Efetivamente, no âmbito dos encontros *Mind and Life*, os diálogos encetados entre duas figuras de proa, o biólogo e neurocientista Francisco Varela e o Dalai Lama, em muito contribuíram para abrir caminho[1] a um grande movimento de discussão e investigação entre várias áreas disciplinares, que podemos apelidar de revolucionário. Este movimento, ao recorrer a diversas abordagens que se pautam pelo maior rigor da investigação científica, tem vindo a edificar um terreno fértil para a mudança de paradigma em torno do entendimento do ser humano e da sua relação com os diversos sistemas com os quais é interdependente (sociais, naturais, planetários, cósmicos).

Essas duas epistemologias, a ciência e as tradições contemplativas, que durante séculos se encontraram de costas voltadas uma para a outra[2], como se de mundos ou vias de conhecimento independentes de tratasse, no final do século XX abriram-se ao diálogo e colaboração, gerando novos caminhos, considerados altamente promissores.

O primeiro programa inteiramente baseado em práticas meditativas que veio a revolucionar as intervenções no domínio da

[1] Na verdade, foi a partir da década de 70 do século passado, que o "diálogo entre poetas (e.g., Octavio Paz), cientistas (e.g., Francisco Varela, Jon Kabat-Zinn), budistas (e.g., Dalai Lama, Chogyam Trungpa), professores de meditação (e.g., Joseph Goldstein) [se tornou] o fermento que desenvolveu um interesse alargado pelas práticas contemplativas e respetivos efeitos a nível do conhecimento, da aprendizagem, das relações sociais e da motivação" (Oliveira & Antunes, 2014, p. 55).

[2] Desde o século XVI com a emergência crescente do movimento científico, que Michel Foucault designou por Momento Cartesiano (Antunes, 2014).

saúde mental foi sistematizado por Kabat-Zinn, em 1979. Nomeado inicialmente por *Stress Reduction & Relaxation Program*, veio a tornar-se internacionalmente conhecido sob a designação de *Mindfulness Based Stress Reduction* (MBSR). Trata-se de um programa "que ensina as pessoas a cuidarem melhor de si próprias" (Kabat-Zinn, 2011, p. 293) e que se baseia no treino intensivo em meditação e yoga. Muito embora tenhamos que ressalvar que o yoga, entendido na sua aceção completa, integra a meditação bem como outras componentes de treino e desenvolvimento humano, é importante destacar que para Kabat-Zinn o objetivo da introdução deste tipo de práticas no contexto da academia nunca foi o de as circunscrever a uma abordagem clínica ou terapêutica, bem pelo contrário. Tal como afirma no seu artigo em que narra a história deste encontro extraordinário entre práticas de tradições milenares e a ciência contemporânea, o propósito da introdução das abordagens baseadas no mindfulness foi o de contribuir para "mover a curva normal da nossa sociedade em direção a maior sanidade e bem-estar [...] o MBSR foi concebido e funciona como uma intervenção de saúde pública, um veículo, tanto para a transformação individual como societal" (p. 282). E, no mesmo artigo, reforçando esta última ideia, acrescenta que o mindfulness "não é mais uma técnica cognitivo-comportamental para ser desenvolvida num paradigma de mudança do comportamento, mas um modo de ser e um modo de ver com profundas implicações para a compreensão da natureza das nossas próprias mentes e corpos" (p. 284).

O significado de mindfulness

Mindfulness é a tradução inglesa do termo *sati* proveniente da língua Pāli (a que foi usada por Buda) e significa, numa primeira aproximação, o ato de prestar atenção ao momento presente.

Porém, como refere Amaro (2015), aquela simples definição capta apenas a função mecanicista de *sati*. Semanticamente, o termo possui um espetro mais completo, correspondendo também à compreensão clara ou à total consciência (*sati-sampajañña*), devendo o objeto alvo de atenção, nesta última aceção, ser considerado no seu contexto (de tempo, lugar e situação) e incorporar preocupações éticas, no sentido de incluir "uma apreciação das atitudes dos praticantes e do impacto que as suas ações têm em si próprios e nos outros" (Amaro, 2015, p. 65). Um terceiro significado do mindfulness (*sati-paññā*) prende-se com "a qualidade que conduz ao total florescimento do bem-estar do ser humano" (p. 65), não sendo por isso indissociável da sabedoria e da necessidade de examinar toda a experiência no seu fluxo permanente.

Nas palavras de Kabat-Zinn (2015), e de forma mais concisa, o mindfulness é "a consciência momento a momento, não ajuizadora, cultivada para prestar atenção de um modo específico, o mais possível de forma não reativa, não ajuizadora e de coração aberto" (p. 1481). Visa assim desenvolver a qualidade da mente que funciona como que um espelho que reflete com clareza o que nele se apresenta. O mindfulness é, portanto, uma qualidade intrínseca da mente que pode ser desenvolvida e refinada com treino sistemático, tratando-se de uma qualidade humana de potencial universal. Como esclarece Kabat-Zinn (2015), "não há nada de particularmente budista em prestar atenção ou a respeito de consciência, nem particularmente oriental ou ocidental [...] a essência do mindfulness é verdadeiramente universal. Tem mais a ver com a natureza da mente humana do que com ideologias, crenças ou com a cultura" (p. 1482).

O mindfulness é sempre do domínio experiencial, não reside no plano concetual, pelo que consiste numa abordagem realista, pragmática, em vez de idealista, relativamente ao desenvolvimento do bem-estar ou à libertação do sofrimento. Consequentemente, como salienta Amaro (2015), ao considerarmos a nossa ação quotidiana, a pergunta *a priori* que deveremos colocar é: *A minha atitude, a minha ação no dia a dia contribuem para construir um mundo melhor para todos, contribuem para aliviar o sofrimento?*, em vez daquela em que tantas vezes nos envolvemos: *A vida e as coisas são como eu acredito que deveriam ser?*

Ao longo dos séculos, esta capacidade humana inata foi investigada, explorada, mapeada, desenvolvida e refinada pelas tradições contemplativas. Segundo Wallace (2007), o termo contemplativo vem do latim *contemplatio*, correspondente à palavra grega *theoria*, referindo-se ambas à busca de "revelar, clarificar e tornar manifesta a natureza da realidade" (p. 1). Encontrando pontes entre a ciência e o conhecimento contemplativo, refere Wallace que, "tal como os cientistas levam a cabo observações e conduzem experimentação com a ajuda de tecnologia, os contemplativos desde há muito fazem as suas próprias observações e experimentações com a ajuda de capacidades atencionais aumentadas e com o envolvimento da imaginação" (p. 2).

Desde o designado Momento Cartesiano, a ciência e as tradições contemplativas divergiram inteiramente no seu caminho como se só o conhecimento erigido sobre a dualidade sujeito-objeto fosse válido – como se só o conhecimento científico de inspiração positivista merecesse o estatuto de credível (Antunes, 2014). Sendo inquestionáveis os avanços que o conhecimento de base científica e tecnológica trouxeram para a melhoria da qualidade de vida das pessoas nas sociedades contemporâneas, nomeadamente os grandes progressos da medicina a nível da melhoria do bem-estar físico, a verdade é que a enorme expansão do conhecimento científico não nos trouxe um aumento correlativo da ética e da virtude; não podemos dizer que a sociedade moderna tenha ficado mais sábia e compassiva. Infelizmente, todos constatamos, como refere Wallace (2007), que "as nossas sociedades se afastaram, de há muito, de prosseguir a felicidade genuína, a verdade e a virtude" (p. 2), tal como era, por exemplo, o desígnio dos estoicos na Grécia Antiga (Reis & Oliveira, 2016).

Porque é que treinarmo-nos em prestar atenção de uma forma muito específica (tal como acima explicitado) e em alcançar uma visão clara, é tão importante para o nosso bem-estar? Como é que o mindfulness origina a transformação da pessoa e do mundo?

Na verdade, treinarmo-nos em 'voltar a casa', i.e., ao momento presente, quebrando padrões de reatividade e automaticidade longamente estabelecidos na pessoa, pode ser uma conquista completamente extraordinária. Trazer de volta a atenção, voluntariamente, uma e outra vez, havia já sido reconhecido por Williams James, na sua obra *Principles of Psychology*, como a educação por excelência: "*the faculty of voluntarily bringing back a wandering attention, over and over again, is the very root of judgment, character, and will. No one is compos sui if he have it not. An education which should improve this faculty would be the education par excellence. But it is easier to define this ideal than to give practical instructions to bringing it about*" (cit. por Kabat-Zinn, 2015, p. 1484).

Intencionalmente trazer de volta a atenção para o momento presente, uma e outra vez, é um tipo de autoeducação que requer compromisso, disciplina e prática. Frequentemente estamos desligados de nós próprios, fora de contacto com a nossa realidade, apresente-se ela sob a forma de sensações, perceções, emoções, pensamentos, imagens, impulsos, ou o que quer que surja no campo atencional. Ou seja, é muito comum estarmos num estado de desatenção, ou como é comumente designado, estado de piloto automático. Sucede, como explica o autor supracitado, que os estados de desatenção levam à desconexão que, por sua vez, conduz à desregulação (afastamento do natural equilíbrio homeostático), a qual resulta na desordem dos tecidos celulares, que, por sua vez, afeta os diversos órgãos e sistemas. Rana (2015), no mesmo sentido, salienta como é tão importante observar atentamente a experiência sem interferir, ou seja, o envolvimento no processo de ganhar distância (recuar um 'passo atrás' ou 'subir acima') e de adoção da atitude fenomenológica para se alcançar mais objetividade sobre as próprias experiências internas e se promover a autorregulação. Por conseguinte, através deste processo, as pessoas tornam-se menos controladas por pensamentos, emoções, sensações, menos fusionadas com os conteúdos da consciência e com os padrões de reatividade habituais.

O mindfulness como via ativadora dos processos de integração

As práticas de mindfulness, pelas características que as qualificam, têm o efeito de contrariar a fragmentação e ativar ou restaurar os processos integrativos, quer em termos físicos quer psicológicos quer ainda sociais. A atenção restabelece e reforça a conexão, levando a maior regulação que, por sua vez, promove um estado de ordem dinâmica, o qual corresponde à assinatura do bem-estar e da saúde (Kabat-Zinn, 2015). Segundo Siegel (2007), quando focamos a atenção nas nossas intenções estamos a estimular as regiões intermédias pré-frontais do córtex cerebral e a favorecer os processos de integração. Esta ativação pré-frontal origina a extensão de fibras axonais que ligam em conjunto várias regiões: o córtex, a zona límbica, o tronco cerebral, o próprio corpo e até o mundo social dos outros cérebros. Do ponto de vista neurofisiológico, "o aumento anatómico das fibras pré-frontais significa que funcionalmente estamos a promover a integração neural" (p. 291).

Assim, o bem-estar físico e mental é criado por um processo de integração que interliga componentes diversas de um sistema num todo funcional. Essas diversas partes integradas formam o designado fluxo FACES: Flexível, Adaptativo, Coerente, Energizado e Estável (Siegel, 2007, p. 288). É o fluxo da experiência que acaba por refletir a qualidade de coerência: conectado, aberto, harmonioso, implicado, recetivo, emergente, noético, compassivo e empático. Para que estas qualidades ocorram, a orientação particular da atenção para a experiência, momento a momento, deve ser caracterizada pela curiosidade, abertura e aceitação – qualidades que definem a visão da mente saudável.

O treino em mindfulness implica, deste modo, que não prendamos, nem rejeitemos a experiência, que cultivemos, antes, a atitude de a acolher, de a observar com curiosidade e interesse, de a aceitar e, neste sentido, pode dizer-se que cultivamos a arte do encontro, da relação. As histórias de encantar do legado de todas as culturas, segundo Kabat-Zin (2003), "lembram-nos que vale a pena procurar o altar aonde as partes fragmentadas e isoladas do nosso ser se possam encontrar e casar, trazendo novos níveis de harmonia e compreensão das nossas vidas, até ao ponto onde possamos realmente viver felizes para sempre, e que significa o intemporal aqui e agora" (p. 88). Observamos os palcos internos e externos através da "perceção silenciosa da realidade" (Wallace, 2007, p. 1).

Mindfulness e envelhecimento

Individual e coletivamente todos desejamos estar envolvidos, o mais possível, em trajetórias de envelhecimento ativo, saudável, com qualidade de vida e feliz. Como favorecer um bom envelhecimento, uma maior satisfação com a vida, a manutenção ou aquisição de níveis ótimos de funcionamento e de qualidade de vida? Como obter mais bem-estar, equilíbrio, saúde, mais sentido na vida? Há múltiplas vias e fatores que a literatura tem vindo a apontar nesse sentido, como mais à frente exploraremos neste capítulo, sendo o treino em mindfulness e compaixão um elemento fundamental a ter presente.

Já se sabia que as práticas de mindfulness, em geral, aumentam a satisfação com a vida, melhoram a capacidade de lidar com a doença, promovem a saúde física e emocional, associam-se a níveis elevados de equanimidade e contentamento, melhoram a concentração e a autoconfiança, bem como promovem a coesão da identidade (e.g., Davidson et al., 2003; Goldin & Gross, 2010), mas o projeto *Resource*, levado a efeito entre 2012-2016 pelo *Max Planck Institute for Human Cognitive and Brain Sciences*, e dirigido por Tania Singer, veio clarificar mais esses efeitos positivos. Sendo considerado um projeto de grande envergadura para testar os efeitos das práticas de mindfulness/meditação e de compaixão a nível da saúde mental e das habilidades sociais da população adulta, os seus principais resultados permitiram constatar melhorias significativas no bem-estar subjetivo, nas funções cerebrais, na saúde (física e emocional) e no comportamento. Após 9 meses de treino foi possível verificar que as diferentes técnicas de meditação[3] têm efeitos diferenciados, mas

[3] O projeto envolveu 229 adultos de meia idade (com média de 41 anos). As práticas estavam organizadas em 3 módulos: Presença (incluindo meditação na respiração e *body scan*), afeto (meditação do amor bondoso) e perspetiva (observação dos pensamentos). Houve 3 grupos, considerados controlos ativos, submetidos a treino de 13 semanas, por componente e por ordem diferente.

também que originam ganhos comuns, nomeadamente, todas contribuírem para aumentar a afetividade positiva, a energia, o foco da atenção no momento presente e a diminuição de distrações (Kok & Singer, 2017).

A regulação emocional é um dos efeitos consistentemente visível com as práticas de mindfulness. A principal conclusão de um estudo recente com homens chineses entre 40 e 70 anos, avaliados em diversas variáveis neurofisiológicas, foi a de que o treino em mindfulness permite operar em fases muito iniciais do processamento afetivo, sendo que cultivar a disposição mindfulness ajuda a reduzir a reatividade habitual da pessoa aos estímulos, sobretudo aos que se associam a eventos considerados negativos (Ho et al., 2015). Também se tem constatado num aumento do comportamento prossocial – evidência que foi reforçada pelos resultados do projeto Resource (Kok & Singer, 2017). As atitudes mais positivas para com os outros explicam-se por se apurarem mudanças no cérebro relacionadas com as áreas habitualmente ativadas quando nos sentimos conectados aos outros (Siegel, 2007).

Precisamente relacionado com o efeito do mindfulness no bem-estar físico e emocional, uma revisão recente que incidiu sobre 15 estudos (Geiger et al., 2016) concluiu que os resultados são consistentes em termos de conduzirem a melhorias no bem-estar emocional (ansiedade, depressão, stress e aceitação da dor). A nível da saúde física, os efeitos revelaram-se inconsistentes (e.g., medidas de inflamação, IL-6, proteína C reativa, respostas imunológicas etc.).

Na esfera cognitiva, a revisão de Gard et al. (2014), que incluiu 12 estudos, apurou melhorias substanciais na atenção (sustentada, menor dispersão) e nas funções executivas.

Os efeitos do treino em mindfulness têm vindo igualmente a ser relacionados com o processo de envelhecimento. Normalmente, à medida que envelhecemos ocorre a diminuição da massa cortical através da perda de neurónios, mas é possível desacelerar o processo de envelhecimento, revelando o treino em mindfulness, neste âmbito, efeitos positivos. A investigação pioneira de Lazar et al. (2005) evidenciou que os meditadores de longa data, comparativamente a grupos de controlo, apresentam um aumento da densidade sinática nos lobos da ínsula e pré-frontal. Os estudos subsequentes dos neurocientistas apontaram no mesmo sentido, ficando bem estabelecido que as práticas repetidas de mindfulness transformam estruturalmente o cérebro em determinadas regiões e promovem a neuroplasticidade (Höltzer et al., 2010, 2011). A investigação de Höltzer et al. (2011), por exemplo, baseada num programa com a duração de 8 semanas (em que os participantes relataram praticar em média 27 minutos por dia), revelou que na segunda semana de treino as imagens de ressonância magnética já evidenciavam um aumento de densidade da matéria cinzenta no hipocampo (estrutura envolvida na aprendizagem e memória) e nas estruturas relacionadas com a autoconsciência, a compaixão e a introspeção. Em contrapartida, verificou-se a diminuição da densidade da amígdala (envolvida nas respostas emocionais como o stress e a ansiedade)[4]. Nenhum dos participantes do grupo de controlo evidenciou essas mudanças.

Na continuidade desta linha de investigação, as investigações mais recentes continuam a evidenciar resultados bastante promissores relativamente ao que acontece no cérebro com treino

[4] Como já havia sido antes destacado, a investigação consistentemente tem vindo a mostrar que as práticas de mindfulness não abrangem apenas circuitos atencionais e mnésicos, mas afetam igualmente os circuitos sociais.

em meditação/mindfulness. Num estudo de grande rigor metodológico (Saggar et al., 2015), baseado em meditação intensiva (que implicou um retiro de 3 meses), em que se compararam dois grupos experimentais com um grupo de controlo, os resultados evidenciaram mudanças longitudinais substanciais nos processos atencionais e cognitivos, especificamente nas dinâmicas cortico-talâmicas, registando-se um retardamento entre elas com treino em meditação *shamata* e amor-bondoso (*loving-kindness*) nos dois grupos experimentais.

A partir de um modelo de estimação da idade do cérebro (*BrainAGE methodology*), testado com 650 sujeitos de idades compreendidas entre os 19 e 86 anos, Luders, Cherbuin e Gaser (2016) comparam 50 meditadores e 50 não meditadores com idades e sexo equivalentes. Estes autores obtiveram um efeito principal do grupo e do sexo, significando que aos 50 anos o cérebro dos meditadores revelava-se em média 7.5 anos mais jovem do que o cérebro dos não meditadores. Quanto ao efeito do sexo, as mulheres apresentavam em média um cérebro 3.4 anos mais jovem do que o dos homens. Para além disso, também detetaram no grupo dos meditadores uma associação significativa entre a idade do cérebro e a idade cronológica, constatando-se que acima dos 50 anos por cada ano de vida, diminuía 1 mês e 22 dias a idade estimada do cérebro.

A nível da doença de Parkinson, num estudo experimental longitudinal constatou-se que o treino em mindfulness (8 semanas, semelhante ao MBSR) apresenta efeitos neurobiológicos positivos nas redes neuronais implicadas na referida doença, tendo-se registado o aumento significativo da densidade da matéria cinzenta nessas redes no grupo experimental (Pickut et al., 2013).

Se considerarmos as mudanças a nível celular, nomeadamente na ação sobre os telómeros[5] e a telomerase, uma revisão recente de 23 artigos (To-Miles, & Backman, 2016), que incidiu sobre o efeito da atividade física e do mindfulness, concluiu, no primeiro caso, que os resultados são inconsistentes, apesar de terem sido encontrados efeitos positivos nalguns estudos, tendo por base a medição dos telómeros. No caso das intervenções baseadas no mindfulness, a evidência é mais consistente, quanto a melhorar a saúde fisiológica, dado que aumentou a atividade da telomerase[6].

Apesar das investigações com abordagens baseadas no mindfulness não estarem ainda muito desenvolvidas em populações idosas, os estudos disponíveis apontam efetivamente para um contributo positivo em termos de melhoria do bem-estar e da qualidade de vida (e.g., Smith, 2004; Sorrell, 2015). Em geral, trata-se de intervenções muito apreciadas, quer porque o mindfulness revela um grande potencial para a compreensão profunda da ligação estreita entre mente e corpo, quer porque muitas pessoas idosas mostram-se bastante interessadas em práticas de reflexão e consciencialização da experiência, do sentido da vida, quer ainda porque constitui uma abordagem muito útil para ajudar a lidar com os problemas associados a doenças crónicas, à depressão, ansiedade, perdas, bem como para aumentar da vitalidade (Rejeski, 2008; Lima, Oliveira, & Godinho, 2011; Oliveira & Cruz, 2015; Kok & Singer, 2017).

---

[5] Os telómeros são as sequências de ADN na extremidade dos cromossomas que os protegem da deterioração e que produzem a enzima telomerase (To-Miles, & Backman, 2016).

[6] É de salientar, contudo, que no caso dos estudos com a atividade física, para além do tipo e da duração da atividade ter sido muito difícil de conhecer, as medidas incidiram no comprimento dos telómeros (cujas alterações só são visíveis após 1 a 6 anos) e não na telomerase (cujas mudanças são detetáveis após algumas semanas de treino).

Num estudo experimental de 8 semanas de duração (MBSR com prática de 30 minutos diários em casa), realizado por Creswell et al., (2012), constatou-se uma redução significativa da solidão e uma diminuição da expressão de genes pró-inflamatórios, a nível de indicadores sanguíneos, nos participantes do grupo experimental.

Foulk et al. (2014), recorrendo a um desenho quase-experimental com pré e pós-teste, envolvendo 50 idosos, constataram que 8 semanas de treino melhoravam significativamente a ansiedade, os pensamentos ruminativos e problemas de sono. Noutro estudo com 40 idosos, de natureza experimental, com 5 semanas de duração, os participantes do grupo experimental (submetidos a treino em mindfulness), em comparação com os do grupo de controlo (submetidos a um programa educativo sobre educação para higiene do sono) não apenas mostraram melhorias significativas no sono como também reduziram a sintomatologia depressiva e a fadiga diurna (Black et al., 2015).

Os diversos ganhos evidenciados pelas investigações a que nos reportámos permitem-nos compreender o título do artigo de Morone et al. (2008), "Senti-me como uma nova pessoa", que se debruçou sobre a dor lombar crónica. Após a participação num curso de mindfulness de 8 semanas, os diários dos participantes revelaram melhorias substanciais na dor e outros ganhos como a capacidade de atenção, a qualidade do sono e qualidade de vida e um sentido acrescido de bem-estar. Estes resultados foram atribuídos à atitude de observação das sensações, sem interferência, e à dissociação das sensações físicas das experiências de dor emocional e cognitiva.

2. Qualidade de vida, bem-estar e envelhecimento

A qualidade de vida e o bem-estar são conceitos inter-relacionados, mas nem sempre consensualmente distinguidos, uma vez que, como refere Simões (2006), há quem os entenda como sinónimos, embora a maioria dos autores os distinga. O construto de qualidade de vida é, em geral, proposto como integrador das condições de vida e das experiências de vida, ou seja, constituído por uma vertente mais objetiva (os recursos materiais e a saúde física, por exemplo) e outra mais subjetiva (a perceção de bem estar que o indivíduo tem da sua vida). Neste sentido, a qualidade de vida é entendida como um conceito abrangente e amplo, que integra fenómenos de ordem física, psicológica e social (Vilar et al., 2016), sendo o bem-estar um indicador da mesma. Acentuando uma dimensão mais subjetiva, mas integrativa de múltiplos fatores, a OMS define a qualidade de vida como "a perceção da pessoa acerca da sua posição na vida, no contexto da cultura e sistema de valores em que se insere e em relação aos seus objetivos, expetativas, padrões e preocupações" (WHOQOL Group, 1995, p. 1405). Daatland (2003) considera que o conceito é, simultaneamente, abrangente e amorfo e que a sua concetualização tem refletido a dominância de duas linhas ou tradições: uma ligada aos padrões de vida, que abarca o rendimento, os bens materiais, as condições de vida da pessoa (também conhecido por nível de vida) e outra ligada às experiências pessoais de vida que integra (à semelhança do que os autores anteriormente citados referem) uma dimensão subjetiva, relacionada com a perceção de bem-estar[7]. A proposta deste último autor é que se integrem as

[7] Relativamente ao bem estar, há construtos específicos que têm vindo a ser estudados na literatura (de teor mais sociológico ou psicológico), tais como os de *bem estar subjetivo* e *bem estar psicológico*, os quais não serão aqui diretamente abordados por estarem fora do escopo deste trabalho, mas deixamos a referência a duas fontes para quem tenha interesse em conhecê-los, respetivamente, de Simões et al. (2000) e Ferreira e Simões (1999).

duas tradições reportadas numa terceira linha agregadora dos fatores objetivos e subjetivos.

Envelhecimento patológico, normal e bem-sucedido

Inscrevendo-se na abordagem positiva acima referida, de há muito que os gerontólogos procuram compreender os fatores que aumentam o período de vida saudável (*healthspan*), sendo o seu objetivo principal o de promover o envelhecimento bem-sucedido (EBS), definido como "a obtenção do máximo de satisfação e felicidade da vida" (Depp, Vahia, & Jeste, 2012, p. 460). Rowe e Khan (1999) foram dos primeiros autores a apresentar uma definição operacional do EBS e a avançar com um estudo de coorte que abrangeu mais de 1000 pessoas idosas. De acordo com estes especialistas, o envelhecimento pode ser caracterizado segundo um *continuum* que vai do envelhecimento patológico até ao envelhecimento bem-sucedido passando pelo envelhecimento normal[8]. Segundo os mesmos autores, a investigação biomédica tinha permitido distinguir o envelhecimento patológico do normal, mas havia que passar a investigar-se também a distinção entre o envelhecimento normal e o bem-sucedido. O modelo do EBS por eles proposto é, por um lado, de grande exigência, uma vez que só contempla os casos em que se verifica, simultaneamente, a inexistência de doenças e incapacidades, um elevado funcionamento cognitivo e físico, e envolvimento social, excluindo de trajetórias de envelhecimento positivo, por exemplo, a adaptação dos idosos à incapacidade e doença. Por

[8] O *envelhecimento normal*, também designado por primário, diz respeito ao processo natural de mudança, à medida que a idade avança, que se caracteriza por ser universal, progressivo e irreversível, não sendo efeito de outro processo nem modificável com tratamento. Por sua vez, o *envelhecimento patológico* (também conhecido por secundário) resulta em processos de degeneração causados por doenças ou por estilos de vida impróprios (Simões, 2006, p. 32).

outro lado, não integra uma dimensão que numerosos estudos foram evidenciando como estando associada a um bom envelhecimento – a dimensão espiritual (Crowther et al., 2002; Simões, 2011). Apesar destes avanços iniciais, continuou a registar-se uma grande falta de acordo entre os investigadores quanto ao que é o EBS e as primeiras investigações consideraram-no sobretudo como sinónimo de não ter incapacidades físicas ou doenças.

Determinantes de um bom envelhecimento

Em contrapartida, os trabalhos mais recentes começaram a debruçar-se sobre os preditores do bom funcionamento cognitivo e da saúde emocional (Labouvie-Vief, 2005; Meadle & Park, 2009; Depp, Vahia & Jeste, 2012; Gutierrez & Isaacson, 2013). Embora o debate continue entre a perspetiva de que o EBS diz respeito a manter os ganhos alcançados anteriormente na vida ou a de que significa a adaptação das pessoas idosas à incapacidade e doença (ou as duas), consideremos o que as investigações mais recentes reportam a respeito do que determina um bom envelhecimento na idade adulta avançada.

Influência genética

Segundo a sistematização do conhecimento realizada por Depp, Vahia e Jeste (2012), embora se assuma que a influência genética diminui com a idade e a correspondente acumulação de experiências de vida, estima-se que os genes e os processos fisiológicos básicos exerçam uma influência entre 20 e 30%. Neste âmbito, destacam-se os telómeros que parecem dar uma boa indicação do envelhecimento celular básico. Tendo recebido o prémio Nobel pela investigação inovadora na área, Epel et al. (2004) mostraram que quanto mais curtos são os telómeros, maior é a aceleração do envelhecimento celular. A telomerase permite manter o comprimento do cromossoma e estabiliza a replicação celular, desacelerando ou acelerando o envelhecimento celular. Sabe-se que a diminuição do tamanho dos telómeros e uma atividade reduzida da telomerase estão associados à morte celular, ao envelhecimento e a uma saúde débil (Epel, 2012). Por exemplo, as pessoas mais velhas, fumadoras, obesas, com artrite reumatoide, cancro, esquizofrenia e demência mostram comprometimentos a nível da saúde dos telómeros (To-Miles, & Backman, 2016). No mesmo sentido, Atzmon et al. (2010), ao estudarem pessoas centenárias, verificaram que uma melhor saúde correspondia a telómeros mais longos, menores índices de diabetes, melhor funcionamento cognitivo e menos incapacidades.

Mestria sobre o ambiente

Um outro fator que a literatura aponta como sendo bastante importante é a *perceção de mestria sobre o ambiente*, correspondendo um sentido mais elevado da mesma a menos sintomas de ansiedade em idosos saudáveis, enquanto que níveis baixos estão associados a mais sintomas depressivos (Deep, Vahia & Jeste, 2012). Um nível elevado de *autoeficácia geral* tem sido associado também a melhor qualidade de vida, menos solidão, e menos stress psicológico em idosos. A mestria e a autoeficácia estão igualmente associadas a um efeito positivo em indicadores biológicos e fisiológicos – menos cortisol e citocinas proinflamatórias, menor risco cardiovascular e sonos REM mais longos (Geiger et al., 2016).

Atitudes para com o envelhecimento

As *atitudes* para com o envelhecimento são outro importante fator a destacar. Há muitas consequências negativas de atitudes desfavoráveis para com o processo de envelhecimento. Mais atitudes negativas significam mais morbilidade (Levy, 2003, cit. por Depp, Vahia & Jeste, 2012). Quando os idosos são expostos a estereótipos positivos acerca do envelhecimento (pessoa sábia, experimentada, com muitas experiências), comparativamente a negativos, o desempenho objetivo em tarefas físicas, tais como caminhar ou escrever, é afetado. Verificou-se efetivamente que os estereótipos têm um efeito no desempenho de tarefas diárias, quando estas são manipuladas experimentalmente. Estes resultados sugerem que os esforços para alterar as atitudes negativas para com o envelhecimento podem modificar positivamente o curso do próprio processo de envelhecimento (Depp, Vahia & Jeste, 2012).

Outros importantes fatores

Considerando particularmente o envelhecimento sem incapacidades e a prevenção da fragilidade cognitiva, para além do que anteriormente mencionámos, merece destaque uma revisão de numerosos estudos, incluindo meta-análises, realizada por Gutierrez e Isaacson (2013). Estes autores identificaram os fatores que mais contribuem para uma boa manutenção das capacidades cognitivas, destacando os seguintes cinco, como os de topo: 1) *a prática regular de exercício físico*, recomendando no mínimo 30 minutos, 3 vezes por semana (caminhadas, exercício aeróbico, jardinagem, golfe, dança, tai chi etc.), sendo que, para efeitos de redução do ritmo de perda de massa cinzenta no cérebro (através da estimulação da neurogénese nas estruturas do hipocampo) e de efeitos positivos nos genes que interferem na estrutura e adaptabilidade das sinapses, propõem a prática de exercício aeróbico, 40 minutos por dia, 5 dias por semana, durante 6 semanas; 2) *a adesão a uma dieta saudável*, aconselhando dietas com restrição clórica e de tipo Mediterrânica, as quais revelaram melhor predizer uma longevidade saudável e sem incapacidades; 3) o *envolvimento em atividades sociais*, sugerindo os autores a manutenção de uma boa rede social e participação em atividades de grupo prazerosas e expressivas (representação, teatro, programas de educação de adultos, dança etc.); 4) *a gestão adequada do stress*, recomendando, tal como vários dos estudos anteriormente mencionados apontaram, a prática de meditação, mindfulness, yoga, biofeedback, pensamento positivo, o estabelecimento de prioridades nas atividades diárias etc.; e 5) *o aumento da atividade cognitiva*, através do envolvimento em aprendizagens novas, desafiantes e do interesse da pessoa, incluindo-se aqui os programas de estimulação cognitiva, o uso de estratégias multissensoriais, a aprendizagem de uma nova língua ou música, a participação em atividades de escrita criativa, de leituras encenadas etc..

Considerações finais

A mudança de perspetiva relativamente à forma como encaramos o envelhecimento, adotando-se uma visão de longo prazo e uma atitude positiva quanto à possibilidade de manter, ou até de aumentar a qualidade de vida, o bem-estar e a vitalidade, impõe-se. Não apenas porque enfrentamos alterações demográficas populacionais profundas e é importante tornar os sistemas de apoio social sustentáveis, mas sobretudo porque, tal como alguns dos visionários da cultura ocidental (desde a Antiguidade Clássica), anteviam e defendiam, é efetivamente possível envelhecer com qualidade de vida e bem-estar. São diversos os fatores que para isso contribuem, salientados por distintas áreas do saber e por desenhos metodológicos igualmente distintos. Mas todos eles, apesar da sua especificidade, apontam no mesmo sentido: pode-se melhorar ou recuperar o bem-estar e a vitalidade e, mesmo, desacelerar o processo de envelhecimento. Sinergicamente, o desafio é o de intervir, a nível pessoal e coletivo, em várias frentes, colocando em prática os significativos saberes que nos vêm não de crenças ou ideologias, mas do cruzamento de muitos dados da investigação científica e de saberes milenares.

Pela multiplicidade dos efeitos benéficos que proporcionam e pelos numerosos mecanismos que mobilizam (atencionais, atitudinais, emocionais, sociais etc.), assumem relevo especial as práticas de mindfulness ou meditação, mas também é importante ter em conta o elevado potencial da atividade física, da perceção de mestria ou de autoeficácia, das relações sociais positivas, do treino cognitivo, da alimentação saudável, o que nos remete para a importância de cultivar um estilo de vida saudável desde as idades mais precoces, como a Estratégia Nacional para o Envelhecimento Ativo e Saudável (2017-2025) acentua (República Portuguesa, 2017), e para a ativação do potencial positivo do ser humano durante toda a vida.

A revisão dos estudos que convocámos para a elaboração deste capítulo não pode ser considerada exaustiva, mas estamos convictas de que a diversidade de fontes consultadas, em torno dos seus tópicos estruturantes, é meritória da nossa melhor atenção, em virtude da qualidade dos trabalhos de investigação que tivemos o gosto de conhecer e rever.

Bibliografia

Amaro, A. (2015). A holistic mindfulness. *Mindfulness, 6*, 63-73. https://doi.org/10.1007/s12671-014-0382-3

Antunes, B. (2014). *The Cartesian moment and the contemplative turn: Reshaping epistemology*. Comunicação apresentada no Summer Research Institute, Mind & Life Europe, 23-29 Agosto, Fraueninsel, Chiemsee.

Atzmon, G. (2010). Genetic variation in human telomerase is associated with telomere length in Ashkenazi centenarians. *PNAS, 107*(1), 1710-1717.

Black, D. S., O'Reilly, G. A., Holmstead, R., Breen, E. C., & Irwin, M. R. (2015). Mindfulness meditation and improvement in sleep quality and daytime impairment among older adults with sleep disturbances. *JAMA International Medicine, 175*(4), 494-501.

Bury, M., & Holme, A. (1991). *Life after ninety*. London. Routledge.

Creswell, J. D. et al. (2012). Mindfulness based stress reduction training reducues loneliness and pro-inflammatory gene expression in older adults: A small randomized control trial. *Brain, Behavior and Immunity, 26,* 1095-1101. doi: 10.1016/j.bbi.2012.07.006

Crowther, M. R. et al. (2002). Rowe and Khan's model of successful aging revisited: Positive spirituality – the forgotten factor. *The Gerontologist, 42*(5), 613-620.

Daatland, S. O. (2003). Quality of life and ageing. In M. L. Johnson (Ed.), *The Cambridge handbook of age and ageing* (pp. 371-377). Cambridge: University Press.

Davidson R. J., Kabat-Zinn J, Schumacher J., et al. (2003). Alterations in brain and immune function produced by mindfulness meditation. *Psychosomatic Medicine, 65*(4), 564-570.

Depp, C. A., Vahia, I. V., & Jeste, D. V. (2012). Successful aging. In S. K. Whithourne, & M. J. Sliwinski (Eds.), *The Wiley-Blackwell handbook of adulthood and aging* (pp. 459-476). Oxford: Wiley-Blackwell.

Direção Geral da Saúde (2015). *A saúde dos portugueses: Perspetiva 2015*. Lisboa: DGS.

Epel, E. S., Blackburn, E. H., Lin J, Dhabhar, F. S., Adler, N. E., Morrow, J. D., & Cawthon, R. M. (2004). Accelerated telomere shortening in response to life stress. *Proceedings of the National Academy of Sciences of the United States of America, 101*, 17312-17315.

Epel, E. (2012). How "reversible" is telomeric aging? *Cancer Prevention Research, 5*, 1163-1168. doi:10.1158/1940-6207.CAPR12-0370

Ferreira, J. A., & Simões, A. (1999). Escalas de bem-estar psicológico (E.B.E.P.). In M. R. Simões, L. S. Almeida & M. M. Gonçalves (Eds.), *Testes e Provas Psicológicas em Portugal* (Vol. 2, pp. 111-121). Braga: Associação dos Psicólogos Portugueses.

Gard, T., Holtzer, B. K., & Lazar, S. W. (2014). The potential effect of meditation on aged-related cognitive decline: A systematic review. *Annals of the New York Academy of Sciences, 1307*, 89-103.

Geiger, P. J., et al. (2016). Mindfulness-based interventions for older adults: A review of the effects on physical and emotional well-being. *Mindfulness, 7*, 296-307.

Goldin P. R., Gross J. J. (2010). Effects of mindfulness-based stress reduction (MBSR) on emotion regulation in social anxiety disorder. *Emotion 10*, 83-91. doi: 10.1037/a0018441

Gutierrez, J., & Isaacson, R. S. (2013). Prevention of cognitive decline. In L. D. Ravdin & H. L. Katzen (eds.), *Handbook of the neuropsychology of aging and dementia* (pp. 167-192). LLC: Springer Science.

Ho, N. S. P., Sun, D., Ting, K-H, Chan, C. C. H., & Lee, T. M. C. (2015). Mindfulness Trait Predicts Neurophysiological Reactivity Associated with Negativity Bias: An ERP Study. *Evidence-Based Complementary and Alternative Medicine, 15*, 1-15.

Hölzel, B. K., Carmody, J., Evans, K. C., Hoge, E. A., Dusek, J. A., Morgan, L., ... Lazar, S. W. (2010). Stress reduction correlates with structural changes in the amygdala. *Social Cognitive and Affective Neuroscience, 5*(1), 11-17.

Hölzel, B. K., Carmody, J., Vangel, M., Congleton, C., Yerramsetti, S. M., Gard, T., & Lazar, S. W. (2011). Mindfulness practice leads to increases in regional brain gray matter density. Psychiatry Research: *Neuroimaging, 191*(1), 36-43.

Kabat-Zinn, J. (1990). *Full catastrophe living: Using the wisdom of your body and mind to face stress, pain and illness.* New York: Delacorte.

Kabat-Zinn, J. (2003). *Onde quer que eu vá* (C. Rodriguez, Trad.). Porto: Oficina do Livro.

Kabat-Zinn, J. (2011). Some reflections on the origins of MBSR, skillful means, and the trouble with maps. *Contemporary Buddhism, 12*(1), 281-306.

Kabat-Zinn, J. (2015). Why paying attention is so supremely important. *Mindfulness, 6*, 1484-1486.

Kok, B. E., & Singer, T. (2017). Phenomenological fingerprints of four meditations: Differential state changes in affect, mind-wandering, meta-cognition, and interoception before and after daily practice across 9 months of training. *Mindfulness, 8*, 218-231. https://doi.org/10.1007/s12671-016-0594-9

Labouvie-Vief, G. (2005). The Psychology of emotions and ageing. In M. Johnson, V. Bengtson, P. Coleman, & T. KirDeath (Eds.), *The Cambridge handbook of age and ageing.* Cambridge: Cambridge University Press (pp. 229-236).

Lazar, S. W. et al. (2005). Meditation experience is associated with increased cortical thickness. *Neuroreport, 16*, 1893-1897.

Lima, M. P., Oliveira, A. L., & Godinho, P. (2011). Promover o bem-estar de idosos institucionalizados: Um estudo exploratório com treino em mindfulness. *Revista Portuguesa de Pedagogia, 45*(1), 165-183. DOI http://dx.doi.org/10.14195/1647-8614_45-1_9f

Luders, E., Cherbuin, N., & Gaser, C. (2016). Estimating brain age using high--resolution pattern recognition: Younger brains in long-term meditation practitioners. *Neuroimage, 134*, 508-513.

Meadle, M. L., & Park, D. C. (2009). Enhancing cognitive function in older adults. Chodzko-Zajko, W. C., Kramer, A. F., & Poon, L. W. *Enhancing cognitive functioning and brain plasticity* (pp. 35-47; vol. 3). Champaign: Human Kinetics.

Morone, N.E., Lynch, C.S., Greco, C.M., Tindle, H.A., & Weiner, D.K. (2008). "I felt like a new person." The effects of mindfulness meditation on older adults with chronic pain: Qualitative narrative analysis of diary entries. *Journal of Pain, 9*, 841-848.

Myers, D. G., & Diener, E. (1995). Who is happy? *Psychological Science, 6*(1), 10-17.

Oliveira, A. L., & Antunes, B. M. G. (2014). A pedagogia contemplativa no ensino superior: Para uma abordagem completa ao que o ser humano convoca. *Revista Portuguesa de Pedagogia, 48*(2), 43-60. http://dx.doi.org/10.14195/1647-8614_48-2_3

Oliveira, A. L., & Cruz, A. C. (2015). O papel do sentido da vida e do mindfulness na compreensão do bem estar de alunos de Universidades Seniores. *Exedra*, 61-78. http://www.exedrajournal.com/wp-content/uploads/2016/02/Cap4.pdf

Pickut, B. A., Hecke, W. V., Kerckhofs, E., Marien, P., Vanneste, S., Cras, P., & Parizel, P. M. (2013). Mindfulness based intervention in Parkinson's disease leads to structural brain challenges on MRI: A randomized control longitudinal trial. *Clinical neurology and neurosurgery, 115*, 2419-2425.

Rana, N. (2015). Mindfulness and loving-kindness meditation: A potential tool for mental health and subjective well-being. *Indian Journal of Positive Psychology, 6*(2), 189-196.

Reis, C. S., & Oliveira, A. L. (2016). Ontological conversion: The place of self--knowledge, the contemplative tradition and contemporary mindfulness in education. *Turkish Online Journal of Educational Technology, Special Issue* for *INTE 2016*, 163-172. http://www.tojet.net/special/2016_12_1.pdf

Rejeski, W. (2008). Mindfulness: Reconnecting the body and the mind in geriatric medicine and gerontology. *The Gerontologist, 48*(2), 135-141.

República Portuguesa (2017). *Estratégia Nacional para o Envelhecimento Ativo e Saudável (2017-2025). Proposta do Grupo de Trabalho Interministerial* (Despacho n.º12427/2016). República Portuguesa: DGS/SNS.

Rowe, J. W., & Kahn, R. L. (1999). *Successful aging*. New York: Dell Publishing.

Saggar, M. et al. (2015). Mean-field thalamocortical modeling of longitudinal EEG acquired during intensive meditation training. *Neuroimage, 114*, 88-104.

Siegel, D. J. (2007). *The Mindful Brain: Reflection and Attunement in the Cultivation of Well-Being*. New York: Norton & Company.

Simões, A. (2006). *A nova velhice: Um novo público a educar*. Lisboa. Âmbar.

Simões, A. (2011). Um modelo mal sucedido de envelhecimento bem sucedido? *Psicologia, Educação e Cultura, XV*(1), 7-28.

Simões, A., Ferreira, J. A., Lima, M. P., Pinheiro, M. R., Vieira, C. M., Matos, A. P., & Oliveira, A. L. (2000). O bem-estar subjectivo: Estado actual dos conhecimentos. *Psicologia, Educação e Cultura, IV*(2), 243-279.

Simões, A., Lima, M. P., Vieira, C. M. C., Oliveira, A. L., Alcoforado, J. L., & Ferreira, J. A. (2009). O sentido da vida: Contexto ideológico e abordagem empírica. *Psychologica, 51*, 101-130. DOI:http://dx.doi.org/10.14195/1647-8606_51_8

Smith, A. (2004). Clinical uses of mindfulness training for older people. *Behavioural and Cognitive Psychotherapy, 32*, 423-430.

Sorrell, J. M. (2015). Meditation for Older Adults: A New Look at an Ancient Intervention for Mental Health. *Journal of Psychosocial Nursing, 53*(5), 15-19.

To-Miles, F. Y. L., & Backman, C. L. (2016). What telomeres say about activity and health: A rapid review. *Canadian Journal of Occupational Therapy, 83*(3) 143-153. DOI: 10.1177/0008417415627345

Vilar, M., Sousa, L., Firmino, H. & Simões, M. (2016). Envelhecimento e qualidade de vida. In H. Firmino; M. Simões & J. Cerejeira (Coords.), *Saúde mental das pessoas mais velhas* (pp. 19-43). Lisboa: Lidel.

Wallace, B. A. (2007). *Contemplative science: Where Buddhism and neuroscience converge*. New York: Columbia University Press.

World Health Organization Quality of Life Group/WHOQOL Group. (1995). The World Health Organization Quality of Life Assessment (WHOQOL): Position paper from the WHO. *Social Science and Medicine, 41*(10), 1403-1409.

CAPÍTULO III

FAMÍLIA E OUTRAS REDES DE SUPORTE SOCIAL NA POPULAÇÃO IDOSA

Sónia Guadalupe

Instituto Superior Miguel Torga – Coimbra

ORCID: https://orcid.org/0000-0003-4898-3942

Henrique Testa Vicente

Instituto Superior Miguel Torga – Coimbra

ORCID: https://orcid.org/0000-0001-5571-9168

Resumo: *Este capítulo aborda as características das redes sociais pessoais na velhice, enquanto recurso fundamental de bem-estar e suporte, problematizando o lugar dos laços familiares na provisão social informal e sua relação com a provisão formal. As redes sociais pessoais são fenómenos dinâmicos, em permanente evolução ao longo do ciclo vital. As perdas normativas na velhice, como as associadas à aposentação ou à diminuição da autonomia, têm sido apontadas como fatores de retração reticular e fragilização dos laços sociais informais. Acresce um conjunto de determinantes sociais associados a condições de vida na idade avançada, assim como*

DOI | https://doi.org/10.14195/978-989-26-1788-6_3

constrangimentos decorrentes das forças estruturais e culturais, que potenciam um quadro de vulnerabilização que tem encontrado resposta na provisão informal, sobretudo das famílias.
O conhecimento acerca das redes sociais na população idosa e da relação entre suporte informal e formal, afigura-se assim relevante e coloca desafios específicos à avaliação e intervenção social.
Palavras-chave: *Redes sociais pessoais; Redes sociais; Suporte social; Família; Laços sociais; Velhice; Envelhecimento; Idosos*

Abstract: *This chapter addresses the characteristics of personal social networks in old age, as a fundamental resource of well-being and support, problematizing the role of family ties in informal social provision and its relationship with formal provision.*
Personal social networks are dynamic phenomena, constantly evolving throughout the life cycle. Normative developmental losses in old age, such as those associated with retirement or decreased autonomy, have been identified as contributing factors to reticular retraction and weakening of informal social ties. In addition, there is a set of social determinants associated with living conditions in old age, as well as constraints resulting from structural and cultural forces, which enhance a situation of vulnerability that has found an answer in informal support provision, especially from families.
Therefore, knowledge about social networks in the elderly population and the relationship between informal and formal support is relevant and poses specific challenges to social assessment and intervention.
Keywords: *Personal social networks; Social networks; Social support; Family; Social ties; Old age; Aging; Elderly*

1. A vitalidade da família no suporte social à pessoa idosa

As famílias, como núcleo central da sociedade que assume a proteção e a promoção do desenvolvimento dos seus membros (Alarcão, 2015; Nunes, 2015), são tidas como as principais responsáveis pelo domínio assistencial e de cuidado às gerações mais jovens e mais velhas (Bazo, 2008; Figueiredo, 2007; Pimentel, 2015; Serapioni, 2005; Sousa, Figueiredo & Cerqueira, 2004; Sousa & Figueiredo, 2007; Williams, 2010).

A centralidade da família como fonte de suporte informal tem-se apresentado como incontornável no contexto sociocultural português (Portugal, 2011, 2014), assente numa premissa axiológica que associa à vinculação parento-filial (ascendente e descendente) e conjugal um conjunto de deveres morais e legais fundados na reciprocidade (Sarti, 2010). Tal tem sido atribuído à forma de organização do bem-estar nos países da Europa do Sul (Ferrera, 2000; Silva, 2002), não se tendo perdido a vitalidade do papel da parentela extensa na contemporaneidade, ainda que a morfologia familiar se tenha democratizado e diversificado numa lógica dominantemente nuclear (Aboim, Vasconcelos & Wall, 2013; Fernandes, 2001; Vasconcelos, 2002).

A família e a sociedade providência, que se materializa nas redes sociais, numa perspetiva mais alargada, constituem o contraponto da proteção social de âmbito estatal. Na relação com a esfera pública, a família é tida como a expressão máxima da vida privada e da intimidade, enquanto nicho afetivo de relações de socialização e de vínculos de inclusão social, estabelecendo-se como condição objetiva e subjetiva de pertença e de filiação (Carvalho, 2010; Castel, 2003). Para além desta dimensão, "a sociabilidade, a entreajuda, o apoio dos conhecidos ou as amizades dos parentes transformaram-se (...) em novas perspetivas para as políticas sociais" (Martin, 1995, p. 62), constituindo uma forte tentação para um Estado-Providência semiperiférico, a lidar com dificuldades financeiras, em recuar nas políticas sociais (Hespanha, 1993; Santos, 1995), ou em combinar recursos e meios em modelos mistos de proteção social, com uma reconfiguração e retração estatal (Marques, 2008; Martin, 1995), redesenhando fronteiras entre a responsabilidade pública e privada (Daatland & Lowenstein, 2005).

As transmutações nos modelos familiares contemporâneos interagem com mudanças demográficas, culturais, económicas e políticas, pluralizando-se configurações e estruturas, com cenários familiares flexíveis e fluídos, associados a um processo de democratização e de diversificação (Bris, 1994; Erera, 2001; Guadalupe & Cardoso, 2018; Pedroso & Branco, 2008; Williams, 2010). No entanto, a família continua a evidenciar uma enorme resistência, adaptação, dinâmica, elasticidade e plasticidade, que se traduz na sua universalidade e ancestralidade (Balandier, 1986), construindo-se e reconstruindo-se no tempo (Leandro, 2011).

Apesar das teses pessimistas que reclamam a dissolução e a extinção da família, o apelo encantatório das solidariedades primárias e familiares tem subsistido (Volpi, 2007; Williams, 2010). "Não admira pois, que (...) a família continue a funcionar como um refúgio onde se espera encontrar protecção e apoio para colmatar as falhas que possam existir a nível de outras formas de laços sociais" (Leandro & Ferreira, 2011, p. 48), particularmente num contexto de austeridade marcado pelo avanço do neoliberalismo, com forte retração na proteção coletiva e na garantia de direitos (Capucha, 2014; Hespanha, Ferreira & Pacheco, 2013). Este movimento de um neoliberalismo familista nas políticas sociais (De Martino, 2001) inscreve-se no debate que opõe processos de desfamilização e de familização (Esping-Andersen, 1999). Os últimos são marcantes no modelo de provisão social do Sul Europeu (Campos & Mioto, 2003), mesmo com o avanço das políticas públicas na área dos idosos, que nunca se revelaram consistentemente substitutivas do papel da família e da solidariedade intergeracional (Daatland & Lowenstein, 2005). A substituibilidade das políticas públicas pelo papel providencial das famílias tem constituido um dos eixos do debate sobre a articulação entre solidariedades familiares e solidariedades públicas. No entanto, alguns autores sublinham as distinções entre as suas naturezas, motivações, regras, princípios, funções e finalidades, não concebendo a possibilidade de substituição mútua (Attias-Donfut & Ogg, 2009; Capucha, 2014; Lesemann & Martin, 1995). Ainda assim, o reconhecido domínio do familismo nas políticas sociais (Andrade, 2009; Campos & Mioto, 2003; Franzoni, 2007; Mioto, 2008; Pereira, 2006) pode ser entendido como uma regressão por metamorfose dos direitos sociais em deveres morais (Yazbeck, 2001), essencialmente a assumir pelos cidadãos e pelas famílias, evidenciando o retorno a um enfoque conservador de política social (Andrade, 2009).

O traço de refamilização reforça a sociedade providencial informal, isto é, a solidariedade provida pelas redes sociais e pelas famílias (Grundy & Tomassini, 2003; Ferreira & Monteiro, 2015; Portugal, 2014; Saraceno, 2010), constituindo a solidariedade intergeracional um factor de estabilidade e de coesão social num contexto de agravada incerteza (Ferreira & Monteiro, 2015; Frade & Coelho, 2015). Paradoxalmente, as condições de vida concretas das famílias, quer pelas transformações que têm vindo a ocorrer nos modelos familiares, quer pelas exigências e limitações de conciliação da vida familiar com a vida laboral, por exemplo, limitam, fragmentam, atomizam ou anulam a sua capacidade potencial para assumir o papel providencial no quotidiano (Capucha, 2014; Guadalupe, 2017; Guadalupe & Cardoso, 2018; Mioto, 2008).

2. O suporte social informal na vulnerabilidade e dependência na população idosa

Sendo cada vez maior a relevância social da população idosa a nível mundial, a formulação das políticas sociais tende a valorizar a sua independência, participação, assistência, autorrealização e dignidade como princípios orientadores (UN, 2015). No entanto, o processo de senescência tende a não favorecer um envelhecimento ativo e autónomo, sendo associados fatores de vulnerabilidade como o declínio da saúde e do rendimento a segmentos populacionais significativamente alargados na idade avançada (UN, 2015). A diversificação, intensificação e agudização da velhice dependente, sobretudo da população de idade mais avançada, constituem enormes desafios sociais, particularmente para os sistemas informais e formais de provisão social, exigindo um estreito laço entre as solidariedades familiares e as políticas sociais (Fernandes, 2001; OCDE, 2012). Quando se efetivam perdas na independência e na autonomia, a solidariedade intergeracional torna-se especialmente relevante, particularmente das gerações mais novas para as mais velhas (Nunes, 2015; Pimentel, 2015). As tendências demográficas, nas famílias e no padrão de envelhecimento, designadamente o tamanho das famílias, a diminuição da natalidade e do número de filhos, o aumento dos divórcios, das famílias monoparentais e isoladas, e dos idosos a viverem sós e sem filhos, conjugam-se numa revolução demográfica que produz diversificação nos modelos, estruturas e papéis intergeracionais (Bazo, 2008), agudizados por outras mudanças societais como o aumento da mobilidade geográfica e da deslocalização, e mudanças no mundo laboral, designadamente a participação da mulher no mercado de trabalho (Bris, 1994; Capucha, 2014). Apesar de as mudanças sociodemográficas trazerem como correlato um aumento das famílias multigeracionais (com maior probabilidade de os indivíduos viverem tempo suficiente para ver nascer netos, bisnetos e trinetos), são menos frequentes os arranjos habitacionais com várias gerações, sendo maior a proporção de idosos que vivem sós e que não têm descendência (Delgado & Wall, 2014; Fernandes, 2001; Vicente & Sousa, 2012). Quando existe descendência, há também maior verticalidade, originando as chamadas "famílias em feijoeiro", caracterizadas pela verticalização das relações, com muitas gerações a coexistir, mas poucos elementos em cada geração (Bengtson, Rosenthal & Burton, 1990; Bengtson, Lowenstein, Putney & Gans, 2003), potenciando-se um quadro em que há um maior número de pessoas para cuidar e um número menor de potenciais cuidadores informais, por períodos cada vez mais longos (Aguirre, 2008; Pimentel, 2015). Assim, se assumirmos que uma trajetória da dependência exige uma trajetória de cuidados (Gil, 2010), vislumbramos a resposta nos sistemas de apoio formais e informais. Apesar das relações entre estes sistemas se assumirem como complementares ou de substituição, os cuidados às pessoas idosas são prestados sobretudo no contexto informal (Gil, 2009, 2010; Pimentel, 2012), havendo uma particular ancoragem familiar.

A família continua a ser valorizada e entendida genericamente como "o lugar primordial das trocas intergeracionais" (Fernandes, 2001, p. 48; Bazo, 2008), sendo centrais os laços de parentesco nas redes de suporte social para a maioria das pessoas na sociedade portuguesa face a necessidades de apoio material e afetivo (Portugal, 2011, 2014; Vasconcelos, 2005). Assim, cuidar dos pais na velhice é tida enquanto retribuição pelas dádivas recebidas, sendo que a norma da reciprocidade na dádiva familiar se estende na cadeia geracional, modificando-se temporalmente os papéis de quem dá e recebe (Attias-Doufut, Lapierre & Segalen, 2002; Portugal, 2011). Quando os pais envelhecem e necessitam de compensar perdas de autonomia, cria-se uma circunstância de reafirmação da norma da reciprocidade (Grundy, 2005; Portugal, 2011), nem sempre ocorrendo simetricamente ou correspondendo às expectativas mútuas (Bazo, 2008; Fernandes, 2001). Quando há autonomia na velhice, as pessoas idosas tendem a apresentar níveis elevados de participação social, constituindo-se como uma forte ajuda na vida familiar (OCDE, 2012), cuidando das gerações mais novas, reiterando assim a horizontalidade na reciprocidade.

Mas a vulnerabilidade social deste segmento populacional, salvaguardando a sua heterogeneidade, não se associa apenas a eventuais limitações na funcionalidade ou a quebras financeiras, mas também ao isolamento social e à solidão, havendo vulnerabilidade acrescida e progressiva quando se conjugam tais determinantes (Castel, 2003; Conselho Económico e Social, 2013). O afastamento e a ausência de laços informais efetivos reitera a fulcral relevância destes mesmos laços, paradoxalmente. As ruturas com as redes de integração primária, nomeadamente com os laços familiares e de pertença comunitária, configuram processos de desafiliação (Castel, 2003; Costa, 2004) e de exclusão social com "privação de tipo relacional, caracterizada pelo isolamento, por vezes associada à falta de auto-suficiência e autonomia pessoal" (Costa, 2004, p. 22). Ora, se as marcas de vulnerabilidade são heterogéneas e se refletem na vida de cada pessoa de forma singular, as formas de enfrentá-las em sociedade têm também de ser plurais e flexíveis.

Se os sistemas informais e de proteção próxima constituem recursos providenciais simples, flexíveis e eficazes na satisfação de necessidades da população idosa em diferentes frentes (Lesemann & Martin, 1995), os sistemas formais não garantem a mesma adaptabilidade e plasticidade. Mas nem sempre os laços familiares existem, estão ativos, disponíveis, próximos, ou nem sempre conseguem reunir condições de vida para cuidar, conciliando esta tarefa com a vida profissional, ou não equacionam apoiar as gerações mais velhas (Guadalupe & Cardoso, 2018).

Na dependência na velhice, apesar da representação social de desresponsabilização familiar e de outros argumentos pessimistas (Bazo, 2008; Pimentel, 2015), a evidência aponta para a persistência das famílias como o pilar dos cuidados, mesmo em países onde o sistema providencial formal garante acesso a amplos apoios e serviços para a população idosa (Bazo, 2008; Bris, 1994; Daatland & Lowenstein, 2005; Gil, 2010; Pimentel, 2015; Serapioni, 2005; Sousa & Figueiredo, 2004). Em Portugal, a proteção social, o bem-estar e o cuidar dependem (e eventualmente vão continuar a depender) fundamentalmente de um modelo assente nas solidariedades familiares e nas transferências transgeracionais (Portugal, 2014).

3. Redes sociais pessoais na velhice

As redes sociais pessoais são mutáveis no tempo e no espaço vivido, num processo contínuo de transações, dinâmicas e escolhas, determinadas pelos atributos pessoais, pelos interesses, necessidades, circunstâncias e constrangimentos nas esferas da vida relacional (Cochran, 1991; Moral, Miguel & Pardo, 2007). As características das redes transformam-se e (re)constroem-se na trajetória do curso de vida por influência dos contextos que exploramos e vivenciamos ou não (Daniel, Ribeiro & Guadalupe, 2011), sendo necessariamente singulares.

Numa perspetiva desenvolvimental e normativa, podemos conceber um mapa diacrónico de evolução das redes pontuado por movimentos expansionistas e contrativos. Apesar de a idade cronológica não ser um indicador a analisar isoladamente, e constituir uma categoria normativa que impõe determinados papéis, identidades e interações sociais, Olsen, Iversen e Sabroe (1991) consideram-na imprescindível na análise das redes, sendo que a sua trajetória pode ser equacionada como um plano inclinado, ora ascendente ora descendente, intercalado com fases de estabilidade nas relações interpessoais (Daniel et al., 2011; Moral et al., 2007; Sluzki, 1996). Assim, "as redes sociais das pessoas modificam-se, expandem-se e contraem-se, à medida que se modificam as suas necessidades físicas, sociais e emocionais" (Specht, 1986, p. 224), sendo apontada genericamente uma associação negativa entre a idade e o tamanho da rede (e.g. Cornwell, Laumann & Schumm, 2008; Schnittker, 2007).

A diversidade de experiências dos sujeitos é marcada por condicionalismos da idade, sendo a heterogeneidade de trajetórias no envelhecimento e na velhice multideterminadas por marcadores sociais inscritos em mudanças ocorridas na passagem do tempo, assim como por processos e acontecimentos de vida relevantes e banais que sucedem ao longo da vida (Arias, 2009; Ginn & Arber, 1996), tanto normativos como acidentais. As redes sociais pessoais na velhice espelham tal heterogeneidade, variando consoante a idade, o sexo e o género, a escolaridade, o estado civil, a composição familiar, as construções, ruturas e reconstruções relacionais, a conjugalidade e a parentalidade, os processos migratórios, o nível de participação social, a ligação a organizações comunitárias, o (des)emprego, o processo de aposentação, a saúde, a autonomia, a autodeterminação, a institucionalização, entre outras variáveis e marcadores sociais.

A metáfora da extinção progressiva da galáxia que Sluzki (1996; 2000) evoca para assinalar o movimento de retração na fase mais tardia da vida, refere-se à desativação e ao desligamento nas relações interpessoais, mas também às perdas que marcam esta etapa, nomeadamente as perdas relacionais, geracionais, perdas nas referências identitárias e perdas de autonomia. Podemos também identificar nesta metáfora o estreitamento da rede resultante de um processo de seletividade socioemocional (Carstensen, Isaacowitz & Charles, 1999) numa fase em que o tempo é perspetivado como limitado, o que se traduz na prioridade dada aos laços fortes e de suporte. Esta ideia é compatível com os estudos centrados nas redes de confiança e de confidentes na idade avançada que nos revelam redes muito restritas no tamanho (Cabral, Ferreira, Silva, Jerónimo & Marques, 2013; McPherson, Smith-Lovin & Brashears, 2006).

Por outro lado, nas fases tardias da vida há também oportunidade para abrir e explorar contextos de interação até então inexplorados e para acrescentar novos vínculos significativos (Arias, 2009). Quando se perde o contacto diário com o mundo do trabalho, na aposentação, potenciam-se quebras relacionais nos laços favorecidos até então pelo contacto quotidiano (Fonseca, 2011; Paúl, 2005a), associando-se uma metamorfose identitária, para muitos, na transição e adaptação à reforma e à velhice (Fonseca, 2011).

No entanto, o desligamento e o afastamento do envolvimento social não são inevitáveis, se tivermos em conta os padrões de interação social adotados ao longo da vida (Fonseca, 2011), até porque "envelhece-se como se vive" (Sande, Dornell & Aguirre, 2011, p. 194). A participação social e as atividades ocupacionais são contextos determinantes para encetar novos vínculos, sendo ainda favorecedores de um envelhecimento ativo e participativo. No entanto, há entraves socioeconómicos à participação, sendo que uma situação de pobreza dificulta as dinâmicas transacionais, impedindo por vezes os contactos frequentes (Phillipson,

Bernard, Phillips & Ogg, 2001). A própria institucionalização pode ser encarada simultaneamente como potenciando quebra e ampliação na esfera relacional, dependendo da situação idiossincrática da pessoa idosa, designadamente da sua saúde e autonomia. Particularmente em situações de isolamento social, a integração institucional pode expandir a rede pelo contacto que proporciona com outras pessoas da mesma geração, assim como com os profissionais que integram as organizações.

Tal representa um paradoxo no envelhecimento populacional: se a maior longevidade traz um potencial favorável de participação na vida social, mais liberto de obrigações, viver mais tempo tem simultaneamente como corolário estar mais exposto ao risco de adoecer e de ficar dependente, o que restringe a disponibilidade para a vida ativa, favorecendo o enfraquecimento e arrefecimento dos laços sociais (Norbert Elias, 2001) que redunda frequentemente em quebra social (Kuyper & Bengtson, 1973) e em isolamento. Os laços sociais são, por outro lado, preditores de uma maior sobrevida e longevidade, tendo o estudo longitudinal de *Alameda County* (EUA, Berkman & Syme, 1979), revelado que as pessoas com menos vínculos sociais e comunitários apresentavam maior risco ou maior probabilidade de morrer mais cedo, mesmo quando eram controladas variáveis como o estado físico da pessoa e o estatuto socioeconómico.

A restrição das redes sociais é, apesar da heterogeneidade de trajetórias, uma ideia dominante associada ao processo natural do envelhecimento (Carstensen, 1992; Sluzki, 1996; Sousa, Figueiredo & Cerqueira, 2004), que representa frequentemente uma sucessão de dificuldades inibidoras da interação social. A fragilização e disrupção das redes na velhice resultam sobretudo de quatro determinantes cumulativos (Sluzki, 1996, 2000): contração do tamanho da rede; diminuição de circunstâncias de renovação dos vínculos; menor energia para ativar, manter e mobilizar os vínculos; perdas geracionais. Sublinhamos como fulcral nesta abordagem deficitária das redes sociais pessoais na população idosa, que o progressivo desaparecimento de pessoas da mesma geração significa perda na estória pessoal e na partilha de memórias comuns (Sluzki, 1996, 2000; Paúl, 2005b). Para além deste determinante, encontramos outros, sobretudo associados a situações de vulnerabilidade pessoal, de existência de barreiras no meio desfavoráveis à manutenção da interação social, de conflitos relacionais acumulados ao longo da vida; à falta de motivação para a participação social; a comportamentos acomodativos; à aposentação; ao declínio das capacidades físicas e cognitivas; à eventual institucionalização; entre outros (Arias, 2009; Daniel et al., 2011; Moral et al., 2007; Paúl, 2005b; Rioseco, Quezada, Ducci & Torres, 2008; Sluzki, 1996; Sousa et al., 2004).

A redução de contactos sociais com vínculos extrafamiliares e a consequente focalização nos laços familiares é também sublinhada como um traço indelével nas redes na velhice (Antonucci & Akiyama, 1987; Cabral et al., 2013; Field & Minkler, 1988; Ham-Chande, Zepeda & Martínez, 2003; Moral et al., 2007; Valle & Garcia, 1994). Vicente (2010, p. 78) usa a analogia do *Big Bang* para referir-se ao universo relacional de um indivíduo que será originário num núcleo denso triangular (pais e filho/a), expandindo-se e diversificando-se ao longo da vida, sendo que no seu final haverá um retorno ao núcleo restrito filio-parental, afirmando que "as relações intergeracionais entre pais e filhos parecem ser o alfa e ómega da rede social pessoal de um indivíduo". O lugar dos laços familiares mais chegados nas redes é também nesta fase da vida um lugar central.

4. Tipos de redes sociais pessoais na velhice

As tipologias dominantes de redes sociais de pessoas idosas são agregações conceptuais e indutivas (a maior parte fundada em análises estatísticas de um vasto número de sujeitos) que conjugam tipos de vínculo, aspetos morfológicos ou estruturais e transacionais nas redes, encontrando-se propostas tipológicas variadas em diferentes contextos socioculturais (Auslander, 1996; Burholt & Dobbs, 2014; Cabral et al., 2013; Doubova, Pérez-Cuevas, Espinosa-Alarcón & Flores-Hernández, 2010; Fiori, Antonucci & Cortina, 2006; Fiori, Smith & Antonucci, 2007; Giannella & Fischer, 2016; Guadalupe & Vicente, 2020; Li & Zhang, 2015; Litwin 1995a, 1995b; Litwin, 1997; Litwin & Landau, 2000; Litwin & Shiovitz-Ezra, 2010; Melkas & Jylha 1996; Mugford & Kendig,1986; Park et al., 2015; Park, Smith & Dunkle, 2014; Wenger, 1991) (Quadro 1). São múltiplos os denominadores comuns entre as tipologias, tanto nas designações como nas características agregadas nos tipos de rede, sendo sobretudo baseadas em variáveis estruturais (tamanho, densidade, composição e distribuição por campos relacionais) e funcionais (tipos de suporte social e reciprocidade), destacando-se ainda os níveis de participação comunitária para a sua configuração (Guadalupe & Vicente, 2019). Neste contexto, importa assinalar que a restrição nas redes está sobretudo associada ao confinamento às relações familiares mais próximas, sendo que o alargamento e a diversificação nas redes se encontram relacionados a um maior envolvimento dos idosos nas organizações da comunidade e em atividades desenvolvidas em contextos que fomentam vínculos extrafamiliares.

Quadro 1. *Tipologias de redes sociais de pessoas idosas*

Mugford & Kendig (1986) Austrália	Wenger (1989; 1991) Norte de Gales (Reino Unido)	Litwin (1995a; 1995b) Israel
Rede atenuada Rede intensa Rede difusa Rede complexa Rede equilibrada	Rede de suporte dependente da família Rede de suporte integrada localmente Rede de suporte local autocontida Rede de suporte focada na comunidade alargada Rede de suporte restrita e privada	Rede de parentela Rede familiar intensa Rede focada nos amigos Rede de laços difusos
Auslander (1996) **Israel**	**Litwin (1997)** **Israel**	**Litwin (2001)** **Israel**
Rede de suporte Rede de substituição Rede tradicional	Rede de suporte diversificada Rede suporte de amizade e vizinhança Rede familiar estreita Rede atenuada Rede de família religiosa Rede de família extensa tradicional	Rede diversificada Rede de amizade Rede de vizinhança Rede familiar Rede restrita
Litwin & Shiovitz-Ezra (2010), USA	**Melkas & Jylhä (1996)** **Finlândia**	**Fiori, Antonucci & Cortina (2006)** **Estados Unidos da América**
Rede diversificada Rede de amizade Rede religiosa Rede familiar Rede restrita	Rede dotada Rede percecionada Rede proactiva Rede familiar intensiva Rede defeituosa	Rede restrita não familiar Rede sem amigos Rede familiar Rede diversa Rede de amigos
Fiori, Smith & Antonucci (2007) **Alemanha**	**Cheng, Lee, Chan, Leung & Lee (2009)** **China**	**(Doubova, Pérez-Cuevas, Espinosa-Alarcón & Flores-Hernández, 2010)** **México**

Rede diversa com suporte Rede focada na família Rede focada na amizade com suporte Rede focada na amizade sem suporte Rede restrita sem amigos insatisfatória Rede restrita sem família e sem suporte	Rede diversa Rede focada na amizade Rede restrita Rede focada na família Rede de família afastada	Rede diversa com participação comunitária Rede diversa sem participação comunitária Rede de viuvez Rede restrita sem amigos Rede restrita sem familiares
Cabral, Ferreira, Silva, Jerónimo & Marques (2013) Portugal	**Burholt & Dobbs (2014) Reino Unido, Índia e Bangladesh**	**Park, Smith & Dunkle (2014) Coreia do Sul**
Rede pequena e predominantemente familiar Rede pequena e predominantemente não-familiar Rede grande e predominantemente familiar Rede grande e predominantemente não-familiar	Agregado multigeracional idoso Agregado multigeracional jovem Rede integrada de família e amigos Rede restrita extra-familiar	Rede restritas Rede conjugal Rede de amizade Rede diversa
Park, Jang, Lee, Ko, Haley & Chiriboga (2015) Estados Unidos da América	**Li & Zhang (2015) China**	**Guadalupe & Vicente (2020) Portugal**
Rede diversa Rede diversa de solteiros Rede de casados Rede familiar Rede restrita de solteiros Rede restrita	Rede diversa Rede de amizade Rede familiar Rede restrita	Rede familiar Rede de amizade Rede de vizinhança Rede institucional

Fontes: Guadalupe (2017, p. 168); Guadalupe & Vicente (2019).

Um estudo recente sobre redes sociais pessoais com idosos portugueses revela quatro tipos de rede: redes familiares, redes de amizade, redes de vizinhança e redes institucionais (Guadalupe, 2017; Guadalupe & Vicente, 2020), numa tipologia tridimensional baseada nas características estruturais, funcionais e relacionais-contextuais da rede. As redes familiares são claramente dominantes, constituídas quase exclusivamente por laços familiares, sublinhando o carácter familista das redes em Portugal nesta população (Guadalupe, 2017; Portugal, 2011, 2014). Seguem-se os outros tipos de rede, todos menos frequentes do que as familiares, e onde os laços com a família também se destacam na composição das redes: as redes de amizade (maiores e mais diversas do que as familiares), as redes de vizinhança (pequenas, mas diversas, com contactos mais frequentes e maior proximidade) e as redes de relações institucionais (menos frequentes, dominadas por laços formais). Os perfis sociográficos revelam que as redes familiares são mais provavelmente detidas por indivíduos *middle-old, casados* ou viúvos e com filhos. As redes de amizade e de vizinhança por sujeitos *young-old* com diferentes estados civis, muitos vivendo sós, havendo maior proporção de homens com redes de amizade. As redes institucionais são sobretudo de sujeitos *old-old* viúvos ou solteiros e sem filhos (Guadalupe, 2017; Guadalupe & Vicente, 2020).

Em Portugal, a tendência familista nas redes sociais (Portugal, 2011) revelada pela sua estruturação em torno dos laços de parentesco, parece dever-se à perspetiva de perenidade destas relações face a outro tipo de relações mais suscetíveis a flutuações e à erosão relacional, sendo encarados como "âncoras instrumentais e afetivas" (Portugal, 2014, p. 208). No entanto, torna-se relevante sublinhar que mesmo noutros contextos socioculturais (distantes do Sul da Europa) a família assume um lugar de destaque, sendo que a maior parte das propostas de tipologia constantes no Quadro 1 incluem um ou mais tipos focados na família, sendo que o peso das relações familiares configura, em múltiplos casos, um fator determinante na construção tipológica.

As redes ancoram indubitavelmente nos laços familiares, mas a diversificação relacional proporciona uma menor densidade nas interconexões e trocas relacionais de recursos mais plurais, o que se associa a redes mais efetivas no suporte e no favorecimento do bem-estar (Sluzki, 1996). Quando os laços familiares se fragilizam, particularmente os laços de conjugalidade e de parentalidade, ora na viuvez ora na ausência ou dispersão da descendência, a compensação surge frequentemente pela via dos laços de amizade, pelas relações de vizinhança ou pelo suporte da rede secundária através de respostas sociais.

5. As redes, as famílias e os desafios na intervenção social

A avaliação do suporte social e o mapeamento da rede são fundamentais na intervenção social com pessoas idosas (Arias, 2009), permitindo antecipar trajetórias de necessidade de apoio e a possibilidade de mobilização de recursos informais. Apesar do traço familista nas redes, consideramos que esta avaliação do suporte disponível e do suporte potencialmente ativável deve ir "além da família" (Dabas, 2006), sob pena de os profissionais reproduzirem o familismo das redes e agudizarem a responsabilidade da provisão familiar, sobrecarregando paradoxalmente as fontes de suporte informais (Guadalupe, 2012). A avaliação diagnóstica das redes deve focar a rede atual, a rede desejada, assim como avaliar retrospetivamente a rede do passado (mais ou menos recente) para identificar vínculos desativados (Nowak, 2001), incluindo as funções genéricas do apoio social, isto é, o apoio emocional, tangível (material ou instrumental) e informativo, mas também funções específicas relevantes para a população idosa, como a companhia social, sendo recomendada a sua avaliação antes e após uma intervenção social (Guadalupe, 2016).

Há vantagens em que a avaliação se baseie na evidência. As tipologias, por exemplo, agregam características das redes a perfis sociodemográficos, constituindo-se como preditoras de trajetórias, possibilitando a prevenção secundária de consequências negativas que afastam os sujeitos de um envelhecimento ativo e salutar. A construção e manutenção de relacionamentos e o apoio social são tidos como determinantes sociais do envelhecimento saudável e ativo (OMS, 2015; UN, 2002). Assim, o conhecimento acerca dos pontos fortes e das fragilidades dos tipos de rede social dos idosos constitui um indicador relevante para o desenho de estratégias, de programas de intervenção social e de medidas de política social (Burholt & Dobbs, 2014; Thiyagarajan, Prince & Webber, 2014).

As intervenções em rede centram-se fundamentalmente em dois eixos: criar e potenciar redes, fomentando novos vínculos ou ativando recursos nos vínculos arrefecidos (Guadalupe, 2016). O primeiro eixo tende a ser mais relevante para casos de isolamento social, sendo que o segundo se constitui como orientador de situações de necessidade de suporte que podem ser respondidas pela própria rede. Mas é fulcral que as redes não sejam meramente equacionadas de forma utilitarista, pois estas também promovem o reconhecimento identitário dos sujeitos com interações imateriais, pelo que a adoção de uma abordagem planeada às redes pode contribuir para aumentar o capital social dos indivíduos e das comunidades (Payne, 2014).

Noutra perspetiva, tendo em consideração a proposta de Cochran (1991) para o desenvolvimento da rede, podemos equacionar estratégias de intervenção focadas em dois âmbitos: 1) aumentar a "*pool*" de potenciais membros, sendo esta influenciada pelos imperativos socioculturais, posição social estrutural do indivíduo, políticas públicas, ecologia da vizinhança e do trabalho, e ideologias pessoais e familiares; 2) fomentar a iniciativa pessoal dos idosos, considerando tanto as experiências educativas, as competências sociocognitivas e as características de personalidade, como o tempo, energia e capacidade disponíveis e as pressões ambientais e desenvolvimentais. As primeiras seriam relevantes nos casos em que se impõem constrangimentos exógenos à iniciativa dos sujeitos; as segundas destacam-se nos casos em que existem potenciais membros disponíveis, mas as limitações se situam no campo intraindividual, podendo concorrer outras barreiras contextuais.

É mais provável que a intervenção ocorra em situações de idosos sem descendência e que vivem sós, pois estes são indicadores de procura de serviços (Albertini & Mencarini, 2011; Bazo, 2008; Brandt & Deindl, 2016; Guadalupe, 2017; Shanas, 1979; Schnettler & Wöhler, 2015; Vikström et al., 2011; Wenger, 2009). A população sem filhos constitui uma minoria cada vez mais alargada, a par da crescente representatividade dos idosos que vivem sozinhos (Delgado & Wall, 2014), sendo a ocorrência simultânea de ambas as circunstâncias cada vez mais provável. Se acrescentarmos condições socioeconómicas desfavoráveis, fragilização da saúde e perdas na autonomia, temos situações-problema que exigem respostas de provisão informal e formal.

Considerações finais

A opção prioritária dos interventores recai quase sempre na família e não nos laços extrafamiliares, onde reconhecemos mais limitações à ativação de laços, inclusive legais. Mas a tendência familista entre os interventores ofusca, por vezes, outras possibilidades interventivas. Esta ampliação deveria ser incentivada, tendo em consideração a diversificação de recursos que encerra, tanto pela intervenção social como por políticas sociais e culturais que promovessem o acesso a contextos favorecedores de construção de novos vínculos, fomentando relações de amizade, de vizinhança e comunitárias.

O alargamento da rede pela via dos laços formais faz-se frequentemente pela existência de dimensões deficitárias na rede informal (Krout, 1985), quer seja pelo isolamento, por lacunas no suporte disponível e efetivo, ou pela incapacidade ou impossibilidade de resposta às necessidades de suporte por parte da rede informal. Aqui também se identifica o desafio de pensar respostas plurais, nomeadamente que assumam formas de apoio direto às redes de suporte informal (laços familiares e extrafamiliares) para que estas assegurem funções equiparadas às assumidas pelas instituições e financiadas pelas políticas públicas. Se o suporte informal responde de forma personalizada às necessidades específicas de cada individuo, as respostas sociais formais são pouco flexíveis e pouco diversas. Neste sentido, é um desafio permanente (re)pensar e equacionar novos serviços que se ajustem e apoiem as situações efetiva e atempadamente (Bloom & Monro, 2015).

Um dos desafios é ultrapassar a representação negativa de "fim de linha" que têm as respostas residenciais (Daniel et al., 2019). Para além de assegurar aos idosos a compensação instrumental e funcional às necessidades da vida quotidiana, deveriam apostar na intensificação de um trabalho de construção relacional. Alley e colaboradores (2007) propõem a criação de comunidades de amigos idosos, com serviços e condições de apoio semelhantes às respostas residenciais existentes, em alternativa a estas, havendo outras possibilidades emergentes para respostas sociais que combinem o apoio formal e informal. Repensar e diversificar as respostas sociais (particularmente as de permanência quotidiana), na sua estruturação e funcionamento, como contextos potenciadores para explorar e para incorporar novos vínculos informais significativos (Sande, Dornell & Aguirre, 2011), representa a aposta no desenvolvimento de novos nichos interpessoais favorecedores de bem-estar para a pessoa idosa.

Os desafios para a intervenção social são assim muitos e múltiplos, focados na criação, recriação, manutenção, ampliação e reforço dos laços sociais.

Bibliografia

Aboim, S., Vasconcelos, P. & Wall, K. (2013). Support, social networks and the family in Portugal: two decades of research, *International Review of Sociology, 23*(1), 47-67, http://dx.doi.org/10.1080/03906701.2013.771050

Aguirre, R. (2008). El Futuro del cuidado. *In* I. Arriagada (ed.), *Futuro para las Familias y Desafíos para las Políticas*. Serie Seminarios y Conferencias. Santiago de Chile: CEPAL.

Alarcão, M. (2015). Família e sistemas envolventes. *In* O. M. Fernandes & C. Maia (coord.). *A família portuguesa no século XXI* (pp.121-132). Lisboa: Parsifal.

Albertini, M. & Mencarini, L. (2011). Childlessness and support networks in later life: a new public welfare demand? Evidence from Italy. *Carlo Alberto Notebooks, 200*. www.carloalberto.org/working_papers

Alley, D., Liebig, P., Pynoos, J., Banerjee, T. & Choi, I.H. (2007). Creating Elder-Friendly Communities: preparations for an aging society. *Journal of Gerontological Social Work, 49*(1-2),1-18, https://doi.org/10.1300/J083v49n01_01

Andrade, F.F. (2009). Desfamiliarização das políticas sociais na América Latina: uma breve análise dos sistemas de proteção social na região. *Barbarói, 31*, 56-71. http://dx.doi.org/10.17058/barbaroi.v2i31.945

Antonucci, T.C. & Akiyama, H. (1987). Social networks in adult life and a preliminary examination of the convoy model. *Journal of Gerontology, 42*(5), 519-527. 10.1093/geronj/42.5.519

Arias, C.J. (2009). La red de apoyo social en la vejez. Aportes para su evaluación. *Revista de Psicologia da IMED, 1*(1), 147-158. https://seer.imed.edu.br/index.php/revistapsico/article/view/20/19

Attias-Donfut, C. & Ogg, J. (2009). Évolution des transferts intergénérationnels: vers un modèle européen?. *Retraite et société, 58*(2), 11-29. https://www.cairn.info/revue-retraite-et-societe1-2009-2-page-11.htm

Attias-Donfut, C. Lapierre, N. & Segalen, M. (2002). *Le nouvel esprit de famille.* Paris: Odile Jacob.

Auslander, G. (1996). The Interpersonal Milieu of Elderly People in Jerusalem. In H. Litwin (editor). *The Social Networks of Older People: A Cross-National Analysis* (pp. 77-97). London: Praeger.

Balandier, G. (1986). *Sens et puissance.* Paris: PUF.

Bazo, M.T. (2008). Personas mayores y solidaridad familiar. *Política y Sociedad, 45*(2), 73-85. https://core.ac.uk/download/pdf/38818777.pdf

Bengtson, V. L., Lowenstein, A., Putney, N. M. & Gans, D. (2003). Global aging and the challenge to families. In V. L. Bengtson & A. Lowenstein (Eds.). *Global aging and challenges to families* (pp. 1-24). New York: Aldine de Gruyter.

Bengtson, V., Rosenthal, C., & Burton, L. (1990). Families and ageing: diversity and heterogeneity. In R. H. Binstock & L. K. George (eds.) *Handbook of aging and the social sciences* (3.ª edição, pp. 263-287). San Diego: Academic Press.

Berkman, L.F. & Syme, S.L. (1979). Social networks, host resistance, and mortality: a nine-year follow-up study of Alameda County residents. *American Journal of Epidemiology, 109*, 186-204, 10.1093/oxfordjournals.aje.a112674

Bloom, M. & Monro, A. (2015). Social Work and the aging family. *The Family Coordinator, 21*(1), 103-115, 10.2307/581791

Brandt, M. & Deindl, C. (2016). Support Networks of Childless Older People in Europe: An Analysis with the Data of the Survey of Health, Ageing and Retirement in Europe (SHARE). http://paa2014.princeton.edu/papers/143298

Bris, H. J-L. (1994). *Responsabilidade familiar pelos dependentes idosos nos países das comunidades europeias*. Lisboa: Conselho Económico e Social. Retrieved from: http://www.ces.pt/download/600/RespFamDep Idosos.pdf

Burholt, V. & Dobbs, C. (2014). A support network typology for application in older populations with a preponderance of multigenerational households. *Ageing & Society, 34*, 1142-1169, 10.1017/S0144686X12001511

Cabral, M.V., Ferreira, P.M., Silva, P.A., Jerónimo, P. & Marques, T. (2013). *Processos de envelhecimento em Portugal – Usos do tempo, redes sociais e condições de vida*. Lisboa: Fundação Francisco Manuel dos Santos. https://www.ffms.pt/FileDownload/b45aa8e7-d89b-4625-ba91-6a6f73f4ecb3/processos-de-envelhecimento-em-portugal

Campos, M.S. & Mioto, R.C.T. (2003). Política de Assistência Social e a posição da família na política social brasileira. *Ser Social, 12*, 165-190, https://doi.org/10.26512/ser_social.v0i12.12932

Capucha, L. (2014). Envelhecimento e políticas sociais em tempos de crise. *Sociologia, Problemas e Práticas, 74*, 113-131. 10.7458/spp2014743203

Carstensen, L.L. (1992). Social and emotional patterns in adulthood: Support for socioemotional selectivity theory. *Psychology and Aging, 7*, 331-338, 10.1037//0882-7974.7.3.331

Carstensen, L.L., Isaacowitz, D.M. & Charles, S.T. (1999). Taking time seriously a theory of socioemotional selectivity. *American Psychologist, 54*(3), 165--18, 10.1037//0003-066x.54.3.165

Carvalho, M.C.B. (2010). Famílias e políticas públicas. *In* A.R. Acosta e M.A.F. Vitale (orgs.). *Família – Redes, laços e políticas públicas* (pp. 267-275, 5.ª edição). São Paulo: CEDPE, PUC-SP: Cortez.

Castel, R. (2003 [1995]). *As metamorfoses da questão social: uma crônica do salário* (4.ª edição). Pertópolis: Vozes.

Cochran, M. (1991). Personal social networks as a focus of support. In D. G. Unger & D. R. Powell (Eds.), *Families as nurturing systems: Support across the life span* (pp. 45-67). New York: The Haworth Press.

Conselho Económico e Social (2013). Parecer de Iniciativa sobre as consequências económicas, sociais e organizacionais decorrentes do envelhecimento da população. Lisboa, Portugal. http://www.ces.pt/download/1335/FINAL_Parecer%20Envelhecimento_aprovado%20em%20Plenario.pdf

Cornwell, B., Laumann, E.O. & Schumm, L.F. (2008). The Social connectedness of older adults: A national profile. *American Sociological Review, 73*(2), 185-203. 10.1177/000312240807300201

Costa, A.B. (2004). *Exclusões sociais*. Lisboa: Gradiva.

Daatland, S.O. & Lowenstein, A. (2005). Intergenerational solidarity and the Family-Welfare State balance. *European Journal of Ageing, 2*, 174-182, 10.1007/s10433-005-0001-1

Dabas, E. (2006). Viviendo redes. *In* E. Dabas (comp.). *Viviendo redes – experiencias y estratégias para fortalecer la trama social* (pp. 23-33). Buenos Aires: FUNDARED, CICCUS.

Daniel, F. C., Brites, A. P., Monteiro, R., & Vicente, H. T. (2019). *De "lar" abominado a estimado (ou tolerado): reconfiguração das representações sobre institucionalização*. Saúde e Sociedade, 28(4), 214-228. http://dx.doi.org/10.1590/s0104-12902019180699

Daniel, F., Ribeiro, A.M. & Guadalupe, S. (2011). Recursos sociais na velhice: um estudo sobre as redes sociais de idosos beneficiários de apoio domiciliário. *In* A.D. Carvalho (coord.), *Solidão e solidariedade: entre os laços e as fracturas sociais* (pp. 73-85). Porto: Edições Afrontamento.

De Martino, M.S. (2001). Políticas sociales y família: estado de bienestar y neoliberalismo familiarista. *Revista Fronteras*, 04, 103-144.

Delgado, A. & Wall, K. (coord.) (2014). *Famílias nos Censos 2011 Diversidade e Mudança*. Lisboa: Instituto Nacional de Estatística e Imprensa de Ciências Sociais. https://www.cig.gov.pt/siic/pdf/2015/FamiliasCensos2011_a.pdf

Doubova, S.V., Pérez-Cuevas, R., Espinosa-Alarcón, P., Flores-Hernández, S. (2010). Social network types and functional dependency in older adults in Mexico. *BMC Public Health, 10*, 104 http://www.biomedcentral.com/1471-2458/10/104

Elias, N. (2001). *A solidão dos moribundos*. Rio de Janeiro: Zahar.

Erera, P.I. (2002). *Family Diversity. Continuity and change in the contemporary family*. Thousand Oaks: Sage Publications.

Esping-Andersen, G. (1999). *Social Foundations of Postindustrial Economies*. Oxford: Oxford University Press.

Fernandes, A.A. (2001). Velhice, solidariedades familiares e política social: itinerário de pesquisa em torno do aumento da esperança de vida. *Sociologia, Problemas e Práticas, 36*, 39-52. http://www.scielo.mec.pt/scielo.php?script=sci_arttext&pid=S0873-65292001000200003

Ferreira, V. & Monteiro, R. (2015). Austeridade, emprego e regime de bem-estar social em Portugal: em processo de refamilização? *Ex aequo, 32*, 49-67, http://www.scielo.mec.pt/scielo.php?script=sci_abstract&pid=S0874--55602015000200005&lng=pt&nrm=iso

Ferrera, M. (2000). *O futuro da Europa social: repensar o trabalho e a protecção social na nova economia*. Lisboa: Celta, Presidência Portuguesa da União Europeia.

Field, D. & Minkler, M. (1988). Continuity and change in social support between young-old and old-old or very-old age. *Journal of Gerontology: Psychological Sciences, 43*(4), 100-106. 10.1093/geronj/43.4.p100

Figueiredo, D. (2007). *Cuidados familiares ao idoso dependente*. Lisboa: Climepsi Editores.

Fiori, K.L., Antonucci, T.C. & Cortina, K.S. (2006). Social Network Typologies and Mental Health Among Older Adults. *Journal of Gerontology, 61B*(1), 25–32. https://doi.org/10.1093/geronb/61.1.P25

Fiori, K.L., Smith, J., & Antonucci, T.C. (2007). Social network types among older adults: a multidimensional approach. *J Gerontol B Psychol Sci Soc Sci., 62*(6), 322-30. 10.1093/geronb/62.6.p322

Fonseca, A.M. (2011). *Reforma e reformados*. Coimbra: Almedina.

Frade, C. & Coelho, L. (2015). Surviving the crisis and austerity: The coping strategies of Portuguese households. *Indiana Journal of Global Legal Studies, 22*(2), 631-664. https://www.jstor.org/stable/10.2979/indjglolegstu.22.2.631

Franzoni, J.M. (2007). *Regímenes del bienestar en América Latina*. Madrid: Fundación Carolina. Retrieved from: http://www.fundacioncarolina.es/wp-content/uploads/2014/08/DT11.pdf

Giannella, E. & Fischer, C.S. (2016). An inductive typology of egocentric networks. *Social Networks*, 47, 15-23. 10.1016/j.socnet.2016.02.003

Gil, A.P.M. (2009). Conciliação entre vida Profissional e vida familiar: o caso da dependência. Lisboa: Núcleo de Estudos e Conhecimento, Instituto de Segurança Social. Retrieved from: http://www.seg-social.pt/documents/10152/135827/conciliacao_vida_profissional_familiar/2d308149-a66d-4075-bbaa-2eb95869c677

Gil, A.P.M. (2010). *Heróis do quotidiano: dinâmicas familiares na dependência*. Lisboa: Fundação Calouste Gulbenkian, Fundação para a Ciência e Tecnologia.

Ginn, J. & Arber, S. (1996). «Mera conexión». Relaciones de género y envejecimiento. *In* S. Arber & J. Ginn (org.). *Relación entre género y envejecimiento. Enfoque sociológico* (pp. 17-34). Madrid: Narcea.

Grundy, E. (2005). Reciprocity in relationships: socio-economic and health influences on intergenerational exchanges between Third Age parents and their adult children in Great Britain. *British Journal of Sociology. 56*(2), 233-55. 10.1111/j.1468-4446.2005.00057.x

Grundy, E. & Tomassini, C. (2003). El apoyo familiar de las personas de edad, en Europa: contrastes e implicaciones. *Notas de Población, 77*, 219-250.

Guadalupe, S. (2016). *Intervenção em rede: Serviço Social, sistémica e redes de suporte social* (2.ª edição). Coimbra: Imprensa da Universidade de Coimbra. http://dx.doi.org/10.14195/978-989-26-0866-2

Guadalupe, S. (2017). *As redes de suporte social informal em Serviço Social: as redes sociais pessoais de idosos portugueses nos processos de avaliação diagnóstica em respostas sociais* [Tese de Doutoramento em Serviço Social]

Lisboa: ISCTE – Instituto Universitário de Lisboa, Escola de Sociologia e Políticas Públicas e CIES, Centro de Investigação e Estudos de Sociologia. https://www.iscte-iul.pt/tese/7228

Guadalupe, S. & Cardoso, J. (2018). As redes de suporte social informal como fontes de provisão social em Portugal: o caso da população idosa. Sociedade & Estado, 33(1), 213-248. https://doi.org/10.1590/s0102-699220183301009

Guadalupe, S. & Vicente, H.T. (2020). Types of personal social networks of older adults in Portugal. *Social Indicators Research*, 1-22, https://doi.org/10.1007/s11205-019-02252-3

Guadalupe, S. & Vicente, H.T. (2019). Social network typologies of older people: A cross-national literature review. *Ciência & Saúde Coletiva* (ahead of print) Retrieved from: http://www.cienciaesaudecoletiva.com.br/artigos/social-network-typologies-of-older-people-a-crossnational-literature-review/17458

Guadalupe, S., Vicente, H.T., & Daniel, F. (2019). Características das redes sociais de pessoas idosas em Portugal. *Redes, Revista Hispana para el análisis de redes sociales, 30*(2), 199-215. https://doi.org/10.5565/rev/redes.816

Ham-Chande, R., Ybáñez-Zepeda, E. & Torres-Martínez, A.L. (2003). Redes de apoyo y arreglos de domicilio de las personas en edades avanzadas en la Ciudad de México. *Notas de Población, 77*, 71-102.

Hespanha, P. (1993). Vers une société-providence simultanément pré- et post-moderne. *Oficina do Centro de Estudos Sociais, 38*. Coimbra: Centro de Estudos Sociais.

Hespanha, P., Ferreira, S., & Pacheco, V. (2013). O Estado social, crise e reformas. *In* Observatório das Crises e das Alternativas (org.). *Anatomia da crise: identificar os problemas para construir alternativas* (pp. 161-249). Coimbra: Centro de Estudos Sociais.

Krout, J.A. (1985). Relationships between informal and formal organizational networks. *In* W.J. Sauer & R.T. Coward (eds.). *Social support networks and the care of the elderly* (pp. 178-195). New York: Springer.

Kuyper, J. & Bengtson, V. (1973). Social breakdown and competence. A model of normal aging. *Human Development. 16*(3), 181-201. 10.1159/000271275

Leandro, M.E. (2011). Laços familiares em questão: antinomias nas sociedades hipermodernas. *In* M.E. Leandro (coord.). *Laços familiares e sociais* (pp. 95-115). Viseu: Psicosoma.

Leandro, M.E. & Ferreira, L. (2011). Os laços sociais em questão. Metamorfoses sociais, metamorfoses de uma noção. *In* M.E. Leandro (coord.), *Laços Familiares e sociais* (pp. 27-57). Viseu: Psicossoma.

Lesemann, F. & Martin, C. (1995). Estado, comunidade e família face à dependência dos idosos. Ao encontro de um "Welfare-Mix". *Sociologia – Problemas e Práticas, 17*, 115-139.

Li, T. & Zhang, Y. (2015). Social network types and the health of older adults: Exploring reciprocal associations. *Social Science & Medicine, 30*(130), 59-68, https://doi.org/10.1016/j.socscimed.2015.02.007

Litwin, H. (1995a). *Unprooted in old age: soviet Jews and their social networks in Israel.* Westport, CT: Praeger.

Litwin, H. (1995b). The social networks of elderly immigrants: An analytic typology. *Journal of Aging Studies, 9*(2), 155-174. 10.1016/0890-4065(95)90009-8

Litwin, H. (1997). The network shifts of elderly immigrants: The case of Soviet Jews in Israel. *Journal of CrossCultural Gerontology, 12*, 45-60. https://link.springer.com/article/10.1023/A:1006593025061

Litwin, H. (2001). Social network type and morale in old age. *The Gerontologist, 41*(4), 516-24. 10.1093/geront/41.4.516 PMID: 11490050

Litwin, H. & Landau, R. (2000). Social network type and social support among the old-old. *Journal of Aging Studies, 14*(2), 213-228. 10.1016/S0890-4065(00)80012-2

Litwin, H. & Shiovitz-Ezra S. (2011). Social network type and subjective well--being in a national sample of older Americans. *Gerontologist, 51*, 379-88. 10.1093/geront/gnq094 PMID: 21097553

Marques, J. (2008). A reconfiguração do estado-providência. *Gestão e Desenvolvimento, 15-16*, 105-119. http://z3950.crb.ucp.pt/Biblioteca/GestaoDesenv/GD15_16/gestaodesenvolvimento15_16_105.pdf

Martin, C. (1995). Os limites da protecção da família – introdução a uma discussão sobre as novas solidariedades na relação Família-Estado.

Revista Crítica de Ciências Sociais, 42, 53-76. https://ces.uc.pt/rccs/rccs.php?id=556&id_lingua=1

McPherson, M. Smith-Lovin, L. & Brashears, M.E. (2006). Social isolation in America: changes in core discussion networks over two decades. American Sociological Review, *71*(3), 353-375. https://www.jstor.org/stable/30038995?seq=1

Melkas, T. & Jylhä, M. (1996). Social network characteristics and social network types among Eldery People in Finland. *In* H. Litwin (Ed.), *The social network of older people: a cross national analysis* (pp. 99-116). Westport, CT: Praeger.

Mioto, R.C.T. (2008). Família e políticas sociais. *In* I. Boschetti, E. Behring, R.C.T. Mioto & S.M.M. Santos (Orgs.). *Política social no capitalismo: tendências contemporâneas* (pp.130-148). São Paulo: Cortez.

Moral, J.C.M., Miguel, J.M.T. & Pardo, E.N. (2007). Análisis de las redes sociales en la vejez através de la entrevista Manheim. *Salud Pública de México, 49*(6), 408-414. http://www.scielo.org.mx/scielo.php?script=sci_arttext&pid=S0036-36342007000600007

Mugford, S. & Kendig, H. (1986). Social relations: Networks and ties. *In* H. Kendig (ed.). Ageing and families: A social networks perspective (pp. 38-59). Sydney: Allen and Unwin.

Nowak, J. (2001). O trabalho social de rede – a aplicação das redes sociais no trabalho social. *In* H. Mouro & D. Simões. *100 Anos de Serviço Social* (pp. 149-184). Coimbra: Quarteto.

Nunes, R. (2015). Ética e família. *In* O.M. Fernandes e C. Maia (coord.). *A Família portuguesa no século XXI* (pp. 39-50). Lisboa: Parfisal.

OCDE/OECD – Organização para a Cooperação e Desenvolvimento Económico (2012). *The Future of Families to 2030*. OECD Publishing. 10.1787/9789264168367-en

Olsen, O., Iversen, L. & Sabroe, S. (1991). Age and the operationalization of social support. *Social Science & Medicine, 32*, 767-771. 10.1016/0277-9536(91)90302-s

OMS – Organização Mundial de Saúde (2015). *Relatório mundial de envelhecimento e saúde*. Genebra: OMS. https://apps.who.int/iris/bitstream/handle/10665/186468/WHO_FWC_ALC_15.01_por.pdf;jsessionid=FEADA8F7E54C5289669E5D5BE95FD5E4?sequence=6

Park, N.S., Jang, Y., Lee, B.S., Ko, J.E., Haley, W.E. & Chiriboga, D.A. (2015). An Empirical Typology of Social Networks and Its Association With Physical and Mental Health: A Study With Older Korean Immigrants. *The Journals of Gerontology Series B: Psychological Sciences and Social Sciences, 70*(1), 67-76. 10.1093/geronb/gbt065

Park, S., Smith, J. & Dunkle, R E. (2014). Social network types and well-being among South Korean older adults. *Aging & Mental Health, 18*(1), 72-80. 10.1080/13607863.2013.801064

Paúl, C. (2005a). A construção de um modelo de envelhecimento humano. *In* Paúl, C. & Fonseca, A. (coords.), *Envelhecer em Portugal: Psicologia, saúde e prestação de cuidados* (pp. 15-41). Lisboa: Climepsi Editores.

Paúl, C. (2005b). Envelhecimento activo e redes de suporte social. *Sociologia, 15*, 275-287. https://ojs.letras.up.pt/index.php/Sociologia/article/view/2392

Payne, M. (2014). Redes sociais em Serviço Social. *In* M.I. Carvalho & C. Pinto (coords.), *Serviço Social: Teorias e práticas* (pp. 181-204). Lisboa: Pactor.

Pedroso, J. & Branco, P. (2008). Mudam-se os tempos, muda-se a família. As mutações do acesso ao direito e à justiça de família e das crianças em Portugal. *Revista Crítica de Ciências Sociais*, 82, 53-83. https://journals.openedition.org/rccs/619

Pereira, P.A.P. (2006). Mudanças estruturais, política social e papel da família: crítica ao pluralismo de bem-estar. *In* A. Mione, M.C. Matos & M.C. Leal (Orgs). *Política social, família e juventude: uma questão de direitos* (pp. 25-42, 2.ª edição). São Paulo: Cortez.

Phillipson, C., Bernard, M., Phillips, J. & Ogg, J. (2001). *The family and community life of older people. Social networks and social support in three urban areas*. London and New York: Routledge.

Pimentel, L. (2012). Cuidar de pessoas idosas dependentes: as interseções entre a esfera pública e a esfera privada. *Rediteia, 45*, 67-77.

Pimentel, L. (2015). As pessoas idosas e os seus contextos familiares: convitea um olhar diferente. *In* O.M. Fernandes & C. Maia (coord.). *A Família portuguesa no século XXI* (pp. 171-178). Lisboa: Parfisal.

Portugal, S. (2011). Dádiva, família e redes sociais. *In* S. Portugal & P.H. Martins (org.). *Cidadania, políticas públicas e redes sociais* (pp. 39-53). Coimbra: Imprensa da Universidade de Coimbra.

Portugal, S. (2014). *Famílias e redes sociais. Ligações fortes na produção de bem-estar.* Coimbra: Almedina.

Rioseco, H.R., Quezada, V.M., Ducci, V.M.E. & Torres, H.M. (2008) Cambio en las redes sociales de adultos mayores beneficiarios de programas de vivienda social en Chile. *Revista Panamericana de Salud Pública, 23*(3), 147--53.

Sande, S., Dornell, T. & Aguirre, M. (2011). Las redes como estrategias en los procesos de intervención ético-política en la vejez. *In Actas III Jornadas Regionales de Trabajo Social*, 1 e 2 de Julho de 2011. Villa María: Universidad Nacional de Villa María. http://biblio.unvm.edu.ar/opac_css/doc_num.php?explnum_id=635

Santos, B.S. (1995). Sociedade-Providência ou autoritarismo social? [editorial]. *Revista Crítica de Ciências Sociais, 42*, i-vii. http://www.boaventuradesousasantos.pt/media/pdfs/Sociedade_Providencia_ou_Autoritarismo_Social_RCCS42.PDF

Saraceno, C. (2010). Social inequalities in facing old-age dependency: a bi--generational perspective. *Journal of European Social Policy, 20*(1), 32–44. 10.1177/0958928709352540

Sarti, C. (2010). Famílias enredadas. *In* A.R. Acosta & M.A.F. Vitale (orgs.). *Família – Redes, laços e políticas públicas* (pp. 22-38, 5.ª edição). São Paulo: CEDPE, PUC-SP, Cortez Editora.

Schnettler, S. e Wöhler, R. (2015). No children in later life, but more and better friends? Substitution mechanisms in the personal and support networks of parents and the childless in Germany. *Ageing and Society, CJO*, 1-25. 10.1017/S0144686X15000197

Schnittker, J. (2007). Look (closely) at all the lonely people: age and the social psychology of social support. *Journal of Aging and Health, 19*, 659-82. https://doi.org/10.1177/0898264307301178

Serapioni, M. (2005). O papel da família e das redes primárias na reestruturação das políticas sociais. *Ciência & Saúde Coletiva, 10(sup)*, 243-253. https://doi.org/10.1590/S1413-81232005000500025

Shanas, E. (1979). The family as a social support system in old age. *The Gerontologist, 19*(2), 169-174. https://doi.org/10.1093/geront/19.2.169

Silva, P.A. (2002). O modelo de welfare da Europa do sul – Reflexões sobre a utilidade do conceito. *Sociologia, Problemas e Práticas. 38*, 25-59. http://www.scielo.mec.pt/pdf/spp/n38/n38a03.pdf

Sluzki, C.E. (1996). *La red social: frontera de la practica sistemica.* Barcelona, Gedisa Editorial.

Sluzki, C.E. (2000). Social network and the elderly: conceptual and clinical issues, and a family consultation. *Family Process, 39*(3), 271-284. 10.1111/j.1545--5300.2000.39302.x

Sousa, L. & Figueiredo, D. (2004). *Services for Supporting Family Carers of Elderly People in Europe: Characteristics, Coverage and Usage.* EUROFAMCARE.

Sousa, L. & Figueiredo, D. (2007). *Supporting family carers of older people in Europe – The national background report for Portugal.* Hamburg: Lit Verlag.

Sousa, L. Figueiredo, D. & Cerqueira, M. (2004). *Envelhecer em família – Os cuidados familiares na velhice.* Porto: Âmbar.

Specht, H. (1986). Social support, social networks, social exchange, and Social Work practice. *Social Service Review, 60*(2), 218-240. https://www.jstor.org/stable/30012339

Thiyagarajan, J.A., Prince, M. & Webber, M. (2014). Social support network typologies and health outcomes of older people in low and middle income countries – A 10/66 Dementia Research Group population--based study. *International Review of Psychiatry, 26*(4): 476-485. 10.3109/09540261.2014.925850

UN – United Nations (2002). *Political Declaration and Madrid International Plan of Action on Ageing*. Second World Assembly on Ageing, Madrid, Spain. New York: UN. http://www.un.org/en/events/pastevents/pdfs/Madrid_plan.pdf

UN – United Nations (2015). *World Population Ageing 2015*. Ney Work: United Nations, Department of Economic and Social Affairs Population Division. http://www.un.org/en/development/desa/population/publications/pdf/ageing/WPA2015_Report.pdf

Valle, J. & Garcia, A. (1994). Redes de apoyo social en usuarios del servicio de ayuda a domicilio de la tercera edad. *Psicothema, 6*(1), 39-47. http://www.psicothema.com/psicothema.asp?id=901

Vasconcelos, P. (2002). Redes de apoio familiar e desigualdade social: estratégias de classe. *Análise Social, XXXVII*(163), 507-544. http://analisesocial.ics.ul.pt/documentos/1218732936N9mRE2wd0Xn17VQ4.pdf

Vasconcelos, P. (2005). Redes sociais de apoio. *In* K. Wall (Org.). *Famílias em Portugal* (pp. 599-631). Lisboa: Imprensa de Ciências Sociais.

Vicente, H.T. (2010). *Família multigeracional e relações intergeracionais: Perspectiva sistémica* [Tese de doutoramento]. Aveiro: Secção Autónoma de Ciências da Saúde, Universidade de Aveiro.

Vicente, H.T. & Sousa, L. (2012). Relações intergeracionais e intrageracionais: A matriz relacional da família multigeracional. *Revista Temática Kairós Gerontologia, 15*(2), 99-117. https://doi.org/10.23925/2176--901X.2012v15iEspecial11p99-117

Vikström, J., Bladh, M., Hammar, M., Marcusson, J., Wressle, E., Sydsjö, G. (2011). The influences of childlessness on the psychological well-being and social network of the oldest old. *BMC Geriatrics, 11*, 78. 10.1186/1471-2318--11-78

Volpi, R. (2007). *La fine della famiglia. La rivoluzione di cui non ci siamo accorti*. Milano: Mondadori.

Wenger, G.C. (1991). A network typology: from theory to practise. *Journal of Aging Studies, 5*(2), 147-162. 10.1016/0890-4065(91)90003-B

Wenger, G.C. (2009). Childlessness at the end of life: evidence from rural Wales. *Ageing & Society, 29*, 1243-1259. 10.1017/S0144686X09008381

Williams, F. (2010). *Repensar as famílias.* Lisboa: Principia.

Yazbeck, M.C. (2001). Pobreza e exclusão social: expressões da questão social. *Temporalis, III*(3), 33-40.

PARTE II

POLÍTICAS E RESPOSTAS SOCIAIS PARA UMA NOVA VELHICE

CAPÍTULO IV

CENTROS DE DIA COMO AGENTES DE ENVELHECIMENTO ATIVO?

Mónica Teixeira

Universidade de Aveiro, Departamento de Educação e Psicologia

ORCID: https://orcid.org/0000-0003-2625-4765

Resumo: *Os Centros de dia são estruturas de semi-institucionalização surgem enquadradas numa política de "envelhecer em casa" e do desenvolvimento do modelo comunitário de intervenção. Os Centros de dia disponibilizam aos utilizadores uma panóplia de serviços, atividades e intervenções em diversas áreas de modo a responder às necessidades e desejos dos idosos e suas famílias.*

O envelhecimento ativo pode definir-se como um processo que tem como objetivo primordial o aumento da qualidade de vida durante o envelhecimento, através da promoção de três níveis: saúde, participação e segurança. O envelhecimento ativo possibilita às pessoas a realização do seu potencial físico, social e participativo e tem em conta a suas diferenças individuais necessidades, desejos e capacidades.

DOI | https://doi.org/10.14195/978-989-26-1788-6_4

O presente capítulo direciona-se na tentativa de integrar conceitos basilares do envelhecimento ativo na intervenção realizada pelos Centros de dia. Destaca-se como pergunta de partida para a análise dos conceitos "Centros de dia como agentes de Envelhecimento Ativo?".

Palavras-chave: *Centros de dia; Envelhecimento ativo; Idosos*

Abstract: *The Day Care Center are structures of semi-institutionalization emerged embedded in a policy of "aging at home" and of the model of community development of intervention. The day care centers provide the users a wide range of services, activities and interventions in many areas in the way to respond to the needs and requests of the elderly and their families.*

Active aging can be defined as a process that has as main objective the increase of quality of life during aging, through the support of three levels: health, involvement and safety. Active aging enables people the fulfilment of their physical capability, social and interactivity and takes into consideration their personal needs, requests and capabilities.

The present chapter is focused on the attempt to integrate basic concepts of active aging in the intervention made by the Day Care Centers. Thereby it stands out as starting question to the analyses of the concepts "Day care Centers as actors of Active Aging?".

Key words: *Day care centers; Active aging; Elderly*

O envelhecimento da população contribuiu para a evolução dos cuidados comunitários de longa duração de modo a conhecer as necessidades de saúde e sociais das pessoas idosas (Baumgarten et al., 2002). O aumento da esperança de vida dos indivíduos, associada à incapacidade, tornou necessário um redimensionamento e uma planificação das políticas públicas para a prestação de cuidados comunitários, uma vez que durante muitos anos os cuidados residenciais terão sido a única alternativa assistencial para idosos com incapacidades (Irigoyen et al., 2002). A implementação de serviços de proximidade também comporta na sua génese a tomada de consciência dos custos excessivos do modelo de internamento definitivo, assim como a ineficiência dessas estruturas de apoio (Pimentel, 2005). Assiste-se deste modo, à integração progressiva da ideologia da "desinstitucionalização" dos cuidados formais (Carvalho, 2006). O modelo comunitário de intervenção emergente defende que as resoluções das dificuldades do indivíduo deverão tendencialmente ser intervencionadas sem que este tenha de ser desenraizado do seu ambiente (Arrazola et al., 2003). Surgem progressivamente novos serviços comunitários suportados pela crise do modelo tradicional e o surgimento das novas políticas direcionadas para o "envelhecer em casa" (Sancho & Rodriguez, 1999); especificamente, o Centro de dia surge como resposta intermédia que veio colmatar a dicotomia existente nos serviços de apoio que, por um lado, se baseavam nos cuidados domiciliários e, por outro lado, nos cuidados residenciais (Arrazola et al., 2003).

O envelhecimento ativo é o processo de otimização das oportunidades de saúde, participação e segurança, tendo como objetivo primordial melhorar a qualidade de vida à medida que as pessoas vão envelhecendo (World Health Organization, 2002). É unânime que a definição de envelhecimento ativo se refere ao processo de otimização do potencial de bem-estar, físico e mental ao longo da vida, a fim de que, na velhice, a pessoa viva de forma ativa e independente (Tamer & Petriz, 2007). Envelhecer ativamente é, segundo Paúl (2005), um processo contínuo que diz respeito a todos, cabendo à sociedade a responsabilidade de desenvolvimento de espaços e de equipamentos sociais diferenciados, seguros e acessíveis, garantindo assim a participação social do idoso.

O presente capítulo pretende problematizar e refletir de que modo os Centros de dia poderão ser agentes de práticas de envelhecimento ativo, a fim de proporcionar aos seus utilizadores mais do que simples serviços básicos assentes sobre rotinas. Este ensaio tenderá a basear-se na experiência prática pretendendo, no entanto, enfatizar uma abordagem para desenvolvimentos futuros. Assim sendo assenta em três pontos: I) Centros de dia e sua caracterização genérica; II) Definição e modelo concetual de envelhecimento ativo; III) Envelhecimento ativo em Centros de dia – reflexão crítica baseada na prática.

Centros de dia-Caraterização

O alicerce dos Centros de dia remonta aos anos 20, na Rússia, onde se deram início aos programas de cuidados diurnos direcionados a portadores de doença mental e cujo objetivo central terá sido encontrar alternativas ao internamento hospitalar (Castiello, 1996). Após a Segunda Guerra Mundial, foi criado o primeiro hospital de dia em Inglaterra com o intuito de perceber quais as necessidades dos idosos (Gaugler et al., 2003). Será na década de 70 que os Centros de dia começam a proliferar pela Europa (Ferrer, 2005) e, nos Estados Unidos, o movimento de "desinstitucionalização" contribui para o crescimento dos cuidados diurnos a idosos (Gaugler & Zarit, 2001).

A delimitação concetual de Centro de dia é complexa, pois existe uma diversidade de modelos de intervenção. Deste modo, misturam-se modelos de intervenção individuais e grupais com modelos de saúde e psicossociais, cuja predominância varia em função do tipo de população a que são dirigidos (Arrazola et al., 2003).

O Centro de dia poderá definir-se como uma estrutura de serviços comunitários (Conrad, & Hughes, 1993) de funcionamento diurno (Cid & Dapia, 2007) dirigido a pessoas idosas. De acordo com Benet (2003), este tipo de estrutura caracteriza-se por ser uma alternativa ao internamento, permitindo assim que o idoso se mantenha o máximo de tempo integrado no seu contexto. Esta resposta multidisciplinar (Manchola, 2000) é direcionada a pessoas que apresentem incapacidades de funcionalidade (Conrad & Hughes, 1993), bem como dificuldades ao nível sócio-assistencial (Manchola, 2000). Esta tipologia de intervenção fomenta a articulação com o cuidador informal através da prestação de apoio aos familiares do cliente (Conrad & Hughes, 1993), proporcionando às famílias o descanso das responsabilidades de cuidar várias horas durante o dia (Gaugler et al., 2003).

A resposta, Centro de dia, tem diversos objetivos entre os quais i) retardar institucionalizações precoces e indesejadas (Castiello, 1996); ii) possibilitar a maximização do nível de recuperação e manutenção da autonomia do idoso de acordo com as suas capacidades individuais (Benet, 2003); iii) prevenir o incremento da dependência através da realização de intervenções reabilitadoras (Castiello, 1996); iv) potenciar as capacidades mentais dos utentes (Irigoyen, 2002); v) promover o companheirismo e a estimulação social (Bilotta et al., 2010); vi) proporcionar a realização de atividades básicas da vida quotidiana favorecendo apoio ao idoso (Castiello, 1996); vii) orientação e aconselhamento dos cuidadores informais (Irigoyen et al., 2002).

As diferenças, na política social e na estrutura reguladora da resposta, podem resultar na organização e produção de programas diferenciados e distintos, decorrentes do modo como a prestação de cuidados se encontra organizada (Jarrott et al., 1998). Assim sendo, as estruturas oferecem serviços de grande variedade, desde as atividades puramente recreativas passando pela intervenção social ou serviços de saúde que englobam a prevenção e a promoção do idoso (Walker et al., 2004). De um modo mais específico, apresentam-se os seguintes tipos de serviços: recreativos, saúde e bem-estar, educacionais, refeições, transportes, artes manuais, programas intergeracionais, grupos de suporte e atividades voluntárias (Martin et al., 2007). É notório que os serviços diurnos para idosos não são apresentados como um conjunto homogéneo de atividades, serviços ou programas, nem tão pouco se destinam à população genérica, existindo uma grande diversidade de acordo com as características da população a ser apoiada (Gaugler & Zarit, 2001).

Salgado e Montalvo (1999) defendem uma categorização tripartida dos serviços prestados pelos Centros de dia, apresentando-os do seguinte modo: serviços básicos, serviços especializados e serviços complementares. Integram os serviços básicos, o transporte dos idosos nas suas diversas deslocações (de e para o domicílio); alimentação e nutrição adequada às necessidades e especificidades individuais; higiene e conforto. Os serviços especiais incluem a realização de atividades diversificadas e simultâneas que integram as áreas da saúde (o acompanhamento a consultas médicas), área ocupacional (os programas de psicomotricidade) e área social (os programas intergeracionais). Por fim, os serviços complementares incluem acesso à biblioteca, podologia e cabeleireiro (Salgado & Montalvo, 1999).

De um modo geral, os Centros de dia proporcionam condições dignas às pessoas idosas dependentes e seus familiares, contribuindo para a manutenção e continuidade do seu modo de vida e maior nível de autonomia (Castiello, 1996). No entanto, as pessoas idosas apresentam-se como um grupo bastante heterogéneo e, como tal, com necessidades distintas, procurando nas instituições respostas diferenciadas e diferenciadoras. Torna-se por isso necessária a implementação de uma multiplicidade de intervenções que permitam aos idosos eleger aquela ou aquelas que melhor se adaptam às suas necessidades individuais (Benet, 2003).

Definição e Modelo Concetual de Envelhecimento Ativo

No pensamento das pessoas, e não só no das pessoas que legislam, mas também na cabeça dos cientistas, clínicos e gerontólogos, existem dois termos que se contradizem: um, habitualmente infere emoções negativas (envelhecimento); o outro, tem um significado positivo ("sucesso", "ótimo", "produtivo", "ativo", entre outros) (Ballesteros, 2008). O envelhecimento ativo pode definir-se como processo que tem como principal fundamento o aumento da qualidade de vida durante o envelhecimento, através da otimização de oportunidades aos diferentes níveis: saúde, participação e segurança (Paúl, 2005). «Ativo» não se refere simplesmente à competência do indivíduo se manter fisicamente ativo ou fazer parte da força de trabalho (World Health Organization, 2002), refere-se igualmente ao envolvimento deste no que diz respeito às questões sociais, económicas e cívicas. Esta conceção implica a substituição das abordagens centradas nas necessidades dos indivíduos por outras que se centram nos direitos das pessoas idosas em todas as dimensões da sua vida (Vallespir & Morey, 2007). Altera, deste modo, a perspetiva baseada na satisfação das necessidades básicas, em que se assume que as pessoas idosas são alvos passivos, para uma ótica que reconhece aos indivíduos direitos a fim de lhes proporcionar igualdade de oportunidade em todos os aspetos da vida e à medida que vão envelhecendo (World Health Organization, 2002). «Ativo» refere-se, segundo Bowling (2008), à contínua participação na sociedade, retendo as capacidades físicas, mentais e sociais para o realizar, mantendo a dignidade, autoeficácia, direitos humanos e ambientes facilitadores de autonomia e independência.

De um modo individual ou grupal, o envelhecimento ativo permite às pessoas realizar o seu potencial físico, social e bem-estar durante a sua trajetória de vida, e a participação na sociedade de acordo com as suas necessidades, desejos e capacidades (World Health Organization, 2002). O envelhecimento ativo inclui, no seu processo, a cultura e o género individuais assim como atribui importância a outros determinantes como sejam os serviços sociais e de saúde, meio físico, caraterísticas comportamentais, sociais e económicas do indivíduo (Paúl, et al., 2005).

O envelhecimento ativo depende de uma diversidade de determinantes entre as pessoas, as famílias e os países em que a World Health Organization (2002) destaca os seguintes: económicos; sociais, pessoais/individuais, ambientais. Resumidamente, os determinantes económicos englobam dois conceitos centrais: o trabalho, através do reconhecimento da necessidade de apoio da contribuição ativa e produtiva dos mais velhos; a proteção social direcionada para os idosos em situação de risco. No que respeita aos determinantes sociais, é consensual que fatores como a solidão, isolamento, a baixa escolaridade e a exposição a situações de violência aumentam os riscos de incapacidade e morte precoce; deste modo, o apoio social, a aprendizagem e a proteção dos indivíduos são fatores essenciais do ambiente social que estimulam a saúde, participação e segurança dos mais velhos. Os determinantes pessoais/individuais evidenciam a importância do sujeito no seu autocuidado, destacando ideias fundamentais como a realização de atividade física adequada, alimentação cuidada, consumo moderado de álcool (entre outros) que contribuem para o atraso do declínio funcional, aumento da longevidade e qualidade de vida; os fatores psicológicos são igualmente incluídos e dos quais se destaca a capacidade cognitiva. Por fim, os determinantes ambientais podem marcar a diferença entre a independência e a dependência das pessoas mais velhas; de na realidade, a insegurança ambiental e as barreiras arquitetónicas são fatores que podem impedir o indivíduo de contactar com os outros, estando, por esse facto, mais propenso ao isolamento.

As medidas de promoção de envelhecimento ativo englobam uma conjugação de estratégias com o objetivo de incrementar a adoção de um papel ativo e determinado, por parte dos idosos, no que respeita à sua própria velhice (Martin et al., 2007). No entanto, sendo o envelhecimento ativo interdependente de uma diversidade de fatores (económicos, sociais e de saúde, comportamentais, ambientais e pessoais), torna-se imperativo a compreensão dos mesmos e das relações que estabelecem entre si neste processo (World Health Organization, 2002).

Os programas e políticas de envelhecimento ativo reconhecem a necessidade de incentivar e equilibrar responsabilidades individuais e pessoais, ambientes favoráveis e solidariedade entre as gerações (World Health Organization, 2002). Estes têm na sua génese razões económicas que conduzem a ações que fomentem o envelhecimento ativo em termos de produtividade e, sobretudo, em termos de gastos sociais e de saúde (Vallespir & Morey, 2007). Quando a saúde, o emprego e mercado de trabalho, a educação e as políticas suportam o envelhecimento ativo haverá potencialmente i)menos mortes prematuras na fase mais produtiva da vida; ii) menos incapacidades e doenças crónicas na velhice; iii) mais pessoas gozam de uma qualidade de vida positiva à medida que vão envelhecendo; iv) mais pessoas participam ativamente, à medida que envelhecem, nos aspetos sociais, culturais, económicos e políticos da sociedade, em papéis pagos e não pagos e na vida doméstica familiar e da comunidade; v) menos custos relacionados com tratamentos médicos e serviços de cuidados (World Health Organization, 2002).

Centros de Dia como Agentes de Envelhecimento Ativo – Uma Abordagem Centrada na Prática

O envelhecimento da população entende-se como um desafio a três dimensões: sociedade, ciência e indivíduo. O aumento da esperança de vida das pessoas não acarretou qualidade de vida, muito pelo contrário, está associada a um aumento progressivo de deficiência em idades mais avançadas tornando-se este um problema para a sociedade. Por outro lado, o desafio coloca-se ao indivíduo na medida em que o envelhecimento individual se torna preocupante, uma vez que o processo é bastante diferenciado entre os indivíduos (uma pessoa mantém-se na sua vida ativa e outra detém enfermidades crónicas que, por vezes, se tornam altamente incapacitantes) (Ballesteros et al., 2005). Parece-nos operacionalizável a promoção de medidas de envelhecimento ativo, quando o idoso mantém preservadas as suas diversas dimensões (física, mental, social...), aspeto que se torna mais complexo à medida que aumentam as incapacidades. De facto, existem diferentes formas de envelhecer que incluem os idosos bem-sucedidos e ativos, e também idosos incapacitados e limitados na sua autonomia pela doença assim como pelo contexto em que estão inseridos (Paúl et al., 2005). A heterogeneidade da população envelhecida só será verdadeiramente apoiada por uma estrutura que disponha de uma ampla gama de serviços e que permite à pessoa escolher aquele que melhor se adapta às suas necessidades (Benet, 2003), como é o caso dos Centros de dia. Esta resposta de parcial institucionalização, na qual os clientes passam um período do seu dia com os seus pares interagindo em atividades organizadas, numa mesma direção institucional (Salari, 2002), poderá assumir-se, no nosso entender, como um agente privilegiado de apoio ao cidadão, uma vez que não possui o peso das palavras "lar de idosos", tornando-se, deste modo, segundo Depalma (2003), uma resposta cada vez mais popular.

A World Health Organization (2002) destaca como apoios basilares do envelhecimento ativo: a saúde, participação e segurança. Através destes conceitos centrais, tentaremos realizar um paralelismo com serviços e programas disponibilizados pelo Centro de dia aos seus utilizadores. Uma das questões importantes a considerar, no que respeita à saúde, é a deteção precoce da patologia e a intervenção nos fatores de risco para o indivíduo que envelhece (Benet, 2003). Para Yuaso & Sguizzato (1996), os idosos residentes na comunidade, apoiados pelo Centro de dia, requerem formas especiais de atenção, no que respeita à saúde, de modo a ser realizada uma intervenção precoce da doença, assim como, um acompanhamento atento da mesma. Para além deste aspeto, a intervenção do Centro de dia direcionada para a saúde torna-se útil na medida em que oferece resposta a pessoas que, por motivos de doença, começam a desenvolver dependências. Assim, o Centro de dia tenta satisfazer as necessidades básicas da pessoa idosa não sendo necessário retirá-la do seu ambiente habitual (Benet, 2003). A nível de saúde, intervém na satisfação das necessidades diárias alimentares e no cuidado nutricional, de modo a contribuir para a diminuição de hábitos dietéticos não apropriados para a saúde em geral, através de um plano estruturado de cuidado (Trinidad, 1996). A promoção da atividade física é também um elemento a considerar num Centro de dia, sendo que o principal objetivo é a manutenção da máxima mobilidade possível do idoso (Alda & Montón, 1996). A fisioterapia é uma intervenção terapêutica relevante, uma vez que possibilita igualmente a manutenção e/ou recuperação da mobilidade dos membros superiores e inferiores, assegurando a independência do idoso (Alda & Montón, 1996), sendo este um serviço necessário ao Centro de dia. A promoção do exercício através da atividade física e fisioterapia começam a ser serviços crescentemente disponibilizados pelos Centros de dia, que permitem aos seus utilizadores a manutenção da sua autonomia. No que concerne à saúde mental, Manchola (2000) salienta como principais patologias presentes nos utilizadores dos Centros de dia: a depressão, a ansiedade e os transtornos comportamentais (tendência para fuga e paranóia). Nesta faixa etária, as intervenções estão muitas vezes direcionadas para a compreensão das aptidões funcionais dos idosos, aceitação dos seus défices e procura de formas de otimização do funcionamento individual, tendo em conta as suas limitações reais (Lima, 2004). Com o mesmo objetivo, os Centros de dia, através da realização de atividades sociais, lúdicas e criativas, tentam potenciar o funcionamento individual do idoso atendendo às suas limitações. Ainda, no que respeita à área da saúde, consideramos que é notória uma clara prevalência de ações preventivas e de manutenção sobre a saúde nos Centros de dia. Do ponto de vista gerontológico, a prevenção é convergente com as premissas de promoção do envelhecimento ativo, caracterizada pela manutenção e preservação das capacidades e potencial do indivíduo e consequente garantia de melhores condições de vida (World Health Organization, 2002).

A participação permite ao idoso fazer parte integrante da sociedade como agente ativo e interventor. Assim, torna-se importante reconhecer nos mais velhos a participação no desenvolvimento económico, no trabalho formal ou informal e em atividades de cariz social, de acordo com as suas preferências, desejos e capacidades. A participação social parece ser um fator importante para o envelhecimento bem-sucedido, no entanto, esta deverá existir em função de um projeto, de uma ação fundamentada, direcionada a um propósito na qual o idoso não seja passivo (Paúl, 2005). Nesta linha de pensamento, os Centros de dia são palco privilegiado para a participação no que respeita a duas ideias principais: aspetos de gestão da organização e aspetos individualizados dos serviços prestados. Ao nível da gestão do Centro de dia, reconhece-se importância da participação dos utentes no que respeita à organização/planificação de atividades lúdicas, culturais, recreativas e religiosas de acordo com os seus gostos; planificação e aferição sobre a qualidade do setor alimentar; a aferição sobre os serviços prestados e a prestar (nomeadamente horários de transportes, necessidade de criação de serviços extra, entre outros). A um nível individual, considera-se importante que o individuo possa ser agente ativo na escolha dos serviços prestados, na sua frequência e tipologia de acordo com as suas necessidades, gostos e preferências. Na realidade, como refere Paúl (2005), o direito à autodeterminação e dignidade fica comprometido sempre que haja uma tentativa de intromissão nas decisões dos idosos, sendo esse facto frequente nas relações com pessoas em idades avançadas. No nosso entender é fundamental que o idoso seja o negociador/decisor na escolha dos serviços a prestar pelo Centro de dia.

Por fim, a segurança retrata a necessidade de assegurar proteção e dignidade às pessoas idosas que se encontram desprotegidas nos seus direitos e necessidades de segurança social, financeira e física (World Health Organization, 2002). Os Centros de dia possibilitam essa proteção, através de esclarecimentos, informações e aconselhamento no que diz respeito aos recursos socioeconómicos existentes e de que forma lhes é possível aceder.

Considerações finais

Em suma, os Centros de dia podem e são aglutinadores de uma diversidade de conceitos, intervenções e programas que estão na base do conceito de envelhecimento ativo, apesar de esta identificação surgir rodeada de inúmeras dificuldades. Em nosso entender, os Centros de dia poderiam ter um papel mais visível no envelhecimento ativo, caso fossem superados diversos obstáculos dos quais destacamos os seguintes: a falta de reconhecimento e valorização da resposta pelo regulador; a dificuldade de implementação da especialização da resposta, sendo na sua maioria Centros de dia de cariz social; no subfinanciamento que dificulta o progresso e modernização do Centro de dia; o processo de implantação e crescimento desordenado tanto ao nível da criação de serviços como da tipologia de utente-alvo; reduzido número de recursos humanos afetos à resposta; e, por fim, a falta de interesse no estudo/investigação sobre o Centro de dia.

A maleabilidade da resposta de Centro de dia permite uma adaptação particularizada às necessidades de cada utilizador, assim como uma atenção individualizada na prestação dos serviços, sendo que estes elementos, no nosso entender, reforçam o seu vínculo ao modelo concetual de envelhecimento ativo.

Acredita-se que esta resposta continuará a ser cada vez mais implementada, no entanto, terá como principais desafios futuros uma reorganização dos serviços conducentes com os "novos idosos", que passará, sobretudo, por um aumento e melhoria da oferta de serviços e uma maior necessidade de formação dos seus técnicos e colaboradores.

No que concerne ao modelo de envelhecimento ativo, o natural envelhecimento populacional tenderá a potenciar a sua implementação e execução, aliviando assim a pressão sobre os sistemas económicos de segurança social, permitindo manter um equilibrado nível de "*well fair*" na população em geral. Por outro lado, urge a necessidade de técnicos e colaboradores que desenvolvem o seu trabalho junto destas respostas terem como missão e visão, no seu trabalho diário, implementar de forma ativa e cuidada os pressupostos do modelo de envelhecimento ativo, bem como, promover as novas políticas de intervenção comunitária que defendem o "envelhecer em casa" e o incentivo e respeito pelo idoso como agente ativo na escolha dos serviços prestados pelo Centro de dia, em detrimento da promoção da dependência e incapacidade das pessoas idosas.

Bibliografia

Alda, J., Montón, J. (1996). Programas de Fisioterapia. *in* Alda, J.; Dompedro, J.; Montalbo, M.; *Centro de Día para Personas Mayores Dependientes*. Madrid: Ministerio de Trabajo y Assuntos Sociales.

Arrazola, F.J., Méndez, A.U. & Lezaun, J J. (2003). *Centros de Día: Atención e Intervención Integral para Personas Mayores Dependientes y con Deterioro Cognitivo*. Gipuzkoa: Departamento de Servicios Sociales: Fundación Matía Gizartekintza.

Baumgarten, M., Lebe, P., Laprise, H., Leclerc, C. & Quinn, C. (2002). Adult Day Care for the Frail Elderly: Outcomes, Satisfaction and Cost. *Journal of Aging and Health*, 14(2), 237-259.

Benet, A.S. (2003). *Los centros de día para personas mayores.* Lleida: Edicions de la Universitat de Lleida.

Bilotta C.,Bergamaschini L., Spreafico S.,Vergani C.(2010).Day care center attendance and quality of live in depressed older adults living in the community. *European Journal of Ageing*, 7(1), 29-35.

Bowling,A.(2008).Enhancing later life :How older people perceive active ageing? *Journal of Aging & Mental Heath*, 12 (3),293-301.

Carvalho, M. I. (2006). *Orientações da política de cuidados às pessoas idosas dependentes – modelo de cuidados em Portugal e nalguns países europeus.* Comunicação no Congresso Internacional sobre Gerontologia, Lisboa.

Castiello, M. T. (1996). *Centro de Día para Personas Dependientes / Centro de Día: Conceptualizacion.* Madrid: Instituto Nacional de Servicios Sociales – Ministerio de Trabajo y Assuntos Sociales.

Cid, X., Dapia, M. (2007).Lazer e tempos livres para as gerações idosas – perspectivas de animação sociocultural e aproximação à realidade Galega. In A. R. Osório & F. C. Pinto (Eds.). *As pessoas idosas – contexto social e intervenção educativa.* Lisboa. Instituto Piaget – Colecção Horizontes Pedagógicos.

Conrad, K. J. & Hughes, S. (1993). Classification of Adult Day Care: a cluster analysis of services and activities. *Journal of Gerontology: Social Sciences*, 48(3), 112-122.

Depalma, J. A. (2003). Positive outcomes and Adult Day Care. *Home, Health, Care, Management & Practice, 15*(4), 342-343.

Fernández-Ballesteros, R., Caprara,J., Iñiguez,J., Garcia,L. (2005). Promoción del envejecimiento active:efectos del programa «Vivir com vitalidad». *Revista Espanhola Geriatria e Gerontologia. 40* (2), 92-102.

Fernández-Ballesteros, R. (2008). Active Aging Promotion Programs. *In* F. Ballesteros *(Ed.) Active Aging.* (133-154). Hogrefe Publishing.

Ferrer, M. T. A. (2005). *Los Centros de Día de Alzheimer y la calidad de vida de los pacientes e sus familiares – Un estudio de caso.* Valência: Colección Interciencias.

Gabinete de Estratégia e Planeamento – Ministério do Trabalho e da Solidariedade Social. (2006). *Carta Social – Rede de Serviços e Equipamentos.* Lisboa: Centro de Informação e Documentação (GEP-CID).

Gaugler, J. E. & Zarit, S. H. (2001). The Effectiveness of Adult day Services for Disabled Older People. *Journal of Aging & Social Policy,* 12(2), 23-47.

Gaugler, J. E., Jarrott, S. E. & Zarit, S. H.(2003). Respite for Dementia Caregivers: the effects of adult day service use on caregiving hours and care demands. *International Psychogeriatrics,* 15(1), 37-58.

Irigoyen, B., Carrasco, M., Hita, J.M. (2002). *Analises* Coste Consecuencia de um Centro de Día Psicogeriátrico. *Revista Española de Geriatria y Gerontologia, 37*(6), 291-297.

Jarrot, S. E., Zarit, S. H., Berg, S. & Johansson, L. (1998). Adult Day Care for Dementia: a Comparison of Programs in Sweden and The United States. *Journal of Cross – Cultural Gerontology,* (13), 99-108.

Lima, M. (2004). *Posso participar? – Actividade de desenvolvimento pessoal para idosos.* Porto: Âmbar

Manchola, E. A. (2000). Recursos sociosanitarios – el Centro de Dia (CD) Psicogeriátrico. *Revista Multidisciplinar de Gerontologia, 10*(2), 105-125.

Martín, I., Gonçalves, D., Paúl, C. & Pinto, F. (2007). Política Sociais para a Terceira Idade. In A. R. Osório & F. C. Pinto (Eds.), *As pessoas idosas – contexto social e intervenção educativa.* Lisboa: Instituto Piaget – Colecção Horizontes Pedagógicos.

Paúl, C. (2005). Envelhecimento activo e redes de suporte social. Retrieved from ler.letras.up.pt/uploads/ficheiros/3732.pdf

Paúl, C., Fonseca A., Martin, I., Amado, J. (2005). Satisfação e qualidade de vida em idosos portugueses. In Paúl e Fonseca (Eds.). *Envelhecer em Portugal, saúde e prestação de cuidados.* Lisboa. Climepsi Editores.

Pimentel, L. (2005). *Lugar do Idoso na Família – Contexto e Trajectória.* Coimbra: Quarteto Editora.

Salari, S. M. (2002). Intergerational Partnership in Adult Day Centers: Importance of age – appropriate environments and behaviors. *The Gerontologist*, 42(3), 321-333.

Salgado, A. A. & Montalvo, J. G. (1999). Centro de Dia para Personas Mayores – Un esquema prático sobre su funcionamento. *Revista Española de Geriatria y Gerontología, 34*(5), 298-305.

Sancho, T. & Rodriguez, P. (1999). *Programas y servicios para personas mayores in Gerontologia conductual – bases para la intervención y âmbitos de la aplicación*. Editorial Síntesis.

Tamer, N., Petriz G. (2007). A Qualidade de vida dos idoso. *In* Osório A. R. Osório & F. C. Pinto (Eds.) *As pessoas idosas – contexto social e intervenção educativa*. Lisboa. Instituto Piaget – Colecção Horizontes Pedagógicos.

Trinidad, D. (1996). *Programas de Salud Física – Centro de Día para Personas Mayores Dependientes*. Madrid: Instituto Nacional de Servicios Sociales – Ministerio de Trabajo y Assuntos Sociales.

Vallespir, J., Morey, M. (2007). A participação dos idosos na sociedade: integração vs segregação. In A. R. Osório & F. C. Pinto (Eds.). *As pessoas idosas – contexto social e intervenção educativa*. Lisboa. Instituto Piaget – Colecção Horizontes Pedagógicos.

Yuaso, D., Sguizzato, G. (1996). Fisioterapia em Pacientes Idosos. *In* M. Netto (Ed). *Gerontologia – A velhice e o envelhecimento em visão globalizada*. S. Paulo. Editora Atheneu.

Walker, J., Porter, R., Flanders, J. (2004). Increasing Prationers Knowledge of Participacion Among Elderly Adults in Senior Center Activities. *Journal Educational Gerontology*. 30, 353-366.

World Health Organization-Dept. of Noncommunicable Disease Prevention and Health Promotion(2002). *Active Ageing: A Policy Framework*. Genebra: Organização Mundial de Saúde.

CAPÍTULO V

AS POLÍTICAS DE SAÚDE COMO DETERMINANTES DA LONGEVIDADE

Joana Vale Guerra
Universidade de Coimbra, Faculdade de Psicologia e Ciências da Educação
Universidade de Coimbra, Centro de Estudos Interdisciplinares (CEIS20)
ORCID: https://orcid.org/0000-0001-7426-5579

Resumo: *As consequências do envelhecimento populacional e individual integram em si uma das principais conquistas da humanidade, mas ao mesmo tempo, colocam novos desafios que necessitam ser conhecidos e ultrapassados para que as pessoas e as sociedades colham os benefícios do "dividendo da longevidade". Não temos a pretensão de sermos nós, neste capítulo, a desvendar todos os limites e potencialidades dessa descoberta. Aqui, privilegiaremos a análise sobre as políticas de saúde relacionadas com o envelhecimento, enquanto chave que poderá ajudar a decifrar melhor a direção do itinerário político, social e económico perante a tendência demográfica e expressiva do aumento da longevidade. Com efeito, não basta considerarmos as políticas de saúde que já tentam responder aos desafios*

DOI | https://doi.org/10.14195/978-989-26-1788-6_5

do duplo envelhecimento demográfico, mas incluir um outro grande desafio, o "triplo envelhecimento". É de esperar que este novo facto demográfico traga implicações para a conceção, implementação e avaliação das políticas de saúde.
Palavras-chave: *Políticas de Saúde; Longevidade; "Triplo envelhecimento"*

Abstract: *The consequences of population ageing at individual or global levels are one of the main achievements of mankind, but at the same time they pose new challenges that need to be known and overcome so that societies reap the benefits of the "longevity dividend". We do not claim to us, to unravel all the limits and potentialities of this discovery. We will focus on health public policies related to ageing, as a key to better decode the direction of the political, social and economic path in the face of the demographic and expressive tendency of increasing longevity. Indeed, it is not enough to reflect about health policies that are already trying to meet the challenges of "twofold ageing", but it is essential to include another major challenge, the "threefold ageing". It is expected that this new demographic element will have effects on the design, implementation and evaluation of health policies.*
Keywords: *Health Policies; Longevity; "Threefold Ageing"*

Nota introdutória

O capítulo que se apresenta constitui-se como um contributo para uma reflexão em torno das políticas de saúde que protegem os cidadãos mais velhos na sociedade atual e uma oportunidade para apresentar uma incursão sobre os principais desafios políticos decorrentes do aumento da longevidade.

De forma preliminar, é importante acentuar duas ideias que criam um melhor entendimento sobre os limites e as potencialidades do debate em torno das políticas de saúde para o envelhecimento. Em primeiro lugar é importante reconhecer que a abundância de referências ao fenómeno do envelhecimento, através de estudos nacionais ou internacionais, através da atualização das projeções demográficas ou mesmo através do crescimento de respostas de apoio social e de saúde dirigidas aos mais velhos, não nos retiram a convicção de se tratar de um dos problemas mais complexos da sociedade contemporânea.

Falamos de um fenómeno novo, em evolução e pode dizer-se que a complexidade surge pelas questões individuais, grupais e societárias colocadas quer pelo processo de envelhecimento individual, quer pelas repercussões sociais do envelhecimento demográfico. Envelhecer no século XXI significa, como diz o relatório do Fundo de População das Nações Unidas (2012), uma celebração e um desafio. Viver mais anos em resultado da melhoria dos estilos de vida, das condições sanitárias, dos avanços da medicina, do acesso aos cuidados de saúde, do bem-estar económico e do acesso ao ensino e à informação é uma vitória da humanidade e do desenvolvimento.

Paradoxalmente, para a generalidade das pessoas, viver mais tempo pode implicar aumentar as oportunidades de experienciar a doença, viver sozinho e necessitar da ajuda de terceiros para a realização das suas atividades no quotidiano. Para as famílias viver mais pode implicar a oportunidade de usufruir da sabedoria, do apoio e da companhia dos seus elementos mais velhos, mas também pode exigir uma readaptação do ambiente familiar, por vezes difícil de concretizar por razões de vária ordem: económicas, laborais, culturais e afetivas. Para a sociedade, o envelhecimento e a longevidade são uma tradução do seu próprio desenvolvimento, mas terá que necessariamente criar condições que colmatem as novas necessidades das pessoas mais velhas e suas famílias. Por exemplo, a criação de melhores respostas na saúde e no apoio social, para que a longevidade seja vivida de forma digna e com qualidade. Compreende-se, no entanto, que a conjuntura económica e social atual coloca em confronto, por um lado, a necessidade de mais e melhores serviços de apoio, mas por outro a escassez de recursos financeiros e orçamentais para a sua implementação. A viabilidade da sua execução num país sob pressão sobre a despesa pública quer do ponto de vista da Segurança Social, quer do sistema de saúde, demonstra dificuldades na construção de uma rede de serviços de qualidade compatível com as exigências que as idades mais avançadas sugerem. A situação torna-se particularmente complexa quando pensamos nas pessoas mais velhas fragilizadas pelo seu contexto individual, familiar e socioeconómico.

É justo afirmar que muito se tem feito para melhorar a compreensão sobre o tema e para dar resposta aos diferentes desafios colocados pelo envelhecimento populacional ou individual. Contudo, a complexidade do fenómeno em si, deixa-nos longe de pensar que encontrámos as soluções necessárias para todos os indivíduos e ao mesmo tempo assegurar a viabilidade do manancial de respostas possíveis para cada forma de viver a longevidade.

Em sequência, acentua-se a segunda ideia preliminar. A complexidade do fenómeno terá, necessariamente, tradução na conceção, implementação e avaliação das políticas de saúde. Para tal, será necessário debater as escolhas que se fazem para a definição de um determinado programa de ação que visa fazer face às necessidades de um país nessa matéria.

Pensar as políticas de saúde para a população mais velha requer uma visão diferenciadora sobre o fenómeno de envelhecer.

Com efeito, basta pensar que envelhecer em contexto de isolamento, com uma reforma baixa, sem retaguarda familiar, com uma rede de apoio formal frágil, sem a acumulação dos benefícios de um estilo de vida saudável é bem diferente de envelhecer com o acesso aos serviços de saúde garantido, com apoio material, informativo, emocional e social. Tanto mais quando se verificam situações de doença crónica, múltipla e incapacitante. O estado de saúde é um fator condicionante, muito relevante, no processo de envelhecimento. Particularmente, quando as capacidades funcionais ficam afetadas temporária ou permanentemente e as pessoas sentem dificuldades em realizar as suas atividades diárias.

Estas duas ideias preliminares sugerem que estudar e criar propostas para os desafios da longevidade não será certamente tarefa fácil. Contudo, sabemos que um melhor funcionamento do setor da saúde e o compromisso societal de proteção sobre os cidadãos mais velhos conduzirá, inequivocamente, a melhorias significativas na qualidade de vida das pessoas. Para tal, é necessário refletir sobre o processo de decisão política que deverá expressar as preferências da sociedade e dos cidadãos.

Como seguidamente se apresenta, o sistema de saúde português integra princípios favoráveis à proteção dos cidadãos independentemente da sua condição perante o trabalho, sexo, idade, religião, profissão ou sistema de proteção. No caso particular dos doentes idosos e dependentes, a criação da Rede Nacional de Cuidados Continuados deu início a um novo paradigma na arte de cuidar, fortalecendo a continuidade ideológica do princípio da universalidade. Contudo, o desafio do "triplo envelhecimento" permanece porque ainda não conhecemos completamente os seus efeitos na dinâmica das sociedades modernas. Como refere a

OMS "*envelhecer no futuro será muito diferente das experiências de gerações anteriores*" (OMS, 2015:10).

O ponto de viragem das políticas de saúde dirigidas aos maiores de 65 anos

A política de saúde em qualquer país traduz a capacidade do Estado ter como critério e medida do próprio desenvolvimento o respeito pelo direito à proteção da saúde da sua população. O compromisso e o consequente investimento na implementação e desenvolvimento de um sistema de saúde organizado, completo e financeiramente equilibrado acolhem benefícios indiscutivelmente positivos e em clara concordância com a fruição do direito à vida. As políticas de saúde são uma pedra angular na definição e reconhecimento da boa governação e traduzem as aspirações do projeto coletivo de uma nação no campo dos direitos humanos e do desenvolvimento social e económico.

A conjuntura política, económica e social dos países desenvolvidos, na segunda metade do século XX, oferecia a garantia de que as decisões políticas atendiam às convicções emanadas da ONU, da OMS e da UE e que cada país edificava o seu sistema de saúde de forma consentânea com o princípio de que a saúde é um direito básico e componente fundamental do desenvolvimento humano e social; e que o estado de saúde da população dependia de uma relação de interdependência dos fatores económicos, sociais, políticos, biológicos, ambientais, culturais e de acessibilidade aos cuidados de saúde; e que os ganhos em saúde eram insofismáveis para o desenvolvimento do indivíduo mas também dos países. A saúde tornou-se, desta forma, num domínio da responsabilidade pública na maior parte dos países europeus.

Em Portugal, o sistema de saúde tem na sua matriz ideológica o modelo beveridgeano. Ou seja, o Estado compromete-se a criar e a organizar os cuidados de saúde acessíveis a toda a população, independentemente do tipo de contrato laboral ou rendimento dos cidadãos ou residentes, financiado a partir da totalidade da riqueza do país, com base nas receitas recolhidas pelo Estado, refletidas no orçamento geral do Estado.

Desde 1976, a conceção e implementação das políticas de saúde centraram-se em princípios de universalidade, equidade e generalidade, que sendo fundamentais, não previam cuidados diferenciados ou integrados para corresponder às necessidades de saúde das pessoas com idade mais avançada. O primeiro passo foi dado em 2006 com a criação da Rede Nacional de Cuidados Continuados Integrados.

Mas pensar as respostas da saúde para um grupo específico não pode ser feito sem reconhecer a natureza das políticas que têm vindo a ser adotadas nos últimos quarenta anos. No computo geral, as transformações mais significativas que foram sendo introduzidas no sistema de saúde português, através da Lei de Bases da Saúde de 1990 e das alterações em 2002, investiram numa conceção ampla do sistema de saúde português (integrando o Serviço Nacional de Saúde, entidades privadas e profissionais liberais), mas sobretudo o entendimento dos cidadãos como os primeiros responsáveis pela própria saúde e a redução do peso do Estado na prestação de atividades de saúde (Simões, 2004).

Cada vez mais, o sentido estratégico das novas políticas de saúde realça o empoderamento do cidadão na saúde e conta ainda com um novo discurso político e económico centrado na gestão e financiamento dos serviços de saúde baseado em resultados de eficiência e eficácia económica associado a uma cultura de privatização.

Não descurando as inúmeras vantagens do ato de se cuidar, é importante questionar as desigualdades implícitas na capacidade de cada um garantir a realização de atividades protetoras de saúde quando os contextos e ambientes de vida podem ser mais importantes na manutenção ou potencialização do *stock* de saúde de cada indivíduo. E tanto mais, porque a idade é um fator naturalmente depreciativo do potencial biológico de cada pessoa. Daqui resultam duas ideias fundamentais, por um lado a garantia de que a idade avançada aumenta as probabilidades de adoecer ou desenvolver algum tipo de dependência, mas por outro, é muito relevante considerarmos que a diversidade de perfis da população com idade mais avançada não é aleatória. Ou seja, o nível de literacia, a qualidade das redes de apoio social, a adequação do rendimento, a perceção de apoio são fatores preditores de um envelhecimento com qualidade (OMS, 2015). Com estes considerandos damos início à ideia de que a composição dos cuidados e serviços de saúde a prestar a uma população envelhecida terá de ser necessariamente diferente. Não pode ser uma resposta generalizada e "mercadorizada" porque o respaldo de cada pessoa varia em função das suas circunstâncias individuais e ambientais muito para além do potencial biológico.

A RNCCI como reforço ideológico do Serviço Nacional de Saúde

A criação da Rede Nacional de Cuidados Continuados Integrados (RNCCI), em 2006, foi o passo dado mais importante para colmatar a enorme lacuna do Serviço Nacional de

Saúde no tratamento dos idosos e dos doentes dependentes. Conforme refere Correia de Campos a "*Rede surgiu da constatação de que a configuração anterior do sistema de saúde não garantia a reabilitação dos seus diferentes grupos-alvo de doentes*" (2008:108). A sua implementação passou a assegurar cuidados de saúde de continuidade através da disponibilização de uma rede de serviços de diferentes tipologias que responde a problemas distintos. São as equipas multidisciplinares com origem nos cuidados diferenciados ou primários que organizam o "processo individual de cuidados continuados" centrado na reabilitação e na promoção da autonomia de cada doente. Os cuidados a prestar são definidos pelas circunstâncias clínicas e sociais das pessoas mais velhas e ou doentes dependentes.

O novo modelo de intervenção introduzido pela RNCCI apresenta na sua génese, por um lado, o reforço sobre a continuidade ideológica do princípio da universalidade do SNS e por outro lado, cria um conjunto de respostas que garantem cuidados de natureza preventiva, curativa, de reabilitação e paliativa.

É bom recordar que a solução anterior respondia às necessidades dos idosos, em situação clínica agravada pela perda de autonomia, através da disponibilização, nos hospitais, de "camas de retaguarda". Esta resposta não tinha outro objetivo senão ganhar tempo até surgir uma resposta familiar ou institucional da comunidade de origem. Quando um doente reunia condições para a alta clínica, mas não reunia condições para regressar ao domicílio, as soluções apresentadas ao doente situavam-se entre: o protelamento da alta em contexto hospitalar, a institucionalização em lar residencial ou o regresso ao domicílio com apoio domiciliário. Note-se que as hipóteses apresentadas não previam qualquer tipo de acompanhamento especializado em função do potencial de recuperação de cada doente. As pessoas ficavam entregues à sua capacidade de pagar serviços de reabilitação e de acompanhamento no dia-a-dia e ainda viam aumentada a probabilidade de agravamento do seu estado de saúde, muitas vezes seguido de uma reentrada no serviço de urgência dos cuidados hospitalares.

Não obstante as dificuldades atuais de ingresso em alguns tipos de unidades de tratamento, nomeadamente pela demora no tempo de espera, a implementação da RNCCI passou a dar resposta e a garantir cuidados de saúde no tempo certo, no local certo e no prestador mais adequado às necessidades das pessoas.

No final de 2016 existiam 8 400 camas em funcionamento e a taxa de ocupação, a nível nacional, revelou, à semelhança dos anos anteriores, valores sempre acima dos 90% em todas as tipologias de unidades (ACSS, 2017).

O número de utentes assistidos em 2016 foi de 52.509. De acordo com o objetivo deste capítulo é importante notar que os utilizadores da RNCCI com idade superior a 65 anos representaram 81,6% do total e a população com idade superior a 80 anos representou 47,4% do total, valor em crescimento ao longo dos anos.

Os dois principais motivos de referenciação para a RNCCI são, a dependência para a realização de atividades da vida diária (90%) e o ensino ao utente ou ao cuidador informal (89%). O destino pós-alta, revela que 10% dos utentes dirigiram-se para respostas sociais. A nível nacional 74% das altas foram para o domicílio e em 73% das altas para o domicílio foi registada necessidade de suporte (ACSS, 2017).

Em conclusão, a ACSS (2017) refere que os utilizadores da RNCCI representam uma população envelhecida, maioritariamente feminina, com baixo nível de escolaridade, carenciada e com elevada incapacidade e dependência.

Do ponto de vista do plano de desenvolvimento da Rede os objetivos passam por aumentar a capacidade das respostas existentes, mas também desenvolver e diversificar novas respostas, como por exemplo os cuidados prestados no domicílio e em ambulatório, o reconhecimento e apoio ao cuidador informal e o desenvolvimento de tecnologias de informação e comunicação relacionado com a criação de Ambientes de Vida Assistida que preveem um aumento da autonomia, autoconfiança e mobilidade dos cidadãos no seu próprio domicílio (ACSS, 2017).

Apesar do modelo inovador introduzido pela RNCCI, pretende-se chamar a atenção de que as características da população, quer do ponto de vista demográfico, social, económico e epidemiológico não se fixam no tempo e vão adquirindo novas configurações e revelando novas necessidades. Neste sentido, o reforço da investigação para uma melhor compreensão sobre o fenómeno do envelhecimento e da longevidade não é intransponível. No próximo ponto abordamos genericamente alguns indicadores que lançam pistas de reflexão para a reconfiguração das políticas de saúde para a atualidade e para o futuro próximo.

A magnitude das necessidades em Portugal

Para melhor compreendermos os desafios do envelhecimento será bom ter em consideração que em Portugal, o peso da população idosa mantem um perfil ascendente, em consequência das tendências de diminuição da fecundidade e de aumento da longevidade. A taxa de fecundidade geral apresenta um decréscimo ininterrupto desde a década de 90 e em 2016, o índice de longevidade foi de 48,7, o nível mais elevado desde 1970. A população com 75 e mais anos é cada vez em maior número. Conforme podemos verificar no gráfico seguinte, não obstante as diferenças entre homens e mulheres, a partir de 2010, as mulheres com 75 ou mais anos são já mais de metade das mulheres com

65 anos. Torna-se inequívoco afirmar que a população idosa está mais envelhecida.

Gráfico 1 – Índice de Longevidade por sexo (1970-2016)

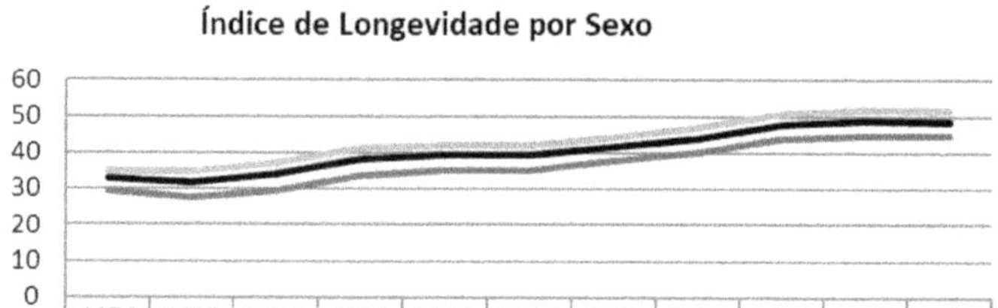

Índice de Longevidade por Sexo

	1970	1975	1980	1985	1990	1995	2000	2005	2010	2015	2016
Homens	29,54	27,64	29,68	33,61	35,21	35,3	37,9	40,1	43,9	44,9	44,6
Mulheres	34,9	34,73	37,27	41,16	42,23	42,1	44,4	47	50,8	51,9	51,6
HM	32,76	31,93	34,19	38,09	39,31	39,3	41,7	44,1	47,9	49	48,7

Fonte dos dados: INE, última atualização a 15 de Junho de 2017

Para o período 2015-2017, a esperança média de vida aos 65 anos situa-se nos 19,45 anos. Ainda que provisório, este valor acompanha uma evolução positiva que se regista de forma significativa desde 1980, altura em que aos 65 anos as pessoas contavam, em média viver, aproximadamente, mais 15 anos. Passados 35 anos, a população portuguesa vive mais quase 20 anos (INE, 2017). Em sequência, a questão que se coloca é sobre o número de anos que uma pessoa com 65 anos pode esperar viver sem sofrer de incapacidades. Os últimos dados indicam que os portugueses podem esperar viver mais sete anos sem incapacidade (OCDE, 2017). Este indicador para além de contribuir para o debate em torno das questões económicas e de produtividade, pode instruir os progressos realizados em termos de acesso, qualidade e sustentabilidade dos cuidados de saúde. No caso de Portugal este indicador revela valores abaixo do expectável e em comparação com os países da UE28. Os europeus contam viver mais 9,4 anos sem perda de funcionalidades após os 65 anos.

Desde 1990, a proporção de indivíduos com 65 e mais anos por 100 residentes com menos de 15 anos (índice de envelhecimento) apresenta uma tendência sistemática de crescimento (72,1 em 1991, 141,3 em 2013 e 150,90 em 2016). Para objetivar, referimos que no ano de 2016 a população residente em Portugal com 65 ou mais anos foi de 2,462,256, ou seja 23,9% da população. De acordo com um estudo da Fundação Francisco Manuel dos Santos o número de pessoas com 65 e mais anos poderá, em 2030, representar quase metade do número de pessoas em idade ativa. Em 2030, a população de Portugal será ainda mais envelhecida, mesmo admitindo um significativo aumento dos níveis de fecundidade (Mendes, 2012).

Para além dos aspetos demográficos indicativos do estado de saúde em idades avançadas, e de acordo com a revisão da literatura, um dos vetores que devemos acompanhar é a disponibilidade de rendimentos adequados ao longo da vida e em particular, após a fase de vida ativa.

O Inquérito às Condições de Vida e Rendimento, realizado em 2017 em Portugal, sobre rendimentos do ano anterior, indica que 17% da população idosa estava em risco de pobreza após as transferências sociais. No caso português é sabido que as transferências relativas a pensões concretizam um importante papel na redução do risco de pobreza monetária. O reconhecimento do direito às pensões é uma forma de garantir a subsistência das pessoas na fase da reforma e permitir que as pessoas mais velhas tenham um nível de vida digno e gozem de independência económica.

Em Portugal, o regime público é a principal fonte de rendimento dos portugueses maiores de 65 anos. As fontes de financiamento do sistema de segurança social são os contribuintes e a administração central. Contudo, as prestações sociais da

Segurança Social, de natureza pecuniária, que visam compensar a perda de remuneração de trabalho ou assegurar valores mínimos de subsistência ao cidadão com 65 ou mais anos revelam dois aspetos que traduzem a fraqueza do debate em torno das políticas sociais de apoio aos idosos. Em primeiro lugar, é de referir que existe um espetro de 70 anos entre a adoção da primeira medida de proteção social, a Pensão de Velhice em 1935 e a última medida criada de combate à pobreza dos idosos, o Complemento Solidário para Idosos em 2005. E em segundo lugar, as compensações monetárias são baixas para o contexto nacional e europeu. Os montantes referenciados colocam em causa por exemplo, entre outras coisas, a capacidade de os reformados adquirirem bens e serviços de saúde fora do sistema público. Por este motivo, todo o investimento efetuado no âmbito do Serviço Nacional de Saúde que tenha como benefício melhorar o acesso dos idosos aos cuidados de saúde será um investimento na qualidade de vida e dignidade das pessoas.

Tabela 1 – Prestações da Segurança Social

Prestações Segurança Social	**Ano de criação**	**Montante**	**Condições de acesso**	**Base de cálculo**	**Regimes**
Pensão de Velhice	1935	264,32 € 382,46 €	Idade	Remunerações registadas	Geral
Pensão Social de Velhice	1980	203,35 € + CES	Idade Residentes	Fixado através do IAS	Não contributivo
Complemento Solidário para Idosos*	2005	Valor máximo 423,69 €	Residente Pensionistas Recursos < ao CSI	Recursos do agregado familiar	Todos os regimes

* Os beneficiários de CSI têm direito, para além do complemento mensal à pensão, a descontos em medicamentos, óculos e próteses dentárias e às tarifas sociais de eletricidade, de gás natural e dos transportes.

Considerações finais

O complemento solidário para idosos e a criação da Rede Nacional de Cuidados Continuados ilustram uma nova geração de políticas sociais públicas. A compreensão sobre o cenário demográfico nacional e o conhecimento sobre a forma como o país organiza os sistemas de saúde e de proteção social contribui para um conhecimento mais aprofundado sobre as necessidades a colmatar no futuro próximo.

Este capítulo reuniu informação que nos permite concluir que há um caminho a percorrer para melhorar a compreensão sobre o processo de envelhecimento com todos os seus determinantes. Apesar da vastidão do tema podemos enunciar quatro tópicos, baseados nas experiências passadas e as projeções demográficas, que podem basear os debates para a reconfiguração das políticas de saúde e de apoio social.

Em primeiro lugar, é necessário criar uma melhor compreensão sobre a diversidade de perfis de envelhecimento de modo a garantirmos a redução das desigualdades na promoção de fatores protetores e na eliminação dos fatores de risco de natureza comportamental e social. Em segundo lugar investir no desenvolvimento do conhecimento sobre os fatores que promovem a melhoria da esperança de vida saudável aos 65 anos. Prolongar os anos de vida saudável, sem perda de funcionalidades é uma vantagem do ponto de vista da produtividade e da qualidade de vida para as pessoas. Em terceiro lugar investir na proteção dos idosos economicamente mais vulneráveis e no reconhecimento de que é fundamental favorecer a melhoria do rendimento, garantindo recursos mínimos e a satisfação de necessidades básicas. Em quarto lugar assegurar que a maior intervenção e participação do sector privado e a tendência para a "mercadorização" dos bens e serviços do setor da saúde revejam uma oportunidade de o Estado português realocar os seus recursos para uma oferta de cuidados a esta população específica, centrando-se nas doenças crónicas, em outras patologias físicas e psicológicas e na medicina preventiva (Rodrigues, 2014). Jamais numa oportunidade para acentuar as desigualdades socioeconómicas que apenas o Serviço Nacional de Saúde de qualidade e eficiente pode fazer face. E por fim, em quinto lugar garantir a manutenção de cuidados continuados integrados e personalizados, procedendo sempre que possível em meio natural.

O envelhecimento está inscrito num complexo sistema de interações onde se procuram equilíbrios que procuram satisfazer as necessidades de cada idoso dentro do quadro dos compromissos societais (Machado, 2017). Será a forma como optamos por solucionar estes novos desafios e a forma de maximizar as oportunidades de uma crescente população idosa que determinará se a sociedade colherá os benefícios do "dividendo da longevidade" (UNFPA, 2012).

Bibliografia

Administração Central do Sistema de Saúde (ACSS) (2017). *Relatório de Monitorização da Rede Nacional de Cuidados Continuados Integrados – 2016*. Acedido a 15 de janeiro de 2108 em: http://www.acss.min-saude.pt/wp-content/uploads/2016/07/RNCCI-Relatorio-Monitorizacao-Anual-2016.pdf

Campos, A.C. (2008). *Reformas da Saúde. O fio condutor*. Coimbra: Almedina.

Direção-Geral da Saúde (2017). *Estratégia Nacional para o Envelhecimento Ativo e Saudável 2017-2025*. Lisboa: DGS.

Gabinete de Estratégia e Planeamento (2017). *Relatório do terceiro ciclo de revisão e avaliação da implementação do Plano Internacional de Ação de Madrid sobre o Envelhecimento*. Lisboa: Ministério do Trabalho, Solidariedade e Segurança Social.

Guerra, J. (2009). Desigualdades completamente indesejáveis em saúde. *Actas do Seminário Combater a Reprodução Intergeracional da Pobreza e da Exclusão Social. Que intervenções?*, Porto: Centro de Investigação em Ciências do Serviço Social, ISSSP, pp. 299-312.

Guerra, J. (2016). Políticas de Saúde em tempo de crise(s). In Albuquerque, C. (coord.) *Políticas Sociais em Tempo de Crise. Perspetivas, tendências e questões críticas*. Lisboa: Pactor, pp:177-192.

Instituto Nacional de Estatística (INE) (2017). Rendimento e Condições de Vida 2017 (dados provisórios). *Destaque*. Lisboa: INE.

Machado, P. (2017). *Isolamento na Velhice: das políticas integradas para a longevidade à construção de um referencial GovInt para o envelhecimento na comunidade*. Lisboa: Fórum para a Governação Integrada.

Mendes, MF e Rosa, MJV (Coords.) (2012). *Projeções 2030 e o Futuro*. Lisboa: Fundação Francisco Manuel dos Santos.

Ministério da Saúde & Ministério do Trabalho, Solidariedade e Segurança Social (2016). *Cuidados Continuados. Saúde e Apoio Social. Plano de Desenvolvimento da RNCCI 2016-2019*. Lisboa. Acedido a 30 de setembro de 2017 em https://www.sns.gov.pt/wp-content/uploads/2016/02/Plano-de-desenvolvimento-da-RNCCI-2016-2019-Ofi%CC%81cial-Anexo-III.pdf

Organização para a Cooperação e Desenvolvimento Económico (OCDE) e OEPSS (2017). *State of Health in the EU. Portugal. Perfil de saúde do país 2017*. Acedido a 15 de janeiro de 2018 em: https://ec.europa.eu/health/sites/health/files/state/docs/chp_pt_portuguese.pdf

Organização Mundial de Saúde (OMS) (2015). *Relatório mundial de envelhecimento e saúde*. Acedido a 30 de setembro de 2017 em: http://apps.who.int/iris/bitstream/10665/186468/6/WHO_FWC_ALC_15.01_por.pdf

Rodrigues, F. (2006). *Políticas Sociais para a Inclusão das Pessoas Idosas*. Comunicação apresentada no 3º Seminário de Gerontologia "A Pessoa Idosa. Momentos de reflexão". Tondela: Gabinete de Ação Social do Município de Tondela.

Rodrigues, TF; Martins, MRO (2014). *Envelhecimento e Saúde. Prioridades Políticas num Portugal em Mudança.* Lisboa: Instituto Hidrográfico.

Simões, J. (2004). *Retrato Político da Saúde. Dependência do percurso e inovação em saúde: da ideologia ao desempenho.* Coimbra: Almedina.

United Nations Population Fund (UNFPA) e HelpAge (2012). *Envelhecimento no Século XXI: Celebração e Desafio.* Acedido a 30 de setembro de 2017 em: https://www.unfpa.org/sites/default/files/pub-pdf/Portuguese-Exec-Summary_0.pdf

CAPÍTULO VI

CIDADANIA ATIVA NUMA SOCIEDADE ENVELHECIDA: O VOLUNTARIADO SÉNIOR

Helena Reis Luz

Universidade de Coimbra, Faculdade de Psicologia e Ciências da Educação, Centro de Estudos Interdisciplinares (CEIS20)

ORCID: https://orcid.org/0000-0003-1592-0953

Isabel Cerca Miguel

Universidade Portucalense Infante D. Henrique – Instituto de Desenvolvimento Humano Portucalense & Departamento de Psicologia e Educação

ORCID: https://orcid.org/0000-0002-5305-7620

Resumo: *O envelhecimento demográfico constitui uma realidade incontornável das sociedades atuais, colocando às sociedades envelhecidas desafios acrescidos que interessa observar. Considerando a cidadania, as concepções capitalistas enfatizam o papel central do trabalho na definição do estatuto de cidadão, destacando a estreita relação entre participação no mercado de trabalho e inclusão social. No entanto, essa abordagem deve ser reequacionada no contexto específico da população idosa, uma vez que na reforma estes encontram vários domínios*

DOI | https://doi.org/10.14195/978-989-26-1788-6_6

significativos de participação social para além do trabalho remunerado. Neste âmbito, o voluntariado sénior afigura-se como um meio proeminente de ativar a participação social dos idosos. O presente capítulo tem como objetivo discutir as fragilidades de uma cidadania baseada no trabalho, enfatizando o voluntariado sénior como uma abordagem de ativação e promoção da participação social dos idosos.
Palavras-chave: *Trabalho; Voluntariado sénior; Ativação dos idosos*

Abstract: *Demographic aging is an inescapable reality of today's societies, placing aged societies great challenges in need of attention. Considering citizenship, capitalist conceptions emphasize the central role of labour in defining a citizen status, and highlight the close relationship between participation in the labour market and social inclusion. In the specific context of the elderly population, this approach must however be pondered, since in retirement elders find several significant domains of social participation beyond paid work. In this context, senior volunteering appears as a prominent means of activating elderly's social participation. This chapter aims to discuss the frailties s of a work-based citizenship, emphasizing senior volunteering as an approach for activating and promoting elderly's social participation.*
Keywords: *Work; Senior volunteering; Activation of the elderly*

Introdução

Nas últimas décadas, o fenómeno de transição demográfica, caracterizado por uma drástica diminuição das taxas de mortalidade e natalidade e consequente aumento da população idosa, trouxe consigo importantes mudanças sociais que impõem desafios consideráveis em variados domínios. A evolução demográfica do mundo ocidental constitui, em simultâneo, um importante desafio tanto para a sociedade e para a ciência, como para os indivíduos. Se, para a sociedade e ciência, os desafios surgidos da evolução demográfica do mundo ocidental se colocam ao nível das diferentes formas de organização social, cultural, económica e política decorrentes do aumento da esperança média de vida e das patologias associadas à vivência da condição de idoso, para o indivíduo estes desafios situam-se na expressão de comportamentos positivos de desenvolvimento, conducentes a um envelhecimento ativo e saudável. É, pois, no quadro de uma nova ordem social associada ao envelhecimento que o estudo dos acontecimentos implicados neste processo tem vindo a adquirir uma expressão nunca antes vista, traduzida num esforço multidisciplinar para o entendimento do fenómeno, e para o qual converge a atenção simultânea das ciências sociais, humanas, económicas e da saúde.

A procura de respostas para os novos problemas e perplexidades decorrentes das modificações na configuração das pirâmides etárias e suas consequentes implicações impõe, deste modo, o aprofundamento de uma visão positiva e original desta etapa da vida que, em larga medida, abandone as conceções tradicionais que associam a velhice a incapacidade e declínio (Miguel, 2014), fomentando estratégias promotoras da qualidade de vida na velhice. Todavia, a redefinição de uma nova imagem do envelhecimento implica que, em articulação com a imputação deste processo à responsabilidade individual, sejam igualmente consideradas dimensões mais latas em termos sociais, institucionais e políticos. É a este nível que, no quadro das políticas sociais, importa (re)equacionar as lógicas conducentes ao exercício da cidadania para a população idosa, reforçando a sua lógica inclusiva e de participação (Levitas, 1998; Lie, Baines & Wheelock,

2009), onde os impulsos de voluntariado sénior assumem uma posição de destaque. Assim, o presente capítulo procura discutir, para os idosos as fragilidades de uma perspetiva de cidadania baseada no trabalho, salientando o voluntariado sénior como um meio de ativação dos idosos, de promoção da participação social e cidadania ativa dos mais idosos.

1. A dimensão construtiva do trabalho: Implicações para a população idosa

É inquestionável o papel central que o trabalho assume na vida humana, sendo o exercício de uma atividade profissional o elemento estruturador em função do qual se organizam e adquirem significado os demais domínios de existência (Kovacs, 2009; Meda, 2007; MOW, 1987; Rapaport & Bailyn, 1998; Reis, 2006), a par da família, grupos de interesse, envolvimento religiosos ou atividade política. Nesta perspetiva, o trabalho assume, para a existência humana, uma importante função promotora de sentido e significado, tanto pessoal como social. O trabalho é mesmo a atividade mais significativa das suas vidas, constituindo-se como uma necessidade existencial para a satisfação das necessidades psicológicas, sociais e económicas (Casey, 1995; Meda, 2007) e, em simultâneo, como um elemento crítico construtor do autoconceito e eixo central da identidade pessoal. Por outro lado, do ponto de vista social, não só o trabalho constitui um organizador de tempo e legitimador social de diferentes fases da vida (estudo, trabalho e reforma), como ainda marca o contributo do indivíduo para os seus contextos envolventes, permitindo afirmar a sua "individualidade como cidadão através dos produtos da sua atividade, o que, por seu turno, lhe permite adquirir uma consciência da sua posição na globalidade das redes de relações estabelecidas em diferentes níveis de interação/participação social e do papel que lhe caberá em termos da definição e (re)formulação das mesmas" (Parada & Coimbra, 1999, p. 94). Com efeito, numa sociedade fortemente marcada por modelos e regras económicas e assente numa lógica de produtividade pelo trabalho, a vida profissional assume tanto uma condição de exigência social como de estatuto pessoal, determinando em larga medida "quem somos" e o "o que significamos socialmente" (Fonseca, 2011). De facto, é através do trabalho, enquanto realidade social, que o cidadão moderno se constrói como tal, e que acede a um conjunto de direitos que, tendencialmente, lhe são específicos, ao mesmo tempo que define a sua existência e identidade social, simbolizada pela profissão e transmitidos através do tipo e quantidade de bens. É, portanto, nesta dupla dimensão constitutiva, a cidadania, por um lado, e a produção de bens e serviços, por outro, que se define e determina em grande medida o modo de vida dos indivíduos, e que o trabalho se constitui um vetor fundamental de organização e coesão social (Kovacs, 2009; Parada & Coimbra, 1999).

Sendo o trabalho um elemento construtor e meio (necessário) para a conquista da cidadania, o desenvolvimento da cidadania surge, deste modo, vinculado a uma lógica capitalista que, pelo exercício de uma atividade profissional, torna real a condição de cidadão. A valorização social do trabalho emerge, nesta ótica, como um meio de afirmar um espaço de participação social, económica, política e cultural, no seio de um contrato social marcadamente inclusivo que, pela vinculação ao mercado de trabalho, promove o acesso a esferas socialmente reconhecidas (Leão & Barros, 2008; Parada & Coimbra, 1999; Polanyi, 1980; Perret, 1993; Castel, 1995). Neste sentido, o exercício de uma cidadania ativa tem sido fortemente relacionado com a participação no mercado de trabalho, consolidada na metáfora de cidadão-trabalhador (Lewis, 2002), enfatizando o trabalho remunerado como meio de expressão de uma cidadania ativa (Levitas, 1998). Todavia, para a população idosa que, após a entrada na reforma, cessa o exercício de uma atividade profissional remunerada, uma cidadania definida em termos da sua participação no mercado de trabalho adquire uma relevância de contornos ambíguos (Amaro da Luz, Miguel & Preto, 2014; Craig, 2004). De facto, considerando que o trabalho não só organiza a vida humana como também ajuda a formar uma imagem pessoal e a definir o lugar de cada um no mundo, a identidade dos indivíduos encontra-se fortemente relacionada com a sua integração social e o seu envolvimento em redes sociais, sendo, portanto, a perda de papéis sociais – como a que, não raro, ocorre na passagem à reforma – um fator cujo impacto negativo se poderá reverter em algum risco de perturbação e mal-estar individual, coincidente com a perda de um sentido de utilidade na vida e oportunidade de integração social (Lazarus & Lazarus, 2006; Victor, 2005).

Deste modo se compreende que "a passagem à reforma pode converter-se num momento particularmente sensível para o bem-estar psicológico e social" (Fonseca, 2011, p. 35), constituindo-se como um acontecimento de vida da maior importância, em face do qual os indivíduos se veem confrontados com a necessidade de proceder a uma adaptação e reorganização dos seus padrões de vida e da sua identidade pessoal, processo este que, não obstante os aspetos positivos que lhe possam ser imputados, não se faz isento de constrangimentos e ambivalências (Atchley, 2000). Não obstante a grande diversidade de padrões individuais e formas de adaptação à reforma (Fonseca, 2011; Han & Moen, 1999; Sterns & Gray, 1999), a ausência de regulação do quotidiano pelo estabelecimento de rotinas e a transição do papel de trabalhador remunerado (condição de "ativo", socialmente bastante determinada) para a situação de "reformado" (que, coincidente com a velhice, se mostra indissociável do processo de envelhecimento) constituem, no geral, os núcleos significantes da reforma (Savishinsky, 2000).

À vivência socialmente pejorativa do envelhecimento, a desvinculação do mercado de trabalho remunerado acrescenta, deste modo, os fundamentos para a negação dos direitos universais de cidadania, frequentemente associados à experiência da pobreza, de discriminações idadistas e da exclusão social por grande parte da população idosa (Higgs, 1995; Marques, 2011; Turner, 1989). Numa alusão específica à população idosa, o risco de exclusão social projeta-se, pois, em vários sentidos, sendo de considerar a exclusão de recursos materiais, das relações sociais, de atividades cívicas, dos serviços básicos e comunitários (Mauritti, 2004).

Scharf, Phillipson e Smith (2005) identificam importantes domínios nos quais os discursos recentes acerca da exclusão podem refletir as particularidades circunstanciais da população idosa: antes de mais, num contexto em que a inclusão social é definida a partir da participação no mercado de trabalho a questão emerge do potencial das pessoas idosas que cessaram o exercício dos seus papéis profissionais para se integrarem na sociedade; depois, e ainda que alguns estudos chamem a atenção para a natureza dinâmica da exclusão social, a situação das pessoas idosas é potencialmente menos suscetível de mudar, sobretudo à luz dos limites mais difusos que o fenómeno reveste para populações mais jovens. Com efeito, a vivência da condição de reformado traz associada a noção de improdutividade decorrente do abandono profissional, enfatizando a ideia dos idosos como dependentes e atuando como restrição ao progresso económico e social, sobretudo devido aos custos com pensões e cuidados necessários e cujos encargos recaem sobre uma porção diminuta da população. A falta de estatuto atribuída aos reformados, resultante das condições de não-trabalhador e economicamente inativo, gera uma polaridade tão inútil quanto imprecisa entre mais novos e mais velhos (Hardill & Baines, 2009), ignorando o papel crítico desempenhado pelos idosos como cidadãos ativos, com o risco de os colocar à margem das sociedades atuais fortemente imbuídas de uma ética de produtividade pelo trabalho (Meda, 2007; Weber, 1930). Com efeito, como refere Silva (2008), "nos atuais espaços de participação sociais e de democracia desvalorizam-se as práticas que são marginais aos objetivos de produtividade, individualismo e consumo, que se protagonizam a par da afirmação de um outro sentido de cidadania, frágil e diminuto" (p. 10).

Pese embora todos os esforços no sentido oposto, a sociedade é ainda dominada por um pensamento baseado em mitos e estereótipos em relação aos idosos (Cerqueira, 2010; Marques, 2011; Miguel, 2014; Sousa & Cerqueira, 2005; Vaz, 2008; WHO, 1999), pelo que permanece a ideia, embora pouco assumida, de que as pessoas idosas, mesmo conservando a sua funcionalidade, são incapazes de se desenvolverem como indivíduos e de exercer a sua cidadania. Cabe, porém, ultrapassar uma visão exclusivamente negativista sobre a experiência do envelhecimento, valorizadora das crises, perdas, incapacidades e declínio, não reproduzindo essa visão de incapacidade individual, especialmente se se considerar o número não negligenciável de idosos que vivem sem doenças e que preservam o seu bem-estar físico e mental, conseguindo manter uma ligação estreita com a família, comunidade e sustentar padrões de elevada qualidade de vida. A superação de muitas das características negativas associadas à idade avançada requer, contudo, uma nova perspetiva do envelhecimento, numa dimensão reconfigurada e moderna, compatível com as atuais circunstâncias sociais, económicas e políticas (Amaro da Luz & Miguel, 2013). No âmbito da definição de uma identidade pessoal e social fortemente baseadas na dimensão profissional, as alterações no contexto de trabalho ou na relação do indivíduo com o mundo do trabalho - da qual a reforma constitui um exemplo da maior importância - poderão traduzir-se em repercussões significativas, tanto a nível pessoal e subjetivo como a nível social, e cuja extensão e intensidade se vai apreendendo aos poucos.

As novas realidades decorrentes da reconfiguração das pirâmides etárias no mundo ocidental suscitam, portanto, a necessidade de construir novas orientações de abordagens aos processos inerentes à "entrada" no envelhecimento, cuja compreensão necessita de ser guiada pela procura de sentidos e significados múltiplos a si inerentes, numa lógica participativa e de inclusão social. Numa etapa da vida em que as dinâmicas de produtividade e consumo não traduzem, como outrora, os elementos centrais da sua participação cívica, impõe-se, pois, o reequacionar de novas formas de expressão de cidadania (Fyfe & Milligan, 2003; Greasly-Adams, 2012; Milligan & Fyfe, 2005). Ainda assim, a posição daqueles que se encontram para além da idade de participação no mercado de trabalho raramente foi examinada quanto à definição da sua cidadania participativa (Craig, 2004; Scharf, Phillipson, & Smith, 2005). Numa lógica de cidadania inclusiva e participada para a população idosa, as posturas que enfatizam o papel central do trabalho na definição do estatuto de cidadão têm vindo a ser criticadas de forma crescente, pela desvalorização que fazem de outras formas de participação social (Amaro da Luz, Miguel, & Preto, 2014; Hank & Erlinghagen, 2010; Miguel, 2016).

Neste sentido, tem vindo a ganhar forma a ideia de que uma perspetiva atual da cidadania exige que esta seja encarada numa lógica inclusiva, interdependente e coletiva que, pela consideração de diferentes formas de exercício de uma cidadania ativa, valorize o amplo contributo social e económico da população idosa. Mais do que a promoção da metáfora do cidadão trabalhador, uma nova abordagem da cidadania na população idosa requer uma aproximação à ideia de uma "cidadania reconquistada", claramente descentrada do estatuto de cidadão definido unicamente a partir da esfera profissional (Miguel, 2016). Neste âmbito, uma abordagem à cidadania que tem atraído renovada atenção nos estudos da cidadania é o de "cidadania republicana" (Dagger, 2002 cit. in Lie et al., 2009), como contrapeso às noções individualistas de cidadania baseadas na reivindicação de direitos. Com foco nos ideais de solidariedade, virtude cívica e "bem comum", esta perspetiva representa uma tradição que acentua a vida cívica como base da cidadania e da comunidade (Delanty, 2002 cit. in Lie et al., 2009).

Ao invés da atividade profissional e económica constituir o cerne da cidadania, o argumento reverte a favor de uma participação social ativa como elemento central da cidadania, impulsionada já não por ganhos económicos individuais, mas por um altruísmo implícito na coesão social. Nesta linha de cidadania inclusiva, Craig (2004) identifica, efetivamente, a "independência e mobilidade, preservação da sua própria identidade e dignidade, escolha e controlo, bem como a capacidade de participar na sociedade tão plenamente quanto possível e de acordo com as suas escolhas" (p. 112) como os fundamentos da cidadania substantiva para as pessoas idosas.

Numa lógica inclusiva, a promoção de um pleno sentido de cidadania por parte dos mais velhos passa, deste modo, pela promoção da sua participação nos mais diversos domínios da vida pessoal e social, respeitando, em simultâneo, as preferências e capacidades individuais. Neste sentido, contrariando as perspetivas que centram no mercado formal de trabalho a sustentação de um sentido identitário e de valorização social, a ênfase deverá ser colocada no domínio da participação total sobre os mais variados domínios económicos, sociais e políticos. Neste pano de fundo, o voluntariado é cada vez mais perspetivado como um instrumento de exercício de uma cidadania ativa potenciadora do bem-estar económico e social dos indivíduos, famílias e comunidades (Robertson, 2012), e as organizações que acolhem estas ações voluntárias, como espaços onde as noções de cidadania podem ser revigoradas (Milligan & Fyfe, 2005). No cenário demográfico atual, o reconhecimento de que a preservação e impulsionamento de espaços democráticos e inclusivos para o voluntariado reveste uma importância primordial para o desenvolvimento de comunidades saudáveis tem, efetivamente, potenciado extensivas ações de promoção da participação ativa por parte dos mais idosos, em especial do voluntariado sénior, no quadro de uma ampla estratégia de envelhecimento ativo (Fernández-Ballasteros, 2008; Ribeiro & Paúl, 2011; Walker, 2002; WHO, 2002).

2. Voluntariado sénior: Uma estratégia de ativação dos idosos

No quadro do envelhecimento ativo (OMS, 2002), ao longo das últimas décadas tem vindo a assistir-se a uma reorientação fundamental nas lógicas de participação social das pessoas idosas, perspetivando-as já não meramente como recetoras de atividades e serviços, mas igualmente – ou sobretudo – como suas prestadoras. Com efeito, esta abordagem protagoniza a participação dos idosos na sociedade como um dos objetivos e compromissos a serem adotados pelos países preocupados em desenvolver uma sociedade inclusiva e para todas as idades (WHO, 2002). As relações estabelecidas com os distintos subsistemas institucionais – família, grupo de pares, comunidade – e o exercício pleno da cidadania emergem, nesta ótica, como palcos incontornáveis da vida social dos indivíduos. Reconhece-se, assim, que a contribuição social dos mais velhos ultrapassa a esfera económica, pois muitos dos seus valiosos contributos têm uma dimensão marcadamente social – tais como os cuidados aos membros da família e a realização de trabalho voluntário na comunidade –, servindo, estas e outras ações, para aumentar e manter o bem-estar pessoal e coletivo.

O voluntariado sénior integra pessoas com idade superior a 65 anos que, por opção, dedicam o seu tempo a atividades sociais não remuneradas e que evidenciam um elevado grau de compromisso. As atividades voluntárias estão, portanto, situadas a meio caminho entre as atividades remuneradas (porque podem constituir um trabalho), entretenimento (porque são escolhidas voluntariamente e ocupam o tempo livre após outras obrigações) e outras atividades sociais (fora da família) (Tomás, Tomás, & Suaréz, 2002). Embora existam diferentes entendimentos sobre o conceito, os estudos convergem nos seus elementos de definição centrais, a saber, a vontade própria, o princípio da gratuidade, a solidariedade/benefício de terceiros (indivíduos, grupos ou organizações) e a sua integração organizacional (Romão, Gaspar, Correia, & Amaro, 2012; Souza & Lautert, 2008).

É expressivo o crescente interesse no voluntariado sénior (Ehlers, Naegele, & Reichert, 2011), devendo-se este impulso a uma dupla combinação de importantes fatores (Smith, 2004). Antes de mais, ao nível microssocial, a preocupação com o bem-estar dos indivíduos levou à consideração da importância que a integração social – a pertença dos indivíduos a um contexto social e relacional específico, de que o voluntariado poderá constituir um importante motor – acarreta importantes benefícios ao longo de todo o ciclo de vida. Depois, ao nível macrossocial, assistiu-se à redução do número de *mulheres domésticas* disponíveis para participação em atividades de voluntariado fruto da sua crescente participação no mercado de trabalho, ao mesmo tempo que se assistiu ao desenvolvimento de um cenário em que um número considerável de idosos tem tempo disponível e capacidade para responder a alguns problemas sociais, através do voluntariado. Muito se ficará a dever à mudança da imagem pública das pessoas idosas, e dos idosos de si próprios, a uma redefinição da natureza e dos méritos do trabalho voluntário, a uma população idosa com maiores níveis educativos e a uma expansão das oportunidades de participação em programas públicos e privados.

O voluntariado é reconhecido como uma importante fonte de satisfação, sociabilidade e autoestima. No entanto, as motivações para o voluntariado podem variar ao longo da vida útil dos indivíduos, de acordo com as diferentes necessidades que são importantes de serem atendidas em cada fase da vida. Tal significa que, nos idosos, as atividades menos importantes possam ser abandonadas (e.g., trabalho remunerado pouco significante) em favor daqueles que implicam um significado mais subjetivo (e.g., solidariedade e ajudar os outros). Este facto explica a relação direta entre a idade e as motivações específicas para o voluntariado, especialmente as baseadas em valores e relações sociais – por exemplo, altruísmo e sentido de pertença –, o desejo de ajudar os outros e o compromisso com a missão da organização, como vários estudos têm vindo a enfatizar (Black & Kovacs, 1999; Greenslade & White, 2005; Hendrick & Curtler, 2004; Martins, 2012). Com efeito, a interrupção do trabalho remunerado é frequentemente associada a uma maior recetividade para o voluntariado e ao aumento das atividades efetivas de voluntariado (Caro & Bass, 1997; Mutchler, Burr, & Caro, 2003), especialmente depois da reforma. Com a desvinculação do mercado de trabalho, as pessoas idosas são desafiadas a criar novas relações sociais fora desse âmbito. Neste sentido, o voluntariado permite manter uma nova perspetiva de vida, um sentido de confiança e utilidade (CES, 2013).

Se, de um modo geral, os contributos de uma sociedade civil ativa são amplamente reconhecidos em termos dos benefícios que produzem tanto para os destinatários dessas ações como para as comunidades nas quais se desenvolvem, a literatura tem vindo a mostrar, de modo consistente, que o envolvimento em ações de voluntariado sénior reverte mais-valias para os próprios. Em particular, um dos importantes benefícios frequentemente associados ao voluntariado é o consequente melhor nível de saúde. Com efeito, a participação em atividades de voluntariado tem-se mostrado associada a melhores níveis percebidos de saúde (Lum & Lightfoot, 2005; Morrow-Howell, Hinterlong, Rozario, & Tang, 2003), maior satisfação com a vida (Nazroo & Mathews 2012; Onyx & Warburton, 2003; Van-Willigen, 2000), diminuída morbilidade e mortalidade (Konrath, Fuhrel-Forbis, Lou, & Brown, 2012; Musick, Herzog, & House, 1999; Onyx & Warburton, 2003), maiores níveis de satisfação e bem-estar (Greenfield & Marks, 2004; Thoits & Hewitt, 2001), menores níveis de dependência funcional e sintomatologia depressiva (Kim & Pai, 2010; Morrow-Howell et al., 2003; Nazroo & Mathews, 2012) e ganhos cognitivos (Cook, 2011). Os benefícios associados ao voluntariado sénior repercutem-se, ainda, no aumento dos contactos sociais e alargamento das redes de apoio social (Misener, Doherty, & Hamm-Kerwin, 2010), contribuindo para a intensificação de sentimentos de pertença e manutenção de uma identidade e auto-estima positivas (Battaglia & Metzer,

2000; Narushima, 2005; Omoto, Snyder, & Martin, 2000; Thoits & Hewitt, 2001), redes essas que não são possíveis de obter a partir do exercício de um papel profissional. Os voluntários séniores tendem, igualmente, a reportar menores níveis de solidão e isolamento social e a encontrar nas ações de voluntariado um apoio, à superação das perdas associadas ao envelhecimento, como a viuvez e à perda de papéis que ocorre com a entrada na reforma (Utz, Carr, Nesse, & Wortmann, 2002).

Para além das vantagens consideráveis para as pessoas mais idosas, são múltiplos os benefícios do voluntariado para as organizações. A este nível, os voluntários séniores mostram-se altamente envolvidos e comprometidos nas suas ações, disponibilizam mais tempo e desenvolvem um voluntariado mais regular no contexto organizacional, constituindo um importante recurso, não só porque permitem minimizar os encargos com a contratação de colaboradores, mas sobretudo porque aportam para os seus papéis de voluntários as competências e sabedoria que adquiriram na sua longa experiência de vida (Chambré, 1993; Zappala & Burrell, 2002).

Embora a coordenação de estratégias, nacionais ou mesmo europeias, de ativação dos mais idosos estejam ainda numa fase emergente, o voluntariado sénior tem vindo a ser reconhecido como uma oportunidade de renovada importância no contexto internacional e, particularmente, no contexto europeu (Hank & Erlinghagen, 2010), manifestamente assumido como o continente mais envelhecido do mundo. Com efeito, o reforço do voluntariado sénior na sociedade atual tem vindo a ser destacado por organizações internacionais. De acordo com a Organização Mundial da Saúde (2002), o voluntariado é um fator importante para manter o bem-estar e a qualidade de vida na velhice, promovendo o envelhecimento ativo e saudável. Através das propostas do Plano de Ação Internacional sobre o Envelhecimento, a Organização das Nações Unidas (2003) salienta a participação dos idosos na sociedade através do trabalho voluntário como um dos compromissos a serem adotados pelos diferentes países, com vista a manter o bem-estar pessoal e coletivo. No contexto europeu, o estabelecimento do Ano Internacional do Voluntariado em 2011 e do Ano Internacional do Envelhecimento Ativo e da Solidariedade entre as Gerações em 2012 acentuaram a maior preocupação com o envelhecimento demográfico e estabeleceram uma conexão clara entre as duas realidades (envelhecimento e voluntariado). Adicionalmente destacaram o papel do voluntariado no alívio das problemáticas associadas à saúde, ao isolamento social e ao envelhecimento.

3. Voluntariado sénior em Portugal: Alguns elementos de diagnóstico

Apesar dos consideráveis benefícios que podem ser identificados e das orientações internacionais relativas ao voluntariado, Portugal apresenta, no contexto europeu, taxas de voluntariado sénior relativamente baixas. Os dados do Eurobarómetro 378 (EC, 2012) revelam que, em Portugal, apenas 6% dos idosos participam de atividades formais de voluntariado sénior, num valor próximo da Turquia (9%) e da Grécia (8%), mas claramente distanciado dos países que a este nível se posicionam no topo, designadamente a Islândia (66%), Suécia (55%) e Holanda (50%).

De forma complementar, os indicadores nacionais (escassos) revelam, a par com uma reduzida taxa de participação em ações de voluntariado formal, uma tendência decrescente de voluntariado (sénior) nos últimos anos. Na verdade, entre 1990 e 1999 houve uma diminuição da taxa de voluntariado formal que passou de 19% para 16% e, em 2001, apenas cerca de 16% dos voluntários eram indivíduos com mais de 65 anos (Delicado, 2002). Num estudo sobre o voluntariado, realizado pelo Observatório do Emprego e Formação Profissional (2008) junto de 8 453 instituições, constatou-se que a percentagem de voluntários (regulares) com idade igual ou superior a 65 anos foi de aproximadamente 11.8%, registando uma maior expressão em instituições como Misericórdias (32.3%), Centros Sociais Paroquiais (31.7%) e Cáritas (26%) (OEFP, 2008). Por seu lado, as estatísticas do Instituto Nacional de Estatística (INE & CASES, 2013) assinalam que, em 2012, 11.5% da população residente e na faixa etária dos 15 anos ou mais participou em, pelo menos, uma atividade formal e/ou voluntariado informal, representando quase 1 milhão e 40 mil voluntários. Novamente, os resultados sugerem que a taxa de voluntariado foi menor em residentes com 65 anos ou mais (7.3%), posicionando-se atrás dos grupos de idade remanescentes considerados (11.6% para o grupo etário de 15-24 anos, 13.1% na faixa dos 25-44 anos e 12.7% para o grupo etário de 45-64 anos). Comparativamente com os restantes países europeus, todos estes indicadores colocam Portugal numa posição pouco expressiva em termos de participação cívica por via do voluntariado, revelando ainda, nos últimos anos, uma dinâmica de involução nas taxas de participação relativas a ações de voluntariado formal, em particular por pessoas de 65 ou mais anos.

Várias são as explicações que têm vindo a ser avançadas para a reduzida incidência do voluntariado em Portugal (Delicado, 2002; Romão, Gaspar, Correia, & Amaro, 2012). Desde logo, as condições políticas e sociais de Portugal, marcadas por uma democracia tardia, vivendo-se ainda as consequências de um longo período de regime autoritário que proibia a maior parte das formas de associativismo, bem como a persistência de uma cultura cívica incipiente, que se traduz, entre outros, na abstenção eleitoral e na indiferença pela política. Depois, os fatores sociodemográficos que se prendem, por um lado, com as baixas habilitações literárias da população portuguesa e diminuta proporção das classes média e média-alta para a participação em algumas ações de voluntariado de base organizacional, qualificado e laico, com novas áreas de motivação e, por outro, as características do mercado laboral português, caracterizado por uma elevada taxa de emprego feminino, na maioria com horários de trabalho a tempo completo. Por fim, a incidência do voluntariado informal – ajuda a familiares e vizinhança – ainda muito frequente no nosso país, e a própria estrutura da economia social em Portugal, que conduz a que grande parte das instituições tenha recursos para contratar funcionários remunerados.

Especificamente no que se refere ao voluntariado sénior, várias têm sido também as barreiras apontadas ao envolvimento por parte dos mais velhos, designadamente, os custos associados à participação voluntária, a localização geográfica, a dificuldade de deslocação e/ou transporte, a inflexibilidade nos horários e na escolha das atividades e funções a serem realizadas pelos voluntários, o requisito de um longo compromisso, e a dificuldade de acesso à informação sobre voluntariado formal (Cook & Sladowski, 2013). No âmbito das organizações, os obstáculos à prática do voluntariado relacionam-se, sobretudo, com a falta de estímulo e apoio à participação dos idosos no trabalho voluntário e com algumas práticas discriminatórias decorrentes da idade. Outros fatores pessoais, como problemas de saúde e barreiras psicológicas decorrentes de falta de confiança e autoestima podem de igual modo contribuir para excluir os idosos das oportunidades de participação a este nível (Warburton, Terry, Rosenman & Shapiro, 2001).

Não obstante, e contrariando a tendência e realidade atual do voluntariado sénior entre nós, as oportunidades e programas de apoio para incentivar os idosos a participar em atividades sociais e cívicas começaram a ser recentemente consideradas. Refira-se, neste contexto, o Programa Nacional de Ação Nacional para o Envelhecimento Ativo (em 2012), a criação de um Programa Nacional de Voluntariado Sénior (desde 2012), o Programa de Ativação Sénior (desde 2014) e várias iniciativas organizacionais provenientes de organizações públicas e de organizações de economia social (e.g., municípios, Fundação de Serralves, Fundação Eugénio de Almeida, Fundação Calouste Gulbenkian). Estas iniciativas destacam, acima de tudo, a relevância do voluntariado como fator importante para manter o bem-estar e a qualidade de vida na velhice, equacionando-o como uma proposta de enorme valia tendente ao envelhecimento ativo (EC, 2012; Ehlers et al., 2011; Lie et al., 2009, Walker, 2002, OMS, 2002).

Um verdadeiro processo de ativação de pessoas idosas reside, portanto, na promoção de sua participação em diversas áreas da vida pessoal e social, promovendo uma cidadania ativa que implica, acima de tudo, uma voz ativa sobre os mais variados domínios económicos, sociais e políticos (Silva, 2008). Tal abre espaços de socialização dos mais idosos, através da criação de ambientes que facilitem a prestação de serviços voluntários em todas as idades, reconhecendo o valor público dessas atividades e facilitando a participação dos idosos, especialmente no caso português (Souza & Lautert, 2008). Saber aproveitar o potencial dos idosos por meio do trabalho voluntário é, pois, não só um mecanismo conservador da saúde e qualidade de vida, mas também uma ferramenta de retirada dos espaços de exclusão social, devolvendo-lhes a autoestima e ajudando-os a exercer sabiamente a sua cidadania, numa etapa da vida em que a exclusão do mercado de trabalho constitui um elemento fragilizador da sua participação social e cívica num contexto societal que atribui à atividade profissional uma génese identitária e um sentido de utilidade (Amaro da Luz, Miguel, & Preto, 2014).

Para esse efeito, várias estratégias têm vindo a ser elencadas com o objetivo de promover a participação ativa da pessoa idosa na sociedade. Por exemplo, a nível cultural, e considerando que os idosos detêm um vasto conhecimento e experiência de vida, argumenta-se a importância da sua contribuição em museus, nomeadamente participando como guias e animadores de museus em visitas de gerações mais jovens, eventos de apoio, bem como a prestação de apoio em tarefas de pesquisa, documentação e inventário (Ribeiro, 2014). A nível social, as práticas de voluntariado podem abranger atividades em centros de idosos, promovendo a participação de membros desses centros (e.g., voluntários séniores com a função de realizar/participar em atividades sociais e culturais nas estruturas residenciais para idosos). As atividades podem, igualmente, integrar projetos com pessoas da mesma faixa etária, organizados com a cooperação de voluntários, numa pequena rede de cooperação doméstica (e.g., atendimento domiciliário a outros idosos). Enfatizando a perspetiva intergeracional, os voluntários séniores podem ser envolvidos na partilha de experiências com grupos de jovens, disponibilizando os conhecimentos e afetando as suas competências a favor de crianças e grupos de jovens (e.g., voluntários seniores que partilham experiências em escolas). Outros tipos de atividades podem incluir o fornecimento de informações a outros idosos, facilitando o acesso aos recursos disponíveis e à prestação de serviços (Martín, Gonçalves, Silva, Paúl & Cabral, 2007).

Considerações finais

O progressivo e rápido envelhecimento demográfico é um fenómeno que carece de atenção nas sociedades ocidentais, considerando os importantes desafios que se colocam a vários níveis. Atendendo às projeções que apontam para o aumento deste fenómeno nas próximas décadas, é claramente importante encetar a procura de soluções inovadoras, que conduzam a novas direções e significados na promoção da qualidade de vida das pessoas mais velhas, cada vez mais vistas não apenas como uma população em crescimento, mas também como uma população com diferentes perfis e necessidades. Perante um número cada vez mais ativo e saudável de idosos, impõe-se repensar novas formas de ativação e participação social, capazes de aumentar o contributo dos idosos com disponibilidade e capacidade para responder a alguns problemas sociais. Se, do ponto de vista sociológico, o tempo de reforma e a cessação da vida profissional foram bem-vindas desde o início do século passado, o abandono de uma atividade profissional remunerada tem, no entanto, conotações sociais e pessoais divergentes. Com efeito, no momento em que o trabalho se tornou uma importante fonte de integração social e de valor social convocando uma participação económica ativa, a entrada na reforma não acontece sem reservas. Nesse sentido, a centralidade do trabalho na discussão de questões relacionadas com o envelhecimento implica relativizar o trabalho como fonte de cidadania e a necessidade de reconfigurar outras formas de participação social. Numa sociedade para todas as idades, o ato social da reforma não pode conduzir à marginalização económica e social das pessoas idosas.

Assim, para muitos idosos, o trabalho voluntário poderá surgir como uma ferramenta que auxilie na maximização do sentimento de atividade e utilidade social, atuando como um mecanismo para a conservação da saúde e qualidade de vida. Além disso, a participação em atividades não remuneradas é uma alternativa em termos de participação social e uma dimensão inquestionável para o exercício de uma cidadania reconquistada, no que se refere à maximização dos contributos individuais e sociais dos idosos. A proposta de ampliar a ativação dos idosos é, portanto, utilizada como base para o desenvolvimento de políticas de envelhecimento multissetorial, a fim de melhorar a saúde e aumentar a participação dos idosos. Para que esse objetivo seja alcançado, é de vital importância facilitar as práticas voluntárias das pessoas mais velhas, criando espaços e ambientes de acolhimento para a realização desta prática. Não só se deve enfatizar que as iniciativas que criam ambientes favoráveis e a promoção de escolhas saudáveis são fundamentais para as políticas que procuram promover a saúde e a qualidade de vida dos adultos mais velhos, mas também é necessária uma reconfiguração das possibilidades de uma plena participação social.

Conforme foi salientado, o voluntariado sénior possui amplas oportunidades de concretização passíveis de influenciar uma reconceptualização da ideia de cidadania sénior. Perante o cenário demográfico atual, mais do que a promoção do "cidadão-trabalhador" e "cidadão-consumidor", uma perspetiva atual da cidadania requer que esta seja encarada numa lógica inclusiva, interdependente e coletiva, que, nomeadamente pelo voluntariado sénior, valorize o amplo contributo social e económico da população idosa. Deste modo, o valor instrumental do voluntariado reverte mais-valias numa dupla dimensão: em primeiro, para os próprios idosos, que veem na sua participação cívica e ativa uma forma de "cidadania reconquistada"; em segundo, para a sociedade em geral que, beneficiária dos contributos dos seus cidadãos de todas as idades, promove uma cidadania coletiva.

É, pois, chegado o momento de consolidar um novo paradigma de velhice, que conceba o idoso como contribuinte ativo e participante de uma sociedade inclusiva, refletindo e promovendo o voluntariado sénior como via de restituição e "reconquista" de uma cidadania ativa e participada, capaz de reverter a imagem de improdutividade que a entrada na reforma (e na velhice) frequentemente acarreta. A superação de uma visão redutora de cidadania, centrada em teses produtivistas, para a consideração de uma perspetiva holística e inclusiva de cidadania beneficiará, deste modo, uma visão societal como espaço alargado de exercício da cidadania.

Bibliografia

Amaro da Luz, M. H., & Miguel, I. (2013). *Empreendedorismo social: Dinâmicas de proximidade territorial a favor de uma cidadania inclusiva*: Manuscript submitted for publication.

Amaro da Luz, H., Miguel, I., & Preto, S. (2014). *Inatividade legitimada: Que alternativas de "trabalho" em período de reforma?* Actas do VIII Congresso Nacional de Sociologia (Portugal).

Atchley, R. C. (2000). *Social forces and aging*. Belmont, CA: Wadsworth.

Atkinson, R., & Davoudi, S. (2000). The concept of social exclusion in the European Union: Context, development and possibilities. *Journal of Common Market Studies, 38*, 427-448.

Baltes, P. B., & Baltes, M. M. (1990). *Successful aging: Perspectives from the behavioral sciences*. Cambridge, UK: Cambridge University Press.

Battaglia, A., & Metzer, J. (2000). Older adults and volunteering: A symbiotic association. *Australian Journal on Volunteering, 5*(1), 5-12.

Black, B. & Kovacs, P. J. (1999). Age-related variation in roles performed by hospice volunteers. *The Journal of Applied Gerontology, 18*(4), 479--497.

Caro, F. & Bass, S. A. (1997). Receptivity to volunteering in the immediate posretiremente period. *Journal of Applied Gerontology, 16*(4), 427-442.

Casey, C. (1995). *Work, self and society: After industrialism*. London: Routledge.

Castel, R. (1995). *Les métamorphoses de la question sociale*. Paris, France:Fayard.

Cerqueira, M. (2010). *Imagens do envelhecimento e da velhice: Um estudo na população portuguesa*. Unpublished Tese de Doutoramento, Universidade de Aveiro, Aveiro.

CES – Centro de Estudos Social da Universidade de Coimbra (2013). Voluntariado em Portugal: Contxetos, atores e práticas. Lisboa: Fundação Eugénio de Almeida.

Chambré, S. M. (1993). Volunteerism by elders: Past trends and future prospects. *The Gerontologist, 33*(2), 221-228.

Cook, S. L. (2011). An exploration of learning through volunteering during retirement. *International Journal of Volunteer Administration, 28*, 9-19.

Cook, S. L., & Sladowski, P. S. (2013). *Volunteering and older adults*: Volunteer Canada Final Report.

Craig, G. (2004). Citizenship, exclusion and older people. *Journal of Social Policy, 33*, 95-114.

Delicado, A. (2002). Caracterização do voluntariado social em Portugal. *Intervenção Social, 25/26*, 127-140.

EC. (2012). *Special Eurobarometer 378: Active ageing*: European Comission: Directorate-General for Employment, Social Affairs and Inclusion.

Ehlers, A., Naegele, G., & Reichert, M. (2011). *Volunteering by older people in the EU*. Brussels: Publications Office of the European Union.

Fernández-Ballasteros, R. (2008). *Active aging: The contribution of psychology*. Cambridge, MA: Hogrefe & Huber Publishers.

Fernández-Ballasteros, R., Caprara, M. G., & García, L. F. (2004). Vivir con vitalidad-M: Un programa europeo multimedia. *Intervención Psicosocial, 13*(1), 63-84.

Fonseca, A. M. (2010). Promoção do desenvolvimento psicológico no envelhecimento. *Contextos Clínicos, 3*(2), 124-131.

Fonseca, A. M. (2011). *Reforma e reformados*. Coimbra: Almedina.

Fyfe, N. R., & Milligan, C. (2003). Space, citizenship, and voluntarism: Critical reflections on the voluntary welfare sector in Glasgow. *Environment and Planning, 35*(11), 2069-2086.

Goss, K. A. (1999). Volunteering and the long civic generation. *Nonprofit and Voluntary Sector Quarterly, 28*(4), 378-415.

Greasly-Adams, C. S. (2012). *Work activities of older people: Beyond paid employment*. Unpublished PhD, University of Stirling, Stirling.

Greenfield, E. A., & Marks, N. F. (2004). Formal volunteering as a protective factor for older adults' psychological well-being. *Journals of Gerontology, 59*, 258-264.

Greenslade, J. H. & White, K. M. (2005). The prediction of above-average participation in volunteerism: A test of the theory of planned behavior and the volunteers function inventory in older Australian adults. *The Journal of Social Psychology*, 145(2), 155-172.

Han, S. K., & Moen, P. (1999). Clocking out: Temporal patterning of retirement. *American Journal of Sociology, 105*(1), 191-236.

Hank, K. & Erlinghagen, M. (2010). Volunteering in "Old" Europe: Patterns, Potentials, Limitations. *Journal of Applied Gerontology, 29*(1), 3-20.

Hardill, I., & Baines, S. (2009). Active citizenship in later life: Older volunteers in a deprived community in England. *The Professional Geographer, 61*(1), 36-45.

Hendrick, J. & Curtler, S. J. (2004). Volunteerism and sociemotional selectivity in later life. *The Journal of Gerontology, 59B*(5), 251-257.

Higgs, P. (1995). Citizenship and old age: The end of the road? *Ageing and Society, 15*, 535-550.

INE & CASES (2013). *Conta Satélite da Economia Social 2010*. Lisboa, Portugal: Instituto Nacional de Estatística.

Kaye, L. W., Butler, S. S., & Webster, N. M. (2003). Toward a productive ageing paradigm for geriatric practice. *Ageing International, 28*(2), 200-213.

Kim, J., & Pai, M. (2010). Volunteering and trajectories of depression. *Journal of Aging & Health Psychology, 22*, 84-105.

Konrath, S., Fuhrel-Forbis, A., Lou, A., & Brown, S. (2012). Motives for volunteering are associated with mortality risk in older adults. *Health Psychology, 31*(1), 87–96.

Kovács, I. (2009). Work and citizenship: crises and alternatives. *Entreprise and work Innovation Studies, 5*(IET), 37-58.

Lazarus, R., & Lazarus, B. (2006). *Coping with aging*. Oxford: Oxford University Press.

Leão, A., & Barros, S. (2008). As representações sociais dos profissionais de saúde mental acerca do modelo de atenção e as possibilidades de inclusão social. *Saúde e Sociedade, 17*(1), 95-106.

Levitas, R. (1998). *The inclusive society? Social exclusion and New Labour.* Basingstoke: Macmillan.

Lewis, J. (2002). Gender and welfare state change. *European Societies, 4*(4), 331-357.

Lie, M., Baines, S., & Wheelock, J. (2009). Citizenship, volunteering and active aging. *Social Policy and Administration, 43*(7), 702-718.

Lum, T. Y., & Lightfoot, E. (2005). The effects of volunteering on physical and mental health of older people. *Research on Aging, 27*, 31-35.

Marques, S. (2011). *Discriminação da terceira idade*. Lisboa, Portugal: Relógio D'Água.

Martín, I., Gonçalves, D., Silva, A., Paúl, C., & Cabral, F. (2007). Políticas Sociais para a Terceira Idade. In A. Osório & F. Pinto (Eds.), *As pessoas idosas: Contexto social e intervenção educativa* (pp. 131-179). Lisboa, Portugal: Instituto Piaget.

Martins, T. (2012). Voluntários/as reformados/as: Práticas de voluntariado na reforma. Dissertação de mestrado, Instituto Superior de Serviço Social do Porto, Senhora da Hora, Portugal

Mauritti, R. (2004). Padrões de vida na velhice. *Análise Social, 39*, 339--363.

Meda, D. (2007). Que sabemos sobre el trabajo? *Revista de Trabajo, 3*(4), 17-32.

Miguel, I. (2014). Envelhecimento e desenvolvimento psicológico: Entre mitos e factos. In H. Amaro da Luz & I. Miguel (Eds.), *Gerontologia Social: Perspetivas de Análise e Intervenção* (pp. 53-67). Instituto Superior Bissaya Barreto. ISBN: 978-989-98952-0-1

Miguel, I. (2016). Mais idades, menos particiapção? Lógicas de "resgate" da cidadania na população idosa. In C. Albuquerque & H. Amaro da Luz (Eds.), *Políticas sociais em tempos de crise: Perspetivas, tendências e questões críticas* (pp. 193-208). Lisboa: Pactor.

Milligan, C., & Fyfe, N. R. (2005). Preserving space for volunteers: Exploring the links between voluntary welfare organisations, volunteering and citizenship. *Urban Studies, 42*(3), 417-433.

Misener, K., Doherty, A., & Hamm-Kerwin, S. (2010). Learning from the experiences of older adult volunteers in sport: A serious leisure perspective. *Journal of Leisure Research, 42*(2), 267-289.

Morrow-Howell, N., Hinterlong, J., Rozario, P. A., & Tang, F. (2003). Effects of volunteering on the well-being of older adults. *Journal of Gerontology, 58*(3), 137-145.

MOW. (1987). *The meaning of work.* New York: Academic Press.

Musick, M. A., Herzog, A. R., & House, J. S. (1999). Volunteering and mortality among older adults: Findings from a national sample. *Journal of Gerontology, 54*, 173-180.

Mutchler, J. E., Burr, J. A., Caro, F. G. (2003). From paid worker to volunteer: Leaving the paid workforce and volunterring in later life. *Social Forces, 81*(4), 1267-1283.

Narushima, M. (2005). 'Payback time': Community volunteering among older adults as a transformative mechanism. *Aging & Society, 25*, 567-584.

Nazroo, J., & Mathews, K. (2012). *The impact of volunteering on well-being in later life.* Cardiff: WRVS.

Observatório do Emprego e Formação Profissional (2008). *Estudo sobre o voluntariado.* Lisboa: Observatório do Emprego e Formação Profissional.

Omoto, A. M., Snyder, M., & Martino, S. C. (2000). Volunteerism and the life course: Investigating age-related agendas for action. *Basic and Applied Social Psychology, 22*(3), 181-197.

Onyx, J., & Warburton, J. (2003). Volunteering and health among older people: A review. *Australasian Journal of Ageing, 22*(2), 65-69.

Parada, F., & Coimbra, J. L. (1999). O trabalho como dimensão de construção da cidadania: Reflexões sobre o papel da escola no processo de formação do indivíduo cidadão/trabalhador. *Inovação, 12*, 93-108.

Polanyi, K. (1980). *A grande transformação – as origens da nossa época*. Rio de Janeiro, Brasil: Editora Campus.

Perret, B. (1993). *L'Avenir du Travail – les démocracies face au chômage*. Paris, France: Éditions du Seuil.

Rapaport, R., & Bailyn, L. (1998). *Rethinking life and work*. Waltham, MA: Pegasus Communications.

Reis, M. H. (2006). *A economia social face às questões do emprego: A função reguladora do terceiro sector no domínio da política económica e social*. Coimbra: Fundação Bissaya Barreto.

Ribeiro, O. (2014). *Práticas do voluntariado nos museus universitários: Contributos para a criação de uma bolsa de voluntários séniores especializados*. Dissertação de mestrado em Museologia, Faculdade de Letras da Universidade do Porto.

Ribeiro, R., & Paúl, C. (2011). *Manual de envelhecimento activo*. Porto: Lidel.

Robertson, G. (2012). *Active ageing and solidarity between generations: The contribution of volunteering and civic engagement in Europe*. London: Volonteurope Secretariat.

Romão, G., Gaspar, V., Correia, T., & Amaro, R. (2012). Estudo de caracterização do voluntariado em Portugal: Trabalho para o Conselho Nacional para a Promoção do Voluntariado. PROACT – Unidade de Investigação e Apoio Técnico ao Desenvolvimento Local, à Valorização do Ambiente e à Luta Contra a Exclusão Social.

Rowe, J. W., & Khan, R. L. (1998). *Successful aging*. New York, NY: Random House.

Savishinsky, J. S. (2000). *Breaking the watch: The meanings of retirement in America*. Ithaca, NY: Cornell University Press.

Scharf, T., Phillipson, C., & Smith, A. (2005). Social exclusion of older people in deprived urban communities of England. *European Journal of Ageing, 2*, 76-87.

Silva, S. M. (2008). *Os espaços de participação das pessoas mais velhas na sociedade portuguesa actual*. Paper presented at the VI Congresso Português de Sociologia: "Mundos Sociais: Saberes e Práticas".

Smith, D. B. (2004). Volunteering in retirement: Perceptions of midlife workers. *Nonprofit and Voluntary Sector Quarterly, 33*(1), 55-73.

Sousa, L., & Cerqueira, M. (2005). As imagens da velhice em diferentes grupos etários: Um estudo exploratório na população portuguesa. *Kairós, 8*(2), 189-206.

Souza, L. M., & Lautert, L. (2008). Trabalho voluntário: Uma alternativa para a promoção da saúde de idosos. *Revista da Escola de Enfermagem USP, 42*(2), 371-376.

Sterns, H. L., & Gray, J. H. (1999). Work, leisure, and retirement. In J. Cavanaugh & S. K. Whitbourne (Eds.), *Gerontology: An interdisciplinary perspective* (pp. 355-390). New York: Oxford University Press.

Thoits, P. A., & Hewitt, L. N. (2001). Volunteer work and well-being. *Journal of Health and Social Behavior, 42*, 115-131.

Tomás, M. S., Tomás, E., & Suaréz, J. (2002). Voluntariado de mayores: Ejemplo de envejecimiento participativo y satisfactorio. *Revista Interuniversitario de Formácion del Professorado, 45*, 107-128.

Turner, B. (1989). Ageing, status politics and social theory. *British Journal of Sociology, 40*, 588-606.

UNO – United Nations Organization (2003). *International Plan of Action on Ageing 2002*. New York: UNO.

Utz, R. L., Carr, D., Nesse, R., & Wortmann, C. B. (2002). The effect of widowhood on older adults' social participation: An evaluation of Activity, Disengagement and Continuity theories. *The Gerontologist, 42*(4), 522--533.

Van-Willigen, M. (2000). Differential benefits of volunteering accross the life course. *Journal of Gerontology, 55*, 308-318.

Vaz, E. (2008). *A velhice na primeira pessoa*. Penafiel, Portugal: Editorial Novembro.

Victor, C. (2005). *The social context of ageing: A textbook of Gerontology*. Oxon: Routledge.

Walker, A. (2002). A strategy for active ageing. *International Social Security Review, 55*(1), 121-138.

Warburton, J., & Cordingley, S. (2004). The contemporary challenges of volunteering in an ageing Australia. *Australian Journal on Volunteering, 9*(2), 67-74.

Warburton, J., Terry, D. J., Rosenman, L., & Shapiro, M. (2001). Differences between older volunteers and non-volunteers: Attitudinal, normative and control beliefs. *Research on Aging, 23*(5), 586-605.

Weber, M. (1930). *The protestant ethic and the spirit of capitalism*. London: Allen and Unwin.

WHO. (1990). *Healthy aging*. Geneva, Switzerland: World Health Organization.

WHO. (1999). *Ageing: Exploring the myths*. Ageing and Health Programme: World Health Organization.

WHO. (2002). *Active aging: A policy framework*. Geneva, Switzerland: World Health Organization.

Zappala, G., & Burrell. (2002). What makes a frequent volunteer? Predicting volunteer commitment in a community services organisation. *Australian Journal on Volunteering, 7*(2), 45-58.

PARTE III

ESPECIFICIDADES NA INTERVENÇÃO SOCIAL COM A POPULAÇÃO IDOSA

CAPÍTULO VII

A INTERVENÇÃO PSICOSSOCIAL AO DOMICÍLIO

Margarida Pedroso de Lima

Universidade de Coimbra, Faculdade de Psicologia e Ciências de Educação

ORCID: https://orcid.org/0000-0002-6239-1137

Resumo: *Edificar uma sociedade que ofereça oportunidades de envelhecer de forma expressiva e saudável ao longo de todo o ciclo de vida e que seja sustentável para todas as idades implica aceitação da nossa enorme variabilidade interpessoal. Neste processo a intervenção como um promotor de bem-estar pessoal e social faz-se figura. Mas para que tal aconteça ela tem que ser variegada de molde a responder às necessidades diversificadas da população das pessoas mais velhas no nosso país. Neste capítulo apresentam-se as características e analisam-se as dificuldades e as vantagens de um tipo de intervenção – a intervenção psicossocial ao domicílio. Mais especificamente do apoio psicológico ou psicoterapêutico ao domicílio.*

Palavras-chave: *Intervenção psicossocial; Apoio psicológico; Psicoterapia; Domicílio; Idade avançada*

DOI | https://doi.org/10.14195/978-989-26-1788-6_7

Abstract: *Building a society that offers opportunities for aging in an expressive and healthy way throughout the life cycle and that is sustainable for all ages implies acceptance of our enormous interpersonal variability. In this process intervention as a promoter of personal and social well-being becomes a figure. But for that to happen, it has to be variegated to respond to the diverse needs of the population of the older people in our country. This chapter presents the characteristics and analyzes the difficulties and advantages of one type of intervention – psychosocial intervention at home. More specifically, psychological or psychotherapeutic support at home.*
Key-words: *Psychosocial intervention; Psychological support; Psychotherapy; Residence; Advanced age.*

Nota introdutória

A população mundial está a envelhecer. Em 2012 havia em todo o planeta 810 milhões de mulheres e homens com 60 anos ou mais. As projeções são para atingirmos 1 bilião em menos de 10 anos aumentando a proporção das pessoas mais velhas comparativamente aos outros grupos etários em todas as regiões do mundo. Estes dados deveriam ser mais do que suficientes para responder à questão '*why older people count?*'. No entanto, como chama atenção o Global AgeWatch (2015), apesar desta tendência a opinião das pessoas mais velhas não é, muitas vezes, incluída em muitas estatísticas. Tal significa que a situação dos mais velhos é frequentemente invisível sendo difícil identificar e documentar padrões de discriminação fortemente arraigados. Quando há dados sobre os mais velhos essa informação nem sempre é desagregada por sexo, pelas diferentes fases que compõem a velhice ou/e pelo local onde estão a residir, sendo que as pessoas a viver na comunidade são muitas vezes negligenciadas. Este capítulo vai ao encontro do espírito do chamamento global das Nações Unidas (2017) de não deixar ninguém para trás "*leave no one behind*". Uma cultura em que as contribuições das pessoas mais velhas para a sua economia e comunidade são reconhecidas e apoiadas é uma cultura que não discrimina com base na idade.

As implicações desta perspectiva são inúmeras nomeadamente a de aceitar a diversidade de formas de envelhecer e de viver a velhice. Como sabemos esta é a variável chave para o bem estar e a qualidade de vida na idade avançada (WHOQOL Group, 1995; Simões et al., 2006). A diversidade e a heterogeneidade, como é sabido, é a característica mais saliente da faixa etária dos mais velhos (Carroll & Nuro, 2002). Consequentemente, uma sociedade preocupada com a qualidade de vida das pessoas mais velhas deve ser desenhada de molde a oferecer opções que respeitem esta variedade de formas de estar nesta fase da vida. Por outro lado, muitas pessoas na velhice têm que lidar com inúmeras mudanças em domínios diversos como o trabalho, a saúde, a cognição, e as relações sociais. Eventualmente, têm de se confrontar com doenças agudas e crónicas, restrições na mobilidade e perda da autonomia. Estes desafios, presentes nas gerações atuais das pessoas mais velhas, alertam-nos para a importância de edificar políticas sociais que vão de encontro às necessidades de serviços e equipamentos que respondam a uma população que está a crescer em número mas também nas suas exigências. Respeitar os mais velhos implica criar respostas que respondam a estas necessidades (Oliveira et al., 2011).

No entanto, neste capítulo, vamos apenas abordar um aspecto deste respeito pela diversidade, ou seja, o direito das pessoas mais velhas fragilizadas continuarem a viver nas suas casas. Para o sucesso desta alternativa o apoio ao domicílio é fundamental. Um dos aspectos menos estudados e mais descurados neste tipo de apoio e intervenção tem sido o apoio psicossocial ao domicílio.

O crescente envelhecimento demográfico e as dificuldades das famílias que procuram apoiar os seus idosos tem-se refletido no aumento das taxas de institucionalização. No futuro que se avizinha, o número de idosos institucionalizados poderá ter uma proporção diretamente relacionada com o envelhecimento demográfico. Porém, os estudos sobre os efeitos da institucionalização apresentam resultados aparentemente contraditórios, fruto das complexidades do fenómeno (Kahana, 2001), a saber, as influências ambientais e as diferenças individuais ao nível da personalidade, história pessoal e motivação. Ora o apoio domiciliário às pessoas idosas pode substituir/adiar a institucionalização, reduzir a hospitalização e promover a qualidade de vida (Brick, 2011; Bernabei et al., 1998) das pessoas que preferem permanecer em sua casa (Mello et al., 2012).

Estas evidências têm contribuído para o aumento do investimento da Europa e das comunidades em profissionais de diferentes áreas da saúde e sociais em serviços e equipamentos de apoio domiciliário. A premissa subjacente é que estes serviços ajudam a manter a independência dos mais velhos, reduzindo incapacidades e fortalecendo as suas escolhas como defendido na política do envelhecimento ativo da WHO (2002). Neste capítulo fazemos uma breve revisão sobre a literatura ainda escassa a este nível e retiramos algumas implicações para a prática.

Desenvolvimento

Envelhecer em casa (*Aging in place*) é uma das respostas mais promissoras para promover a qualidade de vida das pessoas idosas frágeis e doentes e reduzir custos. Efetivamente, de uma perspectiva económica e política envelhecer em casa envolve menores custos financeiros que o cuidado residencial (Brick, 2011) e melhora a saúde dos beneficiários. De acordo com os dados da AARP (2010) 92% das pessoas do segmento de 65+ preferem ficar a residir em suas casas o maior tempo possível, ao invés de serem institucionalizadas (*e.g.*, Mello et al., 2012; Sorocco et al., 2011). Continuar a viver na sua residência habitual de forma independente, perto dos seus amigos e familiares significa manter os seus relacionamentos e estilo de vida (Figueiredo et al., 2012). As casas são particularmente importantes para as pessoas mais velhas, conectando-as com as suas famílias, vizinhos e amigos, a sua história e a sua identidade (Mestheneos, 2011). Assim, as iniciativas que possam contribuir para ir de encontro à aspiração dos mais velhos de *Aging in place* devem ser implementadas.

Muitas das pessoas mais velhas fragilizadas que querem continuar a residir nas suas casas vivem sozinhas sem apoio informal, outras têm apoio por parte de cuidadores informais como os filhos ou esposo(a). Mesmo neste último cenário, cada vez mais é reconhecida a importância de apoiar os cuidadores informais a serem efetivos. Os *Serviços de Descanso no Domicílio* (Portaria n.º 50/2017), os *Serviços de Apoio Domiciliário* (Direção Geral da Ação Social, 1996) e os serviços de saúde para os mais velhos prestados ao domicílio são respostas propostas pelo Governo Português como um complemento ao trabalho desenvolvido pelo cuidador informal em casa. Neste sentido, segundo o Serviço Nacional de Saúde na Área dos Cuidados Continuados Integrados um familiar que cuide de um idoso dependente terá direito não só a horários flexíveis, mas também a certos benefícios fiscais sendo um incentivo para que os idosos fiquem junto da família o máximo de tempo possível. O apoio ao cuidador informal acarreta benefícios como o aumento do bem-estar físico e psicológico e o decréscimo dos níveis de stress, ansiedade, depressão e sentimentos de solidão do cuidador e benefícios consequentes para o idoso dependente.

Estes serviços combinados com outros como centros de dia adiam indiretamente a institucionalização permanente, ou a longo prazo, da pessoa fragilizada ou dependente (Zarit, 2001).

Ao longo da vida, como é sabido o apoio social é um recurso importante para enfrentar desafios e dificuldades, contribuindo para o decréscimo do isolamento social e da solidão (Fernandes, 2014). O sentimento de solidão, definido como o nível de isolamento percebido, é considerado um problema social nesta faixa etária, está relacionado com a depressão (Stella et al., 2002), sendo a sua prevenção uma das premissas do envelhecimento ativo (Paúl, 2005). Devido à importância dada aos relacionamentos interpessoais na promoção do bem-estar e da qualidade de vida (Lubben & Grionda, 2003; Vecchia et al., 2005), a reflexão em torno das respostas sociais de apoio a pessoas idosas que vivam sozinhas, sem cuidador informal ou apenas com cuidador a tempo parcial, tem adquirido uma crescente relevância. O apoio social é, por conseguinte, essencial para promover o bem-estar físico e psicológico e, segundo Vaz Serra (2002), aglutina a rede social, o apoio recebido e o apoio percebido.O apoio recebido é a quantidade de apoio prestado pelos elementos da rede social e o apoio percebido é a avaliação subjetiva que cada um tem do suporte que possuí (Uchino, 2004).

O apoio domiciliário pode ser operacionalizado como qualquer serviço de diagnóstico, terapêutico ou de suporte prestado a casa, abrangendo uma ampla gama de serviços fornecidos por um grupo diversificado de profissionais (Brown et al., 2008).

O objetivo deste tipo de apoio é providenciar cuidados vários, aos utentes e pacientes idosos a residir em suas casas, de modo a promover a sua autossuficiência e independência de molde a poderem permanecer com qualidade nas suas casas (Frade, Barbosa, Cardoso & Nunes, 2015).

Os programas de cuidados de saúde domiciliários geralmente estão bem preparados para satisfazer uma ampla variedade de necessidades de cuidados médicos, contudo muitas vezes são incapazes de fornecer serviços para atender às especificidades dos pacientes idosos que sofrem de uma panóplia diversa de problemas para além dos de saúde, que afetam o seu bem-estar e a sua qualidade de vida. Muito menos estão preparados para ir além do apoio tradicional ao domicílio (*e.g.*, comida, limpeza, serviços médicos) e providenciar serviços como psicoterapia, reabilitação, terapia pela arte, animação sociocultural, desenvolvimento pessoal ou educação.

No entanto, a idade avançada e o estar confinado ao domicílio não é um obstáculo à psicoterapia (Knight, 2004), à reabilitação e às intervenções educacionais e desenvolvimentais. Estes programas, sobretudo se forem focalizados e breves, levam a melhorias ao nível cognitivo, social e emocional e adiam a institucionalização (Rodin, 2009).

A intervenção psicológica e psicossocial (Lima et al., 2014) tem assim como objectivo geral a promoção da qualidade de vida e do bem-estar da pessoa idosa, a prevenção da deterioração (física, neuropsicológica), a redução do *distress* psicológico e a prevenção do surgimento de desordens mentais como a depressão.

No entanto, providenciar serviços de apoio domiciliário personalizados e sustentáveis respondendo a uma panóplia de necessidades de saúde, sociais e educacionais da população das pessoas mais velhas deve partir de uma avaliação rigorosa das suas necessidades. Só assim é possível ter em consideração as especificidades e necessidades individuais.

Nas instituições com uma valência de apoio domiciliário são, geralmente, os técnicos de serviço social que encaminham os utentes para os diversos serviços disponíveis numa determinada área residencial, geralmente, cuidados de alimentação, higiene e de saúde (Sorocco et al., 2011). No entanto, há um crescente reconhecimento da potencialidade das intervenções psicossociais ao domicílio para idosos em situação de fragilidade e que têm problemas de saúde mental (Knight, 2004).

O apoio psicológico ao domicílio pode aliás ser o único meio para que alguns pacientes tenham acesso a apoio social e a cuidados de saúde mental, podendo estes oferecer vantagens em relação aos cuidados hospitalares (Muijen, Marks, Connolly & Audini, 1992).

O apoio psicológico ao domicílio (terapia de suporte) e a psicoterapia individual a casa se for desenhada especialmente para responder às questões da pessoa idosa pode, além de ajudar a aliviar o mal-estar, promover o crescimento psicológico (Lima & Silvério, 2015). Contudo, fornecer apoio a casa apresenta inúmeros benefícios, mas também levanta inúmeras questões clínicas, éticas, legais, de gestão de riscos e segurança (Yang et al., 2009), em comparação com o trabalho realizado em consultório ou ambiente institucional (Yang, Garis, Jackson & McClure, 2009). Assim, para além do desafio emocional espectável da confrontação do terapeuta com a doença, a morte, a solidão ou a ausência de sentido (Lima & Silvério, 2015) ele encontrará um *setting* terapêutico difícil de controlar. Em contexto institucional o ambiente é mais facilmente controlado e previsível, sem interrupções ou presenças indesejadas, o que não é possível em contexto domiciliário, podendo emergir acontecimentos inesperados comprometendo a privacidade das sessões e exigindo uma grande flexibilidade por parte do terapeuta.

No entanto, por outro lado, na psicoterapia ao domicílio, o terapeuta beneficia de fontes de informação diretas privilegiadas, fornecidas pelo próprio ambiente em que o paciente habita. Estas auxiliam o estabelecimento da relação terapêutica, clarificam a situação do cliente e, consequentemente, promovem uma intervenção mais eficaz (Yang et al., 2009).

A menor formalidade presente no contexto domiciliário pode, por um lado, facilitar e aprofundar a relação terapêutica (Maxfield et al., 2008) mas, por outro, dificultar o estabelecimento de limites e a manutenção das fronteiras profissionais. A saber, o idoso ligeiramente pode ver o terapeuta como um amigo ou uma visita em vez de como profissional de saúde (Maxfield et al., 2008) e tentar envolver o terapeuta noutras atividades para além da terapia (Yang et al., 2009).

Para o sucesso da intervenção é necessário que se estabeleçam limites e, para isso, é indispensável relembrar e esclarecer o cliente do propósito das sessões, dos horários e da natureza do relacionamento cliente – terapeuta. Em contrapartida, o terapeuta tem também de estar consciente do direito dos idosos recusarem ou desistirem das sessões (Yang et al., 2009).

Algumas investigações baseadas na evidência providenciam dados sobre a eficácia de serviços e intervenções psicossociais, como a psicoeducação, com adultos seniores fragilizados a viver nas suas casas, levando à redução da depressão. No entanto, poucas são ainda as investigações que relatam o apoio psicológico levado a casa (Lima et al., 2014). Em Portugal temos trabalhos do in-Home Mental Health Services do Home Counselling *Pro-VoluntariU* (Oliveira et al., 2013) e do HIT (Duarte, 2017), mostrando que este tipo de serviços promove a saúde e o bem-estar, bem como, previne a patologia e o isolamento (Lima et al., 2014).

No entanto, a escassa literatura a este nível sublinha que as intervenções devem ter como pano de fundo as orientações internacionais sobre a intervenção com pessoas de idade avançada (Ferreira-Alves, 2010) que estão sintetizadas no modelo CALTAP (*Contextual Adult Life Span Theory for Adapting Psychotherapy*, Knight & Poon, 2008). Este é uma meta modelo desenhado para orientar a abordagem psicoterapêutica com adultos idosos e que evidencia a importância interativa da coorte, da cultura, do contexto imediato e da maturação ao nível cognitivo e emocional na eficácia da intervenção psicoterapêutica (Knight, 2004; Knight & Poon, 2008).

Importante é também que os técnicos tenham formação na área do aconselhamento psicológico e no estabelecimento de uma relação de ajuda baseada nos princípios humanistas da aceitação incondicional, congruência e empatia que possibilitam os sentimentos de auto aceitação e de auto compreensão por parte da pessoa idosa. Apenas neste ambiente é possível clarificar os sentimentos de confusão que eventualmente surjam de molde a promover a mais valia pessoal, bem-estar e sentido de vida (Lima, 2014).

Considerações finais

Neste capítulo apresentamos uma perspetiva mais compreensiva sobre o apoio e o cuidado domiciliário às pessoas mais velhas. O apoio domiciliário psicossocial, complementando os serviços ao domicílio mais tradicionais que intervêm essencialmente nas áreas dos cuidados alimentares, limpeza, higiene e saúde, reconhece que na última fase da vida existem possibilidades significativas de desenvolvimento e de bem-estar pessoal.

O reconhecimento de que as intervenções psicossociais são eficazes na promoção da qualidade de vida contribuiu para, em 1948, a Organização Mundial de Saúde (OMS) mudasse a sua concepção de saúde dando-lhe uma ênfase biopsicossocial. A saúde deixou de se caraterizar pela ausência de doença e incapacidade e passou a ser considerada como um estado de completo bem-estar físico, mental e social (Paúl & Fonseca, 2005).

Para aumentar a expectativa de uma vida saudável e de qualidade, a promoção do 'envelhecimento ativo' como processo de "otimização das possibilidades de saúde, de participação e de segurança" (WHO, 2011) é central. Uma sociedade democrática e socialmente justa deve, consequentemente, prestar atenção às eventuais barreiras que as pessoas podem encontrar neste processo e propor ferramentas diversas para as remover e, assim sendo, ajudar os indivíduos a viverem vidas mais preenchidas e felizes. O apoio ao domicílio contribui na remoção das dificuldades que uma pessoa idosa fragilizada pode experienciar quando quer continuar a participar, aprender, desenvolver-se ou apenas sentir-se melhor. Mas, para que este cenário aconteça é necessário diversificar os serviços disponíveis.

As intervenções psicossociais para terem sucesso devem, no entanto, ser fundadas nos pressupostos da *life span* e na educação ao longo da vida, ser mais estruturadas, acontecerem a um ritmo mais lento, e tendo em atenção a complexa interação entre necessidades físicas, sociais e económicas das pessoas idosas (Fernandes, 2006).

Muitas pessoas idosas estão confinadas à sua casa por um conjunto de vulnerabilidades físicas, psicológicas e sociais e, de modo consequente, uma psicoterapia individual desenhada para responder especificamente às necessidades sentidas pode ajudar a aliviar o mal-estar e promover o crescimento psicológico (Landreville et al., 2001; Areán et al., 2002; Sharpley, 2010). As terapias breves e manualizadas, como é o caso da *Homecounseling Intervention Therapy* (HIT[1]; Lima & Silvério, 2015) oferecem uma boa solução para pessoas de idade avançada sem declínio cognitivo acentuado, pois têm objetivos e tempos delimitados, estando voltadas para a resolução de um conflito ou dificuldade específica. O principal objetivo da HIT é reduzir o mal-estar e a sua prevenção intervindo na relação com cuidadores formais e informais, na adaptação às questões práticas, relacionais e existenciais das vivências nesta situação, na promoção do sentido da vida/propósito positivo e esperança e no empoderamento e desenvolvimento pessoal (Lima & Silvério, 2015).

Ter uma boa qualidade de vida implica ter boas relações interpessoais e sentir prazer de viver numa determinada casa ou bairro e poder aceder a atividades desenvolvimentais e educacionais, participar na sociedade, sentir-se útil, apreciar a vida e ter possibilidades de escolha (Gabriel & Bowling, 2004).

Neste sentido, promover a vida independente e a assistência à população idosa socialmente isolada, nos territórios de baixa densidade e áreas urbanas, através da pesquisa científica e dos

[1] O Atlas, Organização Não-Governamental de Desenvolvimento, iniciou em Coimbra o projeto "Velhos Amigos", com vista à assistência de idosos de Coimbra em situação de carência alimentar e solidão, com apoios ao nível de refeições, serviços de enfermagem e apoio psicológico ao domicilio. O programa HIT surgiu devido à necessidade de melhorar o serviço de apoio psicológico prestado ao domicílio, padronizando procedimentos e fornecendo ferramentas acessíveis, validadas e com eficácia comprovada, com vista a formar psicólogos a trabalhar neste contexto.

cuidados de saúde e sociais é fundamental. Igualmente é a criação de recursos (instrumentos, modelos de intervenção, lições e recomendações, formação de equipas interdisciplinares) para que ação ao nível dos serviços comunitários e de saúde mental seja empiricamente sustentada, e a partilha de saberes (serviços e programas) entre países.

Em casa é mais fácil envelhecer de forma pessoal e criativa quando são dadas as condições para tal. Preservando deste modo o valor da autonomia, dos recursos pessoais, da liberdade e da independência mesmo nas idades mais avançadas (Baltes & Smith, 2003).

Bibliografia

AARP (2010). *Home and Community preferences of the 45+ population*. Retrieved 23 de Novembro 2017. https://assets.aarp.org/rgcenter/general/home-community-services-10.pdf

Areán, P. A., Alvidrez, J., Barrera, A., Robinson, G. S., & Hicks, S. (2002). Would older medical patients use psychological services? *The Gerontologist, 42*(3), 392-398.

Baltes, P. B., & Smith, J. (2003). New frontiers in the future of aging: From successful aging of the young old to the dilemmas of the fourth age. *The Gerontologist, 49*, 123-135.

Bernabei, R., Landi F., Gambassi G., Sgadari A., Zuccala, G., Mor V., Rubenstein, L.Z., & Carbonin, P.U. (1998). Randomized trial of impact of model of integrated care and case management for older people living in the community. *British Medical Journal, 316*, 1348-1351.

Brick, Y. (2011). Aging in Place in Israel. *Global Ageing, 7*(2), 5-16.

Brown, E., Kaiser, R., & Gellis, Z. D. (2008). Screening and assessment of late-life depression in home health care: Issues and challenges. *Annals of Long Term Care, 15*(1), 27-32.

Carroll, K.M., & Nuro, K.F. (2002). One size cannot fit all: a stage model for psychotherapy manual development. *Clinical Psychology: Science and Practice, 9*, 396-406.

Duarte, R. (2017). O impacto da HIT (Homecounseling Intervention Therapy) na Promoção da Qualidade de Vida e na Diminuição da Sintomatologia Depressiva em Pessoas Idosas. Tese de mestrado integrado em Psicologia Clínica (não publicada). FPCE.UC.

Ferreira-Alves, J. (2010). Prática Psicológica com Pessoas Idosas: Uma Leitura Substanciada das orientações da APA. *Revista de Psiquiatria.* CHPL – Hospital Júlio de Matos, *XXII* (3),1-24.

Figueiredo, D., Sousa, L., & Lima, M. (2012). Cuidadores familiares de idosos dependentes: rede social pessoal e satisfação com a vida. *Psicologia, Saúde e Doenças, 13*(2), 47-60.

Fernandes, L. (2006). *Psicoterapias no idoso.* In Firmino, H., Psicogeriatria (pp.133). Coimbra: Psiquiatria Clínica.

Fernandes, M. M. G. C. (2014). *Depressão, qualidade de vida em idosos institucionalizados* (Master's thesis).

Frade, J., Barbosa, P., Cardoso, S., & Nunes, C. (2015). Depression in the elderly: symptoms in institutionalised and non-institutionalised individuals. *Revista de Enfermagem Referência, 4*(4), 41.

Gabriel, Z., & Bowling, A. (2004). Quality of life from the perspectives of older people. *Ageing and Society, 24*, 675-691.

Global AgeWatch (2015). *Global AgeWatch Index.* (www.globalagewatch.org).

Kahana, E. (2001). Institutionalization. In G.L. Maddox (Ed.), *The Enciclopedia of Aging.* New York: Springer Publishing Company.

Knight, B. G. (2004). Psychotherapy with older adults (3rd ed.). New York: Sage.

Knight, B. G., & Poon, C. (2008). Contextual Adult Life Span Theory for Adapting Psychotherapy with Older Adults. *Journal of Rational-Emotional and Cognitive-Behavioral Therapy*, 26, 232-249.

Landreville, P., Landry, J., Baillargeon, L., Guerette, A., & Matteau, E. (2001). Older adults' acceptance of psychological and pharmacological treatments for depression. *Journal of Gerontology , 50B*, P285-P291.

Lima, M. P. & Silvério, L. (2015). *Programa manualizado de prestação de apoio psicológico ao domicílio: Homecounseling Intervention Therapy.* Manuscrito não publicado, Faculdade de Psicologia e Ciências da Educação da Universidade de Coimbra.

Lima, M. P. (2006). Posso participar?: Actividades de desenvolvimento pessoal para idosos. Lisboa: Ambar.

Lima, M. P. (2012). *Intervenção psicoterapêutica com pessoas idosas.* Relatório da disciplina de Intervenção Psicoterapêutica em Adultos Idosos, Faculdade de Psicologia e de Ciências da Educação – Universidade de Coimbra, Portugal.

Lima M.P., Silva, C., Carvalho, M.M., & Fernandes, D. (2014). Providing home counseling for older adults: Benefits and challenges. *Revista de Saúde Pública,* 48 (nº esp.), 75.

Lubben, J. & Grionda, M. (2003). Measuring social networks and assessing their benefits. *In*: Phillipson C, Allan G,& Morgan D (Eds.). *Social networks and social exclusion*: 20-49. Hants, England: Ashgate.

Mello, J.D.A., Durme, T.V., Macq, J., Declerq, A. (2012). Intervention to delay institutionalization of frail older persons: design of longitudinal study in the home care setting. *BMC Public Health, 12,* 615.

Mestheneos, E. (2011). Ageing in place in the European Union. *Global Ageing,* 7(2), 17-23.

Oliveira, A.L., Vieira, C., Lima, M.P., Nogueira, S., Alcoforado, L., Ferreira, J.A., & Zarifis, G. (2011). Developing instruments to improve learning and development of disadvantage seniors in Europe: The paladin project. In Pixel (Ed.), *Conference proceedings of the International Conference The Future of Education* (vol. 1, pp. 268-274). Florence: Simonelli Editore.

Oliveira, A. L. (Coord.), Vieira, C. M., Lima, M. P., Alcoforado, L., Ferreira, S. M., & Ferreira, J. A. (2013), *Promoting conscious and active learning and ageing: How to face current and future challenges?* Coimbra: Imprensa da Universidade de Coimbra. ISBN: 978-989-26-0732-0.

Maxfield, M., & Segal, D. L. (2008). Psychotherapy in nontraditional settings: A case of in-home cognitive-behavioral therapy with a depressed older adult. *Clinical Case Studies, 7*(2), 154-166.

Muijen, M., Marks, I., Connolly, J., & Audini, B. (1992). Home based care and standard hospital care for patients with severe mental illness: a randomised controlled trial. *BMJ, 304*(6829), 749-754.

Nações Unidas (2017). Sustainable Development Goals. https://unstats.un.org/sdgs/report/2016/leaving-no-one-behind.

Paúl, C. & Fonseca, A. M. (2005). *Envelhecer em Portugal: Psicologia, saúde e prestação de cuidados.* Lisboa: Climepsi Editores.

Paúl, C. (2005). Envelhecimento activo e redes de suporte social. *Sociologia, 15.*

Rodin, G. (2009). Individual psychotherapy for the patient with advanced disease. In: *Handbook of Psychiatry in Palliative Medicine*, 2nd Ed. Chochinov, H.M., & Breitbart,W. (eds). Oxford: Oxford University Press.

Sharpley, C. F. (2010). A review of the neurobiological effects of psychotherapy for depression. *Psychotherapy: Theory, research, practice, training, 47*(4), 603.

Simões, A., Ferreira, J., Lima, M., Pinheiro, M., Vieira, C., Matos, A., Oliveira, A., Alcoforado, L., Neto, F., Ruiz, F., Cardoso, A., Felizardo, S., & Sousa, L. (2006). Promover o bem-estar dos idosos: Um estudo experimental. *Psychologica – especial: Envelhecimentos*, 42, 115-131.

Sorocco, K. H., & Lauderdale, S. (2011). *Cognitive behavior therapy with older adults: Innovations across care settings.* New York: Springer Pub. Co, 391-416.

Stella, F., Gobbi, S., Corazza, D. I., & Costa, J. L. R. (2002). Depressão no idoso: diagnóstico, tratamento e benefícios da atividade física. *Motriz, 8*(3), 91--98.

Uchino, B. (2004). *Social Suport & Physical Health. Understanding the health consequences of relationships.* New Haven: Yale University Press.

Vaz Serra, A. (2002). O stress na vida de todos os dias. Coimbra. Ed do autor. ISBN 972-95003-2-0.

Vecchia, R. D., Ruiz, T., Bocchi, S. C. M. & Corrente, J. E. (2005). Qualidade de vida na terceira idade: um conceito subjetivo. *Revista brasileira de epidemiologia*, 246-252.

World Health Organization/WHO (2011). *Active Ageing: Report. Special Eurobarometer. European Commisson.* Genebra: WHO.

WHOQOL Group. (1995). The World Health Organization Quality of Life Assessment: Position paper from the WHO. *Social Science and Medicine, 41*(10), 1403-1409.

Yang, J. A., Garis, J., Jackson, C., & McClure, R. (2009). Providing psychotherapy to older adults in home: Benefits, challenges, and decision-making guidelines. *Clinical Gerontologist, 32*(4), 333-346.

Zarit, S. H., & Leitsch, S. A. (2001). Developing and evaluating community-based intervention programs for Alzheimer's patients and their caregivers. *Aging & Mental Health, 5*(2), 84-98.

CAPÍTULO VIII

EDUCAÇÃO AO LONGO DA VIDA E EDIFICAÇÃO DA SOCIEDADE DO CONHECIMENTO: O CASO DAS "LOJAS DE SABER"

Albertina L. Oliveira
Universidade de Coimbra, Faculdade de Psicologia e Ciências da Educação
Centro de Estudos Interdisciplinares da Universidade de Coimbra (CEIS20)
ORCID: http://orcid.org/0000-0001-9521-528X
Margarida Pedroso de Lima
Universidade de Coimbra, Faculdade de Psicologia e Ciências da Educação
ORCID: https://orcid.org/0000-0002-6239-1137
J.J. Pedroso de Lima
Universidade de Coimbra, Faculdade de Psicologia e Ciências da Educação
ORCID: https://orcid.org/0000-0003-1494-6051

Resumo: *O presente capítulo apresenta as Lojas de Saber, enquanto projeto contextualizado na educação ao longo da vida, destinado a proporcionar oportunidades educativas às pessoas de todas as gerações, a partir dos valiosos saberes de quem se encontra na situação de reformado/a e se revê na missão de educar. Após uma breve análise crítica das forças dominantes na sociedade atual, de*

DOI | https://doi.org/10.14195/978-989-26-1788-6_8

pendor neoliberal, os autores centram-se no contributo das Lojas de Saber para a edificação da sociedade do conhecimento, passando pela necessidade de mobilizar os importantes conhecimentos que as pessoas reformadas desenvolveram nas suas vidas profissionais ao longo de décadas de investimento.Tal contributo acentua a necessidade de se diversificarem as respostas educativas/ formativas e de se promover a igualdade de oportunidades e de acesso à educação, de lutar contra o idadismo bem como de desenvolver a autonomia e empowerment de todos os cidadãos/ãs em todos os espaços e tempos da vida.

Palavras chave: Educação ao longo da vida; Reforma; Sociedade do conhecimento

Abstract: *This chapter presents Lojas de Saber as a project contextualized in lifelong education, designed to provide educational opportunities to people of all generations, based on the valuable knowledge of those who are in the situation of retired and that without payment participate in the mission of the formation of people and communities. After a brief critical analysis of the dominant forces in today's neo-liberal society, the authors focus on the contribution of the Lojas de Saber to the building of the society of knowledge, by mobilizing the important knowledge that the retired people have developed in their professional lives and accumulated over decades of investment. Such a contribution accentuates the need to diversify educational / training responses and to promote equal opportunities and access to education, to fight against discrimination against older people and to develop the autonomy and empowerment of all citizens in all spaces and times of life.*

Key-words: *Lifelong education; Retirement; Society of knowledge*

Introdução

A edificação da sociedade do conhecimento requer um sistema de educação ou longo da vida ou, melhor, de educação permanente, que abranja todos os tempos e espaços da vida dos cidadãos/ãs, de modo a que estes possam promover o seu desenvolvimento e tornar-se mais esclarecidos e socialmente mais participativos. Educar todos, totalmente, em todas as coisas (*omnes, omnia, omnino*) é o desafio que nos vem da maior obra pedagógica do século XVII – a *Pampaedia* – e que ainda hoje não conseguimos concretizar inteiramente, apesar de muito se ter avançado em matéria de educação. Segundo João Amós Comênio existe uma escola para cada idade da vida, sendo a do adulto a escola da vida, tendo por base a própria experiência de vida. Se já no século XVII encontramos esta proposta pioneira, no século XXI ela não poderia ser mais actual. Porém, a educação das pessoas de idade avançada e o aproveitamento dos vastos e ricos saberes que estas desenvolveram e aprimoraram ao longo da vida ainda não são suficientemente valorizados e colocados ao serviço da formação das gerações mais jovens. É neste contexto que surgem as Lojas de Saber, mobilizando os importantes conhecimentos que as pessoas reformadas desenvolveram nas suas vidas profissionais e que acumularam ao longo de 'toda uma vida'. Foram criadas com o objetivo de educar/formar e transmitir voluntariamente informações, saber e experiências não apenas aos que se encontram na última etapa da vida, mas também às novas gerações. Isto é particularmente importante, uma vez que muitos milhares de pensionistas em ótimo estado físico e intelectual estão dispostos a manter uma atividade socialmente útil, podendo contribuir de forma extremamente valiosa para a edificação da sociedade do conhecimento. No presente trabalho, apresentamos as principais características e princípios deste projeto, a vantagem e impacto na qualidade de vida dos seniores envolvidos e na comunidade em geral, no quadro da educação permanente, onde se enfatiza a necessidade de se diversificarem as respostas educativas, a importância de promover o mais possível a igualdade de oportunidades de educação e de acesso e a autonomia ou *empowerment* das pessoas em todo o curso de vida.

Forças dominantes na sociedade contemporânea

A vida na sociedade atual é fortemente marcada pelas forças e movimentos que transformaram e permitiram a passagem da sociedade dita tradicional para a sociedade industrial e pós-industrial. As mudanças que todos temos vindo a presenciar são, simultaneamente, rápidas, profundas e globais e têm revolucionado completamente o modo de vida das pessoas e a organização social. Sem dúvida que a força dominante se deve à influência das transformações científicas e tecnológicas, ou ao que Edgar Faure (1972) denominou, no primeiro relatório internacional da UNESCO sobre educação, a *revolução científico-técnica*, em nada comparável a qualquer período anterior da história humana. Esta dominância tornou-se mais notória a partir da segunda metade do séc. XX, quando se começou a verificar um aumento prodigioso dos conhecimentos e uma aceleração exponencial da mudança.

Efetivamente, estamos, segundo alguns autores, a atravessar um período histórico singular (Melo et al., 1998), correspondente à quarta mudança radical na evolução da humanidade – a era da informação/comunicação – que se constitui como um forte desafio à edificação da sociedade do conhecimento e que resulta de três movimentos conjugados: a sociedade da informação, a globalização e a cultura científica e tecnológica. A sociedade da informação a erigir desafia os sistemas políticos a conceder inteira prioridade à educação, no sentido dos cidadãos e cidadãs saberem avaliar e escolher a informação mais adequada; a globalização, no seu potencial positivo, estimula a que as pessoas se tornem cidadãs do mundo (com horizontes, visão, informações, contactos alargados); a cultura científica e tecnológica apela ao desenvolvimento da capacidade de lidar com o imprevisível, a complexidade, a mudança.

Porém, apesar da informação e do conhecimento serem, incontornavelmente, a matéria-prima das sociedades que se pretendem demarcar pela economia do conhecimento (Delors et al., 1996; Pacheco, 2011), estamos bem longe de viver numa sociedade com cidadãos e cidadãs maioritariamente esclarecidos/as e educados/as, capazes de perceber e questionar o seu papel no mundo, bem como de contribuir ativamente para o bem estar social. O que predomina no mundo atual é a economia de mercado, regida por lógicas inscritas na teoria do capital humano que, como todos sabemos, constitui o berço concetual do neoliberalismo. Estas lógicas, fortemente vinculadas a padrões de eficiência e qualidade, por assentarem na "visão da educação como processo de formação social, orientada para mercados competitivos, clamando que as organizações educativas devem responder a desafios imediatos do mundo económico" (Pacheco, 2011, p. 16), contribuem para gerar fortíssimas desigualdades. 'Descartam' e desvalorizam todos aqueles/as que não estão preparados/as ou que entendem não corresponder ao mundo da 'alta competição'. É neste contexto que se compreende, como veremos mais à frente, porque é que tantas pessoas de idade avançada se sentem excluídas, destratadas e com tão poucas oportunidades.

Voltando à sociedade do conhecimento ou da informação, na proposta de Hutchins (1968, cit. por Jarvis, 2001) da década de 60, a sociedade a edificar, designada por *sociedade de aprendizagem*, seria aquela que contribuiria para instituir a visão de uma 'sociedade boa'[1], suscetível de concretizar os ideais da democracia e do igualitarismo, orientados em função da meta principal de desenvolver o homem completo e de construir sociedades mais justas. Na visão de Hutchins, a sociedade contemporânea reuniria, finalmente, as condições para que esse antigo ideal de Atenas, igualmente visionado por Coménio (com o seu pensamento integrativo, edificante e revolucionário), ao preconizar a trilogia *omnes, omnia, omnino* (educar todos, em todas as coisas e de uma forma total) se pudesse estender a todos os cidadãos e cidadãs (Gomes, 1971), através das grandes conquistas do progresso científico e tecnológico.

Todavia, o que encontramos é uma sociedade movida fundamentalmente pelas forças da globalização em que o conhecimento é visto como recurso económico regulado pelas lógicas de mercado, tendo-lhe subjacente uma conceção instrumental do ser humano e não, como é desejável no quadro da educação permanente ou da educação ao longo da vida, interesses libertadores ou emancipatórios, promotores da inclusão e do desenvolvimento integral e harmonioso das pessoas e sociedades.

---

[1] Encontramos este ideal, pelo menos, na Grécia antiga, onde "a educação não era uma atividade segregada, levada a cabo em certas horas, em certos lugares e num determinado período da vida" (Hutchins, 1968, p. 133, cit. por Candy, 1991, p. 78), mas sim era concebida como o objetivo da própria vida.

A problemática da desvalorização das pessoas idosas

Outra das grandes conquistas das sociedades contemporâneas é o aumento da esperança de vida e da longevidade dos cidadãos e cidadãs, os quais gozam, em geral, de uma boa saúde e de melhores condições de vida, comparativamente a gerações anteriores (e.g., Gondo & Poon, 2007). Como defende Simões (2006), estamos perante uma "nova velhice", uma vez que as pessoas de idade avançada na nossa sociedade são em geral *mais saudáveis, mais longevas e mais instruídas*. Porém, neste cenário, faz todo o sentido perguntarmo-nos, como estamos a tratar os nossos idosos? Que oportunidades estão a ser criadas para as pessoas no pós-reforma, no sentido de se sentirem incluídas, valorizadas e participantes na vida social?

Há toda uma vasta literatura que aponta para a importância de preparar a reforma através, inclusivamente, de programas de intervenção (e.g., Leandro-França, 2016). Subjacente está o reconhecimento da importância dos recursos pessoais e sociais na vivência desta fase (Adams & Taylor, 2015; Dingemans, & Henkens, 2015; Earl, Gerrans, & Halim, 2015). Paradoxalmente, muito embora os discursos e as concetualizações apontem para a inclusão e participação de todos, durante toda a vida, e para a aprendizagem e o desenvolvimento permanente das pessoas, aquele estatuto de que as pessoas de idade avançada gozavam nas sociedades tradicionais, vistas como sábias, conselheiras e alvo de um grande respeito por parte das gerações mais jovens, perdeu-se na sociedade atual, sendo, em contrapartida, muito mais encaradas como um peso e sobrecarga. Ao contrário do que sucedeu no passado, a sociedade em alguns países, incluindo o nosso, desvaloriza e quase completamente desaproveita a sabedoria das pessoas reformadas (Gonçalves & Oliveira, 2013).

A economia de mercado dominante e a visão tecnicista apoiada em modelos de racionalidade técnica e de produção, com os conceitos recentes sobre rendimento e mercado do trabalho, práticas da concorrência, culto de imagem e muitos outros, têm levado à criação e afirmação de 'novos paradigmas' na atitude da sociedade para com a designada terceira idade e o seu saber. Apesar dos novos valores adotados não estarem oficial e explicitamente em conflito com o saber acumulado pelos mais velhos, na prática, verifica-se que todo um rico manancial de conhecimento e experiência, que alicerçou a vida de gerações anteriores, se encontra esquecido ou subaproveitado.

Mesmo nos casos em que há interesse explícito das pessoas idosas em continuarem a servir, graciosamente, a sociedade, com óbvio benefício para esta, não existem mecanismos legais, no nosso país, que o possibilitem facilmente. Em grande parte dos casos não há escolha, sendo as pessoas de idade avançada forçadas a abandonar a sua profissão, mesmo que não o desejem e se encontrem em ótimas condições físicas, cognitivas e emocionais para prosseguirem com ela.

A força 'invisível' dos mecanismos legislativos vigentes leva as pessoas idosas a interiorizarem a sua idade, como se fosse uma culpa ou uma fatalidade envelhecer, e pressiona-as a viverem a esquecer, solitariamente, o que lhes levou décadas a aprender, acabando, não raro, a executarem tarefas menores ou a envolverem-se em meros passatempos em nada estimulantes e recompensadores.

Em sintonia com a cultura prevalecente do desperdício, a sociedade que, ela própria, investiu durante décadas na formação dos seus quadros, impede de forma administrativa o exercício de capacidades a pessoas que, graciosamente, pretendem continuar a servi-la. Como atrás referido, em boa parte graças aos avanços da medicina, a esperança de vida aumentou e muitas pessoas reformadas são cidadãos e cidadãs saudáveis, mental e fisicamente (Depp, Vahia, & Jeste, 2012; Fernández-Ballesteros, 2009, 2013), que têm noção do seu valor intrínseco e que recusam ser tornadas obsoletas por via administrativa. Para estas, a reforma, como existe entre nós, é encarada como uma coação e um desperdício, e não como uma alternativa de vida ou uma escolha.

Neste âmbito, e tendo em conta o que concetualmente se propõe para erigir a sociedade de aprendizagem/conhecimento, ou para implementar o paradigma da educação ao longo da vida, julgamos que, seja qual for a perspetiva que se considere, económica, social, educacional ou ética, a exclusão do conhecimento das pessoas mais velhas é um absurdo. Urge pois recuperar o valor atribuído às pessoas de idade avançada nas sociedades tradicionais para benefício de todos.

A resposta inovadora das Lojas de Saber

É neste contexto, de rutura e de luta contra o *satus quo*, que surge a iniciativa inovadora das Lojas de Saber, propondo-se mobilizar os importantes conhecimentos que as pessoas reformadas desenvolveram nas suas vidas profissionais e que acumularam ao longo de décadas de investimento, de trabalho e de aprendizagem. Foram criadas com o objetivo de educar/formar e transmitir voluntariamente informações, conhecimento e experiências, não apenas aos que se encontram na última etapa da vida, mas também às novas gerações. Isto é particularmente importante, uma vez que muitos milhares de pensionistas em ótimo estado físico e intelectual estão dispostos a manter uma atividade socialmente útil, podendo contribuir de forma extremamente valiosa para a efetiva edificação da sociedade do conhecimento.

As Lojas de Saber constituem assim um projeto que se inscreve no quadro da educação permanente ou da educação ao longo da vida, em que se deseja e enfatiza a necessidade de existirem respostas educativas diversificadas, que possibilitem a igualdade de oportunidades de educação e de acesso a todas as pessoas durante todo o curso de vida (Paixão, Silva, & Oliveira, 2014).

As Lojas de Saber alicerçam-se no pressuposto de que o conhecimento e experiência de milhares de pessoas reformadas, na nossa população, representam um volume de informação precioso e, muitas vezes, insubstituível. Assim, a sociedade não deve ficar indiferente ao desperdício que representa a imobilização ou perda desta informação, desprezando o que levou décadas a ser aprendido e representando, nalguns casos, uma sinopse do conhecimento de gerações passadas.

É justo que a sociedade crie mecanismos que possibilitem um período de vida mais calmo e descansado aos cidadãos e cidadãs de idade avançada, depois de um longo período de trabalho, ajustado sabiamente às potencialidades e limitações humanas. Contudo, não é razoável que o faça, como resultado da mecânica social existente no nosso país, anulando de forma oficial e definitiva as suas capacidades intelectuais e os seus préstimos profissionais, empurrando-os para o esquecimento do que lhes levou décadas a aprender, perdendo-se assim informação de forma inglória.

É urgente consciencializar que há um investimento de toda a sociedade em cada *'cérebro humano'* e que, portanto, na fase da reforma deve haver meios que permitam um retorno voluntário desse investimento à sociedade. Se é certo que muitas pessoas reformadas querem 'começar vida nova' e não estão interessadas no seu passado profissional, muitas outras há que ficariam altamente motivadas se tivessem oportunidade de continuarem a trabalhar naquilo que as valorizou como profissionais. Com efeito, se for criada uma tradição de transmissão de informação de cada pessoa reformada que o deseje, a outros cidadãos e cidadãs, a dez, vinte, sejam quantos forem, ano após ano, terá que haver melhorias na condição global da sociedade e ficaremos mais perto de uma verdadeira sociedade do conhecimento.

Numa outra perspetiva, o crescimento sustentado do país ou, de outro modo, o seu futuro, em muito assenta em ganharmos a batalha da 'qualificação' ou da formação das pessoas. Para atingir esse objetivo, todos os contributos são necessários, e a sabedoria dos seniores é uma enorme reserva de conhecimento passível de ser mobilizado a custo mínimo.

Neste sentido, um conjunto de cidadãos, concluída uma vida profissional que lhes garantiu a aquisição de um importante conjunto de conhecimentos, decidiu pôr à disposição de quem procura melhorar o seu saber, as experiências e conhecimentos adquiridos ao longo da vida, evitando que a sociedade venha a perder, total e escusadamente, estes importantes acervos. Para o efeito, decidiram criar as denominadas "Lojas de Saber", arquitetadas, essencialmente, como locais de transmissão e enriquecimento de conhecimentos e experiências entre gerações. De um ponto de vista social, consideram estar a cumprir um dever de cidadania, devolvendo às novas gerações aquilo que lhes foi transmitido ou que adquiriram ao longo da vida profissional. Procuram também, como durante séculos aconteceu e ainda hoje acontece no domínio familiar, retomar a tradição de passagem de conhecimentos das gerações mais velhas para as mais novas, agora através do trabalho voluntário de reformados.

Desta forma, seguramente que a sociedade ganhará em coesão, em conhecimento e menos erros poderão ser cometidos ou evitados.

A criação de uma prática de transmissão de informação de cada reformado, que o deseje fazer, para outros cidadãos/ãs, traz vantagens recíprocas, para estes e para as pessoas reformadas.

Pretendemos, assim, juntar as potencialidades do saber de pessoas idosas voluntárias de modo a constituir uma força coletiva, dinâmica e útil, que não passe despercebida na nossa sociedade e que, pelo contrário, seja reconhecida, valorizada e reforçada por esta.

Lutamos para que sejam dadas aos cidadãos/ãs mais velhos/as, se assim o desejarem, tarefas responsabilizadoras, úteis, ajustadas aos seus estatutos, e vivificantes do seu amor-próprio e autoestima.

Queremos pessoas reformadas em escolas não só como estudantes, mas como professores voluntários, a ensinarem aquilo em que são especialistas, que aprenderam nas suas vidas. Será uma maneira óbvia de não se perder a experiência e conhecimentos dos mais velhos e de lhes possibilitar a transmissão dessa informação.

Simultaneamente, procuramos contribuir para a difusão do trabalho social voluntário, cujo contributo para a quebra do isolamento e sensação de inutilidade, vivenciados por parte de muitos reformados, se alia a um poderoso auxílio a setores de interesse geral, carenciados em recursos humanos. Uma das palavras-chave deste projeto é, assim, *voluntariado depois da reforma*, com a finalidade de aproveitar as potencialidades especiais adquiridas pelas pessoas nas suas carreiras, ou noutras actividades, implicando o conceito de um retorno voluntário à sociedade do que foi mais valioso na experiência adquirida por cada pessoa.

Temos a convicção de que, no processo de envelhecimento, se a pessoa de idade avançada tiver a certeza da sua utilidade na sociedade, o seu empenho e vontade de viver serão redobrados e a etapa da velhice poderá assim ser, para muitos, uma experiência rica, beneficiando muito mais do que o próprio idoso.

A oportunidade das pessoas idosas servirem a comunidade como voluntárias, se o desejarem, em situações compatíveis com os seus interesses, saberes e competências, em particular ligadas ao seu passado profissional, pode ser um fator importante para um bom envelhecimento, ativo e saudável, tal como pretende a estratégia europeia 2020.

A atividade das Lojas de Saber

Apresentados os pressupostos e princípios em que assentam as Lojas de Saber, passamos agora a dar conta das numerosas atividades já desenvolvidas por estas.

De entre as diversas iniciativas já levadas a efeito destacam-se as mais de cinquenta conferências de cerca de 100 minutos, sobre temas variados, como por exemplo, e considerando apenas as mais recentes: *"Da terra para o céu (Construção de um violino: da árvore até à música)"*; *"Ética em Investigação Científica: faz sentido ensinar?"*; *"Medicina Nuclear. Desenvolvimentos no diagnóstico e no tratamento"*; *"O símbolo da Universidade de Coimbra"*; *"Risco de queda: denominador comum à escala global"*.

Considerando os cursos, estes poderão ser de extensão da formação profissional, elaborados após consulta a instituições das áreas correspondentes, destinados a melhorar insuficiências tradicionais de profissionais no ativo em diversos campos, de formação geral sobre temas com possível interesse no desempenho profissional, ou ter um caráter mais avançado em assuntos muito específicos. Estes cursos não atribuem qualquer reconhecimento formal, podendo ser ministrados uma única vez. Como exemplo, mencionamos os seguintes: "*Como combater o estigma das "doenças mentais"*? (4 horas), "*Saber mais para cuidar bem e reabilitar melhor*" (6 horas); "*Os sons e a vida*" (4 horas).

Para além destas modalidades de formação mais tradicionais, salientamos também a organização de visitas a museus (e.g., Museu Nacional de Machado de Castro; Museu Etnográfico Dr Manuel Lousã Henriques), a promoção de atividades formativas de carácter informal em Lares de Idosos e Centros de Dia (onde os mais capazes e com conhecimentos considerados de interesse alargado são solicitados a ensinar e demonstrar aos seus colegas e a elementos do exterior, os seus saberes e competências).

Integram igualmente as iniciativas das Lojas de Saber, a criação e manutenção de um "*Web site*", com uma secção de divulgação de conselhos e conceitos dos mais velhos, escritos e realizados por estes, com relatórios de entrevistas e uma secção de propostas dos leitores. Prevê-se também, entre outras ações, a publicação de um "*Yearbook of the experience learned secrets of retired people*" com a participação de especialistas nacionais e estrangeiros.

No âmbito de um protocolo estabelecido entre as Lojas de Saber/ Exploratório Centro Ciência Viva e o Diário as Beiras foram já publicados 48 textos subordinados ao tema "Recordar e recrear".

Em síntese, como se depreende pelo leque de atividades mencionadas, a oferta das Lojas de Saber pretende ir para além do campo do cognitivo e desenvolver competências através de demonstrações, trabalhos de grupo, reflexões, comunicações, etc., as quais mobilizam o poder construtivo e a criatividade das pessoas mais velhas. Porém, embora as iniciativas das Lojas de Saber visem, essencialmente, a ação de cidadãos/ãs reformados/as, nada impede que elementos mais jovens colaborem no projeto, materializando o princípio da inclusividade, no caso de constituírem mais valia e em condições idênticas às vigentes para os primeiros. A contribuição de elementos mais novos, em particular da Faculdade de Psicologia e Ciências da Educação de Coimbra, tem sido decisiva.

Considerações finais

Neste trabalho partimos de uma breve análise dos traços dominantes na sociedade contemporânea, fortemente eivada pelo neoliberalismo, ao serviço da economia de mercado, e a consequente instrumentalização e desvalorização dos saberes das pessoas de idade avançada, obtidos depois de uma longa vida de trabalho e investimento. Questionamos uma sociedade que força a abandonar a profissão, mesmo quando tal não é desejado, e defendemos que deverão multiplicar-se as possibilidades de escolha para se ir ao encontro da vasta hetrogeneidade que caracteriza as pessoas seniores: enquanto algumas aspiram a reformar-se cedo, outras, normalmente em ótimas condições físicas, cognitivas e emocionais, ambicionariam continuar a exercer a sua atividade profissional.

Através da caracterização e reflexão em torno do caso das Lojas de Saber, aponta-se para a enorme reserva de conhecimento destas pessoas, fácil de mobilizar, a custo nulo, através de trabalho social voluntário, com enorme contributo para a quebra do isolamento e sensação de inutilidade, podendo constituir, em simultâneo, um agente de agregação intergeracional poderoso e um auxílio a setores carenciados em recursos humanos, no âmbito da formação. A valorização dos saberes das pessoas de idade avançada, seria, ainda, uma forma de ultrapassar a miríade de preconceitos que afetam negativamente a vida das pessoas mais velhas e uma concretização dos objectivos do envelhecimento ativo, participativo e com sentido. Ao serem envolvidas na participação social, as nossas pessoas seniores voltam a sentir-se incluídas. E a inclusão, como todos sabemos, é uma enorme fonte de bem-estar e de justiça social.

Bibliografia

Adams, G. R., & Taylor, E. M. (2015). Friendship and happiness in the third age. In M. Demir (Ed.), *Friendship and happiness: across the Life-Span and Cultures* (pp. 155-169). New York, NY: Springer Netherlands.

Candy. P. C. (1991). *Self-direction for lifelong learning*. San Francisco: Jossey-Bass Publishers.

Delors, J., & Colaboradores (1996). *Educação, um tesouro a descobrir: Relatório para a UNESCO da Comissão Internacional sobre Educação para o século XXI*. Porto: Edições ASA.

Dingemans, E., & Henkens, K. (2015). How do retirement dynamics influence mental well-being in later life? A 10-year panel study. *Scandinavian Journal of Work, Environment & Health, 41*(1), 16-23. Doi:10.5271/sjweh.3464

Earl, J. K., Gerrans, P., & Halim, V. A. (2015). Active and adjusted: Investigating the contribution of leisure, health and psychosocial factors to retirement adjustment. *Leisure Sciences, 37*, 354-372. Doi: 10.1080/01490400.2015.1021881

Depp, C. A., Vahia, I. V., & Jeste, D. V. (2012). Successful aging. In S. K. Whithourne, & M. J. Sliwinski (Eds.), *The Wiley-Blackwell handbook of adulthood and aging* (pp. 459-476). Oxford: Wiley-Blackwell.

Faure, E., Herrera, F., Kaddoura, A-R., Lopes, H., Petrovski, A. V., Rahnema, M., & Ward, F. C. (1972). *Apprendre à être*. Paris: UNESCO.

Fernández-Ballesteros, R. (2009). Jubilación y salud. *Humanitas. Humanidades médicas, 37,* 1-23.

Fernández-Ballesteros, R. (2013). Possibilities and limitations of age. In A. L. Oliveira (Coord.), C. M. Vieira, M. P. Lima, L. Alcoforado, S. M. Ferreira & J. A. Ferreira, *Promoting conscious and active learning and ageing: How to face current and future challenges?* (pp. 25-74). Coimbra: Imprensa da Universidade de Coimbra. http://www.uc.pt/imprensa_uc/catalogo/ebook/E-book_Promoting

Gomes, J. F. (1971). A "Pampaedia" de Coménio. *Revista Portuguesa de Pedagogia, V,* 39-62.

Gondo, Y., & Poon, L.W. (2007). Biopsychosocial approaches to longevity. *Annual Review of Gerontology and Geriatrics,* 129-149.

Gonçalves, C. D., & Oliveira A. L. (2013). Reflections from a study about wisdom with students from a senior university. In A. L. Oliveira et al. (Coord.), *Promoting conscious and active learning and ageing: How to face current and future challenges?* (pp. 113-127). Coimbra: Imprensa da Universidade de Coimbra. http://www.uc.pt/imprensa_uc/catalogo/ebook/E-book_Promoting

Jarvis, P. (2001a). O futuro da educação de adultos na sociedade de aprendizagem. *Revista Portuguesa de Pedagogia, 35*(1), 41-66.

Leandro-França, C., Murta, S. G., Hershey, D. S., & Martins, L. B. (2016). Evaluation of retirement planning programs: A qualitative analysis of methodologies and efficacy. *Educational Gerontology, 42*(7), 497-512. Doi: 10.1080/03601277.2016.1156380

Melo, A. et al. (1998). *Uma aposta educativa na educação para todos. Documento de estratégia para o desenvolvimento da educação de adultos.* Lisboa: Ministério da Educação.

Pacheco, J. A. (2011). *Discursos e lugares das competências em contextos de educação e formação.* Coleção Panorama. Porto: Porto Editora.

Paixão, M. P., Silva, J. T., & Oliveira, A. L. (2014). Perspectives on guidance and counselling as strategic tools to improve lifelong learning in Portugal. In G. K. Zarifis & M. Gravani (Eds.), *Challenging the 'European Area of Lifelong Learning': A critical response* (pp. 167-176). London: Springer. DOI 10.1007/978-94-007-7299-1

Simões, A. (2006). *A nova velhice: Um novo público a educar.* Porto: Ambar.

Simões, A. (2006). Factos e factores do desenvolvimento intelectual do adulto. *Psychologica, 42*, 25-43.

CAPÍTULO FINAL

O IDOSO COMO SUJEITO POLÍTICO

Jacqueline Marques

Universidade Lusófona de Humanidades e Tecnologias

ORCID https://orcid.org/0000-0002-0088-8260

Resumo: *Perante a atual crise da proteção social e, consequente, retração das políticas públicas a população idosa encontra-se numa situação de vulnerabilidade e de perda real de direitos e proteção social. Na presença deste cenário é fundamental que a intervenção social com esta população se transforme para reconquistar os direitos perdidos e para garantir novos que permitam uma melhoria na qualidade de vida dos idosos. Neste capítulo não temos a pretensão de apresentar um guia de ação, e muito menos de boas práticas, mas sim de sugerir alguns contributos para a reflexão sobre a intervenção social com idosos que: i) coloque os idosos como sujeitos políticos, já que estes são sujeitos com capacidade real para conquistar espaços de resistência e luta que lhes garanta a participação na disputa pelos direitos sociais; ii) confira carácter político aos problemas e reivindicações e realize respostas emancipatórias e anti-*

DOI | https://doi.org/10.14195/978-989-26-1788-6_9

opressivas comprometidas com um projeto ético-político, que contribua para a (re)politização do Serviço Social.
Palavras chave: *Envelhecimento; Crise do Estado Social; Sujeito político; Participação; Politização*

Abstract: *Faced with the current crisis of social protection and, consequently, retraction of public policies, the elderly population is in a situation of vulnerability and real loss of rights and social protection. In the presence of this scenario it is fundamental that social intervention with this population is transformed to regain lost rights and to guarantee new ones that allow an improvement in the quality of life of the elderly. In this chapter we do not pretend to present a guide to action, much less good practices, but rather to suggest some contributions to the reflection on social intervention with the elderly that: i) place the elderly as political subjects, since these are subjects with real capacity to conquer spaces of resistance and struggle that guarantees them participation in the dispute for social rights; ii) give political character to the problems and demands and realize emancipatory and anti-oppressive responses committed to an ethical-political project that contributes to the (re) politicization of Social Work.*
Keywords: Aging; Social State crisis; Political subject; Participation; Politization

1. Nota introdutória

A população idosa não constitui um grupo marginal nem minoritário: 11,5% da população global, em 2012, possuía uma idade igual ou superior a 60 anos e essa percentagem poderá chegar aos

22% em 2050 (UNFPA, 2012). Este número é ainda mais expressivo se falarmos na União Europeia onde se espera um aumento de 77% da população com 65 e mais anos e de 174% com mais de 85 anos já em 2050 (Glendinning, 2009, p. 5). Em Portugal, no ano de 2016, 20,9% da população tinha 65 ou mais anos (PORDATA, 2018) e o índice de envelhecimento mais do que duplicará em 2080 (INE, 2017).

Mas nem todos os idosos poderão chegar a essa idade sem apresentar uma situação de fragilidade e dependência, o que se acentua quando se encontram em situações de vulnerabilidade económica. Ser idoso não é sinónimo de ser pobre, mas em Portugal o grupo dos idosos é um dos mais atingidos pela pobreza, como destaca Bruto da Costa no estudo que coordenou (2008, p. 111): *"As pessoas mais idosas (com + 75 anos), embora pouco representadas entre os pobres por razões demográficas, são as mais vulneráveis à pobreza (63% das pessoas desse grupo são pobres)"*. Desengane-se quem considera que tal situação apenas afeta os idosos sem qualquer proteção social ou com uma proteção social mínima (pensão social, CSI, RSI) já que, no mesmo estudo, os autores verificaram que mais de metade dos reformados (51%) do país eram pobres. As causas são variadas e bem conhecidas destacando-se os baixos salários obtidos durante toda a carreira contributiva, bem como as fórmulas de cálculo das pensões.

O envelhecimento traz consigo problemas de saúde, com consequências graves num país que apresenta *"uma população pouco saudável: a esperança de vida livre de doença, em Portugal, era de 55 anos nas mulheres e de 58,2 anos nos homens em 2015, comparativamente com 63,3 anos e 62,6 anos, respetivamente, na União Europeia"* (Mendes et al., 2018, p. 12). O impacto dessa situação é mais marcante e persistente na população pobre[1] que apresentava, em Portugal no ano de 2008, uma incidência de doenças crónicas ou deficiência de 41,6% (Costa et al, 2008). Os idosos pobres acedem menos aos cuidados de saúde quer pela dificuldade em aceder aos cuidados em si, por serem distantes geograficamente ou por apresentarem longas listas de esperas, quer pela dificuldade em comprar medicamentos ou outros tratamentos que lhe sejam prescritos. Isto sem incluir na equação a conhecida perspetiva global de saúde que implicaria falar nas habitações degradadas e húmidas ou na dieta alimentar pouco rica e variada.

Para além desta relação direta entre a falta de saúde, dos cuidados adequados e a falta de recursos sabemos que são outras as dimensões afetadas na vida dos idosos pobres, que acumulam situações de desfavorecimento em diversos domínios: ao nível educativo e cultural, das condições do alojamento, do acesso a bens ou equipamentos que garantam algum conforto, etc. Num estudo comparativo em diversos países da Europa os investigadores Tubeuf e Jusot (2011) concluíram que a saúde dos indivíduos estudados (com 50 e mais anos) aumentava com a capacidade económica dos mesmos. Em Portugal um estudo de Paula Santana (2002) demostrou que os idosos com melhores condições de habitação são mais saudáveis e usam os serviços de saúde de forma mais regular.

Apesar de todos estes ganhos, vivemos hoje uma época de enfraquecimento dos direitos conquistados, de retração de políticas públicas, de privatização de bens e serviços públicos. Paralelemente, vivenciamos uma época de glorificação do

[1] No Reino Unido, país no qual as desigualdades entre os mais pobres e mais ricos não é das mais elevadas, a esperança de vida saudável, livre de incapacidade entre esses dois grupos é de 17 anos (Marmot, 2010 citado em Berhan, 2013, p. 276)

individual, em que se perde ou difunde os sentimentos de pertença que poderiam permitir lutas e conquistas coletivas. Com base nessas preocupações procurei, neste capítulo, apresentar algumas pistas para a reflexão e discussão da intervenção como estratégia que contrarie essa realidade, através do ressurgimento de ação politizada do e na intervenção social.

2. O atual cenário de perdas de direitos

As diversas conquistas do Estado Social através das diferentes políticas públicas, do Sistema Nacional de Saúde, de um sistema de Segurança Social universal permitiram garantir um certo patamar de qualidade de vida na população idosa. Para compreender a importância dos apoios sociais basta verificar os dados sobre as transferências dos apoios sociais – a taxa de risco de pobreza antes de qualquer transferência social era, em 2016, de 45,2%, após as transferências das pensões diminuía para 23,6% e após a transferência sociais para 18,3% (PORDATA, 2018).

No entanto, nas últimas décadas, muitas destas conquistas foram abaladas como consequência da denominada crise do Estado Social. A crise deste tipo de Estado iniciada na década 70 do século passado – mais tarde em Portugal, já que esse tipo de Estado foi implementado (ou quase implementado)[2] apenas no período pós-revolucionário – acabou por se "cristalizar" como um "estado em crise" até aos dias de hoje. São várias as dimensões afetadas, já que se tratou de uma transformação profunda na economia e no modo de regulação social. Um dos

[2] As características *sui generis*, a mescla de diferentes atributos na proteção social, bem como a singularidade da sociedade portuguesa, dão ao país um quadro único que Santos (1990, 1993) denominou de "*Quase Estado-Providência*", ou seja, de uma aplicação parcial deste tipo de Estado.

principais campos de mudança ocorreu no mundo do trabalho, nomeadamente com a generalização do desemprego de longa e de "longuíssima" duração, a precariedade do trabalho, a "institucionalização" de formas atípicas de trabalho, a alternância constante entre trabalho e desemprego, e formas de trabalho financiado e apoiado, entre ouras. Manifestações de uma *"nova questão social"* (Castel, 1998) e de uma *"sociedade de risco"* (Beck, 1994) com impactos (ainda incalculáveis), quer individuais – no bem-estar social e na qualidade de vida dos cidadãos –, quer coletivos – com o questionamento do Estado Social e da relação entre trabalho e proteção social.

As perdas do Estado Social colocaram em causa a perspetiva da idade avançada como um momento em que seria possível reprogramar a vida, com mais tempo para o lazer, para atividades de satisfação e realização pessoal. Com efeito, para Pereirinha et al. (1999) a crise do Estado Social não só coloca em causa a ligação entre trabalho e proteção social, mas configura igualmente uma crise de valores e dos próprios direitos sociais, que deixam de ser promovidos e garantidos pelo Estado. Trata-se de uma *"(...) crise do pensamento igualitário e democrático"* que *"(...) traz no seu bojo propostas reduccionistas na esfera da proteção social"* (Yazbek, 1995, p. 11). Neste cenário, implementa-se um movimento de diminuição, cessação e/ou afunilamento das políticas públicas, nomeadamente, das políticas sociais. Substitui-se políticas sociais universais ou tendencialmente universalizantes, acentuando *"uma tendência para a retração, dando maior ênfase aos deveres, aumentando a seletividade e fornecendo uma maior moralização dos problemas sociais"* (Rodrigues et al., 2005:165).

Esta retração dos patamares de proteção social, com o alastramento da lógica de "condição de recurso" e da terciarização da proteção social, dá lugar a um clima de medo e insegurança, afeta a democracia e a atitude dos cidadãos que ficam na incerteza perante os consecutivos atropelamentos dos direitos sociais. Um ambiente de desconfiança que permitem a transformação da perceção sobre os idosos na sociedade, com diversos discursos que colocam as pensões e os gastos elevados com saúde como fatores fundamentais do risco e insustentabilidade do sistema de Segurança Social e do Sistema de Saúde público. Discursos que potenciam uma luta (mesmo que simbólica) entre pensionista e trabalhadores, entre eles (os idosos) e os outros[3], de modo a fundamentar algumas das soluções drásticas que se operaram com a perda de direitos sociais para essa população.

Esta transformação de universalidade em individualização permite a responsabilização individual pelos problemas e soluções. Assistimos na atualidade a difusão nos meios de comunicação social e nos produtos e serviços "vendidos" aos idosos, a um movimento de privatização da velhice (Debert, 1999) e a responsabilização individual pela mesma. Situação que afeta de forma dramática a população mais vulnerável e pobre, incapaz de garantir durante a sua vida ativa uma proteção individual (monetária) para a velhice e incapaz de com as parcas pensões o garantir quando já é velho.

Trata-se, igualmente, de uma tentativa para fundamentar a ideia que a intervenção do Estado levaria ao desencorajamento da auto-proteção e ao desinvestimento do apoio (e responsabilidade) da família pelos seus membros mais velhos. Transformam uma responsabilidade coletiva em responsabilidades individuais (e familiares) e legitimam a redução de despesas com serviços e apoios sociais e a devolução à comunidade – ou melhor às

[3] Exemplo disso são os idadismo ou etarismo (Centeio, 2006), termo que caracteriza o preconceito e a consequente discriminação a pessoas de mais idade.

famílias nomeadamente às mulheres – da responsabilidade pelo cuidado dos mais velhos. Veja-se o exemplo do programa *community care* que no Reino Unido a partir dos anos 80[4] permitiu uma redução dos gastos públicos com apoio médico e social e favoreceu os cuidados familiares (Martin, 1995).

Esta tendência para o individualismo contemporâneo na qual o indivíduo é forçado a *"criar e gerir, não apenas a sua própria biografia, mas os laços e as redes que a rodeiam"* (Beck; Beck-Gernsheim, 2003, p. 4) põe em perigo a noção de cidadania aliada a um projeto coletivo de bem-estar e justiça social e legitima a ideia do envelhecimento como um perigo social abrindo espaços para leitura moralizadoras (Thompson, 1998) que colocam em causa o contrato social entre gerações.

3. Transformando a intervenção: o idoso como sujeito político

A crise do Estado Social, para além das consequências expressas no ponto anterior, criou o espaço fértil para a alienação e destruição das formas de participação social e política dos trabalhadores e da população em geral. A crise desse tipo de Estado assentou, grosso modo, na crise do regime *fordista*. Um regime de acumulação baseado na produção segundo a conceção *taylorista* e na assimilação maciça por parte dos trabalhadores da lógica de consumo por meio da partilha dos ganhos de produtividade, quer diretamente pelo aumento do salário direto, quer indiretamente através dos benefícios sociais característicos do Estado Social. Trata-se do compromisso social-democrático

[4] Perspetiva alinhada com as ideias hegemónicas que o país vivia na altura com o governo de Margaret Thatcher e a defesa dos ideais neoliberais.

conquistado pela classe trabalhadora que embora tenha permitido a sua integração nos processos sociais e políticos levou, também, ao progressivo abandono da sua capacidade reivindicativa e a uma transformação do campo político, com a gradual alienação das pessoas das diferentes formas de manifestação política e social e de representação democrática (Santos, 1991).

Nota-se a existência de um desinteresse mútuo quer do sistema político e partidário português em relação ao eleitorado, quer do eleitorado que apresenta uma tendência para a abstenção e afastamento da vida política e partidária do país (à exceção das situações de fidelidade eleitoral). Nos últimos 30 anos a abstenção nas eleições em Portugal tem crescido consideravelmente: *"Em 1980, 16% do eleitorado absteve-se. Em 2005, esse valor tinha mais que duplicado, atingindo 36% do eleitorado"* (Torres; Antunes, 2007, p. 98). Estes números aumentam em 2011 para 41,9% e nas eleições de 2015 bateram o recorde de 44,1% (PORDATA, 2015).

Este é terreno fértil para a legitimação das desigualdades sociais e destruição da solidariedade entre gerações, com a revogação de direitos que aparentemente seriam irrevogáveis. Situação que para ser contrariada necessita, entre outros ímpetos, de uma modificação na forma de se ver e fazer a intervenção social, que encare os indivíduos como sujeitos políticos e não como (meros) objetos da intervenção.

Escolhemos a categoria do sujeito pois permite dar aos indivíduos protagonismo, transformá-los em atores políticos, sociais, culturais, em agentes de transformação social, conscientes da sua identidade, do seu papel, da sua história. É um sujeito que faz parte (e sente essa pertence) de uma classe social e, por isso, se transforma em ator coletivo, em sujeito político, capaz de uma ação consciente de luta pelos direitos.

Note-se que a intervenção social baseada nos processos de educação, com inspiração nos métodos de Paulo Freire, centrou-se, no decorrer da década de 90 do século passado, nos processos de empoderamento (empowerment) através do incitamento para o desenvolvimento de potencialidades dos indivíduos e das comunidades para os habilitar para a garantia da sua subsistência. Embora se trate de um processo de enorme importância, o objetivo de incentivar o pensamento crítico e reflexivo dos sujeitos foi-se deslocando para uma ação "habilitadora" de competências de sobrevivência. Por outras palavras, o cerne era um processo de empoderamento do qual emergem sujeitos capazes de garantir a sua (e a do grupo) sobrevivência, de garantir a capacidade dos indivíduos e grupos se integrarem numa sociedade de risco, mas deixava escapar os processos de luta contínua e de indignação.

Uma intervenção social focada nos sujeitos políticos resgata a indignação, a consciencialização de que as mudanças sociais podem ocorrer de baixo para cima, através da organização e da participação da população. Esta forma de trabalho, proposta por Paulo Freire (1979, 1980, 1987), implica de todo o profissional um engajamento social e político com a luta pela justiça e igualdade, com a luta pelos direitos sociais.

Com efeito, a participação como ato político é o caminho para a transformação da intervenção social. Como refere Yazbek (2014, p. 680) *"na política social, a luta contra a pobreza toma o lugar da luta de classes. A perspetiva é de desenvolvimento dos "ativos" dos pobres, desconsiderando os fatores estruturais da pobreza, atribuindo a responsabilidade da pobreza aos próprios pobres"*. Não consciencializar os sujeitos da sua responsabilidade na participação, na reivindicação pelos seus direitos, no protagonismo político que poderão ter é uma forma de despolitização dos direitos sociais. Se a intervenção social não luta por esse espaço ficará, é uma questão de tempo, com funções de gestão de riscos, com uma ação despolitizada isenta de posições e com o objetivo de reproduzir (com base na coesão) as desigualdades sociais existentes (Marques, 2016).

Novamente nas palavras de Yazbek (2014, p. 681) *"compreender que a prática profissional do Serviço Social é necessariamente polarizada pelos interesses das classes sociais em relação, não podendo ser pensada fora dessa trama (...) Trazendo essa tese para o exercício profissional em sua contemporaneidade estamos tratando das disputas políticas no espaço das políticas sociais, mediações centrais no exercício da profissão. Estamos tratando das disputas políticas na esfera pública e nas lutas sociais em seus impactos sobre as relações sociais. Estamos tratando da questão de construção de hegemonia, na condução dos serviços sociais e das necessidades que atendem, bem como dos direitos que asseguram, não apenas como questão técnica, mas como questão essencialmente política, lugar de contradições e resistência".*

Devido à referida falta de participação popular que caracteriza as atuais sociedades é fundamental o desenvolvimento de espaços de consciencialização crítica, na qual os idosos se constituam como sujeitos e não como objeto.

Deste modo são diversos os desafios que se colocam aos profissionais: i) modificar as práticas quotidianas nas instituições de modo a reposicionar o idoso na instituição como sujeito, criando mecanismo de participação real dos idosos na decisão das suas vidas e da vida institucional; ii) apoiar as diversas formas de luta, de participação, de resistência dos idosos, numa articulação estreita com os movimentos sociais e partidos políticos que lutem pelos seus direitos; iii) estimular e consciencializar os idosos da necessidade de integrarem a vida social como sujeitos políticos com voz, com cidadania e com espaço para se afirmarem como tal; iv) reconstruir o Serviço Social como profissão que assume coletivamente as lutas e resistências das populações, em especial dos mais vulneráveis.

Considerações finais

Como nos lembra Faleiros (2013, p. 44) o idoso deve ser concebido como um "*sujeito político de direitos implicado na e pela restruturação económica, familiar, social, política e nas dimensões pessoais e biológicas*", para o efeito o autor considera "*fundamental que as pessoas idosas se organizem e se manifestem socialmente, tanto na esfera dos partidos políticos e dos movimentos sociais, como na reivindicação do seu lugar social*".

Procura-se uma ação capaz de denunciar a mercantilização dos direitos sociais dos idosos e a propaganda que evoca a responsabilização individual dos idosos pelos riscos associados ao envelhecimento. Deste modo, para além de travar a perda de direitos e a retração do Estado Social[5], esta ação poderá contribuir para a conquista de direitos que permitam uma melhor qualidade de vida aos idosos de hoje e das gerações futuras. Toda esta ação resultará, também, numa melhoria da autoimagem e do reconhecimento social dos idosos, pelo acesso à expressão pública e sua, consequente, capacidade de representação.

[5] Essa situação possui um impacto que ultrapassa os idosos, afetando também a sua família já que, desde o adensar da crise de 2008, muitas famílias foram obrigados a viver na dependência económica (e em muitos casos habitacional) dos seus pais (idosos) que assumem o sustento da família mais alargada com as suas pensões. Os idosos que até aí recebiam apoio dos descendentes tornam-se um porto de abrigo para muitas famílias e foram, em muitos casos, os ascendentes a garantir a subsistência das famílias.

Esta forma de intervenção social poderá contribuir para a desconstrução da imagem dos idosos como uma massa individual de indivíduos e para colaborar com a construção de um coletivo organizado que participa na luta pelos seus direitos.

O processo que vivenciamos de despolitização do social – que retira a legitimidade de grupos que pretendem denunciar a violação dos direitos sociais e as desigualdades sociais – é acompanhado, como referimos, pela *"desresponsabilização social"* do Estado (Faleiros, 1999), legitimando intervenções sociais despolitizadas, moralizadora e individualizada. É, assim, essencial retomar a politização do coletivo e das políticas sociais e o Serviço Social assume um papel essencial nessa politização, através de práticas emancipatórias e anti-opressivas, integradas num projeto ético-político que permita repolitizar o próprio Serviço Social. Para Faleiros esta perspetiva para além de fortalecer a cidadania permite uma *"(...) articulação da dimensão política com a dimensão de serviços, não se reduzindo o Serviço Social, nem a relações psicológicas nem a relações burocráticas para acesso a determinados benefícios"* (1999, p. 169).

Com esta aproximação ao tema espero ter deixado alguns pontos de reflexão e discussão, que permitam ao Serviço Social construir uma prática que, como refere Gadotti (2014, p. 56) numa explicação sobre as práticas de Paulo Freire, transforme *"(...) "massa" amorfa em "povo" participante".*

Bibliografia

Beck, U. (1994). The Reinvention of Politics: Towards a theory of reflexive modernization. In U. Beck, A. Giddens, S. Lash (orgs.). *Reflexive modernization.*

Politics, tradition and aesthetics in the modern social order. Oxford: Blackwell Publishers, pp. 1-55.

Beck, U; Beck-Gernsheim, E. (2003). *La individualización – El individualismo institucionalizado y sus consecuencias sociales y políticas*. Barcelona: Paidos.

Berhan, M. (2013) "Ser Pobre, Ser-se Pobre. Reflexão critica sobre os números da pobreza". In Raquel Varela (coord.). A Segurança Social é sustentável. Lisboa: Bertrand Editora. pp. 275-287.

Castel, R. (1998). *As Metamorfoses da Questão Social: Uma Crónica do Salário*. Petrópolis: Ed. Vozes.

Centeio, L. (2006). "Envelhecimento e Barreiras da Idade no Emprego". In *Protecção Social, Cadernos Sociedade e Trabalho*, nº VII. pp. 179--198.

Costa, B. (coord.) (2008). *Um Olhar sobre a Pobreza. Vulnerabilidade e Exclusão Social no Portugal Contemporâneo*. Lisboa: Gradiva.

Debert, G. (1999). *A reinvenção da velhice*. São Paulo: Edusp/Faspesp.

Faleiros, V. P. (1999). "Os desafios do serviço social na era da globalização". In *Serviço Social & Sociedade, v. 20, nº 61*, p. 153-187.

Faleiros, V. P. (2013). "Autonomia relacional e cidadania protegida: Paradigma para envelhecer bem". In Maria Irene Carvalho (coord.). *Serviço Social no Envelhecimento*. Lisboa: Pactor.

Freire, P. (1979). *Educação e Mudança*. São Paulo: Paz e Terra.

Freire, P. (1980). *Conscientização: Teoria e Prática da Libertação – Uma introdução ao pensamento de Paulo Freire*. São Paulo: Moraes.

Freire, P. (1987). *Pedagogia do Oprimido*. 17ª edição. Rio de Janeiro: Paz e Terra.

Gadotti, M. (2014). "Alfabetizar e politizar. Angicos, 50 anos depois". In *Foro de Educación*, v.12, n.16, pp. 51-70.

Glendinning, C. (2009). *Combining choice, quality and equity in social services. Synthesis report: Peer Review in Social Protection and Social Inclusion*. Denmark. EU.

INE (2017). Projeções de População Residente 2015-2080. *Destaque*. Lisboa: INE.

Marques, J. (2016). *Itinerário de uma política : olhares sobre o rendimento social de inserção no concelho de Aveiro*. Tese de Doutoramento em Serviço Social, Instituto Superior de Serviço Social de Lisboa da Faculdade de Ciências Humanas e Sociais da Universidade Lusíada de Lisboa.

Mendes, F.; Duarte-Ramos, F.; Barros, H.; Ferreira, P.; Gaspar, R. Santana, R. (2018). "Meio caminho andado". *Relatório Primavera 2018*. Lisboa: Observatório Português dos Sistemas de Saúde. Disponível em http://opss.pt/relatorios/relatorio-de-primavera-2018/.

Pereirinha, J. et al. (1999). *Exclusão Social em Portugal: Estudo de Situações e Processos de Avaliação das Políticas Sociais: Relatório de Investigação*. Lisboa: CISEP/ CESIS.

Rodrigues, F.; Constantin, T.; Hoven, R.; Nunes, M. (2005). *Pobreza e Perpectivas Europeias*. Frankfurt: Peter Lang

Santana, P. (2002). "Poverty, social exclusion and health in Portugal". Social Science & Medicine, Volume 55, Issue 1, pp. 33-45. Disponível em https://www.sciencedirect.com/science/article/pii/S0277953601002180

Santos, B. (1990). O Estado, a Sociedade e as Políticas Sociais: O caso das Políticas de Saúde. In. Boaventura Sousa Santos. *O Estado e a Sociedade em Portugal (1974-1998)*. Porto: Afrontamento. pp. 193-266.

Santos, B. (1991). "Subjectividade, Cidadania e Emancipação". In *Revista Critica de Ciencias Sociais n.º 32*. pp. 135-191.

Santos, B. (1993). O Estado, as Relações Salariais e o Bem-estar Social na Semi-periferia: O caso português. In Boaventura Sousa Santos (org.). *Portugal: Um Retrato Singular*. Porto: Afrontamento. pp. 15-56.

Thompson, E. (1998). *Costumes em comum: estudos sobre a cultura popular tradicional*. São Paulo: Companhia das letras.

Torres, A.; Antunes, M. (2007). *O Regresso dos Partidos*. Lisboa: Âncora Editora.

Tubeuf, S.; Jusot, F. (2011). "Social health inequalities among older Europeans: the contribution of social and family background". The European Journal of Health Economics, vol. 12(1), pp. 61-77. Disponível em https://ideas.repec.org/a/spr/eujhec/v12y2011i1p61-77.html.

UNFPA e HelpAge (2012). *Envelhecimento no Século XXI: Celebração e Desafio.* Disponível em https://www.unfpa.org/sites/default/files/pub-pdf/Portuguese-Exec-Summary_0.pdf.

Yazbeck, M. (1995). A Política Social brasileira nos anos 90: refilantropização da Questão Social. In *Cadernos ABONG nº 11.*

Yazbeck, M. (2014). "A dimensão política do trabalho do assistente social". In *Serviço Social & Sociedade.* nº 120, pp. 677-693.

BIBLIOGRAFIA GERAL

AARP (2010). Home and Community preferences of the 45+ population. Retrieved 23 de Novembro 2017. https://assets.aarp.org/rgcenter/general/home--community-services-10.pdf

Adams, G. R., & Taylor, E. M. (2015). Friendship and happiness in the third age. In: M. Demir (Ed.), Friendship and happiness: across the Life-Span and Cultures (pp. 155-169). New York, NY: Springer Netherlands.

Aguirre, R. (2008). El Futuro del cuidado. In I. Arriagada (ed.), Futuro para las Familias y Desafíos para las Políticas. Serie Seminarios y Conferencias. Santiago de Chile: CEPAL.

Alarcão, M. (2015). Família e sistemas envolventes. In O. M. Fernandes e C. Maia (coord.). A família portuguesa no século XXI (pp.121-132). Lisboa: Parsifal.

Albertini, M. e Mencarini, L. (2011). Childlessness and support networks in later life: a new public welfare demand? Evidence from Italy. Carlo Alberto Notebooks, 200. Online in: www.carloalberto.org/working_papers.

Alley, D., Liebig, P., Pynoos, J., Banerjee, T. e Choi, I.H. (2007). Creating Elder-Friendly Communities: preparations for an aging society. Journal of Gerontological Social Work, 49(1-2),1-18.

Amaro, A. (2015). A holistic mindfulness. Mindfulness, 6, 63-73. https://doi.org/10.1007/s12671-014-0382-3

Andrade, F.F. (2009). Desfamiliarização das políticas sociais na América Latina: uma breve análise dos sistemas de proteção social na região. Barbarói, 31, 56-71.

Antonucci, T.C. e Akiyama, H. (1987). Social networks in adult life and a preliminary examination of the convoy model. Journal of Gerontology, 42(5), 519-527.

Antunes, B. (2014). The Cartesian moment and the contemplative turn: Reshaping epistemology. Comunicação apresentada no Summer Research Institute, Mind & Life Europe, 23-29 Agosto, Fraueninsel, Chiemsee.

Areán, P. A., Alvidrez, J., Barrera, A., Robinson, G. S., & Hicks, S. (2002). Would older medical patients use psychological services?. The Gerontologist, 42(3), 392-398.

Arias, C.J. (2009). La red de apoyo social en la vejez. Aportes para su evaluación. Revista de Psicologia da IMED, 1(1), 147-158.

Attias-Donfut, C. e Ogg, J. (2009). Évolution des transferts intergénérationnels: vers un modèle européen?. Retraite et société, 58, 11-29.

Attias-Donfut, C. Lapierre, N. e Segalen, M. (2002). Le nouvel esprit de famille. Paris: Odile Jacob.

Atzmon, G. (2010). Genetic variation in human telomerase is associated with telomere length in Ashkenazi centenarians. PNAS, 107(1), 1710--1717.

Balandier, G. (1986). Sens et puissance. Paris: PUF.

Baltes, P. B., & Smith, J. (2003). New frontiers in the future of aging: From successful aging of the young old to the dilemmas of the fourth age. The Gerontologist, 49, 123-135.

Bazo, M.T. (2008). Personas mayores y solidaridad familiar. Política y Sociedad, 45(2), 73-85.

Bengtson, V. L., Lowenstein, A., Putney, N. M., & Gans, D. (2003). Global aging and the challenge to families. In V. L. Bengtson & A. Lowenstein (Eds.). Global aging and challenges to families (pp. 1-24). New York: Aldine de Gruyter.

Bengtson, V., Rosenthal, C., & Burton, L. (1990). Families and ageing: diversity and heterogeneity. In R. H. Binstock & L. K. George (eds.) Handbook of aging and the social sciences (3.ª edição, pp. 263-287). San Diego: Academic Press.

Berkman, L.F. e Syme, S.L. (1979). Social networks, host resistance, and mortality: a nine-year follow-up study of Alameda County residents. American Journal of Epidemiology, 109, 186-204.

Bernabei, R., Landi F., Gambassi G., Sgadari A., Zuccala, G., Mor V., Rubenstein, L.Z., & Carbonin, P.U. (1998). Randomized trial of impact of model of integrated care and case management for older people living in the community. British Medical Journal, 316, 1348-1351.

Black, D. S., O'Reilly, G. A., Holmstead, R., Breen, E. C., & Irwin, M. R. (2015). Mindfulness meditation and improvement in sleep quality and daytime impairment among older adults with sleep disturbances. *JAMA International Medicine, 175*(4), 494-501.

Bloom, M. e Monro, A. (2015). Social Work and the aging family. *The Family Coordinator, 21*(1), 103-115.

Brandt, M. e Deindl, C. (2016). Support Networks of Childless Older People in Europe: An Analysis with the Data of the Survey of Health, Ageing and Retirement in Europe (SHARE). *Online in*: http://paa2014.princeton.edu/papers/143298.

Brick, Y. (2011). Aging in Place in Israel. *Global Ageing, 7*(2), 5-16.

Bris, H. J-L. (1994). *Responsabilidade familiar pelos dependentes idosos nos países das comunidades europeias*. Lisboa: Conselho Económico e Social. *Online In* http://www.ces.pt/download/600/ RespFamDepIdosos.pdf.

Brown, E., Kaiser, R., & Gellis, Z. D. (2008). Screening and assessment of late-life depression in home health care: Issues and challenges. *Annals of Long Term Care, 15*(1), 27-32.

Burholt, V. e Dobbs, C. (2014). A support network typology for application in older populations with a preponderance of multigenerational households. *Ageing & Society, 34*, 1142-1169.

Bury, M., & Holme, A. (1991). *Life after ninety*. London. Routledge.

Cabral, M.V. (coord.), Ferreira, P.M., Silva, P.A., Jerónimo, P. e Marques, T. (2013). *Processos de envelhecimento em Portugal – Usos do tempo, redes sociais e condições de vida*. Lisboa: Fundação Francisco Manuel dos Santos.

Cagnin, S. (2009). Neuropsicologia Cognitiva e Psicologia Cognitiva: O que o estuda dacognição deficitária pode nos dizer sobre o funcionamento cognitivo normal? *Psicologia em Pesquisa, 3*(1), 16-30.

Campos, M.S. e Mioto, R.C.T. (2003). Política de Assistência Social e a posição da família na política social brasileira. *Ser Social, 12*, 165-190.

Candy. P. C. (1991). *Self-direction for lifelong learning*. San Francisco: Jossey-Bass Publishers.

Capucha, L. (2014). Envelhecimento e políticas sociais em tempos de crise. *Sociologia, Problemas e Práticas, 74*, 113-131. DOI:10.7458/spp2014743203.

Carroll, K.M., & Nuro, K.F. (2002). One size cannot fit all: a stage model for psychotherapy manual development. *Clinical Psychology: Science and Practice*, 9, 396-406.

Carstensen, L.L. (1992). Social and emotional patterns in adulthood: Support for socioemotional selectivity theory. *Psychology and Aging, 7*, 331-338.

Carstensen, L.L., Isaacowitz, D.M. e Charles, S.T. (1999). Taking time seriously a theory of socioemotional selectivity. *American Psychologist, 54*(3), 165-18.

Carvalho, M.C.B. (2010). Famílias e políticas públicas. *In* A.R. Acosta e M.A.F. Vitale (orgs.). *Família – Redes, laços e políticas públicas* (pp. 267-275, 5.ª edição). São Paulo: CEDPE, PUC-SP: Cortez.

Castel, R. (2003 [1995]). *As metamorfoses da questão social: uma crônica do salário* (4.ª edição]. Pertópolis: Vozes.

Cochran, M. (1991). Personal social networks as a focus of support. In D. G. Unger & D. R. Powell (Eds.), *Families as nurturing systems: Support across the life span* (pp. 45-67).New York: The Haworth Press.

Conselho Económico e Social (2013). Parecer de Iniciativa sobre as consequências económicas, sociais e organizacionais decorrentes do envelhecimento da população. Lisboa, Portugal. *Online in:* http://www.ces.pt/download/1335/FINAL_Parecer%20Envelhecimento_aprovado%20em%20Plenario.pdf.

Cornwell, B., Laumann, E. O., e Schumm, L. F. (2008). The Social connectedness of older adults: A national profile. *American Sociological Review, 73*(2), 185-203.

Costa, A.B. (2004). *Exclusões sociais*. Lisboa: Gradiva.

Creswell, J. D. et al. (2012). Mindfulness based stress reduction training reducues loneliness and pro-inflammatory gene expression in older adults: A small randomized control trial. *Brain, Behavior and Immunity, 26,* 1095-1101. doi: 10.1016/j.bbi.2012.07.006

Crowther, M. R. et al. (2002). Rowe and Khan's model of successful aging revisited: Positive spirituality – the forgotten factor. *The Gerontologist, 42*(5), 613-620.

Daatland, S. O. (2003). Quality of life and ageing. In M. L. Johnson (Ed.), *The Cambridge handbook of age and ageing* (pp. 371-377). Cambridge: University Press.

Daatland, S.O. e Lowenstein, A. (2005). Intergenerational solidarity and the Family-Welfare State balance. *European Journal of Ageing, 2*, 174-182. DOI: 10.1007/s10433-005-0001-1.

Dabas, E. (2006). Viviendo redes. *In* E. Dabas (comp.). *Viviendo redes – experiencias y estratégias para fortalecer la trama social* (pp. 23-33). Buenos Aires: FUNDARED, CICCUS.

Daniel, F., Ribeiro, A.M., e Guadalupe, S. (2011). Recursos sociais na velhice: um estudo sobre as redes sociais de idosos beneficiários de apoio domiciliário. *In* A.D. Carvalho (coord.), *Solidão e solidariedade: entre os laços e as fracturas sociais* (pp. 73-85). Porto: Edições Afrontamento.

Davidson R. J., Kabat-Zinn J, Schumacher J., et al. (2003). Alterations in brain and immune function produced by mindfulness meditation. *Psychosomatic Medicine, 65*(4), 564-570.

De Martino, M. S. (2001). Políticas sociales y família: estado de bienestar y neoliberalismo familiarista. *Revista Fronteras*, n. 04, p.103-144.

Delgado, A. e Wall, K. (coord.) (2014). *Famílias nos Censos 2011 Diversidade e Mudança*. Lisboa: Instituto Nacional de Estatística e Imprensa de Ciências Sociais.

Delors, J., & Colaboradores (1996). *Educação, um tesouro a descobrir: Relatório para a UNESCO da Comissão Internacional sobre Educação para o século XXI*. Porto: Edições ASA.

Depp, C. A., Vahia, I. V., & Jeste, D. V. (2012). Successful aging. In S. K. Whithourne, & M. J. Sliwinski (Eds.), *The Wiley-Blackwell handbook of adulthood and aging* (pp. 459-476). Oxford: Wiley-Blackwell.
Depp, C. A., Vahia, I. V., & Jeste, D. V. (2012). Successful aging. In S. K. Whithourne, & M. J. Sliwinski (Eds.), *The Wiley-Blackwell handbook of adulthood and aging* (pp. 459-476). Oxford: Wiley-Blackwell.
Díez-Cirarda M, Ojeda N, Peña J, Cabrera-Zubizarreta A, Gómez-Beldarrain MÁ, Gómez-Esteban JC, Ibarretxe-Bilbao N. (2015). Neuroanatomical Correlates of Theory of Mind Deficit in Parkinson's Disease: A Multimodal Imaging Study. PLoS One.10(11):e0142234. doi: 10.1371/journal.pone.0142234. eCollection 2015.

Dingemans, E., & Henkens, K. (2015). How do retirement dynamics influence mental well-being in later life? A 10-year panel study. *Scandinavian Journal of Work, Environment & Health, 41*(1), 16-23. Doi:10.5271/sjweh.3464

Direção Geral da Saúde (2015). *A saúde dos portugueses: Perspetiva 2015.* Lisboa: DGS.

Doubova, S.V., Pérez-Cuevas, R., Espinosa-Alarcón, P., Flores-Hernández, S. (2010). Social network types and functional dependency in older adults in Mexico. *BMC Public Health, 10,* 104 http://www.biomedcentral.com/1471-2458/10/104.

Duarte, R. (2017). O impacto da HIT (Homecounseling Intervention Therapy) na Promoção da Qualidade de Vida e na Diminuição da Sintomatologia Depressiva em Pessoas Idosas. Tese de mestrado integrado em Psicologia Clínica (não publicada). FPCE.UC.

Earl, J. K., Gerrans, P., & Halim, V. A. (2015). Active and adjusted: Investigating the contribution of leisure, health and psychosocial factors to retirement adjustment. *Leisure Sciences, 37,* 354-372. Doi: 10.1080/01490400.2015.1021881

Elias, N. (2001). *A solidão dos moribundos.* Rio de Janeiro: Zahar.

Epel, E. (2012). How "reversible" is telomeric aging? *Cancer Prevention Research, 5,* 1163-1168. doi:10.1158/1940-6207.CAPR12-0370

Epel, E. S., Blackburn, E. H., Lin J, Dhabhar, F. S., Adler, N. E., Morrow, J. D., & Cawthon, R. M. (2004). Accelerated telomere shortening in response to life stress. *Proceedings of the National Academy of Sciences of the United States of America, 101,* 17312-17315.

Erera, P.I. (2002). *Family Diversity. Continuity and change in the contemporary family.* Thousand Oaks: Sage Publications.

Esping-Andersen, G. (1999). *Social Foundations of Postindustrial Economies.* Oxford: Oxford University Press.

Faure, E., Herrera, F., Kaddoura, A-R., Lopes, H., Petrovski, A. V., Rahnema, M., & Ward, F. C. (1972). *Apprendre à être.* Paris: UNESCO.

Fernandes, A.A. (2001). Velhice, solidariedades familiares e política social: itinerário de pesquisa em torno do aumento da esperança de vida. *Sociologia, Problemas e Práticas, 36,* 39-52.

Fernandes, L. (2006). *Psicoterapias no idoso.* In Firmino, H., Psicogeriatria (pp.133). Coimbra: Psiquiatria Clínica.

Fernandes, M. M. G. C. (2014). *Depressão, qualidade de vida em idosos institucionalizados* (Master's thesis).

Fernández-Ballesteros, R. (2009). Jubilación y salud. *Humanitas. Humanidades médicas, 37,* 1-23.

Fernández-Ballesteros, R. (2013). Possibilities and limitations of age. In A. L. Oliveira (Coord.), C. M. Vieira, M. P. Lima, L. Alcoforado, S. M. Ferreira & J. A. Ferreira, *Promoting conscious and active learning and ageing: How to face current and future challenges?* (pp. 25-74). Coimbra: Imprensa da Universidade de Coimbra. ISBN: 978-989-26-0732-0 http://www.uc.pt/imprensa_uc/catalogo/ebook/E-book_Promoting

Ferreira, J. A., & Simões, A. (1999). Escalas de bem-estar psicológico (E.B.E.P.). In M. R. Simões, L. S. Almeida & M. M. Gonçalves (Eds.), *Testes e Provas Psicológicas em Portugal* (Vol. 2, pp. 111-121). Braga: Associação dos Psicólogos Portugueses.

Ferreira-Alves, J. (2010). Prática Psicológica com Pessoas Idosas: Uma Leitura Substanciada das orientações da APA. *Revista de Psiquiatria.* CHPL – Hospital Júlio de Matos, *XXII,* (3),1-24.

Ferrera, M. (2000). *O futuro da Europa social: repensar o trabalho e a protecção social na nova economia*. Lisboa: Celta, Presidência Portuguesa da União Europeia.

Field, D. e Minkler, M. (1988). Continuity and change in social support between young-old and old-old or very-old age. *Journal of Gerontology: Psychological Sciences, 43*(4), 100-106.

Figueira, A. P. & Paixão., R. (in press). REHACOG. https://dialnet.unirioja.es/servlet/autor?codigo=3741233

Figueiredo, D., Sousa, L., & Lima, M. (2012). Cuidadores familiares de idosos dependentes: rede social pessoal e satisfação com a vida. *Psicologia, Saúde e Doenças, 13* (2), 47-60.

Fonseca, A.M. (2011). *Reforma e reformados*. Coimbra: Almedina.

Frade, C. e Coelho, L. (2015). Surviving the crisis and austerity: The coping strategies of Portuguese households. *Indiana Journal of Global Legal Studies, 22*(2), 631-664.

Frade, J., Barbosa, P., Cardoso, S., & Nunes, C. (2015). Depression in the elderly: symptoms in institutionalised and non-institutionalised individuals. *Revista de Enfermagem Referência, 4*(4), 41.

Franzoni, J.M. (2007). *Regímenes del bienestar en América Latina*. Madrid: Fundación Carolina. *Online in*: http://www.fundacioncarolina.es/wp-content/uploads/2014/08/DT11.pdf.

Gabriel, Z., & Bowling, A. (2004). Quality of life from the perspectives of older people. *Ageing and Society, 24*, 675-691.

Gard, T., Holtzer, B. K., & Lazar, S. W. (2014). The potential effect of meditation on aged-related cognitive decline: A systematic review. *Annals of the New York Academy of Sciences, 1307*, 89-103.

Geiger, P. J., et al. (2016). Mindfulness-based interventions for older adults: A review of the effects on physical and emotional well-being. *Mindfulness, 7*, 296-307.

Giannella, E. e Fischer, C.S. (2016). An inductive typology of egocentric networks. *Social Networks*, 47, 15-23. DOI: 10.1016/j.socnet.2016.02.003.

Gil, A.P.M. (2009). Conciliação entre vida Profissional e vida familiar: o caso da dependência. Lisboa: Núcleo de Estudos e Conhecimento, Instituto de Segurança Social. *Online in*: http://www.seg-social.pt/documents/10152/135827/conciliacao_vida_profissional_familiar/2d308149-a66d-4075-bbaa-2eb95869c677.

Gil, A.P.M. (2010). *Heróis do quotidiano: dinâmicas familiares na dependência*. Lisboa: Fundação Calouste Gulbenkian, Fundação para a Ciência e Tecnologia.

Ginn, J. e Arber, S. (1996). «Mera conexión». Relaciones de género y envejecimiento. *In* S. Arber e J. Ginn (org.). *Relación entre género y envejecimiento. Enfoque sociológico* (pp. 17-34). Madrid: Narcea.

Global AgeWatch (2015). *Global AgeWatch Index*. (www.globalagewatch.org).

Goldin P. R., Gross J. J. (2010). Effects of mindfulness-based stress reduction (MBSR) on emotion regulation in social anxiety disorder. *Emotion 10,* 83-91. doi: 10.1037/a0018441

Gomes, J. F. (1971). A "Pampaedia" de Coménio. *Revista Portuguesa de Pedagogia, V,* 39-62.

Gonçalves, C. D., & Oliveira A. L. (2013). Reflections from a study about wisdom with students from a senior university. In A. L. Oliveira et al. (Coord.), *Promoting conscious and active learning and ageing: How to face current and future challenges?* (pp. 113-127). Coimbra: Imprensa da Universidade de Coimbra. ISBN: 978-989-26-0732-0 http://www.uc.pt/imprensa_uc/catalogo/ebook/E-book_Promoting

Gondo, Y., & Poon, L.W. (2007). Biopsychosocial approaches to longevity. *Annual Review of Gerontology and Geriatrics,* 129-149.

Grundy, E. (2005). Reciprocity in relationships: socio-economic and health influences on intergenerational exchanges between Third Age parents and their adult children in Great Britain. *British Journal of Sociology. 56*(2), 233-55. DOI: 10.1111/j.1468-4446.2005.00057.x.

Grundy, E. e Tomassini, C. (2003). El apoyo familiar de las personas de edad, en Europa: contrastes e implicaciones. *Notas de Población, 77,* 219-250.

Guadalupe, S. (2016). *Intervenção em rede: Serviço Social, sistêmica e redes de suporte social* (2.ª edição). Coimbra: Imprensa da Universidade de Coimbra.

Guadalupe, S. (2017). "As redes de suporte social informal em Serviço Social: as redes sociais pessoais de idosos portugueses nos processos de avaliação diagnóstica em respostas sociais" [Tese de Doutoramento em Serviço Social] Lisboa: ISCTE – Instituto Universitário de Lisboa, Escola de Sociologia e Políticas Públicas e CIES, Centro de Investigação e Estudos de Sociologia.

Gutierrez, J., & Isaacson, R. S. (2013). Prevention of cognitive decline. In L. D. Ravdin & H. L. Katzen (eds.), *Handbook of the neuropsychology of aging and dementia* (pp. 167-192). LLC: Springer Science.

Ham-Chande, R., Ybáñez-Zepeda, E. e Torres-Martínez, A.L. (2003). Redes de apoyo y arreglos de domicilio de las personas en edades avanzadas en la Ciudad de México. *Notas de Población, 77*, 71-102.

Hespanha, P. (1993). Vers une société-providence simultanément pré- et post-moderne. *Oficina do Centro de Estudos Sociais, 38*. Coimbra: Centro de Estudos Sociais.

Hespanha, P., Ferreira, S., e Pacheco, V. (2013). O Estado social, crise e reformas. *In* Observatório das Crises e das Alternativas (org.). *Anatomia da crise: identificar os problemas para construir alternativas* (pp. 161-249). Coimbra: Centro de Estudos Sociais.

Ho, N. S. P., Sun, D., Ting, K-H, Chan, C. C. H., & Lee, T. M. C. (2015). Mindfulness Trait Predicts Neurophysiological Reactivity Associated with Negativity Bias: An ERP Study. *Evidence-Based Complementary and Alternative Medicine, 15,* 1-15.

Hölzel, B. K., Carmody, J., Evans, K. C., Hoge, E. A., Dusek, J. A., Morgan, L., ... Lazar, S. W. (2010). Stress reduction correlates with structural changes in the amygdala. *Social Cognitive and Affective Neuroscience, 5*(1), 11-17.

Hölzel, B. K., Carmody, J., Vangel, M., Congleton, C., Yerramsetti, S. M., Gard, T., & Lazar, S. W. (2011). Mindfulness practice leads to increases in regional brain gray matter density. Psychiatry Research: *Neuroimaging, 191*(1), 36-43.

Jarvis, P. (2001a). O futuro da educação de adultos na sociedade de aprendizagem. *Revista Portuguesa de Pedagogia, 35*(1), 41-66.

Kabat-Zinn, J. (1990). *Full catastrophe living: Using the wisdom of your body and mind to face stress, pain and illness.* New York: Delacorte.

Kabat-Zinn, J. (2003). *Onde quer que eu vá* (C. Rodriguez, Trad.). Porto: Oficina do Livro.

Kabat-Zinn, J. (2011). Some reflections on the origins of MBSR, skillful means, and the trouble with maps. *Contemporary Buddhism, 12*(1), 281-306.

Kabat-Zinn, J. (2015). Why paying attention is so supremely important. *Mindfulness, 6,* 1484-1486.

Kahana, E. (2001). Institutionalization. In G.L. Maddox (Ed.), *The Enciclopedia of Aging.* New York: Springer Publishing Company.

Knight, B. G. (2004). Psychotherapy with older adults (3rd ed.). New York: Sage.

Knight, B. G., & Poon, C. (2008). Contextual Adult Life Span Theory for Adapting Psychotherapy with Older Adults. *Journal of Rational-Emotional and Cognitive-Behavioral Therapy*, 26, 232-249.

Kok, B. E., & Singer, T. (2017). Phenomenological fingerprints of four meditations: Differential state changes in affect, mind-wandering, meta-cognition, and interoception before and after daily practice across 9 months of training. *Mindfulness, 8,* 218-231. https://doi.org/10.1007/s12671-016-0594-9

Krout, J.A. (1985). Relationships between informal and formal organizational networks. *In* W.J. Sauer e R.T. Coward (eds.). *Social support networks and the care of the elderly* (pp. 178-195). New York: Springer.

Kuyper, J. e Bengtson, V. (1973). Social breakdown and competence. A model of normal aging. *Human Development. 16*(3), 181-201.

Labouvie-Vief, G. (2005). The Psychology of emotions and ageing. In M. Johnson, V. Bengtson, P. Coleman, & T. KirDeath (Eds.), *The Cambridge handbook of age and ageing.* Cambridge: Cambridge University Press (pp. 229--236).

Landreville, P., Landry, J., Baillargeon, L., Guerette, A., & Matteau, E. (2001). Older adults' acceptance of psychological and pharmacological treatments for depression. *Journal of Gerontology , 50B* , P285-P291.

Lazar, S. W. et al. (2005). Meditation experience is associated with increased cortical thickness. *Neuroreport, 16,* 1893-1897.

Leandro, M.E. (2011). Laços familiares em questão: antinomias nas sociedades hipermodernas. *In* M.E. Leandro (coord.). *Laços familiares e sociais* (pp. 95-115). Viseu: Psicosoma.

Leandro, M.E. e Ferreira, L. (2011). Os laços sociais em questão. Metamorfoses sociais, metamorfoses de uma noção. *In* M.E. Leandro (coord.), *Laços Familiares e sociais* (pp. 27-57). Viseu: Psicossoma.

Leandro-França, C., Murta, S. G., Hershey, D. S., & Martins, L. B. (2016). Evaluation of retirement planning programs: A qualitative analysis of methodologies and efficacy. *Educational Gerontology, 42* (7), 497-512. Doi: 10.1080/03601277.2016.1156380

Lesemann, F. e Martin, C. (1995). Estado, comunidade e família face à dependência dos idosos. Ao encontro de um "Welfare-Mix". *Sociologia – Problemas e Práticas, 17,* 115-139.

Li, T e Zhang, Y. (2015). Social network types and the health of older adults: Exploring reciprocal associations. *Social Science & Medicine, 30*(130), 59-68.

Lima M.P., Silva, C., Carvalho, M.M., & Fernandes, D. (2014). Providing home counseling for older adults: Benefits and challenges. *Revista de Saúde Pública,* 48 (nº esp.), 75.

Lima, M. P. & Silvério, L. (2015). *Programa manualizado de prestação de apoio psicológico ao domicílio: Homecounseling Intervention Therapy.* Manuscrito não publicado, Faculdade de Psicologia e Ciências da Educação da Universidade de Coimbra.

Lima, M. P. (2006). Posso participar?: Actividades de desenvolvimento pessoal para idosos. Lisboa: Ambar.

Lima, M. P. (2012). *Intervenção psicoterapêutica com pessoas idosas.* Relatório da disciplina de Intervenção Psicoterapêutica em Adultos Idosos, Faculdade de Psicologia e de Ciências da Educação – Universidade de Coimbra, Portugal.

Lima, M. P., Oliveira, A. L., & Godinho, P. (2011). Promover o bem-estar de idosos institucionalizados: Um estudo exploratório com treino em mindfulness. *Revista Portuguesa de Pedagogia, 45*(1), 165-183. DOI http://dx.doi.org/10.14195/1647-8614_45-1_9f

Litwin, H. (1995a). *Unprooted in old age: soviet Jews and their social networks in Israel.* Westport, CT: Praeger.

Litwin, H. (1995b). The social networks of elderly immigrants: An analytic typology. *Journal of Aging Studies, 9*(2), 155-174.

Litwin, H. (1997). The network shifts of elderly immigrants: The case of Soviet Jews in Israel. *Journal of CrossCultural Gerontology, 12*, 45-60.

Litwin, H. (2001). Social network type and morale in old age. *The Gerontologist, 41*(4), 516-24. DOI: 10. 1093/geront/41.4.516 PMID: 11490050.

Litwin, H. e Landau, R. (2000). Social network type and social support among the old-old. *Journal of Aging Studies, 14*(2), 213-228.

Litwin, H. e Shiovitz-Ezra S. (2011). Social network type and subjective well-being in a national sample of older Americans. *Gerontologist, 51*, 379-88. DOI: 10.1093/geront/gnq094 PMID: 21097553.

Lopes, J. A., Cruz, M. C., Mathur, S. R., Quinn, M. M. & RutherFord Jr., R. B. (2006). *Competências Sociais, Aspectos comportamentais, emocionais e de aprendizagem*. Braga: Psiquilibrios.

Lubben, J. & Grionda, M. (2003). Measuring social networks and assessing their benefits. *In*: Phillipson C, Allan G,& Morgan D (Eds.). *Social networks and social exclusion*: 20-49. Hants, England: Ashgate.

Luders, E., Cherbuin, N., & Gaser, C. (2016). Estimating brain age using high-resolution pattern recognition: Younger brains in long-term meditation practitioners. *Neuroimage, 134,* 508-513.

Maia, L. Correia, C. & Leite, R. (2007). *Manual Prático de Avaliação & Intervenção Neuropsicológica – Estudos de casos e instrumentos.* Covilhã: Éditos Prometaicos.

Marques, J. (2008). A reconfiguração do estado-providência. *Gestão e Desenvolvimento, 15-16*, 105-119.

Martin, C. (1995). Os limites da protecção da família – introdução a uma discussão sobre as novas solidariedades na relação Família-Estado. *Revista Crítica de Ciências Sociais, 42*, 53-76.

Maxfield, M., & Segal, D. L. (2008). Psychotherapy in nontraditional settings: A case of in-home cognitive-behavioral therapy with a depressed older adult. *Clinical Case Studies*, *7*(2), 154-166.

McPherson, M. Smith-Lovin, L. e Brashears, M.E. (2006). Social isolation in America: changes in core discussion networks over two decades. American Sociological Review, *71*(3), 353-375.

Meadle, M. L., & Park, D. C. (2009). Enhancing cognitive function in older adults. Chodzko-Zajko, W. C., Kramer, A. F., & Poon, L. W. *Enhancing cognitive functioning and brain plasticity* (pp. 35-47; vol. 3). Champaign: Human Kinetics.

Melkas, T. e Jylhä, M. (1996). Social network characteristics and social network types among Eldery People in Finland. *In* H. Litwin (Ed.), *The social network of older people: a cross national analysis* (pp. 99-116). Westport, CT: Praeger.

Mello, J.D.A., Durme, T.V., Macq, J., Declerq, A. (2012). Intervention to delay institutionalization of frail older persons: design of longitudinal study in the home care setting. *BMC Public Health, 12*, 615.

Melo, A. et al. (1998). *Uma aposta educativa na educação para todos. Documento de estratégia para o desenvolvimento da educação de adultos.* Lisboa: Ministério da Educação.

Mestheneos, E. (2011). Ageing in place in the European Union. *Global Ageing, 7*(2), 17-23.

Mioto, R.C.T. (2008). Família e políticas sociais. *In* I. Boschetti, E. Behring, R.C.T. Mioto e S.M.M. Santos (Orgs.). *Política social no capitalismo: tendências contemporâneas* (pp.130-148). São Paulo: Cortez.

Moral, J.C.M., Miguel, J.M.T. e Pardo, E.N. (2007). Análisis de las redes sociales en la vejez através de la entrevista Manheim. *Salud Pública de México, 49*(6), 408-414.

Morone, N.E., Lynch, C.S., Greco, C.M., Tindle, H.A., & Weiner, D.K. (2008). "I felt like a new person." The effects of mindfulness meditation on older adults with chronic pain: Qualitative narrative analysis of diary entries. *Journal of Pain, 9,* 841-848.

Mugford, S. e Kendig, H. (1986). Social relations: Networks and ties. *In* H. Kendig (ed.). Ageing and families: A social networks perspective (pp. 38-59). Sydney: Allen and Unwin.

Muijen, M., Marks, I., Connolly, J., & Audini, B. (1992). Home based care and standard hospital care for patients with severe mental illness: a randomised controlled trial. *BMJ, 304*(6829), 749-754.

Myers, D. G., & Diener, E. (1995). Who is happy? *Psychological Science, 6*(1), 10-17.

Nações Unidas (2017). Sustainable Development Goals. https://unstats.un.org/sdgs/report/2016/leaving-no-one-behind.

NeuroLab-Neurologia de los Trastornos Medicos Severos, REHACOP: Programa integral de rehabilitación cognitiva en Psicosis, Acedido em: http://neurolab.deusto.es/rehacop-programa-integral-derehabilitacion-cognitiva-en-psicosis/

Nunes, R. (2015). Ética e família. *In* O.M. Fernandes e C. Maia (coord.). *A Família portuguesa no século XXI* (pp. 39-50). Lisboa: Parfisal.

OCDE/OECD – Organização para a Cooperação e Desenvolvimento Económico (2012). *The Future of Families to 2030.* OECD Publishing. DOI:10.1787/9789264168367-en.

Ojeda N., Peña J., Bengoetxea E, García A., Sánchez P., Segarra R. C, Ezcurra J., Gutiérrez Fraile M., Eguíluz J.I. C. (2012). REHACOP: Programa de Rehabilitación Cognitiva en Psicosis. *Revista de Neurología, 54*(6), 337-342.

Oliveira, A. L. (Coord.), Vieira, C. M., Lima, M. P., Alcoforado, L., Ferreira, S. M., & Ferreira, J. A. (2013), *Promoting conscious and active learning and ageing: How to face current and future challenges?* Coimbra: Imprensa da Universidade de Coimbra. ISBN: 978-989-26-0732-0.

Oliveira, A. L., & Antunes, B. M. G. (2014). A pedagogia contemplativa no ensino superior: Para uma abordagem completa ao que o ser humano convoca. *Revista Portuguesa de Pedagogia, 48*(2), 43-60. http://dx.doi.org/10.14195/1647-8614_48-2_3

Oliveira, A. L., & Cruz, A. C. (2015). O papel do sentido da vida e do mindfulness na compreensão do bem estar de alunos de Universidades Seniores. *Exedra*, 61-78. http://www.exedrajournal.com/wp-content/uploads/2016/02/Cap4.pdf

Oliveira, A.L., Vieira, C., Lima, M.P., Nogueira, S., Alcoforado, L., Ferreira, J.A., & Zarifis, G. (2011). Developing instruments to improve learning and development of disadvantage seniors in Europe: The paladin project. In Pixel (Ed.), *Conference proceedings of the International Conference The Future of Education* (vol. 1, pp. 268-274). Florence: Simonelli Editore.

Olsen, O., Iversen, L. e Sabroe, S. (1991). Age and the operationalization of social support. *Social Science & Medicine, 32*, 767-771.

OMS – Organização Mundial de Saúde (2015). *Relatório mundial de envelhecimento e saúde*. Genebra: OMS.

Pacheco, J. A. (2011). *Discursos e lugares das competências em contextos de educação e formação*. Coleção Panorama. Porto: Porto Editora.

Paixão, M. P., Silva, J. T., & Oliveira, A. L. (2014). Perspectives on guidance and counselling as strategic tools to improve lifelong learning in Portugal. In G. K. Zarifis & M. Gravani (Eds.), *Challenging the 'European Area of Lifelong Learning': A critical response* (pp. 167-176). London: Springer. ISBN 978-94-007-7298-4; DOI 10.1007/978-94-007-7299-1

Park, S. Smith, J. e Dunkle, R E. (2014). Social network types and well-being among South Korean older adults. *Aging & Mental Health, 18*(1), 72-80. DOI: 10.1080/13607863.2013.801064.

Paúl, C. & Fonseca, A. M. (2005). *Envelhecer em Portugal: Psicologia, saúde e prestação de cuidados*. Lisboa: Climepsi Editores.

Paúl, C. (2005). Envelhecimento activo e redes de suporte social. *Sociologia, 15*.

Paúl, C. (2005a). A construção de um modelo de envelhecimento humano. *In* Paúl, C. e Fonseca, A. (coords.), *Envelhecer em Portugal: Psicologia, saúde e prestação de cuidados* (pp. 15-41). Lisboa: Climepsi Editores.

Paúl, C. (2005b). Envelhecimento activo e redes de suporte social. *Sociologia, 15*, 275-287.

Payne, M. (2014). Redes sociais em Serviço Social. *In* M.I. Carvalho e C. Pinto (coords.), *Serviço Social: Teorias e práticas* (pp. 181-204). Lisboa: Pactor.

Pedroso, J. e Branco, P. (2008). Mudam-se os tempos, muda-se a família. As mutações do acesso ao direito e à justiça de família e das crianças em Portugal. *Revista Crítica de Ciências Sociais*, 82, 53-83.

Peña, J., Ibarretxe-Bilbao, N., García-Gorostiaga, I., Gomez-Beldarrain, M. A., Díez-Cirarda, M., & Ojeda, N. (2014). Improving functional disability and cognition in Parkinson disease Randomized controlled trial. *Neurology, 83*(23), 2167-2174.

Pereira, M. (in press). Transtornos Neurocognitivos – Leve e Maior. *Manual de formação para educadores, módulo do programa REHACOG*. Tese de Mestrado Integrado a apresentar à FPCE.UC.

Phillipson, C., Bernard, M., Phillips, J. e Ogg, J. (2001). *The family and community life of older people. Social networks and social support in three urban areas*. London and New York: Routledge.

Pickut, B. A., Hecke, W. V., Kerckhofs, E., Marien, P., Vanneste, S., Cras, P., & Parizel, P. M. (2013). Mindfulness based intervention in Parkinson's disease leads to structural brain challenges on MRI: A randomized control longitudinal trial. *Clinical neurology and neurosurgery, 115*, 2419--2425.

Pimentel, L. (2012). Cuidar de pessoas idosas dependentes: as interseções entre a esfera pública e a esfera privada. *Rediteia, 45*, 67-77.

Pimentel, L. (2015). As pessoas idosas e os seus contextos familiares: convitea um olhar diferente. *In* O.M. Fernandes e C. Maia (coord.). *A Família portuguesa no século XXI* (pp. 171-178). Lisboa: Parfisal.

Portugal, S. (2011). Dádiva, família e redes sociais. *In* S. Portugal e P.H. Martins (org.). *Cidadania, políticas públicas e redes sociais* (pp. 39-53). Coimbra: Imprensa da Universidade de Coimbra.

Portugal, S. (2014). *Famílias e redes sociais. Ligações fortes na produção de bem-estar*. Coimbra: Almedina.

Rana, N. (2015). Mindfulness and loving-kindness meditation: A potential tool for mental health and subjective well-being. *Indian Journal of Positive Psychology, 6*(2), 189-196.

REHACOP.https://www.google.pt/search?biw=1366&bih=651&q=rehacop&oq=rehacop&gs_l=psyab.3..35i39k1l2j0i30k1l2.5025.7741.0.8227.3.3.0.0.0.0.123.353.0j3.3.0....0...1.1.64.psy-ab..0.3.350...0j0i10k1.I78Qd5BFrJU

Rejeski, W. (2008). Mindfulness: Reconnecting the body and the mind in geriatric medicine and gerontology. *The Gerontologist, 48*(2), 135-141.

República Portuguesa (2017). *Estratégia Nacional para o Envelhecimento Ativo e Saudável (2017-2025). Proposta do Grupo de Trabalho Interministerial* (Despacho n.º12427/2016). República Portuguesa: DGS/SNS.

Rios, L., & Fraguela, J. A. G. (2007). *La psicologia en la intervención social.* Madrid: Editorial Síntesis.

Rioseco, H.R., Quezada, V.M., Ducci, V.M.E. e Torres, H.M. (2008) Cambio en las redes sociales de adultos mayores beneficiarios de programas de vivienda social en Chile. *Revista Panamericana de Salud Pública, 23*(3), 147-53.

Rodin, G. (2009). Individual psychotherapy for the patient with advanced disease. In: *Handbook of Psychiatry in Palliative Medicine*, 2nd Ed. Chochinov, H.M., & Breitbart,W. (eds). Oxford: Oxford University Press.

Rowe, J. W., & Kahn, R. L. (1999). *Successful aging.* New York: Dell Publishing.

Saggar, M. et al. (2015). Mean-field thalamocortical modeling of longitudinal EEG acquired during intensive meditation training. *Neuroimage, 114,* 88--104.

Sánchez, P., Peña, J., Bengoetxea, E., Ojeda, N., Elizagárate, E., Ezcurra, J., Gutiérrez, M. (2014) Improvement in negative symptoms and functional outcome after new generation cognitive remediation program (REHACOP): A randomized controlled trial. *Schizophrenia Bulletin, 40*(3), 707-15.

Santos, B.S. (1995). Sociedade-Providência ou autoritarismo social? [editorial]. *Revista Crítica de Ciências Sociais, 42,* i-vii.

Saraceno, C. (2010). Social inequalities in facing old-age dependency: a bi-generational perspective. *Journal of European Social Policy, 20*(1), 32-44. DOI: 10.1177/0958928709352540.

Sarti, C. (2010). Famílias enredadas. *In* A.R. Acosta e M.A.F. Vitale (orgs.). *Família – Redes, laços e políticas públicas* (pp. 22-38, 5.ª edição). São Paulo: CEDPE, PUC-SP, Cortez Editora.

Schnettler, S. e Wöhler, R. (2015). No children in later life, but more and better friends? Substitution mechanisms in the personal and support networks of parents and the childless in Germany. *Ageing and Society, CJO*, 1-25. DOI:10.1017/S0144686X15000197.

Schnittker, J. (2007). Look (closely) at all the lonely people: age and the social psychology of social support. *Journal of Aging and Health, 19*, 659-82.

Serapioni, M. (2005). O papel da família e das redes primárias na reestruturação das políticas sociais. *Ciência & Saúde Coletiva, 10(sup)*, 243-253.

Shanas, E. (1979). The family as a social support system in old age. *The Gerontologist, 19*(2), 169-174.

Sharpley, C. F. (2010). A review of the neurobiological effects of psychotherapy for depression. *Psychotherapy: Theory, research, practice, training, 47*(4), 603.

Siegel, D. J. (2007). *The Mindful Brain: Reflection and Attunement in the Cultivation of Well-Being*. New York: Norton & Company.

Silva, P.A. (2002). O modelo de welfare da Europa do sul – Reflexões sobre a utilidade do conceito. *Sociologia, Problemas e Práticas. 38*, 25-59.

Simões, A. (2006). *A nova velhice: Um novo público a educar.* Porto: Ambar.

Simões, A. (2006). Factos e factores do desenvolvimento intelectual do adulto. *Psychologica, 42,* 25-43.

Simões, A. (2011). Um modelo mal sucedido de envelhecimento bem sucedido? *Psicologia, Educação e Cultura, XV*(1), 7-28.

Simões, A., Ferreira, J. A., Lima, M. P., Pinheiro, M. R., Vieira, C. M., Matos, A. P., & Oliveira, A. L. (2000). O bem-estar subjectivo: Estado actual dos conhecimentos. *Psicologia, Educação e Cultura, IV*(2), 243-279.

Simões, A., Ferreira, J., Lima, M., Pinheiro, M., Vieira, C., Matos, A., Oliveira, A., Alcoforado, L., Neto, F., Ruiz, F., Cardoso, A., Felizardo, S., & Sousa, L. (2006). Promover o bem-estar dos idosos: Um estudo experimental. *Psychologica — especial: Envelhecimentos*, 42, 115-131.

Simões, A., Lima, M. P., Vieira, C. M. C., Oliveira, A. L., Alcoforado, J. L., & Ferreira, J. A. (2009). O sentido da vida: Contexto ideológico e abordagem empírica. *Psychologica, 51,* 101-130. DOI:http://dx.doi.org/10.14195/1647-8606_51_8

Sluzki, C.E. (1996). *La red social: frontera de la practica sistemica.* Barcelona, Gedisa Editorial.

Sluzki, C.E. (2000). Social network and the elderly: conceptual and clinical issues, and a family consultation. *Family Process, 39*(3), 271-284.

Smith, A. (2004). Clinical uses of mindfulness training for older people. *Behavioural and Cognitive Psychotherapy, 32,* 423-430.

Sorocco, K. H., & Lauderdale, S. (2011). *Cognitive behavior therapy with older adults: Innovations across care settings.* New York: Springer Pub. Co, 391-416.

Sorrell, J. M. (2015). Meditation for Older Adults: A New Look at an Ancient Intervention for Mental Health. *Journal of Psychosocial Nursing, 53*(5), 15-19.

Sousa, L. e Figueiredo, D. (2004). *Services for Supporting Family Carers of Elderly People in Europe: Characteristics, Coverage and Usage.* EUROFAMCARE.

Sousa, L. e Figueiredo, D. (2007). *Supporting family carers of older people in Europe – The national background report for Portugal.* Hamburg: Lit Verlag.

Sousa, L. Figueiredo, D. e Cerqueira, M. (2004). *Envelhecer em família – Os cuidados familiares na velhice.* Porto: Âmbar.

Specht, H. (1986). Social support, social networks, social exchange, and Social Work practice. *Social Service Review, 60*(2), 218-240.

Stella, F., Gobbi, S., Corazza, D. I., & Costa, J. L. R. (2002). Depressão no idoso: diagnóstico, tratamento e benefícios da atividade física. *Motriz, 8*(3), 91-98.

Thiyagarajan, J.A., Prince, M. e Webber, M. (2014). Social support network typologies and health outcomes of older people in low and middle income countries – A 10/66 Dementia Research Group population-based study. *International Review of Psychiatry, 26*(4): 476-485.

To-Miles, F. Y. L., & Backman, C. L. (2016). What telomeres say about activity and health: A rapid review. *Canadian Journal of Occupational Therapy, 83*(3) 143-153. DOI: 10.1177/0008417415627345

Uchino, B. (2004). *Social Suport & Physical Health. Understanding the health consequences of relationships*. New Haven: Yale University Press.

UN - United Nations (2002). *Political Declaration and Madrid International Plan of Action on Ageing*. Second World Assembly on Ageing, Madrid, Spain. New York: UN. *Online in*: http://www.un.org/en/events/pastevents/pdfs/Madrid_plan.pdf.

UN - United Nations (2015). *World Population Ageing 2015*. Ney Work: United Nations, Department of Economic and Social Affairs Population Division. *Online in*: http://www.un.org/en/development/desa/ population/publications/pdf/ageing/WPA2015_Report.pdf.

Valle, J. e Garcia, A. (1994). Redes de apoyo social en usuarios del servicio de ayuda a domicilio de la tercera edad. *Psicothema*, *6*(1), 39-47.

Vasconcelos, P. (2002). Redes de apoio familiar e desigualdade social: estratégias de classe. *Análise Social*, *XXXVII*(163), 507-544.

Vasconcelos, P. (2005). Redes sociais de apoio. *In* K. Wall (Org.). *Famílias em Portugal* (pp. 599-631). Lisboa: Imprensa de Ciências Sociais.

Vaz Serra, A. (2002). O stress na vida de todos os dias. Coimbra. Ed do autor. ISBN 972-95003-2-0.

Vecchia, R. D., Ruiz, T., Bocchi, S. C. M. & Corrente, J. E. (2005). Qualidade de vida na terceira idade: um conceito subjetivo. *Revista brasileira de epidemiologia*, 246-252.

Vicente, H.T. (2010). *Família multigeracional e relações intergeracionais: Perspectiva sistémica* [Tese de doutoramento]. Aveiro: Secção Autónoma de Ciências da Saúde, Universidade de Aveiro.

Vicente, H.T. e Sousa, L. (2012). Relações intergeracionais e intrageracionais: A matriz relacional da família multigeracional. *Revista Temática Kairós Gerontologia*, *15*(2), 99-117.

Vieira, D. (2017). Perturbações do Espetro do Autismo – Módulo de psicoeducação para o programa REHACOG. Tese de Mestrado Integrado apresentada à FPCE.UC.

Vilar, M., Sousa, L., Firmino, H. & Simões, M. (2016). Envelhecimento e qualidade de vida. In H. Firmino; M. Simões & J. Cerejeira (Coords.), *Saúde mental das pessoas mais velhas* (pp. 19-43). Lisboa: Lidel.

Volpi, R. (2007). *La fine della famiglia. La rivoluzione di cui non ci siamo accorti*. Milano: Mondadori.

Wallace, B. A. (2007). *Contemplative science: Where Buddhism and neuroscience converge*. New York: Columbia University Press.

Wenger, G.C. (1991). A network typology: from theory to practise. *Journal of Aging Studies, 5*(2), 147-162.

Wenger, G.C. (2009). Childlessness at the end of life: evidence from rural Wales. *Ageing & Society, 29*, 1243-1259. DOI: 10.1017/S0144686X09008381.

WHOQOL Group. (1995). The World Health Organization Quality of Life Assessment: Position paper from the WHO. *Social Science and Medicine, 41*(10), 1403-1409.

Williams, F. (2010). *Repensar as famílias*. Lisboa: Principia.

World Health Organization Quality of Life Group/WHOQOL Group. (1995). The World Health Organization Quality of Life Assessment (WHOQOL): Position paper from the WHO. *Social Science and Medicine, 41*(10), 1403-1409.

World Health Organization/WHO (2011). *Active Ageing: Report. Special Eurobarometer. European Commisson*. Genebra: WHO.

Yang, J. A., Garis, J., Jackson, C., & McClure, R. (2009). Providing psychotherapy to older adults in home: Benefits, challenges, and decision-making guidelines. *Clinical Gerontologist, 32*(4), 333-346.

Yazbeck, M.C. (2001). Pobreza e exclusão social: expressões da questão social. *Temporalis, III*(3), 33-40.

Zarit, S. H., & Leitsch, S. A. (2001). Developing and evaluating community based intervention programs for Alzheimer's patients and their caregivers. *Aging & Mental Health, 5*(2), 84-98.

NOTA BIOGRÁFICA DOS AUTORES:

Albertina Lima de Oliveira

Professora Auxiliar da Faculdade de Psicologia e Ciências da Educação da Universidade de Coimbra e investigadora do Centro de Estudos Interdisciplinares (CEIS20) da mesma universidade. Psicóloga e doutorada em Ciências da Educação. Autora de diversos livros, artigos e capítulos de livros sobre educação, potenciação do bem-estar e qualidade de vida de pessoas adultas e idosas. Formadora certificada de Mindfulness pelo Centre for Mindfulness Research and Practice da University of Bangor (Reino Unido). Introduziu a disciplina de Educação para o Mindfulness no plano de estudos da Licenciatura em Ciências da Educação da Universidade de Coimbra, na última revisão curricular.

Clara Cruz Santos

Clara Cruz Santos (P.hD) é professora Auxiliar na Faculdade de Psicologia e Ciências de Educação da Universidade de Coimbra. É coordenadora do mestrado em intervenção social, Inovação e empreendedorismo (MISIE) e da Licenciatura em Serviço Social da mesma instituição. Possui publicações e trabalho desenvolvido nas áreas de Risco Social e de Serviço Social. É investigadora em projetos de investigação nacionais e internacionais focados na análise das políticas e práticas sociais nas áreas da Proteção Social e Pensamento Critico. É coordenadora do Observatório de Cidadania e Intervenção Social (OCIS) da FPCEUC.

Helena Reis Amaro da Luz

Professora Auxiliar Convidada na Faculdade de Psicologia e Ciências da Educação da Universidade de Coimbra. Doutora em Economia pelo Instituto Superior de Economia e Gestão da Universidade de Lisboa (ISEG-UL). Doutoranda em Serviço Social no Instituto Superior de Ciências do Trabalho e da Empresa do Instituto Universitário de Lisboa (ISCTE-IUL). Mestre em Economia e Política Social pelo ISEG-UL e Licenciada em Serviço Social pelo Instituto Superior de Serviço Social de Coimbra. Investigadora associada do Centro Interdisciplinar de Ciências Sociais da Universidade Nova de Lisboa (CICS.NOVA). Investigadora do Observatório da Cidadania e da Intervenção Social (OCIS/FPCEUC). Investigadora e autora de publicações e artigos científicos nos domínios da Economia Social, Intervenção Social, Gerontologia e Políticas Sociais, sendo cocoordenadora da recente obra publicada pela PACTOR, intitulada *Políticas Sociais em Tempos de Crise: Perspetivas, Tendências e Questões Críticas*.

Henrique Testa Vicente

Psicólogo Clínico e Psicoterapeuta. Licenciado em Psicologia pela Faculdade de Psicologia e de Ciências da Educação da Universidade de Coimbra (2002). Doutorado em Ciências da Saúde pela Universidade de Aveiro (2010). Professor Auxiliar convidado no Instituto Superior Miguel Torga (ISMT), onde leciona unidades curriculares da Licenciatura em Psicologia e do Mestrado em Psicologia Clínica (área de especialidade em Psicoterapia Psicodinâmica). Coordenador do Grupo de Investigação "Bem-estar, Saúde e Envelhecimento" do Centro de Estudos da População, Economia e Sociedade (CEPESE). Coordenador do Núcleo João dos Santos do ISMT.

Isabel Cerca Miguel

Professora Auxiliar no Departamento de Psicologia e Educação da Universidade Portucalense Infante D. Henrique. É, desde 2010, Doutorada em Psicologia Social pela Faculdade de Psicologia e Ciências da Educação da Universidade de Coimbra. Docente nos domínios da psicologia social, psicologia do desenvolvimento do adulto e do idoso, psicogerontologia, intervenção familiar, intervenção para o envelhecimento ativo. É autora de várias publicações e artigos científicos nacionais e internacionais e é revisora convidada de revistas científicas nacionais e internacionais. É membro do Instituto de Desenvolvimento Humano Portucalense.

Jacqueline Ferreira Marques

Assistente Social, Professora Auxiliar na Universidade Lusófona de Humanidades e Tecnologias. Doutorada em Serviço Social pelo Instituto de Superior de Serviço Social de Lisboa – Universidade Lusíada de Lisboa, Mestre e licenciada em Serviço Social, em 2003 e 1997 respetivamente. Leciona nos domínios da Política Social, Serviço Social e Políticas Sociais e metodologias de intervenção, com publicações e trabalho desenvolvido nas áreas das Políticas Sociais Públicas, medidas de combate a pobreza e exclusão social e governança do território.

Joana Guerra

Professora Auxiliar convidada na Faculdade de Psicologia e Ciências da Educação da Universidade de Coimbra. Doutorada em Serviço Social pela Universidade Católica Portuguesa. Mestre em Saúde Pública pela Faculdade de Medicina da Universidade de Coimbra. Leciona diversas disciplinas nos domínios do Serviço Social e Política Social, com publicações e trabalho desenvolvido nas áreas de intervenção das políticas de saúde. É investigadora em projetos de investigação nacionais e internacionais focados na análise das políticas sociais públicas nas áreas da Saúde, Proteção Social e Educação.

Margarida Pedroso de Lima

Psicóloga, Mestre em Psicologia da Educação e Doutorada em Psicologia do Desenvolvimento, exerce funções de Professora Associada na Faculdade de Psicologia e de Ciências da Educação da Universidade de Coimbra, onde lecciona disciplinas de psicologia do desenvolvimento e de intervenção na idade adulta.

As suas áreas de interesse vão para a intervenção desenvolvimental e terapêutica com grupos e para a investigação sobre os factores promotores de bem-estar na idade adulta avançada. É autora de vários livros como o 'Posso Participar' e 'Envelhecimentos' e de programas de intervenção como o HIT.

Mónica Alexandra Vidal Teixeira

Professora Auxiliar Convidada do Instituto Superior de Serviço Social do Porto onde leciona na área da Gerontologia Social.

Licenciada em Serviço Social pelo Instituto Miguel Torga, possui formação especializada, mestrado e doutoramento na área da Gerontologia realizados na Universidade de Aveiro. Em 2017 obteve o título de especialista na área de Trabalho Social e Orientação pelo Instituto Politécnico de Leiria.

Frequenta o Pós-doutoramento em Gerontologia, no Departamento de Educação e Psicologia da Universidade de Aveiro, sob a orientação da Profª Doutora Liliana Sousa e cuja temática é "Modelos colaborativos com pessoas Idosas".

Na sua experiência como formadora e docente destacam-se os seguintes momentos: Entre 2000 e 2006 foi formadora no Instituto Superior Miguel Torga no qual lecionou diversas formações sobre idosos e entre 2005 e 2007 foi docente das Pós-Graduações de Gerontologia Social. Na Universidade de Aveiro colaborou nas unidades curriculares de Estágio em Gerontologia, Prática Profissional em Gerontologia e na unidade curricular Atividade Física e Mental, na categoria de Assistente Convidada entre 2005 a 2014.

Sónia Guadalupe

Assistente social. Licenciada em Serviço Social pelo Instituto Superior de Serviço Social de Coimbra (1995). É mestre em Família e Sistemas Sociais pelo Instituto Superior Miguel Torga (2000), doutorada em Saúde Mental pela Universidade do Porto (2009) e doutorada em Serviço Social pelo ISCTE-IUL (2017).

Professora Auxiliar no Instituto Superior Miguel Torga, em Coimbra. Coordenadora da licenciatura (1.º ciclo) em Serviço Social no ISMT. Professora na licenciatura e mestrado em Serviço Social, e no ramo de especialidade em Terapias Familiares e Sistémicas do mestrado em Psicologia Clínica.

Foi presidente da Associação de Estudantes do Instituto Superior de Serviço Social de Coimbra (1993/1994). Foi vice-presidente da Associação dos Profissionais de Serviço Social (2008-2015).

Editora-chefe da Revista Portuguesa de Investigação Comportamental e Social.

Investigadora sobre a temática das redes de suporte social em populações vulneráveis. Investigadora do CEISUC (Centro de Estudos e Investigação em Saúde da Universidade de Coimbra) e colaboradora do CEPESE (Centro de Estudos da População, Economia e Sociedade).

www.ingramcontent.com/pod-product-compliance
Ingram Content Group UK Ltd.
Pitfield, Milton Keynes, MK11 3LW, UK
UKHW022026190726
13853UKWH00005B/2132